MW01054586

NEUROIMAGING PERSONALITY, SOCIAL COGNITION, AND CHARACTER

http://booksite.elsevier.com/9780128009352

Neuroimaging Personality, Social Cognition and Character

by John R. Absher, Editor and Jasmin Cloutier, Co-Editor

For supplementary materials, please access the companion web site for this volume.

NEUROIMAGING PERSONALITY, SOCIAL COGNITION, AND CHARACTER

Editor

JOHN R. ABSHER
University of South Carolina, Greenville, SC, USA

Co-Editor

JASMIN CLOUTIER
University of Chicago, Chicago, IL, USA

AMSTERDAM • BOSTON • HEIDELBERG • LONDON
NEW YORK • OXFORD • PARIS • SAN DIEGO
SAN FRANCISCO • SINGAPORE • SYDNEY • TOKYO
Academic Press is an imprint of Elsevier

Academic Press is an imprint of Elsevier
125 London Wall, London EC2Y 5AS, UK
525 B Street, Suite 1800, San Diego, CA 92101-4495, USA
50 Hampshire Street, Cambridge, MA 02139, USA
The Boulevard, Langford Lane, Kidlington, Oxford OX5 1GB, UK

ISBN: 978-0-12-800935-2

British Library Cataloguing-in-Publication Data
A catalogue record for this book is available from the British Library

Library of Congress Cataloging-in-Publication Data
A catalog record for this book is available from the Library of Congress

For information on all Academic Press publications
visit our website at www.elsevier.com

Senior Acquisition Editor: Natalie Farra
Senior Editorial Project Manager: Kristi Anderson
Production Project Manager: Karen East and Kirsty Halterman
Designer: Matthew Limbert

Typeset by TNQ Books and Journals
www.tnq.co.in

Printed and bound in the United States of America

Front Cover:

Background image: From pixabay.com http://pixabay.com/en/brain-intelligence-human-science-311522/ CC0 Public Domain License. Modified by Tianyi Li.

DTI Image: *Courtesy of the Laboratory of Neuro Imaging and Martinos Center for Biomedical Imaging, Consortium of the Human Connectome Project* - www.humanconnectomeproject.org

Top right brain image: Created by Carlos Cardenas-Iniguez.

Bottom right brain image: Created by Ivo Gyurovski.

Contents

16. Honesty

FRANCESCA MAMELI, GIUSEPPE SARTORI, CRISTINA SCARPAZZA, ANDREA ZANGROSSI, PIETRO PIETRINI, MANUELA FUMAGALLI AND ALBERTO PRIORI

VI
BRAIN IMAGING AND SOCIETY

17. The Henchman's Brain: Neuropsychological Implications of Authoritarianism and Prejudice

KELSEY WARNER, DANIEL TRANEL AND ERIK ASP

18. The Neural Mechanisms of Prejudice Intervention

KEITH B. SENHOLZI AND JENNIFER T. KUBOTA

19. Political Neuroscience

INGRID J. HAAS

20. Science in Society: Neuroscience and Lay Understandings of Self and Identity

CLIODHNA O'CONNOR

21. Toward a Neuroscience of Wisdom

PATRICK B. WILLIAMS AND HOWARD C. NUSBAUM

List of Contributors

John R. Absher Department of Neuropsychiatry and Behavioral Science, University of South Carolina School of Medicine, Greenville, SC, USA; Via College of Osteopathic Medicine, Spartanburg, SC, USA; Alliance for Neuro Research, LLC, Greenville, SC, USA

Reginald B. Adams Jr. Department of Psychology, The Pennsylvania State University, University Park, PA, USA

Daniel N. Albohn Department of Psychology, The Pennsylvania State University, University Park, PA, USA

Andrey P. Anokhin Department of Psychiatry, Washington University School of Medicine, St. Louis, MO, USA

Erik Asp Department of Psychiatry, University of Iowa, Iowa City, IA, USA; Department of Psychology, Hamline University, St. Paul, MN, USA

Michael J. Banissy Department of Psychology, Goldsmiths, University of London, London, UK

Anam Barakzai Department of Psychology, University of Chicago, Chicago, IL, USA

Andrea Bari Department of Brain and Cognitive Sciences, The Picower Institute for Learning and Memory, Massachusetts Institute of Technology, Cambridge, MA, USA

Taylor S. Bolt Department of Psychology, Wake Forest University, Winston-Salem, NC, USA

Carlos Cardenas-Iniguez Department of Psychology, University of Chicago, Chicago, IL, USA

Jasmin Cloutier Department of Psychology, University of Chicago, Chicago, IL, USA

Philip J. Corr Department of Psychology, City University London, London, UK

Dale Dagenbach Department of Psychology, Wake Forest University, Winston-Salem, NC, USA

Colin G. DeYoung Department of Psychology, University of Minnesota, Minneapolis, MN, USA

Håkan Fischer Department of Psychology, Stockholm University, Stockholm, Sweden

William Fleeson Department of Psychology, Wake Forest University, Winston-Salem, NC, USA

Jonathan B. Freeman Department of Psychology, New York University, New York, NY, USA

Manuela Fumagalli Centro Clinico per la Neurostimolazione, le Neurotecnologie ed i Disordini del Movimento, Fondazione IRCCS Ca' Granda, Ospedale Maggiore Policlinico, Milan, Italy

R. Michael Furr Department of Psychology, Wake Forest University, Winston-Salem, NC, USA

Ivo Gyurovski Department of Psychology, University of Chicago, Chicago, IL, USA

Ingrid J. Haas Department of Political Science and Center for Brain, Biology, and Behavior, University of Nebraska-Lincoln, Lincoln, NE, USA

Ryan S. Hampton Department of Psychology, Arizona State University, Tempe, AZ, USA

Tanja S. Kellermann Department of Neurosurgery, Medical University of South Carolina, Charleston, SC, USA

Jennifer T. Kubota Department of Psychology, University of Chicago, Chicago, IL, USA; Center for the Study of Race, Politics, and Culture, University of Chicago, Chicago, IL, USA

Paul J. Laurienti Department of Radiology, Wake Forest University School of Medicine, Winston-Salem, NC, USA

Tianyi Li Department of Psychology, University of Chicago, Chicago, IL, USA

Caroline Di Bernardi Luft Department of Psychology, Goldsmiths, University of London, London, UK

Francesca Mameli Centro Clinico per la Neurostimolazione, le Neurotecnologie ed i Disordini del Movimento, Fondazione IRCCS Ca' Granda, Ospedale Maggiore Policlinico, Milan, Italy

Sebastian Markett Department of Psychology and Center for Economics and Neuroscience, University of Bonn, Bonn, Germany

Neil McNaughton Department of Psychology, University of Otago, Dunedin, New Zealand

Christian Montag Department of Psychology, University of Ulm, Ulm, Germany

Joseph M. Moran Center for Brain Science, Harvard University, Cambridge, MA, USA; US Army Natick Soldier Research, Development, and Engineering Center, Natick, MA, USA

Howard C. Nusbaum The University of Chicago, Chicago, IL, USA

Cliodhna O'Connor Department of Psychology, Maynooth University, Maynooth Co Kildare, Ireland

Andreas Olsson Emotion Lab at Karolinska Institutet, Department of Clinical Neuroscience, Division of Psychology, Stockholm, Sweden

Pietro Pietrini Università degli Studi di Pisa, Pisa, Italy

Alberto Priori Centro Clinico per la Neurostimolazione, le Neurotecnologie ed i Disordini del Movimento, Fondazione IRCCS Ca' Granda, Ospedale Maggiore Policlinico, Milan, Italy; Università degli Studi di Milano, Milan, Italy

Martin Reuter Department of Psychology and Center for Economics and Neuroscience, University of Bonn, Bonn, Germany

Giuseppe Sartori Università degli Studi di Padova, Padua, Italy

Cristina Scarpazza Università degli Studi di Padova, Padua, Italy

Keith B. Senholzi Harvard Medical School, Boston, MA, USA

Ryan M. Stolier Department of Psychology, New York University, New York, NY, USA

Bettina Studer Department of Neurology, St. Mauritius Therapieklinik, Meerbusch, Germany

Rongxiang Tang Department of Psychology, Washington University, St. Louis, MO, USA

Yi-Yuan Tang Presidential Endowed Chair in Neuroscience, Department of Psychological Sciences, Texas Tech University, Lubbock, TX, USA

Nicholas M. Thompson Department of Psychology, Goldsmiths, University of London, London, UK

Daniel Tranel Department of Neurology, University of Iowa, Iowa City, IA, USA; Department of Psychology, University of Iowa, Iowa City, IA, USA

Irem Undeger Emotion Lab at Karolinska Institutet, Department of Clinical Neuroscience, Division of Psychology, Stockholm, Sweden

Kelsey Warner Department of Neurology, University of Iowa, Iowa City, IA, USA; Department of Psychiatry, University of Iowa, Iowa City, IA, USA

Patrick B. Williams The University of Chicago, Chicago, IL, USA

Jonathan Yi Emotion Lab at Karolinska Institutet, Department of Clinical Neuroscience, Division of Psychology, Stockholm, Sweden

Rongjun Yu Department of Psychology, National University of Singapore, Singapore

Andrea Zangrossi Università degli Studi di Pisa, Pisa, Italy

Maryam Ziaei School of Psychology, University of Queensland, St. Lucia, QLD, Australia

Preface

Fascination with Neuroimaging Personality, Social Cognition, and Character (NPSCC) is ubiquitous and has been around since long before brain imaging was invented. Since then, its importance in popular culture and in neuroscience has grown tremendously, and its impact will continue to broaden.

Science fiction nicely captures our fascination with NPSCC. Several sci-fi movies highlight the implications of understanding people through NPSCC. Only a few pertinent examples are discussed here. For example, *Minority Report* explores a world in which it is possible to predict criminal behavior based partly on brain imaging. *Lucy* suggests that genetics can unlock a brain's potential; brain imaging reveals just how vast this potential may be. *Star Trek* (several depictions) demonstrates the importance of Theory of Mind, as characterized by the Vulcan mind meld. *Coherence*, *Brainscans*, and, most recently, *Hector and the Search for Happiness* have capitalized on the popularity of NPSCC. In *Hector*, real-time brain imaging is used to study rapidly changing human emotions such as love, misery, anger, and joy. The examples of PSCC movies (those without neuroimaging depictions) are, well, almost *Limitless*. In *Limitless*, it is arguably the protagonist's social cognitive capacity that is more fascinating than his raw intellect.

These science fiction movies illustrate the ubiquity of NPSCC depictions in popular culture, but its popularity is evident in the nonfiction realm as well. The idea for the current volume, for example, was the offspring of a Wall Street Journal article on altruism[a] and a popular book about optimism[b] mixed with my own fascination with NPSCC. What makes people altruistic, optimistic, honest, or courageous? What do we know about these phenomena based on brain imaging and neuroscience? How can we know whether someone is a PSCC "outlier" by choice, or by virtue of underlying neuroscientific imperatives? Where do we draw the line when there is no clear demarcation between choice and biology? Almost every time a science journal or news report commented on neuroscience, I became a little more optimistic about the possibility that neuroscience might help us answer such questions one day. Does such optimism bias the conclusions drawn from NPSCC research or the manner in which those results are presented in the media (as discussed later in this volume)? We are surely a long way off from definitive answers, but neuroscience is progressing faster than I imagined at the outset of this worthy effort.

When Elsevier requested book ideas, optimism probably motivated me (far more than altruism) to offer the idea for NPSCC, a relatively new realm of neuroscience. After initially searching for chapter authors to write about optimism and altruism, many other enticing topics emerged. Nonfiction works offered valuable perspectives on moral decision-making, automatic versus controlled cognitive and emotional processing mechanisms, and several additional topics addressed in this volume. The field is growing and diversifying rapidly. Current research in the field reinforces the ascendance of NPSCC and its increasingly important societal implications.

I hope that this volume will stimulate awareness of personality neuroscience, social neuroscience, and character/morality neuroscience throughout a broad array of fields, from anthropology to cell biology and many others. A truly interdisciplinary effort will be required to understand the most complex object in the universe, the human brain, and NPSCC delivers valuable assistance in this collaborative and important research enterprise. Readers of this volume will find ample support for these conclusions.

References

1. Svoboda E. Hard-wired for giving: contrary to conventional wisdom that humans are essentially selfish, scientists are finding that the brain is built for generosity. *Wall Str J*. August 31, 2013.
2. Sharot T. *The Optimism Bias: A Tour of the Irrational Positive Brain*. New York: Vintage Books; 2011.
3. *Minority Report* (movie).
4. *Coherence* (movie).
5. *Lucy* (movie).
6. *Brainscans* (movie).
7. *Star Trek* (movie series).
8. *Hector and the Search for Happiness* (movie).
9. *Limitless* (movie).

Acknowledgments

Many people contributed to the completion of this volume, and this acknowledgment conveys sincere gratitude for these efforts.

First, and foremost, Elsevier was exceptionally supportive. Natalie Farra stimulated the process from its original conceptualization, and Kristi Anderson provided ongoing assistance and encouragement throughout the entire process of securing chapter author commitments, collecting chapter submissions, completing necessary revisions, and verifying that each manuscript and its accompanying material were ready for the production team project managers, Karen East and Kirsty Halterman, who methodically put the finishing touches on the volume. Elsevier also provided temporary access to ScienceDirect to facilitate the identification of chapter authors.

Second, I appreciate the support systems upon which the editors and authors relied. Specifically, I wish to thank my wife, Laureen, who endured countless hours of preoccupation and distraction over these many months; I am grateful for her patience and understanding, as I am sure my colleagues are grateful for the key members of their personal support system.

Third, I greatly appreciate Dr. Jasmin Cloutier (Jas), for co-editing this volume for his insightful contributions throughout the entire process, his relationships with leaders in this field, and his outstanding chapter contribution. I am also thankful to Tianyi Li for producing an attractive cover design.

Fourth, I am most grateful to the authors and co-authors of this work for enduring seemingly endless and at times nit-picky editorial remarks, and for diligently refining their chapter submissions to their current superb levels. The readers of this volume are also appreciated for valuing NPSCC research, and its potential ramifications for so many aspects of neuroscience and culture. I am grateful to all of the above for the opportunity to bring NPSCC to fruition, because this is a topic that will increase in importance in the coming decades, and not just for the science fiction writers around the world who are adept at promoting it.

SECTION I

INTRODUCTION

C H A P T E R

1

Hypersexuality and Neuroimaging Personality, Social Cognition, and Character

John R. Absher[1,2,3]

[1]Department of Neuropsychiatry and Behavioral Science, University of South Carolina School of Medicine, Greenville, SC, USA; [2]Via College of Osteopathic Medicine, Spartanburg, SC, USA; [3]Alliance for Neuro Research, LLC, Greenville, SC, USA

O U T L I N E

1. INTRODUCTORY QUESTIONS AND DEFINITIONS

1.1 Questions

Can neuroimaging improve our understanding of personality, social cognition, and character (NPSCC)? The human brain determines the best and worst aspects of humanity. This deterministic view presupposes that a person may misbehave because there is something wrong with the brain, and they may behave magnificently due to something special happening in the brain. NPSCC is an important vantage point for the study of how the brain permits us to be unique, creative, intelligent, loving, generous, and socially adept creatures. What does neuroimaging reveal about the neural underpinnings of our personalities, social cognition, and character? How does personality relate to social behavior and character, and how do brain structure and function determine these phenomena across the lifespan in health and disease? Are there neuroimaging techniques that clearly differentiate personality, social cognition, and character (PSCC) from other basic mental operations related to cognition? What biological processes explain the differences between a transient state (e.g., sadness) and a persistently depressed temperament or trait? To what extent do free will, belief, emotional regulation, and cognitive capacity enable (for example) fear to be overcome by determination, lust to be subdued by moral beliefs, or anxiety to be calmed by self-reflection? Such questions are motivation for this volume.

Some of the more epistemological questions may extend beyond NPSCC. For example, what are the criteria that define the penetrance or expressivity of deterministic influences, the boundaries of free will, the aspects of PSCC that are malleable, and those that are unalterably imbedded into the genetic code, brain organization, and chemoarchitecture? Can neuroimaging unambiguously

Neuroimaging Personality, Social Cognition, and Character
http://dx.doi.org/10.1016/B978-0-12-800935-2.00001-4

distinguish the immutable from the malleable aspects of brain function? Neuroimaging is one of a broad array of tools and academic disciplines that must be utilized to address such profound questions, and NPSCC may, therefore, be of great interest to scholars from a broad array of intellectual disciplines,[1] even if its perspectives offer a limited or partial perspective.

This introductory chapter is focused on several specific aims:

1. To define some of the key NPSCC terms used throughout this volume
2. To provide a brief overview of hypersexuality as a structure within which major NPSCC themes are presented (as an exemplar of NPSCC research), because
   a. Personality, social cognition, and character (PSCC) all impact hypersexuality
   b. Brain lesions that produce hypersexuality may also influence PSCC
   c. The brain networks and structures implicated in hypersexuality are involved in many aspects of PSCC
   d. The complexities of hypersexuality research mirror those encountered in other aspects of NPSCC
3. To introduce the limitations of narrow localizationism and present the three most important, overlapping, integrated, and complementary brain networks related to hypersexuality
4. To summarize the structural and functional imaging approaches to NPSCC, focusing on contributions from lesion studies of hypersexuality to provide a historical backdrop for NPSCC
5. To summarize some of the complexities and challenges specific to neuroimaging
6. To revisit hypersexuality in the closing section and recapitulate relevant findings from the literature that directly relate to PSCC themes, some of which are developed in great detail in this volume

1.2 Definitions

Neuroimaging is a term used to capture a broad array of structural and functional imaging technologies. **Structural neuroimaging** includes all scales of anatomical imaging, from gross sectional anatomy (e.g., the Visible Human Project)[2] to the subcellular level.[3,4] Cranial computerized tomography (CT), and magnetic resonance imaging (MRI) are the primary modalities used for structural neuroimaging, as applied in NPSCC research, and may provide measures in one dimension (e.g., cortical thickness),[5–28] two dimensions (surface area),[5–7,9,10,20,29–34] three dimensions (volume),[5,6,8–11,13,15–17,24,29,34–151] or even four dimensions (e.g., change in volume over time in the same individuals).[152–165] Many studies examine volume changes over time in groups, often comparing one subgroup to another. **Voxel-based morphometry** (VBM) enables statistical analyses among images transformed to stereotaxic space, both within subjects over time and between subjects.[166] **Diffusion tensor imaging** (DTI)[167] and **diffusion spectrum imaging** (DSI)[168] map white matter tracts using diffusion MRI. These techniques are the backbone of brain morphometry: length or thickness, surface area, volume, changes in volume over time, or differences in volume among study groups. Anatomical connectivity may be defined by DTI and DSI, and effective or functional connectivity can also be defined using MRI. In addition, chemoarchitecture, cytoarchitecture, receptor type and density, and dendritic spine density and organization may be defined using many neuroscientific techniques that complement structural mapping and analysis. For pertinent overviews, see the ENIGMA Consortium,[169] the Human Connectome Project,[170] Dennis and Thompson,[171] or Catani and Thiebaut de Schotten.[172]

Functional neuroimaging techniques measure aspects of brain function, such as glucose or oxygen metabolism, global or regional cerebral blood flow, electrical activity, neurotransmitter receptor or transporter activity, or tissue concentrations of specific chemical moieties. These techniques reveal both structural and functional information. Electroencephalography (EEG), magnetoencephalography (MEG), and evoked potentials (EP) may be mapped to the cortical surface.[173] Single photon emission computed tomography (SPECT), positron emission tomography (PET), functional MRI (fMRI),[174] MR spectroscopy, and perfusion CT scanning are all mapped in three dimensions throughout the brain. New tools and techniques continue to expand this list, such as functional near infrared spectroscopy (fNIRS), which may reveal "between-brain connectivity during dynamic interactions between two subjects."[175]

These functional neuroimaging tools can quantify and localize many types of brain activity. Depending on the techniques employed, they may reveal information on the millisecond time scale, or in some cases, minutes to hours. For example, EEG and EP can reveal changes in electrical activity at the millisecond level, while mapping glucose metabolism with PET may require 30–60 min.[174] Many clinical, experimental, demographic, genetic, and other variables may be linked statistically in time and space to brain function using such techniques to reveal new and important insights into the complexities of personality, social cognition, and character, as emphasized in this volume.

Stanley and Adolphs summarize additional strategies for relating structure to function in social neuroscience research,[176] such as computational neuroscience, optogenetics (primarily for animal studies), and others. Recent developments may permit "**Optogenetics**, a technology that allows scientists to control brain activity by shining light on neurons, [and] relies on light-sensitive proteins that can suppress or stimulate electrical signals within cells," to be applied to human studies.[177] Such techniques promise tantalizing options for functional imaging approaches to NPSCC research. For example, deactivating specific elements of a network using optogenetics may enable scientists to test the effects of localized cortical failure, in real time, in individuals with otherwise perfectly healthy brains, and to extend the progress gained through techniques such as transcranial direct current stimulation (tDCS),[177b] transcranial magnetic stimulation (TMS),[177c] and human intracranial electrophysiology (HIE).[177d]

Personality has been defined as "an individual's unique variation on the general evolutionary design for human nature, expressed as a developing pattern of dispositional traits, characteristic adaptations, and integrative life stories, complexly and differentially situated in culture."[178] Statistical analyses of personality traits support "a hierarchical model of traits based on the Big Five dimensions: extraversion, neuroticism, agreeableness, conscientiousness, and openness/intellect."[179] Three of these Big Five dimensions (neuroticism, agreeableness, and conscientiousness) are encompassed by the "metatrait," stability, and the other two (extraversion and openness/intellect) by the metatrait, plasticity.[179] There are several aspects of each Big Five dimension, and many facets of each aspect.[179] A major challenge of **personality neuroscience** is to relate all of these facets, aspects, dimensions, and metatraits to specific neurobiological processes.

The concept of **social cognition** has recently been defined as follows: "Social cognition is the study of how people make sense of other people and themselves. It focuses on how ordinary people think and feel about people—and how they think they think and feel about people."[180] **Social neuroscience** "encompasses all levels of biological analysis (genetic polymorphisms, neurotransmitters, circuits and systems, as well as collective behavior in groups) and stages of processing (sensory systems, perception, judgment, regulation, decision-making, action)."[176] Social cognition has been one focus of social neuroscience, which developed over the last 35 years, with **Theory of Mind (ToM)** research as one of its early topics of investigation.[181] Studies of ToM have investigated brain mechanisms for **mental state attribution (MSA)** and **shared representations** formed by individuals (**perceivers**) as they attempt to understand the minds of others (**targets**) and themselves.[182] These studies are hampered by the complexity of social interactions, which depend upon cognitive synthesis of multimodal information that is dynamic and contextually embedded and difficult to study in a naturalistic setting.[182] Understanding others requires the ability to read behavioral cues to gain insight into a target's personality and character, processing a vast and dynamic stream of contextually embedded multimodal information and developing beliefs derived from these cues, processes which may be affected by the stage of brain development, acquired or congenital deficits (e.g., deafness), and other factors.[183] To the extent that personality and character are defined in relation to an individual's interactions with others in social situations, social neuroscience also encompasses these components of NPSCC.

An attempt to define human **character traits** is offered by Miller:

> "(a) They [character traits] are personality trait dispositions which manifest as beliefs, desires, and/or actions of a certain sort appropriate to that trait, as a result of being stimulated in a way appropriate to that trait.
>
> (b) They are those personality traits for which a person can be appropriately held responsible and/or be normatively assessed.
>
> (c) They are metaphysically grounded in the interrelated mental state dispositions specific to the given trait such that necessarily, if these dispositions obtain in a person, then the character trait disposition obtains as well.
>
> (d) Some of them are virtues and vices, but perhaps there are character traits which are neither virtues nor vices."[184]

Character traits may have moral valence, such as honest/dishonest and altruistic/selfish. Other character traits, such as impulsive/reserved, may not necessarily connote moral valence or moral character.

1.3 Definition of Hypersexuality and Rationale for Inclusion

As mentioned earlier, hypersexuality is presented in this chapter to illustrate how NPSCC applies to a specific clinical condition. Hypersexuality is not just a convenient topic that structures the discussion of how neuroimaging informs understanding of complex, PSCC-related clinical phenomena. In addition, hypersexuality is a topic this author has had an opportunity to study with colleagues in a multidimensional assessment that included advanced neuroimaging methods (discussed in detail later in this chapter). This background provided the context within which the richness of hypersexuality as a topic appropriate for NPSCC was realized. For example, the topic nicely illustrates the failure of strict localizationism because so many different brain areas have been implicated in hypersexuality. Also, the challenges posed by coexisting clinical symptoms (e.g., disinhibition and dysexecutive phenomena), anatomical variability, age effects such as brain atrophy, concomitant medications, and many other factors are challenges the author experienced in studying hypersexuality (and other topics) that have direct relevance to other clinical phenomena of interest to scientists in this field. For these reasons, this clinical phenomenon is an appropriate introduction to a largely nonclinical text.

The proposed definition of **hypersexuality** is offered and then deconstructed to introduce these points, which are elaborated in later sections:

A. A minimum of six months of "recurrent and intense sexual fantasies, sexual urges, or sexual behaviors [FUB] in association with three or more of the following five criteria:
  A1. Time consumed by sexual [FUB] repetitively interferes with other important (nonsexual) goals, activities, and obligations.
  A2. Repetitively engaging in sexual [FUB] in response to dysphoric mood states (e.g., anxiety, depression, boredom, or irritability).
  A3. Repetitively engaging in sexual [FUB] in response to stressful life events.
  A4. Repetitive but unsuccessful efforts to control or significantly reduce these sexual [FUB].
  A5. Repetitively engaging in sexual behaviors while disregarding the risk for physical or emotional harm to self or others.

B. There is clinically significant personal distress or impairment in social, occupational, or other important areas of functioning associated with the frequency and intensity of these sexual [FUB].

C. These sexual [FUB] are not due to the direct physiological effect of an exogenous substance."[185]

These definitions highlight the relationships of hypersexuality to personality, social cognition, and character. Hypersexuality is defined as disruptive (A1), causes social or occupational dysfunction (B), and may relate directly to emotional status (A2) or stress (A3). Emotional predilection and regulation and the degree to which stress has an influence on a person are related to aspects of personality. The character definition mentions behavior for which one may be held accountable, and the definition of hypersexuality requires that the behavior be "disruptive," both indicating that hypersexuality overlaps the character domain. Impulsive and compulsive tendencies (A4) and a lack of inhibitory control for risk avoidance (A5) are personality-related components of the hypersexuality definition. While many drugs are thought to provoke hypersexuality, the definition (C) excludes cases due to exogenous substances. For the purposes of this chapter, it is assumed that brain-damaged patients may also fulfill this definition, assuming all other criteria are met.

1.4 Hypothetical Hypersexuality Scenario

A brief hypothetical scenario is presented to structure the discussion of hypersexuality and its many analytical complexities. Assume a hypersexual (HS) individual encounters a potential target (T). Under normal circumstances, both HS and T may be expected to perceive and process the other's physical characteristics, including facial features, body HS and size and shape, coloration, odor, hair length, distinguishing anatomical features, and race. Behavioral information would also be perceived and processed, including such things as body movements or mannerisms, speech rate/volume/variation/melodic intonation, facial expressiveness, affective/emotional expression, gestures, eye contact, and the denoted or connoted meanings of verbal and nonverbal communication. Both HS and T may be expected to form an understanding of the other and to assimilate and interpret these various physical, behavioral, cognitive, and social characteristics. Some of this information may be automatically and unconsciously processed. Additional information from top-down processing is also likely to impact these interpretations. Based on a unique history, traits and dispositions, character adaptations, life stories, and culture, HS and T may develop favorable, unfavorable, or neutral multifaceted impressions of each other and improve their knowledge of each other; the multiple facets of such impressions may include some that are positive (e.g., sexual attractiveness), some that are negative (e.g., eye contact and communication skills), and some facets that are neutral (e.g., intellect). The valence of these facets could change in another context (e.g., intellect may be important during courting rituals, but probably not for HS).

Over the course of their evolving relationship, HS and T are likely to utilize basic perceptual processes, such as smell, sight, touch, hearing, and taste, and cognitive mechanisms, such as memory, language, visual perception, executive function, judgment, and decision making. Their social cognition, character, and personality are assumed to rely on these several subordinate cognitive mechanisms, as well as influences within the social group and culture. Can we point to one specific aspect of this complex process and identify the cause of HS's hypersexuality? Are there multiple, independent, and equally effective causes of hypersexuality? Are there always several contributing mechanisms? Is it possible that disruption of bottom-up, top-down, or regulatory influences alone can account for hypersexuality? Brain lesion cases offer a unique opportunity to study the various aspects of brain function that may contribute to hypersexuality; in non-lesion cases, hypersexuality could be postulated to arise from a wide variety of brain dysfunctions. Just as our understanding of cognitive function and neuropsychology has been enhanced by studying individuals with brain lesions, our understanding of personality, social cognition, and character may be informed by studying brain lesion cases. In the following section, a concise literature review pertaining to both non-brain lesion and lesion-induced hypersexuality subjects will be presented.

2. HYPERSEXUALITY WITHOUT AND WITH STRUCTURAL BRAIN LESIONS

2.1 PSCC in Non-Brain-Lesion HS Subjects

Aspects of personality, social cognition, and character (PSCC) may malfunction and contribute to hypersexuality. Several studies are informative in this regard. Reid and colleagues studied 95 men in one study[186] and 47 in another study that also involved 31 women,[187] using the NEO (Neuroticism, Extraversion and Openness) Personality Inventory–Revised.[188] In the first study, a path analysis revealed that shame indirectly impacted hypersexuality through the "mechanism" of neuroticism, with one facet of neuroticism (impulsivity) being a "particular salient facet."[186] The second study compared men and women[187] and found these same factors to be important. In addition, "The prevalence of less trust, competence, and dutifulness was significantly higher among the women patients. A greater percentage of the women also struggled with stress proneness (vulnerability) and exhibited excitement-seeking tendencies."[187] Both studies support the possibility that "maladaptive shame coping" indirectly contributes to hypersexuality in individuals with these personality characteristics, but there are clear differences in the underlying personality structures that predispose them to hypersexuality, as well as the degree to which hypersexuality becomes manifested.

In a large, population-based sample, hypersexuality was found to be "organized along a continuum of increasing sexual frequency and preoccupation, with clinical cases of hypersexuality falling in the upper end of the continuum or dimension."[189] Two independent analyses were presented, including 2101 men and women selected from the general population, and 716 male sex offenders, and the conclusions were the same in both groups. The authors cite other disorders that also seem to be organized along a continuum (e.g., attention deficit/hyperactivity disorder) and argue that the extent of hypersexuality is more important in defining the disorder than the fulfillment of various taxonomic criteria. A compelling and relevant part of their discussion concerns the conditions under which hypersexuality may be considered a mental disorder:

> A disorder comprises two parts: harmfulness, a value term based on social norms, and dysfunction, a scientific term based on evolutionary principles. If hypersexuality is nothing more than variation in normal sexual functioning, then its status as a disorder is brought into question by the absence of genuine dysfunction; and if hypersexuality is not a disorder, then it does not fit into the diagnostic nomenclature.[189]

A primary implication of these studies is the suggestion that hypersexuality in individuals without brain lesions occurs along a continuum; although some facets of neuroticism seem particularly important in its genesis, it is likely that non-brain-lesion HS "is usually the result of the additive effect of many small individual influences."[188,189]

A second important implication of these studies is that they illustrate the difficulty we have defining a clear demarcation for "disorder." For example, there is a continuum (of sorts) of disruptive sexual behavior extending from infidelity and promiscuity in monogamous relationships, on the one hand, to a dangerous form of promiscuity that produces a public health threat (e.g., the famous case of "patient zero" who contributed to the spread of acquired immune deficiency syndrome), on the other.[190] When does non-lesion-induced hypersexuality cross the line where we define it as a disorder? If we abandon criteria-based classifications of disease in favor of drawing a demarcation along disease continua, then it will be critical to establish such boundaries.

A third important implication of these studies is that they illustrate the difficulty we have defining "dysfunction" in non-lesion-induced hypersexuality; even if we find metabolic, physiological, or structural distinctions in affected individuals, on what basis are brain differences defined as dysfunction? Must we correct the aberrant phenomena and see if the behavior is normalized in order to know if it represents "genuine dysfunction," as defined by Walters et al.? If so, how can we accomplish this in non-brain-lesion hypersexuality cases?

To some extent, the literature presented in this section contributes to our understanding of the reported relationships between hypersexuality and neuroticism, impulsivity, shame, vulnerability, dependence, and other PSCC phenomena. In the next section, some of these same phenomena are reviewed as they occur in association with localized brain damage.

2.2 Brain Lesions and Hypersexuality

Acquired hypersexuality due to brain damage probably differs mechanistically from that seen in non-brain-lesion hypersexuality. One key advantage of studying hypersexuality due to brain lesions is that cerebral dysfunction is easy to assume, given that there is a structural lesion, and there may be an abrupt change in behavior. Thus, lesion-induced hypersexuality cases offer an opportunity to study the neural underpinnings of a dysfunction attributable to the damaged brain area and may provide clues regarding how brain damage and dysfunction leads to disorder/harmfulness, as determined by social and cultural norms of behavior; also, the "disorder," required for the clinical diagnosis of hypersexuality, may be easier to detect when it begins abruptly in lesion-induced hypersexuality. Furthermore, brain lesions may occur in isolation, without the confounders seen in many other forms of hypersexuality (e.g., neurochemical alterations of mania, coexisting epileptic pathology in seizure-related or interictal hypersexuality, microscopic pathology of Parkinson's disease seen in deep brain stimulation cases of hypersexuality, etc.). For these reasons, the study of lesion-induced hypersexuality can be informative in ways that studying non-brain-lesion hypersexuality may not. In this section, the focus will be to briefly introduce lesion-induced hypersexuality as a tool to demonstrate how NPSCC can be used to understand complex neuropsychiatric syndromes. For a recent review of human lesion studies and human sexual behavior, see Baird et al.[191]

That brain damage can lead to dramatic and abrupt alterations in PSCC is hardly a matter of debate. After Phineas Gage was reported by Harlowe, Bigelow, and Jackson in the later part of the nineteenth century, it was evident that frontal lobe destruction could impair social cognition, comportment, and willingness to conform to societal norms.[192–194] As more brain lesion cases were reported, it became obvious that what was true of devastating frontal damage could also be true of small lesions in many different parts of the brain.

Although narrow localizationism led to clinically useful observations, mechanistic understanding continued to challenge behavioral neuroscience. For example, many neurobehavioral symptoms had an anomalous localization, and the strict localizationist views necessarily had to be abandoned.[195] Disconnectionism helped usher in an era of network or associationist theory, because both were superior to narrow localizationism in explaining how damage to separate anatomical areas may produce similar cognitive and behavioral phenomena.[111,195] Damasio and Damasio noted that aphasia, apraxia, visual agnosia, and other neuropsychological functions may result from a wide variety of lesion locations,[196] and the same phenomenon has been observed for depression, executive dysfunction, psychosis, personality change, emotional dysregulation, and hypersexuality. Several brain areas interact in neuroanatomically extensive functional networks that orchestrate cognitive and behavioral phenomena. No longer is it sufficient to attach a particular cognition or behavior to a specific brain area, because more likely than not, other parts of the brain are implicated in the same cognition or behavior.

Network/associationist constructs of brain organization will be relied upon to illustrate aspects of NPSCC that are pertinent to the study of hypersexuality, referencing later chapters in this volume when possible. It is clear from lesion-induced hypersexuality[191] that there are multiple networks involved, including a frontal, temporal, and subcortical/limbic network. These networks participate in the three classical phases of "the human sexual response cycle as defined by Kaplan… (1) sexual desire, (2) excitement, and (3) orgasm."[191] Sexual desire or drive is "mediated by subcortical structures, specifically the hypothalamus, ansa lenticularis, and pallidum. The temporal lobes, specifically the amygdalae, also play a significant role in this initial phase. Sexual excitement (phase 2) is mediated by cortical structures, namely the parietal and frontal lobes, controlling genital sensation and the motor aspects of sexual response, respectively, sustaining sexual activity until progression to orgasm (phase 3) for which the septal region has been implicated."[191] To oversimplify, the subcortical/limbic network creates the sexual urges, the frontal lobes constrain or permit sexual behaviors, the parietal region mediates sexual sensations, and the septal area is most closely linked to orgasm. All three networks relate to the human sexual response, and lesion-induced hypersexuality may relate to damage in any of these three important networks.

2.2.1 Frontal Lobes

An exhaustive review of frontal lobe lesions and their relation to PSCC is beyond the scope of this chapter. Instead, a focused review of key frontal lobe regions and the dysfunctions related to hypersexuality is presented. This same approach is taken throughout this section on lesion-induced hypersexuality.

Dorsolateral prefrontal cortex (dlPFC) lesions may impair response inhibition, cognitive estimation, executive functions, prosody (specifically, affective expression), and abstract thinking.[111] Medial prefrontal cortex (mPFC) lesions may impair drive, motivation, and level of interest.[111] Manifestations of medial frontal lobe lesions can include apathy, blunted affect, amotivation, and in extreme cases, akinetic mutism.[111] More precisely, within mPFC, anterior cingulate cortex (ACC) lesions can cause not only akinetic mutism, but also autonomic dysregulation (for a review see Damasio and Van Hoesen[197]). The ACC has been related to numerous relevant clinical phenomena, including penile erection, reward systems (e.g., self-stimulation), inappropriate social behavior, disinhibition, compulsive behavior, and hypersexuality (see Devinsky et al.[198] for review).

To the extent that dlPFC lesions impair self-regulatory mechanisms or estimations of the likelihood of reward resulting from hypersexual advances, lesions of the dlPFC could contribute to lesion-induced hypersexuality. Similarly, the mPFC activates in response to sexual stimuli, at least in some studies, though the activation may occur on the right, the left, or bilaterally.[199] ACC lesions could participate in the genesis of lesion-induced hypersexuality through several of the clinical phenomena noted above (and others), as the ACC serves as an integrative region for many frontal, autonomic, and regulatory functions of the frontal lobe.

Damage to the orbitofrontal cortex (OFC) can also lead to disinhibition, antisocial or inappropriate personality, and character changes.[111] Obsessive-compulsive behavior[113,200–202] and Witzelsucht have been linked to damage within the OFC, the uncinate fasciculus, or other fiber tracts interconnecting subcortical nuclei.[203–205] Impaired judgment and inappropriate risk-taking, as measured by the Iowa Gambling Task, are also associated with OFC lesions.[80,206–211] Antisocial personality disorder after left OFC damage has been reported with loss of libido.[212] Given the theoretical similarity between obsessive compulsive disorder (OCD) and hypersexuality ("compulsive sexual behavior"), Miner and colleagues performed a diffusion tensor imaging (DTI) study to compare eight hypersexual patients with eight controls and found that the impulsivity, compulsive tendencies, and greater negative affect (anxiety and depression) correlated with superior frontal fractional anisotropy (FA; "a measure of the extent to which water diffusion is directionally restricted").[213] No *inferior* frontal lobe differences were found for FA, or mean diffusivity (MD; "a measure of overall diffusivity in the tissue") in the eight hypersexual subjects.[213] In addition, compulsive sexual behavior severity was negatively correlated with superior frontal MD.[213] Although the study was small, used a normal control group, and did not include nonhypersexual controls with impulse control disorders, the findings suggest that reduction of superior frontal lobe white matter integrity may predispose to hypersexuality, and there may be anatomical distinctions between the impulse control disorders and patients with hypersexuality. It is unclear whether disturbances of white matter integrity impacted dlPFC or mPFC connections to functional networks implicated in hypersexuality, but such a possibility cannot be ignored.

Ventromedial prefrontal cortex (vmPFC) damage is often associated with personality alterations such as antisocial behavior.[214–216] Affective ToM has also been linked to vmPFC damage.[150] The authors note that the rich connections of vmPFC with "the anterior insula, temporal pole, inferior parietal region, and amygdala place it in a position to evaluate and regulate incoming limbic information which can consequently be used to inhibit behavior, regulate emotions and empathize with the experience of others."[150] Individuals with vmPFC damage and affective ToM deficits tend to be "particularly impaired on tasks that involve integration of emotion and cognition, such as affective mental state attribution."[150] One recent review of vmPFC function, for example, offers "a meaning-centered view of vmPFC [function that] predicts that vmPFC and its subcortical connections are not essential for simple forms of affect, valuation, and affective learning, but are essential when conceptual information drives affective physiological and behavioral responses."[216b] This view of vmPFC function is similar to that proposed by Eslinger and Damasio in the EVR case, in which vmPFC damage disconnected dlPFC and limbic systems structures, thus impairing "two kinds of regulatory activity: (1) 'modulation' of innate hypothalamic drives that are informed of environmental rules and contingencies; and (2) activation of higher cortices by basic drives and tendencies."[217] The three interrelated networks responsible for the human sexual response (discussed above) may depend on vmPFC to mediate the interactions between excessive drive (bottom-up) and socially defined norms for restraining hypersexuality behaviors (top-down).

In addition to searching for published cases of lesion-induced hypersexuality, it is also informative to look at clinical phenomena that have been strongly correlated with hypersexuality and that also have been strongly linked to frontal lobe lesions. In such cases, a direct relation to hypersexuality cannot be established, because the lesion-induced neurobehavioral disorder may be the direct cause of hypersexuality. This concern has led Baird et al.[191] to distinguish hypersexuality due to disinhibition from a more "true" hypersexuality that results from excess sexual drive. This concern applies to many clinical phenomena, such as autism spectrum disorders or attention deficit disorders, which are associated with several symptoms and linked to a variety of underlying etiologies. NPSCC might play a role in unraveling the multifaceted nature of such complex clinical disorders.

For example, mania is associated with an increase in various psychic tones, drives, and tendencies, such as hypersexuality, hypertalkativeness, and psychomotor agitation.[218] Starkstein and colleagues report a consecutive series of 12 patients with mania attributed to brain injury.[219] Seven of the 12 subjects also had hypersexuality. While selection bias or circular reasoning (i.e., the hypersexuality may be considered to be a symptom of mania) may account for this high prevalence of hypersexuality in acquired mania, all of the hypersexual cases had either right (3/7) or bilateral brain damage, and 6/7 included frontal lobe damage. This report nicely reviews the early literature on organic causes of mania but unfortunately does not specifically address the hypersexuality issue, even in their table of manic symptoms, or precisely define the anatomical boundaries of these frontal lobe lesions.

Disinhibition has often been cited as a risk factor for hypersexuality.[220,221] Frontal leucotomy and frontal lobotomy have both been reported to produce disinhibition and hypersexuality.[191] Bancroft reviewed the neuroscience of central inhibition of sexual response in the male and implicated the frontal lobe as part of a broad network for central inhibition, including "the hypothalamus, amygdala, septum, and ventral striatum, with additional cortical components in the hippocampus, cingulate gyrus, and orbital parts of the frontal cortex."[222] Damage to this inhibitory system leads to disinhibition. However, evidence of expected frontal/executive dysfunction is not always evident in hypersexuality, leading to speculation that impulsivity, poor judgment, and risky behaviors may be "domain specific."[223] Many other personality and emotional factors have been related to hypersexuality, such as impulsivity, emotional dysregulation, shame, neuroticism, stress-proneness, and a tendency to distrust others.[186,187,189]

Further research in lesion-induced hypersexuality will be required to clarify the potency of such phenomena, individually or collectively, in producing hypersexuality.

The foregoing discussion of hypersexuality and frontal lobe damage is necessarily selective. Nevertheless, some tentative conclusions can be made. First, the frontal lobes are large, diverse structures with myriad functions, and damage to a number of different frontal lobe structures can contribute to lesion-induced hypersexuality. Second, hypersexuality is related in some respects to impulsivity, obsessive compulsive behaviors, mood regulation disorders, and disinhibition. Third, dysexecutive and disinhibited phenomena are probably not necessary or sufficient for the manifestation of hypersexuality; indeed, some authors consider such cases unrelated to true lesion-induced hypersexuality.[191] Fourth, depression, anxiety, shame, and other aspects of negative affect are often seen in subjects with secondary (lesion-induced) mania who also manifest hypersexuality. Fifth, the other structures implicated in lesion-induced hypersexuality (see following sections) all have connections to the frontal lobe. Sixth, important frontal lobe structures in the underlying networks implicated in hypersexuality are the ACC, OFC, and vmPFC, which may allow for the influence of experience, beliefs, emotion/affect, and affective ToM and other regulatory influences to interact with limbic centers, such as the hypothalamus, which drives sexual urges and desires. Catani et al. list the Brodmann areas and white matter tracts implicated in a variety of such frontal lobe syndromes (see Figure 1, used with permission) and describe the key anatomical structures connected by these tracts.[111]

This brief discussion is not intended to clarify the complex role of the frontal lobes in hypersexuality or to exhaustively review frontal lesion studies seen in relation to hypersexuality. Instead, the summary identifies some of the frontal lobe areas and clinical phenomena that are potentially important for future neuroimaging studies of lesion-induced and non-lesion-induced hypersexuality.

2.2.2 *Temporal Lobes*

Temporal lobe damage may impair auditory and prosodic comprehension, facial recognition and emotional decoding, naming, and verbal fluency, among other things.[111] Hypersexuality following temporal lobe damage has been associated with rage, passivity, apathy, indifference, aphasia, and aggression. In patients with bilateral temporal lobe damage, elements of the human Kluver-Bucy syndrome (KBS; visual agnosia, hyperorality (such as excessive eating or mouthing of nonfood objects), hypermetamorphosis ("an irresistible impulse to touch, loss of normal anger, and fear responses")) are often evident.[224–226] Disorders of social cognition may also result from temporal lobe dysfunction, as discussed in several chapters in this volume.

This section on the relation of the temporal lobe to hypersexuality, like the preceding section, will illustrate only a selection of the most important clinical phenomena and temporal lobe regions thought to relate to hypersexual behavior. Catani et al. (Figure 2) depict some of the key anatomical connections in the temporal lobe network and review some of these various clinical manifestations of disturbances in the temporal lobe network.[111]

Bilateral temporal lobe destruction, including the amygdala, has long been recognized to reduce sexual inhibitions and provoke hypersexuality.[114] The KBS is perhaps the most extensively studied acquired human hypersexuality syndrome. *Hypo*sexuality also may follow temporal lobe damage, particularly that resulting from epilepsy.[191,227] Mania has been seen following right basotemporal strokes, and is seen more often following right than left sided lesions.[205,228] Braun et al. performed a "multiple single-case literature review analysis" of subjects

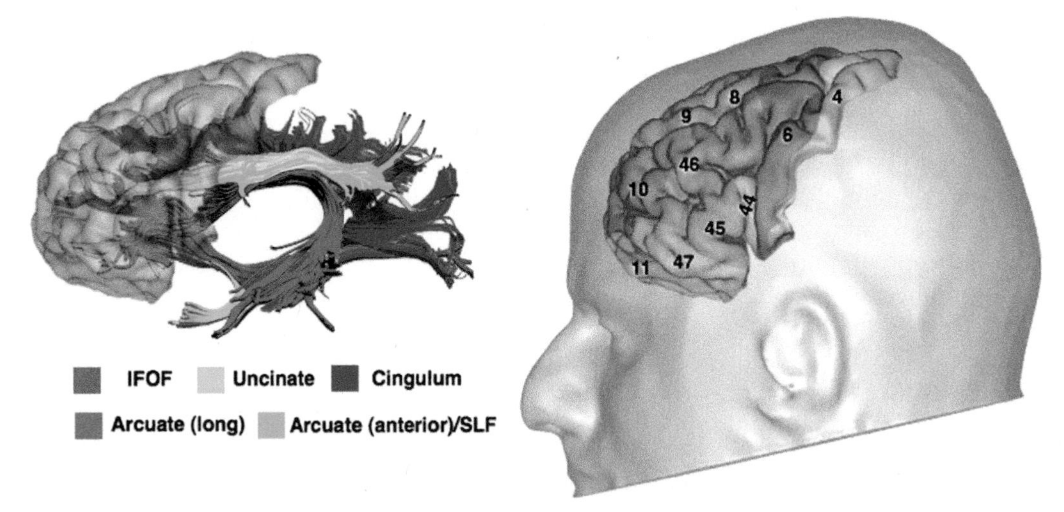

FIGURE 1 Frontal lobe and its main associative connections.[111] Brodmann areas are numbered, and major tracts are color coded, as shown. IFOF is inferior fronto-occipital fasciculus. SLF is superior longitudinal fasciculus.

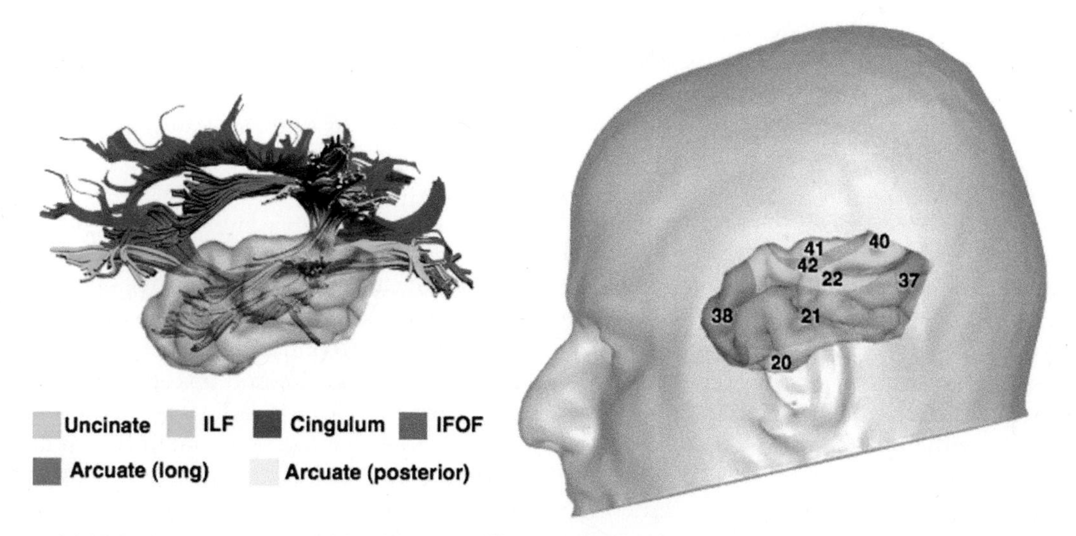

FIGURE 2 Temporal lobe and its main associative connections.[111] Brodmann areas are numbered, and major tracts are color coded, as shown. The inferior longitudinal fasciculus is ILF, and the inferior fronto-occipital fasciculus is IFOF.

with brain lesions seen in association with either hyposexuality or hypersexuality and found that 8/11 lesions among those with hypersexuality were located in the right hemisphere.[218] Left hemisphere lesions were reported more often in cases of hyposexuality, and both hyper- and hyposexuality were most often related to lesions within the temporal lobe.[218] They propose a hemispheric specialization for "psychic tone [which] includes motor, moral, emotional, language, and sexual dimensions, and these are reflected in… intellectual representations" through the mechanism of "approach-avoidance disposition."[218]

One large series of published cases of hypersexuality following unilateral brain lesions (n = 33) replicated the higher frequency of right hemispheric damage (26/33)[138] and compared Braun's model with the theory of "contralesional release."[114,138] Sexual tension was proposed to relate to left hemispheric functioning, and orgasm—representing a release from sexual tension—was proposed to be a function of the right hemisphere.[138] Central autonomic mechanisms were suspected as an underlying basis for the hemispheric specialization, with contrasting (left hemispheric) parasympathetic and (right) sympathetic systems.[138] The amygdala was suggested as a likely site within the temporal lobes, supporting this hemispheric specialization for contrasting aspects (tension vs. release) of sexual function.[138]

The role of the amygdala in hypersexuality may relate to its relationship to emotion, and particularly negative emotion, as negative affect, and emotion is prominent in both lesion-induced and non-lesion-induced hypersexuality. In patients with hypersexuality following temporal lobectomy, contralesional amygdalar volume correlates with the degree of postoperative hypersexuality.[229] Further, isolated lesions of the amygdala, sparing the surrounding temporal lobe, do not reproduce the KBS in monkeys.[230] Recognition of emotional facial expressions, such as fear, involves the amygdala, and the amygdala is highly interconnected to brain areas implicated in hypersexuality, including the septum, basal forebrain, and anterior temporal cortices.[231] The amygdala's well-known contributions to approach-avoidance (see Chapter 2), reward mechanisms,[123] and emotional salience detection could also partly explain its importance in lesion-induced hypersexuality. The amygdala may also contribute to moral and ethical decision making,[220] as will be discussed later in this volume. These observations suggest a definite, although indirect, role of the amygdala in lesion-induced hypersexuality.

Hypersexuality may manifest in many ways, including autoerotic behaviors, homosexuality, promiscuity, paraphilia, compulsive viewing of pornography, and alterations/switches in sexual preference, behaviors that have all been reported in patients with temporal lobe damage. Case reports also implicate the temporal lobes in pedophilic hypersexuality.[126] Devinsky et al. report a patient with pedophilic hypersexuality following right temporal lobe surgery who was convicted of possessing child pornography.[126] Such cases underscore the multifaceted nature of hypersexuality. A complex human behavior, like hypersexuality, typically has a wide variety of manifestations, influences, causes, and associated features. This premise is likely true for most complex human behaviors, because our neuroscientific understanding is usually not sophisticated enough to deconstruct these multifaceted clinical phenomena into their most fundamental elements.

2.2.3 Other Brain Areas

Several reviews of sexual psychobiology have been published.[191,232,233] According to one account, there are three subcortical and three cortical structures primarily involved in the human sexual response: the septal area, hypothalamus, the ansa lenticularis and pallidum, the frontal lobes, the parietal lobes, and the temporal lobes.[191] The frontal and temporal lobe contributions are highlighted above. The remaining areas are discussed in this section.

The hypothalamus is thought to relate to the neuroendocrine and autonomic aspects of sexual function.[191] Klein Levin syndrome is a rare disorder related to hypothalamic damage, characterized by hypersomnolence, hyperphagia, and hypersexuality.[234] One of the difficulties in studying hypothalamic lesions is the close proximity of other important structures, such as the median forebrain bundle, thalamus, and fornix, and the possibility that different hypothalamic subnuclei may exert contrasting effects.[191]

Heath reported a case in which a chemical stimulation probe was inserted into the septal nuclei, and orgasm was produced each time the area was stimulated.[235] Heath also found "pleasurable responses" in the majority of 54 subjects who underwent either electrical or chemical stimulation of the septal region.[191] The septal area is the only brain area where stimulation is consistently associated with orgasm in humans.

The medial thalamus is in close anatomical proximity to the hypothalamus and has also been implicated in hypersexuality.[203,236] Bilateral globus pallidus lesions due to carbon monoxide toxicity were also implicated in a case of hypersexuality.[237] Even damage to the occipital lobe has been reported to cause hypersexuality.[238] One VBM study of pedophilic men demonstrated that "two major fiber bundles: the superior fronto-occipital fasciculus and the right arcuate fasciculus" (see Figure 1) demonstrated lower frontal lobe white matter volumes in comparison to men with nonsexual offenses, suggesting that these pathways connect "a network for recognizing sexually relevant stimuli."[55] Major white matter connections exist among many of the brain areas discussed in this section, and the involvement of so many structures supports the notion that several functional networks are implicated in lesion-induced hypersexuality, particularly the limbic network (see Figure 3).

In one case of stroke-related hypersexuality, functional neuroimaging demonstrated abnormal frontal, temporal, and parietal lobe glucose metabolism following damage to the subthalamic nucleus (STN).[239] Absher et al. speculate that damage to the medial forebrain bundle (MFB) may contribute to the lesion-induced hypersexuality in this case and also may contribute to some of the functional manifestations that were observed on PET.[239] However, the thalamic radiation is also situated anatomically close to the area of damage,[201,240] and the authors did not address the possibility that functional changes related to damage to this structure, rather than the MFB. Stimulation of the STN resulted in hypersexuality behavior in two subjects reported by Doshi and Bhargava.[241] The infrequent occurrence of hypersexuality following STN stimulation might relate to subtle differences in the site of stimulation, the effects of Parkinson's disease, medications, or other factors. Nevertheless, damage to the STN or nearby structures may produce hypersexuality.

In summary, the clinical interpretation of these lesion studies is complex, yet the data strongly suggest that lesion-induced hypersexuality may emerge following damage to widely separated neuroanatomical structures. These anatomical areas are connected in several networks assumed to become anatomically or physiologically disrupted by the damage. The lesions that contribute to hypersexuality may affect all aspects of the human sexual response cycle, including the drive/motivation phase, the excitement phase, and the orgasm phase. In addition, brain lesions may impact

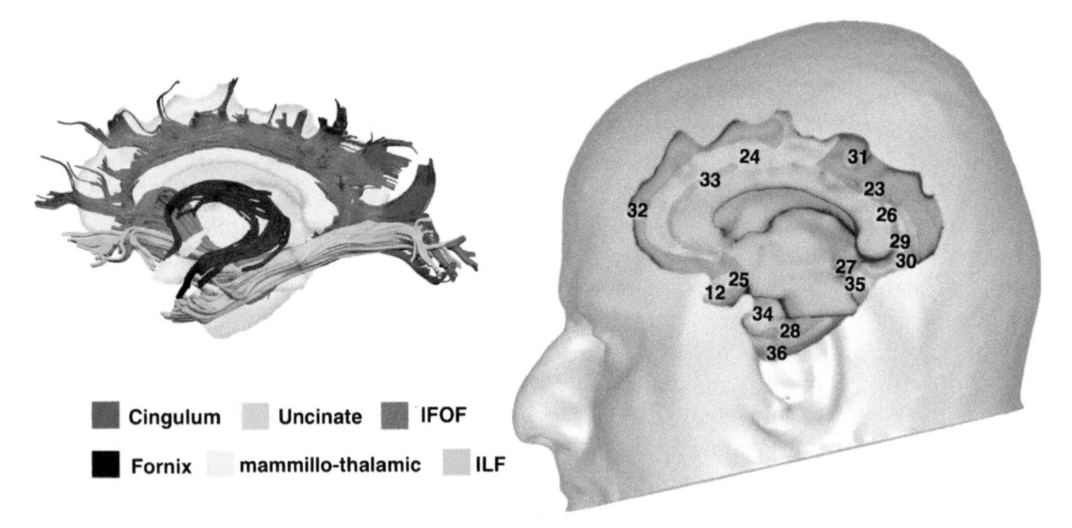

FIGURE 3 Limbic lobe and its main associative connections.[111] Brodmann areas are numbered, and major tracts are color coded, as shown. ILF is inferior longitudinal fasciculus, and IFOF is inferior fronto-occipital fasciculus.

many aspects of PSCC, including impulsivity, emotional regulation, inhibitory mechanisms, theory of mind, moral decision making, etc. In our efforts to understand complex human behaviors such as hypersexuality, it is tempting to revert to narrow localizationism to simplify the task. For example, the septum does orgasm, the hypothalamus does "drive," and the amygdala does the emotion piece. The purpose of the foregoing discussion was to introduce the absurdity of such narrow localizationism and introduce the three overlapping, integrated, and complementary brain networks, which are likely important in hypersexuality. Lesion studies can help define these networks and refine the role of each network in the integration of bottom-up, top-down, and regulatory/modulatory influences on behavior.

3. THE MANY CHALLENGES OF NPSCC WITH HYPERSEXUALITY AS AN EXAMPLE

3.1 The Lesion Method

Systematic analytic techniques for studying patients with brain lesions evolved largely in an attempt to make sense of interesting cases, like Phineas Gage.[242] Norman Geschwind facilitated the development of the lesion method and subsequent network/associationist theories of brain function by emphasizing the importance of disconnection syndromes in his landmark 1965 papers.[243,244] Later, he summarized the many factors that can produce behavioral and functional change following brain lesions,[245] including an extensive list of biochemical, structural, physiological, psychological, and delayed effects. Since these early days, the use of brain-lesion cases to improve understanding of complex human behaviors has met tremendous success, and yet there has been plenty of time to reveal the weaknesses of this approach. In this section, key aspects of the lesion method are summarized.

"The essence of **the lesion method** is the establishment of a correlation between a circumscribed region of damaged brain and changes in some aspect of an experimentally controlled behavioral performance."[196] The structure-function correlations proposed using the lesion method require the following: (1) clinical, behavioral, or experimental variables to which the lesions are related—the functions, (2) underlying structure-function theories utilized to make the correlations—hypotheses, (3) sophisticated understanding of the underlying brain tissues impacted by the lesion—structural specifications, and (4) sufficient anatomical resolution and analytic technique—structural measurements.[196] The lesion method was extended from single cases to case series and continues to provide a "unique window into brain function."[246]

Understanding a complex phenomenon such as hypersexuality requires complex conceptions of multiple, closely integrated brain networks, just as other "complex psychological functions must emerge from the cooperation of multiple components of integrated networks."[196] If there are dozens of cortical centers for visual processing, how many neuroanatomical structures might be implicated in the network dysfunctions responsible for hypersexuality? We have reviewed many brain areas implicated by the lesion literature, and surely there are more. Lesion studies by themselves are limited in their ability to define these multiple, closely integrated brain networks, but may become more informative when multiple cases with the same clinical phenomena (e.g., lesion-induced hypersexuality) are studied, when the results of ancillary data such as neuropsychological or neurophysiological measures all implicate a specific network, and when the lesion data are combined with functional neuroimaging.

In the case reported by Absher et al.[239] the subthalamic nucleus (STN) was damaged (possibly also involving the medial forebrain bundle and/or thalamic fasciculus). This small lesion produced glucose metabolism alterations in the OFC, mPFC, dlPFC, and the temporal and parietal lobes, as demonstrated on PET scanning. A wide array of cognitive alterations were noted, as well as disinhibition and excessive jocularity. The complexity of the underlying neuroanatomical, cognitive, behavioral, and functional substrates of even a small lesion in a single subject is substantial. If a clinical phenomenon such as disinhibition or mania can provoke hypersexuality, how are we to fully control the wide array of confounders in lesion studies? When we combine data from different lesion subjects in a "similar" location, how are we to mitigate the effect of various confounders that differ between subjects, some of which may influence the behavior under study?

Thus, a number of conceptual issues can limit the applicability of the lesion method. For instance, the risks of drawing false conclusions from "reverse inference" (drawing conclusions about brain functions that have not been directly tested) are well-known.[247,248] The idea that functions influenced by a single brain region may be "subtracted" from a normally functioning "integrated network" to draw conclusions about how a normal and intact network would function is perhaps the greatest weakness seen in the underlying theoretical constructs applied to the lesion method, a weakness that Rorden refers to as the "modularity assumption."[246] Due to its inherent requirement for sophisticated hypotheses, both for study design and result interpretation, the lesion method is susceptible to bias introduced by circular reasoning; prior experience with a lesion in a particular area may lead to the hypothesis that this area is important for a particular function. This process is similar to the mania/hypersexuality circularity mentioned in a prior section. Such post hoc hypotheses are almost unavoidable. Another weakness of the lesion method is that single-subject studies are rarely controlled (i.e., there is no "normal" comparison), unless the lesion is a preplanned neurosurgical intervention (thalamotomy, temporal lobectomy, subthalamic deep brain stimulation, cingulotomy, etc.). These intentional human lesions are not performed in individuals with normal brain functioning, and underlying disease introduces an unavoidable confound into the analysis of such cases. Group studies may compare lesion patients lacking a symptom of interest (the normal comparison or control group) to lesion subjects manifesting a symptom of interest. Multiple variables can be controlled using approaches such as Statistical Parametric Mapping (http://www.fil.ion.ucl.ac.uk/spm/) to map the probability that a specific structure or function correlates with the dependent variable(s) under study. The success of such mapping approaches depends largely on the a priori hypotheses being explored, knowledge of which variables to control, the underlying models or conceptions of brain function relied upon to develop the hypotheses, and numerous other factors.

The confounding effects of different clinical phenomena (e.g., mania and disinhibition may provoke hypersexuality) that may impact the behavior under study (e.g., lesion-induced hypersexuality) greatly complicate hypothesis generation in lesion studies, as well as the interpretation of results. In addition, the enormous anatomical and functional variability from person to person must be considered.[249,250] The lesion method has its place, but its role in facilitating our understanding of brain function is increasingly supplemented and supplanted by new tools that may be applied to NPSCC.

3.2 Newer Imaging Techniques

Many new approaches have been developed that will improve NPSCC by reducing the impact of neuroanatomical variability. The size, location, and pathways of cerebral cortical sulci vary from person to person, change during the course of normal development and aging, and may be irrevocably altered as a consequence of disease, trauma, and even substance abuse. The cytoarchitecture of any specific region varies between and within individuals, and the boundaries between various cytoarchitectonic zones are difficult to define.[249,250] This neuroanatomical and functional complexity has led to new neuroimaging structural/functional parcellation schemes, although universal acceptance of these new mapping techniques has not yet been achieved.

Neurotransmitter systems likewise differ among individuals.[251] Even if we are to find the perfect method of mapping both the cytoarchitecture and anatomy in neuroimaging studies, can we be sure we are correctly mapping the chemoarchitecture and brain functions dependent upon these underlying structures? The neurotransmitters, receptor types and distributions, receptor densities, and number of connections between various brain areas are among the many facets of chemoarchitecture that can influence function. Functional brain mapping tools, such as resting-state fMRI,[252] and effective connectivity models[253] are being deployed to circumvent some of the limitations, and a call for a "best practices" approach has been issued.[254] None of the available structural imaging techniques, other than perhaps microscopy/histology, enable perfect delineation of cortical cytoarchitectonic zones from one another, leading some authors to propose that the best way to accurately define cortical neuroanatomical structure is in fact to rely on functional imaging characteristics.[250] These efforts will continue to improve our ability to manage the structural and functional differences among individuals and to predict aspects of NPSCC based on these structural and functional differences.[255] The application of models of brain organization and function (e.g., multiregional retroactivation)[256] is also necessary to make sense of the data and to reveal when our understanding of neuroanatomy is deficient.[257]

Due to the complexity of the tracts connecting a given cortical area with other nearby and distant brain areas, even a small degree of error in neuroanatomical or functional localization may implicate the wrong network hubs or core functional areas.[168] Even with precise delineation of the neuroanatomy and functional zones at all levels (e.g., tracts, cortical cytoarchitecture, chemoarchitecture, and functional/effective connectivity), drawing conclusions about the function of individual brain areas in NPSCC is challenging due to our imperfect models of overall brain functioning.

Combining data from multiple subjects requires controlling multiple variables. The power of this approach has been well demonstrated, as reviewed here and elsewhere in this volume. It should also be noted that small studies (even single case studies) with fMRI, ERP, tDCS, TMS, optogenetics, DTI, and other techniques can be informative, especially when specific hypotheses are being explored. Studying subjects with lesion-induced changes in PSCC can also be informative to develop hypotheses about the structures and networks that might be explored through larger studies.

This brief overview of the lesion method and newer imaging techniques introduces some of the challenges investigators might wish to consider in NPSCC research. Regardless of its limitations, the lesion method will continue to be important and will likely continue to complement the newer neuroimaging and analytic techniques.[250,258]

4. SUMMARY

An overview of the complexities of neuroimaging of personality, social cognition, and character is provided in this introductory chapter, using hypersexuality as an example. In Section 1, many questions are posed for which there are no easy answers. Definitions for many of the terms used throughout this volume are introduced. The issue of hypersexuality is described to provide an organizing construct that facilitates the discussion of NPSCC in general. We return to this construct in this Summary Section. In Section 2, a selective review of hypersexuality is presented, covering both non-brain-lesion and lesion-induced hypersexuality. The main brain areas likely to relate to hypersexuality are summarized. Localizationist and network/associationist theories of brain organization are presented to explain current models of hypersexuality in relation to three key brain networks implicated by the research (Figures 1–3), within which specific brain areas seem to exert their most potent influences on hypersexual behavior. Finally, in Section 3, the lesion method and newer neuroimaging tools are briefly introduced. Many of the well-known challenges to conclusive NPSCC research are summarized. The goal of these sections is not to provide a definitive review of hypersexuality or the many conceptual and methodologic challenges faced in the study of hypersexuality. Instead, a major goal of these sections is to reveal stumbling blocks encountered in the study of one clinical phenomenon (hypersexuality) that might be considered in future NPSCC research.

Another major goal of this chapter has been to summarize the neuroanatomy, clinical phenomenology, and existing literature pertaining to a disorder that epitomizes NPSCC. As emphasized in the earlier sections, hypersexuality research includes findings of great relevance to personality neuroscience, social cognition, and character research. By discussing a single phenomenon, hypersexuality, this discussion has given us a scaffolding upon which to build a discussion of several ideas that are developed in the chapters to follow.

For example, in our hypothetical scenario, HS and T may experience approach-avoidance conflict. Hypersexuality is described in one popular model[250] as strictly an excess drive, which would be an exaggerated or excessive consummatory motivation/goal. The involvement of the amygdala in regulating hypothalamic drive, and Braun et al.'s[218] conception of left-right hemispheric contributions to drive both support approach-avoidance mechanisms as an important feature of hypersexuality.

Studies using path analysis[187,188] have revealed relationships between hypersexuality and the Big Five dimension, neuroticism, and one of its facets, impulsivity. Mood disorders relate to hypersexuality via brain damage leading to mania, and non-brain-lesion hypersexuality is associated with anxiety, depression, and negative affect. The ability to regulate sexual urges and drives is thought to be a key feature of hypersexuality, and the lesion-induced hypersexuality studies implicate key structures for emotion regulation (e.g., vmPFC).

The preceding discussion supports the premise that the study of hypersexuality can be viewed as a microcosm of NPSCC research. Personality, social cognition, and character issues are vitally important to human sexual behavior, and alterations in all three of these domains are seen in the hypersexuality literature. NPSCC research is applicable to the study of normal brain function, and its application to the study of complex human disorders such as hypersexuality can also be informative.

References

1. Cacioppo JR, Berntson GG, Sheridan JF, McClintock MK. Multilevel integrative analysis of human behavior: social neuroscience and the complementing nature of social and biological approaches. *Psychol Bull.* 2000;126(6):829–843.
2. The National Library of Medicines Visible Human Project. Available at: http://www.nlm.nih.gov/research/visible/visible_human.html; Accessed 06.12.14.
3. Tadrous PJ. Subcellular microanatomy by 3D deconvolution brightfield microscopy: method and analysis using human chromatin in the interphase nucleus. *Anat Res Int.* 2012;2012:848707. http://dx.doi.org/10.1155/2012/848707.
4. Yoo TS, Bliss D, Lowekamp BC, Chen DT, Murphy GE, Narayan K, Hartnell LM, Do T, Subramaniam S. Visualizing cells and humans in 3D: biomedical image analysis at nanometer and meter scales. *IEEE Comput Graph Appl.* 2012:39–49.
5. Hyatt CJ, Haney-Caron E, Stevens MC. Cortical thickness and folding deficits in conduct-disordered adolescents. *Biol Psychiatry.* 2012;72(3):207–214. http://dx.doi.org/10.1016/j.biopsych.2011.11.017.
6. Ameis SH, Ducharme S, Albaugh MD, et al. Cortical thickness, cortico-amygdalar networks, and externalizing behaviors in healthy children. *Biol Psychiatry.* 2014;75(1):65–72. http://dx.doi.org/10.1016/j.biopsych.2013.06.008.
7. Schilling C, Kühn S, Romanowski A, Schubert F, Kathmann N, Gallinat J. Cortical thickness correlates with impulsiveness in healthy adults. *Neuroimage.* 2012;59(1):824–830. http://dx.doi.org/10.1016/j.neuroimage.2011.07.058.
8. Frick A, Howner K, Fischer H, Eskildsen SF, Kristiansson M, Furmark T. Cortical thickness alterations in social anxiety disorder. *Neurosci Lett.* 2013;536:52–55. http://dx.doi.org/10.1016/j.neulet.2012.12.060.
9. Bjørnebekk A, Fjell AM, Walhovd KB, Grydeland H, Torgersen S, Westlye LT. Neuronal correlates of the five factor model (FFM) of human personality: multimodal imaging in a large healthy sample. *Neuroimage.* 2013;65:194–208. http://dx.doi.org/10.1016/j.neuroimage.2012.10.009.
10. Lewis GJ, Panizzon MS, Eyler L, et al. Heritable influences on amygdala and orbitofrontal cortex contribute to genetic variation in core dimensions of personality. *Neuroimage.* 2014;103C:309–315. http://dx.doi.org/10.1016/j.neuroimage.2014.09.043.

11. Howner K, Eskildsen SF, Fischer H, et al. Thinner cortex in the frontal lobes in mentally disordered offenders. *Psychiatry Res*. 2012;203 (2–3):126–131. http://dx.doi.org/10.1016/j.pscychresns.2011.12.011.
12. Hayashi A, Abe N, Fujii T, et al. Dissociable neural systems for moral judgment of anti- and pro-social lying. *Brain Res*. 2014;1556:46–56. http://dx.doi.org/10.1016/j.brainres.2014.02.011.
13. Forsman LJ, de Manzano O, Karabanov A, Madison G, Ullén F. Differences in regional brain volume related to the extraversion-introversion dimension–a voxel based morphometry study. *Neurosci Res*. 2012;72(1):59–67. http://dx.doi.org/10.1016/j.neures.2011.10.001.
14. Poletti M, Enrici I, Adenzato M. Cognitive and affective theory of mind in neurodegenerative diseases: neuropsychological, neuroanatomical and neurochemical levels. *Neurosci Biobehav Rev*. 2012;36(9):2147–2164. http://dx.doi.org/10.1016/j.neubiorev.2012.07.004.
15. Hecht D. Depression and the hyperactive right-hemisphere. *Neurosci Res*. 2010;68(2):77–87. http://dx.doi.org/10.1016/j.neures.2010.06.013.
16. Savitz JB, Price JL, Drevets WC. Neuropathological and neuromorphometric abnormalities in bipolar disorder: view from the medial prefrontal cortical network. *Neurosci Biobehav Rev*. 2014;42C:132–147. http://dx.doi.org/10.1016/j.neubiorev.2014.02.008.
17. Fabian JM. Neuropsychological and neurological correlates in violent and homicidal offenders: a legal and neuroscience perspective. *Aggress Violent Behav*. 2010;15(3):209–223. http://dx.doi.org/10.1016/j.avb.2009.12.004.
18. Anderson NE, Kiehl KA. The psychopath magnetized: insights from brain imaging. *Trends Cogn Sci*. 2012;16(1). http://dx.doi.org/10.1016/j.tics.2011.11.008.
19. Babiloni C, Buffo P, Vecchio F, et al. Brains "in concert": frontal oscillatory alpha rhythms and empathy in professional musicians. *Neuroimage*. 2012;60(1):105–116. http://dx.doi.org/10.1016/j.neuroimage.2011.12.008.
20. Menon V. Large-scale brain networks and psychopathology: a unifying triple network model. *Trends Cogn Sci*. 2011;15(10):483–506. http://dx.doi.org/10.1016/j.tics.2011.08.003.
21. Müller JL, Gänssbauer S, Sommer M, et al. Gray matter changes in right superior temporal gyrus in criminal psychopaths. Evidence from voxel-based morphometry. *Psychiatry Res*. 2008;163(3):213–222. http://dx.doi.org/10.1016/j.pscychresns.2007.08.010.
22. Dennis M, Simic N, Bigler ED, et al. Cognitive, affective, and conative theory of mind (ToM) in children with traumatic brain injury. *Dev Cogn Neurosci*. 2013;5:25–39. http://dx.doi.org/10.1016/j.dcn.2012.11.006.
23. Hyde LW, Shaw DS, Hariri AR. Understanding youth antisocial behavior using neuroscience through a developmental psychopathology lens: review, integration, and directions for research. *Dev Rev*. 2013;33(3):168–223. http://dx.doi.org/10.1016/j.dr.2013.06.001.
24. Hu X, Erb M, Ackermann H, Martin JA, Grodd W, Reiterer SM. Voxel-based morphometry studies of personality: issue of statistical model specification–effect of nuisance covariates. *Neuroimage*. 2011;54(3):1994–2005. http://dx.doi.org/10.1016/j.neuroimage.2010.10.024.
25. Waldman DA, Balthazard PA, Peterson SJ. Social cognitive neuroscience and leadership. *Leadersh Q*. 2011;22(6):1092–1106. http://dx.doi.org/10.1016/j.leaqua.2011.09.005.
26. Simkin DR, Thatcher RW, Lubar J. Quantitative EEG and neurofeedback in children and adolescents: anxiety disorders, depressive disorders, comorbid addiction and attention-deficit/hyperactivity disorder, and brain injury. *Child Adolesc Psychiatr Clin N Am*. 2014;23(3):427–464. http://dx.doi.org/10.1016/j.chc.2014.03.001.
27. Frodl T, Skokauskas N. Neuroimaging of externalizing behaviors and borderline traits. *Biol Psychiatry*. 2014;75(1):7–8. http://dx.doi.org/10.1016/j.biopsych.2013.10.015.
28. Hecht D. An inter-hemispheric imbalance in the psychopath's brain. *Pers Individ Dif*. 2011;51(1):3–10. http://dx.doi.org/10.1016/j.paid.2011.02.032.
29. Wallace GL, White SF, Robustelli B, et al. Cortical and subcortical abnormalities in youths with conduct disorder and elevated callous-unemotional traits. *J Am Acad Child Adolesc Psychiatry*. 2014;53(4):456–465. http://dx.doi.org/10.1016/j.jaac.2013.12.008.
30. Elliott R, Deakin B. Role of the orbitofrontal cortex in reinforcement processing and inhibitory control: evidence from functional magnetic resonance imaging studies in healthy human subjects. *Int Rev Neurobiol*. 2005;65(04):89–116. http://dx.doi.org/10.1016/S0074-7742(04)65004-5.
31. Nagai M, Kishi K, Kato S. Insular cortex and neuropsychiatric disorders: a review of recent literature. *Eur Psychiatry*. 2007;22(6):387–394. http://dx.doi.org/10.1016/j.eurpsy.2007.02.006.
32. Yamasue H, Abe O, Suga M, et al. Gender-common and -specific neuroanatomical basis of human anxiety-related personality traits. *Cereb Cortex*. 2008;18(1):46–52. http://dx.doi.org/10.1093/cercor/bhm030.
33. Coenen VA, Panksepp J, Hurwitz TA, Urbach H, Mädler B. Human medial forebrain bundle (MFB) and anterior thalamic radiation (ATR): imaging of two major subcortical pathways and the dynamic balance of opposite affects in understanding depression. *J Neuropsychiatry Clin Neurosci*. 2012:223–236.
34. Goerlich-Dobre KS, Bruce L, Martens S, Aleman A, Hooker CI. Distinct associations of insula and cingulate volume with the cognitive and affective dimensions of alexithymia. *Neuropsychologia*. 2014;53:284–292. http://dx.doi.org/10.1016/j.neuropsychologia.2013.12.006.
35. Lu F, Huo Y, Li M, et al. Relationship between personality and gray matter volume in healthy young adults: a voxel-based morphometric study. *PLoS One*. 2014;9(2):1–8. http://dx.doi.org/10.1371/journal.pone.0088763.
36. Pardini DA, Raine A, Erickson K, Loeber R. Lower amygdala volume in men is associated with childhood aggression, early psychopathic traits, and future violence. *Biol Psychiatry*. 2014;75(1):73–80. http://dx.doi.org/10.1016/j.biopsych.2013.04.003.
37. Andari E, Schneider FC, Mottolese R, Vindras P, Sirigu A. Oxytocin's fingerprint in personality traits and regional brain volume. *Cereb Cortex*. February 2014;24:479–486. http://dx.doi.org/10.1093/cercor/bhs328.
38. Van Schuerbeek P, Baeken C, De Raedt R, De Mey J, Luypaert R. Individual differences in local gray and white matter volumes reflect differences in temperament and character: a voxel-based morphometry study in healthy young females. *Brain Res*. 2011;1371:32–42. http://dx.doi.org/10.1016/j.brainres.2010.11.073.
39. Montag C, Schoene-Bake J-C, Wagner J, et al. Volumetric hemispheric ratio as a useful tool in personality psychology. *Neurosci Res*. 2013;75(2):157–159. http://dx.doi.org/10.1016/j.neures.2012.11.004.
40. Lewis PA, Rezaie R, Brown R, Roberts N, Dunbar RIM. Ventromedial prefrontal volume predicts understanding of others and social network size. *Neuroimage*. 2011;57(4):1624–1629. http://dx.doi.org/10.1016/j.neuroimage.2011.05.030.
41. Powell JL, Lewis PA, Dunbar RIM, García-Fiñana M, Roberts N. Orbital prefrontal cortex volume correlates with social cognitive competence. *Neuropsychologia*. 2010;48(12):3554–3562. http://dx.doi.org/10.1016/j.neuropsychologia.2010.08.004.
42. Zhang X, Yao S, Zhu X, Wang X, Zhu X, Zhong M. Gray matter volume abnormalities in individuals with cognitive vulnerability to depression: a voxel-based morphometry study. *J Affect Disord*. 2012;136(3):443–452. http://dx.doi.org/10.1016/j.jad.2011.11.005.

43. Matsui MM, Yoneyama E, Sumiyoshi T, et al. Lack of self-control as assessed by a personality inventory is related to reduced volume of supplementary motor area. *Psychiatry Res Neuroimaging*. 2002; 116(1–2):53–61. http://dx.doi.org/10.1016/S0925-4927(02)00070-7.
44. Alvarenga PG, do Rosário MC, Batistuzzo MC, et al. Obsessive-compulsive symptom dimensions correlate to specific gray matter volumes in treatment-naïve patients. *J Psychiatr Res*. 2012;46(12): 1635–1642. http://dx.doi.org/10.1016/j.jpsychires.2012.09.002.
45. Kaasinen V, Maguire RP, Kurki T, Brück A, Rinne JO. Mapping brain structure and personality in late adulthood. *Neuroimage*. 2005;24(2): 315–322. http://dx.doi.org/10.1016/j.neuroimage.2004.09.017.
46. Schumann CM, Barnes CC, Lord C, Courchesne E. Amygdala enlargement in toddlers with autism related to severity of social and communication impairments. *Biol Psychiatry*. 2009;66(10): 942–949. http://dx.doi.org/10.1016/j.biopsych.2009.07.007.
47. Li Y, Qiao L, Sun J, et al. Gender-specific neuroanatomical basis of behavioral inhibition/approach systems (BIS/BAS) in a large sample of young adults: a voxel-based morphometric investigation. *Behav Brain Res*. 2014;274:400–408. http://dx.doi.org/10.1016/j.bbr.2014.08.041.
48. Structures RD. Brain pathology in pedophilic offenders. *Arch Gen Psychiatry*. June 2007;64:737–746.
49. Gardini S, Cloninger CR, Venneri A. Individual differences in personality traits reflect structural variance in specific brain regions. *Brain Res Bull*. 2009;79(5):265–270. http://dx.doi.org/10.1016/j.brainresbull.2009.03.005.
50. Shefer G, Marcus Y, Stern N. Is obesity a brain disease? *Neurosci Biobehav Rev*. 2013;37(10 Pt 2):2489–2503. http://dx.doi.org/10.1016/j.neubiorev.2013.07.015.
51. Evans DW, Lazar SM, Boomer KB, Mitchel AD, Michael AM, Moore GJ. Social cognition and brain morphology: implications for developmental brain dysfunction. *Brain Imaging Behav*. 2014. http://dx.doi.org/10.1007/s11682-014-9304-1.
52. de Oliveira-Souza R, Hare RD, Bramati IE, et al. Psychopathy as a disorder of the moral brain: fronto-temporo-limbic grey matter reductions demonstrated by voxel-based morphometry. *Neuroimage*. 2008;40(3):1202–1213. http://dx.doi.org/10.1016/j.neuroimage.2007.12.054.
53. Gopal A, Clark E, Allgair A, et al. Dorsal/ventral parcellation of the amygdala: relevance to impulsivity and aggression. *Psychiatry Res*. 2013;211(1):24–30. http://dx.doi.org/10.1016/j.pscychresns.2012.10.010.
54. Yang J, Wei D, Wang K, Qiu J. Gray matter correlates of dispositional optimism: a voxel-based morphometry study. *Neurosci Lett*. 2013;553:201–205. http://dx.doi.org/10.1016/j.neulet.2013.08.032.
55. Cantor JM, Kabani N, Christensen BK, et al. Cerebral white matter deficiencies in pedophilic men. *J Psychiatr Res*. 2008;42(3):167–183. http://dx.doi.org/10.1016/j.jpsychires.2007.10.013.
56. Koelsch S, Skouras S, Jentschke S. Neural correlates of emotional Personality : a structural and functional magnetic resonance imaging study. *PLoS One*. 2013;8(11):1–17. http://dx.doi.org/10.1371/journal.pone.0077196.
57. Schiffer B, Peschel T, Paul T, et al. Structural brain abnormalities in the frontostriatal system and cerebellum in pedophilia. *J Psychiatr Res*. 2007;41(9):753–762. http://dx.doi.org/10.1016/j.jpsychires.2006.06.003.
58. Boccardi M, Bocchetta M, Aronen HJ, et al. Atypical nucleus accumbens morphology in psychopathy: another limbic piece in the puzzle. *Int J Law Psychiatry*. 2013;36(2):157–167. http://dx.doi.org/10.1016/j.ijlp.2013.01.008.
59. Dickey CC, Vu M-AT, Voglmaier MM, Niznikiewicz MA, McCarley RW, Panych LP. Prosodic abnormalities in schizotypal personality disorder. *Schizophr Res*. 2012;142(1–3):20–30. http://dx.doi.org/10.1016/j.schres.2012.09.006.
60. Scharmüller W, Schienle A. Voxel-based morphometry of disgust proneness. *Neurosci Lett*. 2012;529(2):172–174. http://dx.doi.org/10.1016/j.neulet.2012.09.004.
61. Newsome MR, Scheibel RS, Chu Z, et al. The relationship of resting cerebral blood flow and brain activation during a social cognition task in adolescents with chronic moderate to severe traumatic brain injury: a preliminary investigation. *Int J Dev Neurosci*. 2012;30(3):255–266. http://dx.doi.org/10.1016/j.ijdevneu.2011.10.008.
62. Li H, Li W, Wei D, et al. Examining brain structures associated with perceived stress in a large sample of young adults via voxel-based morphometry. *Neuroimage*. 2014;92:1–7. http://dx.doi.org/10.1016/j.neuroimage.2014.01.044.
63. Banissy MJ, Kanai R, Walsh V, Rees G. Inter-individual differences in empathy are reflected in human brain structure. *Neuroimage*. 2012;62(3):2034–2039. http://dx.doi.org/10.1016/j.neuroimage.2012.05.081.
64. Baumgartner T, Schiller B, Hill C, Knoch D. Impartiality in humans is predicted by brain structure of dorsomedial prefrontal cortex. *Neuroimage*. 2013;81:317–324. http://dx.doi.org/10.1016/j.neuroimage.2013.05.047.
65. Poeppl TB, Nitschke J, Santtila P, et al. Association between brain structure and phenotypic characteristics in pedophilia. *J Psychiatr Res*. 2013;47(5):678–685. http://dx.doi.org/10.1016/j.jpsychires.2013.01.003.
66. Lackner HK, Weiss EM, Schulter G, Hinghofer-Szalkay H, Samson AC, Papousek I. I got it! Transient cardiovascular response to the perception of humor. *Biol Psychol*. 2013;93(1):33–40. http://dx.doi.org/10.1016/j.biopsycho.2013.01.014.
67. Mohnke S, Müller S, Amelung T, et al. Brain alterations in paedophilia: a critical review. *Prog Neurobiol*. 2014. http://dx.doi.org/10.1016/j.pneurobio.2014.07.005.
68. Kapogiannis D, Sutin A, Davatzikos C, Costa P, Resnick S. The five factors of personality and regional cortical variability in the Baltimore longitudinal study of aging. *Hum Brain Mapp*. 2013;34(11):2829–2840. http://dx.doi.org/10.1002/hbm.22108.
69. Gansler DA, Lee AK, Emerton BC, et al. Prefrontal regional correlates of self-control in male psychiatric patients: impulsivity facets and aggression. *Psychiatry Res*. 2011;191(1):16–23. http://dx.doi.org/10.1016/j.pscychresns.2010.09.003.
70. Barrett LF, Satpute AB. Large-scale brain networks in affective and social neuroscience: towards an integrative functional architecture of the brain. *Curr Opin Neurobiol*. 2013;23(3):361–372. http://dx.doi.org/10.1016/j.conb.2012.12.012.
71. Barrett LF, Bliss-moreau E, Duncan SL, Rauch SL, Wright CI. The amygdala and the experience of affect. *Soc Cogn Affect Neurosci*. 2007:73–83. http://dx.doi.org/10.1093/scan/nsl042 (January 2001).
72. Becker B, Mihov Y, Scheele D, et al. Fear processing and social networking in the absence of a functional amygdala. *Biol Psychiatry*. 2012;72(1):70–77. http://dx.doi.org/10.1016/j.biopsych.2011.11.024.
73. Shany-Ur T, Poorzand P, Grossman SN, et al. Comprehension of insincere communication in neurodegenerative disease: lies, sarcasm, and theory of mind. *Cortex*. 2011;48(10):1329–1341. http://dx.doi.org/10.1016/j.cortex.2011.08.003.
74. Starcke K, Brand M. Decision making under stress: a selective review. *Neurosci Biobehav Rev*. 2012;36(4):1228–1248. http://dx.doi.org/10.1016/j.neubiorev.2012.02.003.
75. Lauterbach EC, Cummings JL, Kuppuswamy PS. Toward a more precise, clinically–informed pathophysiology of pathological laughing and crying. *Neurosci Biobehav Rev*. 2013;37(8):1893–1916. http://dx.doi.org/10.1016/j.neubiorev.2013.03.002.
76. Spampinato MV, Wood JN, De Simone V, Grafman J. Neural correlates of anxiety in healthy volunteers : a voxel- based morphometry study. *J Neuropsychiatry Clin Neurosci*. 2009:199–205.

77. Xu J, Potenza MN. White matter integrity and five-factor personality measures in healthy adults. *Neuroimage*. 2012;59(1):800–807. http://dx.doi.org/10.1016/j.neuroimage.2011.07.040.
78. Van Overwalle F, Baetens K, Mariën P, Vandekerckhove M. Social cognition and the cerebellum: a meta-analysis of over 350 fMRI studies. *Neuroimage*. 2013;86:554–572. http://dx.doi.org/10.1016/j.neuroimage.2013.09.033.
79. Domschke K, Dannlowski U. Imaging genetics of anxiety disorders. *Neuroimage*. 2010;53(3):822–831. http://dx.doi.org/10.1016/j.neuroimage.2009.11.042.
80. Willis ML, Palermo R, Burke D, McGrillen K, Miller L. Orbitofrontal cortex lesions result in abnormal social judgements to emotional faces. *Neuropsychologia*. 2010;48(7):2182–2187. http://dx.doi.org/10.1016/j.neuropsychologia.2010.04.010.
81. Baumgartner T, Gianotti LRR, Knoch D. Who is honest and why: baseline activation in anterior insula predicts inter-individual differences in deceptive behavior. *Biol Psychol*. 2013;94(1):192–197. http://dx.doi.org/10.1016/j.biopsycho.2013.05.018.
82. Dunn BD, Dalgleish T, Lawrence AD. The somatic marker hypothesis: a critical evaluation. *Neurosci Biobehav Rev*. 2006;30(2):239–271. http://dx.doi.org/10.1016/j.neubiorev.2005.07.001.
83. Rabin JS, Braverman A, Gilboa A, Stuss DT, Rosenbaum RS. Theory of mind development can withstand compromised episodic memory development. *Neuropsychologia*. 2012;50(14):3781–3785. http://dx.doi.org/10.1016/j.neuropsychologia.2012.10.016.
84. Péron J, El Tamer S, Grandjean D, et al. Major depressive disorder skews the recognition of emotional prosody. *Prog Neuropsychopharmacol Biol Psychiatry*. 2011;35(4):987–996. http://dx.doi.org/10.1016/j.pnpbp.2011.01.019.
85. Feinstein A, DeLuca J, Baune BT, Filippi M, Lassman H. Cognitive and neuropsychiatric disease manifestations in MS. *Mult Scler Relat Disord*. 2013;2(1):4–12. http://dx.doi.org/10.1016/j.msard.2012.08.001.
86. Lee H, Heller AS, van Reekum CM, Nelson B, Davidson RJ. Amygdala-prefrontal coupling underlies individual differences in emotion regulation. *Neuroimage*. 2012;62(3):1575–1581. http://dx.doi.org/10.1016/j.neuroimage.2012.05.044.
87. Seitz RJ, Franz M, Azari NP. Value judgments and self-control of action: the role of the medial frontal cortex. *Brain Res Rev*. 2009;60(2):368–378. http://dx.doi.org/10.1016/j.brainresrev.2009.02.003.
88. Thoma P, Friedmann C, Suchan B. Empathy and social problem solving in alcohol dependence, mood disorders and selected personality disorders. *Neurosci Biobehav Rev*. 2013;37(3):448–470. http://dx.doi.org/10.1016/j.neubiorev.2013.01.024.
89. Miller LE, Saygin AP. Individual differences in the perception of biological motion: links to social cognition and motor imagery. *Cognition*. 2013;128(2):140–148. http://dx.doi.org/10.1016/j.cognition.2013.03.013.
90. Morris PL, Robinson RG, de Carvalho ML, et al. Lesion characteristics and depressed mood in the stroke data bank study. *J Neuropsychiatry Clin Neurosci*. Spring 1996;8(2):153–159.
91. Montag C, Reuter M. Disentangling the molecular genetic basis of personality: from monoamines to neuropeptides. *Neurosci Biobehav Rev*. 2014;43:228–239. http://dx.doi.org/10.1016/j.neubiorev.2014.04.006.
92. Brüne M, Brüne-Cohrs U. Theory of mind–evolution, ontogeny, brain mechanisms and psychopathology. *Neurosci Biobehav Rev*. 2006;30(4):437–455. http://dx.doi.org/10.1016/j.neubiorev.2005.08.001.
93. Coutinho JF, Sampaio A, Ferreira M, Soares JM, Gonçalves OF. Brain correlates of pro-social personality traits: a voxel-based morphometry study. *Brain Imaging Behav*. 2013;7(3):293–299. http://dx.doi.org/10.1007/s11682-013-9227-2.
94. Koelsch S. Towards a neural basis of music-evoked emotions. *Trends Cogn Sci*. 2010;14(3):131–137. http://dx.doi.org/10.1016/j.tics.2010.01.002.
95. Barbey AK, Colom R, Paul EJ, Chau A, Solomon J, Grafman JH. Lesion mapping of social problem solving. *Brain*. 2014. http://dx.doi.org/10.1093/brain/awu207.
96. Schiffer B, Paul T, Gizewski E, et al. Functional brain correlates of heterosexual paedophilia. *Neuroimage*. 2008;41(1):80–91. http://dx.doi.org/10.1016/j.neuroimage.2008.02.008.
97. Marsh AA, Finger EC, Fowler KA, et al. Reduced amygdala-orbitofrontal connectivity during moral judgments in youths with disruptive behavior disorders and psychopathic traits. *Psychiatry Res*. 2011;194(3):279–286. http://dx.doi.org/10.1016/j.pscychresns.2011.07.008.
98. Spinella M. The role of prefrontal systems in sexual behavior. *Int J Neurosci*. 2007;117(3):369–385. http://dx.doi.org/10.1080/00207450600588980.
99. Fabian JM. Neuropsychology, neuroscience, volitional impairment and sexually violent predators: a review of the literature and the law and their application to civil commitment proceedings. *Aggress Violent Behav*. 2012;17(1):1–15. http://dx.doi.org/10.1016/j.avb.2011.07.002.
100. Sundram F, Deeley Q, Sarkar S, et al. White matter microstructural abnormalities in the frontal lobe of adults with antisocial personality disorder. *Cortex*. 2012;48(2):216–229. http://dx.doi.org/10.1016/j.cortex.2011.06.005.
101. Declerck CH, Boone C, Emonds G. When do people cooperate? The neuroeconomics of prosocial decision making. *Brain Cogn*. 2013;81(1):95–117. http://dx.doi.org/10.1016/j.bandc.2012.09.009.
102. Savic I, Garcia-Falgueras A, Swaab DF. Sexual differentiation of the human brain in relation to gender identity and sexual orientation. *Prog Brain Res*. 2010:41–62. http://dx.doi.org/10.1016/B978-0-444-53630-3.00004-X. Elsevier B.V.
103. Bergvall ÅH, Nilsson T, Hansen S. Exploring the link between character, personality disorder, and neuropsychological function. *Eur Psychiatry*. 2003;18(7):334–344. http://dx.doi.org/10.1016/j.eurpsy.2003.03.008.
104. Kelly C, Toro R, Di Martino A, et al. A convergent functional architecture of the insula emerges across imaging modalities. *Neuroimage*. 2012;61(4):1129–1142. http://dx.doi.org/10.1016/j.neuroimage.2012.03.021.
105. Chacko RC, Corbin MA, Harper RG. Acquired obsessive- compulsive disorder associated with basal ganglia lesions. *J Neuropsychiatry Clin Neurosci*. Spring 2000;12(2):269–272.
106. Müller JL. Are sadomasochism and hypersexuality in autism linked to amygdalohippocampal lesion? *J Sex Med*. 2011;8(11):3241–3249. http://dx.doi.org/10.1111/j.1743-6109.2009.01485.x.
107. Figee M, Wielaard I, Mazaheri A, Denys D. Neurosurgical targets for compulsivity: what can we learn from acquired brain lesions? *Neurosci Biobehav Rev*. 2013;37(3):328–339. http://dx.doi.org/10.1016/j.neubiorev.2013.01.005.
108. Schaefer M, Heinze H-J, Rotte M. Touch and personality: extraversion predicts somatosensory brain response. *Neuroimage*. 2012;62(1):432–438. http://dx.doi.org/10.1016/j.neuroimage.2012.05.004.
109. Tost H, Bilek E, Meyer-Lindenberg A. Brain connectivity in psychiatric imaging genetics. *Neuroimage*. 2012;62(4):2250–2260. http://dx.doi.org/10.1016/j.neuroimage.2011.11.007.
110. Järvinen-Pasley A, Vines BW, Hill KJ, et al. Cross-modal influences of affect across social and non-social domains in individuals with Williams syndrome. *Neuropsychologia*. 2010;48(2):456–466. http://dx.doi.org/10.1016/j.neuropsychologia.2009.10.003.
111. Catani M, Dell'acqua F, Bizzi A, et al. Beyond cortical localization in clinico-anatomical correlation. *Cortex*. 2012;48(10):1262–1287. http://dx.doi.org/10.1016/j.cortex.2012.07.001.
112. Penney S. Impulse control and criminal responsibility: lessons from neuroscience. *Int J Law Psychiatry*. 2012;35(2):99–103. http://dx.doi.org/10.1016/j.ijlp.2011.12.004.
113. Friedlander L, Desrocher M. Neuroimaging studies of obsessive-compulsive disorder in adults and children. *Clin Psychol Rev*. 2006;26(1):32–49. http://dx.doi.org/10.1016/j.cpr.2005.06.010.

114. Devinsky J, Sacks O, Devinsky O. Kluver-Bucy syndrome, hypersexuality, and the law. *Neurocase*. 2010;16(2):140–145. http://dx.doi.org/10.1080/13554790903329182.
115. Swaab DF. Sexual differentiation of the brain and behavior. *Best Pract Res Clin Endocrinol Metab*. 2007;21(3):431–444. http://dx.doi.org/10.1016/j.beem.2007.04.003.
116. Boccardi M, Frisoni GB, Hare RD, et al. Cortex and amygdala morphology in psychopathy. *Psychiatry Res*. 2011;193(2):85–92. http://dx.doi.org/10.1016/j.pscychresns.2010.12.013.
117. Grèzes J, Dezecache G. How do shared-representations and emotional processes cooperate in response to social threat signals? *Neuropsychologia*. 2013:1–10. http://dx.doi.org/10.1016/j.neuropsychologia.2013.09.019.
118. Beauregard M. Mind does really matter: evidence from neuroimaging studies of emotional self-regulation, psychotherapy, and placebo effect. *Prog Neurobiol*. 2007;81(4):218–236. http://dx.doi.org/10.1016/j.pneurobio.2007.01.005.
119. De Pascalis V, Cozzuto G, Caprara GV, Alessandri G. Relations among EEG-alpha asymmetry, BIS/BAS, and dispositional optimism. *Biol Psychol*. 2013;94(1):198–209. http://dx.doi.org/10.1016/j.biopsycho.2013.05.016.
120. Adenzato M, Cavallo M, Enrici I. Theory of mind ability in the behavioural variant of frontotemporal dementia: an analysis of the neural, cognitive, and social levels. *Neuropsychologia*. 2010;48(1):2–12. http://dx.doi.org/10.1016/j.neuropsychologia.2009.08.001.
121. Markett S, Reuter M. Neuroeconomics: individual differences in behavioral decision making. *Pers Individ Dif*. 2014;60:S10. http://dx.doi.org/10.1016/j.paid.2013.07.181.
122. Platek SM, Keenan JP, Gallup GG, Mohamed FB. Where am I? The neurological correlates of self and other. *Brain Res Cogn Brain Res*. 2004;19(2):114–122. http://dx.doi.org/10.1016/j.cogbrainres.2003.11.014.
123. Georgiadis JR, Kringelbach ML. The human sexual response cycle: brain imaging evidence linking sex to other pleasures. *Prog Neurobiol*. 2012;98(1):49–81. http://dx.doi.org/10.1016/j.pneurobio.2012.05.004.
124. Waterhouse L. The social brain is a super-network Chapter 3, In: Rethinking Autism: *Variation and Complexity*. Elsevier: Amsterdam, 2013:97–155. http://dx.doi.org/10.1016/B978-0-12-415961-7.00003-4.
125. Gonzalez-Liencres C, Shamay-Tsoory SG, Brüne M. Towards a neuroscience of empathy: ontogeny, phylogeny, brain mechanisms, context and psychopathology. *Neurosci Biobehav Rev*. 2013;37(8):1537–1548. http://dx.doi.org/10.1016/j.neubiorev.2013.05.001.
126. Mendez M, Chow T, Ringman J, Twitchell G, Hinkin CH. Pedophilia and temporal lobe disturbances. *J Neuropsychiatry Clin Neurosci*. 2000;12(1):71–76.
127. Gasquoine PG. Localization of function in anterior cingulate cortex: from psychosurgery to functional neuroimaging. *Neurosci Biobehav Rev*. 2013;37(3):340–348. http://dx.doi.org/10.1016/j.neubiorev.2013.01.002.
128. Bao A-M, Swaab DF. Sex differences in the brain, behavior, and neuropsychiatric disorders. *Neuroscientist*. 2010;16(5):550–565. http://dx.doi.org/10.1177/1073858410377005.
129. Kirsch HE. Social cognition and epilepsy surgery. *Epilepsy Behav*. 2006;8(1):71–80. http://dx.doi.org/10.1016/j.yebeh.2005.09.002.
130. Dal Monte O, Schintu S, Pardini M, et al. The left inferior frontal gyrus is crucial for reading the mind in the eyes: brain lesion evidence. *Cortex*. 2014;58C:9–17. http://dx.doi.org/10.1016/j.cortex.2014.05.002.
131. Neumann D, Zupan B, Babbage DR, et al. Affect recognition, empathy, and dysosmia after traumatic brain injury. *Arch Phys Med Rehabil*. 2012;93(8):1414–1420. http://dx.doi.org/10.1016/j.apmr.2012.03.009.
132. Robbins TW, Gillan CM, Smith DG, de Wit S, Ersche KD. Neurocognitive endophenotypes of impulsivity and compulsivity: towards dimensional psychiatry. *Trends Cogn Sci*. 2012;16(1):81–91. http://dx.doi.org/10.1016/j.tics.2011.11.009.
133. Volkow ND, Baler RD. Addiction science: uncovering neurobiological complexity. *Neuropharmacology*. 2014;76(Pt B):235–249. http://dx.doi.org/10.1016/j.neuropharm.2013.05.007.
134. Motofei IG. A dual physiological character for sexual function: libido and sexual pheromones. *BJU Int*. 2009;104(11):1702–1708. http://dx.doi.org/10.1111/j.1464-410X.2009.08610.x.
135. Kim Y-T, Kwon D-H, Chang Y. Impairments of facial emotion recognition and theory of mind in methamphetamine abusers. *Psychiatry Res*. 2011;186(1):80–84. http://dx.doi.org/10.1016/j.psychres.2010.06.027.
136. Vaske J, Galyean K, Cullen FT. Toward a biosocial theory of offender rehabiltiation: why does cognitive-behavioral therapy work? *J Crim Justice*. 2011;39(1):90–102. http://dx.doi.org/10.1016/j.jcrimjus.2010.12.006.
137. Pawliczek CM, Derntl B, Kellermann T, Kohn N, Gur RC, Habel U. Inhibitory control and trait aggression: neural and behavioral insights using the emotional stop signal task. *Neuroimage*. 2013;79:264–274. http://dx.doi.org/10.1016/j.neuroimage.2013.04.104.
138. Suffren S, Braun CMJ, Guimond A, Devinsky O. Opposed hemispheric specializations for human hypersexuality and orgasm? *Epilepsy Behav*. 2011;21(1):12–19. http://dx.doi.org/10.1016/j.yebeh.2011.01.023.
139. Sexton CE, Mackay CE, Ebmeier KP. A systematic review of diffusion tensor imaging studies in affective disorders. *Biol Psychiatry*. 2009;66(9):814–823. http://dx.doi.org/10.1016/j.biopsych.2009.05.024.
140. Mitchell IJ, Beech AR. Towards a neurobiological model of offending. *Clin Psychol Rev*. 2011;31(5):872–882. http://dx.doi.org/10.1016/j.cpr.2011.04.001.
141. Willner P, Scheel-Krüger J, Belzung C. The neurobiology of depression and antidepressant action. *Neurosci Biobehav Rev*. 2013;37(10 Pt 1):2331–2371. http://dx.doi.org/10.1016/j.neubiorev.2012.12.007.
142. Limbrick-Oldfield EH, van Holst RJ, Clark L. Fronto-striatal dysregulation in drug addiction and pathological gambling: consistent inconsistencies? *NeuroImage Clin*. 2013;2:385–393. http://dx.doi.org/10.1016/j.nicl.2013.02.005.
143. Markowitsch HJ, Staniloiu A. Amygdala in action: relaying biological and social significance to autobiographical memory. *Neuropsychologia*. 2011;49(4):718–733. http://dx.doi.org/10.1016/j.neuropsychologia.2010.10.007.
144. Fischer H, Nyberg L, Bäckman L. Age-related differences in brain regions supporting successful encoding of emotional faces. *Cortex*. 2010;46(4):490–497. http://dx.doi.org/10.1016/j.cortex.2009.05.011.
145. Spence SA, Kaylor-Hughes C, Farrow TFD, Wilkinson ID. Speaking of secrets and lies: the contribution of ventrolateral prefrontal cortex to vocal deception. *Neuroimage*. 2008;40(3):1411–1418. http://dx.doi.org/10.1016/j.neuroimage.2008.01.035.
146. Yamasue H, Kuwabara H, Kawakubo Y, Kasai K. Oxytocin, sexually dimorphic features of the social brain, and autism. *Psychiatry Clin Neurosci*. 2009;63(2):129–140. http://dx.doi.org/10.1111/j.1440-1819.2009.01944.x.
147. Nishijo H, Hori E, Tazumi T, et al. Role of the monkey amygdala in social cognition. Cognition and emotion in the brain. Selected topics of the International Symposium on Limbic and Association Cortical Systems. *Int Congr Ser*. 2003;1250:295–310. http://dx.doi.org/10.1016/S0531-5131(03)01070-7.
148. Bickart KC, Dickerson BC, Feldman Barrett L. The amygdala as a hub in brain networks that support social life. *Neuropsychologia*. 2014;63:235–248. http://dx.doi.org/10.1016/j.neuropsychologia.2014.08.013.
149. Zhan S, Liu W, Li D, et al. Long-term follow-up of bilateral anterior capsulotomy in patients with refractory obsessive-compulsive disorder. *Clin Neurol Neurosurg*. 2014;119:91–95. http://dx.doi.org/10.1016/j.clineuro.2014.01.009.

150. Shamay-Tsoory SG, Aharon-Peretz J, Perry D. Two systems for empathy: a double dissociation between emotional and cognitive empathy in inferior frontal gyrus versus ventromedial prefrontal lesions. *Brain*. 2009;132(Pt 3):617–627. http://dx.doi.org/10.1093/brain/awn279.
151. Davis KL, Panksepp J. The brain's emotional foundations of human personality and the affective neuroscience personality scales. *Neurosci Biobehav Rev*. 2011;35(9):1946–1958. http://dx.doi.org/10.1016/j.neubiorev.2011.04.004.
152. Fonov V, Evans AC, Botteron K, Almli CR, McKinstry RC, Collins DL. Unbiased average age-appropriate atlases for pediatric studies. *Neuroimage*. 2011;54(1):313–327. http://dx.doi.org/10.1016/j.neuroimage.2010.07.033.
153. Kuklisova-Murgasova M, Aljabar P, Srinivasan L, et al. A dynamic 4D probabilistic atlas of the developing brain. *Neuroimage*. 2011; 54(4):2750–2763.http://dx.doi.org/10.1016/j.neuroimage.2010.10.019.
154. Lebel C, Walker L, Leemans A, Phillips L, Beaulieu C. Microstructural maturation of the human brain from childhood to adulthood. *Neuroimage*. 2008;40(3):1044–1055. http://dx.doi.org/10.1016/j.neuroimage.2007.12.053.
155. Raz N, Schmiedek F, Rodrigue KM, Kennedy KM, Lindenberger U, Lövdén M. Differential brain shrinkage over 6 months shows limited association with cognitive practice. *Brain Cogn*. 2013;82(2): 171–180. http://dx.doi.org/10.1016/j.bandc.2013.04.002.
156. Sullivan EV, Pfefferbaum A, Rohlfing T, Baker FC, Padilla ML, Colrain IM. Developmental change in regional brain structure over 7 months in early adolescence: comparison of approaches for longitudinal atlas-based parcellation. *Neuroimage*. 2011;57(1): 214–224. http://dx.doi.org/10.1016/j.neuroimage.2011.04.003.
157. Taki Y, Thyreau B, Kinomura S, et al. A longitudinal study of the relationship between personality traits and the annual rate of volume changes in regional gray matter in healthy adults. *Hum Brain Mapp*. 2013;34(12):3347–3353. http://dx.doi.org/10.1002/hbm.22145.
158. Thompson PM, Mega MS, Vidal C, Rapoport JL, Toga AW. Detecting disease-specific patterns of brain structure using cortical pattern matching and a population-based probabilistic brain atlas. *Inf Process Med Imaging*. 2001;2082:488–501. http://dx.doi.org/10.1007/3-540-45729-1_52.
159. Whitford TJ, Grieve SM, Farrow TF, et al. Progressive grey matter atrophy over the first 2–3 years of illness in first-episode schizophrenia: a tensor-based morphometry study. *Neuroimage*. 2006;32(2): 511–519. http://dx.doi.org/10.1016/j.neuroimage.2006.03.041.
160. Wu G, Wang Q, Shen D. Registration of longitudinal brain image sequences with implicit template and spatial-temporal heuristics. *Neuroimage*. 2012;59(1):404–421. http://dx.doi.org/10.1016/j.neuroimage.2011.07.026.
161. Brain Development Cooperative Group. Total and regional brain volumes in a population-based normative sample from 4 to 18 years: the NIH MRI study of normal brain development. *Cereb Cortex*. 2012;22(1):1–12. http://dx.doi.org/10.1093/cercor/bhr018.
162. Dennison M, Whittle S, Yücel M, et al. Mapping subcortical brain maturation during adolescence: evidence of hemisphere- and sex-specific longitudinal changes. *Dev Sci*. 2013;16(5):772–791. http://dx.doi.org/10.1111/desc.12057.
163. Zipparo L, Whitford TJ, Redoblado Hodge MA, et al. Investigating the neuropsychological and neuroanatomical changes that occur over the first 2–3 years of illness in patients with first-episode schizophrenia. *Prog Neuropsychopharmacol Biol Psychiatry*. 2008;32(2): 531–538. http://dx.doi.org/10.1016/j.pnpbp.2007.10.011.
164. Rohrer JD, Ridgway GR, Modat M, et al. Distinct profiles of brain atrophy in frontotemporal lobar degeneration caused by progranulin and tau mutations. *Neuroimage*. 2010;53(3):1070–1076. http://dx.doi.org/10.1016/j.neuroimage.2009.12.088.
165. Farrow TFD, Thiyagesh SN, Wilkinson ID, Parks RW, Ingram L, Woodruff PWR. Fronto-temporal-lobe atrophy in early-stage Alzheimer's disease identified using an improved detection methodology. *Psychiatry Res*. 2007;155(1):11–19. http://dx.doi.org/10.1016/j.pscychresns.2006.12.013.
166. Ashburner J, Friston KJ. Voxel-based morphometry–the methods. *Neuroimage*. 2000;11(6 Pt 1):805–821. http://dx.doi.org/10.1006/nimg.2000.0582.
167. Phillips JS, Greenberg AS, Pyles JA, et al. Co-analysis of brain structure and function using fMRI and diffusion-weighted imaging. *J Vis Exp*. 2012;69:1–11. http://dx.doi.org/10.3791/4125.
168. Hagmann P, Cammoun L, Gigandet X, et al. Mapping the structural core of the human cerebral cortex. *PLoS Biol*. 2008;6(7): 1479–1493.
169. Thompson PM, Stein JL, Medland SE, et al. The ENIGMA Consortium: large-scale collaborative analyses of neuroimaging and genetic data. *Brain Imaging Behav*. 2014. http://dx.doi.org/10.1007/s11682-013-9269-5.
170 Van Essen DC, Smith SM, Barch DM, Behrens TEJ, Yacoub E, Ugurbil K. The WU-Minn Human Connectome Project: an overview. *Neuroimage*. 2013;80:62–79. http://dx.doi.org/10.1016/j.neuroimage.2013.05.041.
171. Dennis EL, Thompson PM. Mapping connectivity in the developing brain. *Int J Dev Neurosci*. 2013;31(7):525–542. http://dx.doi.org/10.1016/j.ijdevneu.2013.05.007.
172. Catani M, Thiebaut de Schotten M. A diffusion tensor imaging tractography atlas for virtual in vivo dissections. *Cortex*. 2008;44(8): 1105–1132. http://dx.doi.org/10.1016/j.cortex.2008.05.004.
173. Malmivuo J. Comparison of the properties of EEG and MEG in detecting the electric activity of the brain. *Brain Topogr*. 2012;25(1): 1–19. http://dx.doi.org/10.1007/s10548-011-0202-1.
174. Meyer PT, Rijntjes M, Weiller C. Neuroimaging: functional neuroimaging: functional magnetic resonance imaging, positron emission tomography, and single-photon emission computed tomography. In: *Bradley's Neurology in Clinical Practice*. 6th ed. Elsevier; 2012: 529–548.e5. http://dx.doi.org/10.1016/B978-1-4377-0434-1.00037-2.
175. Holper L, Scholkmann F, Wolf M. Between-brain connectivity during imitation measured by fNIRS. *Neuroimage*. 2012;63(1): 212–222. http://dx.doi.org/10.1016/j.neuroimage.2012.06.028.
176. Stanley DA, Adolphs R. Toward a neural basis for social behavior. *Neuron*. 2013;80:816–826.
177. a. Chuong AS, Miri ML, Busskamp V, et al. Noninvasive optical inhibition with a red-shifted microbial rhodopsin. *Nat Neurosci*. 2014;17:1123–1129. http://dx.doi.org/10.1038/nn.3752.
b. Mameli F, Mrakic-Sposta S, Vergari M, Fumagalli M, Macis M, Ferrucci R, et al. Dorsolateral prefrontal cortex specifically processes general – but not personal – knowledge deception: multiple brain networks for lying. *Behav Brain Res*. 2010;211(2):164–168. http://dx.doi.org/10.1016/j.bbr.2010.03.024.
c. Kalbe E, Schlegel M, Sack AT, et al. Dissociating cognitive from affective theory of mind: a TMS study. *Cortex*. 2010;46(6):769–780. http://dx.doi.org/10.1016/j.cortex.2009.07.010.
d. Guillory SA, Bujarski KA. Exploring emotions using invasive methods: review of 60 years of human intracranial electrophysiology. *Soc Cogn Affect Neurosci*. 2014. http://dx.doi.org/10.1093/scan/nsu002.
178. McAdams DP, Pals JL. A new Big Five: fundamental principles for an integrative science of personality. *Am Psychol*. 2006;61(3): 204–217.
179. DeYoung CG. Personality neuroscience and the biology of traits. *Soc Pers Psychol Compass*. 2010;4(12):1165–1180.
180. Fiske ST, Taylor SE. *Social Cognition from Brains to Culture*. Los Angeles: Sage; 2013; 1.
181. Premack DG, Woodruff G. Does the chimpanzee have a theory of mind? *Behav Brain Sci*. 1978;1(4):515–526.
182. Zaki J, Ochsner K. The need for a cognitive neuroscience of naturalistic social cognition. *Ann NY Acad Sci*. 2009;1167:16–30. http://dx.doi.org/10.1111/j.1749-6632.2009.04601.x.
183. Saxe R, Carey S, Kanwisher N. Understanding other minds: linking developmental psychology and functional neuroimaging. *Annu Rev Psychol*. 2004;55:87–124. http://dx.doi.org/10.1146/annurev.psych.55.090902.142044.

184. Miller CB. *Character and Moral Psychology*. Oxford: Oxford; 2014. 35–36.
185. Kafka MP. Hypersexual disorder: a proposed diagnosis for DSM-V. *Arch Sex Behav*. 2010;39(2):377–400. http://dx.doi.org/10.1007/s10508-009-9574-7.
186. Reid RC, Stein JA, Carpenter BN. Understanding the roles of shame and neuroticism in a patient sample of hypersexual men. *J Nerv Ment Dis*. 2011;199(4):263–267. http://dx.doi.org/10.1097/NMD.0b013e3182125b96.
187. Reid RC, Dhuffar MK, Parhami I, Fong TW. Exploring facets of personality in a patient sample of hypersexual women compared with hypersexual men. *J Psychiatr Pract*. 2012;18(4):262–268. http://dx.doi.org/10.1097/01.pra.0000416016.37968.eb.
188. Costa PT, McCrae RR. *Revised NEO Personality Inventory (NEO-PIR) and NEO Five-Factor Inventory Professional Manual*. Odessa, FL: Psychological Assessment Resources; 1992 (cited in 206).
189. Walters GD, Knight RA, Långström N. Is hypersexuality dimensional? Evidence for the DSM-5 from general population and clinical samples. *Arch Sex Behav*. 2011;40(6):1309–1321. http://dx.doi.org/10.1007/s10508-010-9719-8.
190. Auerbach DM, Darrow WW, Jaffe HW, Curran JW. Cluster of cases of the acquired immune deficiency syndrome. Patients linked by sexual contact. *Am J Med*. 1984;76(3): 487–492. http://dx.doi.org/10.1016/0002-9343(84)90668-5. PMID: 6608269.
191. Baird AD, Wilson SJ, Bladin PF, Saling MM, Reutens DC. Neurological control of human sexual behaviour: insights from lesion studies. *J Neurol Neurosurg Psychiatry*. 2007;78(10):1042–1049. http://dx.doi.org/10.1136/jnnp.2006.107193.
192. Bigelow HJ, Barnard J. *Phineas Gage*. 2002;3:843–857.
193. Filley CM. Chapter 35: the frontal lobes. In: 3rd ed. *Handbook of Clinical Neurology*. vol. 95. Elsevier B.V; 2010:557–570. http://dx.doi.org/10.1016/S0072-9752(08)02135-0.
194. Damasio H, Grabowski T, Frank R, Galaburda AM, Damasio AR. The return of Phineas Gage: clues about the brain from the skull of a famous patient. *Science*. 1994;264:1102–1105. http://dx.doi.org/10.1126/science.8178168.
195. Absher JR, Benson DF. Disconnection syndromes: an overview of Geschwind's contributions. *Neurology*. 1993;43(5):862–867. http://dx.doi.org/10.1212/WNL.43.5.862.
196. Damasio H, Damasio AR. *Lesion Analysis in Neuropsychology*. New York: Oxford; 1989.
197. Damasio AR, Van Hoesen GW. Emotional disturbance associated with focal lesions of the limbic frontal lobe. Chapter 4. In: Keilman K, Sats P, eds. *Neuropsychology of Human Emotion*. The Guilford Press; 1983:85–110.
198. Devinsky O, Morell MJ, Vogt BA. Contributions of anterior cingulate cortex to behavior. *Brain*. 1995;118:279–306.
199. Stoléru S, Fonteille V, Cornélis C, Joyal C, Moulier V. Functional neuroimaging studies of sexual arousal and orgasm in healthy men and women: a review and meta-analysis. *Neurosci Biobehav Rev*. 2012;36(6): 1481–1509. http://doi.org/10.1016/j.neubiorev.2012.03.006.
200. Salinas C, Davila G, Berthier ML, Green C, Lara JP. Lesions in prefrontal- subcortical circuits. *J Neuropsychiatry Clin Neurosci*. 2009;21:332–334.
201. Coenen VA, Panksepp J, Hurwitz TA, Urbach H, Mädler B. Anterior thalamic radiation (ATR): imaging of two major subcortical pathways and the dynamic balance of opposite affects in understanding depression. *J Neuropsychiatry Clin Neurosci*. 2012;24(2):223–236.
202. Berthier ML, Kulisevsky J, Gironell A, Heras JA. Obsessive-compulsive disorder associated with brain lesions: clinical phenomenology, cognitive function, and anatomic correlates. *Neurology*. 1996;47:353–361.
203. Spinella M. Hypersexuality and dysexecutive syndrome after a thalamic infarct. *Int J Neurosci*. 2004;114(12):1581–1590. http://dx.doi.org/10.1080/00207450490509339.
204. Chen Y-C, Tseng C-Y, Pai M-C. Witzelsucht after right putaminal hemorrhage: a case report. *Acta Neurol Taiwan*. 2005;14(4): 195–200. Available at: http://www.ncbi.nlm.nih.gov/pubmed/16425547.
205. Braun CMJ, Daigneault R, Gaudelet S, Guimond A. Diagnostic and Statistical Manual of Mental Disorders, Fourth Edition symptoms of mania: which one(s) result(s) more often from right than left hemisphere lesions? *Compr Psychiatry*. 2008;49(5):441–459. http://dx.doi.org/10.1016/j.comppsych.2008.02.001.
206. Shamay-Tsoory SG, Harari H, Aharon-Peretz J, Levkovitz Y. The role of the orbitofrontal cortex in affective theory of mind deficits in criminal offenders with psychopathic tendencies. *Cortex*. 2010;46(5): 668–677. http://dx.doi.org/10.1016/j.cortex.2009.04.008.
207. Szczepanski SM, Knight RT. Insights into human behavior from lesions to the prefrontal cortex. *Neuron*. 2014;83(5):1002–1018. http://dx.doi.org/10.1016/j.neuron.2014.08.011.
208. Webb CA, DelDonno S, Killgore WDS. The role of cognitive versus emotional intelligence in Iowa Gambling Task performance: what's emotion got to do with it? *Intelligence*. 2014;44:112–119. http://dx.doi.org/10.1016/j.intell.2014.03.008.
209. Croft KE, Duff MC, Kovach CK, Anderson SW, Adolphs R, Tranel D. Detestable or marvelous? Neuroanatomical correlates of character judgments. *Neuropsychologia*. 2010;48(6):1789–1801. http://dx.doi.org/10.1016/j.neuropsychologia.2010.0.001.
210. Mah LW, Arnold MC, Grafman J. Deficits in social knowledge following damage to ventromedial prefrontal cortex. *J Neuropsychiatry Clin Neurosci*. 2005 Winter;17(1):66–74.
211. Damasio AR. *Descartes' Error: Emotion, Reason, and the Human Brain*. New York, NY: Putnam Publishing; 1994.
212. Meyers CA, Berman SA, Scheibel RS, Hayman A. Case report: acquired antisocial personality disorder associated with unilateral left orbital frontal lobe damage. *J Psychiatr Neurosci*. 1992;17(3):121–125.
213. Miner MH, Raymond N, Mueller BA, Lloyd M, Lim KO. Preliminary investigation of the impulsive and neuroanatomical characteristics of compulsive sexual behavior. *Psychiatr Res Neuroimaging*. 2009;174:146–151.
214. Damasio a R, Tranel D, Damasio H. Individuals with sociopathic behavior caused by frontal damage fail to respond autonomically to social stimuli. *Behav Brain Res*. 1990;41(2):81–94. Available at: http://www.ncbi.nlm.nih.gov/pubmed/2288668.
215. Mercadillo RE, Luis Diaz J, Barrios FA. Neurobiology of moral emotions. *Salud Ment*. 2007;30:1–11. Available at: <Go to ISI>://WOS:000248342100001.
216. a. Contreras-Rodríguez O, Pujol J, Batalla I, Harrison BJ, Soriano-Mas C, Deus J, et al. Functional connectivity bias in the prefrontal cortex of psychopaths. *Biol Psychiatry*. 2014:1–8. http://dx.doi.org/10.1016/j.biopsych.2014.03.007.
b. Roy M, Shohamy D, Wager TD. Ventromedial prefrontal-subcortical systems and the generation of affective meaning. *Trends Cogn Sci*. 2012;16(3):147–156. http://dx.doi.org/10.1016/j.tics.2012.01.005. Epub 2012 February 5.
217. Eslinger PJ, Damasio AR. Severe disturbance of higher cognition after bilateral frontal lobe ablation: patient EVR. *Neurology*. 1985;35:1731–1741.
218. Braun CMJ, Dumont M, Duval J, Hamel I, Godbout L. Opposed left and right brain hemisphere contributions to sexual drive: a multiple lesion case analysis. *Behav Neurol*. 2003;14(1–2): 55–61. Available at: http://www.ncbi.nlm.nih.gov/pubmed/12719639.
219. Starkstein SE, Boston JD, Robinson RG. Mechanisms of mania after brain injury: 12 case reports and review of the literature. *J Nerv Ment Dis*. 1988;176(2):87–100.
220. Moll J, de Oliveira-Souza R, Bramati IE, Grafman J. Functional networks in emotional moral and nonmoral social judgments. *Neuroimage*. 2002;16(3):696–703. http://dx.doi.org/10.1006/nimg.2002.1118.

221. Moll J, Zahn R, de Oliveira-Souza R, et al. Impairment of prosocial sentiments is associated with frontopolar and septal damage in frontotemporal dementia. *Neuroimage*. 2011;54(2):1735–1742. http://dx.doi.org/10.1016/j.neuroimage.2010.08.026.
222. Bancroft J. Central inhibition of sexual response in the male: a theoretical perspective. *Neurosci Biobehav Rev*. 1999;23(6):763–784. http://dx.doi.org/10.1016/S0149-7634(99)00019-6.
223. Reid RC, Garos S, Carpenter BN, Coleman E. A surprising finding related to executive control in a patient sample of hypersexual men. *J Sex Med*. 2011;8(8):2227–2236. http://dx.doi.org/10.1111/j.1743-6109.2011.02314.x.
224. Lilly R, Cummings JL, Benson EF, Frankel M. The human Kluver-Bucy syndrome. *Neurology*. 1983;33:1141–1145.
225. Pilleri G. The Kluver-Bucy syndrome in man: a clinico-anatomical contribution to the function of the medial temporal lobe structures. *Psychiatr Neurol (Basel)*. 1966;152:65–103.
226. Terzian H, Ore GD. Syndrome of Kluver and Bucy reproduced in man by bilateral removal of the temporal lobes. *Neurology*. 1955;5(6):373–380.
227. Devinsky J, Schachter S. Norman Geschwind's contribution to the understanding of behavioral changes in temporal lobe epilepsy: the February 1974 lecture. *Epilepsy Behav*. 2009;15(4):417–424. http://dx.doi.org/10.1016/j.yebeh.2009.06.006.
228. Cummings JL. Neuropsychiatric manifestations of right hemisphere lesions. *Brain Lang*. 1997;57:22–37.
229. Baird AD, Wilson SJ, Bladin PF, Saling MM, Reutens DC. The amygdala and sexual drive: insights from temporal lobe surgery. *Ann Neurol*. 2004;55:87–96.
230. Adolphs R, Spezio M. Role of the amygdala in processing visual social stimuli. *Prog Brain Res*. 2006;156(06):363–378. http://dx.doi.org/10.1016/S0079-6123(06)56020-0.
231. Satterthwaite TD, Wolf DH, Pinkham AE, et al. Opposing amygdala and ventral striatum connectivity during emotion identification. *Brain Cogn*. 2011;76(3):353–363. http://dx.doi.org/10.1016/j.bandc.2011.04.005.
232. Byne W, Kemmether E. The sexual brain. In: *Biological Psychiatry*. 2000:59–86.
233. Agmo A. Chapter 6: neural control of sexual behavior. In: *Functional and Dysfunctional Sexual Behavior: A Synthesis of Neuroscience and Comparative Psychology*. ; 2011:231–256.
234. Levin M. Periodic somnolence and morbid hunger: a new syndrome. *Brain*. 1936;59:494–504.
235. Gorman DG, Cummings JL. Hypersexuality following septal injury. *Arch Neurol*. 1992;49:208–310.
236. Inzelberg R, Nisipeanu P, Joel D, Sarkantyus M, Carasso RL. Acute mania and hemichorea. *Clin Neuropharmacol*. 2001;24:300–303.
237. Starkstein SE, Berthier ML, Leiguarda R. Psychic akinesia following bilateral pallidal lesions. *Int J Psychiatry Med*. 1989;19(2):155–164.
238. Cao Y, Zhu Z, Wang R, Wang S, Zhao J. Hypersexuality from resection of left occipital arteriovenous malformation. *Neurosurg Rev*. 2010;33(1):107–114. http://dx.doi.org/10.1007/s10143-009-0232-2.
239. Absher JR, Vogt BA, Clark DG, et al. Hypersexuality and hemiballism due to subthalamic infarction. *Neuropsychiatry Neuropsychol Behav Neurol*. 2000;13:220–229.
240. Cavazos JE, Wang C, Sitoh Y, Ng SES, Tien RD. Anatomy and pathology of the septal region. *Neuroim Clin N Am*. 1997;7(1):67–78.
241. Doshi P, Bhargava P. Hypersexuality following subthalamic nucleus stimulation for Parkinson's disease. *Neurol India*. 2008;56(4):474–477.
242. Fleischman J. *Phineas Gage: A Gruesome but True Story about Brain Science*. Boston: Houghton Mifflin Company; 2002.
243. Geschwind N. Disconnexion syndromes in animals and man, Part I. *Brain*. 1965;88:237–294.
244. Geschwind N. Disconnexion syndromes in animals and man, Part II. *Brain*. 1965;88:585–644.
245. Geschwind N. Mechanisms of change after brain lesions. *Ann NY Acad Sci*. 1985;457:1–11.
246. Rorden C, Karnath H-O. Using human brain lesions to infer function: a relic from a past era in the fMRI age? *Nat Rev Neurosci*. 2004;5(10):813–819. http://dx.doi.org/10.1038/nrn1521.
247. Poldrack RA. Inferring mental states from neuroimaging data: from reverse inference to large-scale decoding. *Neuron*. 2011;72(5):692–697. http://dx.doi.org/10.1016/j.neuron.2011.11.001.
248. Adolphs R. Conceptual challenges and directions for social neuroscience. *Neuron*. 2010;65(6):752–767. http://dx.doi.org/10.1016/j.neuron.2010.03.006.
249. Van Essen DC. Cartography and connectomes. *Neuron*. 2013;80(3):775–790. http://dx.doi.org/10.1016/j.neuron.2013.10.027.
250. Amunts K, Schleicher A, Zilles K. Cytoarchitecture of the cerebral cortex–more than localization. *Neuroimage*. 2007;37(4):1061–1065. http://dx.doi.org/10.1016/j.neuroimage.2007.02.037. Discussion 1066–1068.
251. Eickhoff SB, Schleicher A, Scheperjans F, Palomero-Gallagher N, Zilles K. Analysis of neurotransmitter receptor distribution patterns in the cerebral cortex. *Neuroimage*. 2007;34(4):1317–1330. http://dx.doi.org/10.1016/j.neuroimage.2006.11.016.
252. Smith SM, Vidaurre D, Beckmann CF, et al. Functional connectomics from resting-state fMRI. *Trends Cogn Sci*. 2013;17(12):666–682. http://dx.doi.org/10.1016/j.tics.2013.09.016.
253. Woolrich MW, Stephan KE. Biophysical network models and the human connectome. *Neuroimage*. 2013;80:330–338. http://dx.doi.org/10.1016/j.neuroimage.2013.03.059.
254. Devlin JT, Poldrack RA. In praise of tedious anatomy. *Neuroimage*. 2007;37(4):1033–1041. http://dx.doi.org/10.1016/j.neuroimage.2006.09.055. Discussion 1050–1058.
255. Mueller S, Wang D, Fox MD, et al. Individual variability in functional connectivity architecture of the human brain. *Neuron*. 2013;77(3):586–595. http://dx.doi.org/10.1016/j.neuron.2012.12.028.
256. Damasio AR. Time-locked multiregional retroactivation: a systems-level proposal for the neural substrates of recall and recognition. *Cognition*. 1989;33:25–62.
257. Man K, Kaplan J, Damasio H, Damasio A. Neural convergence and divergence in the mammalian cerebral cortex: from experimental neuroanatomy to functional neuroimaging. *J Comp Neurol*. 2013;521:4097–4111. http://dx.doi.org/10.1002/cne.23408.
258. Mah Y-H, Husain M, Rees G, Nachev P. Human brain lesion-deficit inference remapped. *Brain*. 2014:2522–2531. http://dx.doi.org/10.1093/brain/awu164.

SECTION II

PERSPECTIVES ON THE NEURAL BASIS OF PERSONALITY AND DISPOSITIONS

CHAPTER

2

Approach/Avoidance

Neil McNaughton[1], *Colin G. DeYoung*[2], *Philip J. Corr*[3]

[1]Department of Psychology, University of Otago, Dunedin, New Zealand; [2]Department of Psychology, University of Minnesota, Minneapolis, MN, USA; [3]Department of Psychology, City University London, London, UK

OUTLINE

1. BASICS OF APPROACH/AVOIDANCE—BEHAVIOR AND BRAIN

Approach and avoidance behaviors are fundamental to survival. As such, they depend on phylogenetically old systems with many conserved features. For this reason, the basic human brain systems for approach and avoidance have much in common with those of other species. Through study, we know more about the neurobiology of these systems and related traits than we do about most others, and this has translated into progress in human neuroimaging research. In this chapter, we lay out basic principles for understanding the processes of approach and avoidance, and then we briefly discuss neuroimaging research on the states related to them before discussing the progress in research on related personality traits.

Importantly, we define personality traits in terms of longer-term stabilities in patterns of states. That is, the level of a trait reflects the likelihood of being in a particular type of state, given a particular set of eliciting stimuli. The activation of approach and avoidance systems in any given situation requires careful long-term control of its precise intensity for any given input, and this long-term trait control of levels of activation is influenced by genes, developmental processes, and life events. These two systems and their associated traits can be seen as providing a foundation for the more complex processes from which mind and personality emerge.

Both the specific states that result from the activation of approach and avoidance systems and the longer-term sensitivities that tune these activations to match current functional requirements can be assessed indirectly through many techniques, including self-report and behavioral data. But increasingly the more direct measurements of neuroimaging are affording new insights. Detailed analysis of neuroimaging specific aspects of approach/avoidance systems is provided

Neuroimaging Personality, Social Cognition, and Character
http://dx.doi.org/10.1016/B978-0-12-800935-2.00002-6

in chapters 3, 5 and 6 of the book. Here, we provide a more general overview of the fundamental nature of approach and avoidance, the systems that control them, details of a third system that resolves conflicts between goals, and the range of resultant states and traits that should be open to analysis by neuroimaging. It is important to note that within this chapter, we will use the simple term "avoidance" to refer to active avoidance (often termed withdrawal) and the terms "goal conflict processing" and "behavioral inhibition" to refer to passive avoidance. This is an important functional distinction, as these two forms of avoidance are mediated by different, and partially opposing, systems of behavior regulation.[1,2]

It is important to distinguish more general, positively motivated, goal-directed, behavior from object-specific consummatory behaviors (e.g., eating, drinking, and mating). Likewise, direct or very close contacts with specific affectively negative objects require specific defensive behaviors (e.g., attacking an enemy or avoiding contact with fluids from an Ebola corpse). There can be individual differences in these object-specific systems (with extreme sensitivity seen in clinical conditions such as uncontrollable aggression and stimulus-specific phobia). However, once there is even a moderate distance between the organism and the object, the adaptive requirements for approach or avoidance become essentially independent of the specific object—allowing us to talk about more general systems of approach and avoidance that are separate from specific consummatory and defensive reactions. Evolution, therefore, has shaped what can be seen as two general systems dedicated to approach and avoidance, respectively; reflecting the fundamental nature of these systems, they are represented in the major traits of personality.

1.1 Positive and Negative Goals

An important concept in dealing with mammalian approach and avoidance systems is the idea that they process goal representations. The nature of this internal representation needs some explanation and should be kept completely separate from the "goals" that people often attribute to behaviors in terms of external function (obtaining the food at the end of the runway) or evolutionary explanations (achieving survival).

The simplest approach behaviors can be controlled by the detection of gradients rather than by goal representations. A bacterium will approach food through the detection of, for example, chemical gradients in its immediate environment. It has receptors that can detect the strength of a signal (you can think of this as a smell or taste with chemicals) and move in the direction of increasing signal strength. Similarly, a simple multicelled organism can essentially scan gradients of physical stimuli by taking a twisting path through its environment or, when it is close to the source, by wagging its head or body. It then heads in the direction where the signal is strongest, which should ultimately lead it to its food even though it has no information as to the particular location at which it will ultimately arrive. Avoidance behavior can also be governed by gradients. For many organisms, being in the light is dangerous. So even simple detection of light strength allows the organism to move in the opposite direction and find safety (Figure 1).

These kinds of movement, controlled by a local gradient, are called "taxes"[3] (pronounced *tack-seize*). Taxes are often taken to involve a reaction to only very simple stimulus aspects—so that light intensity (as in Figure 1) would be included but visual stimuli that depend on form would not.[3,4] Critically, taxes are not goal-directed in the sense that the series of individual behavioral steps are not determined by a single internal representation of their endpoint. Although each behavioral step can be viewed with the goal of reducing the current light level, the final point at which the animal comes to a halt is simply the point at which such a sequence self-terminates and is not represented internally. That is, the maggot follows a path determined by the local gradients even if that is a circuitous route and does not terminate in the darkest place in the environment, whereas a rat controlled by a goal-representation will often take a straight-line path through a strongly lit area to reach the darkest area. So, as external observers, we can often see taxes as causing an organism to reach a "goal" (in a functional and/or evolutionary sense), but the behavior of the organism itself is not driven by an internal representation of the end state of the sequence of behaviors.

So, what is a goal for an organism? What kind of internal representation is a goal? Imagine you are a hungry rat and you have just been placed in a T maze by an experimenter. If you are hungry and you know from previous experience that there is food and it is at the end of one arm, then that particular food-bearing arm will be a goal for you. The other arm does not have food and so is not a goal for you at the moment. Likewise, if the food is moved to the other arm (and you know that it has been moved), that arm will now become your goal. However, if you are not hungry, even the arm with food will not be a goal, and you will probably decide to curl up in the end of the start arm and have a snooze (experimenters often test rats in the day time, which is the time they normally sleep). A goal, therefore, has both cognitive/identifying and motivational/consummatory properties. Its cognitive properties distinguish it from other places, times, or combinations of stimuli that may have the same motivational properties (we will refer to each specific set of such cognitive properties as a "situation"). Its motivational properties derive from the organism's current need to acquire some specific stimulus—food, drink, etc.—and

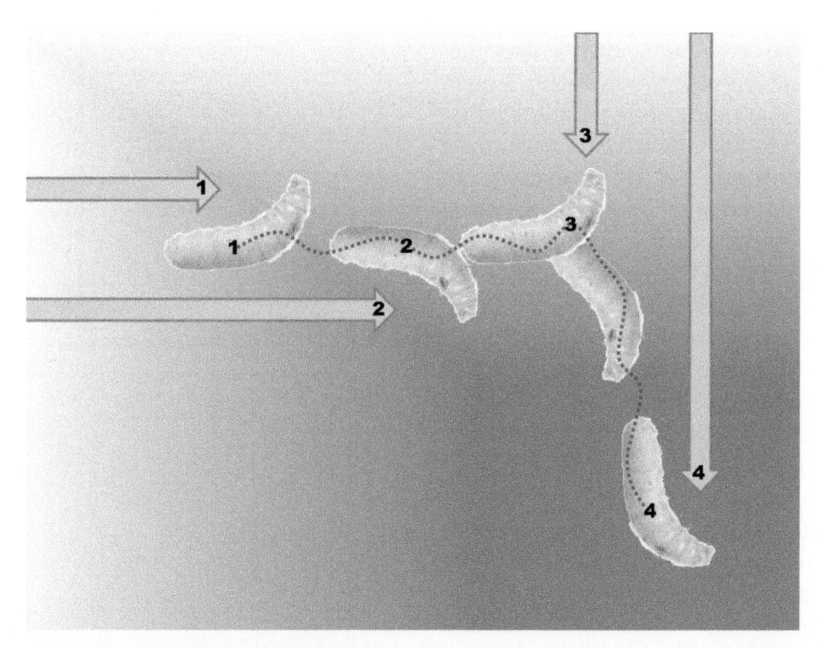

FIGURE 1 After feeding, maggots head for the dark. If light comes from the left, it is detected by head receptors when the maggot turns its body up (point 1), making it turn away (point 2). Successive turns away from the light take the maggot to the right (dotted track). At point 3, the light instead comes from the top. Turns away from the light now move the maggot downwards. *Based on Figure 7.3 in Hinde.*[4]

the presence of the relevant motivational stimulus in the situation. Importantly, neither the "situation" nor the "motivation" by themselves will generate goal-directed behavior. It is their compound "goal representation" that does so. In contrast, while we can discern "goals" for behavior in an evolutionary sense (e.g., the goal of random exploration could be said to be the discovery of food), this behavior is not goal-directed in the sense that behavior is controlled by an internal representation of the end point of the behavioral sequence.

To say that an animal has a goal needs some justification. How do we know that a rat's behavior is not simply the result of taxes or, in the case of complex, learned running in a maze, simply a long sequence of stimulus-response reactions that act like a string of taxes? In an experiment where a tone predicted a shock, it was found that a sheep lifted its leg off the pad that delivered the shock. Had the sheep learned a simple stimulus–response relationship? If the sheep was turned over so its head was on the pad, it lifted its head rather than its leg in response to the tone. Even in this very simple case, the sheep has clearly learned the relationship between the two stimuli and has not learned a fixed "conditioned response."[5] Or, suppose you inject a rat with enough anesthetic to seriously affect its coordination; it will still immediately escape from a box where it has been shocked using new movements that can include rolling out of the box.[6] Or, if you lesion its motor control systems, it can navigate a complex maze perfectly accurately using quite new movements. "The essential point here is that the new movements are not stereotyped, but selected from variable patterns in such a manner as to bring the animal nearer the goal. Furthermore, the new patterns are directly and efficiently substituted without any random activity."[4] In all these cases, the animals are demonstrating control of behavior by an internal representation that is a compound of an identifying situation and a motivating consummatory stimulus, which calls forth whatever behaviors are available to the animal, given the situation, to achieve consummation.

We also need to be clear that, in this sense of situation–motivation compound, there can be two very different types of goals relating to approach or avoidance, which can conflict with, or reinforce, each other in the control of behavior. The word "goal" is typically used in English to signify something we want to achieve in the positive sense. As we have already seen, there can be positive and negative taxes (with the organism moving to increase or decrease the relevant signal, moving up or down its gradient, respectively). Likewise, there can be positive and negative goals, with the goal creating what might be called a "cognitive gradient" that then determines the animal's specific behavior. A positive goal is an attractor, something to be included in the desired future state; a negative goal is a repulsor, something to be excluded or avoided in the desired future state.

In the top left-hand panel of Figure 2, a shock or some other unpleasant event or object is represented in the bottom left-hand corner. If a rat is aware of this danger, it will run directly away—shown by the arrows in the figure—and the tendency to run will decrease as the rat moves away from the object.[7] The object is a repulsor. The red zone represents all the paths that will be taken by any rat, and its grading represents the change in running strength, resulting from the cognitive gradient created by the rat's internal representation of the danger. Conversely, in the top right-hand panel is the equivalent representation of the paths taken by a rat running to food, or some other attractor, that is in the top right hand corner of the rectangular box. The only substantive difference between the shock and the food is that the direction of movement, relative to the gradient, is the opposite—as with positive and negative taxes.

Whether behavior is controlled by a positive or negative goal can be difficult to determine without careful analysis. It might seem obvious that danger is negative. However, when faced with danger (such as a cat), a rat may be motivated to seek safety (its home burrow), which constitutes a positive goal. In many threatening situations then, both avoidance of danger and approach to safety can occur. Typically, when close to danger, the rat will avoid it (moving directly away from the danger because any other path will take it closer to the cat and increase the chance of being caught), and when close to its burrow, it will head straight for that. The result (Figure 2, bottom panel) will be a curvilinear path in the simplest cases with the initial running being avoidance of the negative goal of danger and the later running being approach to the positive goal of safety. With several examples of the same situation and with an animal starting in different positions, we can determine the nature of the controlling goal from the set of trajectories. If they diverge from a point (as in Figure 2, top left), then they are controlled by a negative goal and if they converge (top right), then they are controlled by a positive goal.

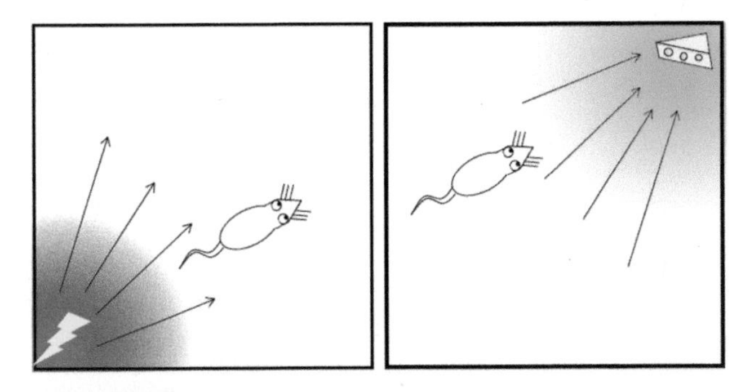

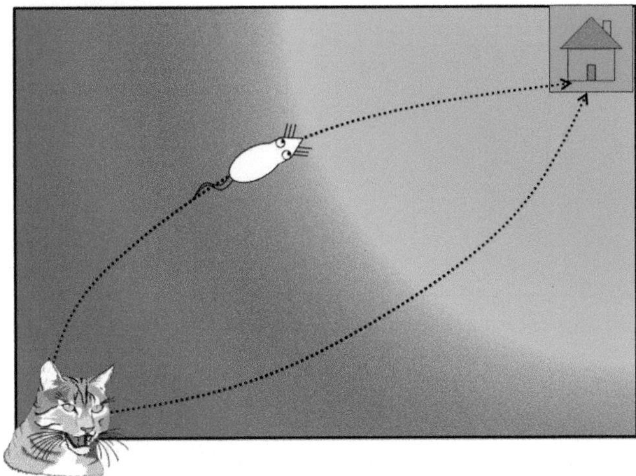

FIGURE 2 Diagrammatic representation of cognitive gradients created by shock (top left), by food (top right), and by the combination of danger and safety (bottom). The solid arrows represent the direction of movement of a rat located at the base of the arrow. The dashed curves represent the path taken by a rat from danger to safety (see text).

1.2 Valuation versus Motivation

The positive or negative nature of a goal is not determined just by whether the stimulus generating the situation is itself positive or negative. In the example that we just considered of the rat fleeing the cat, the presence of the cat generates a negative goal at one point in space, but the absence of the cat (guaranteed by the nature of the burrow) generates a positive goal in the burrow entrance. (If there is no burrow, then the rat will simply run directly away from the negative goal—as in the top left panel of Figure 2—until it reaches the limits of the apparatus since there is no safe place to attract it.)

The capacity of a single class of motivational stimuli to give rise to opposite goals in different circumstances is most obvious with consummatory stimuli (like food and water) and in economic experiments. The presentation of a positive stimulus produces positive goals, but the omission of expected food, omission of expected water, loss of money, and any other negative contingencies of positive events generate the aversive state of frustration.[8–10] The situation linked to omission or loss will therefore be associated with negative motivation, and so their compound will be a negative goal. When the same situation occurs in the future, it will then generate avoidance, thereby reducing exposure to frustration. An important point here is that the immediate experience of frustration produces escape,[11] fighting,[8] learned avoidance, and many other responses that are also typical of the immediate experience of pain.[12] In general, the omission of negative and positive events can be treated as having the same effects as the presentation of positive and negative events, respectively.

The idea that omission of a positive event creates a negative goal in exactly the same way as presentation of a negative event requires one caveat—the two outcomes do not have the same value. For an event to affect behavior, it must first be valued. This value, for any given object, will vary with both time and the particular individual. Gorgonzola cheese will have a high positive value for many hungry adults but will usually have a high negative value for young children. Likewise, a rat that is not hungry will not value food highly (i.e., will not work hard to obtain it), and a rat that has undergone taste aversion conditioning for a particular flavor will not value a food with that flavor but will value other food. You might think that a specific object (like a dollar)

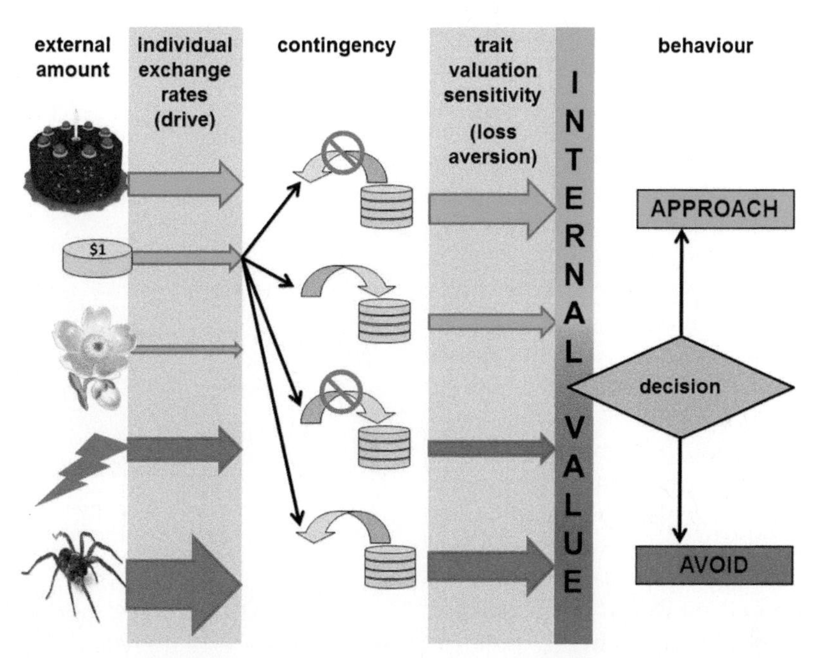

FIGURE 3 Relations between external amount, contingency, and value. An external item will have a specific amount (e.g., one entire cake) that, together with the current level of drive (which acts like a currency exchange rate) for that kind of item for that person, determines its primary internal value (thickness of arrows in first column). As shown for the case of $1, this interacts with whether the item will be gained or lost to determine the direction and size of its internal value as ultimately measured by the effect on behavior. The direction of this effect is reversed if the gain or loss is omitted. Loss (removal from a store of items) is most easily controlled with money but will also occur when, for example, one rat steals the food from another rat.

will have the same value, whether it is being gained or lost—after all, it is the most fungible of all stimuli. But it turns out that, in a wide range of situations, a lost dollar is treated as having greater value than a gained dollar. That is, someone will work harder to avoid a dollar loss than they will to make a dollar gain. This important and very general phenomenon, discovered in behavioral economics,[13–15] is termed loss aversion.

Individual differences in valuation are often not important for scores obtained in experiments in which some manipulation affects approach or avoidance. For example, individual variation in hunger drive among a group of rats will simply increase variation in the running speed within a group and will not, in the absence of sampling biases, change the difference between the means of the groups. However, there are times when we may want a full understanding of the effects of individual variations in approach and avoidance tendencies. There are often times, in particular, when we want to assess the long-term sensitivities of approach and avoidance systems (i.e., individual personality traits). If so, we will want to take account of both the valuation of specific objects (via their specific exchange rates in relation to the single internal currency on which choices are based) and whether the situation involves a positive or negative contingency with the occurrence of the object (Figure 3). If we use only positive objects to assess trait approach, our measures will be confounded by the variation in trait positive valuation. It should also be noted that valuation, as we have used it here, involves an essential interaction between "wanting" and "liking" components of a positively valued object that have different neural correlates.[16,17]

1.3 Goal Interactions, Gradients, and Goal Conflict

Approach and avoidance behavior are fairly simple to understand when there is only one goal. However, when more than one goal is available, we need to consider the way goal gradients interact and the special effects that occur when goals are in conflict with each other.

We have already seen one simple example of interaction between two goals. The rat first fleeing from the cat and then racing toward safety (Figure 2) has two compatible goals. Which goal is in control depends on the rat's position and the fact that the effect of a goal has a gradient (i.e., a decrease in the strength of the effect of the goal on behavior as distance increases). These gradients are represented by the fading of the colors with distance from the points at which each goal is located. In one sense, it is obvious that a rat should first run directly away from a cat, as this means it is least likely to get caught. However, more mechanistically, we can say that, at a very short distance, the effect of the cat is strong and so produces avoidance, while the effect of very distant safety is weak and so produces minimal approach. The reverse is true at the other end of the rat's trajectory: at intermediate distances, we see a balance of push and pull in operation, with both tendencies generating much the same running movements.

The notion of the diminution of the strength of a goal with distance is intuitively obvious. But it has also been demonstrated in a range of experiments[18] that test, for example, how strongly a rat fitted with a harness will pull to move toward a positive goal or away from a negative one.[19] A similar diminution is seen with delays between action and the achievement of a goal—a phenomenon referred to in behavioral economics as "temporal discounting," which shows a gain/loss asymmetry[20] similar to that shown with simple value.

A more problematic interaction, from the point of view of both the organism and the experimenter, is the interaction between incompatible goals. The theoretically simplest example is what is called approach-avoidance conflict. For example, if a hungry rat is placed in one end of a straight alley and knows there is food in the goal box at the other end, it will run to this positive goal. However, if we also arrange it so that it will receive a shock in the goal box, the resultant behavior is not simple. With a weak shock, it will run slower but still reach the goal box. With a moderate shock, however, it will start to run, slow down as it gets closer to the goal box, and then dither to-and-fro. None of this can be explained by a simple economic calculation that subtracts the intrinsic value of the shock from that of the food, which would result in the rat either not running at all (receiving neither food nor shock) or always running all the way to the goal box (receiving shock but also food).

To understand approach-avoidance conflict, we need to look at the nature and interaction of the positive and negative goal gradients affecting the rat. The experiments with rats in harnesses[19] demonstrated that the fall-off with distance of the power of a goal is much greater for a negative one than a positive one. These gradients and their summative interaction[7,21,22] are shown in Figure 4. Initially, because its gradient is shallow, the positive goal (the memory of food) attracts the rat. In the absence of shock, the rat would run progressively faster as it got nearer to the goal box. But, part way down the runway, the negative goal (the memory of shock) begins to affect the rat, slowing it down and making its path less direct. If the shock is strong enough, so that the negative goal is more highly valued than the positive goal when the rat is in the goal box, then at a point before the goal box is reached, the approach tendency and the avoidance tendency will be equal, and the rat will not reach the goal box.

Approach-avoidance conflict does not simply make the rat stop running. The positive and negative values do not just cancel out, leaving the rat unmotivated. Instead, at the balance point, the rat will dither between approach and avoidance, turning first away and then back toward the goal box (dashed path in Figure 4).

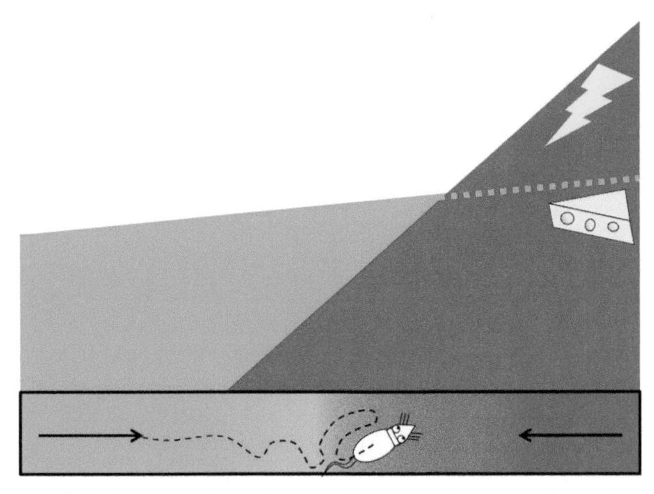

FIGURE 4 Interaction of approach and avoidance gradients. Positive goals (e.g., cheese for a hungry rat) have an effect on behavior that decreases slowly with distance (green). Negative ones (e.g., shock) have a steeper gradient (red). Their interaction (graded color in bottom panel) means a rat will initially run toward food but then will stop part way if there is a shock (see text).

(We also dither, experiencing strong emotion, as we wonder "should I stay or should I go," etc.) Approach-avoidance conflict will also often produce what appears to be completely irrelevant behavior, such as grooming. This is technically termed "displacement activity,"[3] and you are likely to have experienced this in yourself: chewing your nails as you worry about what to do or pacing up and down as you wait for a challenging interview.

The novel behavioral patterns elicited by approach-avoidance conflict and the effects of antianxiety drugs on them, but not in simple avoidance,[23] show that a third system, beyond the approach and avoidance systems, is involved. Termed the "Behavioral Inhibition System" (BIS), and described in considerable detail by Gray,[1,18,24] this system has outputs that inhibit the behavior that would be generated by the positive and negative goals (without reducing the activation of the goals themselves), increases arousal and attention (generating exploration and displacement activities), and increases the strength of avoidance tendencies (i.e., increases fear and risk aversion). Increased avoidance during goal conflict is adaptive since, faced with risk, failing to obtain food or some other positive goal is likely to be easy to make up at another time, but experiencing danger could have severe consequences. This increase in aversion produced by goal conflict is sensitive to anxiolytic drugs. So, if the rat shown in Figure 4 is treated with an anxiolytic drug, it will no longer dither and will approach closer to, and sometimes reach, the goal box[25,26]—since the drug affects the passive avoidance generated by goal conflict but not basic approach or basic active avoidance tendencies.[23]

As with approach and active avoidance, the functional requirements of approach-avoidance conflict are sufficiently fundamental that passive avoidance appears early in phylogenetic terms, being present in coelenterates[27] with anxiolytic benzodiazepine receptors appearing in primitive vertebrates, such as the lungfish, and being present in fish, amphibians, reptiles, birds, and mammals.[28]

1.4 Hierarchical Control

An important feature of neural systems is that they are hierarchically organized. Both in terms of evolution and development, neural systems must fulfill preexisting adaptive requirements while adding the machinery for more sophisticated functions. As a result, higher order circuits are overlaid on lower order ones, and whether behavior is controlled by a quick and dirty, or slow and sophisticated, circuit can depend on time pressure.[29] For example (Figure 5), incoming information can be evaluated quickly, but only sparsely, in the thalamus. If an important stimulus (e.g., a potential danger) is detected, a signal can be sent directly (and so immediately) to the amygdala, which can start taking action. The incoming information is then passed to the cortex for more detailed (and so slower) processing. If the cortex confirms the thalamic evaluation, action (e.g., avoidance) is continued; if it disconfirms, then different action can be initiated. Critically, in the bulk of situations, raising a false alarm has few consequences, while a slow response to a real threat can be fatal; this is a variant of the "Life-Dinner Principle," namely that it is better to sacrifice one's dinner than one's life.[30]

With the generation of either approach or avoidance behavior, the control of simple motor acts and of larger scale actions is essentially independent of the type of goal. Running through a doorway involves the same basic perceptual-motor requirements whether you are attempting to leave one room because there is a snake in it or enter the next room to get the last cookie before your friends reach it. There is, therefore, a simple hierarchical organization of control systems, from act to action to goal[31–33] (Figure 6), which means that neuroimaging can focus on particular levels of the system when asking specific levels of questions.

The most immediate control is of what we will call "acts." These are selected in more posterior parts of frontal cortex close to the primary motor strip–supplementary motor area/SMA and Area 8 (including frontal eye control fields). (For details of the numbered cytoarchitectonic cortical areas defined by Brodmann, see the Figures provided in most neuropsychology text books, e.g., Kolb & Wishaw[34] Figures 1–10, Bear, Connors and Paradiso[35] Figure 7.26.) Sets of acts make up "actions," which require deeper levels of processing (i.e., requiring more computational layers or more recursive cycles between a fixed set of layers). These are selected in parts of dorsolateral and lateral orbitofrontal

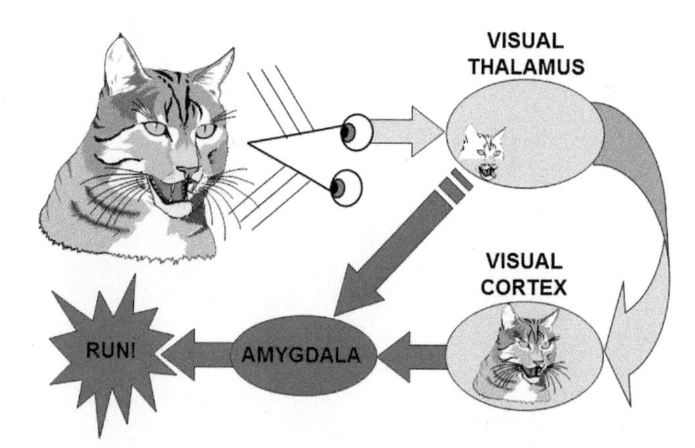

FIGURE 5 Quick and dirty perceptual processing in parallel with slow and sophisticated. Initial partial processing by the thalamus produces a quick, potentially wrong, response. Slower processing through the visual cortex will be more accurate but not desirable if survival depends on speed. *Based on Ledoux.*[29]

cortex (OFC)—among the many other functions of OFC. Sets of actions, in turn, are shaped by goals, which not only require even deeper processing but also include a motivational component, distinguishing between positive and negative valence. Goals are selected by limbic areas (anterior cingulate/ACC, prelimbic, infralimbic, and medial OFC). These act/action/goal control systems can be viewed as hierarchically organized, not only because of the progressive ordering of their functions, but also because they are interconnected and progress from the most recent isocortex through to the oldest allocortex and because they all have the same fundamental pattern of connections with subcortical structures (Figure 6). Analysis of approach/avoidance (as opposed to generic motor control independent of the goal) is therefore analysis of the way activity in limbic systems controls other aspects of the brain.

A final important aspect of hierarchical control—the matching of psychological to neural hierarchy—has been most studied in relation to avoidance and behavioral inhibition. Careful analysis of the behaviors of rats faced with cats in the laboratory determined that specific avoidance-related behaviors occurred when there were different distances between the rat and the cat.[37–40] Defensive distance reflects both a negative goal gradient of the type we have already discussed and a hierarchy of behavioral responses ranging from quick and dirty to slow and sophisticated. Importantly for neuroimaging, this behavioral hierarchy maps to a similar hierarchy of neural structures ranging from caudal (and phylogenetically old) to rostral (and phylogenetically recent).[41,42] We have suggested[2] that the systems controlling avoidance and behavioral inhibition can be seen as having a parallel hierarchical organization of this type (Figure 7(A)), where the position on the gradient of defensive distance determines the neural level that will be maximally

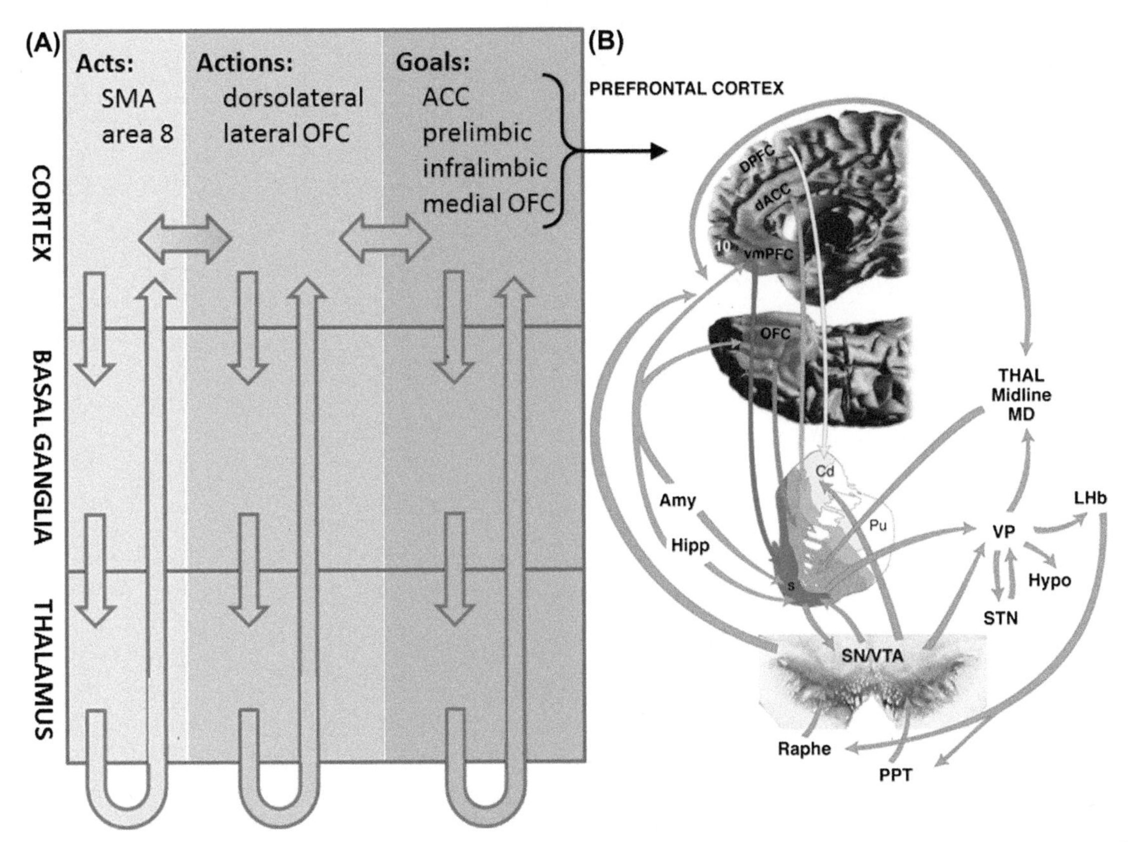

FIGURE 6 (A) Acts, actions, and goals are processed by different, interacting levels of the frontal cortex. These can be seen as parallel systems and have similar topographic relationships with the same subcortical areas.[31–33] Abbreviations: ACC = anterior cingulate cortex; OFC = orbital frontal cortex; SMA = supplementary motor area. (B) A more detailed picture of goal processing areas that process reward-related information, showing the retention of topographic organization (yellow-red arrows) of the direct connections of the various cortical areas to the basal ganglia, specifically the ventral striatum. They are also less directly connected via a range of other areas (brown arrows). *Reprinted by permission from Macmillan Publishers Ltd: Neuropsychopharmacology,[36] copyright (2009).*

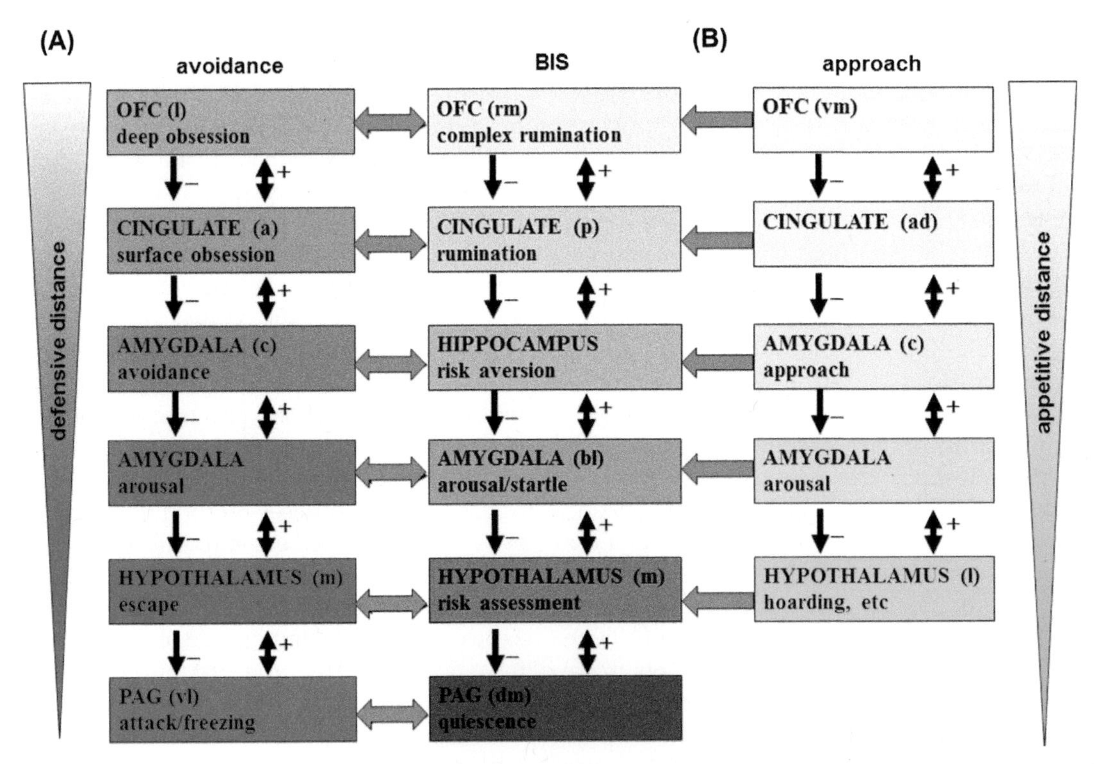

FIGURE 7 (A) Hierarchical organization of avoidance and behavioral inhibition (BIS) in terms of behavior and neural level. Lower levels process small defensive distances; higher levels process greater ones (i.e., negative events that are more distant in space or time). Activation tends to spread through the whole system (double-headed black arrows) but strong activation of a higher level (e.g., avoidance) inhibits (single-headed arrows) the behavioral output from (but not the activation of) lower levels (e.g., escape). *(Adapted from McNaughton and Corr.[2])* (B) Postulated equivalent organization of approach.[36,43,44] Abbreviations: PAG = periaqueductal gray; OFC = orbital frontal cortex.

activated and in control of behavior, and the presence of a sufficiently strong conflicting approach tendency will switch control from the avoidance system to the BIS. Although this has been less explicitly elaborated, it appears that the approach system (often referred to as the behavioral approach system, or BAS) has a similar hierarchical organization[16,36] (Figure 7(B)).

As with the motor control system, appropriate inputs activate multiple modules (anatomically localized processing units) within these systems both via direct input and via reciprocal connections between the modules. The selection of a particular module for the control of behavior essentially involves a release from inhibition (allowing control to pass quickly between modules that are already primed to produce output). Where a higher level module controlling more sophisticated responses (e.g., simple active avoidance) is highly activated, it will inhibit the specific behavioral outputs from lower level modules controlling quicker and dirtier responses (e.g., undirected escape). However, to permit fast switching between such behaviors and because the behaviors share common autonomic requirements, activation of the higher order module will not reduce (and can increase) the activation of lower order modules. As a result, although behavioral output from the lower order modules is blocked, other outputs to the sympathetic nervous system will still occur.

1.5 From System Architecture to Neuroimaging

On the face of it, the neuroimaging of approach and avoidance behaviors could not be simpler: we would first develop a task that motivated simple approach and simple avoidance, and then we would observe blood-oxygen-level dependent (BOLD) or electroencephalogram (EEG) signals during the performance of these behaviors. We would create a contrast with an appropriate control condition to ensure that we were measuring motivation-specific processes. However, our description of the complexities of the generation of approach and avoidance behavior should warn us against this seductively simple approach.

Specifically, it can be seen from our overview so far that the neuroimaging of approach/avoidance states and traits needs to take primary account of three basic systems: controlling approach, avoidance, and goal (e.g., approach-avoidance) conflict (the BIS). Neuroimaging that is intended to be specific to approach, avoidance, or conflict also needs to take into account the valuation of affective stimuli, which will differ for a gain and a loss of the stimulus as well as being transient (e.g., the value of food will depend on the current level of hunger). Simple actions (like running) can fulfill both approach and avoidance goals, and so a substantial part of the frontal cortex, basal ganglia, and thalamus can be ignored (Figure 6) if we are interested in *differences* between approach and avoidance systems, which we expect to involve limbic goal control areas. (That said, the strength of activation of the approach, avoidance, and goal conflict systems could be assessed via variation in their output to act and action systems in the same way as it can be assessed by the strength of behavioral output, e.g., speed of running.)

At the state level, for each of approach and avoidance, we should expect activation of specific modules of distinct hierarchically organized systems spanning the caudal subcortex through to the rostral cortex. The specific module (and system) momentarily controlling behavior will vary with the location of the animal within the current goal gradient(s), as well as with variations in the external amount of the motivating stimulus and with the internal valuation of that amount. However, because of the common inputs from objects in the world to different modules and because of reciprocal interconnections between the modules (Figure 7), many modules of these systems will be active simultaneously and so, to an approximation, we can also expect to detect some activation of the system as a whole.

This holistic view of approach and avoidance systems is even more appropriate at the trait level. Not only, as we have noted, are the systems strongly interconnected, but there are both evolutionary and pharmacological reasons for expecting trait modulation to often act on each system as a whole rather than targeting one specific module. This has important implications when we consider the neuroimaging of basic systems of personality.

First, in adaptive terms, we would expect sensitivity to motivationally significant stimuli to impact on the systems as a whole. In many cases, for example, an increase in the intensity of a threat is equivalent to (if not caused by) a decrease in defensive distance. Thus any trait adjustment (genetically or through development or learning) of the general strength of approach or avoidance responses will act primarily to determine which module of a system is activated at any point in time rather than altering only the intensity of activation of a single module. More module-specific trait sensitivities would affect the probability of a particular class of output. For example, there could be a selective increase in the probability of panic or of obsession to any given level of threat input without a change in more general threat sensitivity (fearfulness).

Second, pharmacologically, the systems can be viewed as relatively homogenous. This is clearest in relation to the role of serotonin, noradrenaline, and endogenous benzodiazepine ligands in avoidance and behavioral inhibition but is likely to be similar in relation to the role of dopamine and endogenous opiates in approach. As shown in Figure 8, the relatively small

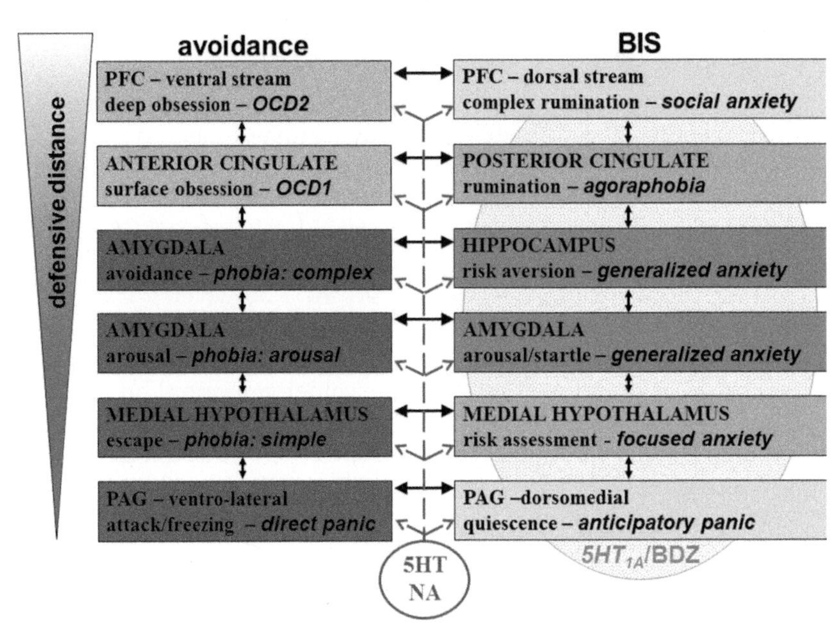

FIGURE 8 Pharmacology of avoidance and behavioral inhibition (BIS). Anxiolytic drugs—$5HT_{1A}$ agonists, BDZ—act (blue shading) on structures of the BIS to reduce the effects of goal conflict but do not affect active avoidance behaviors. More complex forms of anxiety tend to be less affected (lighter shading). Panicolytic drugs—via 5HT and NA transporter and breakdown systems—act (purple shading) on structures of the avoidance system to reduce panic and avoidance and also on structures of the BIS (where increased 5HT leads to increased $5HT_{1A}$ activation) to reduce goal conflict. Because of variations in the 5HT transporter systems, only some panicolytic drugs also reduce obsessions (lighter shading). Abbreviations: 5HT = serotonin; $5HT_{1A}$ = serotonin 1A receptor; BDZ = benzodiazepine; NA = noradrenaline; OCD = obsessive compulsive disorder; PAG = periaqueductal gray; PFC = prefrontal cortex. *Adapted from McNaughton.*[2,45,46]

numbers of serotonin and noradrenaline cells in the raphe and locus coeruleus send multiple collaterals that innervate essentially the whole of the systems controlling avoidance and behavioral inhibition. Genetic, developmental, or situation-related variation in synthesis and release would tend to affect the two systems as a whole—essentially altering defensive distance for both avoidance and behavioral inhibition. Similarly, although there appears to be some variation (see blue shading in Figure 8) in the density or effectiveness of benzodiazepine receptors among brain areas, endogenous benzodiazepines can essentially impact the whole of the BIS, altering goal conflict sensitivity independent of defensive distance. Likewise, relatively small numbers of dopamine neurons send collateral innervation to vast areas of the frontal cortex and basal ganglia and so are able to modulate approach systems quite generally. (Note, however, that the dopamine system does not merely control approach behavior; rather, it more generally facilitates flexibility in behavior to learn from and take advantage of possibilities that arise from failures of prediction.)[47]

Module-specific traits could result from more local changes in the same systems. For example, pharmaceutical companies have created compounds that target obsession more than panic, and vice versa. They take advantage of differences in the transporter molecules existing in different neural areas to produce area-specific alterations in the level of transmitters in the synaptic cleft for any given released amount. Genetic variation in transmitter uptake among these areas could, therefore, give rise to specific traits. Likewise, changes in receptor subtype or density within an area would change the postsynaptic effect achieved by any particular level of transmitter.

We can expect the state imaging of approach/avoidance to show changes in both the focus and magnitude of activation within distinct approach and avoidance goal-processing systems (Section 1.1). These changes in activation should depend both on changes in valuation affected by exchange rate factors, including loss aversion (Section 1.2), and on the location of the individual along spatial and temporal gradients (Section 1.3). When approach and avoidance goals are concurrently and similarly activated (i.e., the individual is in the range of intersection of their opposing goal gradients), the behavioral inhibition system will also be activated, in addition to the approach and avoidance systems (Section 1.3). With trait imaging, we can expect variation to be evident more globally across each system (and even between systems) and to be dependent on more global (e.g., hormonal) biological factors. However, the hierarchical organization of the systems means that the detection of trait variations may be as much a matter of detecting changes in the typical neural focus of activation within a system as it is a matter of detecting changes in the level of activation of a specific module within a system.

2. STATE NEUROIMAGING OF APPROACH, AVOIDANCE, AND GOAL CONFLICT

There is a vast amount of literature looking at the details of neural reactions to motivationally significant stimuli, choices, and responses. We will provide in this section only a brief, high-level overview, placing it in the context of the behavioral and neural foundations of goal-directed approach and avoidance provided in Section 1. We will focus on more global, systemic issues as a transition between the details of the key neural systems that we have already reviewed and attempt to assess their trait sensitivities, i.e., approach-prone and avoidance-prone personalities, which we will review in the next section.

Human imaging has not generally focused on approach or avoidance behavior as such. It has more often focused on "reward" or "punishment," which usually blend valuation and learning with approach or avoidance behavior. Human imaging has also seldom combined both negative and positive events with both negative and positive response contingencies. Further, in many cases, the neural response to different stimuli has been measured without any requirement to generate behavior. Where behavior is generated, it tends to be limited to pressing buttons, as there are strong technical reasons for limiting movement during functional magnetic resonance imaging (fMRI) and, although free-moving radio-transmitted EEG recording is available, the bulk of evoked potential and rhythmic EEG recording limits movement to reduce artifacts. Current imaging paradigms therefore make it difficult to identify signals that are specific to goal-directed approach or avoidance and are not confounded by other factors.

A further complication is that brain activation in relation to a motivationally significant event can reflect several different aspects of that event. Specifically,[48] valence (positive/neutral/negative) may be signaled independently of amount, salience (increasing with amount) may be signaled independently of valence, and value (valence x amount) may be signaled selectively for one valence with no value variation for the other valence.

The summary provided below largely ignores these complications. It lumps gain/reward with approach and loss/punishment with avoidance. It blurs the processing of upcoming goals with the evaluation of the outcomes of responding. Conversely, it focuses on differences in neural localization, or attempting to distinguish approach, avoidance, and goal conflict systems. This ignores the complication that there may be strong neural overlap between these different types of goal processing (with, e.g., lateral OFC, anterior insular cortex, and ACC showing valence-independent activations).[48] But there is also the likelihood that particular subregions may differentiate between approach, avoidance, and goal conflict in a way that generates a combined, blurred signal with current and relatively poorly localized imaging methods.

2.1 State Neuroimaging of Approach/Reward

As shown in Figure 6, goal-processing areas control behavior via links to the striatum,[36] and this is a major locus where the release of dopamine can alter future behavior. It will be significant for Section 3 that alterations in the dopamine system affect a broad range of motivated behaviors,[16] and so here we will focus first on the striatum and dopamine and then consider the goal-processing areas (Figure 7) on the output of which striatal dopamine can impact.

Imaging studies[36,49] show dopamine release linked to responses for secondary incentives as well as primary incentives, such as pleasant sounds or the simple presentation of food items to hungry participants. Striatal activation is observed prior to monetary choice and is maintained by gain more than loss (with value affecting dorsal but not ventral striatum and the latter preferentially reacting to gain as compared to loss outcomes). However, the striatal response is not mediated by the simple delivery of a positive or negative outcome but rather reflects whether the associated action is being reinforced. Additional data "strongly suggest that the human dorsal striatum is involved in reward processing, specifically learning and updating actions that lead to reward, rather than representing and identifying rewards, a function postulated to occur in frontal cortex."[49] The same appears to be true of the ventral striatum, which may sometimes code "reward prediction error."[36] Although reinforcement is important for determining the specifics of approach (and avoidance) behavior, these data provide no evidence that the basal ganglia are part of the specific systems controlling approach, to which we turn next.

The lowest level at which positive goals are known to be controlled is the lateral hypothalamus (Figure 7(B)). Imaging of the hypothalamus[50] via scalp EEG is impossible and via fMRI is difficult. It is less than 10 mm across, close to the sinuses, and contains different closely adjacent nuclei that may show opposite responses to a situation and therefore cancel out each other's signals. However, it has been studied to a small extent in the context of sexual arousal[51] and appetite control with results that suggest, consistent with Figure 7(B), that it "acts as a central gateway modulating homeostatic and nonhomeostatic drives."[50] Its activity is reduced by glucose (but not by either artificial sweetener or a nonsweet calorific solution) and this response is reduced in obese people and absent in type 2 diabetics. Its activity appears to be increased by pictures of fattening food and is positively correlated to caloric intake.[50]

The amygdala is activated by food images in hungry participants (but not by nonfood images or in satiated participants) and by high-calorie items as opposed to low calorie items, with a stronger response and weaker amygdalar modulation of other areas to high calorie items in obese people.[50] (The hippocampus is also activated by food images—this will be dealt with in more detail in the section on goal conflict, below.) The "hunger hormone," ghrelin, also activates the amygdala.[50,52] The amygdala is also activated[36,53] by potential rewards, and its response decreases with reward devaluation. Compared with the ventral striatum, the response of the amygdala to reward shows rapid habituation, and the extent of activation does not differ from that produced by potential punishments when arousal level is controlled. It is also activated during sexual arousal.[51] These data are consistent with the view that the amygdala is important for the processing of positive goal stimuli and controls arousal for all motivational systems, both positive and negative.

The dorsal anterior cingulate cortex (dACC), as with the other areas we have considered, is activated by food relative to nonfood items; is more activated by high calorie items in obese people; and, together with other aspects of ACC, is involved in sexual arousal.[51] However, dACC (and dorsal prefrontal cortex/PFC) also engage in reward processing that is not directly linked to valuation (for example, supporting working memory for incentives that can then be used for outcome evaluation).[36] Indeed, "it has been proposed that the overall function of the dACC might involve the use of outcome, and particularly reward-related, information to guide action selection."[54] This outcome evaluation process, however, produces particularly clear activations (with fMRI and, particularly, EEG rhythmicity) when there is outcome conflict (i.e., when an expected reward is not delivered). Thus a major function of dACC (not inconsistent with simple outcome evaluation) appears to be outcome conflict monitoring.[54] As we discussed earlier, this outcome conflict (which should not be confused with goal conflict, see below) will generate a negative goal and will have all the effects expected of an explicit punishment. Interpretation of imaging results from dACC, therefore, has similar complications to the hypothalamus and amygdala, with both positive and negative information being processed in what appears to be closely adjacent areas.

OFC, as with the other areas we have considered above, is activated by visual food as opposed to nonfood images and shows a particularly clear differentiation in this response between fasted and sated states. It shows a high level of functional connectivity with the hypothalamus, increased activity in response to ghrelin,[52] and in general can be considered to be the most significant node in the networks encoding the rewarding value of stimuli.[50] Its posterior part is also activated during sexual arousal.[51,55] Its more anterior portions are activated by more abstract rewards, such as monetary gain.[55] Whereas medial regions respond to rewards (both in terms of magnitude and probability and adjusted for temporal discounting), more caudolateral regions respond to punishments and more rostromedial lateral regions appear to be involved in behavioral inhibition.[36,48] Medial OFC, together with adjacent medial frontal cortex, appears to be activated by the expectation of reward and is not activated by habitual stimulus-response learning.[53]

2.2 State Neuroimaging of Avoidance/Punishment

Some imaging studies focusing directly on defensive behavior have been undertaken and are considered in the following paragraph in relation to the entire system shown in Figure 7(A). Studies that investigate loss and punishment are then briefly discussed in relation to the studies of gain and reward reviewed in Section 2.1.

Volunteers in a virtual maze with a virtual predator (which could capture them and produce real pain via electric shock to a finger) showed activation of ventromedial PFC, rostral ACC and medial OFC, basolateral amygdala, central amygdala, and periaqueductal gray (PAG). This activation was strongest in more rostral structures to distal threat and more caudal ones to proximal threat—consistent with Figure 7(A).[56,57] Similarly, with simple aversive conditioning of a shock with a simple stimulus or a contextual stimulus, there was stimulus-related activation in the amygdala and hippocampus, respectively,[58] whereas with a virtual reality context, both amygdala and hippocampal activation were detected[59]—also consistent with Figure 7(A) (see also Section 2.3). In experimentally induced panic attacks, the PAG (and a range of other parts of the upper brain stem) and hypothalamus are activated, while medial PFC is not, and the ACC can become deactivated. This is consistent not only with the rostral-caudal shifts observed with decreasing defensive distance, but also with the idea that the strong activation of one level of the system will tend to reduce the involvement of other levels.[42] A PAG-hypothalamus-(amygdala)-premotor cortex network is activated by video clips of threatening actions,[60] while the anterior midcingulate responds to a range of stimuli that generate an intense negative effect, ranging from simple pain, through pain anticipation, the closeness of a spider to a foot, and emotionally charged words.[61] These results are all consistent with the notion of a hierarchically organized active avoidance system presented in Figure 7(A). Also consistent with the distinction between active avoidance (anxiolytic insensitive) and BIS (anxiolytic sensitive), is the finding that benzodiazepine administration does not reduce pain-related activations in the brainstem, ACC, anterior and posterior insula, PFC, and other areas.[62]

The bulk of studies investigating loss or punishment in the same types of experiments as they investigate gain or reward involve the omission of the loss/punishment under conditions where it is difficult to distinguish the avoidance of danger from the approach to safety (Section 1). With this caveat, there is evidence for affectively negative activation in the amygdala, ACC, and more caudolateral aspects of OFC.[36,48] Explicit manipulation of the strength of unconditioned frustration (theoretically equivalent to punishment) activates PAG, amygdala, dACC and insula.[63] However, the areas controlling approach and avoidance may be difficult to distinguish.[48] With rewarding and punishing outcomes (as opposed to goal-oriented anticipation), activation can be virtually ubiquitous both with very little distinction between them and a capacity "for positive and negative outcomes to directly influence neural processing throughout nearly the entire brain."[64]

2.3 State Neuroimaging of Goal Conflict

Goal conflict is, in principle, more complex than simple approach or avoidance. But it has been subjected to much more extensive theoretical and neuropsychological analysis (Section 1.3), and critically, the actions of anxiolytic drugs have distinguished the BIS from simple approach and avoidance systems and so can be used to validate behavioral tests across species and dissect BIS-related (conflict) activations from those produced by a simple threat. However, it should be noted that one of the effects of BIS activation is to amplify avoidance tendencies.[1,2] Areas that are activated both by conflict and in simple avoidance tasks will clearly be involved in generating BIS-related output but cannot be definitively assigned as parts of the BIS itself, unlike those activated solely by conflict.

A recent, 2014, particularly explicit assessment of approach-avoidance conflict activation in humans used monetary tokens to generate approach and a virtual predator capable of causing the loss of a large number of tokens. Consistent with their BIS-theory-based hypotheses (see Figures 7 and 8), there was a linear correlation between the level (probability of loss) of imposed threat and the size of BOLD response in the left anterior hippocampus. There may also have been activation in the adjacent amygdala. Reliable (whole brain corrected) activations were also seen in right inferior frontal gyrus/insula, bilateral parahippocampal gyrus, and right fusiform gyrus. Interestingly, they confirmed the functional status of the hippocampal activations by showing that patients with a hippocampal lesion showed reduced approach-avoidance conflict in the same task.[65] Animal work has also linked contextual but not simple cued fear conditioning to the BIS. Consistent with this, the right amygdala, ACC, and insula (see also Section 2.4) were activated by a simple cue, as well as by the context during conditioning with a shock to the forearm as an unconditional stimulus. In contrast, the left hippocampus was activated only in the context condition.[58] Contextual conditioning (but without a simple cue as a contrast) with foot shock has also been reported to activate the *right* anterior hippocampus (as well as a range of other structures).[59] A potential predator threat has also been contrasted with direct predator interaction in a task with a virtual predator delivering delayed shocks. This contrast should specifically assess areas involved in the BIS, as it controls for more general threat detection and avoidance. Potential rather than direct predator interaction involved posterior cingulate cortex, bilateral hippocampus, hypothalamus, and amygdala (all consistent with Figure 7), as well as ventromedial PFC and subgenual ACC.[57]

Go/No-Go is a simple test of response inhibition with clear "involvement of a right prefrontal region, comprising the posterior part of the inferior frontal gyrus (IFG) and the adjacent part of the middle frontal gyrus (MFG) in the inhibitory process. ...The anterior cingulate gyrus is also commonly activated ... but has been attributed a more generic role of selective, executive attention and performance monitoring, which is consistent with the finding of its activation in particular during failed inhibition trials."[66] In contrast to more lateral regions of lateral OFC (Section 3), more rostromedial regions of lateral OFC appear to be involved in behavioral inhibition generally.[36,48]

Anxiolytic drug effects are generally consistent with the picture provided by simple conflict-related activations. Benzodiazepines, specific serotonin reuptake inhibitors (SSRIs), and, to some extent, pregabalin reduce threat-related activation in the amygdala, insula (see Section 2.4), and medial PFC.[67] Benzodiazepine reduced the activation generated by the anticipation of pain clearly in the right insula/inferior frontal gyrus/superior temporal gyrus but only marginally and nonsignificantly in the ACC, while not reducing any activations generated by pain itself.[62] Delta-9 tetrahydrocannabinol reduces Go/No-Go activation differences in the right inferior frontal gyrus and ACC.[66]

The most theoretically driven experiments attempting to image the BIS have involved the use of EEG. The detailed neuropsychological theory of the BIS[1,2] has, at its core, the fact that, without any false positive or negatives so far, all clinically effective anxiolytic drugs (including those that have no effect on panic or depression) reduce the frequency of hippocampal rhythmical slow activity (RSA, 5–12 Hz in the rat, likely somewhat lower in humans).[68]

"We developed a human homologue of rat RSA as a biomarker for BIS hyper-reactivity[/anxiety]. Hippocampal depth recording is impractical for assessing [BIS activity] in humans. However, in rats, rhythmicity in frontal cortex becomes coherent (phase-locked) with hippocampal RSA during risk assessment behaviours.[69] Since the hippocampus itself shows RSA even when it is not in control of behaviour, this outflow of RSA to PFC should be more predictive of BIS functional output and act as a better biomarker than hippocampal recording. We therefore searched for rhythmicity in human frontal cortex that was generated by goal (approach-avoidance) conflict and sensitive to anxiolytic drugs.

We measured human scalp EEG during approach, conflict, and avoidance, subtracting the average power in approach and avoidance from conflict to measure *goal – conflict-specific rhythmicity* (GCSR). We found GCSR at a right frontal cortex site (F8).[70,71] Right frontal cortex (particularly the inferior frontal gyrus) controls stopping[72–75] (a major output of the BIS) in the Stop Signal Task (SST)[76]. ...We used the SST to extract GCSR from F8 and found that this correlated positively with both trait anxiety and neuroticism.[77] Critically, we later showed that F8 GCSR was reduced by both benzodiazepine and 5HT1A drugs[78] that share, in the clinic, only BIS and not [avoidance] or antidepressant actions. So, right frontal GCSR elicited in the SST task in humans is pharmacologically homologous to RSA elicited by electrical stimulation in rats." McNaughton[45], p. 140

2.4 "A Link between the Systems"—State Neuroimaging of the Insula

In the previous sections, we have mentioned only in passing one structure that is routinely activated in tasks involving goals:[48,53] the insula. The previous sections reviewed systems that are likely to be specifically involved in goal-directed approach, avoidance, or conflict and that are likely to be distinguished either in terms of the large-scale structures involved (e.g., hippocampus) or in terms of the specific nuclei involved within a structure (e.g., within the amygdala). The insula is a "distinct, but entirely hidden lobe... (mostly) reciprocally connected with the amygdala, and with many limbic and association cortical areas, and is implicated in an astonishingly large number of widely different functions, ranging from pain perception and speed production to the processing of social emotions."[79] It acts as a major network hub,[80,81] which can be viewed as a "limbic integration cortex."[82] Rather than specifically supporting approach, avoidance, conflict, or any more complex aspect of goal processing, the anterior insula in particular[81] appears to act as a "link between the systems,"[83] allowing a mixed readout from the motivational activations of all of them.

The anterior insula appears to be involved in at least the initial aspects of goal processing, particularly valuation. Ghrelin (see also Section 2.1) activates anterior insula.[52] Reward anticipation activates the anterior insula more than reward outcome.[84] The anterior insula has activations related to the subjective value of rewards independent of type,[55] to sexual arousal,[85] to the values of losses,[48] and to the proximity of threat.[86] Variations in loss aversion between people are mirrored by valuation-related differences in activation to gain and loss in the insula.[20] Risk, in many different forms, activates the anterior insula,[87] and risk averse people show stronger anterior insula responses in anticipation of high risk gambles.[88]

Overall, the anterior insula, despite some parcelation and differentiation, appears to be "instrumental in integrating disparate functional systems involved in processing affect, sensory-motor processing and general cognition and is well suited to provide an interface between feelings, emotion and cognition."[89] Suggested integrative functions include the following: (a) "mediating dynamic interactions between other large-scale brain networks involved in externally oriented attention and internally oriented or self-related cognitions... [so as to mark salient] events for additional processing and initiate appropriate control signals ... to guide behavior";[90] and (b) integrating "different qualities into a coherent experience of the world and setting the context for thoughts and actions."[83] It appears, then, to be an area where the outputs of approach, avoidance, and conflict systems can become integrated rather than being a part of any one of those systems or having separate zones within it dedicated to each.

Despite its apparent role in integrating across the motivational systems, there may be reason to see the insula as particularly important for anxiety in general (the BIS in particular). Its likely role in monitoring higher levels of the motivational systems and "initiating appropriate control signals" clearly includes involvement in anxious anticipation,[86] and the equivalent of the conflict-monitoring and resolution functions of the BIS. Like the BIS, its dysfunction has been specifically linked to phobic and anxious disorders.[91] It is also closely associated with the right inferior frontal gyrus, which makes a major contribution to response inhibition[92] in distinction to mesial PFC, which is more involved in error detection,[93] and to ACC, which is more involved in outcome conflict monitoring.[54,94]

The insula also contains both benzodiazepine[95] and 5HT1A[96] receptors. An indication of their likely function is given by the fact that the benzodiazepine, midazolam, reduces anterior insula activation by anticipated pain but not by pain itself.[62] Likewise, benzodiazepines and SSRIs (and to some extent pregabalin) reduce threat-related activation in the insula.[67] The presence of direct targets for different types of anxiolytic in the insula raises the possibility that it may contain frontal components of the BIS. (The neuropsychology of the BIS, see Figure 7, has not been worked out in detail for the frontal cortex.)

However, the insula appears to go beyond simple goal conflict even in its involvement in inhibitory control,[32] and it has been proposed as part of "a ventral system, including the amygdala, insula, ventral striatum, ventral ACC, and PFC, for identification of the emotional significance of a stimulus, production of affective states, and automatic regulation of emotional responses [and not part of] a dorsal system, including the hippocampus, dACC, and PFC, for the effortful regulation of affective states and subsequent behavior."[97,98] On this view, it would be quite distinct from the BIS, which has the hippocampus as its most important node.

3. TRAIT NEUROIMAGING OF APPROACH, AVOIDANCE, AND GOAL CONFLICT

Psychological traits describe a variation in the likelihood of being in particular classes of states, and much of this variation is assumed to be due to differences in the sensitivities or strengths of the systems that generate the relevant states in response to appropriate eliciting stimuli.[99] Traits that reflect variation specifically in approach or avoidance behavior, therefore, are hypothesized to stem from variation in the systems described above as underlying approach- and avoidance-related states. Investigating this hypothesis requires testing whether traits are associated with relevant neurobiological parameters.

For three reasons, trait research is more difficult than research designed to understand the nature of approach and avoidance systems as such. First, much larger sample sizes are required in order to have sufficient statistical power to test associations with individual differences, relative to the samples required for comparing the operation of a particular brain system under different conditions in the same individuals. Second, an additional type of measure must be included in the research—that is, measures of traits, which must be validated as measuring a reasonably stable pattern of behavior over an extended period of time. Third, neurobiological parameters that can adequately explain trait variation must themselves be reasonably stable over time. This requires proof that measures of those parameters are sufficiently trait-like.

The field of personality neuroscience is sufficiently new that at least two of those three requirements have not been met in much of the existing scientific literature. First, many MRI and positron emission tomography (PET) studies have used such small samples that their findings are of little evidentiary value. This is because underpowered samples not only increase the likelihood of false negatives, in which a real effect cannot be detected as statistically significant, they also increase the proportion of significant findings that are false positives, in which an effect that has been detected as significant does not in fact exist. The latter problem is a direct result of sampling variability—the smaller the sample, the more likely it is to be so unrepresentative of the general population as to yield, by chance, a parameter value large enough to be significant when the true value is, in fact, close to zero (or even far from zero in the opposite direction).

Unfortunately for the study of individual differences, correlations are particularly susceptible to outliers, and close to 200 participants are necessary to achieve 80% power to detect the average effect in personality research ($r=0.21$,[100] which is similar to the typical effect size in psychology more generally).[101] In the last few years, sample sizes in personality neuroscience have been increasing, but there is still relatively little trustworthy existing research to review. The median sample size in a random sample of 241 neuroimaging papers published after 2007 was 15.[102] Many of these were studies of within-person effects (though even for within-person research, 15 is typically inappropriately small), but many correlational studies have also been published with samples smaller than 20, and these should not be trusted.

Additionally, most neural variables have not been examined in terms of their test-retest reliability. If a measure is not reliable as an index of a stable parameter, then it is not trait-like and, therefore, probably cannot be systematically linked to any given trait, even if the system being measured is genuinely related to that trait. Some research has begun on the test-retest reliability of MRI assessments, and additional work in this vein will be crucial for the advancement of personality neuroscience.[103–105]

In contrast to neuroimaging variables, trait measures, especially questionnaire measures, are much more likely to have been validated as having sufficient test-retest reliability. Still, questionnaire measures have limitations of their own. Various biases may influence the way that people respond to questionnaires, and the best questionnaire assessment can be achieved by including peer ratings in addition to self-ratings.[106,107] Other methods of trait assessment exist, using decision-making or behavioral tasks instead of questionnaires, but these have been much less extensively developed. One example of a task that has been validated as having trait stability is a decision-based assessment of temporal discounting.[108,109]

One obvious limitation of all psychological trait measures is that they do not reveal the neurobiological systems underlying the behavior they assess. For this reason, labeling questionnaires with the names of neural systems is potentially misleading. The widely used BIS/BAS scales,[110] for example, do not measure and have not been tested against the sensitivity of BIS (see Section 2.3) and BAS as neural systems. They assess patterns of behavior and emotional experience that their authors hypothesized to be linked to BIS and BAS sensitivity. At this early stage of personality neuroscience, we should not assume the very hypotheses that need to be tested. The question of how the approach, avoidance, behavioral inhibition, and other neural systems are responsible for the variation in traits is precisely what the field must strive to discover.

Neuroimaging research to test whether various traits are related to these neural systems has typically been carried out using questionnaire measures of traits. Two major approaches have been used to develop personality questionnaires: the theoretical approach and the empirical approach. In the former, one starts with a trait construct identified through observation and theory and attempts to develop items rationally. These items are then typically winnowed, through psychometrics, so as to be sufficiently unidimensional, and the scale is often validated by showing convergent and discriminant validity with other scales and by demonstrating that it predicts some hypothetically relevant behavior. This is the approach that was used to develop the BIS/BAS scales, for example.[110] The major limitation of this approach is that one's theory and intuitions about what items will be good indicators of the construct in question may be wrong. One may develop a scale that is reliable psychometrically but lacks validity (i.e., that does not adequately measure the construct intended) and predicts seemingly relevant behaviors for reasons other than those dictated by the guiding theory. For example, the Autonomic Perception Questionnaire does not relate to people's actual individual capacity to perceive their heart rate.[111]

In empirical questionnaire development, in contrast, one starts with a broad pool of variables (items or scales), without a priori hypotheses regarding exactly what trait dimensions are measured. Factor analysis is then used to determine the major dimensions of covariation among the variables. With a sufficiently broad and unbiased pool of variables (difficult to achieve in practice), this approach is theoretically capable of addressing one of the central questions of personality research, namely which traits tend to manifest together in the same people. This approach is crucial for understanding the structure of personality traits, and it has led to the widely used Five-Factor Model or Big Five,[112,113] sometimes extended to include a sixth factor.[114] Such models are important because they identify the major dimensions of personality, but they have the serious limitation that they reveal nothing about the sources of those dimensions. Considerable research at lower (e.g., neural) levels is subsequently needed to understand what causes the traits in each dimension to covary.[115,116]

A third approach, which has so far been little used in personality neuroscience but is a promising alternative to the standard theoretical and empirical approaches, is the criterion approach. This approach would start with a well-validated biomarker (a trait-like neural parameter) and then identify questionnaire items most strongly associated with that variable. This is already possible with the BIS.[45,117] A scale developed on this approach would not only provide new insights regarding the behavioral and experiential correlates of the biomarker, it would also allow the best possible questionnaire measurement of a neural parameter. Importantly, it is theoretically possible that some key, stable, biological source of individual differences could be identified first, and the nature of its emergent psychological construct determined only later.

3.1 Neuroimaging and Approach Traits

Both theoretically and empirically derived scales have been used in neuroimaging research. Although approach and avoidance tendencies have not been measured by questionnaires in such a way as to disentangle motivation from valuation, we can nonetheless identify traits related to these motivational factors. To begin with those related to approach, the most commonly studied traits are those in the extraversion family. Extraversion is one of the Big Five dimensions, reflecting the shared variance of more specific traits, such as gregariousness, assertiveness, enthusiasm, talkativeness, activity level, and excitement-seeking. Though often expressed in the social domain, extraversion has been hypothesized to reflect sensitivity to reward more generally, with the tendency to approach positive goals as an important component of the trait.[115,116,118] The BAS scale[110] shows reasonable convergent validity with extraversion and can be included in this family of traits.[119,120]

Research on extraversion is covered in more depth in Chapter 6, but we will briefly review some of the studies that have linked extraversion and related traits to components of the BAS. The most compelling evidence comes from studies showing that people high or low on extraversion respond differently to pharmacological manipulation of the dopaminergic system.[121–129] A number of fMRI studies have reported that extraversion or the BAS scale is associated with neural activation in response to emotionally positive or rewarding stimuli, often in approach-related brain regions, including those in Figure 7.[130–133] However, all of the studies just cited used samples smaller than 20, meaning that they are of little evidentiary value. In the future, higher-powered fMRI studies should be one of the most powerful methods for testing the hypothesis that variation in the approach system underlies traits in the extraversion family.

Structural MRI studies have begun to use larger sample sizes, and several of them have found that extraversion is associated with volume in the ventromedial

PFC/OFC.[134–137] Other studies, however, have not replicated these findings.[104,138–140] However, these studies have varied in their methods and populations studied, which could account for some of the inconsistencies. More large primary studies, as well as meta-analyses, will be needed to provide convincing tests of this and other effects in personality neuroscience. Associations of extraversion with other brain regions have been even more inconsistent in structural MRI studies. One study worth mentioning because of its large sample size ($N=486$) reported that an extraversion-related scale (positive emotionality from the Multidimensional Personality Questionnaire) was positively associated with left amygdala volume.[141]

Resting EEG hemispheric asymmetry, in which the one frontal lobe of the brain is more active than the other, is another neural parameter that has been linked to motivation.[142] Controversy exists, however, regarding whether left versus right bias corresponds to approach versus avoidance or to goal-directed activity (including both approach and active avoidance) versus goal conflict and passive avoidance.[143] A number of studies have found that extraversion is related to left-dominant asymmetry, but failures to replicate have been reported as well, and a meta-analysis found no evidence for the effect.[144] Nonetheless, although EEG frontal asymmetry at rest may not be generally related to trait approach, it is possible that frontal asymmetry is related to approach-related traits in contexts that evoke approach motivation. In an all-male sample, the BAS scale was found to predict resting-state asymmetry only for participants interacting with a female experimenter whom they rated as attractive[129]—such findings point to further complexities in the neuroimaging laboratory that should not be ignored. Another much smaller study found a similar effect: a trait measure of approach-related positive affect was associated with asymmetry only under the condition of a positive mood manipulation, but not in negative or neutral-mood conditions.[145] The possibility that extraversion is associated with asymmetry only in certain contexts would be consistent with the definition of traits as tendencies to respond in particular ways to particular classes of stimuli. Without the presence of a relevant stimulus, the trait may not be manifested, and individual differences in behavior or neural activity may not be apparent.

Gray[24] originally hypothesized that the trait most associated with BAS sensitivity is impulsivity, but more recent research suggests extraversion is probably more specifically related to BAS sensitivity.[118,120,146,147] Nonetheless, some forms of impulsivity (particularly those related to extraversion) appear to be linked to the neural systems involved in approach. Impulsivity is "the tendency to act on immediate urges, either before consideration of possible negative consequences or despite consideration of likely negative consequences."[148] Because any instance of impulsivity requires both the presence of an impulse to act and a failure to constrain that impulse, variation in both bottom-up impulse systems related to approach and avoidance, as well as top-down constraint systems in the PFC, may lead to different types of impulsivity.

Two types of impulsivity appear to be importantly related to approach behavior: (1) the tendency to act quickly to approach potential reward, with little deliberation or premeditation and (2) the tendency to take risks for the sake of excitement or novel experience. Both of these traits have been linked to dopaminergic function, using PET imaging to show that they reduced D2 binding in the midbrain, which in turn predicts greater dopaminergic release in the striatum in response to amphetamine.[149,150] A lack of deliberation has also been shown, in fMRI, to predict increased ventral striatal activity in response to cues of reward.[151,152] (For further discussion of impulsivity, see Chapter 8.)

3.2 Neuroimaging and Avoidance and Goal Conflict Traits

Personality research has tended not to distinguish clearly between traits reflecting individual differences in active avoidance versus those that reflect differences in processing goal conflict. This is probably due to the fact that activation of the BIS leads to increased arousal of active avoidance systems and biases motivation toward avoidance, as well as to the fact that certain neuromodulators, including serotonin and noradrenaline, influence both systems. In psychometric research, trait measures of anxiety, depression, and other traits related to passive avoidance covary strongly with trait measures of fear, panic, irritability, and other emotional forms of active avoidance. Together, the tendency to experience all of these negative emotional states (and related cognitive, motivational, and behavioral states) constitutes the broad Big Five dimension labeled neuroticism. Neuroticism is the major risk factor for psychopathology.[153]

Many theoretically derived trait measures fall within the neuroticism family, including Cloninger's Harm Avoidance,[154] various measures of trait anxiety,[155] and Carver and White's BIS scale.[110] (For further review of neuroimaging related to traits in this family, see Chapter 7.) The fact that a scale is labeled an "anxiety" scale does not mean it measures anxiety in the specific sense, related to the BIS, we have used in this chapter, and, due to the ambiguity of the concept, it may not measure "anxiety" in a more general sense either. Neuroticism appears to have two major subfactors, one reflecting anxiety, depression, and other internalizing problems, the other reflecting irritability, anger, emotional lability, and the tendency to get upset easily.[156] Scales in the neuroticism family tend to measure either the first factor or a blend of both factors. Only rarely do they target more specific facets capable of distinguishing, for example, between anxiety and depression. A recent attempt to develop separate scales reflecting approach, avoidance, and conflict sensitivities, using the theoretical method, was carried out by Corr and Cooper.[157]

Gray and McNaughton[1] posited that neuroticism reflects the joint sensitivity of behavioral inhibition and avoidance systems, implying that it should be influenced by the neuromodulators serotonin and noradrenaline. This hypothesis has been supported for serotonin using a variety of methods, including genetics and pharmacological manipulation, but also neuroimaging:[115] several PET studies have found that neuroticism predicts variation in serotonin receptor or transporter binding.[158–160] Only the most recent of these used a sample large enough to be particularly informative, however, and additional studies are necessary. Less evidence exists linking noradrenaline to neuroticism, but it remains a promising hypothesis.[161,162]

A number of fMRI studies have reported that neuroticism predicts neural responses to aversive stimuli in relevant brain areas from Figure 7, but most of these have used samples so small as to preclude confidence in their results. Of 21 samples in a recent meta-analysis of these effects,[163] only 7 of them were larger than 25, and only 1 was larger than 60. Unfortunately, meta-analysis is not a panacea for the problems created by underpowered samples, as meta-analytic conclusions are likely to be biased by their inclusion. Nonetheless, it is worth noting that the meta-analysis in question found neuroticism to be associated with neural activity only in aversive relative to neutral conditions and not in positive relative to neutral conditions, which is encouraging for the theory that neuroticism reflects the major manifestation of both active and passive avoidance motivation in personality; however, it does little to throw much light on the nature of this association.

Many theoretical accounts of the neurobiology of neuroticism highlight a central role for the amygdala, which is unsurprising given its role in the mobilization of anxiety and fear. Although the meta-analysis by Servaas and colleagues[163] did not implicate the amygdala, some larger fMRI studies have found associations between avoidance traits and amygdala reactivity to aversive stimuli. Unfortunately, these studies have used a variety of different methods and experimental paradigms, so it remains difficult to draw any firm conclusions. Neuroticism or trait anxiety has been found to predict: (1) a slower decrease in amygdala activity after viewing aversive images;[164] (2) greater amygdala activity in response to aversive images, but only in people generally lacking in social support;[165] (3) reduced synchrony between amygdala and other limbic regions, especially in the PFC;[166] and (4) reduced synchrony between the left amygdala and the medial PFC when viewing negative versus neutral emotion faces, but increased synchrony between these structures in the right hemisphere.[167] The last two findings in this list raise the possibility that functional interactions between the amygdala and regions of the frontal cortex may be important in determining levels of neuroticism.

One fMRI study reporting a link between neuroticism and neural activity in the amygdala is worth mentioning, despite having a sample of only 17, because it used an innovative method to distinguish between valuation and motivation.[168] Participants viewed positive, negative, and neutral images and were required either to approach them (by pressing a button that enlarged the image, creating the illusion of approach) or to avoid them (by pressing a button that reduced the image in size). The study found that one of the two major subfactors of neuroticism (related to anxiety and depression) predicted amygdala reactivity to approach relative to avoidance (regardless of stimulus valence), whereas the other (related to anger and lability) predicted amygdala reactivity to negative relative to neutral and positive stimuli (regardless of motivational direction).

The apparent importance of the amygdala for neuroticism in terms of brain function is consistent with structural neuroimaging studies that have shown traits in the neuroticism family to be related to increased amygdala volume.[169,170]. As with approach-related traits, however, there has been little consistency in studies of the association of traits in the neuroticism family with the volume of specific brain regions, even as sample sizes have increased, and other studies have not found amygdala volume to be associated with these traits.[135,171,172] Some of the inconsistencies here may reflect methodological differences, given differences in the questionnaires used, and differences in MRI analysis. Refreshingly, in this case, a nearly definitive study has been carried out in a sample of over 1000 people, which found that neuroticism scores (based on the average of several commonly used questionnaire measures) were indeed positively correlated with amygdala volume (after controlling for age, sex, and total brain volume), albeit very weakly ($r=0.1$, i.e., accounting for only 1% of the variance[103]). Only one other subcortical structure, the hippocampus, was also significantly correlated with neuroticism ($r=0.1$). These findings are important both because they confirm that neuroticism is associated with the two subcortical structures most strongly theoretically implicated in the trait, and also because they suggest an explanation of inconsistent findings. The structural effects studied are weak enough that even samples that are very large by neuroimaging standards may be underpowered to detect them.

In addition to demonstrating the positive association of neuroticism with both amygdala and hippocampal volume, Holmes and colleagues[103] found that neuroticism was negatively associated with the thickness of a region of the rostral ACC and the adjacent medial PFC. Further, among the individuals scoring highest in neuroticism (more than one standard deviation above the mean), cortical thickness in this region was negatively correlated with amygdala volume (whereas they were unrelated in the rest of the sample). These additional findings are consistent with the theory that neuroticism results not only from functional sensitivity of subcortical structures involved in avoidance, but also from impaired regulation of these structures by higher-level control systems. Also consistent with this theory are diffusion tensor imaging studies (which assess the structure of white matter tracts) showing that traits in the neuroticism family are associated with reduced integrity of tracts connecting cortical and subcortical regions.[138,173–175]

One additional broad pattern has appeared in structural neuroimaging research on neuroticism: the trait appears to be negatively related to global measures of brain volume, such as volume of cerebral gray matter, ratio of brain volume to intracranial volume, and total brain volume.[138,140,176,177] This association of neuroticism with reduced brain volume has been hypothesized to reflect cell death due to the potentiation of excitotoxic processes by the stress hormone cortisol.[177] Several studies have shown that neuroticism is associated with elevated basal levels of cortisol.[178–181]

The last finding we will consider in relation to avoidance traits is the complement to the finding that extraversion may be associated with hemispheric asymmetry in EEG. Considerable research has shown that traits in the neuroticism family predict greater right relative to left neural activity in the frontal lobes when viewing stimuli or while at rest.[182,183] Similarly, a recent study used near-infrared reflection spectroscopy, a technique that uses light to measure regional cerebral oxygenated hemoglobin, to demonstrate that cerebral blood flow is increased in the right frontal lobe in individuals high on neuroticism during anticipation of a shock.[184] It would be a mistake, however, to think that all traits in the neuroticism family are associated with right-dominant frontal asymmetry. The effect appears to be limited to traits involving passive avoidance, such as anxiety and depression. In contrast, traits of anger-proneness and hostility are associated with greater left-dominant frontal asymmetry,[185–187] further supporting the hypothesis that the right hemisphere is specialized for processing goal conflict, rather than for all avoidance-related processes.

3.3 Future Directions for Trait Research

There are a number of aspects of current trait research on the neuroimaging of approach, avoidance, and conflict that are clearly unsatisfactory. Most obvious is the fact that, particularly in the strict senses defined in Section 1, there has been no neuroimaging that can be unambiguously linked to traits specifically reflecting approach, or avoidance, or conflict. A conflict-specific biomarker has recently been identified (and correlates moderately with neuroticism and trait anxiety)[77] but has not yet been used to identify specific related trait components. Trait scales related to positive affect (extraversion, etc.) and negative affect (neuroticism, harm avoidance, etc.) have been shown to have neural correlates, but none of these are pure measures of approach, avoidance, or conflict sensitivity. It is clear that neuroticism is not pure avoidance sensitivity or pure conflict sensitivity, although it appears linked to both of these factors and to others (e.g., depression) as well. Similarly, extraversion encompasses behaviors reflecting not only the tendency to approach but also positive emotional tendencies having to do with the enjoyment of rewards after they are received.

We believe there are five main steps that need to be taken to improve research on the functional neuroimaging of approach and avoidance traits: (1) to take account of the detailed knowledge derived from both animal and human work about the approach, avoidance, and conflict systems (Section 1.1–1.3) to develop more specifically targeted experimental tasks, using carefully designed contrasts between sets of conditions; (2) to test more focused anatomical hypotheses specified by theory, using carefully designed regions of interest (Section 1.4–1.5); (3) to take advantage of related state research (Section 2) to develop appropriate anchoring biomarkers; (4) to collect samples large enough for good research on individual differences—over 100 at a minimum, preferably over 200 (while recognizing that such samples may still be too small to detect some effects of interest); and (5) to ensure that the appropriate level in the trait hierarchy is being matched to the appropriate aspect of neural organization, taking into consideration that an observed association with one trait might, in reality, be either more specific (with a facet of the trait in question) or more general (with a trait at a higher level of the hierarchy).

4. FROM BASICS TO STATES AND TRAITS: ASSESSING APPROACH, AVOIDANCE, AND GOAL CONFLICT

Our analysis of the basics of approach, avoidance, and goal conflict shows that care must be exercised when using complex combinations of motivational stimuli and complex paradigms. Variations in valuation, such as loss aversion, differential effects of approach and avoidance gradients, direct interactions between approach and avoidance systems, and the asymmetric impact of goal conflict on avoidance relative to approach, must all be taken into account when interpreting many of the paradigms currently used. However, in principle, state analysis of these systems is straightforward.

One simplifying step is to use money as the source of motivation. Organizations that find work for students and other casual workers can supply participants with a hunger for money sufficient to make them willing to work for the local minimum wage. Importantly, loss of money from an existing store can then be used as a motivator, with the knowledge that its external value is the same as the gain of the same amount of money used as a positive motivator. As shown in Figure 3, gain and loss can be presented or omitted to generate approach or avoidance. The amounts of gain and loss can then be varied parametrically to allow mathematical extraction, separately, of the contribution of gain/loss sensitivity differences and of approach/avoidance sensitivity differences. Using these methods, loss aversion and approach preference have been demonstrated.[188]

For neuroimaging, it is also important to use designs that allow the calculation of appropriate contrasts. If one wishes to image goal conflict activation, one must accept that gain, loss, approach, avoidance, and other systems will all necessarily be activated when approach-avoidance conflict is being generated. To deal with this requires the use of at least three conditions. For example, with conditions that deliver two alternatives with a 50% probability on any trial, one could have: (1) net gain (−10c, +20c); (2) conflict (−15c, +15c); and (3) net loss (−20c, +10c). A contrast of neuroimaging activation in condition 2 against the average of condition 1 and condition 3 would assess goal conflict-specific activation while eliminating the effects of external value (15c = (10c+20c)/2) and controlling for effects of factors such as risk. In practice, because of loss aversion, to statistically eliminate the effects of gain, loss, approach, and avoidance, when assessing conflict, one would need the ratio of gain/loss amounts tailored to each individual's degree of loss aversion. Additional conditions would allow the separation of the effects of gain from the effects of loss and effects of approach from the effects of avoidance.[188]

For those interested in goal gradients (Figures 2 and 4), existing virtual reality maze paradigms (see Section 2.2) or even simpler runway analogues could be used. These have already demonstrated effects related to distance from a "predator," as well as differences between simple anticipation of shock and the response to actual shock delivery. Combined with the presentation of money (to selected money-hungry participants), these virtual reality paradigms allow manipulation of the full gamut of parameters that have previously been used in animal behavior tests.

It is tempting, in the imaging of personality, to select questionnaires that have been designed, in theory, to tap into specific neurobiological functions (e.g., scales purporting to measure Gray's Behavioral Inhibition System) but that have not in fact been neurobiologically validated. However, as we noted earlier, the nascent neuroscience of personality should not assume the very hypotheses that need to be tested. Psychologists' presuppositions about which neural systems are responsible for any given trait, as measured by a questionnaire, may well be wrong. With approach, avoidance, and conflict, we are dealing with primordial biological systems whose elements have evolved to fulfill system-specific purposes. The state activation of these systems can be, and has been, assessed directly, with specific components extractable through appropriate contrasts. These specific components of neural state activation provide, we would argue, the best basis both for assessing personality-related variation in activation and for deriving questionnaire scales or other measures of approach, avoidance, and goal conflict traits, using the criterion approach described in Section 3. How the sensitivities of the approach, avoidance, behavioral inhibition, and other neural systems give rise to variation in traits is the key question that the field must strive to solve. A genuinely neuroscientific approach will provide a solid basis for future attempts to understand the contribution of these fundamental neural systems to traits such as extraversion, neuroticism, impulsivity, and others.

References

1. Gray JA, McNaughton N. *The Neuropsychology of Anxiety: An Enquiry into the Functions of the Septo-hippocampal System*. 2nd ed. Oxford: Oxford University Press; 2000.
2. McNaughton N, Corr PJ. A two-dimensional neuropsychology of defense: fear/anxiety and defensive distance. *Neurosci Biobehav Rev*. 2004;28:285–305.
3. McFarland D. *The Oxford Companion to Animal Behaviour*. Oxford: Oxford University Press; 1987.
4. Hinde RA. *Animal Behaviour*. New York: McGraw-Hill Book Company; 1966.
5. Cahill L, McGaugh JL, Weinberger NM. The neurobiology of learning and memory: some reminders to remember. *Trends Neurosci*. 2001;24(10):578–581.
6. Towe AL, Luschei ES. Preface. In: Towe AL, Luschei ES, eds. *Motor Coordination*. New York: Plenum Press; 1981:vii–viii.
7. Miller NE. Experimental studies of conflict. In: Hunt JM, ed. *Personality and the Behavioural Disorders*. New York: Ronald Press; 1944:431–465. [Cited by Kimble, 1961].
8. Amsel A. Frustrative nonreward in partial reinforcement and discrimination learning: some recent history and a theoretical extension. *Psychol Rev*. 1962;69:306–328.
9. Amsel A. *Frustration Theory: An Analysis of Dispositional Learning and Memory*. Cambridge: Cambridge University Press; 1992.
10. Amsel A, Roussel J. Motivational properties of frustration. I. Effect on a running response of the addition of frustration to the motivational complex. *J Exp Child Psychol*. 1952;43:363–368.
11. Adelman HM, Maatsch JL. Learning and extinction based upon frustration, food reward, and exploratory tendency. *J Exp Psychol*. 1956;52:311–315.

12. Gray JA. *The Psychology of Fear and Stress*. London: Cambridge University Press; 1987.
13. Kahneman D, Tversky A. Prospect theory: an analysis of decision under risk. *Econometrica*. 1979;47:263–291.
14. Novemsky N, Kahneman D. The boundaries of loss aversion. *J Mark Res*. 2005;42:119–128.
15. Tversky A, Kahneman D. Loss aversion in riskless choice: a reference dependent model. *Q J Econ*. 1991;106:1039–1061.
16. Berridge KC. Motivation concepts in behavioral neuroscience. *Physiol Behav*. 2004;81:179–209.
17. Berridge KC. Food reward: brain substrates of wanting and liking. *Neurosci Biobehav Rev*. 1996;20:1–25.
18. Gray JA. *Elements of a Two-process Theory of Learning*. London: Academic Press; 1975.
19. Brown JS. Gradients of approach and avoidance responses and their relation to level of motivation. *J Comp Physiol Psychol*. 1948;41:450–465.
20. Tanaka SC, Yamada K, Yoneda H, Ohtake F. Neural mechanisms of gain–loss asymmetry in temporal discounting. *J Neurosci*. April 16, 2014;34(16):5595–5602.
21. Miller NE. Liberalization of basic S-R concepts: extensions to conflict behaviour, motivation and social learning. In: Koch S, ed. *Psychology: A Study of a Science*. New York: Wiley; 1959:196–292. [Cited by Gray, 1975].
22. Kimble GA. *Hilgard and Marquis' Conditioning and Learning*. New York: Appleton-Century-Crofts; 1961.
23. Gray JA. Drug effects on fear and frustration: possible limbic site of action of minor tranquilizers. In: Iversen LL, Iversen SD, Snyder SH, eds. *Handbook of Psychopharmacology Vol. 8. Drugs, Neurotransmitters and Behaviour*. Vol. 8. New York: Plenum Press; 1977:433–529.
24. Gray JA. *The Neuropsychology of Anxiety: An Enquiry in to the Functions of the Septo-hippocampal System*. Oxford: Oxford University Press; 1982.
25. Bailey CJ, Miller NE. The effect of sodium amytal on an approach-avoidance conflict in cats. *J Comp Physiol Psychol*. 1952;45(4):205–208.
26. Barry H, Miller NE. Effects of drugs on approach-avoidance conflict tested repeatedly by means of a telescope alley. *J Comp Physiol Psychol*. 1962;55(2):201–210.
27. Razrin G. *Mind in Evolution*. Boston: Houghton Mifflin; 1971. [cited by Gray, 1975].
28. Hebebrand J, Friedl W, Breidenbach B, Propping P. Phylogenetic comparison of the photoaffinity-labeled benzodiazepine receptor subunits. *J Neurochem*. 1987;48:1103–1108.
29. LeDoux JE. Emotion, memory and the brain. *Sci Am*. 1994;270: 50–59.
30. Dawkins R, Krebs JR. Arms races between and within species. *Proc R Soc Lond B Biol Sci*. 1979;205:489–511.
31. Haegelen C, Rouaud T, Darnault P, Morandi X. The subthalamic nucleus is a key-structure of limbic basal ganglia functions. *Med Hypotheses*. 2009;72:421–426.
32. Chambers CD, Garavan H, Bellgrove MA. Insights into the neural basis of response inhibition from cognitive and clinical neuroscience. *Neurosci Biobehav Rev*. 2009;33:631–646.
33. Haber SN, Calzavara R. The cortico-basal ganglia integrative network: the role of the thalamus. *Brain Res Bull*. 2009;78(2–3): 69–74.
34. Kolb B, Whishaw IQ. *Fundamentals of Human Neuropsychology*. San Francisco: W. H. Freeman; 1980.
35. Bear MF, Connors BW, Paradiso MA. *Neuroscience: Exploring the Brain*. Baltimore: Williams and Wilkins; 1996.
36. Haber SN, Knutson B. The reward circuit: linking primate anatomy and human imaging. *Neuropsychopharmacology*. January 2010;35(1):4–26.
37. Blanchard RJ, Flannelly KJ, Blanchard DC. Defensive behaviors of laboratory and wild *Rattus norvegicus*. *J Comp Psychol*. 1986;100(2):101–107.
38. Blanchard RJ, Blanchard DC. Antipredator defensive behaviors in a visible burrow system. *J Comp Psychol*. 1989;103(1):70–82.
39. Blanchard RJ, Blanchard DC. An ethoexperimental analysis of defense, fear and anxiety. In: McNaughton N, Andrews G, eds. *Anxiety*. Dunedin: Otago University Press; 1990:124–133.
40. Blanchard RJ, Blanchard DC. Anti-predator defense as models of animal fear and anxiety. In: Brain PF, Parmigiani S, Blanchard RJ, Mainardi D, eds. *Fear and Defence*. Chur: Harwood Acad. Pub.; 1990:89–108.
41. Graeff FG. Neuroanatomy and neurotransmitter regulation of defensive behaviors and related emotions in mammals. *Braz J Med Biol Res*. 1994;27:811–829.
42. Graeff FG, Del-Ben CM. Neurobiology of panic disorder: from animal models to brain neuroimaging. *Neurosci Biobehav Rev*. 2008;32:1326–1335.
43. Shizgal P, Fulton S, Woodside B. Brain reward circuitry and the regulation of energy balance. *Int J Obes*. 2001;25:S17–S21.
44. Mahler SV, Berridge KC. Which cue to "want?" Central amygdala opioid activation enhances and focuses incentive salience on a prepotent reward cue. *J Neurosci*. May 20, 2009;29(20): 6500–6513.
45. McNaughton N. Development of a theoretically-derived human anxiety syndrome biomarker. *Trans Neuro*. 2014;5(2):137–146.
46. McNaughton N. Aminergic transmitter systems. In: D'Haenen H, Den Boer JA, Westenberg H, Willner P, eds. *Textbook of Biological Psychiatry*. John Wiley & Sons; 2002:895–914.
47. DeYoung CG. The neuromodulator of exploration: a unifying theory of the role of dopamine in personality. *Front Hum Neurosci*. 2013;7:762.
48. Jessup RK, O'Doherty JP. Distinguishing informational from value-related encoding of rewarding and punishing outcomes in the human brain. *Eur J Neurosci*. June 2014;39(11):2014–2026.
49. Delgado MR. Reward-related responses in the human striatum. *Ann N Y Acad Sci*. May 2007;1104:70–88.
50. De Silva A, Salem V, Matthews PM, Dhillo WS. The use of functional MRI to study appetite control in the CNS. *Exp Diabet Res*. 2012;2012:A764017.
51. Bishop SJ. Trait anxiety and impoverished prefrontal control of attention. *Nat Neurosci*. 2009;12(1):92–98.
52. Malik S, McGlone F, Bedrossian D, Dagher A. Ghrelin modulates brain activity in areas that control appetitive behavior. *Cell Metab*. May 2008;7(5):400–409.
53. O'Doherty JP. Neural mechanisms underlying reward and punishment learning in the human brain: insights from fMRI. In: Vartanian O, Mandel DR, eds. *Neuroscience of Decision Making*. New York: Psychology Press; 2011:173–198.
54. Botvinick M. Conflict monitoring and decision making: reconciling two perspectives on anterior cingulate function. *Cogn Affect Behav Neurosci*. 2007;7(4):356–366.
55. Sescousse G, Redoute J, Dreher J-C. The architecture of reward value coding in the human orbitofrontal cortex. *J Neurosci*. September 29, 2010;30(39):13095–13104.
56. Mobbs D, Petrovic P, Marchant JL, et al. When fear is near: threat imminence elicits prefrontal-periaqueductal gray shifts in humans. *Science*. 2007;317:1079–1083.
57. Mobbs D, Marchant JL, Hassabis D, et al. From threat to fear: the neural organization of defensive fear systems in humans. *J Neurosci*. September 30, 2009;29(39):12236–12243.
58. Marschner A, Kalisch R, Vervliet B, Vansteenwegen D, Buchel C. Dissociable roles for the hippocampus and the amygdala in human cued versus context fear conditioning. *J Neurosci*. September 2008; 28(36):9030–9036.

59. Alvarez RP, Biggs A, Chen G, Pine DS, Grillon C. Contextual fear conditioning in humans: cortical-hippocampal and amygdala contributions. *J Neurosci*. June 2008;28(24):6211–6219.
60. Pichon S, de Gelder B, Grèzes J. Threat prompts defensive brain responses independently of attentional control. *Cereb Cortex*. February 1, 2012;22(2):274–285.
61. Shackman AJ, Salomons TV, Slagter HA, Fox AS, Winter JJ, Davidson RJ. The integration of negative affect, pain and cognitive control in the cingulate cortex. *Nature Rev Neurosci*. 2011;12:154–167.
62. Wise RG, Lujan BJ, Schweinhardt P, Peskett GD, Rogers R, Tracey I. The anxiolytic effects of midazolam during anticipation to pain revealed using fMRI. *Magn Reson Imaging*. July 2007;25(6):801–810.
63. Yu R, Mobbs D, Seymour B, Rowe JB, Calder AJ. The neural signature of escalating frustration in humans. *Cortex*. May 2014; 54:165–178.
64. Vickery TJ, Chun MM, Lee D. Ubiquity and specificity of reinforcement signals throughout the human brain. *Neuron*. 2011;72(1):166–177.
65. Bach DR, Guitart-Masip M, Packard PA, et al. Human hippocampus arbitrates approach-avoidance conflict. *Curr Biol*. February 18, 2014;24:541–547.
66. Borgwardt SJ, Allen P, Bhattacharyya S, et al. Neural basis of Delta-9-tetrahydrocannabinol and cannabidiol: effects during response inhibition. *Biol Psychiatry*. December 2008;64(11):966–973.
67. Aupperle RL, Tankersley D, Ravindran LN, et al. Pregabalin effects on neural response to emotional faces. *Front Hum Neurosci*. 2012;6. Article 42.
68. McNaughton N, Kocsis B, Hajós M. Elicited hippocampal theta rhythm: a screen for anxiolytic and pro-cognitive drugs through changes in hippocampal function? *Behav Pharmacol*. 2007;18:329–346.
69. Young CK, McNaughton N. Coupling of theta oscillations between anterior and posterior midline cortex and with the hippocampus in freely behaving rats. *Cereb Cortex*. January 1, 2009;19(1):24–40.
70. Neo PSH. *Theta Activations Associated with Goal-conflict Processing: Evidence for the Revised Behavioural Inhibition System* [Ph.D. thesis]. Dunedin: Department of Psychology, University of Otago; 2008.
71. Neo PSH, McNaughton N. Frontal theta power linked to neuroticism and avoidance. *Cogn Affect Behav Neurosci*. 2011;11:396–403.
72. Aron AR, Poldrack RA. Cortical and subcortical contributions to stop signal response inhibition: role of subthalamic nucleus. *J Neurosci*. 2006;26:2424–2433.
73. Aron AR, Fletcher PC, Bullmore ET, Sahakian BJ, Robbins TW. Stop-signal inhibition disrupted by damage to right inferior frontal gyrus in humans. *Nat Neurosci*. 2003;6:115–116.
74. Aron AR. The neural basis of inhibition in cognitive control. *Neuroscientist*. 2007;13:214–228.
75. Floden D, Stuss DT. Inhibitory control is slowed in patients with right superior medial frontal damage. *J Cogn Neurosci*. 2006;18:1843–1849.
76. Logan GD, Cowan WB, Davis KA. On the ability to inhibit simple and choice reaction-time responses–a model and a method. *J Exp Psychol Hum Percept Perform*. 1984;10(2):276–291.
77. Neo PSH, Thurlow J, McNaughton N. Stopping, goal-conflict, trait anxiety and frontal rhythmic power in the stop-signal task. *Cogn Affect Behav Neurosci*. 2011;11:485–493.
78. McNaughton N, Swart C, Neo PSH, Bates V, Glue P. Anti-anxiety drugs reduce conflict-specific "theta" – a possible human anxiety-specific biomarker. *J Affect Disord*. 2013;148:104–111.
79. Nieuwenhuys R. The insular cortex: a review. *Prog Brain Res*. 2012;195:123–163.
80. Liang X, Zou Q, He Y, Yang Y. Coupling of functional connectivity and regional cerebral blood flow reveals a physiological basis for network hubs of the human brain. *Proc Natl Acad Sci USA*. January 14, 2013;110(5):1929–1934.
81. Cauda F, D'Agata F, Sacco K, Duca S, Geminiani G, Vercelli A. Functional connectivity of the insula in the resting brain. *Neuroimage*. March 1, 2011;55(1):8–23.
82. Augustine JR. Circuitry and functional aspects of the insular lobe in primates including humans. *Brain Res Rev*. October 1996;22(3):229–244.
83. Kurth F, Zilles K, Fox PT, Laird AR, Eickhoff SB. A link between the systems: functional differentiation and integration within the human insula revealed by meta-analysis. *Brain Struct Func*. June 2010;214(5–6):519–534.
84. Liu X, Hairston J, Schrier M, Fan J. Common and distinct networks underlying reward valence and processing stages: a meta-analysis of functional neuroimaging studies. *Neurosci Biobehav Rev*. 2011;35(5):1219–1236.
85. Stoléru S, Fonteille V, Cornélis C, Joyal CC, Moulier V. Functional neuroimaging studies of sexual arousal and orgasm in healthy men and women: a review and meta-analysis. *Neurosci Biobehav Rev*. 2012;36(6):1481–1509.
86. Carlson JM, Greenberg T, Rubin D, Mujica-Parodi LR. Feeling anxious: anticipatory amygdalo-insular response predicts the feeling of anxious anticipation. *Soc Cog Affect Neurosci*. January 2011;6(1):74–81.
87. Mohr PNC, Biele G, Heekeren HR. Neural processing of risk. *J Neurosci*. May 12, 2010;30(19):6613–6619.
88. Rudorf S, Preuschoff K, Weber B. Neural correlates of anticipation risk reflect risk preferences. *J Neurosci*. November 21, 2012;32(47):16683–16692.
89. Chang LJ, Yarkoni T, Khaw MW, Sanfey AG. Decoding the role of the insula in human cognition: functional parcellation and large-scale reverse inference. *Cereb Cortex*. March 1, 2013;23(3):739–749.
90. Menon V, Uddin LQ. Saliency, switching, attention and control: a network model of insula function. *Brain Struct Func*. June 1, 2010;214(5–6):655–667.
91. Paulus MP, Stein MB. An insular view of anxiety. *Biol Psychiatry*. August 15, 2006;60(4):383–387.
92. Aron AR, Robbins TW, Poldrack RA. Inhibition and the right inferior frontal cortex: one decade on. *Trends Cogn Sci*. April 2014;18(4):177–185.
93. Rubia K, Smith AB, Brammer MJ, Taylor E. Right inferior prefrontal cortex mediates response inhibition while mesial prefrontal cortex is responsible for error detection. *Neuroimage*. 2003;20:351–358.
94. Botvinick M, Nystrom LE, Fissell K, Carter CS, Cohen JD. Conflict monitoring versus selection-for-action in anterior cingulate cortex. *Nature*. November 11, 1999;402(6758):179–181.
95. Geuze E, van Berckel BNM, Lammertsma AA, et al. Reduced GABAA benzodiazepine receptor binding in veterans with post-traumatic stress disorder. *Mol Psychiatry*. July 31, 2007 online;13(1):74–83.
96. Moller M, Jakobsen S, Gjedde A. Parametric and regional maps of free serotonin 5HT1A receptor sites in human brain as function of age in healthy humans. *Neuropsychopharmacology*. August 2007;32(8):1707–1714.
97. Phillips ML, Drevets WC, Rauch SL, Lane R. Neurobiology of emotion perception I: the neural basis of normal emotion perception. *Biol Psychiatry*. September 1, 2003;54(5):504–514.
98. Phillips ML, Drevets WC, Rauch SL, Lane R. Neurobiology of emotion perception II: implications for major psychiatric disorders. *Biol Psychiatry*. September 1, 2003;54(5):515–528.
99. Corr PJ, DeYoung CG, McNaughton N. Motivation and personality: a neuropsychological perspective. *Soc Pers Psych Compass*. 2013;7:158–175.
100. Richard FD, Bond Jr CF, Stokes-Zoota JJ. One hundred years of social psychology quantitatively described. *Rev Gen Psych*. 2003;7:331–363.

101. Hemphill JF. Interpreting the magnitudes of correlation coefficients. *Am Psychol.* 2003;58:78–80.
102. Carp J. The secret lives of experiments: methods reporting in the fMRI literature. *Neuroimage.* 2012;63:289–300.
103. Holmes AJ, Lee PH, Hollinshead MO, et al. Individual differences in amygdala-medial prefrontal anatomy link negative affect, impaired social functioning, and polygenic depression risk. *J Neurosci.* 2012;32:18087–18100.
104. Kapogiannis D, Sutin A, Davatzikos C, Costa P, Resnick S. The five factors of personality and regional cortical variability in the Baltimore longitudinal study of aging. *Hum Brain Mapp.* 2013;34:2829–2840.
105. Poppe A, Barch D, Carter C, Ragland D, Silverstein S, MacDonald A. Test-retest reliability of GLM and ICA in schizophrenia patients and healthy controls. *Biol Psychiatry.* 2014;75:373S–374S.
106. Connelly BS, Ones DS. An other perspective on personality: meta-analytic integration of observers' accuracy and predictive validity. *Psychol Bull.* 2010;136:1092–1122.
107. Vazire S. Who knows what about a person? The self–other knowledge asymmetry (SOKA) model. *J Pers Soc Psychol.* 2010;98:281–300.
108. Kirby KN. One-year temporal stability of delay-discount rates. *Psychon Bull Rev.* 2009;16:457–462.
109. Ohmura Y, Takahashi T, Kitamura N, Wehr P. Three-month stability of delay and probability discounting measures. *Exp Clin Psychopharm.* 2006;14:318–328.
110. Carver CS, White TL. Behavioral inhibition, behavioral activation, and affective responses to impending reward and punishment: the BIS/BAS scales. *J Pers Soc Psychol.* 1994;67:319–333.
111. Schandry R. Heart beat perception and the emotional experience. *Psychophysiology.* 1981;18:483–488.
112. John OP, Naumann LP, Soto CJ. Paradigm shift to the integrative Big Five trait taxonomy: history: measurement, and conceptual issues. In: John OP, Robins RW, Pervin LA, eds. *Handbook of Personality: Theory and Research.* New York: Guilford Press; 2008:114–158.
113. Markon KE, Krueger RF, Watson D. Delineating the structure of normal and abnormal personality: an integrative hierarchical approach. *J Pers Soc Psychol.* 2005;88:139–157.
114. Saucier G. Recurrent personality dimensions in inclusive lexical studies: indications for a big six structure. *J Pers Soc Psychol.* 2009;77:1577–1614.
115. DeYoung CG. Personality neuroscience and the biology of traits. *Soc Pers Psych Compass.* 2010;4:1165–1180.
116. DeYoung CG. Cybernetic Big Five Theory. *J Res Pers.* 2014;56:33–58.
117. McNaughton N, Corr PJ. Approach, avoidance, and their conflict: the problem of anchoring. *Front Syst Neurosci.* 2014;8.
118. Depue RA, Collins PF. Neurobiology of the structure of personality: Dopamine, facilitation of incentive motivation and extraversion. *Behav Brain Sci.* 1999;22:491–569.
119. Quilty LC, DeYoung CG, Oakman JM, Bagby RM. Extraversion and behavioural activation: integrating the components of approach. *J Pers Assess.* 2014;96:87–94.
120. Wacker J, Mueller EM, Hennig J, Stemmler G. How to consistently link extraversion and intelligence to the catechol-o-methyltransferase (COMT) gene: on defining and measuring psychological phenotypes in neurogenetic research. *J Pers Soc Psychol.* 2012;102: 427–444.
121. Chavanon ML, Wacker J, Stemmler G. Paradoxical dopaminergic drug effects in extraversion: dose-and time-dependent effects of sulpiride on EEG theta activity. *Front Hum Neurosci.* 2013;7:A117.
122. Depue RA, Luciana M, Arbisi P, Collins P, Leon A. Dopamine and the structure of personality: relation of agonist-induced dopamine activity to positive emotionality. *J Pers Soc Psychol.* 1994;67:485–498.
123. Depue RA, Fu Y. On the nature of extraversion: variation in conditioned contextual activation of dopamine-facilitated affective, cognitive, and motor processes. *Front Hum Neurosci.* 2013;7:A288.
124. Mueller EM, Burgdorf C, Chavanon ML, Schweiger D, Wacker J, Stemmler G. Dopamine modulates frontomedial failure processing of agentic introverts versus extraverts in incentive contexts. *Cogn Affect Behav Neurosci.* June 2014;14(2):756–768.
125. Rammsayer TH. Extraversion and dopamine: individual differences in response to changes in dopaminergic activity as a possible biological basis of extraversion. *Eur Psychol.* 1998;3:37–50.
126. Rammsayer TH, Netter P, Vogel WH. A neurochemical model underlying differences in reaction times between introverts and extraverts. *Pers Individ Differ.* 1993;14:701–712.
127. Wacker J, Stemmler G. Agentic extraversion modulates the cardiovascular effects of the dopamine D2 agonist bromocriptine. *Psychophysiology.* 2006;43:372–381.
128. Wacker J, Chavanon ML, Stemmler G. Investigating the dopaminergic basis of extraversion in humans: a multilevel approach. *J Pers Soc Psychol.* 2006;91:171–187.
129. Wacker J, Mueller EM, Pizzagalli DA, Hennig J, Stemmler G. Dopamine-D2-receptor blockade reverses the association between trait approach motivation and frontal asymmetry in an approach-motivation context. *Psychol Sci.* 2013;24:489–497.
130. Canli T, Sivers H, Whitfield SL, Gotlib IH, Gabrieli JDE. Amygdala response to happy faces as a function of extraversion. *Science.* 2002;296(5576):2191.
131. Canli T, Zhao Z, Desmond JE, Kang E, Gross J, Gabrieli JDE. An fMRI study of personality influences on brain reactivity to emotional stimuli. *Behav Neurosci.* 2001;115(1):33.
132. Cohen MX, Young JAT, Baek J-M, Kessler C, Ranganath C. Individual differences in extraversion and dopamine genetics predict neural reward responses. *Cogn Brain Res.* 2005;25(3): 851–861.
133. Mobbs D, Hagan CC, Azim E, Menon V, Reiss AL. Personality predicts activity in reward and emotional regions associated with humor. *Proc Natl Acad Sci USA.* 2005;102(45):16502–16506.
134. Cremers H, van Tol MJ, Roelofs K, et al. Extraversion is linked to volume of the orbitofrontal cortex and amygdala. *PLoS One.* 2011;6:e28421.
135. DeYoung CG, Hirsh JB, Shane MS, Papademetris X, Rajeevan N, Gray JR. Testing predictions from personality neuroscience. *Psychol Sci.* June 2010;21(6):820–828.
136. Li Y, Qiao L, Sun J, et al. Gender-specific neuroanatomical basis of behavioral inhibition/approach systems (BIS/BAS) in a large sample of young adults: a voxel-based morphometric investigation. *Behav Brain Res.* November 1, 2014;274:400–408.
137. Omura K, Constable RT, Canli T. Amygdala gray matter concentration is associated with extraversion and neuroticism. *Neuroreport.* 2005;16:1905–1908.
138. Bjørnebekk A, Fjell AM, Walhovd KB, Grydeland H, Torgersen S, Westlye LT. Neuronal correlates of the five factor model (FFM) of human personality: multimodal imaging in a large healthy sample. *Neuroimage.* 2013;65:194–208.
139. Hu X, Erb M, Ackermann H, Martin JA, Grodd W, Reiterer SM. Voxel-based morphometry studies of personality: issue of statistical model specification—effect of nuisance covariates. *Neuroimage.* 2011;54:1994–2005.
140. Liu WY, Weber B, Reuter M, Markett S, Chu WC, Montag C. The Big Five of Personality and structural imaging revisited: a VBM–DARTEL study. *Neuroreport.* 2013;24:375–380.
141. Lewis GJ, Panizzon MS, Eyler L, et al. Heritable influences on amygdala and orbitofrontal cortex contribute to genetic variation in core dimensions of personality. *Neuroimage.* September 28, 2014;103C:309–315.

142. Harmon-Jones E, Gable PA, Peterson CK. The role of asymmetric frontal cortical activity in emotion-related phenomena: a review and update. *Biol Psychol*. 2010;84:451–462.
143. Wacker J, Chavanon ML, Leue A, Stemmler G. Is running away right? The behavioral activation-behavioral inhibition model of anterior asymmetry. *Emotion*. April 2008;8(2):232–249.
144. Wacker J, Chavanon ML, Stemmler G. Resting EEG signatures of agentic extraversion: new results and meta-analytic integration. *J Res Pers*. 2010;44:167–179.
145. Coan JA, Allen JJ, McKnight PE. A capability model of individual differences in frontal EEG asymmetry. *Biol Psychol*. 2006;72: 198–207.
146. Pickering AD. The neuropsychology of impulsive antisocial sensation seeking personality traits: from dopamine to hippocampal function? In: Stelmack RM, ed. *On the Psychobiology of Personality: Essays in Honour of Marvin Zuckerman*. Oxford: Elsevier; 2004:453–457.
147. Smillie LD, Pickering AD, Jackson CJ. The new reinforcement sensitivity theory: implications for personality measurement. *Pers Soc Psychol Rev*. 2006;10:320–325.
148. DeYoung CG. Impulsivity as a personality trait. In: Vohs KD, Baumeister RF, eds. *Handbook of Self-Regulation: Research, Theory, and Applications*. 2nd ed. New York: Guilford Press; 2010:485–502.
149. Buckholtz JW, Treadway MT, Cowan RL, et al. Mesolimbic dopamine reward system hypersensitivity in individuals with psychopathic traits. *Nat Neurosci*. 2010;13:419–421.
150. Buckholtz JW, Treadway MT, Cowan RL, et al. Dopaminergic network differences in human impulsivity. *Science*. 2010;329:532.
151. Forbes EE, Brown SM, Kimak M, Ferrell RE, Manuck SB, Hariri AR. Genetic variation in components of dopamine neurotransmission impacts ventral striatal reactivity associated with impulsivity. *Mol Psychiatry*. 2007;14:60–70.
152. Forbes EE, Brown SM, Kimak M, Ferrell RE, Manuck SB, Hariri AR. Genetic variation in components of dopamine neurotransmission impacts ventral striatal reactivity associated with impulsivity. *Mol Psychiatry*. January 2009;14(1):60–70.
153. Lahey BB. Public health significance of neuroticism. *Am Psychol*. 2009;64:241–256.
154. Cloninger CR, Svrakic DM, Przybecky TR. A psychobiological model of temperament and character. *Arch Gen Psychiatry*. 1993;50:975–990.
155. Spielberger CD, Gorusch RL, Lushene R, Vagg PR, Jacobs GA. *Manual for the State-Trait Anxiety Inventory (Form Y)* CA94306. Palo Alto: Consulting Psychologists Press; 1983.
156. DeYoung CG, Quilty LC, Peterson JB. Between facets and domains: ten aspects of the Big Five. *J Pers Soc Psychol*. 2007;93:880–896.
157. Corr PJ, Cooper A. The Reinforcement Sensitivity Theory of Personality Questionnaire (RST-PQ): Development and Validation, submitted.
158. Frokjaer VG, Mortensen EL, Nielsen FÅ, et al. Frontolimbic serotonin 2A receptor binding in healthy subjects is associated with personality risk factors for affective disorder. *Biol Psychiatry*. 2008;63:569–576.
159. Takano A, Arakawa R, Hayashi M, Takahashi H, Ito H, Suhara T. Relationship between neuroticism personality trait and serotonin transporter binding. *Biol Psychiatry*. 2007;62:588–592.
160. Tauscher J, Bagby RM, Javanmard M, Christensen BK, Kasper S, Kapur S. Inverse relationship between serotonin 5-HT_{1A} receptor binding and anxiety: a [^{11}C]WAY-100635 PET investigation in healthy volunteers. *Am J Psychiatry*. 2001;158(8):1326–1328.
161. Hennig J. Personality, serotonin, and noradrenaline. In: Stelmack RM, ed. *On the Psychobiology of Personality: Essays in Honor of Marvin Zuckerman*. New York: Elsevier; 2004:379–395.
162. White TL, Depue RA. Differential association of traits of fear and anxiety with norepinephrine- and dark-induced pupil reactivity. *J Pers Soc Psychol*. 1999;77:863–877.
163. Servaas MN, van der Velde J, Costafreda SG, et al. Neuroticism and the brain: a quantitative meta-analysis of neuroimaging studies investigating emotion processing. *Neurosci Biobehav Rev*. 2013;37:1518–1529.
164. Schuyler BS, Kral TR, Jacquart J, et al. Temporal dynamics of emotional responding: amygdala recovery predicts emotional traits. *Soc Cog Affect Neurosci*. February 2014;9(2):176–181.
165. Hyde LW, Gorka A, Manuck SB, Hariri AR. Perceived social support moderates the link between threat-related amygdala reactivity and trait anxiety. *Neuropsychologia*. March 2011;49(4): 651–656.
166. Mujica-Parodi LR, Korgaonkar M, Ravindranath B, et al. Limbic dysregulation is associated with lowered heart rate variability and increased trait anxiety in healthy adults. *Hum Brain Mapp*. January 2009;30(1):47–58.
167. Cremers HR, Demenescu LR, Aleman A, et al. Neuroticism modulates amygdala-prefrontal connectivity in response to negative emotional facial expressions. *Neuroimage*. January 1, 2010;49(1):963–970.
168. Cunningham WA, Arbuckle NL, Jahn A, Mowrer SM, Abduljalil AM. Aspects of neuroticism and the amygdala: chronic tuning from motivational styles. *Neuropsychologia*. 2010;48:3399–3404.
169. Iidaka T, Matsumoto A, Ozaki N, et al. Volume of left amygdala subregion predicted temperamental trait of harm avoidance in female young subjects: a voxel-based morphometry study. *Brain Res*. 2006;1125:85–93.
170. Koelsch S, Skouras S, Jentschke S. Neural correlates of emotional personality: a structural and functional magnetic resonance imaging study. *PLoS One*. 2013;8:e77196.
171. Cherbuin N, Windsor TD, Anstey KJ, Maller JJ, Meslin C, Sachdev PS. Hippocampal volume is positively associated with behavioural inhibition (BIS) in a large community-based sample of mid-life adults: the PATH through life study. *Soc Cog Affect Neurosci*. 2008;3(3):262–269.
172. Fuentes P, Barrós-Loscertales A, Bustamante JC, Rosell P, Costumero V, Ávila C. Individual differences in the Behavioral Inhibition System are associated with orbitofrontal cortex and precuneus gray matter volume. *Cogn Affect Behav Neurosci*. 2012;12:491–498.
173. Taddei M, Tettamanti M, Zanoni A, Cappa S, Battaglia M. Brain white matter organisation in adolescence is related to childhood cerebral responses to facial expressions and harm avoidance. *Neuroimage*. 2012;61:1394–1401.
174. Westlye LT, Bjørnebekk A, Grydeland H, Fjell AM, Walhovd KB. Linking an anxiety-related personality trait to brain white matter microstructure: diffusion tensor imaging and harm avoidance. *Arch Gen Psychiatry*. 2011;68:369–377.
175. Xu J, Potenza MN. White matter integrity and five-factor personality measures in healthy adults. *Neuroimage*. 2012;59: 800–807.
176. Jackson J, Balota DA, Head D. Exploring the relationship between personality and regional brain volume in healthy aging. *Neurobiol Aging*. 2011;32:2162–2171.
177. Knutson B, Momenan R, Rawlings RR, Fong GW, Hommer D. Negative association of neuroticism with brain volume ratio in healthy humans. *Biol Psychiatry*. 2001;50:685–690.
178. Garcia-Banda G, Chellew K, Fornes J, Perez G, Servera M, Evans P. Neuroticism and cortisol: pinning down an expected effect. *Int J Psychophysiol*. 2014;91:132–138.
179. Miller GE, Cohen S, Rabin BS, Skoner DP, Doyle WJ. Personality and tonic cardiovascular, neuroendocrine, and immune parameters. *Brain Behav Immun*. 1999;13:109–123.
180. Nater UM, Hoppmann C, Klumb PL. Neuroticism and conscientiousness are associated with cortisol diurnal profiles in adults—role of positive and negative affect. *Psychoneuroendocrinology*. 2010;35:1573–1577.

181. Polk DE, Cohen S, Doyle WJ, Skoner DP, Kirschbaum C. State and trait affect as predictors of salivary cortisol in healthy adults. *Psychoneuroendocrinology*. 2005;30:261–272.
182. Shackman AJ, McMenamin BW, Maxwell JS, Greischar LL, Davidson RJ. Right dorsolateral prefrontal cortical activity and behavioral inhibition. *Psychol Sci*. 2009;20:1500–1506.
183. Sutton SK, Davidson RJ. Prefrontal brain asymmetry: a biological substrate of the behavioral approach and inhibition systems. *Psychol Sci*. May 1997;8(3):204–210.
184. Morinaga K, Akiyoshi J, Matsushita H, et al. Anticipatory anxiety-induced changes in human lateral prefrontal cortex activity. *Biol Psychol*. 2007;74:34–38.
185. Everhart DE, Demaree HA, Harrison DW. The influence of hostility on electroencephalographic activity and memory functioning during an affective memory task. *Clin Neurophysiol*. 2008;119:134–143.
186. Harmon-Jones E, Allen JJ. Anger and frontal brain activity: EEG asymmetry consistent with approach motivation despite negative affective valence. *J Pers Soc Psychol*. 1998;74:1310–1316.
187. Harmon–Jones E. On the relationship of frontal brain activity and anger: examining the role of attitude toward anger. *Cognit Emotion*. 2004;18:337–361.
188. Hall PJ, Chong W, McNaughton N, Corr PJ. A economic perspective on the reinforcement sensitivity theory of personality. *Pers Individ Differ*. 2011;51:242–247.

CHAPTER

3

Integrating Personality/Character Neuroscience with Network Analysis

Taylor S. Bolt[1], *Ryan S. Hampton*[2], *R. Michael Furr*[1], *William Fleeson*[1], *Paul J. Laurienti*[3], *Dale Dagenbach*[1]

[1]Department of Psychology, Wake Forest University, Winston-Salem, NC, USA; [2]Department of Psychology, Arizona State University, Tempe, AZ, USA; [3]Department of Radiology, Wake Forest University School of Medicine, Winston-Salem, NC, USA

OUTLINE

When first approached by one of the editors of this volume about contributing a chapter on network science and personality and character, we responded with some trepidation. One of us had actually proposed working on this topic to several of the other authors previously, but we had not really followed through on it. There were many reasons for that: busy people with too many other good research questions to address, the difficulties of bringing people together across campuses, etc... But there also was the daunting awareness of the extreme intractability of the problem. First, there is the lingering failure of the last extreme effort to do this in the pseudoscience of phrenology in the nineteenth century. The effort to localize characteristics such as conscientiousness to a particular brain area seemed to be fundamentally misguided. Phrenology failed because both its brain measures (convex and concave areas on the scalp) and its concepts of the kinds of functions that would be localized were flawed. We currently have much improved brain measures (functional Magnetic Resonance Imaging (fMRI), Positron Emission Tomography (PET), Diffusion Tensor Imaging (DTI), evoked potentials, etc...) but

Neuroimaging Personality, Social Cognition, and Character
http://dx.doi.org/10.1016/B978-0-12-800935-2.00003-8

still wrestle with the question of how to think about the kinds of functions that we should use this high-powered technology to assess. And, to be fair, skeptics of contemporary localization of function research employing our new techniques for measuring the brain's activity over the past 20 years have sometimes been heard to comment on it as "the new phrenology," at least in conference hallways, if not in print. Fortunately, the other chapters in this volume illustrate that some advances can be made in addressing this issue for personality and character.

This chapter diverges from them by moving away from typical approaches to localization of function. From 2005 to 2015, an alternative approach to understanding brain functioning has been developed that emphasizes functional connectivity between brain areas using the tools of graph theory and network science. Rather than trying to isolate functions to particular areas, network analyses try to understand brain functioning as a complex system of interacting nodes and their connections. This begs the question of how best to characterize a complex network of interacting parts. In addition, and more problematically for something like trying to understand brain networks and personality or character, it also raises the question of how to compare or contrast brain networks. Fundamentally, this chapter is about how we might do that for personality. There are a few pioneering studies that have begun to do this that we will review. More importantly, however, in this chapter we try to lay out a research program that acknowledges the issues to be addressed and the problems to be avoided. In addition, we provide a preliminary illustration of how network science might inform personality research (and perhaps vice versa) using behavioral data and network science methods. We do not attempt to characterize similar efforts applied to other interesting topics, such as character or social cognition.

A reasonable start might be to ask the following questions. What aspects of personality should be used to try to gain traction on this problem? Does it make sense to focus on aspects of personality that might have a stronger biological basis? Does it make sense to try to capture network differences using an extreme contrast approach—high conscientiousness versus low conscientiousness? Or to treat personality components as continuous variables? In addition, does it make sense to try to understand a personality characteristic in isolation? Are brain networks associated with conscientiousness modified by the presence or absence of another personality characteristic, such as extroversion? Is it possible to combine the complexity of network science applied to the brain and the complexity of a large set of personality characteristics? Finally, under what conditions should the attempt to understand brain networks and personality data be collected? The vast majority of network science brain analyses have looked at resting state data in which the participant is either doing nothing with both eyes shut or is looking at a fixation point. But personality is not a fixed quality: certain personality characteristics are activated by particular situational factors. Would one need to have the situational factors that activate that characteristic present during the brain data collection period in order to fully understand the neural underpinnings of that characteristic? In one sense, these are all empirical questions that time and programmatic research might answer.

In the next three sections, we try to expand upon these questions by (1) characterizing network science and its applications to the brain, including the smattering of studies that have tried to combine this with measures of personality; (2) considering contemporary accounts of personality, how they have been derived, and how these might best be used to devise a research program that can be integrated with network science analyses; and (3) using network science to study personality using behavioral data to illustrate commonalities and differences between these fields of study. In addition, throughout, we try to consider what additional insights the application of network science to personality might provide.

1. WHAT IS NETWORK SCIENCE?

The field of network science can be said to have started after the publication of two key manuscripts in 1998.[1,2] The main focus of network science is the study of how systems are connected. A network is a system of things (nodes) connected by relationships (edges). Systems science shifts the focus from individual elements of a system to the interactions between the elements. A practical example is warranted here, and we will refrain from using the brain until the next section. Consider the airport system. It is composed of many airports across the world. Each airport has flights that go to other airports. The airport network is then a system of nodes (airports) that are connected by flights. The power of network science is that one can evaluate the airport network as a complete interconnected system. The tools and methods that have been developed over 15 years enable researchers to begin to understand the complex organization, or topology, of the network. The topology of a network governs the flow in the network. In order to understand such a system, one should not focus exclusively on the node—the connections matter and the position of the node within the network matters. A node with many connections, often called a hub, can have a greater impact on the system than a node with few connections. This difference in the number of connections requires that the research focus on the relationships between nodes, as well as the nodes themselves. When a single airport closes due to weather, it does not only affect the flights to and from that airport. The system as a whole is impacted, and flight schedules across the world can change. The flow of people from New York to London can be altered by a closure or delay of flights in Chicago, a highly connected hub, but may not be influenced at all by the closure of the airport in Norman, Oklahoma.

Network science is based in graph theory, which describes and analyzes graphs of information represented by the nodes and edges of a system. By applying graph theory to real-world systems, one can characterize the connectivity of the system in a theoretically grounded and relatively consistent manner. Network science already has been applied widely from studies of the Internet[3] to an assortment of social and biological systems[4,5a] and even to complex metabolic networks.[6]

Years of research in network science have led to identifying a set of characteristics that appear to be common in naturally occurring networks. Two fundamental forms of networks are random graphs, in which all of the connections between nodes are assigned at random, and lattices, in which all of the nodes connect with some regular tiling. Many of the naturally occurring complex networks studied to date fall somewhere in between these two extremes. This is a concept known as the small-world phenomenon. This is perhaps most well-known because of Travers and Milgram's classic experiment, in which they had participants in Nebraska and Boston try to get a letter through acquaintances to a target person in Massachusetts. They noted that the number of intermediaries required was surprisingly low at an average of 5.2 and that those in Boston, who were geographically closer to the target person, required fewer intermediaries than the more distant Nebraskans.[7] The important concept derived from this experiment is that despite the scale and complexity of social networks in the entire United States, the shortest path from any one node in these networks to any other node is relatively short. This is true despite the fact that social networks are clustered like a lattice, with friends of any given node having a high likelihood of also being friends. In summary, small-world networks can be characterized as having the high clustering of lattices and the short, typical path lengths of random networks.[1]

The degree distribution of a network summarizes the number of connections that individual nodes possess. Random networks exhibit a binomial distribution, whereas all nodes in a lattice have the exact same degree, resulting in a point function. Though it was originally assumed that the Internet was similar to a random graph, Barabási summarized the Internet as a "scale-free" network.[3] Instead of exhibiting the binomial distribution of a random graph, the Internet exhibited a power law distribution of a few highly connected nodes and a vast majority of lower-degree nodes. The high-degree nodes were often several orders of magnitude higher than the low-degree nodes, and thus the network was termed scale-free. Other social and biological networks also tend to show varying degrees of this scale-free connectivity.[4,5a]

1.1 Network Metrics

At present, there is a wide array of metrics used to describe and classify networks. These metrics are grounded both in the mathematical properties of graph theory and in successful application to known networks, such as those mentioned previously. Network science is a field in which new measures are created and updated as our understanding of networks changes.

One way of characterizing a network is to describe its nodes. Perhaps the simplest characteristic of a node is its number of connections, or degree (k). A node i with $k_i = 20$ is one which connects to 20 other nodes within the network. The clustering coefficient (C) is a measure of interconnectedness of nodes in a cluster. It describes the proportion of extant connections between neighbors of a node to the potential number of connections between those same neighbors. The path length (L) of a node is the average number of edges that must be used to travel the shortest functional distance to any other node in the network. Related to the concepts of C and L are those of local and global efficiency, respectively.[5b] Local efficiency (E_{loc}) is meant to describe the overall efficiency of clusters in a network and, as Latora and Marchiori describe, reveals fault tolerance, or how efficient communication between the neighbors of node i is when i is removed. Global efficiency (E_{glob}) describes the overall efficiency of communication for the whole network and is simply the reciprocal of L, allowing for the inclusion of disconnected nodes and scaling similar to that of E_{loc} from 0 to 1. Betweenness centrality identifies important nodes in a network based on the number of shortest links from all noted to all other nodes linked to the original node, and thus provides a potential measure of a given node's importance in communication in the overall network.[8]

Another commonly evaluated metric is assortativity.[9] Assortativity describes the tendency of nodes with similar degree k to be connected to each other. Assortativity, a simple correlation of the degree of nodes at each end of an edge, ranges from −1 (highly disassortative) to 1 (highly assortative). Thus, an assortative network will have high/low degree nodes connecting with other high/low degree nodes, and a less assortative or disassortative network will contain high degree nodes that connect primarily with lower degree nodes, and a network with an assortativity of 0 would not have any sort of preferential connectivity in regards to degree.

More complex analyses are also available. K-core decomposition identifies the core nodes of a network that are of a high degree and are highly interconnected, while also giving an index of how quickly the core of a network fragments.[10,11] Modularity and scaled inclusivity are used to identify the community structure of a network and the extent to which this community structure changes over time or between different networks.[4,5a,12,13] An analysis of network "hubs" using functional cartography is further meant to assess communication within and between these communities.[6,14] A hub is a node that is among the highest degree nodes in the network, and these typically either organize connections within communities (provincial hubs) or connect different communities within the overall network (connector hubs).

1.2 Network Science and Neuroscience

Insights from network science have brought about an amazing transformation in neuroscience since 2005. Network science has revolutionized how researchers are examining the human brain. The discovery of small-world networks is particularly important for neuroscience, as it is commonly accepted that the brain exhibits regional specificity (neighborhoods), as well as distributed processing (high interconnectivity). Small-world networks have been the focus of considerable research, and many critical methods necessary for evaluating the network properties have been developed.[1,2,9,15] The first functional brain network analysis for humans was performed in 2005.[16] Since that first publication, the field has literally been transformed. This change in technique has been accompanied with a change in perspective. Rather than looking for a specific brain area that controls a cognitive function, investigators are trying to understand how interactions between brain regions result in high-order brain function.

A key feature of brain networks compared to traditional brain imaging is that the data are multivariate. That is, the network analyses focus on the relationships between areas rather than the activity within each area. The quantitative metrics that can be used to assess the organization of the networks are all multivariate in nature. For example, path length, or the number of steps it takes to get from node i to all other nodes in the network, is not an intrinsic quality of node i but rather reflects the connectivity of node i within the context of the entire network. A change in connection between other nodes, say nodes k and l, can have dramatic effects on the path length of node i. The methods needed to statistically assess brain networks are just beginning to be developed.[17–20]

There are notable methodological limitations in many network science studies of the brain. For example, the vast majority of ongoing work looking at functional connectivity continues to use massively univariate statistics developed for traditional imaging or more qualitative assessments. Most of these methods rely on the consistency between subjects to identify differences between populations. In other words, the majority of analyses are focused on finding consistency in network organization across predefined populations of people. The resultant outcomes focus on specific topological properties of the network that are common and exhibit limited variability across people. The weakness of this strategy is that the complexity inherent in the brain networks is likely rooted in the more variable aspects of the connectivity rather than in the structure that is highly consistent.

Studies of complex brain networks have clearly demonstrated that brain circuits are modulated by cognitive task,[21–24] experimental condition,[24–26] drugs,[27,28] and disease states.[29–33]

The default mode network (DMN) is a brain circuit that has received considerable attention. This circuit is highly interconnected at rest and becomes disengaged when an individual is performing a cognitive task. As the DMN disengages, other task-relevant circuits become more interconnected. For example, when one compares brain connectivity at rest to connectivity during the performance of a working memory task, there is a clear shift in connectivity from the DMN to the working memory circuit.[22,23] Brain network analyses hold the promise for us to gain a much more fundamental understanding of how interactions between brain regions result in cognitive processes, clinical disorders, and possibly personality. Bullmore and Sporns provide an excellent review of network science applications to both structural and functional brain networks, as well as a thorough introduction to the basic concepts applied.[34]

1.3 Network Science, Neuroscience, and Personality

A very small number of studies have begun to use network science to understand the neural correlates of personality. The ones reviewed below focus on personality characteristics rather than personality disorders, although preliminary work on the latter also has begun. Additionally, other studies have examined the relationship between personality and resting state fMRI data[35,36] but have not focused on functional connectivity per se and therefore are outside the purview of this chapter. As noted earlier, other possibly related topics, such as social cognition, are also not covered.

Gao et al. constructed resting state brain networks using fMRI data and then related both whole brain network metrics and network metrics for more localized regions to different aspects of personality.[37] Eysenck's model of personality guided their approach, and therefore, they focused on the personality characteristics of extraversion and neuroticism. For global measures of network functioning, only extraversion and the clustering coefficient correlated significantly. For analyses of network metrics within smaller regions of interest selected on the basis of prior research, betweenness centrality correlated with extroversion in the left insula and the bilateral middle temporal gyrus (MTG). Betweenness centrality provides a measure of how influential a particular node is for information transfer in a network, assuming that such transfer follows the shortest possible path lengths. Nodes with high betweenness centrality lie on the shortest path between other nodes. Higher extraversion was associated with higher betweenness centrality in the insula, while the opposite pattern was observed in MTG. No global network measures correlated with neuroticism, but neuroticism was positively correlated with betweenness centrality in the right precentral gyrus, right olfactory cortex, right caudate nucleus, and bilateral amygdala. It was suggested that the correlation between extroversion and clustering coefficient could be related to a higher arousal threshold and a higher tolerance of arousal in more extroverted individuals. The correlations between neuroticism and the amygdala and the right caudate nucleus would be consistent with limbic structures playing a higher role in network communication in more neurotic individuals.

Kong et al. used resting state fMRI data to try to understand the neural mechanisms by which personality characteristics might mediate eudonic well-being (EWB).[38] EWB was assessed using a 42-item Scales of Psychological Well-being Questionnaire,[39] and personality traits were measured using the Revised NEO Personality Inventory, yielding measures posited by the five-factor theory of personality.[40] Analyses of the resting state fMRI data first focused on normalized low frequency fluctuations in the blood oxygen dependent level (BOLD) signal that has previously been found to correlate with conditions such as attention deficit hyperactivity disorder, Alzheimer's dementia, and major depressive disorder. These analyses were used to identify core brain regions that correlated significantly with EWB. Spherical seeds were then inserted into these regions of interest and resting state functional connectivity between the spheres and other regions in the network was assessed. The core regions identified were in the right superior temporal gyrus and an area primarily in the left thalamus but extending to part of the right thalamus. Subsequent functional connectivity analyses first looked at how EWB was affected by the connections between these core regions and other locations. EWB was negatively correlated with the strength of the connection between the thalamus and the right insula. Further analyses correlated the personality traits from the Revised NEO Personality Inventory and the normalized low-frequency fluctuations and resting state functional connectivity. Neuroticism was correlated with the normalized low frequency fluctuations in the superior temporal gyrus and the thalamus, and with thalamus-insula connectivity.

Resting state functional connectivity, rather than whole-brain network analyses, also has been used to explore the neural correlates of the five-factor personality domains. In general, anterior cingulate cortex (ACC) based resting state functional connectivity to medial prefrontal gyrus, paracingulate gyrus, and anterior/central precuneus was related to personality. Similarly, precuneus-based resting state functional connectivity correlates of personality were in and around the precuneus, the dorsomedial prefrontal cortex, the posterior cingulate gyrus, and primary motor and primary visual cortices. As an illustration of how functional connectivity analyses may inform personality, consider this study's interpretation of neuroticism: individuals high on neuroticism showed more local connectivity within a region of the precuneus associated with social and emotional functioning that was targeted with a seed-based analysis, while also showing more connections between this region and other areas of the precuneus more associated with cognitive processes. This connectivity could be linked to the integration of social and emotional information and perhaps increased sensitivity to social-emotional and cognitive conflicts. The same initial region also connected to the dorsomedial prefrontal cortex, which is involved in self-evaluation and in social interaction. Thus, this pattern of connectivity may show a relationship between cognitive processes of self-evaluation and the integration of social and emotional information consistent with the characteristic of neuroticism but manifesting in the single set of analyses from resting state data rather than piecemealed through many studies of individual characteristics.[41]

These preliminary studies of brain networks and their relationship to personality are doubtless the first of many more to come. We would suggest that the potential of this approach is manifested more in studies that arrive at a greater understanding of a personality characteristic by considering the interplay between various neural regions and how that might map onto personality characteristics rather than just correlations between areas and personality characteristics.

2. PERSONALITY AND PROGRESS BUILT UPON FACTOR ANALYSIS

The previous sections focused on network science and its application to neuroscience. This section deals more extensively with personality. A review of the contemporary personality field seems warranted before trying to understand possible connections between personality and network science. Two main points of this section are the following: (1) network analysis is a technique that may produce equivalent results to factor analysis, but nonetheless provokes different theories and conceptualizations of the underlying phenomenon; and (2) personality psychology could benefit from the reconceptualization attendant upon this new technique for accomplishing one of its main goals.

This is not because personality has not made considerable progress already on this goal. This main goal of personality psychology is to organize the many ways individuals differ from each other into a coherent structure. Personality psychology deals with many variables. Variables are content on which individuals differ from each other. For example, optimism is a personality psychology variable because individuals differ in how optimistic they are. Friendliness is a personality psychology variable because individuals differ in how friendly they are. Need for cognition is a personality variable because individuals differ in how much they enjoy effortful cognitive endeavors. So are guilt proneness, grit, hope, psychological mindfulness, alexithymia, arrogance, achievement motivation, warmth, generosity, politeness, and curiosity. This volume investigates many more.

In fact, the number of such variables numbers in the tens of thousands. There are over 17,000 single words referring to personality characteristics in the English language alone.[42] For example, just considering trait words that start with the letter a, there are at least the following: artistic, arresting, alert, attentive, argumentative, antsy, antisocial, asinine, alluring, aloof, assertive, aggressive, ardent, anal, arrogant, artistic, and so on. Naturally, one of the main goals of personality psychology is to provide some order to this state of affairs.

This state of affairs creates two main problems. First, identifying the important traits requires comprehensiveness, that is, a system that includes all the important traits. Second, the potential traits are not organized, creating an unwieldy collection of traits, in which it is unknown how the different traits relate to each other, whether they are redundant or unique, and which ones are central or most important.

Since these are variables, the natural approach to providing order is to investigate how they covary. Covariances among the variables will reveal patterns of relationships among individual differences. The idea is that the patterns of individual differences will reveal the systematic organization of traits, as well as the traits that are the most important and that capture the nature of the person. Investigation of patterns of covariance is an activity ripe for network analysis.

However, personality psychologists have used factor analysis as their analytic technique of choice (with wild success). John, Naumann, and Soto describe the brilliant two-step development of the Big Five.[43] It started with the lexical hypothesis, which claimed that a separate word existed in the language for each and every important difference between individuals' behavior.[42] Thus, to obtain a comprehensive system, the only requirement was to include all of the trait words in a given language's dictionary; every important difference, in fact every difference, between people in their behavior would be included on such a list. But such a list turned out to be too large and too disorganized. The second stage of the solution was to use factor analysis to reduce redundancy in the list. What was especially creative about this step was that it allowed people, as their personalities actually were, to determine what the redundancies were—factor analysis designates redundancy precisely in those cases where levels of two traits co-occur naturally in people. Participants rated themselves on the traits, and if individuals' scores on one trait were similar to their scores on another trait, then those two traits were designated as redundant with each other.

It turned out that redundancies were large among the 17,000 words, and redundancies only became small when the 17,000 traits were reduced to about five traits, a remarkably small number. Suddenly, the problems had a viable solution: all important traits were included, there was little redundancy, and traits were organized relative to each other. The five traits were the by-now-familiar extraversion, agreeableness, conscientiousness, emotional stability, and intellect. Extraversion, for example, was one trait, because any person who was relatively talkative was likely to be also relatively energetic, relatively assertive, and relatively bold (among other related characteristics).

2.1 What Is Known about the Big Five

Personality psychologists have not been content with the simple five-trait view of personality but rather have created a hierarchical tree structure with levels both higher and lower than the Big Five. Even the middle level has been revised, as it now appears that adding a sixth trait, honesty, may create a better representation.[43,44] Moving up the hierarchy eliminates even the slight redundancies among the Big Five and results in two super traits: growth (extraversion/openness) and citizenship (agreeableness, conscientiousness, and emotional stability).[45] Moving down the hierarchy results in smaller, subcomponent traits. For example, reliability, orderliness, impulse control, decisiveness, punctuality, formalness, conventionality, and industriousness may be subcomponents or facets of conscientiousness.[46] It is not clear yet whether there are 10,[47] 30,[40] or another number of these smaller traits.[48] Filling out this hierarchy not only adds sophistication and subtlety to traits, but it provides additional organization and characterization to the various traits. In particular, as one moves higher in the organization, traits become broader and more widely predictive. Redundancy within a trait, as well as redundancy between traits goes down. As one moves down the hierarchy, traits become narrower and more specific, although perhaps stronger, in their predictions.[49] Furthermore, redundancy within a trait and between traits increases.

2.2 Widespread Impact on Life Outcomes

One of the most startling findings to emerge this decade was that traits powerfully predict important life outcomes, both concurrently and predictively over decades.[50,51] Extraversion and neuroticism have been known to predict positive affect and negative affect at about the 0.3 to 0.4 level for some time.[52a,b] This prediction level is greater than any other variable's ability to predict affect, and it has now been replicated across multiple labs, age groups, and cultures.[52a] It is even true within introverts, as Fleeson, Malanos, and Achille showed that both extraverts and introverts were happier when they acted extraverted, and McNiel and Fleeson verified that even manipulating state extraversion induced state positive affect.[53,54]

Although happiness may be a subjective phenomenon, conscientiousness predicts work and school performance about equally as strongly as does intelligence, regardless of the type of measure of work performance.[55,56] Even physical health and death are influenced by personality: four of the Big Five traits have been shown to predict lifespan. People higher in conscientiousness, extraversion, emotional stability, and agreeableness live longer. These associations are strong, at least as strong as those connecting socioeconomic status (SES) and intelligence quotient (IQ) with mortality. Personality traits are also directly related to numerous highly prevalent mental disorders with substantial documented social costs (e.g., mood, anxiety, substance use, and antisocial disorders).

2.3 Predictive Validity

There have been a wide variety of unrelated studies predicting a variety of outcomes from traits. Some of this research focused on predicting validation criteria, such as behavior, to demonstrate that traits really do exist.[57,58] Other research focused on what it is like to be someone with a given trait: what their living spaces look like,[59] how likable and normal they are considered to be,[60,61] what sort of e-mail names or Web pages they have,[62,63] their music preferences,[64] and their relationship behaviors.[65] One creative development in this research was honing in on cultural trends, such as speed dating, the Internet, and Web and e-mail presences.[66]

2.4 Target and Observer Agreement

The accuracy of personality impressions returned to favor in the last 10–20 years. Results show that targets and observers agree in their judgments about the targets' personalities, including both close others and brand new acquaintances.[67,68] Even people who have never interacted with the participant agree about his or her personality: "thin slices" of the targets depicted in photographs or very brief video clips can lead to agreement in impressions of a variety of personality traits.[69–71]

2.5 Equivalent Statistical Techniques but Nonequivalent Conceptual Prompts

The preceding review of the field of personality highlighted numerous insights. This tremendous success in personality psychology is largely attendant upon the use of factor analysis. Although we will argue that factor analysis and network analysis produce essentially equivalent results, we believe their differing assumptions prompt differing conceptual directions. We will describe the differing assumptions and their proposed impact on conceptual leanings.

2.6 Assumption of Factor Analysis

The key relevant assumption of factor analysis is that the covariance among the variables is being caused by a latent variable (or a small number of latent variables). In the case of traits, the assumption would be that the smaller subcomponent traits covary with each other because they are all caused by the same latent Big Five trait. For example, assertiveness, boldness, and dominance all covary with each other because they are each caused by latent extraversion. (Although the unique variances of each subcomponent could relate to each other, often they do not in standard factor analysis.)

This assumption prompts the researcher to particular conceptual leanings. Importantly, it does not in any way require these leanings, and personality psychologists are uniformly capable of resisting those leanings. However, making an assumption that the causal structure is a certain way does prioritize thinking that the causal structure is that way. First, making this assumption may prioritize identifying the nature of the latent variable. For example, it may prompt a search for the neural anatomy corresponding to the latent variable of extraversion. Several chapters in this volume demonstrate this approach. Note that these chapters also demonstrate the effectiveness of this approach.

Second, making the assumption of a single latent cause may prompt the conceptual leaning towards believing the causal machinery of traits is relatively monolithic. Because it is assumed that one or a small number of latent causes contribute to the subcomponent traits and their covariance, the theoretical assumption that the causal machinery consists of one or a small number of latent causes is prioritized. For example, explaining extraversion might begin with a search for the single neurochemical system that is responsible for individual differences in extraversion or for the primary motivation that is responsible.

2.7 Assumption of Network Analysis

An alternative technique for investigating the pattern of covariation among many variables is network analysis. We will compare factor analysis and network analysis in detail in the next section; for now, we are concentrating on the basic assumptions because of their prompting of certain conceptual directions. In contrast to factor analysis, network analysis assumes that variables covary with each other because of their multiple bivariate connections to each other. It attempts to characterize the many bivariate associations in a digestible format.

2.8 Factor Analysis and Network Analysis

Although these two techniques may ultimately be very similar, they may nonetheless prompt different conceptual developments. Given that factor analysis has been dominant in personality psychology, whereas network analysis has been barely present, we believe this possibility could lead to significantly new conceptual developments in personality psychology.

First, the assumption that subcomponents of traits are directly linked to each other may prompt the search for causal mechanisms that explain direct connections between subcomponents. For example, personality psychologists may search for the ways in which assertiveness causes talkativeness or talkativeness causes assertiveness rather than for the common cause of both. Second, the sorts of causal connections that are investigated may be different than those investigated as part of a latent, joint cause. For example, the kinds of connections between talkativeness and assertiveness may be relatively cognitive or social in nature.

Third, theoretical models of traits may come to emphasize multiple causal connections among subcomponent traits rather than a small number of causal connections acting on traits. For example, the causal connections between talkativeness and assertiveness may be different from the causal connections between sociability and courage, even though they are all in the trait of extraversion. Fourth, the explanation for the emergence of the Big Five may focus on webs of interconnections rather than from sharing a similar latent cause. Different traits emerge because some clusters of subcomponent traits are more tightly integrated than are others. Explanations of the emergence of the Big Five may focus on these interconnections rather than on the different latent structures.

2.9 Examples of Network-Like Models of Personality

Although factor analysis is dominant in personality psychology, there are at least two examples of personality models that share these conceptual leanings with network analysis.

Whole trait theory is one model of personality that shares these conceptual leanings.[72,73] In whole trait theory, subcomponents of traits represent ways of reacting to situations and of pursuing goals. For example, talkativeness (a subcomponent of extraversion) is a reaction to situations welcoming talkativeness and a pursuit of goals involving conveying information to others.[74,75] Being organized (a subcomponent of conscientiousness) is a reaction to a situation demanding completing a task, and a pursuit of the goal of using time effectively.[76]

Furthermore, traits and their subcomponents are proposed to have both trait and state aspects. States have the same content as traits and subcomponents but describe the way the person is for a short time rather than the way the person is in general. For example, the content of talkativeness—speaking a lot—can describe the way a person is for a short time or can be descriptive of the way a person is in general. This state aspect of subcomponents allows them to be closely tied to the immediate needs of the situation and the individual, rather than being only expressions of underlying, stable latent forces. That is, states can be activated when the situation demands it, the person needs it to accomplish a goal, or other subcomponents facilitate it.[73]

This characterization of subcomponents leads to proposing causal structures among the subcomponents and rooting those structures in the subcomponents themselves rather than in latent, underlying variables. For example, being assertive (one subcomponent state) is often called upon when a person wants to be talkative (another subcomponent state) but others are talking. Whole trait theory proposes that generalization, the learning of abstract principles, direct causal connections, and logical, biological, and cultural processes all lead to causal interconnections among the subcomponents.[73,77]

Cramer et al. presented what may be the only explicitly network-based model of personality.[78] They use network analysis as a statistical technique, with the explicit goal of characterizing personality dimensions in terms of networks of interrelationships among affective, cognitive, and behavioral components. They explicitly reject the assumption of a latent causal entity responsible for the covariations among these elements. In particular, they propose that direct causal, homeostatic, or logical connections among the components give rise to the covariation among components. However, they also accept that some smaller-sized latent variables may contribute to some of those covariances.

The Cramer et al. model is a good example of how this analytic technique can guide readers toward different conceptual leanings. Their article generated many commentaries because of its novelty. Some of these commentaries disagreed with the Cramer et al. characterization of traits. Thus, new perspectives were developed as a result of their proposal demonstrating the potential of network theory to lead to new insights in personality.

3. NETWORK SCIENCE APPLIED TO BEHAVIORAL DATA

As mentioned above, the tools of network science have been applied to a striking variety of phenomena. Yet, these tools are currently used minimally, if at all, by personality researchers. Even the elegant Cramer model described in the previous section, while strongly influenced by network theory, failed to utilize many of the metrics developed by network science in other contexts. In this section, we articulate and demonstrate fundamental connections between some of the analytic tools frequently used by network science and personality science, and we begin exploring the differences.

Exploring these connections and differences can enhance each discipline's familiarity with the concepts and tools of the other, and this has at least two potential benefits. First, it can help foster cross talk and collaboration by giving each discipline a deeper understanding of each other's common techniques and language. Second, it has the potential to generate new questions and insights, as concepts and techniques from one discipline's analytic toolbox might provide novel ideas among researchers in the other discipline.

With these goals in mind, we focus on exploratory factor analysis (an analytic tool very familiar to most personality psychologists) and network analysis. We address three specific issues. First, we briefly explain the basic logic of both tools, revealing their common basis and distinctive languages. Greater depth and technical discussions of the two tools are available elsewhere. Second, we apply both analytic tools to the same data set, highlighting the similarity in the core results. Third, we provide initial comments about the degree to which the two tools provide different, and potentially complementary, information.

3.1 Conceptual Overview of Both Tools

At a general level, both exploratory factor analysis and network science describe interrelationships among a set of variables. That is, both approaches are intended to provide insight into the structure of a set of bivariate associations. However, the two techniques conceptualize and carry out this description in radically different ways.

3.1.1 Factor Analysis

Factor analysis conceptualizes the structure of associations in terms of latent variables or "factors" that give rise to observed, manifested, or measured variables. Factor analysis (and the closely-related principal components analysis) accomplishes this by identifying sets of observed variables that have more in common with each other than with other observed variables in the analysis. Factor analysis begins with a correlation matrix of bivariate associations among observed variables. Conceptually, factor analysis scans the matrix to identify which observed variables go together. It searches for clusters of observed variables that are strongly correlated with each other and that are weakly correlated with observed variables in other clusters. More technically, it extracts factors that account for as much variation in the observed variables as possible.

Exploratory factor analysis can be seen as steps that are often conducted in an iterative, back-and-forth manner: extraction, selection of a number of factors, rotation, and examination of factor loadings and (potentially) factor correlations.[79] The first step involves applying an "extraction method" that identifies combinations of observed variables, and these combinations are called factors. There are several types of extraction methods, but principal axis factor analysis and principal components analysis are the most frequently used. Extraction produces one *eigenvalue* for each potential factor, with as many potential factors as there are observed variables. A factor's eigenvalue can be seen as the amount of variance in the observed variables explained by the factor.

In the second step, researchers decide on the number of factors that adequately summarize the relationships between the original variables. The "appropriate" number of factors can be ambiguous, but there are rules-of-thumb to aid in the process.[80] The rules-of-thumb generally depend on the relative magnitudes of the eigenvalues, but information from subsequent steps can be used to inform the decision (e.g., clarity of the factor loadings, see step 4).

In the third step, researchers usually use a "rotation" to clarify the psychological meaning of the factors. Rotation is intended to produce *simple structure*, a pattern of associations in which each observed variable associates strongly with (i.e., "loads on") one factor and only one factor. There are two general types of rotation: orthogonal rotation generates factors that are uncorrelated, and oblique rotation generates factors that can be correlated with each other.

Fourth, researchers draw psychological conclusions based on key statistical outcomes, primarily factor loadings and (if relevant) interfactor correlations. Factor loadings are values representing associations between each observed variable and each factor. By noting which observed variables are most strongly associated with each factor, researchers can interpret the psychological meaning of the factors. There are several types of factor loadings that might be produced, but they are all roughly or literally on a correlational metric of −1 to +1, with values closer to −1 or +1 representing strong associations, and values close to 0 indicating no connection between an observed variable and a factor. Interfactor correlations are obtained when researchers extract more than one factor and implement an oblique rotation, and they reveal the degree to which the dimensions underlying the observed variables are themselves associated with each other.

3.1.2 Network Analysis

Network analysis can also be illustrated in a series of steps: choosing a threshold, applying the threshold to a correlation matrix to produce an adjacency matrix, and producing the network from the adjacency matrix. Like factor analysis, network analysis can begin with a correlation matrix of associations among a set of observed variables.

In the first and second step, a threshold is chosen and applied to "binarize" or "dichotomize" the correlation matrix, creating an adjacency matrix. Correlations with an absolute value above the threshold are given a "1" and those below are given a "0." (The binarization process is optional; an alternative, although a computationally more complex option, is to construct a weighted network). From the adjacency matrix, a network can be straightforwardly constructed: each observed variable is represented as a "node" in the network, and any pair of nodes with a "1" in the adjacency matrix is given an "edge" or connection between them. Note that the choice of threshold is a controversial one and could have a significant effect on the structure of the resultant network.[81] The choice of threshold may depend on several factors: the size of the sample from which the data was drawn, the choice of type I error rate, the density of the resulting network, and the domain from which the data was drawn. Network metrics should ideally be applied across a range of thresholds to demonstrate the result is not based on an arbitrary threshold determination. Fortunately, most network scientists are sensitive to this issue, and many networks have been observed to have robust community structure across a range of thresholds.[82]

Although there are many network science metrics, as described earlier, this section assesses one particular measure related to *community structure*. In networks derived from real world data, it is often observed that networks can be partitioned into groups of nodes that are more interconnected among themselves than with nodes outside those groups. Groups of nodes are called "communities" or "modules," and a network is partitioned into communities through the operation of community detection algorithms. There are many such algorithms, each with advantages and disadvantages.[83] Perhaps the most popular kind of community detection algorithm among network researchers is *modularity maximization algorithms*. These algorithms operate by searching through possible partitions of a network to find those partitions with the highest modularity value (Q). *Modularity* is a measure designed to quantify community structure. More specifically, it is designed to measure the quality of a particular partition of a network into *modules* or groups. Computationally, modularity (often referred to as *Q*) reflects the number of links between nodes within a module minus what would be expected given a random distribution of links between all nodes regardless of modules.[5b] This value varies from 0 to 1, with a higher value reflecting stronger community structure. It is also important to note that this algorithm does not allow for any overlap in its assignment of nodes to communities, meaning that nodes are placed into only one community. In most cases this is desirable, but there are also circumstances in which overlapping communities are more appropriate (e.g., social networks).[84] For example, in a friendship network, it would be expected that certain nodes (i.e., persons) cluster into multiple modules (i.e., friend groups), and an accurate community detection algorithm would partition the network into overlapping communities.

3.2 Analytic Comparison in Behavioral Data Set

Considering both the similarity in the conceptual purposes of factor analysis and network analysis and the analytic dissimilarity between the two, our goal is to examine their practical convergences and divergences. To what degree do they produce similar results?

For this comparison, we apply both techniques to a well-known personality inventory—the 44-item Big Five Inventory (BFI).[85] A fundamental question in the development and evaluation of personality inventories (or any psychological test) is dimensionality. Particularly, how many meaningful dimensions are reflected in the items in a personality inventory, and what exactly are those dimensions? The goal is to identify sets of items that go together (i.e., that measure similar things), which has important conceptual and practical implications.[79]

The data for this example was collected through 115 introductory psychology students who completed the BFI as part of a course requirement. The BFI is a self-report questionnaire designed to measure the Big Five dimensions: extraversion, agreeableness, openness to experience, conscientiousness, and neuroticism. It consists of 44 items, with 8 items measuring extraversion,

9 items measuring agreeableness, 10 items measuring openness to experience, 9 items measuring conscientiousness, and 8 items measuring neuroticism.

3.2.1 *Factor Analysis of Behavioral Data Set*

Factor analysis is the tool used by most personality researchers to examine dimensionality. From this perspective, one would expect to identify five factors that correspond to the Big Five dimensions. For example, we would expect to find good evidence of five factors, and we would expect the "agreeableness" items to load together strongly on a single factor and load weakly on any other factors. Similarly, we would expect the extraversion items to load strongly on one factor and weakly on the other factors.

As mentioned above, the start of the factor analysis procedure is a correlation matrix. Table 1 presents a correlation matrix of the first five BFI items (test items are prefixed with letters representing the first letter of their intended Big Five dimension). Of course, the complete correlation matrix includes all correlations among the 44 items.

A principal axis factor extraction method was used to extract factors from the full correlation matrix. Table 2 displays the first 10 extracted factors and their eigenvalues (variance explained by each factor). As researchers scan (either in graphical form or tabular form) the descending levels of eigenvalues, they hope to see a clear point at which the differences in eigenvalues becomes minimal.

In the current case, the eigenvalues do not provide perfectly clear insight, but they might indicate a six-factor solution. However, examination of a six-factor structure revealed factor loadings (see step 4, above) that were significantly problematic. The first five factors strongly resembled the Big Five traits, but the sixth factor included only a single item that loaded more strongly on it than on any other factor. The latter finding is generally seen as nonoptimal, and it motivates consideration of alternative structures. Indeed, exploratory factor analysis is often a back-and-forth process informed by both statistical and conceptual clarity.

Therefore, we extracted five factors, conducted an oblique rotation (*direct oblimin rotation*) and examined the solution. Table 3 displays the factor loadings for each item (each item is prefixed by the starting letter of the dimension it is designed to measure; items are sorted by the size of their factor loadings on each factor, and the strongest factor loading for each item is bolded). Examination of these factor loadings revealed reasonable simple structure, with nearly all items loading strongly on one and only one factor.

The pattern of factor loadings allows researchers to interpret the factors, with the items loading most strongly on each factor revealing the meaning of each factor. For

TABLE 2 Eigenvalues, or the Variance Explained in the Observed Variables for Each Factor

Total Variance Explained			
	Initial eigenvalues		
Factor	**Total**	**% of Variance**	**Cumulative %**
1	6.588	14.973	14.973
2	5.020	11.410	26.383
3	3.975	9.033	35.416
4	3.066	6.967	42.383
5	2.778	6.315	48.698
6	2.021	4.594	53.292
7	1.459	3.315	56.607
8	1.339	3.044	59.651
9	1.303	2.961	62.611
10	1.112	2.527	65.138

Each factor is in descending order based on the size of its associated eigenvalue. Column three is the percentage of variance explained in the observed variables, and column four is the cumulative percentage of variance explained in the observed variables.

TABLE 1 Correlation Matrix Displaying the Bivariate Correlations between the First Five Items of the BFI Personality Test

	Is talkative (E1)	**Tends to find fault with others (R) (A1)**	**Does a thorough job (C1)**	**Is depressed, blue (N1)**	**Is original, comes up with new ideas (01)**
Is talkative (E1)	1	−0.285**	0.101	−0.052	0.171
Tends to find fault with others (R) (A1)	−0.285**	1	0.003	−0.264**	−0.108
Does a thorough job (CI)	0.101	0.003	1	−0.246**	0.081
Is depressed, blue (N1)	−0.052	−0.264**	−0.246**	1	0.085
Is original, comes up with new ideas (01)	0.171	−0.108	0.081	0.085	1

Test items are labeled with a letter (E, A, C, N, & O) and a number, referring to the Big Five Dimension most associated with that item and the order in which they appear in the test, respectively.
**$p < .01$

TABLE 3 Factor Loadings of Each Item for the Five-Factor Solution

	Factor				
	1	2	3	4	5
Is considerate and kind to almost everyone (A7)	**0.677**	0.112	−0.033	0.071	0.150
Starts quarrels with others (R) (A3)	**0.676**	0.029	0.293	0.111	−0.052
Is sometimes rude to others (R) (A8)	**0.666**	−0.071	0.160	−0.022	−0.214
Can be cold and aloof (R) (A6)	**0.634**	−0.265	−0.288	−0.061	−0.090
Is helpful and unselfish with others (A2)	**0.612**	0.153	0.058	0.181	0.126
Has a forgiving nature (A4)	**0.576**	0.108	0.014	0.016	0.154
Likes to cooperate with others (A9)	**0.512**	0.064	−0.232	0.159	0.036
Tends to find fault with others (R) (A1)	**0.452**	−0.165	0.143	−0.016	−0.298
Is generally trusting (A5)	**0.419**	−0.112	−0.156	0.083	−0.104
Is emotionally stable, not easily upset (R) (N5)	**−0.286**	−0.118	0.017	−0.264	0.272
Likes to reflect, play with ideas (O8)	0.042	**0.712**	0.127	0.067	−0.094
Is inventive (O5)	0.077	**0.664**	−0.078	−0.051	−0.085
Has an active imagination (O4)	0.038	**0.642**	−0.026	−0.022	0.104
Is original, comes up with new ideas (O1)	0.014	**0.616**	−0.161	−0.019	−0.067
Is sophisticated in art, music, or literature (O10)	0.056	**0.563**	0.073	−0.018	0.019
Values artistic, aesthetic experiences (O6)	−0.013	**0.547**	0.009	0.063	0.144
Is ingenious, a deep thinker (O3)	−0.227	**0.545**	0.046	0.189	−0.024
Is curious about many different things (O2)	−0.024	**0.509**	−0.098	0.044	−0.140
Has few artistic interests (R) (O9)	0.070	**0.470**	0.028	−0.122	−0.081
Prefers work that is routine (R) (O7)	−0.119	**0.409**	−0.059	−0.292	−0.043
Is outgoing, sociable (E8)	0.215	0.095	**−0.788**	0.076	0.028
Tends to be quiet (R) (E5)	−0.130	−0.200	**−0.784**	−0.122	−0.127
Is sometimes shy, inhibited (R) (E7)	−0.189	−0.070	**−0.675**	0.020	−0.326
Is talkative (E1)	−0.047	0.017	**−0.675**	−0.042	0.196
Is full of energy (E3)	0.097	0.167	**−0.606**	0.211	−0.020
Generates a lot of enthusiasm (E4)	0.149	0.156	**−0.564**	0.160	0.048
Is reserved (R) (E2)	0.010	−0.064	**−0.555**	−0.362	−0.178
Has an assertive personality (E6)	−0.105	0.218	**−0.366**	0.254	0.162
Does things efficiently (C7)	0.126	0.069	−0.153	**0.702**	0.073
Does a thorough job (C1)	0.086	0.085	−0.019	**0.701**	0.198
Perseveres until the task is finished (C6)	−0.100	0.067	−0.054	**0.664**	0.036
Is a reliable worker (C3)	0.218	0.028	0.105	**0.631**	0.129
Tends to be disorganized (R) (C4)	−0.054	−0.215	0.060	**0.584**	−0.188
Can be somewhat careless (R) (C2)	0.107	0.054	0.083	**0.530**	−0.087
Tends to be lazy (R) (C5)	0.073	−0.031	−0.050	**0.497**	−0.199
Makes plans and follows through with them (C8)	0.075	−0.093	−0.078	**0.459**	−0.003
Can be tense (N3)	−0.001	0.046	0.024	0.023	**0.612**

TABLE 3 Factor Loadings of Each Item for the Five-Factor Solution—cont'd

	Factor				
	1	2	3	4	5
Gets nervous easily (N8)	0.059	−0.217	0.230	0.000	**0.585**
Worries a lot (N4)	−0.051	−0.011	−0.058	0.044	**0.572**
Is relaxed, handles stress well (R) (N2)	−0.013	−0.114	0.044	0.020	**0.504**
Is easily distracted (R) (C9)	−0.213	−0.126	0.173	0.405	**−0.459**
Can be moody (N6)	−0.392	−0.023	−0.080	0.174	**0.446**
Remains calm in tense situations (R) (N7)	0.038	−0.270	0.089	−0.233	**0.435**
Is depressed, blue (N1)	−0.335	0.148	0.193	−0.161	**0.402**

Each column represents one of the five factors and contains every item's factor loadings on that factor. Items are ordered by factor loading size. The patterns of factor loadings on each factor allow the researcher to "interpret" the meaning of each factor. Notice that all items accurately load on their intended factors, with the exception of items N5 and C9.

TABLE 4 Correlation Matrix Containing the Interfactor Correlations between Each Factor

Factor	1	2	3	4	5
1	1.000	−0.001	−0.027	0.221	−0.141
2	−0.001	1.000	−0.160	0.088	0.049
3	−0.027	−0.160	1.000	−0.048	0.081
4	0.221	0.088	−0.048	1.000	−0.066
5	−0.141	0.049	0.081	−0.066	1.000

example, the nine items intended to reflect agreeableness all load robustly on factor one and load weakly on the other four factors. Similarly, the 10 items intended to reflect openness load robustly on factor two and load weakly on the other four factors. All items loaded most strongly on their predicted factor with the exception of two items, N5 and C9. Examining the pattern of loadings for the entire loading matrix clearly suggests that factor one represents the Big Five dimension of agreeableness, factor two represents openness to experience, factor three represents extraversion, factor four represents conscientiousness, and factor five represents neuroticism. Thus, factor analysis "captures" the intended five-factor structure of the BFI quite closely. Another relevant factor analysis output is the relationships between the five factors. Because an oblique rotation was used, the factors are allowed to correlate with each other. Table 4 presents the interfactor correlations among the five factors. The factors are relatively independent of each other (only one coefficient reaches an effect size of 0.20). However, there is slight positive correlation between the "agreeableness" factor (factor one) and the "conscientiousness" factor (factor four), a negative correlation between the "openness to experience" factor (factor two) and the "extraversion" factor (factor three), and a negative correlation between the "agreeableness" factor (factor one) and the "neuroticism" factor (factor five).

3.2.2 *Network Analysis of Behavioral Data Set*

Factor analysis seems to produce a relatively clear psychological interpretation of the data, but does network science produce comparable results? To answer this question, the procedure described above was followed. First, the correlation matrix containing the correlations between all 44 items in the BFI was dichotomized with a threshold coefficient of $r = \pm 0.20$. This resulted in an adjacency matrix representing dichotomized associations between each item in the personality test. Table 5 presents the adjacency matrix of the first five items represented in the correlation matrix above (those item pairs labeled "1" are highlighted: ignoring the diagonal of the matrix.) When applied to the entire 44-item correlation matrix, the adjacency matrix reflects a network of 44 nodes with 218 edges between them (i.e., 218 correlation > the 0.20 threshold).

The application of a modularity maximization algorithm (*Louvain method*)[83] for community detection produced a partition of the network into five modules or communities. The Q of this partition was 0.344, falling within the common range of 0.3–0.7 observed in most other networks.[5b] In other words, the algorithm produced a partition with a moderately strong community structure.

Because the choice of a threshold can affect the overall network structure, we examined networks derived from alternative thresholds. Community structure was relatively consistent across thresholds near 0.20. For example, a higher threshold of $r = \pm 0.3$ produced the same five-module network, though it was a less densely connected network. However, raising the threshold to $r = 0.4$ produced a 14-module network with a less interpretable structure. The relative consistency of community structure depends on various factors, such as the size of the network and the number of within-module connections.

TABLE 5 The Adjacency Matrix of the First Five Items, Produced with a Threshold Correlation Coefficient or $r = \pm 2$

	Is talkative (E1)	Tends to find fault with others (R) (A1)	Does a thorough job (C1)	Is depressed, blue (N1)	Is original, comes up with new ideas (O1)
Is talkative (E1)	1				
Tends to find fault with others (R) (A1)	1	1			
Does a thorough job (C1)	0	0	1		
Is depressed, blue (N1)	0	1	1	1	
Is original, comes up with new ideas (O1)	0	0	0	0	1

The adjacency matrix represents the dichotomized associations between each item in the personality test. Those item pairs labeled with "1" are highlighted, ignoring the diagonal of the matrix. when applied to the entire 44-item correlation matrix, the adjacency matrix reflects a network of 44 nodes with 290 edges between them (i.e., 218 correlations > the ±0.2 threshold).

Figure 1 displays the results of the analysis based on the $r = \pm 0.20$ threshold. In this figure, nodes represent items of the personality tests, and they are prefixed with letters representing the first letter of the Big Five dimension they are designed to measure (A = agreeableness, E = extraversion, C = conscientiousness, O = openness to experience, and N = neuroticism). In order to facilitate visualization of the network, a force-directed algorithm was also applied to the network. The operation of this particular force-directed algorithm[86] causes nodes to repel each other like magnets, while edges attract the nodes they connect like springs. Node size is determined by the degree of that node. Nodes with larger degrees (i.e., larger number of connections) were larger in size relative to nodes with smaller degrees. Finally, the five modules discovered in the community detection algorithm were differentiated by color.

Clearly, the modularity algorithm reproduced the five dimensions of the Big Five in the context of a network. The algorithm converged on a structure that included five modules, and all nodes are grouped in modules only with the nodes they were designed to be associated with (in terms of BFI content). For example, all the neuroticism "nodes" were placed together in a single module, as were all of the agreeableness items, and so on.

'There is a striking similarity between the results of the factor analysis and the network analysis. First, they both produced a fairly clear structure that included five factors or modules. Of course, this depends on some analytic choices (e.g., a larger threshold in network analysis produced more modules, and the eigenvalues in factor analysis ambiguously suggested six factors), but the current analytic choices are fairly standard. Second, the "same" factors or modules were obtained. That is, the five modules produced by the network analysis had the same content as the five factors produced by the factor analysis. Notice also that item C9 ("is easily distracted (R)") was placed in the "neuroticism" module, corresponding to the observation in the factor loading matrix that item C9 loaded most strongly on the "neuroticism" factor. More specifically, both procedures "captured" the Big Five, consistent with the intended structure of the BFI.

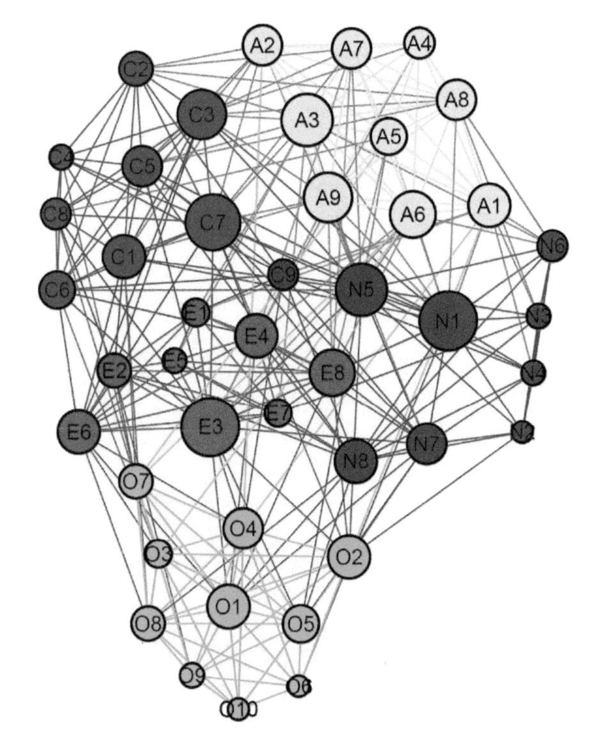

FIGURE 1 Network of 44 BFI items. Nodes represent items of the personality test (labeled by item). A force-directed algorithm was applied to the network to facilitate observation of the network. The algorithm operates by distancing the space between the nodes, while bringing those nodes sharing edges closer together. Node size is scaled by degree of node. Modules are partitioned by color (G = "openness to experience" module, P = "neuroticism" module, R = "extraversion" module, B = "conscientiousness" module, and Y = "agreeableness" module). Notice all nodes are placed in their intended module, with the exception of node C9.

In sum, the two tools achieved significant similarity in their results, despite radical differences in the procedures of both.

It is worth noting that network analysis converges with the *rotated* structure obtained via factor analysis. In other words, both analyses produced structures that approximated simple structure. Network analysis produced a structure in which each item was placed into the module containing the items it was most interconnected with, and each module is clearly delineated (despite a number of interconnections between modules). This corresponded to the results produced by factor analysis that had been rotated to simple structure. The unrotated factor structure does not reflect the Big Five, and it lacks simple structure in that few items load robustly on any factor and many items load moderately on multiple factors. This correspondence underscores the fact that the community-detection algorithm used here (and many other community-detection algorithms) is designed to produce a partition with the strongest community structure, or highest Q. And a strong community structure is defined roughly as a network consisting of modules that have high within-module connections relative to connections between modules. A network with a strong community structure consists of nodes that tightly interconnect within their own group and do not connect with other nodes outside their group.

With this in mind, modularity is essentially a metric of simple structure, and its maximization produces networks with the strongest possible community structure. In the context of this example, the partition of the network with the highest modularity value was the partition consisting of five modules. The five-module partition achieved the strongest community structure relative to other potential partitions. In other words, the five-module partition of the BFI network, with each node accurately placed into its designed module, achieved the highest degree of simple structure.

3.3 Differences and Additional Thoughts on Network Analysis

Despite these strong convergences between factor analysis and network analysis, there are important differences, and they might provide complementary information.

3.3.1 *Conceptual and Statistical Basis*

The procedures themselves are based upon different assumptions and statistical models. Factor analysis is a form of latent variable modeling that assumes that the observed variables (e.g., items of a personality test) are indicators of one's score on a latent variable. The items in a factor analysis are important, in so far as they are indicators of more general, underlying factors. For example, from a latent variable modeling perspective, the scores on the extraversion items above are assumed to be indicators or "manifest" variables of a more general, underlying extraversion "factor," or what is often referred to as a trait. The goals of factor analysis are to "extract" factors (i.e., linear weighted combinations of the original variables) that explain as much variance as possible in the common variance among the original variables and to have factors that are interpretable. Interpretable factors consist of items that are weighted (i.e., load) strongly on one factor and weighted weakly on the rest (i.e., simple structure). In contrast, the network science approach does not conceive of items as indicators of more general underlying factors or dimensions. Modules or communities are composed of the nodes they have because of the interconnections among those nodes. Interconnections among the nodes of a module are not conceived as deriving from a more general, underlying structure. In the network approach, all items form part of the network and are just as real as the modules of which they are a component of. These theoretical differences have important implications in the interpretation of data collected from a personality test.[80]

3.3.2 *Modularity*

The modularity metric might offer a new and insightful index for factor analytic types of questions. As noted earlier, Q is an important metric in the quantification of community structure. An important application of this metric for network researchers is the comparison of the strength of community structures across networks or the same network over time. For example, comparing the Q across the friendship networks of two individuals quantifies the relative segregation between groups of friends (i.e., modules) for each individual.

In the context of personality research, a concept like modularity might be useful, for example, in test construction. For example, one might compare the modularity between five-factor and six-factor structures of a given test, with modularity revealing the relative strength of one structure over the other. The structure producing the higher Q (i.e., stronger community structure) would reflect the clearer structure. This could provide a relatively concrete and objective index of "simple structuredness" that could be used to evaluate the appropriateness of different factor analytic structures.

3.3.2.1 Participation Coefficient

Another network statistic, known as the participation coefficient (*Pi*), may offer another important measure to the personality researcher. In contrast to modularity, the participant coefficient is calculated at the item level, with a participant coefficient calculated for each item. The participant coefficient compares the number of links of a node to other modules with its number of links to its own modules.[6] Conceptually, the *Pi* value for each node can be thought of as how interconnected the node is with other modules relative to its own module. A node with a *Pi* of 0 is completely interconnected with its own module, while a node with a *Pi* that approaches 1 has a nearly uniform distribution of connections among all of the modules. For the personality researcher, there is an important application of this item-level statistic. From a factor analysis perspective, the value of a node's participant coefficient quantifies the "simple structuredness" of each node. In other words, the metric quantifies how related the item is to its own module. Examining the *Pi* for each node in the module informs the test designer of how well each item satisfies simple structure. Those nodes with low *Pi* values are more interconnected with their own module, while those nodes with high *Pi* are more uniformly connected to all of the modules in the network. Again, this concrete objective information can help determine which items may need to be removed, if any, to achieve a test with simpler structure.

The *Pi* can be illustrated with the BFI network. Calculating a *Pi* for each node (items E1–E8) of the "extraversion" module (items E1–E8) gives the degree to which each node connects to items within the extraversion module relative to nodes in the other four Big Five modules. Items E1, E5, and E7 have the lowest item-level *Pi* values. Examining the connections of each item reveals that the majority of each item's connections are within its own module ("extraversion"), with no more than three connections outside its own module. In other words, all three of these items are almost completely interconnected within their own module. The only node in the network completely interconnected with its own module is item O10 ("is sophisticated in art, music, or literature") of the "openness to experience" module, revealed by a *Pi* of 0. The nodes in the BFI network with the largest number of connections, E3 ("is full of energy") and N1 ("is depressed, blue"), tend to have a sizable amount of connections with other modules and thus have some of the highest *Pi* values. Figure 2 illustrates these observations. To emphasize those nodes that have a high degree of connections within their own module, node size is scaled by 1-*Pi* for each node. Notice that the nodes with the smallest amount of connections tend to have more connections within their own module and are thus more emphasized in the network (by node size).

Caution should be exercised before conceiving of this metric as corresponding to an item's factor loadings from the factor analysis output. In the BFI network, the number of connections of each item (i.e., degree) is determined by the correlations between that item and every other item in the personality test. The *Pi* is determined by the distribution of these links within and between modules. On the other hand, factor loadings of an item from a factor analysis output represent the degree of association between that item and an underlying *factor* (latent variable). Thus, factor loadings are not comparing the degree to which each item is associated with other individual items, but the degree to which that item is associated with each extracted factor. An item with simple structure from a factor analysis perspective is an item that has a strong factor loading with one factor and loads weakly on the rest of the extracted factors. An item with simple structure, as defined from the perspective of network science, is an item with a large number of within-module connections relative to between-module connections with other *items* in the network.

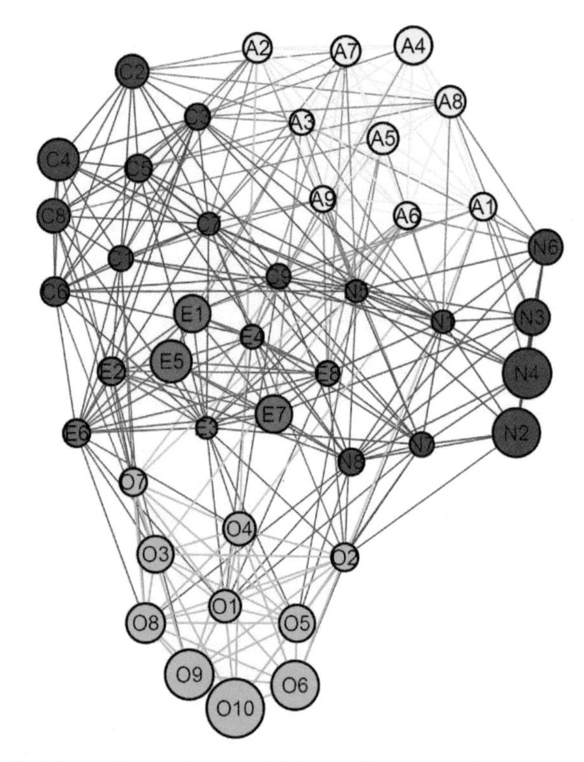

FIGURE 2 Network of 44 BFI items. This network was generated in the same way as Figure 1, except node size is scaled by 1-*Pi*, rather than node degree. Notice that those nodes more interconnected in their own module are larger, while those with interconnections outside their module are smaller.

3.3.2.2 Visualization

As illustrated in Figure 2, a potentially powerful and common function of network science is the availability of tools for the *visualization* of a system. It is possible that such network-based visualizations may be valuable communicative supplements to factor analysis. That is, factor analysts might be able to use such graphics to help present and describe complex factorial structures. However, one potential objection to this use is that two-dimensional network visualizations miss the complexity of factor analytic structures with greater dimensionality (e.g., three-dimensional structures). Thus, network science offers the ease of a two-dimensional representation of the data, but caution should be exercised before making any strong conclusions about the structure of the personality test from this representation. One preliminary observation of the BFI network is the extensive connectivity among separate modules. Despite the relatively tight within-module structure, there are a large number of between-module connections. This illustrates the large amount of associations among the items from separate dimensions or factors: an observation that may be overlooked by the examination of solely factor structure and factor loadings.

3.3.2.3 Additional Metrics

Another important category of metrics applied to networks are *centrality* metrics. Measures of centrality attempt to identify nodes that are highly central to the structure of the network. There are a variety of centrality metrics that highlight nodes important to the network in different ways (e.g., *degree centrality, eigenvector centrality, and betweenness centrality*); for an extensive review of different centrality measures see Joyce et al.[14] More research is needed to determine whether highly central nodes, as defined by these metrics, in networks models based on personality test items identify important items in the personality test. An important property of most network metrics is that they can be applied at the level of a single item, a module, or the entire network. Most network metrics are computed at the level of each item (e.g., *degree*) and summed/averaged to attain module or network level values of these metrics (e.g., *mean degree)*. Thus, networks can be analyzed at several different levels, and each level may offer insights into the network not offered by others. Consider networks derived from personality tests; analysis can be focused at the level of item, communities of items, and all items considered together. Further research is needed to determine what insights each metric offers, at each level of analysis, in the examination of data collected from personality tests. Factor analysis also offers analysis at the level of item and factors. In particular, factor loadings can be examined to determine each item's association with each extracted factor, and interfactor correlations can be examined to determine each factor's association with the other extracted factors.

Of course, there are additional components of network analyses not present in standard personality approaches, such as factor analysis. One is the possibility of grounding some parts of the network, in particular brain areas, and doing so in ways that factor analytic approaches might not suggest. For example, instead of looking for the locus of extraversion in the brain, which classic factor analysis might direct one's attention to, one might look for structures involved in connecting various components of extraversion to each other. Of course, how to integrate the network results from the personality assessments with neuroimaging data remains an open question. Several challenges must be overcome to make this possible. In many cases, the network analysis of personality data is based on a sample of individuals using correlations between the various personality scores. In these cases, there is no one personality network that can be extracted for each person that can be associated with that individual's brain imaging data, but it may be possible to associate sample-based personality networks with neuroimaging markers extracted from the same sample. It would be relatively simple to associate weights of the edges in personality networks to measures extracted from neuroimaging data. Such an analysis could help identify where variation in a particular brain region is associated with variation in the topology of the personality networks. It would also be possible to perform a more complex analysis that integrates functional brain networks with the personality networks. While the methods are beyond the scope of this chapter, the concept is as follows. A new (tripartite) network is created with individual study participants being one node type, brain network hubs being another node type, and personality characteristics being the final node type. In a tripartite network, there can only be connections between different node types rather than within node types. Each study participant is connected to the brain regions that are hubs in their brain weighted by the node degree. Each person is also connected to the personality traits weighted by their score. Now a brain personality trait network can be extracted where brain regions are connected to personality traits, based on being shared by individual study subjects. This type of network has the potential to help us understand the relationships between the personality traits and the organization of individuals' brain networks. These possibilities are just that, possibilities, as we have not tested any of these methods. However, the combining of personality networks with anatomical or functional brain imaging is a ripe and exciting open area for future research.

The goal of this brief overview using behavioral data was to introduce the basic concepts of network analysis in the context of factor analysis, a tool very familiar to many personality psychologists. Network analysis shows meaningful convergence with factor analysis, illustrating a common purpose and, to some degree, a common analytic strategy. Network analysis offers many metrics, some of which might act as a potential supplement of factor analysis. However, the utility of some metrics might depend heavily on the phenomenon being examined. Thus, the application of network tools to this sort of data is preliminary, and further research is needed to determine whether these tools offer insights into the data overlooked by factor analysis. We hope that further research illuminates the differences and similarities among these tools and the capabilities of each, and we hope that this overview is a first step in that direction.

4. CONCLUSIONS

The preceding reviews of network science, network science applied to neuroscience, contemporary personality theory and research, and the illustration of both standard personality and network science methods being applied to the same data suggests the potential for further fruitful analyses. With respect to the integration of neuroscience using network science measures and personality, the earlier noted promise of greater understanding based on the consideration of the interactions of various neural regions and how those may map onto personality characteristics seems especially promising. Cautions are warranted, but it seems that the addition of the network science tool kit to the methodology of personality researchers has the potential to generate new and important insights.

References

1. Watts DJ, Strogatz SH. Collective dynamics of 'small-world' networks. *Nature*. 1998;393:440–442.
2. Barabasi AL, Albert R. Emergence of scaling in random networks. *Science*. 1999;286:509–512.
3. Barabasi AL. Network science. *Phil Trans R Soc A*. 2013;371:1–3.
4. Girvan M, Newman MEJ. Community structure in social and biological networks. *Proc Natl Acad Sci USA*. 2002;99:7821–7826.
5. a. Newman MEJ, Girvan M. Finding and evaluating community structure in networks. *Phys Rev E*. 2004;69:02611.
   b. Latora V, Machiori M. Efficient behavior of small-world networks. *Phys Rev Lett*. 2001;87:198701.
6. Guimera R, Amaral LAN. Functional cartography of complex metabolic networks. *Nature*. 2005;433:895–900.
7. Travers J, Milgram S. An experimental study of the small world problem. *Sociometry*. 1969;32:425–443.
8. Freeman LC. Centrality in social networks: I. Conceptual clarification. *Soc Netw*. 1979;1:215–239.
9. Newman MEJ. Assortative mixing in networks. *Phys Rev Lett*. 2002;89(20):208701.
10. Alvarez-Hamelin JI, Vespignani A, Dall'Asta A. K-core decomposition of internet graphs: hierarchies, self-similarity, and measurement biases. *Netw Heterog Media*. 2008;3:371–393.
11. Hagmann P, Cammoun L, Gigander X, et al. Mapping the structural core of the human cerebral cortex. *PLoS Biol*. 2008;6:1479–1493.
12. Moussa MN, Steen MR, Laurienti PJ, Hayasaka S. Consistency of network modules in resting-state fMRI connectome data. *PLoS One*. 2012;7(8):e44428.
13. Steen MR, Hayasaka S, Joyce KE, Laurienti PJ. Assessing the consistency of community structure in complex networks. *Phys Rev E*. 2011;84:016111.
14. Joyce KE, Laurienti PJ, Burdette JH, Hayasaka S. A new measure of centrality for brain networks. *PLoS One*. 2010;5(8):e12200.
15. Newman MEJ. Mathematics of networks. In: Blume LE, Durlauf SN, eds. *The New Palgrave Encyclopedia of Economics*. Basingstoke: Macmillan; 2008.
16. Eguiluz VM, Chialvo DR, Cecchi GA, Baliki M, Apkarian AV. Scale-free brain functional networks. *Phys Rev Lett*. 2005;94(1):018102.
17. Simpson SL, Bowman FD, Laurienti PJ. Analyzing complex functional brain networks: fusing statistics and network science to understand the brain. *Stat Surv*. 2013;7:1–36.
18. Simpson SL, Lyday RG, Hayasaka S, Marsh AP, Laurienti PJ. A permutation testing framework to compare groups of brain networks. *Front Comput Neurosci*. 2013;25(7):171.
19. Zalesky A, Fornito A, Bullmore ET. Network-based statistic: identifying differences in brain networks. *Neuroimage*. 2010;53:1197–1207.
20. Smith SM, Vidaurre D, Beckmann CF, et al. Functional connectomics from resting-state fMRI. *Trends Cogn Sci*. 2013;17(12):666–682.
21. Moussa MN, Vechlekar CD, Burdette JH, et al. Changes in cognitive state alter human functional brain networks. *Front Hum Neurosci*. 2011;5:1–15.
22. Stanley ML, Dagenbach D, Lyday RG, Burdette JH, Laurienti PJ. Changes in global and regional modularity associated with increasing working memory load. *Front Hum Neurosci*. December 1, 2014;8.
23. Rzucidlo JK, Roseman PL, Laurienti PJ, Dagenbach D. Stability of whole brain and regional network topology within and between resting and cognitive states. *PLoS One*. August 5, 2013;8(8):e70275.
24. Pessoa L. Understanding brain networks and brain organization. *Phys Life Rev*. 2014;11(3):400–435.
25. Paolini BM, Laurienti PJ, Norris J, Rejeski WJ. Meal replacement: calming the hot-state brain network of appetite. *Front Psychol*. March 2014;5.
26. Bullins J, Laurient PJ, Morgan AR, et al. Drive for consumption, craving, and connectivity in the visual cortex during the imagery of desired food. *Front Aging Neurosci*. 2013;5:77.
27. Telesford QK, Laurienti PJ, Friedman DP, Kraft RA, Daunais JB. The effects of alcohol on the nonhuman primate brain: a network science approach to neuroimaging. *Alcohol Clin Exp Res*. 2013;37:1891–1900.
28. Yuan KW, Qin J, Liu Q, et al. Altered small-world brain functional networks and duration of heroin use in male abstinent heroin-dependent individuals. *Neurosci Lett*. 2010;477:37–42.
29. Brier MR, Thomas JB, Fagan AM, et al. Functional connectivity and graph theory in preclinical Alzheimer's disease. *Neurobiol Aging*. 2014;35:757–768.
30. Hugenschmidt CE, Burdette JH, Morgan AR, et al. Graph theory analysis of functional brain networks and mobility disability in older adults. *J Gerontol Med Sci*. 2014;69:1399–1406.
31. Ma S, Calhoun VD, Eichele T, Du W, Adali T. Modulations of functional connectivity in the healthy and schizophrenia groups during task and rest. *Neuroimage*. 2012;62:1694–1704.
32. Yao ZY, Zhang L, Lin Y, et al. Abnormal cortical networks in mild cognitive impairment and Alzheimer's disease. *PLoS Comput Biol*. 2010;6:e1001006.
33. Micheloyannis S, Pachou E, Stam CJ, et al. Small-world networks and disturbed functional connectivity in schizophrenia. *Schizophr Res*. 2006;87:60–66.
34. Bullmore E, Sporns O. Complex brain networks: graph theoretical analysis of structural and functional systems. *Nat Rev Neuro*. 2009;10:186–198.
35. Kunisato Y, Okamoto Y, Okada G, et al. Personality traits and the amplitude of spontaneous low-frequency oscillations during resting state. *Neurosci Lett*. 2011;492:109–111.
36. Hahn T, Dresler T, Ehlis AC, et al. Randomness of resting state brain oscillations encodes Gray's personality trait. *Neuroimage*. 2012;59:1842–1845.
37. Gao Q, Qiang X, Duan X, et al. Extraversion and neuroticism relate to topological properties of resting state brain networks. *Front Hum Neurosci*. 2013;7:257.
38. Kong F, Liu L, Wang X, et al. Different neural pathways linking personality traits and eudemonic well-being: a resting state functional magnetic resonance imaging study. *Cogn Affect Behav Neurosci*. November 2014.
39. Ryff C, Almeida DM, Ayanian JS, et al. *Midlife Development in the United States (MIDUS II), 2004–2006*. Ann Arbor, MI: Inter-university Consortium for Political and Social Research (ICPSR); 2007.
40. Costa PT, McCrae RR. *Professional manual: revised NEO personality inventory (NEO-PI-R) and NEO five-factor inventory (NEO-FFI)*. Odessa, FL: Psychological Assessment Resources; 1992.
41. Adelstein JS, Shehzad Z, Mennes M, et al. Personality is reflected in the brain's intrinsic functional architecture. *PLoS One*. 2011;6:e27633.
42. Allport GW, Odbert HS. Traitnames: a psycho-lexical study. *Psychol Mongr*. 1936;47:1–171.

43. John OP, Naumann LP, Coto CJ. Paradigm shift to the integrative Big-Five trait taxonomy: history, measurement, and conceptual issues. In: John OP, Robins RW, Pervin LA, eds. *Handbook of Personality: Theory and Research*. New York, NY: Guilford Press; 2008:114–158.
44. Lee K, Ashton MC. The HEXACO personality factors in the indigenous personality lexicons of English and 11 other languages. *J Pers*. 2008;76:1001–1054.
45. Saucier G. Recurrent personality dimensions in inclusive lexical studies: indications for a big six structure. *J Pers*. 2009;77: 1577–1614.
46. Roberts BW, Pomerantz EM. On traits, situations, and their integration: a developmental perspective. *Pers Soc Psychol Rev*. 2004;8:402–416.
47. DeYoung CG, Quilty LC, Peterson JB. Between facets and domains: 10 aspects of the Big Five. *J Pers Soc Psychol*. 2007;93:880–896.
48. Soto CJ, John OP. Using the California psychological inventory to assess the Big Five personality domains: a hierarchical approach. *J Res Pers*. 2009;43:25–38.
49. Paunonen SV, Ashton MC. Big Five factors and facets and the prediction of behavior. *J Pers Soc Psychol*. 2001;81:24–539.
50. Ozer DJ, Benet-Martinez V. Personality and the prediction of consequences. *Ann Rev Psychol*. 2006;57:401–421.
51. Roberts BW, Kuncel NR, Shiner R, Caspi A, Goldberg LR. The power of personality: the comparative validity of personality traits, socioeconomic status, and cognitive ability for predicting important life outcomes. *Perspect Psychol Sci*. 2007;2:313–345.
52. a. Lucas RE, Fujita F. Factors influencing the relation between extraversion and pleasant affect. *J Pers Soc Psychol*. 2000;79: 1039–1056.
b. Costa Jr PT, McCrae RR. *Neo PI-R Professional Manual*. Odessa, FL: Psychological Assessment Resources, Inc; 1992.
53. Fleeson W, Malanos AB, Achille NM. An intraindividual process approach to the relationship between extraversion and positive affect: is acting extraverted as "good" as being extraverted? *J Pers Soc Psychol*. 2002;83:1409–1422.
54. McNiel JM, Fleeson W. The causal effects of extraversion on positive affect and neuroticism on negative affect: manipulating state extraversion and state neuroticism in an experimental approach. *J Res Pers*. 2006;40:529–550.
55. Noftle EE, Robins RW. Personality predictors of academic outcomes: Big Five correlates of GPA and SAT scores. *J Pers Soc Psychol*. 2007;93:116–130.
56. Ones DS, Viswesvaran C, Schmidt FL. Comprehensive meta-analysis of integrity test validities: findings and implications for personnel selection and theories of job performance. *J Appl Psychol*. 1993;78: 679–703.
57. Fleeson W, Gallagher P. The implications of Big-Five standing for the distribution of trait manifestation in behavior: fifteen experience-sampling studies and a meta-analysis. *J Pers Soc Psychol*. 2009;97:1097–1114.
58. Back MD, Schmukle SC, Egloff B. Predicting actual behavior from the explicit and implicit self-concept of personality. *J Pers Soc Psychol*. 2009;97:533–548.
59. Gosling SD, Ko SJ, Mannarelli T, Morris ME. A room with a cue: judgments of personality based on offices and bedrooms. *J Pers Soc Psychol*. 2002;82:379–398.
60. Wood D, Gosling SD, Potter J. Normality evaluations and their relation to personality traits and well-being. *J Pers Soc Psychol*. 2007;93:861–879.
61. Wortman J, Wood D. The personality traits of liked people. *J Res Pers*. 2011;45:519–528.
62. Heisler JM, Crabill SL. Who are "stinkybug" and "Packerfan4"? Email pseudonyms and participants' perceptions of demography, productivity, and personality. *J Comput Mediat Commun*. 2006;12:114–135.
63. lucasVazire S, Gosling SD. e-Perceptions: personality impressions based on personal websites. *J Pers Soc Psychol*. 2004;87:123–132.
64. Rentfrow PJ, Gosling SD. The do re mi's of everyday life: the structure and personality correlates of music preferences. *J Pers Soc Psychol*. 2003;84:1236–1256.
65. Asendorf JB, Wilpers S. Personality effects on social relationships. *J Pers Soc Psychol*. 1998;74:1531–1544.
66. Luo S, Zhang G. What is attractive: similarity, reciprocity, security or beauty? Evidence from a speed-dating study. *J Pers*. 2009;77: 933–964.
67. Hall JA, Andrzejewski SA, Murphy NA, Schmid Mast M, Feinstein BA. Accuracy of judging others' traits and states: can accuracy levels be compared across tests? *J Res Pers*. 2008;2008(42):1476–1489.
68. Connely BS, Ones DS. Meta-analytic integration of observers' accuracy and predictive validity. *Psychol Bull*. 2010;136:1092–1122.
69. Borkenau P, Mauer N, Riemann R, Spinath FM, Angleitner A. Thin slices of behavior as cues of personality and intelligence. *J Pers Soc Psychol*. 2004;86:599–614.
70. Ames DR, Kammrath LK, Suppes A, Bolger N. Not so fast: the (not-quite-complete) dissociation between accuracy and confidence in thin-slice impressions. *Pers Soc Psychol Bull*. 2010;36:264–277.
71. Ambady N, Bernieri F, Richeson J. Towards a histology of social behavior: judgmental accuracy from thin slices of behavior. In: Zanna MP, ed. *Advances in Experimental Social Psychology*. San Diego, CA: Academic Press; 2000:201–272.
72. Fleeson W. Perspectives on the person: rapid growth and opportunities for integration. In: Deaux K, Snyder M, eds. *The Oxford Handbook of Personality and Social Psychology*. New York: Oxford University Press; 2012:33–63.
73. Fleeson W, Jayawickreme E. Whole trait theory. *J Res Pers*. 2014. http://dx.doi.org/10.1016/j.jrp.2014.10.009.
74. Fleeson W. Situation-based contingencies underlying trait-content manifestation in behavior. *J Pers*. 2007;75:825–862.
75. McCabe KO, Fleeson W. What is extraversion for? Integrating trait and motivational perspectives and identifying the purpose of extraversion. *Psychol Sci*. 2012;73:1498–1505.
76. McCabe KO, Fleeson W. Are traits useful? Explaining trait manifestations as tools in the pursuit of goals. *J Pers Soc Psychol*, in press.
77. Allport GW. *Personality: A Psychological Interpretation*. Oxford, England: Holt; 1937.
78. Cramer AOJ, van der Sluis S, Noordhof A, et al. Dimensions of normal personality as networks in search of equilibrium: you can't like parties if you don't like people. *Eur J Pers*. 2012;26:414–431.
79. Furr RM, Bacharach VR. *Psychometrics: An Introduction*. 2nd ed. London, UK: Sage; 2014. http://dx.doi.org/10.1111/j.1083-6101.2006.00317.
80. Fabrigar LR, Wegner DT, MacCallum RC, Strahan EJ. Evaluating the use of exploratory factor analysis in psychological research. *Psychol Meth*. 1999;4:272–299.
81. van Wijk BC, Stam CJ, Daffertshofer A. Comparing brain networks of different size and connectivity density using graph theory. *PLoS One*. October 28, 2010;5(10):e13701.
82. Achard S, Salvador R, Whitcher B, Suckling J, Bullmore E. A resilient, low-frequency, small-world human brain functional network with highly connected association cortical hubs. *J Neurosci*. 2006;26:63–72.
83. Newman MEJ. Modularity and community structure in networks. *Proc Natl Acad Sci USA*. 2006;103:8577–8582.
84. Mokken RJ. Cliques, clubs, and clans. *Qual Quant*. 1979;13:61–75.
85. John OP, Srivastava S. The Big Five trait taxonomy: history, measurement, and theoretical perspectives. In: Pervin LA, John OP, eds. *Handbook of Personality: Theory and Research*. 2nd ed. New York: Guilford; 1999:102–138.
86. Jacomy M, Venturini T, Heymann S, Bastian M. ForceAtlas2, a continuous graph layout algorithm for handy network visualization designed for the Gephi software. *PLoS One*. 2014:e98679.

CHAPTER

4

Genetics, Brain, and Personality: Searching for Intermediate Phenotypes

Andrey P. Anokhin

Department of Psychiatry, Washington University School of Medicine, St. Louis, MO, USA

OUTLINE

1. INTRODUCTION: SEARCHING FOR THE BIOLOGICAL ROOTS OF PERSONALITY AND INDIVIDUAL DIFFERENCES

Recent theoretical and methodological advances in human genetics and neuroscience offer novel opportunities for understanding biological bases of personality. Building a unifying, biologically based theory of personality will require the elucidation of complex links between genes, brain, and personality traits. Accordingly, in this chapter we review, summarize, and critically evaluate three related lines of research: genetic research on personality, investigation of the relationships between individual variability in the brain and personality traits, and genetic studies of brain structure and function. We also demonstrate how these research directions can be integrated using a genetically informative research design in order to determine how genetic influences on personality are mediated by variability in brain structure and function. In conclusion, we identify critical issues that should be addressed in future studies.

Personality emerges as a result of a complex dynamic interplay between genetic and environmental factors in the course of individual development. However, there are significant gaps in our knowledge related to the three research directions specified above. The structure of personality has been extensively studied, and major dimensions have been identified and described, but this research mainly relied on statistical analyses of self-reports that

Neuroimaging Personality, Social Cognition, and Character
http://dx.doi.org/10.1016/B978-0-12-800935-2.00004-X

may be biased in a variety of ways. Research on the etiology of individual differences typically took a top-down approach by attempting to identify genetic determinants and neurobiological correlates of personality dimensions derived by self-report. A major limitation of this behavioral genetics approach is that it relies on the assumption of congruency between the structure of personality derived statistically from self-reported data and the structure of the underlying neurobiological variation.

An alternative, bottom-up approach (personality neuroscience) would start with distinct, objectively measured biobehavioral dimensions of individual differences and determine their correlates in self-reported behaviors. Genetic contributions to these individual brain variations may also be elucidated. Such a biologically-based approach may lead to the emergence of a novel taxonomy of personality rooted in neuroscience and genetics. It is important to note that the top-down and bottom-up approaches are complementary, and personality research will benefit from their integration.

Although the search for biological roots of personality has a very long history dating back to the ancient Greek time (e.g., Hippocrates' and Galen's humoral theories of temperament), the progress over the centuries has been relatively modest. Recent methodological advances in genetics and neuroscience present unique opportunities for personality researchers and can lead to a revolutionary change in our understanding of the natural history of personality and individual differences. This goal can be achieved by the integration of three major directions of research (Figure 1): first, we need a better understanding of the genetic underpinnings of personality; second, we need a better understanding of the neural substrates and mechanisms underlying personality traits; third, we need to understand the genetic determinants of these neurobiological variables. Ideally, all three aspects of the biological research on personality should be integrated in a single study; however, bridging the knowledge obtained in studies addressing each of these aspects individually also can significantly advance our understanding of the biological bases of personality.

The purpose of this chapter is to summarize and critically evaluate recent advances in the three directions of research outlined above, identify problems and pitfalls, and outline promising directions for future research. The chapter is structured according to these three major directions: first, we discuss the current state of the genetic research on personality; then we review and summarize the most important evidence linking personality traits to individual variability in brain structure and function; next, we discuss the evidence for genetic influences on structural and functional variability in the brain. Finally, we discuss possible ways of integrating these three lines of research and illustrate them with examples from recent literature and our ongoing studies.

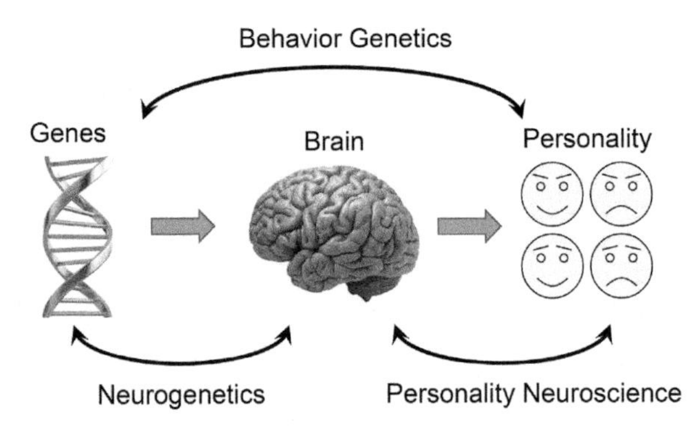

FIGURE 1 Genetics, brain, and personality: an integrative perspective. Behavior Genetics of personality can assess heritability of personality traits, estimate the degree of genetic overlap among different traits, reveal gene × environment interactions, and identify specific genetic variants contributing to heritability. However, exactly *how* genes affect personality traits remains beyond the reach of the traditional behavioral genetics approaches. To address this fundamental issue, we need Personality Neuroscience that will elucidate the neural substrates of personality and Neurogenetics that will establish genetic factors underlying this neurobiological variation. Through the integration of all three approaches, we hope to achieve the "full picture" of genes→brain→personality relationships.

2. GENETIC DETERMINANTS OF PERSONALITY

2.1 Personality "Phenotypes"

Building a biologically informed model of personality requires a better understanding of the genetic and environmental factors that shape individual differences and give rise to distinct personality dimensions. Genetic studies of personality can be subdivided into two groups: (1) assessment of heritability and genetic covariances among personality dimensions using genetically informative samples, such as twins and families, and (2) identification of specific genes influencing personality dimensions using genetic association studies. A central issue in genetic studies of complex traits is the choice and definition of the phenotype. The majority of genetic studies on personality focused on three widely adopted taxonomies of personality traits.

Eysenck's model[1] was the first to gain a wide acceptance due to its purported neurobiological foundations that fit well into the leading behavioral neuroscience theories of that time and its operationalization in the Eysenck Personality Questionnaire (EPQ). This model construes personality as composed of two independent dimensions of temperament: neuroticism (N) and extraversion (E)— and a third dimension, psychoticism (Psy).

The second is Cloninger's model[2,3] psychobiological theory of personality that was based on evidence from genetic studies, neuropharmacology, neuroanatomy, and behavioral neuroscience research. In its present form, it includes four temperament traits—harm avoidance (HA), novelty seeking (NS), reward dependence (RD), and persistence (P)—and three character traits—self-directedness (SD), cooperativeness (C), and self-transcendence (ST). These dimensions are operationalized in the Temperament and Character Inventory (TCI). Earlier version of this model included only NS, HA, and RD (Tridimensional Personality Questionnaire, TPQ), with persistence added later as a separate dimension, based on factor analyses of accumulated data.[3]

The third influential taxonomy of personality is the Five-Factor Model (FFM), or the "Big Five."[4] Unlike Eysenck's and Cloninger's models, which were derived from biosocial theories of animal and human behavior, the Costa and McCrae model was largely atheoretical and derived from factor analyses of correlations among adjectives used for self-descriptions of personality. The main assumption behind the construction of this taxonomy was that descriptors of personality used in the everyday language capture the true, objectively existing structure of personality. These analyses yielded a robust factor structure including five dimensions: openness (O), conscientiousness (C), extraversion (E), agreeableness (A), and neuroticism (N) included in the NEO personality inventory.[5] This five-factor structure has been replicated in multiple cultures and languages, indicating the construct validity of the NEO. Below I summarize the main findings of genetic studies using these three personality models.

2.2 Assessment of Heritability: Genetic and Environmental Influences on Personality Traits

Heritability is the proportion of the total variance of the trait that can be explained by genetic variation, and it can be expressed in percentage units from 0% to 100%. The remainder of the variance in the trait is caused by nongenetic (environmental) factors that can be further subdivided into two categories: shared environmental factors representing those aspects of the environment that are common to co-twins (e.g., ethnicity, culture, family, neighborhood) and therefore tend to increase their similarity, and nonshared, or individual, environmental factors and experiences that are unique to each of the co-twins and therefore tend to decrease twins' similarity.[6,7] Importantly, monozygotic (MZ) twins share 100% of their segregating genes, whereas dizygotic (DZ) twins share only 50% on the average, the same as non-twin siblings. However, both MZ and DZ twins reared together share their environment to the same extent (one of the key assumptions of the twin method). Fitting structural equation models to the observed twin data provides tests of different models that explain the variance in the trait by some combination of genetic, shared environmental, and individual environmental factors and parameter estimates for the best-fitting model, including heritability. Importantly, nonshared environmental variance also includes the measurement error.[7]

There are a few important attributes of heritability that should be taken into account when interpreting the results of twin studies. First, heritability is a characteristic of a population and cannot be applied to an individual. Second, heritability applies to a given population at a given moment in time, although personality traits show a good convergence of heritability estimates obtained in different populations at different times.

Multivariate analysis of twin data can be used to estimate *genetic correlations* between different traits that show the extent to which two personality traits are influenced by the same or different genetic factors, that is, the degree of "genetic overlap." For a more detailed overview of the twin method and its various extensions, the reader is referred to methodological reviews.[6,8,9]

Several large-scale twin studies investigated heritability of the scores on scales of Cloninger's TPQ. Heath and colleagues[10] conducted genetic analysis of data from 2680 adult Australian twin pairs and found significant heritability of HA, NS, and RD scales. Genetic factors accounted for between 54% and 61% of the stable variation in these traits. Furthermore, this study also collected data from the same twins on the revised EPQ and investigated genetic overlap between the scales of the two instruments by computing genetic correlations. This analysis allowed the authors to test the hypothesis that TPQ and EPQ provide alternative descriptions of the same underlying heritable dimensions of personality. However, the results showed only partial overlap of genetic influences on TPQ and EPQ, suggesting that each of the instruments provides only partial descriptions of the underlying structure of heritable personality differences. This finding has an important implication for the strategy of the search for neuroanatomical correlates of personality: personality traits derived from multiple assessment instruments using genetically informative twin samples may be more genetically homogenous, which may increase the likelihood of mapping them onto the variability in brain structure that has been shown to be highly heritable (reviewed below in greater detail).

Other studies of Cloninger's model involved samples of diverse age, ranging from adolescents to the elderly. The first twin study of Cloninger's personality dimensions in adolescence that included 1851 twins between the ages of 11 and 18 years found moderate heritability of NS, HA, and RD ranging from 0.28 to 0.36 but no evidence for genetic influences on PS. There was no evidence for gender differences in genetic influences on the TPQ scales.[11] Another study[12] focused on the other end of the lifespan. In a sample including 2420 women and 870 men aged 50–96, phenotypic factor analysis supported the four-factor structure in both genders, and genetic factor structure (assessed using genetic correlations) turned out to be gender-specific: in women, four genetic factors emerged, while in men, the genetic covariance among the TPQ dimensions could be explained by only three genetic factors.

The Big Five personality traits of openness (O), conscientiousness (C), extraversion (E), agreeableness (A), and neuroticism (N) have also been shown to be significantly influenced by genetic factors. Using data pooled from several twin studies in different countries, Distel et al.[13] assembled a combined twin sample that involved 4403 MZ twins, 4425 DZ twins, and 1661 siblings from 6140 families. Analyses of this combined sample showed significant heritability of all personality dimensions in the FFM: genetic factors accounted for 43%, 36%, 43%, 47%, and 54% of interindividual variability in N, A, C, E, and O, respectively.

In summary, twin studies indicate that 30–60% of interindividual variation in personality traits can be attributed to genetic differences among individuals. Second, studies have shown that genetic influences are significant across the lifespan, from adolescence to older age. Finally, these studies indicate that there is a partial genetic overlap between different personality scales, suggesting that genetically-based dimensions of personality may be distinct from the dimensions based on phenotypic correlations.

2.3 Finding Specific Genes Using Genetic Linkage and Association Methods

The classical twin method provides important information about the genetic and environmental origin of individual differences, as well as commonality versus specificity of genetic influences on different phenotypes, but it does not specify genes influencing the trait. The latter goal can be achieved by genetic linkage and association studies. Genetic association refers to the co-occurrence of a certain allele of a genetic marker and the phenotype of interest in the same individuals at above-chance level.[14] Association studies fall into two broad categories: candidate gene association studies and genome-wide association studies (GWAS).

Candidate gene studies focus on genetic polymorphisms selected by their biological relevance to the studied phenotype. Usually, these are functional polymorphisms, that is, their selection is based on the evidence that they produce functional effects at the molecular and cellular level, such as changes in gene expression, enzyme activity, or receptor characteristics. The candidate gene approach has obvious strengths: it is hypothesis driven, utilizes genetic variants that are likely to be causal variants, and therefore has a strong potential to provide a mechanistic explanation for the observed association. It also involves a limited number of statistical tests, thus mitigating the multiple comparisons problem.[15] Consequently, this analysis does not require very large samples, which is an important consideration for phenotypes that are difficult and costly to measure, such as neuroimaging phenotypes. However, in recent years, candidate gene studies, as well as candidate gene–environment interaction studies, have drawn much criticism for their inherently restrictive nature (i.e., limiting the search for genes involved in the determination of a complex phenotype to a handful of *apriori*-selected variants while neglecting the rest of the genome).[16–18]

Candidate gene studies of personality yielded mixed results. This is unsurprising, given that most of these studies relied on small samples. A recent review of 369 genetic studies of personality[19] concluded that results of candidate gene studies have been mixed, even when meta-analyses were conducted. Historically, following the first candidate gene study that found an association between neuroticism and a polymorphism in the serotonin transporter gene,[20] a large portion of candidate gene studies focused on the serotonergic system, in particular on the serotonin transporter gene polymorphism described by Lesch et al.[20–22]

In the past 15 years, the advent of molecular-genetic technologies and the identification of numerous SNPs throughout the genome paved the way for GWAS of personality. In contrast to the focused and restricted candidate gene approach, the GWAS approach[23] is largely atheoretical exploratory approach that does not require an *apriori* hypothesis. It is based on genome-wide "scanning" for association of thousands and even millions of SNPs for association with the phenotype of interest. To mitigate the problem of multiple testing and thus increased probability of false positive findings (Type I errors) while still retaining the ability to detect small effects, such studies have to be based on very large samples, usually of the order of thousands. The main advantage of the GWAS approach is that it is not confined to a specific hypothesis and is thus "unbiased"; its main disadvantage is the requirement of very large sample sizes. An analogy can be drawn between these two analytical strategies in genetics and similar approaches in neuroimaging: the candidate gene approach is analogous to the *apriori* region-of-interest analysis in MRI studies, whereas the GWAS approach is analogous to a voxel-wise whole-brain scan.

The results of GWAS of personality have largely been disappointing. Thus, the first GWAS study of Cloninger's temperament scales involved a sample of 5117 individuals and 1,252,387 genetic markers.[24] However, no genetic variants that significantly contribute to personality variation were identified, although the sample afforded ample statistical power to detect single genetic variants that explain only 1% of the trait variance.[24] A recent genome-wide meta-analysis of TCI scales included several large samples with a total number exceeding 11,000 subjects.[25] Although this study employed a sample of unprecedented size and combined various approaches to facilitate gene finding, such as meta-analysis, gene-based tests, and pathway analysis, its results were negative, with no SNPs, genes, or pathways showing a significant association with the four temperament dimensions after correcting for multiple testing. The authors concluded that identification of genetic variants significantly associated with temperament and personality might require even larger samples and/or a more refined phenotype.[25]

The first whole-genome study of Eysenck's Neuroticism scale used the DNA pooling approach (measuring allele frequencies using DNA pooled from groups of individuals selected by personality phenotype, rather than genotyping individual samples) in a sample of 2054 individuals selected on extremes of neuroticism scores.[26] A second group of 1534 individuals was used as a replication sample. Although a few interesting associations emerged, they did not survive replication in independent samples. A more recent study of neuroticism performed by the Genetics of Personality Consortium used a sample of an unprecedented size, including 63,661 participants from 29 discovery cohorts and 9786 participants from a replication cohort. An association meta-analysis yielded a genome-wide significant SNP in MAGI1 gene previously implicated in schizophrenia and bipolar disorder, but this association was not reproduced in the replication sample.

Several large-scale association studies focused on the Big Five dimensions of personality. De Moor and colleagues[27] have undertaken one of the largest analyses of personality ever by combining results from 10 GWAS studies, including 17,375 individuals. Furthermore, five additional samples consisting of 3294 individuals served as replication samples for any genome-wide significant SNP findings. Genotyping data included over two million SNPs. Significant associations with several SNPs were found for openness to experience (O) and conscientiousness (C); however, these associations failed to reach significance in replication samples, except that the direction of the effect of the KATNAL2 gene on 18q21.1 on C showed the same trend in all replication samples. A meta-analysis performed by Amin et al.[28] included 6149 individuals from multiple extended families and families with sibships from whom data on the NEO were available. This analysis yielded a significant association between O and an SNP on chromosome 11q24, and KCNJ1 gene was identified as a possible candidate. However, this region has not been implicated in other GWAS studies using NEO.[27]

A number of important conclusions can be drawn from candidate gene and GWAS studies of personality. First, GWAS studies failed to confirm many earlier positive findings from candidate gene studies (see a critical review by Munafo et al.[29]). Given that the latter have typically been based on much smaller samples, it is reasonable to conclude that most of these previous findings were false positives. Second, GWAS studies of personality generally failed to generate any new robust and replicable findings. One possible explanation of these disappointing outcomes is that the effects of individual genes are extremely small, and even larger-scale studies may be needed to detect these tiny effects.

Yet another possible explanation is that the effects of genes contributing to personality variation are nonadditive, including both intralocus (dominance) and interlocus (epistasis) allelic interaction. In the presence of nonadditive genetic effects, individual phenotype is affected not only by the presence or absence of specific alleles in the genotype, but also by the combination of different alleles, such that effects of one allele may be attenuated or enhanced by the presence of other alleles in the individual's genotype. Indeed, most GWAS analyses assessed only additive effects of genes contributing to personality dimensions. However, given the complexity and the interactive nature of the putative neurobiological pathways potentially influencing personality traits, this assumption may not be true. Most candidate gene studies conducted to date focused on neurotransmitter systems (e.g., serotonergic, dopaminergic, and noradrenergic). However, pathway analysis using rapidly developing bioinformatics tools[30] will likely yield novel candidate gene systems related to other aspects of brain functioning, such as neurodevelopment and hormonal modulation. Finally, a gene by environment interaction may also play a significant role. Both genetic nonadditivity and gene by environment interactions have been largely ignored by most GWAS analyses of personality published so far.

If both candidate gene approach and the genome-wide search failed to provide consistent and replicable results, what strategy should succeed in future studies? As often is the case, the truth may lie in the middle. Approaches based on the knowledge of biological pathways may provide a better coverage of genomic variation than traditional candidate gene studies, but at the same time, they are more focused and biologically informed compared to a fully "blind" genome-wide search.

Finally, increasing evidence suggests that epigenetic modifications of gene expression may play a particularly important role in brain functioning.[31,32] Unfortunately, due to the tissue-specific nature of gene expression and its modification, direct investigation of the relationship between epigenetic variation and personality is not possible due to the lack of access to brain tissue in living humans (for these and other limitations of human behavioral epigenetics research, see Ref. 33). However, translational neuroscience research using animal models of personality can help to fill this gap.

3. PERSONALITY AND THE BRAIN: TOWARD INTERMEDIATE PHENOTYPES (ENDOPHENOTYPES)

3.1 Personality and Individual Differences in Brain Structure

It has been known for a long time from postmortem brain morphology studies that the human brain shows enormous individual variability in its structural characteristics, such as the overall volume, relative size of cortical areas and subcortical structures, sulcal pattern, cytoarchitectonic structure, shape and orientation of different types of cells, dendritic arborization, and other micro- and macroanatomical features. However, establishing associations between individual characteristics of brain structure assessed postmortem and personality traits was hardly possible. The emergence of noninvasive imaging of the living brain opened an exciting opportunity for the investigation of brain-personality relationships.

Structural variability of the brain can be assessed in living humans using magnetic resonance imaging (MRI). MRI signal varies as a function of tissue type (e.g., gray and white matter, ventriculae), which allows for a qualitative and quantitative description of shape, size, and volume of cortical areas and subcortical brain structures. Most widely used methods for the analysis of structural MRI data include volumetry—counting the number of voxels within a brain structure delineated either manually or using an automatic segmentation approach—and surface-based analysis. The most representative structural MRI studies of major personality taxonomies (Big Five, Cloninger's TCI, and BIS/BAS) are summarized in Table 1.

In one of the largest (n = 265) studies of brain-personality relations using the NEO Personality Inventory, Bjornebekk et al.[34] investigated total and regional brain volumes, regional cortical thickness and arealization, and diffusion tensor imaging (DTI) indices of white matter (WM) microstructure. Of the five NEO scales, neuroticism showed the most consistent relationship with brain structure. Higher scores on neuroticism were associated with smaller total brain volume, a decrease in WM microstructure, and smaller frontotemporal surface area. Extraversion was inversely associated with the thickness of inferior frontal gyrus, while conscientiousness was inversely associated with arealization of the temporoparietal junction. Agreeableness and openness did not show any consistent associations with brain structure.

In another structural imaging study of NEO scales in 116 healthy adults,[35] a higher neuroticism score was associated with smaller volume in the dorsomedial prefrontal cortex (PFC) and a part of the left medial temporal lobe, including the posterior hippocampus, and larger volumes in the midcingulate gyrus. According to the authors, these findings were generally consistent with biologically based model of the Big Five. In particular, it was noted that neuroticism showed associations with volumes of brain regions implicated in threat, punishment, and negative affect; extraversion correlated with volume of the medial orbitofrontal cortex, a brain region involved in processing reward information; agreeableness covaried with volume in regions that process information about the intentions and mental states of other individuals; and conscientiousness was associated with volume in lateral PFC, a region involved in planning and the voluntary control of behavior.[35]

Cremers et al.,[36] using data from 65 healthy participants, investigated the relationship between the Big Five scales and "affective" brain regions, including the amygdala, orbitofrontal cortex, and the anterior cingulate cortex (ACC). Contrary to their expectation, the authors did not find any significant correlation with neuroticism. Instead, they found a positive correlation between extraversion and regional brain volume in the medial orbitofrontal cortex (OFC) and centromedial amygdala, as well as total gray matter (GM) volume, suggesting that increased volumes of medial OFC and amygdala may play a role in the increased sensitivity to reward and thus the propensity to experience positive affect. This study also found a sex by extraversion interaction in the ACC, with males showing a positive correlation between ACC volume and extraversion, and females showing a negative correlation. Interestingly, this finding was consistent with an earlier study in adolescents that also found a sex by volume interaction effect for the medial prefrontal gyrus.[37]

Another study of structural brain correlates of NEO scales in 62 healthy subjects placed a strong emphasis on the analysis of covariates that might modulate brain–personality relationships.[38] Correlations between regional brain volumes and personality scales were strongly dependent on specific combinations of covariates included in the model, such as gender, age, total gray and white matter volumes, etc. This study tested different combinations of "nuisance covariates" and found significant correlations with personality traits only with some of these combinations but not the others. A study of elderly participants[39] that compared morphometric brain measures and personality scores assessed at two time points two years apart found an inverse correlation between neuroticism and right orbitofrontal and dorsolateral PFCs and rolandic operculum. Extraversion was positively associated with larger left temporal, dorsolateral prefrontal, and ACCs. Openness correlated

TABLE 1 Structural Brain Correlates of Personality

Author	Sample, n (n females)	Age	Personality assessment	Key findings
Bjornebekk et al.[34]	265 (150)	20–85	NEO-PI-R	N: ↓ total brain volume, fronto-temporal surface area, and WM microstructure; E: ↓inferior frontal gyrus thickness; C: ↓ arealization of the temporoparietal junction; A,O: no reliable findings
Cremers et al.[36]	65 (0)	21–56	NEO-FFI	N: n.s.; E: ↑ r. medial OFC, incl. subgenual cingulate gyrus; ↑ total GM volume
Xu[49]	62 (31)	20–40	NEO-FFI	N, E: n.s.; A: ↓ regional cerebellar GM volume; O: ↑ l. superior OFC
Kapogiannis et al.[39]	87 (42)	72+/−7.7	NEO-PI-R	N: r. lateral OFC, r. dorsolateral PFC, ↑ ventral visual stream areas; E: ↑ l. AC, dorsolateral PFC, temporal regions; O: ↓ r. medial OFC, l. insula, ↑ fronto-polar cortex; A: r. OFC, ↓ dorsomedial PFC; C: ↑ sensorimotor areas involved in motor planning (BA3,5,6)
DeYoung et al.[35]	116 (58)	18–40	NEO-PI-R	N: ↓ r. dorsomedial PFC, ↓ l. medial temporal lobe incl. posterior hippocampus; basal ganglia, ↑ middle ACC, middle temporal gyrus, cerebellum, ↓ r. precentral gyrus; E: ↑ medial OFC; A: ↑ posterior CC, fusiform gyrus, ↓ superior temporal sulcus; C: ↑ left middle frontal gyrus; ↓ posterior fusiform gyrus; O: n.s.
Joffe et al.[115]	113 (48%)	36.8+/−13.3	NEO-FFI	N: ↓ hippocampus in met allele carriers but not Val/Val homozygotes of the BDNF Val66Met polymorphism
Yamasue et al.[40]	183 (66)	21–40	TCI	HA: ↓ r. hippocampus in both sexes; ↓ l. anterior PFC in women only
Iidaka et al.[42]	56 (26)	22.4	TCI	HA: ↑ l. amygdala in women only; NS: ↑ l. medial frontal gyrus; RD: ↑ r. Caudate (tail)
Pujol et al.[41]	100 (50)	20–40	TCI	HA: ↑ r. anterior cingulated gyrus; NS: ↑ l. posterior cingulate region
Fuentes et al.[43]	114 (0)	18–53	BIS/BAS	BIS: ↓ r. and medial OFC, precuneus; BAS: Not reported

Notes: The table includes only studies with a total sample size of n > 50 using any of the following personality assessment instruments: NEO-FFI is NEO Five Factor Inventory (N, neuroticism; E, extraversion; A, agreeableness; O, openness; C, conscientiousness); NEO-PI-R is the Revised NEO personality inventory (same scales); TCI is Cloninger's Temperament and Character Inventory (HA, harm avoidance; NS, novelty seeking; RD, raward dependence); and BIS/BAS is Behavioral Activation/Behavioral Inhibition Systems scales. ↓ and ↑ denote decreased and increased size of a brain region in individuals with higher scores on a given personality scale; (l. and r.) denote left and right hemisphere structures, respectively; GM, gray matter; PFC, prefrontal cortex; OFC, orbitofrontal cortex; n.s. indicates that no significant associations with brain structure were found. Designation of cortical areas and subcortical structures and regions is kept most closely to the original sources.

with larger right fronto-polar and smaller orbitofrontal and insular cortices, and agreeableness was associated with a larger right orbitofrontal cortex. Finally, individuals with higher conscientiousness scores were characterized by larger dorsolateral prefrontal and smaller fronto-polar cortices.

Although the majority of structural imaging studies was based on the FFM of personality, several studies used Cloninger's taxonomy implemented in TPQ and, later, TCI questionnaires. Using a large sample of 183 participants, Yamasue et al.[40] found that higher scores on HA were associated with smaller regional GM volume in the right hippocampus in both men and women. This finding suggests that a smaller right hippocampus may represent a neuroanatomical correlate of anxiety-related traits. In another study of 100 participants,[41] HA correlated positively with the anterior cingulate gyrus volume, while higher NS scores were associated with larger left posterior cingulate volume. Another study found positive correlation between HA and left amygdala volume in women but not in men.[42] In addition, NS was associated with increased medial frontal gyrus, and RD correlated positively with the volume of the tail of the right caudate nucleus.

A study by Fuentes and colleagues[43] focused on a Behavioral Inhibition System (BIS) scale from the

BIS/BAS questionnaire.[44] Analysis of regional brain volumes in 114 participants showed that higher BIS scores were associated with reduced volume of the right and medial orbitofrontal cortices and the precuneus, suggesting that[45] anxiety-related personality traits may be associated with reduced brain volume in brain structures implicated in emotional regulation.

Finally, a study by Montag and colleagues stands out in that they focused their analysis on hemispheric asymmetry of gray and white matter volume, rather than on absolute volumes.[46] Using a measure of the so-called volumetric hemispheric ratio in a fairly large sample (n=267) of healthy participants, they found that men, but not women, with greater GM volume in the left, rather than right, hemisphere score higher on extraversion assessed using the EPQ-R. This finding underscores the potential importance of relative, rather than absolute, structural brain measures for understanding the biology of personality. This approach can be extended to other structural measures (area, thickness, structural connectivity) and other "ratios," such as anterior to posterior or cortical to subcortical.

More recently, studies have begun to investigate the relationships between personality and structural connectivity in the brain. WM fiber tracts play a fundamental role in the integrative brain function because they represent the neuroanatomical substrate for communication among distant brain regions and their integration into a coherent functional network supporting complex behaviors. The main tool for the investigation of structural connectivity in the human brain is DTI that provides measures of WM integrity based on water diffusivities in parallel and perpendicular directions relative to axons. Two measures derived from DTI data, fractional anisotropy (FA) and mean diffusivity (MD), are differentially sensitive to the direction of diffusivity, with higher FA and lower MD indicating greater WM integrity.[47]

Recent DTI studies have found significant correlation between structural brain connectivity and personality traits; however, the reports are somewhat conflicting. Montag and coauthors[48] examined correlations between a single factor of "trait anxiety" extracted from three relevant personality scales (BIS from the BIS/BAS, HA from the TCI, and N from the EPQ-R) and DTI measures in 110 healthy young adults. Significant correlations were observed between FA values and trait anxiety in males but not females. The largest correlation ($r=0.49$) was observed for the WM tracts linking the hippocampus with the posterior cingulum, suggesting that in males, nearly 25% of the variance in trait anxiety was accounted for by this DTI measure. In a smaller (n=51) DTI study of the Big Five scales,[49] higher neuroticism scores were associated with higher MD values, indicating reduced integrity of WM in multiple fiber tracts, including corpus callosum, corona radiata, inferior frontal occipital fasciculus, and superior longitudinal fasciculus. In particular, higher neuroticism was associated with reduced integrity of WM interconnecting the PFC and amygdala (the anterior cingulum and uncinate fasciculus), consistent with the notion of decreased top-down emotion regulation from the PFC among persons with high neuroticism. In contrast, higher scores on openness correlated with better integrity of WM tracts interconnecting extensive cortical and subcortical structures, including the dorsolateral PFC. However, in contrast to Montag et al.,[48] this study did not find significant correlations between neuroticism scores and FA measures. E and C scores did not show significant correlations with any DTI parameters.

What inferences can be drawn from structural imaging studies of personality traits? The first and most important conclusion is that there is overall poor consistency of findings across studies. An overview of Table 2 shows that there is a great deal of disagreement, even among studies that used the same personality measures and were based on reasonably large samples ($n > 50$). Variability of findings across studies could be attributed to two major sources: sample characteristics and structural imaging procedures. Table 2 shows that samples differed substantially with respect to age; moreover, in some studies the age range was very broad (e.g.,Refs 34,36,43). When participants' ages range from adolescents to the elderly within a single study, including age as a covariate may not fully exclude the confounding effects of age that can lead to both false positive and false negative findings. This may happen if age-related changes of brain structure and/or personality are nonlinear, or

TABLE 2 Comparison of Functional Neuroimaging Methods

	EEG/ERP	fMRI
Relation to neuronal activity	Direct	Indirect
Temporal resolution	High (1/1000 s)	Limited (1–5 s)
Spatial resolution	Limited (20–30 mm)	High (3–4 mm)
Cost	Low	High

EEG, electroencephalogram; ERP, event-related brain potentials; fMRI, functional magnetic resonance imaging.

if brain structure is differentially related to personality traits in different periods of the life span. Furthermore, there is substantial variability across studies with respect to the sources of the samples (patients, college students, volunteers responding to advertisements) and exclusion criteria applied. In particular, exclusion versus inclusion of participants with the history of psychopathology and substance use or abuse may have contributed to variability of the results. Studies also differed with respect to scanner type, image acquisition parameters and sequences, preprocessing pipelines, and quantification of the structural measures, such as the use of different segmentation and normalization methods and accounting for total brain volume. All these factors, separately and in combination, may have contributed to the variability of findings across studies.

Finally, studies differed with respect to the degree of control over the number of statistical tests performed. The problem of excessive rate of false-positive findings has been a subject of heated debate in the neuroimaging field.[50,51] As far as neuroimaging of personality is concerned, replicability of findings is the most important issue to be addressed in future studies. This can be done, in particular, using multisite collaborative studies with a systematic ascertainment of population-representative samples, harmonization of acquisition and analysis, and collection of data on potentially important covariates. Such studies can provide samples with the sufficient statistical power required for comprehensive analyses involving important covariates and, at the same time, an adequate control over multiple testing.

On the bright side, some relatively consistent findings are emerging. For example, the relationship between anxiety- and negative affect-related traits (neuroticism, harm avoidance) and reduced hippocampal volumes has been reported by at least two relatively large studies.[35,40] Another example is the positive association between extraversion and orbitofrontal cortex volume that has been found in at least two larger studies [35,36] and two smaller studies.[52,53]

3.2 Methods for the Assessment of Brain Function: Strengths and Weaknesses

Before summarizing the evidence for association between personality and individual differences in brain function, we provide a brief comparative overview of the most relevant methods for functional brain imaging.

Historically, neural substrates of personality have been studied using brain electrophysiology techniques long before the advent of hemodynamic imaging methods. Electroencephalography (EEG) is a recording of neuronal electric activity using small censors (electrodes) attached to the scalp. Due to the "blurring" of the brain activity by tissues located between the cortical surface and the recording electrode, spatial resolution of EEG is rather coarse but still shows topographical specificity, which is sufficient for distinguishing between major cortical regions. The resting EEG represents synchronized activity of a large number of neurons (primarily, gradual postsynaptic potentials), which in the resting state manifests itself as oscillations at different frequencies. Event-Related Potentials (ERPs) represent EEG changes associated with specific events, such as stimuli or responses. The ERP is obtained by averaging over multiple trials of EEG fragments time-locked to a specific event. In the process of averaging, ongoing activity that is not systematically related to the event is canceled out, while activity synchronized with the event becomes apparent due to increased signal-to-noise ratio. ERP waveforms contain a number of peaks with different polarity and latency corresponding to specific stages of the information processing in the brain.

Over the past two decades, functional MRI (fMRI) has become another commonly used method for investigation of brain function, including studies of personality and individual differences. fMRI measures changes of the local oxygenation of blood, which has been shown to be systematically related to neuronal activity. The raw blood oxygenation level-dependent (BOLD) signal needs to be analyzed to determine how the time course of a BOLD signal at each voxel is associated with specific task events. This analysis results in a statistical map that shows the strength of the association between the observed BOLD signal and its changes predicted using the parameters of a "standard" hemodynamic response and the timing of specific events in the task.

Each of these methods has its strengths and weaknesses that are summarized in Table 1. EEG/ERP is a direct, real-time measure of neuronal activity, whereas the BOLD signal is not. There is strong evidence that the BOLD signal correlates with neural activity, but the exact nature of that activity reflected by the BOLD signal is not fully understood, and correlations are not always perfect,[54] in particular, the relative contribution of inhibitory (IPSP) and excitatory (EPSP) postsynaptic potentials to the BOLD signal remains unclear.[55] Next, due to their excellent temporal resolution, electrophysiology methods capture real-time neural dynamics, such as distinct stages of information processing in the brain (stimulus detection, categorization, response selection, error detection, etc.) that unfold within a period of less than one second. In contrast, the BOLD signal reaches its peak only about five seconds after the stimulus and takes 10 more seconds to return to baseline.[56] The main weakness of electrophysiology methods is their limited spatial resolution due to volume conduction effects and "blurring" of the electrical signal by the tissues between the cerebral cortex and the recording sensor on the scalp. Although the use of dense electrode arrays and novel analytical approaches has substantially improved spatial resolution of neuroelectric imaging, reliable source localization is mostly limited to cortical activity because the contribution of subcortical structures to the scalp-recorded EEG is relatively weak and difficult to isolate from the cortical activity, except specific forms of pathological activity. In contrast, the main advantage of fMRI is its high spatial resolution, allowing for accurate localization of task-related changes in the BOLD signal, including both the cerebral cortex and subcortical structures.

Two other methodological considerations are specifically important for the study of individual differences. First, the studies of the functional neural correlates of personality normally assume that measures of brain activity represent stable, trait-like characteristics. Evidence available to date indicates high reliability of resting-state EEG measures (test-retest correlations of the order of 0.7–0.9),[57] moderate to high reliability of ERP measures (0.4–0.7),[58,59] and low to moderate reliability of resting-state and task-related fMRI measures (0.3–0.5).[60] Second, because studies of individual differences require larger samples than studies of within-subject effects, another important consideration is the cost of assessments, which is modest for EEG/ERP (provided the equipment is available) and rather high for fMRI. Importantly, the above two issues are related: the lower the test-retest reliability of a brain-based measure is, the larger sample is needed to detect a significant correlation with personality.

In summary, a researcher choosing a method for the assessment of brain function faces a dilemma: either to choose inexpensive and reliable EEG/ERP assessments that provide little glimpse into the role of specific brain structures, or rather go with fMRI that may require substantial investments and large samples to achieve robust and reproducible results. Ideally though, these two methods should be combined to achieve the highest temporal and spatial resolution in the assessment of task-related neural activity.

3.3 Resting-State EEG and Personality

EEG studies of personality and individual differences have identified a number of neuroelectric correlates of personality traits and symptoms of psychopathology. The great bulk of these studies concerned frontal EEG asymmetry (FA-EEG), which is expressed as the difference in alpha-band power between the left and right anterior scalp regions. Spectral band powers of the resting EEG reflect basic characteristics of neuronal oscillatory activity. Overall, there is a relative consensus that abundant alpha-band (8–13 Hz) oscillations reflect cortical deactivation, whereas scarce or absent alpha oscillation indicates increased level of cortical arousal. Davidson et al. suggested that the direction of FA-EEG is associated with "affective style" such that greater left frontal alpha-band power presumably indicating *lower* left than right level of prefrontal activation is associated with stronger withdrawal motivation (avoidance) and increased vulnerability to depression, while the opposite pattern of asymmetry is associated with stronger approach motivation and low risk for depression.[61–63] FA-EEG shows good test-retest reliability[64] and modest but significant heritability.[65] Although this attractive hypothesis has generated an extensive literature, evidence for the association between FA-EEG, depression, and relevant personality traits and behavioral measures remains somewhat mixed.[62,66–71] Nevertheless, FA-EEG is still considered by many researchers as an indicator of affective style and risk for internalizing psychopathology.[66,71–73] In particular, developmental studies converge to suggest that relatively greater right frontal activation is associated with anxious and withdrawn temperaments in infants and children,[61,74,75] while the opposite pattern of frontal asymmetry (i.e., greater relative left frontal EEG activity) in infants has been associated with a higher risk of externalizing behaviors in toddlerhood.[76] Together, personality correlates of FA-EEG assessed in developmental and adult samples suggests that greater left cortical activation is associated with approach-related traits, including reward-seeking, pleasure, and aggression,[77] while greater right than left frontal activation is associated with withdrawal-related (avoidance) traits and behaviors, such as sadness, fear, and inhibition.

3.4 Event-Related Brain Potentials (ERPs) and Personality

Compared to resting-state EEG, ERPs have a more straightforward functional interpretation because they are elicited in tasks that are designed to probe specific cognitive and emotional processes. It is important that neural substrates of some of the ERP phenotypes have been relatively well established, in particular, using multimodal imaging studies combining ERP and fMRI recordings. Such neuroanatomically validated electrophysiological phenotypes are particularly valuable because they combine high temporal resolution of ERPs with the knowledge of underlying neural substrates and provide an affordable instrument for the studies of individual differences, including genetic studies, that normally require large samples, a condition which is often cost-prohibitive to be met by fMRI studies.

One such ERP phenotype that attracted much attention in recent research is Error-Related Negativity (ERN), a neurophysiological marker of error monitoring, a fundamental mechanism of behavioral self-regulation that involves automatic, preconscious detection of the mismatch between the intended and actually executed action.[78,79] Converging evidence from studies using ERP source localization analyses, multimodal imaging (EEG and fMRI), single unit recording, and studies of patients with brain lesions indicates that the main anatomical source of ERN is the ACC.[80–85] A previous study in our laboratory has demonstrated significant heritability of individual differences in ERN, suggesting that this neuroelectric marker can serve as an endophenotype for personality traits and disorders characterized by self-regulation deficits.[86] In recent years, ERN has been increasingly used in the investigation of neurocognitive mechanisms underlying individual differences in personality and psychopathology (reviewed in Refs 78,87,88). This evidence converges to suggest that increased ERN, presumably indicating an abnormally overactive error monitoring system, is associated with higher scores on behavioral inhibition, withdrawal, and negative affect, as well as with obsessive-compulsive, depressive, and anxiety-spectrum symptomatology,[89] whereas reduced ERN is associated with personality traits indicating impulsivity, low socialization, and externalizing symptoms in children and adults, presumably resulting from reduced sensitivity to errors and, hence, poor learning from errors.[90–93] These correlations with personality and psychopathology are also consistent with the notion that ERN reflects not only cognitive but also emotional processing of errors.[94,95]

3.5 fMRI and Personality

A number of studies used task-related changes in the BOLD signal to make inferences about individual differences in regional brain activity that could be associated with personality. Compared to structural MRI studies, these studies are more difficult to summarize because they typically used diverse experimental designs. However, a number of convergent findings can be identified. We refer the reader to a recent comprehensive review by Kennis et al.,[96] as well as other chapters in the volume.

In brief, personality traits related to behavioral activation and reward seeking showed relatively consistent positive correlations with the activation of the ventral and dorsal striatum and orbitofrontal cortex, subgenual and ventral ACC, in response to positive stimuli, consistent with the evidence for the role of these cortical regions in reward processing, including both reward expectancy and reward receipt.[96] Personality traits related to withdrawal-avoidance behaviors were positively correlated to amygdala activation in response to negative stimuli, as well as decreased functional connectivity of the amygdala with the PFC, ACC, and hippocampus.[96] Resting-state fMRI investigations indicate that higher scores on avoidance-withdrawal-related personality traits are associated with increased functional connectivity between the cingulate gyrus and other brain areas.[96] Interestingly, in most studies, personality traits correlated with bilateral activation patterns, and only few studies reported hemispheric asymmetry in the correlations.[96]

A recent meta-analysis using a parametric coordinate-based approach has focused on functional neuroimaging studies of neuroticism.[97] To identify brain regions that are consistently associated with neuroticism, the authors selected 18 studies using the contrasts "negative > neutral" and "positive > neutral" in emotion-processing tasks. It should be noted that the majority of studies included in the meta-analysis were based on very small samples ($n < 20$). Significant correlations in multiple regions emerged only for negative > neutral contrast. Neuroticism was associated with decreased activation in the ACC, thalamus, hippocampus/parahippocampus, striatum, and several temporal, parietal, and occipital brain areas, as well as increased activation in the hippocampus/parahippocampus and frontal and cingulate regions. Interestingly, different types of studies tended to contribute to negative versus positive correlations between neuroticism and negative > neutral activation. Negative correlations were mostly reported by studies using experimental paradigms with anticipation of aversive stimuli, whereas studies using fear conditioning and general emotion processing tended to report positive correlations. One unexpected but notable negative finding of this meta-analysis was the lack of significant correlation between neuroticism and higher amygdala reactivity.

Importantly, emerging evidence from studies combining neuroimaging of personality with the "imaging genetics" approach (i.e., candidate gene association studies of imaging phenotypes) suggests that relationships between task-related BOLD activation and personality traits can depend on the genotype, including even the direction of correlation. For example, the correlation between self-reported sensitivity to punishment measured using Torrubia's Senstivity to Punishment and Reward Questionnaire (SPRQ)[98] and functional connectivity between the amygdala and hippocampus during anticipation of monetary loss has been shown to be moderated by functional polymorphisms of two genes involved in serotonergic neurotransmission, hydroxylase-2 gene (TPH2) and serotonin transporter gene (5-HTTLPR). The correlation between BIS and amygdala–hippocampus connectivity was positive in individuals homozygous for the TPH2 G-allele and negative in carriers of the T-allele. Similarly, the correlation was positive in homozygotes for the 5-HTTLPR L_A variant, while carriers of the S/L_G allele showed a trend toward a negative association.[99]

In summary, there is growing evidence that individual differences in task-related BOLD signal changes are systematically associated with personality traits. However, despite some degree of convergence, this evidence remains quite disparate, presumably to the fact that most published studies of fMRI correlates of personality relied on small samples that may be insufficient for reliable estimation of correlations with personality, especially if one takes into account the need for multiple comparison correction (typically, analyses were not confined to a single *apriori* region of interest) and for potentially important covariates such as age, gender, psychopathology, and other factors. For example, among 76 studies reviewed by Kennis et al.,[96] in only 24 studies (less than one-third), sample size exceeded 30 participants; in only seven studies, samples were over 50; and just two studies were based on samples including over 100 participants. The problem of low statistical power may be further exacerbated by the fact that test-retest reliability of task-related fMRI measures is not very high (reviewed in Ref. 60) and may vary across tasks and among brain regions within a given task.[100] Studies searching for brain correlates of personality hinge on the assumption (explicit or implicit) that imaging phenotypes used in such analyses represent stable, trait-like individual differences. Surprisingly, the validity of this assumption is very rarely addressed, which may be one of the reasons for the observed variability of results across studies.

Based on the issues discussed above, several recommendations can be made regarding the selection of functional imaging phenotypes for the study of personality. First, studies of individual differences employing fMRI may be better served if test-retest reliability of "candidate" imaging phenotypes is established first, because focusing on a limited number of reliable and functionally interpretable patterns of regional activation can drastically reduce the number of statistical tests to be performed and thus the need for strict correction for multiple comparisons. Another approach that can be recommended is using regions of interest (ROI) that show overlap across different tasks tapping into the same theoretical construct. Using reliability and cross-task convergence as criterions for ROI selection will maximize the chances that these imaging measures represent valid and reliable neurophenotypes. It is important to note, however, that while the ROI approach helps to mitigate the multiple testing problem, it also carries the risk of ignoring other regions that may play a role in personality differences. Finally, developmental stability of individual differences in measures of brain function is another important criterion, since it can increase the generalizability of findings across the lifespan.

4. THE GENETICS OF POTENTIAL NEUROBIOLOGICAL ENDOPHENOTYPES FOR PERSONALITY TRAITS

Twin studies using structural MRI have provided consistent evidence for strong genetic influences on brain structure.[101,102] Both the total brain volume and regional volumes show consistently high heritability from childhood to old age, with genetic factors accounting for 60% to over 90% of the observed variance.[101,102] However, the degree of genetic influences varies of as a function of brain region, with frontal lobe volumes showing higher heritability (over 90%) than the hippocampus (around 50%), while several medial brain areas were influenced primarily by environmental factors. A recent meta-analysis of published twin studies by Blockland et al.[102] indicates substantial heritability of intracranial volume, total brain volume, and total and regional GM and WM volumes, cerebellar volumes, subcortical structures, and area of the corpus callosum. This meta-analysis also showed higher heritabilities for larger brain structures than smaller structures and higher heritabilities for WM volumes than GM volumes.

Most of the genetic studies of brain structure focused on volumetric measures. However, cortical volume is the product of two other structural measures, cortical thickness and surface area. Using structural MRI data collected in the Vietnam Era Twin Study of Aging (VETSA), Kremen and colleagues[103] demonstrated, using multivariate genetic analysis, that neocortical thickness and surface area are influenced by largely distinct genetic factors. Furthermore, thickness and area measures showed a distinct structure of genetic correlations: while the pattern of genetic correlations for the cortical surface area was mostly spatially contiguous (i.e., neighboring areas showed the highest correlations), cortical thickness was characterized by high genetic correlations among spatially disconnected cortical-thickness clusters in different lobes. Overall, these results suggest that cortical thickness and surface area are influenced by different genetic mechanisms and characterized by distinct time courses during neurodevelopment.[104] The finding of distinct genetic influences on cortical thickness and area has important implications for the investigation of structural imaging correlates of personality: if personality dimensions are differentially related to thickness and area, then GM volume may not be an optimal measure for the search of personality correlates.

Another important issue is the extent to which genetic factors operate at the global versus local level. Heritability of global area and thickness measures was high, while heritability of regional area and thickness measures was substantially lower (about 50%), and it was further reduced after accounting for the global measures.[103] This pattern of findings suggests that genetic influences on specific cortical regions are largely accounted for by heritability of the global area, thickness, and volume. Based on these analyses, the first genetically based map of cortical ROIs was created, with regions determined based on distinct genetic influences.[103]

In other analyses of the same sample, Kremen and colleagues[103] also investigated heritability and the structure of genetic relationships among 19 subcortical volumetric

ROI. Using factor analysis, they identified four factors based on shared genetic influences: the basal ganglia/thalamus factor, including putamen, pallidum, thalamus, and the caudate; the ventricles factor; the limbic factor (hippocampus, amygdala); and a separate factor represented by nucleus accumbens. It is important that, in contrast to the cortical measures that were strongly affected by respective global factors, no single genetic factor was found for subcortical structures.

In contrast to genetic studies of the brain structure, far fewer studies investigated heritability of functional MRI measures. The results are mixed, which is unsurprising, given a broad range of task designs employed and typically small sample sizes in most of these studies.[105–111] In the largest twin study using fMRI, Blokland et al.[106] conducted voxel-wise analysis of BOLD response in a working memory task and found significant heritability of task-related changes in BOLD response in inferior, middle, and superior frontal gyri, left supplementary motor area, precentral and postcentral gyri, middle cingulate cortex, superior medial gyrus, angular gyrus, superior parietal lobule (including precuneus), and superior occipital gyri. In these regions, genetic factors accounted up to 65% of the phenotypic variance in the task-related response.

Another (much smaller) study using an n-back working memory task[108] found significant heritability of task-related BOLD signal change in a number of areas, including frontal, anterior cingulate, temporoparietal, and visual areas, as well as the visual cortex, but only in the distraction period of the task and not during encoding and retrieval, which makes the interpretation of these findings in the context of genetic influences on the neural mechanisms of working memory rather problematic. Another small-sample study reported a 38% heritability of BOLD response in the dorsal ACC in a response inhibition task, whereas heritability was zero in other ROI analyzed in this study. However, even the reported heritability of 38% was nonsignificant because the confidence interval included zero.[109] A substantially larger study[107] assessed BOLD response to subjective experience of sadness induced by film clips in eight-year-old twins and failed to detect any significant genetic effects. A study of neural correlates of response inhibition using an antisaccade task in twins[112] found significant heritability only in the left thalamus but not in ROI typically associated with response inhibition.

Overall, evidence for genetic influences on brain function measured using task-related changes of the BOLD response is far less consistent compared with brain structure and variation in brain function measured using EEG/ERP methods. The reasons for this inconsistency can be twofold. First, heritability studies of both resting-state and task-related fMRI measures are still very scarce, and many of them were based on small samples that are insufficiently powered to detect small or even moderate heritability (under 50%). Therefore, larger scale and hence better-powered studies are needed to clarify the issue of genetic influences on brain function measured by fMRI. The second reason is related to test-retest reliability of fMRI measures. Because only stable, trait-like individual differences can be heritable, test-retest reliability can be viewed as the theoretical upper limit for heritability. Available evidence (much of which is summarized in a review by Bennett and Miller[60]) suggests that, overall, test-retest reliability of fMRI measures is not very high, although it appears to vary as a function of task design. Recent reliability studies of resting-state connectivity fMRI measures also suggest a modest stability of individual differences (test-retest intraclass correlations of 0.2–0.4 with a typical scan length—up to 10 min—but somewhat increasing with longer scan times).[113] Thus, it is hard to expect high heritability of these measures, which necessitates the use of sufficiently large samples (of the order of hundreds of twin pairs) to detect significant genetic influences. The problem is further compounded by the fact that fMRI analyses involve numerous statistical tests and require a stringent protection against Type I error, which results in the need of even larger samples.

5. LINKING GENETICS, BRAIN, AND PERSONALITY

In the previous sections, we discussed evidence for genetic influences on personality, relationships between personality and the brain, and genetic influences on the brain. Together, this evidence suggests that genetic influences on personality traits can be mediated by genetic influences on brain structure and function. Testing this hypothesis requires a conjoint analysis of all three domains (genetics, brain, and personality) in a single study. So far, only a handful of studies combined data on neuroimaging and personality in genetically informative samples.

A recent study used structural MRI and personality data from a twin study[114] to investigate structural variability in the amygdala and medial orbitofrontal cortex. Modest phenotypic correlations were observed between left amygdala volume and positive emotionality ($r = 0.16$) and between left medial orbitofrontal cortex thickness and negative emotionality ($r = 0.34$). Importantly, low but significant genetic correlations also emerged between neuroanatomical and personality measures, suggesting that at least a modest portion of genetic influences on personality may be mediated by genetic influences on brain structure.

Other studies have suggested the role for specific genetic variants in mediating the relationship between brain structure and personality traits. A study by[115] suggests that the Val66Met polymorphism of the gene for the brain-derived neurotrophic factor (BDNF) can modulate the relationship between trait neuroticism and hippocampal volume: in Met allele carriers, higher neuroticism and trait depression and stress were associated with lower total hippocampal GM volume. However, no such associations were observed in Val homozygotes. The interpretation of these findings offered by the authors is that Met carriers who also have elevated depression may be vulnerable to hippocampal GM loss, while Val/Val homozygotes may be resistant to such loss even in the presence of higher depression. In other words, the Met allele may confer increased susceptibility to hippocampal GM loss in depressed individuals. Yet another interpretation of these findings can be offered: Met allele confers increased risk for depression, but only in individuals with reduced hippocampus, whereas a large hippocampus acts as a protective factor against the risk associated with Met allele. However, because the study is cross-sectional, it is difficult to test these alternative hypotheses. Longitudinal studies or co-twin control studies will be needed to resolve the direction of causality in the relationships between neuroticism, depression, hippocampal volumes, and BDNF polymorphism.

The only published study of this kind sheds some light on the causal relationships between genetic liability, environmental influences such as stressful life experience, and hippocampal tissue loss. The relationship between exposure to severe stress and reduced hippocampal volumes has been well documented by both animal and human studies; in particular, hippocampus reduction has been consistently reported in individuals with posttraumatic stress disorder (PTSD). However, the direction of causality in this relationship was unclear because reduced hippocampal volume in PTSD patients could be a consequence of stress-induced neurotoxicity or, conversely, smaller hippocampi could be a preexisting condition that is associated with an increased risk for developing PTSD, given traumatic exposure. In a study of MZ twins discordant for trauma exposure, Gilbertson et al.[116] found evidence that smaller hippocampi indeed might constitute a risk factor for the development of stress-related psychopathology. The severity of PTSD symptoms in the PTSD patients from these pairs was negatively correlated not only with their own hippocampal volumes, but also with the hippocampal volumes of their unaffected and nontrauma-exposed co-twins. Furthermore, both twins from PTSD-discordant pairs had significantly smaller hippocampi than twins in non-PTSD pairs. Because MZ twins are genetically identical, this pattern of findings strongly suggests that smaller hippocampal volume constitutes a preexisting vulnerability factor for developing PTSD after trauma exposure, rather than a result of trauma-induced neurotoxicity.

A significant reduction of hippocampal volumes has been reported not only in psychopathological conditions such as PTSD and anxiety disorders, but also in individuals scoring high on personality traits associated with behavioral inhibition and withdrawal behaviors such as harm avoidance and neuroticism [35,40]. The role of genetic factors in this association remains unclear.

Below we illustrate how this problem can be approached using genetically informative samples using data from our recent neuroimaging study of a cohort of monozygotic and DZ twins. MRI and personality data were collected from 66 twins (36 female) at the age of 18 years. MRI images were acquired using Siemens Trio 3T scanner, and T1-weighted volumes were analyzed using standard FreeSurfer pipeline[117] to perform subcortical segmentation and measure the volume of subcortical GM structures, as well as to perform cortical surface reconstruction and automatic parcellation of the cortex. A previous study[40] reported an inverse correlation between HA and the right hippocampal volume in a large sample of young adults. Since HA and N capture similar behavioral tendencies (anxiety, negative affect, and withdrawal-avoidance behaviors), we expected that individuals with elevated scores on the N scale would show reduced right hippocampal volumes. This expectation was confirmed by the analysis of correlations between MRI and personality data: the neuroticism score was significantly and inversely correlated with both the right ($r=-0.366$, $p=0.001$) and left ($r=-0.271$, $p=0.015$) hippocampal volumes.

Next, we asked the question whether this observed (phenotypic) association can be accounted for by common genetic and environmental factors that influence both traits. To test this hypothesis, we computed within-pair correlation between neuroticism in one first twin of a pair and hippocampal volume in the other twin (cross-twin, cross-trait correlation). Since the two variables (neuroticism and hippocampal volume) included in this analysis were measured in different individuals, any significant correlation can arise only due to their familial relatedness, including both genetic and shared environmental factors (a similar correlation in randomly assembled pairs of unrelated individuals is always expected to be zero).

Furthermore, higher MZ than DZ correlation in this cross-trait, cross-twin analysis would indicate the contribution of shared genetic factors, whereas equal size of MZ and DZ correlations is expected if the correlation arises due to common environmental factors (that are assumed to be shared to the same degree by MZ and DZ twins). Finally, if the observed phenotypic association between neuroticism and hippocampal volume arises due to individual-specific experiences, such as stressful or traumatic life events, there should be no cross-trait, cross-twin correlation within twin pairs.

The results of this analysis (Figure 2) clearly support the genetic hypothesis: cross-trait, cross-twin correlation

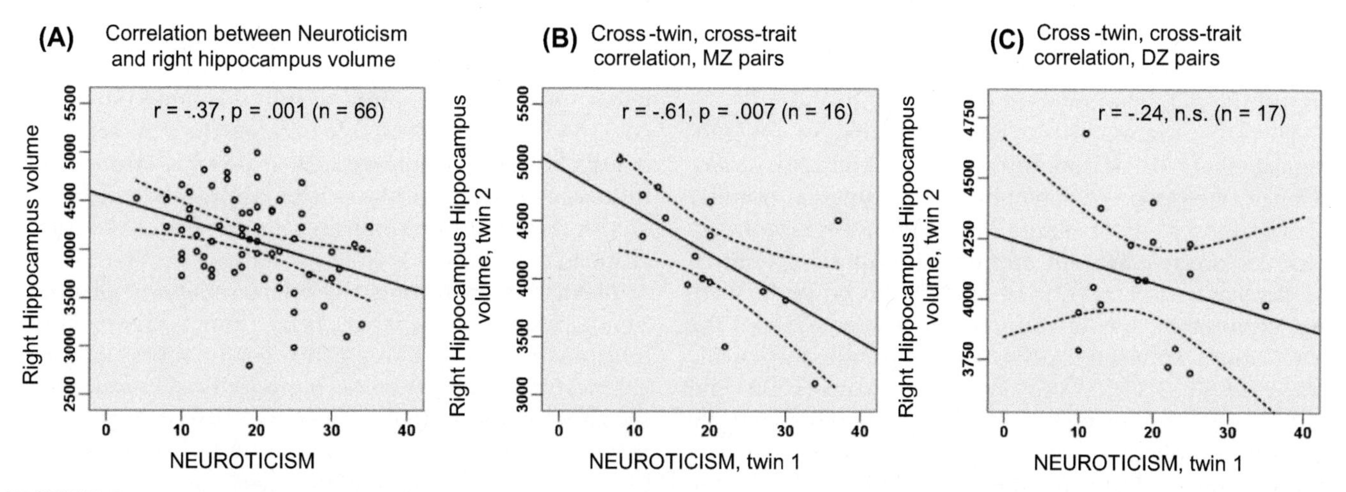

FIGURE 2 Neuroticism and right hippocampal volume. (A) Correlation across subjects in the entire sample. (B) Within-pair correlation between neuroticism in the first twin of a pair and hippocampal volume in the second twin, in MZ twins (cross-twin, cross-trait correlation). (C) Same correlation as in (B), but in DZ twins. Regression line (solid) and its 95% confidence intervals (broken lines) are shown.

was significant in MZ pairs ($r=-0.61$, $p=0.007$, $n=16$) but showed only a nonsignificant trend in DZ pairs ($r=-0.24$, n.s., $n=17$). In other words, hippocampal volume of one twin significantly predicted neuroticism in the other twin, and this was true for MZ but not for DZ twins. Although the sample was too small for a multivariate genetic analysis, this pattern of results provides preliminary support for the notion that the relationship between smaller hippocampal volumes and elevated scores on neuroticism is mediated by common genetic influences, rather than shared or individually specific environmental exposures. This pattern of results converges with the study of PTSD-discordant twins discussed above,[116] suggesting that smaller hippocampal volume constitutes a preexisting vulnerability factor for developing PTSD after trauma exposure, rather than a result of trauma-induced neurotoxicity. These preliminary findings provide a proof of concept for future larger-scale genetically informative investigation of the relationships between hippocampal volume reduction, neuroticism, and risk for internalizing spectrum psychopathology. More broadly, findings discussed in this section indicate that imaging studies in twin samples are a promising approach to disentangling genetic and environmental sources of covariance between individual variability in brain structure and personality.

6. SUMMARY AND FUTURE DIRECTIONS: IN SEARCH OF A UNIFYING, BIOLOGICALLY INFORMED MODEL OF PERSONALITY

In previous sections, we have discussed the progress in the understanding of the biological bases of personality using genetic and neuroscience methods and have identified some issues and pitfalls associated with this research. In summary, studies indicate high heritability of brain structure and moderate heritability of personality traits, and evidence starts to emerge suggesting that at least some of the genetic influences on personality can be mediated by genetically transmitted variation in brain structure. On the other hand, evidence for heritability of functional imaging phenotypes remains scarce and mixed. Studies investigating the relationships between individual variability in the brain and personality traits are abundant, but their findings are often disparate for the same personality traits.

This variability of findings can be attributed to the fact that many of these studies relied on small samples that are prone to producing false positive findings, especially when multiple statistical tests are made, as is typical in research involving neuroimaging data. There is a great hope that emerging large, multisite collaborative studies, such as ENIGMA Enhancing (Enhancing Neuroimaging Genetics through Meta-Analysis; http://enigma.loni.ucla.edu),[118] will clarify the picture.

Will neuroimaging studies on personality follow the fate of genetic studies? As sample sizes grow and large-sample studies emerge, will most of the published findings in this field be refuted, similar to how small-sample candidate gene findings in personality research were refuted by subsequent large-scale genome-wide analyses? Although it may be premature to draw such analogies, the following predictions regarding the future development of the field seem reasonable. First, it is likely that many previous findings of high correlations between personality and brain-based measures will be challenged. Second, the effect sizes of significant findings will be diminishing as the sample sizes grow. Third, novel and often unexpected associations between neural circuitry and personality will be discovered in large-sample studies. Finally, similar to genetics research, we will see a shift from one-to-one and few-to-one models of relationships between brain-based measures and personality traits to many-to-many models, positing that numerous and relatively distinct neurobiological factors contribute to the variation in personality traits.

Further progress in the understanding of the biological bases of personality using genetic and neuroscience methods will require addressing a number of challenges. First, large, population-representative samples are needed. An appropriate sample size must be determined based on the analyses to be performed. An exploratory, whole brain, voxel-wise analysis of the correlations between a personality measure and BOLD response in a task condition (e.g., extraversion and BOLD response during reward anticipation) may require a larger sample than a more focused, hypothesis-driven analysis restricted to *apriori* selected ROI, assuming the same expected effects size. It is likely that the required sample size for adequately powered analyses will be in the hundreds or even thousands of participants. Since collecting samples of this size is impractical for most research groups, multisite collaborative studies will be necessary. One of such initiatives, the ENIGMA network, is an international effort to combine data obtained by different research groups in order to achieve sufficiently large samples necessary to detect the modest gene effect sizes on neuroimaging phenotypes, including MRI, DTI, fMRI, and EEG. However, increasing sample sizes alone will not solve the problem of multiple testing. It is already a great challenge in GWAS and neuroimaging studies taken separately, and the combination of these two approaches (genome-wide, whole-brain search for association) will multiply the problem. Therefore, the application of multivariate statistical methods for data reduction is extremely important.[119]

Another important sample consideration is its representativeness of the general population. Results obtained using convenience samples, such as university students or patients of a clinic, may not be fully generalizable to the population at large. Furthermore, potential clinical significance of such findings may be limited due to various biases such samples can introduce. Therefore, it is imperative for a well-designed study to ensure that the sample is well-representative of the general population. Relevant approaches have been long used in epidemiology research, and comprehensive investigation of the biological bases of personality might require an integration of cognitive neuroscience and epidemiology.

An important condition for further understanding of the links between personality and brain function is designing appropriate experimental paradigms tapping into the hypothesized biobehavioral processes and mechanisms. Many previous studies were not originally designed to elucidate the neurobiological bases of personality. Rather, personality questionnaires were administered as ancillary measures, and analyses were performed post hoc using data from tasks that were designed for a different purpose, such as finding neural correlates of a psychiatric disorder. Furthermore, using multiple tasks tapping into the same personality construct is highly desirable because the convergence of results across multiple tasks will provide stronger and more conclusive evidence. This approach will help to ensure that the observed associations are not idiosyncratic to a particular task.

Next, demonstrating test-retest reliability of neurobiological measures is essential from both theoretical and practical perspective. Since personality dimensions are normally construed as temporally stable, trait-like measures of individual differences, their neurobiological underpinnings must be represented by temporally stable, trait-like individual differences in brain functioning. It is all the more important because current evidence for test-retest reliability of fMRI measures is rather mixed. Therefore, before a given task can be used for the investigation of biological bases of personality, test-retest reliability of measures derived from that task should be demonstrated first. This is also important from a practical perspective because such a prescreening of tasks and specific measures from these tasks will permit restricting analyses to the most important and promising measures, thus mitigating the multiple comparisons problem.

Collecting data on potential confounders is also very important. One of the crucial components of the investigation of biological bases of personality is ruling out potential spurious correlations between personality and neurobiological measures. Such spurious correlations may arise, for example, when the sample is heterogeneous due to admixture of different subsamples, such as pooling together data from college students' samples, clinical samples, and population-based samples. Finally, gender and ethnicity can confound associations between personality dimensions and neurobiological responses if distinct groups constituting the sample differ on both variables.

In this chapter, we have reviewed genetic research on personality, studies of the brain-personality relationships, and genetic studies of brain structure and function. Although substantial progress has been made over recent years in these distinct areas, building a unifying biological model of personality will require a much better integration of these research directions, which should become an important priority for future studies.

Glossary

Candidate gene Genetic polymorphisms selected by their biological relevance to the studied phenotype (i.e., a hypothesis linking the gene's known or presumed function and the biological mechanisms underlying the phenotype). For example, the role of dopamine in the processing of reward has been well established. Therefore, genetic variants known to alter dopaminergic neurotransmission may be plausible candidates for personality traits describing individual differences in reward responsivity.

Electroencephalography (EEG) A recording of neuronal electric activity using small censors (electrodes) attached to the scalp. The resting EEG represents synchronized activity of a large number of neurons (primarily, gradual postsynaptic potentials), which in the resting state manifests itself as oscillations at different frequencies. In research settings, EEG is typically converted from the time domain into the frequency domain using spectral analysis to measure the power of different frequency bands that have been linked to distinct aspects of cortical information processing.

Event-Related Potentials (ERP) EEG changes associated with specific task events, such as stimuli or responses. The ERP is obtained by averaging over multiple trials of EEG fragments time-locked to a specific event. In the process of averaging, any ongoing oscillatory activity that is not systematically related to the event is canceled out, while event-related activity synchronized with the event becomes apparent due to increased signal-to-noise ratio. ERP waveforms contain a number of peaks with different polarity and latency corresponding to specific stages of the information processing in the brain.

Functional magnetic resonance imaging or functional MRI (fMRI) A functional neuroimaging procedure that measures changes in BOLD signal over time. Resting state fMRI measures spontaneous fluctuations in BOLD signal and correlations between these fluctuations in different brain regions. Task-related fMRI measures a statistical association between fluctuation in BOLD signal and specific task periods (block design) or events such as stimuli or responses (event-related design). For example, a significant association of the BOLD signal with given stimulus in a specific brain region indicates that the region is "activated" by the stimulus. The relationship of the BOLD signal to actual neural activity is not perfect and is based on a number of assumptions.

Genetic association Co-occurrence of a certain allele of a genetic marker and the phenotype of interest in the same individuals at above-chance level. Association studies fall into two broad categories: candidate gene association studies and genome-wide association studies (GWAS).

Genome-Wide Association Study (GWAS) A genetic association study using a large number of SNPs (hundreds of thousands to millions) for scanning the entire genome for possible association with a given phenotype. In contrast to candidate gene studies that are typically based on an *apriori* biological hypothesis, GWA studies are purely exploratory and aim at the discovery of novel associations.

Heritability The proportion of the total phenotypic (observed) variance of a trait that can be attributed to genetic factors. For example, a 40% heritability of a personality trait means that 40% of the variance in that trait can be explained by genetic differences among individuals, while the other 60% are explained by nongenetic factors, including measurement error.

Single nucleotide polymorphism (SNP) Variation in the DNA sequence resulting from a substitution of a single nucleotide with another one. Because SNPs occur in very large numbers throughout the genome and their chromosomal location is typically well established, they can serve as markers of specific chromosomal regions. Some SNPs may be "functional polymorphisms," that is, occur within the gene of interest and affect that gene's function and are thus regarded as candidate genes, while others just mark a chromosomal region where another, yet to be determined, functional mutation may be located.

References

1. Eysenck JH. Principles and methods of personality description, classification and diagnosis. *Br J Psychol.* August 1964;55:284–294.
2. Cloninger CR, Svrakic DM. Integrative psychobiological approach to psychiatric assessment and treatment. *Psychiatry.* Summer 1997;60(2):120–141.
3. Cloninger CR, Svrakic DM, Przybeck TR. A psychobiological model of temperament and character. *Arch Gen Psychiatry.* December 1993;50(12):975–990.
4. McCrae RR, Costa Jr PT. The five factor theory of personality. In: John OP, Robins RW, Pervin LA, eds. *Handbook of Personality: Theory and Research.* Third ed. New York: Guilford; 2008:159–181.
5. McCrae RR, Yang J, Costa PT, et al. Personality profiles and the prediction of categorical personality disorders. *J Pers.* April 2001;69(2):155–174.
6. Boomsma D, Busjahn A, Peltonen L. Classical twin studies and beyond. *Nat Rev.* November 2002;3(11):872–882.
7. Posthuma D, Beem AL, de Geus EJ, et al. Theory and practice in quantitative genetics. *Twin Res.* October 2003;6(5):361–376.
8. Rijsdijk FV, Sham PC. Analytic approaches to twin data using structural equation models. *Brief Bioinform.* 2002;3(2):119–133.
9. van Dongen J, Slagboom PE, Draisma HH, Martin NG, Boomsma DI. The continuing value of twin studies in the omics era. *Nat Rev.* September 2012;13(9):640–653.
10. Heath AC, Cloninger CR, Martin NG. Testing a model for the genetic structure of personality: a comparison of the personality systems of Cloninger and Eysenck. *J Pers Soc Psychol.* April 1994;66(4):762–775.
11. Heiman N, Stallings MC, Young SE, Hewitt JK. Investigating the genetic and environmental structure of Cloninger's personality dimensions in adolescence. *Twin Res.* October 2004;7(5):462–470.
12. Stallings MC, Hewitt JK, Cloninger CR, Heath AC, Eaves LJ. Genetic and environmental structure of the Tridimensional Personality Questionnaire: three or four temperament dimensions? *J Pers Soc Psychol.* January 1996;70(1):127–140.
13. Distel MA, Trull TJ, Willemsen G, et al. The five-factor model of personality and borderline personality disorder: a genetic analysis of comorbidity. *Biol Psychiatry.* December 15, 2009;66(12):1131–1138.
14. Cordell HJ, Clayton DG. Genetic association studies. *Lancet.* September 24–30, 2005;366(9491):1121–1131.
15. Kwon JM, Goate AM. The candidate gene approach. *Alcohol Res Health.* 2000;24(3):164–168.
16. Dick DM, Agrawal A, Keller MC, et al. Candidate gene-environment interaction research: reflections and recommendations. *Perspect Psychol Sci.* January 2015;10(1):37–59.
17. Duncan LE, Keller MC. A critical review of the first 10 years of candidate gene-by-environment interaction research in psychiatry. *Am J Psychiatry.* October 2011;168(10):1041–1049.
18. Flint J, Munafo MR. Candidate and non-candidate genes in behavior genetics. *Curr Opin Neurobiol.* August 6, 2012.
19. Balestri M, Calati R, Serretti A, De Ronchi D. Genetic modulation of personality traits: a systematic review of the literature. *Int Clin Psychopharmacol.* January 2014;29(1):1–15.
20. Lesch KP, Bengel D, Heils A, et al. Association of anxiety-related traits with a polymorphism in the serotonin transporter gene regulatory region. *Science.* November 29, 1996;274(5292):1527–1531.
21. Ficks CA, Waldman ID. Candidate genes for aggression and antisocial behavior: a meta-analysis of association studies of the 5HTTLPR and MAOA-uVNTR. *Behav Genet.* September 2014;44(5):427–444.
22. Murphy DL, Maile MS, Vogt NM. 5HTTLPR: white knight or dark blight? *ACS Chem Neurosci.* January 16, 2013;4(1):13–15.
23. Visscher PM, Brown MA, McCarthy MI, Yang J. Five years of GWAS discovery. *Am J Hum Genet.* January 13, 2012;90(1):7–24.
24. Verweij KJ, Zietsch BP, Medland SE, et al. A genome-wide association study of Cloninger's temperament scales: implications for the evolutionary genetics of personality. *Biol Psychol.* October 2010;85(2):306–317.
25. Service SK, Verweij KJ, Lahti J, et al. A genome-wide meta-analysis of association studies of Cloninger's Temperament Scales. *Transl Psychiatry.* 2012;2:e116.

26. Shifman S, Bhomra A, Smiley S, et al. A whole genome association study of neuroticism using DNA pooling. *Mol Psychiatry*. March 2008;13(3):302–312.
27. de Moor MH, Costa PT, Terracciano A, et al. Meta-analysis of genome-wide association studies for personality. *Mol Psychiatry*. March 2012;17(3):337–349.
28. Amin N, Hottenga JJ, Hansell NK, et al. Refining genome-wide linkage intervals using a meta-analysis of genome-wide association studies identifies loci influencing personality dimensions. *Eur J Hum Genet*. August 2013;21(8):876–882.
29. Munafo MR, Flint J. Dissecting the genetic architecture of human personality. *Trends Cogn Sci*. September 2011;15(9):395–400.
30. Mooney SD, Krishnan VG, Evani US. Bioinformatic tools for identifying disease gene and SNP candidates. *Methods Mol Biol*. 2010;628:307–319.
31. Mehler MF. Epigenetic principles and mechanisms underlying nervous system functions in health and disease. *Prog Neurobiol*. December 11, 2008;86(4):305–341.
32. Petronis A. Epigenetics as a unifying principle in the aetiology of complex traits and diseases. *Nature*. June 10, 2010;465(7299):721–727.
33. Miller G. Epigenetics. The seductive allure of behavioral epigenetics. *Science*. July 2, 2010;329(5987):24–27.
34. Bjornebekk A, Fjell AM, Walhovd KB, Grydeland H, Torgersen S, Westlye LT. Neuronal correlates of the five factor model (FFM) of human personality: multimodal imaging in a large healthy sample. *NeuroImage*. January 15, 2013;65:194–208.
35. DeYoung CG, Hirsh JB, Shane MS, Papademetris X, Rajeevan N, Gray JR. Testing predictions from personality neuroscience. Brain structure and the big five. *Psychol Sci*. June 2010;21(6):820–828.
36. Cremers H, van Tol MJ, Roelofs K, et al. Extraversion is linked to volume of the orbitofrontal cortex and amygdala. *PLoS One*. 2011;6(12):e28421.
37. Blankstein U, Chen JY, Mincic AM, McGrath PA, Davis KD. The complex minds of teenagers: neuroanatomy of personality differs between sexes. *Neuropsychologia*. January 2009;47(2):599–603.
38. Hu X, Erb M, Ackermann H, Martin JA, Grodd W, Reiterer SM. Voxel-based morphometry studies of personality: issue of statistical model specification–effect of nuisance covariates. *NeuroImage*. February 1, 2011;54(3):1994–2005.
39. Kapogiannis D, Sutin A, Davatzikos C, Costa Jr P, Resnick S. The five factors of personality and regional cortical variability in the Baltimore longitudinal study of aging. *Hum Brain Mapp*. November 2013;34(11):2829–2840.
40. Yamasue H, Abe O, Suga M, et al. Gender-common and -specific neuroanatomical basis of human anxiety-related personality traits. *Cereb Cortex*. January 2008;18(1):46–52.
41. Pujol J, Lopez A, Deus J, et al. Anatomical variability of the anterior cingulate gyrus and basic dimensions of human personality. *NeuroImage*. April 2002;15(4):847–855.
42. Iidaka T, Matsumoto A, Ozaki N, et al. Volume of left amygdala subregion predicted temperamental trait of harm avoidance in female young subjects. A voxel-based morphometry study. *Brain Res*. December 13, 2006;1125(1):85–93.
43. Fuentes P, Barros-Loscertales A, Bustamante JC, Rosell P, Costumero V, Avila C. Individual differences in the Behavioral Inhibition System are associated with orbitofrontal cortex and precuneus gray matter volume. *Cogn Affect Behav Neurosci*. September 2012;12(3):491–498.
44. Carver CS, White TL. Behavioral inhibition, behavioral activation, and affective responses to impending reward and punishment: the BIS/BAS Scales. *J Pers Soc Psychol*. August 1994;67(2):319–333.
45. Weyandt L, Swentosky A, Gudmundsdottir BG. Neuroimaging and ADHD: fMRI, PET, DTI findings, and methodological limitations. *Dev Neuropsychol*. 2013;38(4):211–225.
46. Montag C, Schoene-Bake JC, Wagner J, et al. Volumetric hemispheric ratio as a useful tool in personality psychology. *Neurosci Res*. February 2013;75(2):157–159.
47. Griffa A, Baumann PS, Thiran JP, Hagmann P. Structural connectomics in brain diseases. *NeuroImage*. October 15, 2013;80:515–526.
48. Montag C, Reuter M, Weber B, Markett S, Schoene-Bake JC. Individual differences in trait anxiety are associated with white matter tract integrity in the left temporal lobe in healthy males but not females. *Neuroscience*. August 16, 2012;217:77–83.
49. Xu J, Potenza MN. White matter integrity and five-factor personality measures in healthy adults. *NeuroImage*. January 2, 2012;59(1):800–807.
50. Bennett CM, Wolford GL, Miller MB. The principled control of false positives in neuroimaging. *Soc Cogn Affect Neurosci*. December 2009;4(4):417–422.
51. Vul E, Pashler H. Voodoo and circularity errors. *NeuroImage*. August 15, 2012;62(2):945–948.
52. Omura K, Todd Constable R, Canli T. Amygdala gray matter concentration is associated with extraversion and neuroticism. *Neuroreport*. November 28, 2005;16(17):1905–1908.
53. Rauch SL, Milad MR, Orr SP, Quinn BT, Fischl B, Pitman RK. Orbitofrontal thickness, retention of fear extinction, and extraversion. *Neuroreport*. November 28, 2005;16(17):1909–1912.
54. Swettenham JB, Muthukumaraswamy SD, Singh KD. BOLD responses in human primary visual cortex are insensitive to substantial changes in neural activity. *Front Hum Neurosci*. 2013;7:76.
55. Logothetis NK. What we can do and what we cannot do with fMRI. *Nature*. June 12, 2008;453(7197):869–878.
56. Magri C, Schridde U, Murayama Y, Panzeri S, Logothetis NK. The amplitude and timing of the BOLD signal reflects the relationship between local field potential power at different frequencies. *J Neurosci*. January 25, 2012;32(4):1395–1407.
57. Napflin M, Wildi M, Sarnthein J. Test-retest reliability of resting EEG spectra validates a statistical signature of persons. *Clin Neurophysiol*. November 2007;118(11):2519–2524.
58. Olvet DM, Hajcak G. Reliability of error-related brain activity. *Brain Res*. August 11, 2009;1284:89–99.
59. Brunner JF, Hansen TI, Olsen A, Skandsen T, Haberg A, Kropotov J. Long-term test-retest reliability of the P3 NoGo wave and two independent components decomposed from the P3 NoGo wave in a visual Go/NoGo task. *Int J Psychophysiol*. July 2013;89(1):106–114.
60. Bennett CM, Miller MB. How reliable are the results from functional magnetic resonance imaging? *Ann N Y Acad Sci*. 2010;1191:133–155.
61. Davidson RJ, Fox NA. Asymmetrical brain activity discriminates between positive and negative affective stimuli in human infants. *Science*. December 17, 1982;218(4578):1235–1237.
62. Davidson RJ. Anterior electrophysiological asymmetries, emotion, and depression: conceptual and methodological conundrums. *Psychophysiology*. September 1998;35(5):607–614.
63. Wheeler RE, Davidson RJ, Tomarken AJ. Frontal brain asymmetry and emotional reactivity: a biological substrate of affective style. *Psychophysiology*. January 1993;30(1):82–89.
64. Tomarken AJ, Davidson RJ, Wheeler RE, Kinney L. Psychometric properties of resting anterior EEG asymmetry: temporal stability and internal consistency. *Psychophysiology*. September 1992;29(5):576–592.
65. Anokhin AP, Heath AC, Myers E. Genetic and environmental influences on frontal EEG asymmetry: a twin study. *Biol Psychol*. March 2006;71(3):289–295.
66. Allen JJ, Kline JP. Frontal EEG asymmetry, emotion, and psychopathology: the first, and the next 25 years. *Biol Psychol*. October 2004;67(1–2):1–5.
67. Hagemann D. Individual differences in anterior EEG asymmetry: methodological problems and solutions. *Biol Psychol*. October 2004;67(1–2):157–182.

68. Allen JJ, Urry HL, Hitt SK, Coan JA. The stability of resting frontal electroencephalographic asymmetry in depression. *Psychophysiology*. March 2004;41(2):269–280.
69. Coan JA, Allen JJ. Frontal EEG asymmetry and the behavioral activation and inhibition systems. *Psychophysiology*. January 2003;40(1):106–114.
70. Harmon-Jones E, Gable PA, Peterson CK. The role of asymmetric frontal cortical activity in emotion-related phenomena: a review and update. *Biol Psychol*. July 2010;84(3):451–462.
71. De Pascalis V, Cozzuto G, Caprara GV, Alessandri G. Relations among EEG-alpha asymmetry, BIS/BAS, and dispositional optimism. *Biol Psychol*. September 2013;94(1):198–209.
72. Allen JJ, Coan JA, Nazarian M. Issues and assumptions on the road from raw signals to metrics of frontal EEG asymmetry in emotion. *Biol Psychol*. October 2004;67(1–2):183–218.
73. Gatzke-Kopp LM, Jetha MK, Segalowitz SJ. The role of resting frontal EEG asymmetry in psychopathology: afferent or efferent filter? *Dev Psychobiol*. November 20, 2012.
74. Fox NA, Rubin KH, Calkins SD, et al. Frontal activation asymmetry and social competence at four years of age. *Child Dev*. December 1995;66(6):1770–1784.
75. Davidson RJ, Fox NA. Frontal brain asymmetry predicts infants' response to maternal separation. *J Abnorm Psychol*. May 1989;98(2):127–131.
76. Smith CL, Bell MA. Stability in infant frontal asymmetry as a predictor of toddlerhood internalizing and externalizing behaviors. *Dev Psychobiol*. March 2010;52(2):158–167.
77. Carver CS, Harmon-Jones E. Anger is an approach-related affect: evidence and implications. *Psychol Bull*. March 2009;135(2):183–204.
78. van Noordt SJ, Segalowitz SJ. Performance monitoring and the medial prefrontal cortex: a review of individual differences and context effects as a window on self-regulation. *Front Hum Neurosci*. 2012;6:197.
79. Segalowitz SJ, Dywan J. Individual differences and developmental change in the ERN response: implications for models of ACC function. *Psychol Res*. November 2009;73(6):857–870.
80. Ridderinkhof KR, Ullsperger M, Crone EA, Nieuwenhuis S. The role of the medial frontal cortex in cognitive control. *Science*. October 15, 2004;306(5695):443–447.
81. Mathalon DH, Whitfield SL, Ford JM. Anatomy of an error: ERP and fMRI. *Biol Psychol*. October 2003;64(1–2):119–141.
82. Debener S, Ullsperger M, Siegel M, Fiehler K, von Cramon DY, Engel AK. Trial-by-trial coupling of concurrent electroencephalogram and functional magnetic resonance imaging identifies the dynamics of performance monitoring. *J Neurosci*. December 14, 2005;25(50):11730–11737.
83. Herrmann MJ, Rommler J, Ehlis AC, Heidrich A, Fallgatter AJ. Source localization (LORETA) of the error-related-negativity (ERN/Ne) and positivity (Pe). *Brain Res Cogn Brain Res*. July 2004;20(2):294–299.
84. Miltner WH, Lemke U, Weiss T, Holroyd C, Scheffers MK, Coles MG. Implementation of error-processing in the human anterior cingulate cortex: a source analysis of the magnetic equivalent of the error-related negativity. *Biol Psychol*. October 2003;64(1–2):157–166.
85. Iannaccone R, Hauser TU, Staempfli P, Walitza S, Brandeis D, Brem S. Conflict monitoring and error processing: new insights from simultaneous EEG-fMRI. *NeuroImage*. January 15, 2015;105:395–407.
86. Anokhin AP, Golosheykin S, Heath AC. Heritability of frontal brain function related to action monitoring. *Psychophysiology*. July 2008;45(4):524–534.
87. Olvet DM, Hajcak G. The error-related negativity (ERN) and psychopathology: toward an endophenotype. *Clin Psychol Rev*. December 2008;28(8):1343–1354.
88. Moser JS, Moran TP, Schroder HS, Donnellan MB, Yeung N. On the relationship between anxiety and error monitoring: a meta-analysis and conceptual framework. *Front Hum Neurosci*. 2013;7:466.
89. Aarts K, Vanderhasselt MA, Otte G, Baeken C, Pourtois G. Electrical brain imaging reveals the expression and timing of altered error monitoring functions in major depression. *J Abnorm Psychol*. November 2013;122(4):939–950.
90. Dikman ZV, Allen JJ. Error monitoring during reward and avoidance learning in high- and low-socialized individuals. *Psychophysiology*. January 2000;37(1):43–54.
91. Santesso DL, Segalowitz SJ, Schmidt LA. ERP correlates of error monitoring in 10-year olds are related to socialization. *Biol Psychol*. October 2005;70(2):79–87.
92. Stieben J, Lewis MD, Granic I, Zelazo PD, Segalowitz S, Pepler D. Neurophysiological mechanisms of emotion regulation for subtypes of externalizing children. *Dev Psychopathol*. Spring 2007; 19(2):455–480.
93. Hall JR, Bernat EM, Patrick CJ. Externalizing psychopathology and the error-related negativity. *Psychol Sci*. April 2007;18(4):326–333.
94. Aarts K, De Houwer J, Pourtois G. Erroneous and correct actions have a different affective valence: evidence from ERPs. *Emotion*. October 2013;13(5):960–973.
95. Koban L, Pourtois G. Brain systems underlying the affective and social monitoring of actions: an integrative review. *Neurosci Biobehav Rev*. March 26, 2014.
96. Kennis M, Rademaker AR, Geuze E. Neural correlates of personality: an integrative review. *Neurosci Biobehav Rev*. January 2013;37(1):73–95.
97. Servaas MN, van der Velde J, Costafreda SG, et al. Neuroticism and the brain: a quantitative meta-analysis of neuroimaging studies investigating emotion processing. *Neurosci Biobehav Rev*. September 2013;37(8):1518–1529.
98. Torrubia R, Ávila C, Moltó J, Caseras X. The Sensitivity to Punishment and Sensitivity to Reward Questionnaire (SPSRQ) as a measure of Gray's anxiety and impulsivity dimensions. *Pers Indiv Differ*. 10/15/2001;31(6):837–862.
99. Hahn T, Heinzel S, Notebaert K, et al. The tricks of the trait: neural implementation of personality varies with genotype-dependent serotonin levels. *NeuroImage*. November 1, 2013;81:393–399.
100. Bennett CM, Miller MB. fMRI reliability: influences of task and experimental design. *Cogn Affect Behav Neurosci*. December 2013;13(4):690–702.
101. Peper JS, Brouwer RM, Boomsma DI, Kahn RS, Hulshoff Pol HE. Genetic influences on human brain structure: a review of brain imaging studies in twins. *Hum Brain Mapp*. June 2007;28(6):464–473.
102. Blokland GA, de Zubicaray GI, McMahon KL, Wright MJ. Genetic and environmental influences on neuroimaging phenotypes: a meta-analytical perspective on twin imaging studies. *Twin Res Hum Genet*. June 2012;15(3):351–371.
103. Kremen WS, Fennema-Notestine C, Eyler LT, et al. Genetics of brain structure: contributions from the Vietnam Era Twin Study of Aging. *Am J Med Genet B Neuropsychiatr Genet*. October 2013;162b(7):751–761.
104. Chen CH, Fiecas M, Gutierrez ED, et al. Genetic topography of brain morphology. *Proc Natl Acad Sci USA*. October 15, 2013;110(42):17089–17094.
105. Blokland GA, McMahon KL, Hoffman J, et al. Quantifying the heritability of task-related brain activation and performance during the N-back working memory task: a twin fMRI study. *Biol Psychol*. September 2008;79(1):70–79.
106. Blokland GA, McMahon KL, Thompson PM, Martin NG, de Zubicaray GI, Wright MJ. Heritability of working memory brain activation. *J Neurosci*. July 27, 2011;31(30):10882–10890.

107. Cote C, Beauregard M, Girard A, Mensour B, Mancini-Marie A, Perusse D. Individual variation in neural correlates of sadness in children: a twin fMRI study. *Hum Brain Mapp*. June 2007;28(6):482–487.
108. Koten JW, Wood G, Hagoort P, et al. Genetic contribution to variation in cognitive function: an fMRI study in twins. *Science*. March 2009;323(5922):1737–1740.
109. Matthews SC, Simmons AN, Strigo I, Jang K, Stein MB, Paulus MP. Heritability of anterior cingulate response to conflict: an fMRI study in female twins. *NeuroImage*. October 15, 2007;38(1):223–227.
110. Park J, Shedden K, Polk TA. Correlation and heritability in neuroimaging datasets: a spatial decomposition approach with application to an fMRI study of twins. *NeuroImage*. January 16, 2012;59(2):1132–1142.
111. Polk TA, Park J, Smith MR, Park DC. Nature versus nurture in ventral visual cortex: a functional magnetic resonance imaging study of twins. *J Neurosci*. December 19, 2007;27(51):13921–13925.
112. Macare C, Meindl T, Nenadic I, Rujescu D, Ettinger U. Preliminary findings on the heritability of the neural correlates of response inhibition. *Biol Psychol*. December 2014;103:19–23.
113. Birn RM, Molloy EK, Patriat R, et al. The effect of scan length on the reliability of resting-state fMRI connectivity estimates. *NeuroImage*. December 2013;83:550–558.
114. Lewis GJ, Panizzon MS, Eyler L, et al. Heritable influences on amygdala and orbitofrontal cortex contribute to genetic variation in core dimensions of personality. *NeuroImage*. September 28, 2014;103c:309–315.
115. Joffe RT, Gatt JM, Kemp AH, et al. Brain derived neurotrophic factor Val66Met polymorphism, the five factor model of personality and hippocampal volume: Implications for depressive illness. *Hum Brain Mapp*. April 2009;30(4):1246–1256.
116. Gilbertson MW, Shenton ME, Ciszewski A, et al. Smaller hippocampal volume predicts pathologic vulnerability to psychological trauma. *Nat Neurosci*. November 2002;5(11):1242–1247.
117. Fischl B, van der Kouwe A, Desrieux C, et al. Automatically parcellating the human cerebral cortex. *Cereb Cortex*. 2004;14(1):11–22.
118. Thompson PM, Stein JL, Medland SE, et al. The ENIGMA Consortium: large-scale collaborative analyses of neuroimaging and genetic data. *Brain Imaging Behav*. June 2014;8(2):153–182.
119. Hibar DP, Kohannim O, Stein JL, Chiang MC, Thompson PM. Multilocus genetic analysis of brain images. *Front Genet*. 2011;2:73.

CHAPTER

5

Anxiety and Harm Avoidance

Sebastian Markett[1], *Christian Montag*[2], *Martin Reuter*[1]

[1]Department of Psychology and Center for Economics and Neuroscience, University of Bonn, Bonn, Germany;
[2]Department of Psychology, University of Ulm, Ulm, Germany

OUTLINE

In the present chapter, we are going to review and integrate neuroimaging findings on one aspect of human personality, called harm avoidance. The term harm avoidance has been coined by Cloninger[1] in his biosocial theory of personality. Harm avoidance is a temperament that disposes people to worry in anticipation, to fear uncertainty, to be shy toward strangers, and to become tired very quickly. All these behaviors are related to "anxiety" and indeed, the harm avoidance trait is conceptualized as a personality disposition. Before we start to gather empirical findings from the neuroimaging literature, we begin with a brief synopsis on personality in general, the theoretical considerations that led to the establishment of the harm avoidance construct in personality neuroscience, and a brief discussion of temperament and character in a biopsychological framework.

1. WHAT IS PERSONALITY?

Personality describes the entirety of stable characteristics of a person over time and situations that account for individual differences in behavior, cognition, and emotion. Most taxonomies of personality, however, focus exclusively on the emotional and motivational attributes of a person, such as approach and avoidance motivation, anxiety, or conscientiousness. A prominent definition holds that personality can be viewed as the organization of psychophysical systems that determine an individual's unique adjustment to his or her environment.[2] According to this definition, personality is composed of a set of psychophysical systems that are organized in a certain way. There is broad consensus in academic psychology that the best way to characterize these systems is the trait concept. Traits are enduring dispositions that

Neuroimaging Personality, Social Cognition, and Character
http://dx.doi.org/10.1016/B978-0-12-800935-2.00005-1

are thought to be dimensional in nature. According to this view, each individual can be characterized between the dimension's two poles. Trait anxiety, for example, describes such a dimensional trait: anxiety is an emotion characterized by worrying, rumination, and pessimism, and is most prominently elicited by uncertainty or unfamiliar situations. A person with a high level of trait anxiety will experience a higher level of anxiety in a given situation and will experience anxiety more often than a person with low levels of trait anxiety. A negative definition can give a more clear insight what the trait concept stands for: traits can be distinguished from states, which are of a more transient nature. States focus on the given moment and on the situation at hand and describe the current emotional experience of a person ("Right now, I am worried that I will fail the exam"), while traits focus on the stable aspect of such experiences ("I am always worried that I might fail"). Even though anxious states can be experienced independently from the individual trait level (certain situations might frighten everybody, no matter how anxious or brave the person is), there is a certain overlap between the two constructs, because people with high trait anxiety levels will experience anxious states more often than a person with low trait anxiety.[3]

But how many traits are there (or in other words, how many traits do we need to consider in order to fully characterize a person's personality?), what is their content, and how do they relate to each other? The most parsimonious models assume independent (i.e., orthogonal) traits that account for a substantial proportion of individual differences in behavior and behavioral disposition. From the numerous possible traits, those traits that capture and explain the most variance can be extracted by means of factor analysis. Factor analysis describes a set of statistical methods to reduce the dimensionality of data by extracting an underlying factor structure. The input data for this approach can be clinical data, such as symptoms or diagnoses, or adjectives from a dictionary that can be used to characterize a person. Factor analysis can reveal the number of factors (which are interpreted as traits) that are needed to parsimoniously describe the data set. The content of the factors can be derived by analyzing factor loadings, i.e., the covariation between the extracted factors and single variables of the input data. The derived factors constitute latent dimensions that influence the measured variables of the input data. The factor analytic approach has been widely applied in personality research, but even though results converge toward solutions of five plus minus two factors, there is still some debate going on regarding how the input data should look, how many factors need to be extracted, how they should be rotated during factor analysis, and most importantly, what their content is. The solution most widely agreed upon is the five-factor model of the "Big Five" personality traits "neuroticism," "extraversion," "openness to experience," "agreeableness," and "conscientiousness."[4] The merit of this taxonomy lies in its robustness: it has been derived from self-report data, observations, and peer-descriptions, as well as in different cultures and age groups.[5] Still, the Big Five model has been questioned, and numerous alternative taxonomies have been proposed.[6,7] This unsatisfactory state of affairs arises mainly from the descriptive and data-driven nature of the factor-analytic approach that is blind to the underlying causal, neurobiological structures leading to individual differences in behavior. Regarding the first two traits from the Big Five model, neuroticism and extraversion, for example, it has been shown that certain types of anxiolytic drugs affect behaviors associated with both traits, even though both traits are factor-analytically conceptualized as independent.[8] The biosocial theory of personality by Cloninger takes up this critique and proposes a trait model where fundamental traits are conceptualized based on genetic, rather than statistical, similarity. In the following section, we will briefly outline this model because our review of the neuroimaging literature will be based on one of its traits.

2. TEMPERAMENT AND CHARACTER: THE BIOSOCIAL THEORY OF PERSONALITY

The biosocial theory of personality originates from the work by Robert J. Cloninger.[9–11] The model differentiates between seven traits to describe personality. A crucial distinction is made between temperament and character traits. Temperaments are congenital emotional dispositions and refer to automatic and associative responses to emotional stimuli. Temperaments have a strong heritable basis, are developmentally stable, emotion based, and are unaffected by sociocultural learning.[11] Temperaments can be best understood in terms of behavioristic learning theory and describe individual differences in the way people react to reinforcement or punishment. Characters have only a weak genetic basis and develop later in life. Crucial for this development is the interaction between temperament dispositions and sociocultural learning. In contrast to earlier personality taxonomies, the biosocial theory of personality therefore distinguishes between two qualitatively different classes of traits, which are—at least with respect to their ontogenetic development—hierarchically organized. From a biological point of view, the temperament dimensions are of special interest because of their strong genetic basis (about 50% additive genetic effects, temperaments are claimed to be genetically independent from each other[11]) and their hypothesized implementation in neural circuits devoted to the processing of reinforcing and punishing stimuli. Of note, the notion that temperament traits are more influenced by genetics compared to character traits has been challenged by twin studies showing equal heritability estimates.[12]

As outlined above, typical stimuli associated with temperaments are stimuli of reward, stimuli of punishment, and the absence of such stimuli. The biosocial theory of personality distinguishes between four different temperaments:[1,10] *novelty seeking* (NS) is the temperament dimension associated with responding strongly toward stimuli of reward, novelty, or the potential relief of punishment or monotony. The response triggered by such stimuli is behavioral activation, which includes behaviors such as exploratory pursuit, appetitive approach, or active avoidance (to escape stimuli signaling punishment or nonreward). In terms of behavioral learning theory, novelty seeking mediates the significance of positive and negative reinforcement. The second temperament dimension, *harm avoidance* (HA), is associated with stimuli[a] of punishment and frustrating nonreward. The reaction triggered by such stimuli is passive avoidance or rapid extinction[b] of a certain behavior in case of frustration. The third and fourth temperaments are called *reward dependence* and *persistence* and are both triggered by conditioned and unconditioned signals for reward or for the relief of punishment. While harm avoidance and novelty seeking describe behavioral inhibition (HA) and behavioral activation (NS) to a stimulus at hand, reward dependence and persistence both deal with the maintenance of behavior and the resistance to extinction in the case of reward omission or punishment. The stimuli particularly relevant for reward dependence are (verbal) signals of social approval. The typical behavior triggered by such stimuli is the maintenance of behavior that has been socially rewarded in the past with the intention to increase the possibility of getting the approval of others also in the future. Persistence, on the contrary, describes the tendency to perseverate a certain behavior, despite fatigue and frustration. In the initial formulation of the theory,[a] persistence was considered an element of reward dependence (thus, only three temperaments were distinguished), but empirical studies have led to the refinement to consider both facets of reward dependence as separable temperaments. The main difference is that reward dependence has a strong social component and describes behavioral maintenance in order to reach (social) reward, while persistence lacks the social component and describes behavioral maintenance despite frustration.

Besides the four temperament dimensions, the biosocial theory of personality distinguishes three character dimensions. Character is developed later in life through the conceptual organization of perception, insight learning, and the reorganization of self-concepts. Character dimensions include the following: *self-directedness*, which reflects the tendency to act responsibly and resourcefully to initiate and organize steps toward goal achievement; *cooperativeness*, which comprises social tolerance, empathy, helpfulness, and compassion and opposes selfishness and hostile revengefulness; and *self-transcendence*, which reflects the degree of spirituality, religion, and the feeling of participation in one's surroundings as a unitive whole.

In order to quantify temperament and character traits, self-report measures have been developed. Questions were formulated in order to capture the essence of temperaments and characters by asking for trait-relevant behavior. The initial questionnaire measure was the tridimensional personality questionnaire (TPQ[92]), aiming at the three temperaments—novelty seeking, harm avoidance, and reward dependence (including persistence). In response to empirical findings and theoretical refinement, a further questionnaire was developed: the temperament and character inventory (TCI[93]). The TCI measures all seven hypothesized traits (four temperaments and three characters). Even though the TCI is based on the more recent and empirically valid version of the theory, the TPQ is still used in some research settings to solely assess individual differences in temperament dimensions. When reviewing the literature, we will therefore explicitly state which measure was used in each study.

3. ANXIETY AND HARM AVOIDANCE

All personality theories, no matter whether biologically informed or factor-analytically derived, include a trait that captures anxiety. These traits have been labeled neuroticism,[14] fear,[15] behavioral inhibition system,[13] or HA. It is interesting to see that self-report measures of the different conceptualizations are highly intercorrelated.[16] This great overlap between personality theories points toward trait anxiety as an integral part of human nature. This centrality might reflect the immense consequences of excessively high scores on measures of trait anxiety: clinical diagnoses such as major depression and anxiety disorders (general anxiety disorder, social

[a] The initial conceptualization by Cloninger highlights conditioned stimuli as relevant for the anxiety response. The idea that only conditioned stimuli are relevant for anxiety originates from the work by Jeffrey Gray and the initial formulation of his Reward Sensitivity Theory.[8] The revised Reward Sensitivity Theory[13] states that conditioned but also unconditioned stimuli can trigger anxiety, as long as the stimulus has a conflicting or unfamiliar component. A similar revision for the harm avoidance construct would be appropriate.

[b] Extinction is a concept from learning theory and describes the observation that no longer reinforced behaviors gradually stop occurring over time. The notion that learning relates to an innate temperament is not necessarily at odds: innate factors do not affect the processing of a given stimulus but the underlying neural circuits that are devoted to learning and extinction.

phobia, specific phobia, panic disorders, and posttraumatic stress disorder) are characterized by high scores on trait anxiety in a way that suggests trait anxiety to be at least a strong vulnerability factor for affective disorders, if not an endophenotype.[17,18]

Psychiatric disorders place a considerable burden on the inflicted individual and their families, but also on health care systems and society as a whole.[19] There is reason to believe that the detection of biomarkers of disorders will not only help to diagnose and treat, but also prevent psychiatric disease.[20] Biomarkers are measurable indicators of a healthy state of a person, but also of disease or disorder.[21] Understanding the neural network underlying trait anxiety is an important step toward establishing such biomarkers by creating a reference scheme to evaluate clinical abnormalities against the normal anxiety response or individual differences in the nonpathological range. Modern neuroimaging provides tools to study neural correlates of trait anxiety. A promising strategy toward this goal is to quantify the individual levels of trait anxiety with tools provided by personality psychology (e.g., psychometric self-report measures) and relate these scores to variation in brain structure and function. This approach will be particularly useful if personality measures are used that are conceptualized with a clear link to their biological implementation. Harm avoidance, as conceptualized in the biosocial theory of personality, represents such a measure and is therefore an ideal candidate in the pursuit of deciphering the neural basis of individual differences in anxiety.

Harm avoidance is additively composed of four different facets that can be measures by subscales from the TCI (HA1–HA4) that are characterized by two poles, each describing either a prototypically high versus low level of the trait: anticipatory worry versus uninhibited optimism (HA1), fear of uncertainty versus confidence (HA2), shyness with strangers versus gregariousness (HA3), and fatigability versus vigor (HA4). People scoring high on harm avoidance (and average on other temperaments) are cautious, apprehensive, fatigable, and inhibited, while people scoring low on harm avoidance are described as fearless, carefree, sociable, and energetic. For a more detailed introduction, please see Kose.[22]

4. NEUROIMAGING OF TEMPERAMENT TRAITS

From very early on, trait theorists have argued that traits are thought to reflect neuropsychological systems that operate to make different stimuli functionally equivalent and to initiate a consistent and adaptive response.[2] With respect to anxiety, functional equivalence means that entirely different situations and stimuli might elicit an anxiety response, as long as they bear uncertainty or the prospect of punishment or nonreward. Such situations can be open fields, the odor of a predator who is not in sight, the cracking of a branch in the park at night, a public speech situation, or an important job interview. All situations bear different contexts and even convey information via different senses. But still, they all bear uncertainty and potential danger, which will require assessment and a closer examination.[c] It is thought to be mainly the trait that extracts this information and initiates an orientation and exploration response. This definition of a trait also implies that the initiated response is consistent across time and situations. This means that the behavioral adaptation is consistent (i.e., the already-described anxiety response), and a person with higher trait anxiety will react with a consistently stronger anxiety response across all mentioned situations than a person with low trait anxiety.

You may have noted that this definition of a trait is routed in behaviorism: a trait is thought to bridge between a stimulus and the behavioral response toward it. In contrast to a stimulus from the environment and overt behavioral responses, traits cannot be directly observed but only inferred by aggregating observations across situations and time. Importantly, internal stimuli can also trigger anxiety. Free floating anxiety is a situation where negative emotions are not attributable to a specific situation or object (also called rumination behavior or cognitive anxiety). While the behaviorist cannot investigate the trait (it is in the "black box"), modern neuroimaging fortunately provides tools to peek into the "black box."

The notion that traits are neuropsychological systems implies that traits should be reflected in the structural and functional properties of the brain. At least some aspects of a trait should be reflected in the brain independently from external stimulation. Properties can either be localized in a circumscribed structure or a distributed network and should covary with the individual degree of the trait. Structural properties may include local gray or white matter concentrations or the integrity of white matter tracts across the brain. Functional properties, on the contrary, may include intrinsically generated activity and spontaneous functional connectivity between brain regions. Furthermore, implicated brain regions or networks should respond in a consistent manner to external stimulation. This means that brain regions and networks implicated in anxiety should be activated by all kinds of

[c] Uncertainty of a situation, including risk assessment (am I really in a dangerous situation or not?), reflects the heart of anxious behavior, which has also been conceptualized by Gray and McNaughton.[13] In contrast, fear is elicited by a clear, unambiguous stimulus threatening a person. For example, being confronted with a person pointing a gun at your head clearly would elicit fear and not anxiety.

stimuli that convey threat and uncertainty. Again, activation strength in these brain regions should covary as a function of individual differences in trait levels. Lastly, traits should bias various stages of information processing in order to extract relevant information that is needed to make various stimuli from different modalities functionally equivalent and to initiate an appropriate response. That is, a trait can also manifest itself on attentional, perceptual, and even motor processes, which have no direct relationship with emotional and affective processing that are described in the trait definition. In the following, we are going to review various imaging findings regarding the harm avoidance construct.

5. REVIEW OF EMPIRICAL STUDIES

To our knowledge, a total of 41 imaging studies have been published on the neural correlates of harm avoidance so far.[d] These studies encompass the whole array of neuroimaging techniques and target variables: morphological analyses of local gray and white matter volume, assessment of structural and functional connectivity, analyses of resting state activity and metabolism, and tracing of the density of receptor molecules using positron emission tomography (PET), as well as experimental approaches which examine the influence of harm avoidance on task evoked neural activity using blood oxygen level dependent functional magnetic resonance imaging (BOLD fMRI), near infrared spectroscopy (NIRS), and event-related potentials. Each neuroimaging technique capitalizes on partly complementary information and the synthesis of all published findings points toward a complex network that underlies the harm avoidance temperament. Several brain areas appear consistently across studies and seem to be of special relevance for the harm avoidance construct: first, there is a temporal complex, encompassing parts of the temporal lobe, the hippocampus formation, and the amygdala (12 studies). These results are consistent with the work by Joseph LeDoux,[23] who identified the amygdala as a central processing hub for fear and anxiety, and the work by Jeffrey Gray,[13] who identified the septo-hippocampal system as the most crucial part of the anxiety-axis in the mammalian brain. Second, the basal ganglia and, most prominently, the striatum have been linked to harm avoidance in five studies. The basal ganglia are part of the "salience network,"[24,25] which also encompasses the anterior insula and the anterior cingulate. Functional and structural properties of these two structures were found in nine studies to correlate with harm avoidance. And third, the prefrontal cortex (PFC), especially its medial part, shows consistent association to harm avoidance across studies (15 studies). The medial PFC and the dorsal anterior cingulate cortex (ACC) are also the core structures involved in fear extinction by suppressing the activity of the amygdala.[26,27] The prefrontal cortices have traditionally been related to executive control, but a new strain of research on the brain's behavior outside of attention-demanding tasks has identified the medial PFC as a central hub of this network devoted to self-referential thought.[28] This network has been labeled "the brain's default mode network."[29] The second hub region of the default mode network is the posterior cingulate, and some evidence (two studies so far) point toward an involvement of this structure in harm avoidance as well. In the recent past, network analytic techniques have confirmed that these brain regions form a network, and properties of this network relate to harm avoidance. We have visualized the brain regions implicated in the "harm avoidance network" in Figure 1. Please note that this figure is merely an illustration based on a narrative review of the literature (i.e., no empirical or meta-analytic data). Emphasizing the network view on brain function, Montag et al.[30] have proposed that

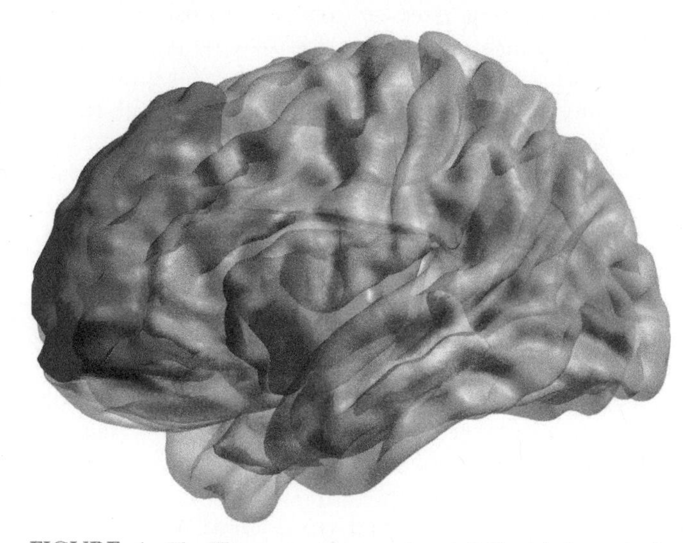

FIGURE 1 The "harm avoidance network." Depicted are brain regions whose functional and structural properties are most consistently found to correlate with harm avoidance: a set of brain regions in the medial prefrontal cortex (red); a set of brain areas comprising the striatum, the insula, and the anterior cingulate (yellow); and the temporal complex with the hippocampus, parahippocampal gyrus, and amygdala (green). The figure was created based on the automatic anatomical labeling atlas as implemented in the WFU pick atlas[90] and BrainNetViewer[91] solely for illustrative purposes.

[d] We searched Pubmed.com and Google Scholar in October 2014 with combinations of the following keywords: harm avoidance, electroencephalogram, event-related potentials, near infrared spectography, positron emission tomography, diffusion tensor imaging, voxel based morphometry, structural MRI, functional MRI, and resting-state fMRI. Given how unlikely it is to find all relevant studies in this broad research field, we apologize for omissions.

the interplay between prefrontal and subcortical areas is crucial for trait anxiety, and improper checks and balances between ascending affective signaling and top-down control lead to high levels of trait anxiety.

After this brief synopsis of the empirical findings, we are going to review the studies in depth. Following the theoretical considerations, we will first review studies that describe the morphological and structural correlates of harm avoidance and resting state studies looking at activity, metabolism, and functional connectivity in the absence of a clear behavioral task. We will then review experimental approaches using task fMRI, task NIRS, and event-related potentials. The latter studies give especially valuable insights into the influence of the harm avoidance temperament on information processing.

5.1 VBM and Morphological Studies

A total of seven morphological studies on harm avoidance have been published so far. The first study[31] focused on the cingulate gyrus. In 100 healthy participants (equal gender distribution), the surfaces of the anterior and posterior cingulate were manually traced bilaterally and correlated with the harm avoidance scores obtained from the TCI. The right but not the left anterior cingulate surface correlated positively with harm avoidance, accounting for 24% of the variance. This effect could be observed in both gender groups. No such effect was present for the posterior cingulate. An analysis of the four subscales revealed that all four harm avoidance facets correlated significantly with the surface measurement of the right ACC. The largest effects were reported for the facet fatigability ($r=0.40$) and anticipatory worry ($r=0.35$). The study by Pujol and colleagues was also the first to highlight the role of the anterior cingulate in harm avoidance, a result confirmed by many studies to follow. The anterior cingulate might be of special relevance for anxiety, because it is well-known that this region is triggered by conflicting stimuli.[32] As mentioned above, both uncertainty and conflicting situations are able to trigger anxiety.

The second study[33] used voxel based morphometry (VBM) to study the relationship between local gray and white matter volume and TCI dimensions in the elderly (N=42 healthy participants, 18 men, 24 women, mean age 60 years). Both global and regional volumes of gray matter, white matter, and cerebro-spinal fluid were analyzed. Even though the authors report associations between their outcome variables and character dimensions, no significant association with harm avoidance was found, neither with regional or global volumes of the different tissue classes.

Three other VBM studies, however, were able to detect significant associations between regional gray matter volume and harm avoidance. Iidaka et al.[34] analyzed data from N=56 healthy participants (30 men and 26 women) who completed the TCI. Regional gray matter volume in the left amygdala was positively correlated with harm avoidance, yielding a higher volume in participants scoring high on harm avoidance. Even though this effect was present in the entire sample, a gender-specific analysis revealed that this relationship was only present in females. In the entire sample, trends for positive relationships between harm avoidance and gray matter volume in the right middle temporal and angular gyri and orbitofrontal cortex were also reported.

Yamasue et al.[35] carried out VBM on gray matter segmentations in 183 healthy participants (177 males and 66 females) who had filled in the TCI. In the entire sample, a robust negative correlation was found between harm avoidance and the regional volume in the right hippocampus. Higher harm avoidance scores related to lower gray matter volume. Gender specific analyses revealed reduced gray matter volume in the left anterior prefrontal cortex in females scoring high on harm avoidance. No such relationship was present in males. The hippocampus finding is very interesting, because smaller hippocampus volumes are discussed to be of high relevance for a better understanding of depression.[36]

Gardini et al.[37] carried out VBM on gray matter segments in 85 healthy participants (58 men and 27 women) who had filled in the TPQ. The authors reported negative correlations between harm avoidance and gray matter in the right cuneus and inferior parietal lobule, the left precuneus, the middle occipital and the middle frontal gyri, and the bilateral inferior frontal gyri. No gender specific effects were reported.

In a more recent VBM study, van Schuerbeck and colleagues[38] investigated gray and white matter volume in 68 healthy female participants. Most results in this study confirmed a relationship between harm avoidance and gray matter volume: in line with the results by Gardini et al.,[37] a negative correlation between harm avoidance and regional volumes of the inferior frontal gyri was reported. This fits also with findings from Montag et al.,[16] who observed a negative correlation between gray matter volumes and neuroticism (a trait being closely linked to harm avoidance). Furthermore, participants scoring high on harm avoidance had a reduced regional volume in the anterior cingulate and the cerebellar tonsils and an increased volume in superior frontal gyri and in the lingual gyrus. The negative relationship between harm avoidance and anterior cingulate volume is at odds with the study by Pujol et al.,[31] which might be attributable to the differences in methodological approach. The positive correlation of gray matter volume in the lingual gyrus and harm avoidance was paralleled by a similar relationship between regional white matter volume and harm avoidance.

The result regarding the cerebellum was further analyzed by Larrichiuta et al.,[39] who quantified gray and white matter volume of the entire cerebellum. They also reported a negative relationship between cerebellar volume and harm avoidance, but only in males and not in females. Van Schuerbeck et al.[38] had only studied female participants. Besides this inconsistency, both studies point toward the cerebellum as an interesting target structure for harm avoidance. The cerebellum has long been neglected in behavioral neuroscience research, but recent results have demonstrated a rich connectivity pattern of this structure with neocortical association cortices.[40] The two studies reviewed here are the only studies that report cerebellar involvement in harm avoidance. Clearly, further studies are needed to characterize this relationship in depth, but given the recent findings in other branches of the field, this will surely be a promising endeavor.

5.2 Diffusion Tensor Imaging Studies: Structural Connectivity

Five studies have used diffusion tensor imaging to look at white matter characteristics as a function of harm avoidance. The first study by Westlye and colleagues[41] utilized the rather data-driven approach of tract-based spatial statistics (TBSS) to assess differences in white matter integrity in high and low harm avoidant participants. TBSS[42] capitalizes on fractional anisotropy (FA), the most commonly derived measure from diffusion imaging data. FA quantifies how strongly directional a local white matter tract structure is. Based on the directionality of water diffusion measured in a given voxel, a scalar ranging from 0 to 1 is assigned to the voxel, where 0 means entirely unrestricted diffusion, while 1 indicates strong directionality in diffusivity. In the large fluid-filled ventricles of the brain, FA values range close to 0, while in the thick white matter bundles of the corpus callosum, FA values can be as high as 0.85. TBSS allows for a voxel-based group analysis of FA values quite similar to local morphological analyses using VBM. Subject-specific FA maps are projected onto a mean reference template, and subsequently, voxelwise statistics can be applied to test for group differences in white matter structure. Westlye et al.[41] tested 263 healthy participants (150 women and 113 men in an age range from 20 to 85) who had filled in the items from the TPQ's harm avoidance scale. The whole brain analysis revealed very widespread associations, with FA in 42% of all voxels being negatively related to harm avoidance. That is, subjects scoring high on harm avoidance showed lower integrity of white matter fiber tracts as indicated by lower FA values throughout the brain. The mean FA value across all voxels in the reference image accounted for 4% of the variance in harm avoidance scores. The global whole brain analysis was complemented by a set of tract-specific analyses based on atlas-derived regions of interest. These analyses revealed strong associations between harm avoidance and FA in corticolimbic circuitry, especially in the anterior thalamic radiation, connecting the thalamus with the frontal lobes, and the uncinate fasiculus, connecting mid-temporal structures (including the hippocampus and amygdala) with orbitofrontal areas. Further results included an association between FA in the subgenual anterior cingulate and harm avoidance. The results from this first diffusion tensor imaging study on the harm avoidance temperament substantiate the network aspect of harm avoidance. All three clusters of brain regions that have been implicated in harm avoidance in previous work (i.e., frontal cluster, salience cluster including the ACC, and temporal cluster) are structurally connected, and the integrity of these connections relates to harm avoidance.

The role of the uncinate fasiculus that connects limbic midtemporal regions with orbitofrontal cortices has been further analyzed by Taddei et al.[43] In line with the findings reported by Westlye and colleagues, a negative relationship between FA in the right uncinate fasiculus and harm avoidance was observed. Again, this negative correlation indicates less structural connectivity in highly harm avoidant participants. Taddei and colleagues found this relationship in a sample of 20 adolescents (mean age 15 years, 11 males and 9 females) whose harm avoidance scores had been rated by their teachers on a 12-item scale derived from the Junior Temperament and Character Inventory on a previous visit to the laboratory about six years earlier. Aside from the already mentioned negative relationship between harm avoidance and FA in the uncinate fasiculus, a whole brain TBSS analysis revealed also a significant negative relationship between mean FA and harm avoidance. Again, this finding is in line with the global TBSS result from Westlye et al.

The other three diffusion tensor imaging studies relied on tractography or region of interest analyses to assess the relationship between white matter integrity and harm avoidance. Montag et al.[44] focused on mean FA values in atlas-derived fiber tracts in the temporal lobes. In this study, the TCI was administered along with several other personality inventories to extract a higher order factor called "negative emotionality" by means of principal component analysis. The study included 110 healthy participants (44 men and 66 women). Results confirmed positive relationships between negative emotionality (including harm avoidance) and fiber tracts in the temporal lobe. The strongest relationship was observed in the fiber tract connecting the hippocampus with the cingulate. Other results included a positive relationship between harm avoidance and the inferior fronto-occipital fasiculus, the superior longitudinal fasiculus, and the uncinate fasiculus. In general, higher harm avoidance scores went along with higher FA values, a result that contradicts the two previous studies that found negative relationships exclusively. The results reported by Montag et al., however, showed a strong sex dimorphism: results were only significant in the male subsample. According to the Montag et al. data, anxiety might be a result of strong bursts of subcortical anxiety signals that are transferred via intact white matter tracts (although there is still a huge debate what biological processes actually underlie the diffusion weighted MRI signal[45]).

The two more recent studies by Lei et al.[46] and Laricchiuta et al.[47] focused on the basal ganglia. Lei et al.[46] used the striatum as a seed for the probabilistic tracking of fiber projections to nine target regions in frontal, cingulate, and temporal regions. As the main outcome measure, the proportional ratio of the number of fiber projections to the target region to the total number of fiber projections was calculated for every voxel in a given fiber tract. In a sample of 50 healthy, all-male participants, positive correlations between harm avoidance and structural connectivity of the striatum with the posterior cingulate and the striatum with the dorsolateral prefrontal cortex were obtained.

Laricchiuta et al.[47] focused on the entire basal ganglia by segmenting the putamen, the caudate body, and the pallidum in high-resolution structural images and extracted diffusivity parameters from coregistered diffusion tensor imaging (DTI) scans. In a sample of 125 healthy participants (52 men and 73 women), mean diffusivity (MD) in the right and left putamen correlated positively with harm avoidance scores. This result was obtained while controlling for age, gender, and total volume of the structures. Mean diffusivity is a diffusion measure that reflects the total amount of diffusion in a voxel (both directed and undirected diffusion). Therefore, a higher total diffusion in the putamen seems to be associated with a higher harm avoidance score.

5.3 Molecular Imaging: PET

Three studies used PET as a technique to image the activity or availability of molecules in the brain and the relationship of this data to harm avoidance. Please note that these studies are not the only PET studies on harm avoidance. We have decided to split the review of PET studies into two paragraphs because of a fundamental difference: the three studies reviewed here have not acquired information on neural activity in terms of blood flow or oxygen and glucose metabolism and therefore reveal insights into structural rather than functional properties.

The first molecular imaging study by Kaasinen et al.[48] used [^{18}F]-fluoro-L-dopa to probe striatal dopaminergic functioning. This radioligand tracks DOPA decarboxylase, the enzyme catalyzing the synthesis of dopamine from its direct precursor L-dopa. Therefore, the metabolic rate of the radioligand allows one to draw conclusions on the individual's dopamine synthesis capacity. Twenty-five healthy elderly volunteers (mean age 60 years, 13 men and 12 women) participated in the study. All participants had filled in the TCI. The analysis was confined to regions of interest (ROIs) in the striatum (head of caudate nucleus and putamen) and an occipital control region. No robust association with harm avoidance and the metabolic rate of [^{18}F]-fluoro-L-dopa was observed. This negative finding indicates that harm avoidance was not related to striatal dopamine synthesis capacity in these elderly subjects.

The second molecular imaging study on harm avoidance focused also on the dopaminergic system. Kim et al.[49] used [^{11}C]raclopride to track the availability of dopamine D2 and D3 receptors. Twenty-one healthy participants (13 men and 8 women) who had filled in the TCI participated in the study. Five manually drawn ROIs in the striatum were analyzed, with the cerebellum as region of (control) reference. Harm avoidance correlated negatively with receptor availability in the precommissural dorsal caudate and the postcommissural caudate. No correlations were found in the putamen. The results by Kim and colleagues are the first to show a molecular correlate of harm avoidance in the harm avoidance network. Harm avoidance seems to be characterized by a low availability of D2/D3 dopamine receptor in the caudate, a part of the striatum and basal ganglia formation.

The third molecular imaging study traced mu-opioid receptor availability with [^{11}C]-carfentanil as radioligand.[50] A total of 22 middle-aged volunteers (mean age 38 years, 11 men and 11 women) who were preselected according to their TCI harm avoidance scores from a large data base (lower and upper quartile of the distribution) participated in the study. Harm avoidance affected receptor availability in the anterior part of the harm avoidance network: participants scoring high on harm avoidance had higher receptor availability in the ventromedial and dorsolateral prefrontal cortex, as well as in the anterior cingulate and anterior insula. The effects were mainly driven by the shyness and fatigability subscales. Mu-opioid receptors have a high affinity for endogenous beta-endorphins and enkephalin. Both transmitter molecules play an important role in nociception.

5.4 Resting-State Functional Studies: PET, SPECT, fMRI, and NIRS

A total of 10 studies looking for harm avoidance correlates in cerebral metabolism and blood flow in task-free conditions have been published so far. These studies include resting-state fMRI studies that investigate functional connectivity in BOLD time series, positron emission tomography studies imaging cerebral blood flow and glucose metabolism, single photon emission computed tomography (SPECT) studies of cerebral blood flow, and studies using near infrared spectrography (NIRS). We decided to group these studies together despite dissimilarities in the applied imaging techniques, because all studies focus on functional properties of the human brain with high spatial resolution in a task-free resting state.[51] We will begin by reviewing PET studies, then switch over to fMRI studies, and finally discuss findings from NIRS studies.

5.4.1 Positron Emission Tomography

Three resting-state PET studies tracking regional glucose metabolism and one SPECT study tracking cerebral blood flow have been published on harm avoidance. The first study by Youn et al.[52] used [^{18}F]-fluorodeoxyglucose as a radioligand to track glucose metabolism in 19 healthy subjects (13 men and 6 women) who had filled in the TCI. Negative correlations between harm avoidance and regional blood flow in temporal and cingulate regions were found: participants scoring high on harm avoidance had lower regional glucose metabolism in bilateral inferior temporal gyrus, left middle temporal gyrus, and the right anterior cingulate. No positive relationships between harm avoidance and metabolism were observed. All results were obtained while statistically controlling for the other three temperaments. The study by Youn et al. is the first to confirm that baseline brain function is also altered in temporal and cingulate regions that have been implicated in harm avoidance using structural imaging techniques.

The second study by Hakamata et al.[53] used the same radioligand as Youn and colleagues to track resting-state metabolism in 31 healthy middle-aged subjects (21 men and 10 women, mean age of 54 years, age range 34–73 years) who had filled in a 60-item selection of the TCI on the four temperaments, including harm avoidance. The rationale of the study was to extend the findings by Youn et al. in a larger sample. The authors were successful in replicating associations of harm avoidance with neural structures from the previous study. Harm avoidance was negatively correlated with glucose metabolism in the bilateral middle temporal gyrus. Other temporal regions that showed a negative correlation with harm avoidance included left parahippocampal and left fusiform gyri. No correlation with the anterior cingulate could be observed; however, two clusters in the posterior cingulate showed a negative relationship with harm avoidance. Furthermore, a negative correlation with the left middle frontal gyrus and a positive correlation with the right thalamus were reported. All results were obtained while controlling for the other three temperament dimensions. Besides the replication of Youn et al., these results point toward a functional role of the prefrontal cortex in harm avoidance, another region detected in structural imaging studies. Furthermore, the study by Hakamata et al. is one of the two studies that point toward involvement of the default mode network in harm avoidance.

The third PET study published by Nishio et al.[54] also used [^{18}F]-fluorodeoxyglucose to track glucose metabolism in the resting-state of the human brain. Their study included 102 healthy middle aged participants (65 males and 37 females, mean age 52.7 years, age range 34–73 years) who had answered all temperament items from the TCI. The authors did not find any positive correlation between resting-state glucose metabolism and harm avoidance. However, they report one large cluster in the right medial frontal gyrus (Brodman area 10) that showed a negative correlation: participants scoring high on harm avoidance had a reduced glucose metabolism. This effect was found after controlling for age, gender, and the remaining three temperament dimensions. Gender-specific analyses revealed no significant association between harm avoidance and resting-state metabolism in men but a negative association between harm avoidance and metabolism in the left anterior medial prefrontal cortex (mPFC) in women. The anterior mPFC is the second hub region of the default mode network, which would be in line with the finding by Hakamata et al.[53] This also fits with the findings by Yamasue et al.[35] showing reduced gray matter volume of the left anterior prefrontal cortex in females. What is still rather surprising is that the associations between the temporal areas found in the two previous PET studies could not be detected by Nishio, even though their sample was about twice as large as those in earlier studies.

In their SPECT study, Turner et al.[55] used 99mTC-HMPAO (Technetium 99m-hexamethylpropylene amine oxime) as a radioligand to track variations in regional cerebral blood flow (rCBF) in 20 healthy male volunteers. Prior to SPECT scanning, participants had filled in the TCI. Based on the TCI harm avoidance scores, participants were grouped into four quartile groups, and pairwise comparisons between the four groups were conducted. Two contrasts remained significant after controlling for multiple comparisons: the third quartile group showed increased rCBF compared to the lowest quartile group in a cluster comprising the right fusiform gyrus, right middle and superior temporal gyri, as well as the parahippocampal gyrus. Graphically plotting the results indicated an increase in rCBF from quartile group one to three and an asymptotic plateau afterward (no difference between group three and four). Furthermore, the third quartile group showed increased rCBF in occipital regions and cuneus compared to the second quartile group. Both rCBF and glucose metabolism are correlates of neural activity. Therefore, the positive correlation between rCBF and harm avoidance is surprising, given the negative correlations in temporal regions found in the first two PET studies.

5.4.2 Functional Magnetic Resonance Imaging

Four studies have used BOLD fMRI to study functional connectivity during rest and its relationship to harm avoidance. Three of these studies have made use of a seed-based approach. In a seed-based study, a priori ROI is chosen based on theoretical considerations of previous work. This ROI then serves as a seed for the analysis: pairwise correlations are computed between the mean BOLD time series from the seed region and the BOLD time series in all others voxels across the brain. This results in a functional connectivity map (fcMap) where every voxel carries a scalar that corresponds to its correlation with the seed region. For individual differences analyses, the correlation values are usually transformed to approximate a normal distribution and then regressed onto the individual difference measure of interest, in our case harm avoidance. In the study by Li et al.,[56] different nuclei of the amygdala served as seed regions. The basolateral, centromedial, and superficial amygdala subregions were located using a probabilistic cytoarchitectonic atlas. A large sample of 291 healthy young adults (153 women and 138 men) who had filled in the TPQ was investigated. The functional connectivity of all three nuclei was modulated by harm avoidance. Higher harm avoidance scores went along with (1) lower functional connectivities between the basolateral amygdala and temporal regions, especially inferior temporal gyri and the temporal parietal junction; (2) lower functional connectivity between the centromedial amygdala and frontal regions, especially ventromedial PFC; (3) lower functional connectivity between the superficial amygdala and temporal and prefrontal regions; and (4) higher functional connectivity between the superficial amygdala and the ventral striatum. All results were obtained while controlling for age and a depression screening (Beck Depression Inventory[94]) and while masking out voxels that did not show any functional connectivity with the amygdala at the group level. Besides the results that were found in the whole sample, gender-specific results were also reported. In general, females scoring high on harm avoidance had stronger negative functional coupling of the basolateral seed region across target regions, while males scoring high on harm avoidance had stronger negative coupling of the centromedial seed. The superficial amygdala's connectivity was mostly unaffected by gender. In general, the results by Li et al. confirm many brain regions that have previously been related to harm avoidance: the temporal complex, prefrontal cortex (including its medial subregion), and the basal ganglia. Interestingly, all brain regions revealed themselves in an analysis that was focused on the amygdala. No functional connectivities between the target regions (i.e., temporal, frontal, and striatal regions) were computed and it therefore remains speculative if the amygdala qualifies as a central processing hub in the harm avoidance network (this would be the case if the connections to the amygdala were stronger or more central than other connections in the network). The broad role of the amygdala in affective processing and the widespread modulation of its functional connectivity by harm avoidance scores give us reason to believe so.

Baeken et al.[57] also looked at functional connectivity of an amygdala seed. Their study included 60 female participants who had filled in the TCI. In contrast to the study by Li et al., no different nuclei of the amygdala were investigated. Instead, two spheres with a 6mm radius were centered around atlas-based coordinates in the left and the right amygdala, respectively. Coordinates were selected from the literature. The authors report a significant influence of harm avoidance on functional connectivities of the amygdalae and widespread cortical regions. High harm avoidant participants showed stronger functional connectivity between the left amygdala and occipital and temporal target regions and weaker functional connectivity between the left amygdala and prefrontal areas. Apart from that, both stronger and weaker functional connectivity to cerebellar regions were observed. The findings regarding the cerebellum fit with the structural imaging work by Larichiutta et al.[39] and van Schuerbeck et al.[38] reviewed above. Both studies have reported an association between structural properties of the cerebellum and individual differences in harm avoidance. Besides the left amygdala, Baeken et al. also report modulatory influences of harm avoidance on right amygdala connectivity that are more widespread than those of the left amygdala. Participants scoring high on harm avoidance showed a stronger coupling of the right amygdalar seed with parietal, temporal, limbic, and prefrontal sites and a weaker coupling between the right amygdala and the anterior cingulate and the cerebellum. In contrast to the first resting state connectivity study, more positive correlations between harm avoidance and functional connectivity were observed. This might be attributable to the differences in seed definition.

The third seed-based study by Markett et al.[58] focused on the insular network. The insular salience network is a resting-state network spanning the anterior insula, the anterior cingulate, parts of the basal ganglia, and cortical regions along the operculum. In a sample of 24 female participants who had filled in the TCI, the insular network was characterized by two seeds in the right and left anterior insulae. The seed regions were 12mm spheres around atlas coordinates selected from previous work. Participants scoring high in harm avoidance showed a stronger functional coupling between the right anterior insula and the dorsal anterior cingulate and between the right anterior insula and a cluster in the dorsolateral prefrontal cortex. No effect was found for the left anterior insula. The analysis of the four harm avoidance subscales revealed that the right lateralized effects were most strongly driven by the shyness facets, albeit all subscales showed an association with functional connectivity between the anterior insula and the anterior cingulate. This study was the first to confirm that functional connectivity in the resting brain apart from the amygdala nuclei contribute to individual differences in harm avoidance. The findings that functional connectivity in the right hemisphere relate more strongly to harm avoidance is also consistent with the findings by Baeken et al.,[57] who found stronger and more widespread associations between harm avoidance and right, rather than left, amygdala connectivity.

The fourth resting-state fMRI study used a different methodological approach. Kyeong et al.[59] applied graph analysis to multiple regions of interest. Graph analysis is a mathematical modeling technique to represent a network as a set of nodes (brain regions in the case of brain networks) and edges (pairwise functional connectivities between all brain regions). The resulting connectivity matrix (number of nodes × number of nodes) can then be subjected to graph analytical algorithms that provide measures of network organization, network efficiency, and importance of single nodes to the network architecture. A total of 40 healthy volunteers were included in the analysis. All participants had filled in the TCI. In this sample, harm avoidance was strongly intercorrelated with the novelty-seeking temperament dimension (−0.56). Intercorrelations between harm avoidance and novelty seeking are not uncommon, even though both temperaments are conceptualized as orthogonal. The inverse relationship, however, is usually much weaker (e.g., $r = -0.34$[44]). Because of this confound, Kyeong et al.[59] used a clustering algorithm to split their sample in two groups: 19 participants were grouped into a high harm avoidance/low novelty seeking (HA+/NS−) group and 21 participants were grouped into a low harm avoidance/high novelty seeking (HA−/NS+) group. In both groups, mean connectivity matrices were computed. This was done by averaging individual matrices obtained from a functional connectivity analysis of BOLD time series extracted from 90 cortical and subcortical regions, as segmented in the automatic anatomical labeling (AA[95]) atlas. These two group connectivity matrices (one per group) represent the entire brain as one network. By means of a graph theoretical modularity estimation algorithm, the entire brain network can be decomposed into nonoverlapping subnetworks. In both temperament groups, the algorithm found five distinct functional communities, with a high degree of between-group similarity: a frontotemporal, a sensorimotor, and a visual module were identical in both groups. The remaining two modules, however, differed between temperament groups: in the HA+/NS− group, prefrontal and cingulate regions were clustered together with limbic regions (amygdala, hippocampus, parahippocampal gyrus) in the fourth module, while the basal ganglia and thalamus formed the fifth. In the HA−/NS+ group, to the contrary, prefrontal and cingulate regions grouped themselves with the basal ganglia and the thalamus in the fourth module, while the limbic regions formed the fifth. That is, in the highly harm avoidant group, prefrontal and cingulate regions aligned with limbic areas, while in the low harm avoidant group, prefrontal and cingulate regions pledged their allegiance to a basal ganglia/thalamus complex. These results are the first to describe qualitative (rather than quantitative) differences in brain network architecture in relationship to individual differences in basic temperaments. All former studies had demonstrated either local functional and structural differences or individual differences in regional connectivity strength as correlates of harm avoidance. Implicated brain regions that cluster differently into networks contain regions that have been previously described as relevant for harm avoidance: prefrontal, anterior cingulate, striatal, limbic, and temporal. The main finding was complemented by two further analyses. The density of functional connections (fraction of present connections to possible connections) was computed for the three complexes that showed different clustering results. The connection density between prefrontal and limbic regions was increased in the HA+/NS− group compared to the HA−/NS+ group. Although the correlative nature of the data does not allow for conclusions on directionality, the data is in line with the idea that stronger signaling from subcortical to frontal structures underlies trait anxiety.[30] As a further analysis, regional gray matter volume was calculated based on structural T1-weighted images acquired from the same participants. Gray matter volume was extracted for each of the three functional complexes and then compared between the two temperament groups. Between-group differences were found for the basal ganglia/thalamus and for the limbic complex with HA+/NS− showing larger local gray matter volume. This recent resting-state fMRI study surely is a hallmark toward deciphering the biological implementation of the harm avoidance temperament in the brain. It will be important to see if these findings replicate in females and how the often described sex differences contribute to the functional clustering of brain regions. Furthermore, it will be important to replicate the finding in a sample that shows a more usual relationship between different temperaments.

5.4.3 NIRS

Only one study on harm avoidance has been published so far that relies on NIRS to study metabolic correlates of neural activity in the resting-state. In this study, Nakao et al.[60] investigated a sample of 22 healthy participants (12 men and 10 women) who had filled in the TCI. NIRS sensor probes were positioned on the scalp above frontal and temporal lobes. For the study, however, only signals from 15 channels over medial PFC were used for data analysis. From a total of 10 minutes (eyes open and eyes closed), the band-limited power was calculated for two different frequency bands: low-frequency fluctuations in the range of 0.04–0.15 Hz and very slow oscillations in the range of 0.02–0.04 Hz. The power in this very low frequency range was negatively related to harm avoidance. The finding was constrained to one electrode over the medial part of Brodman area 9 (medial prefrontal cortex) and to the condition during which the participants were asked to keep their eyes closed. The clear advantage of NIRS over fMRI is a much faster sampling rate and therefore a finer temporal resolution, making spectral power analysis feasible. Again, the medial prefrontal cortex, a central region in the brain's default mode network, was related to individual differences in harm avoidance.

5.5 Task-Based Studies

In the following, we are going to review task-based activation studies that investigate the influence of harm avoidance on neural correlates of observable behavior. In these studies, research participants engage in a task, in contrast to the studies reviewed so far where non-functional or task-free resting-state correlates of harm avoidance were investigated. Task studies allow us to observe how brain areas that have been implicated in harm avoidance change their activity in response to task demands and external stimulation. We will start by reviewing five fMRI studies, followed by three NIRS studies, and six electroencephalogram (EEG) studies. The latter technique has millisecond accuracy and has been often used to study temporal dynamics in response to stimuli (often with no direct relevance to anxiety). Therefore, these studies provide important evidence that individual differences in a temperament can affect a large range of cognitive and affective processes.

5.5.1 *Functional Magnetic Resonance Imaging*

The first fMRI study used a risky-gains task to investigate neural correlates of risk processing.[61] A total of 17 healthy participants (11 males and 6 females) participated in the study. Questionnaire data (TCI) was available from 15 participants. In the task, participants gambled for points: possible gains of ascending magnitude appeared subsequently on-screen, and participants could either decide to claim a given gain or to wait for a higher gain to show up. With higher possible gains, however, a chance of losing an equal amount of points was introduced. That is, with every number appearing on screen (20, 40, or 80 points), participants were to choose a safe bet or a risky gamble. Unknown to the participant, the probabilities of winning or losing were manipulated in a way that every strategy was equally successful in getting high scores. Paulus et al. constructed five different events of interest for their analysis: trials where the participant chose the safe 20 points option, trials where the participant chose the 40 points option (and won), trials where the participant chose the 80 points option (and won), trials where the participant lost 40 points, and trials where the participant lost 80 points. This allowed the authors to isolate neural activity correlated with risky responses (selecting 40 or 80 points over 20 points) and neural activity correlated with punishment (losing 40 points or losing 80 points compared to baseline). The central result from the group analysis identified the right anterior insula as a key structure: the insula was activated during risky choice but also during punishment. Furthermore, activation strength in punished trials was positively correlated with the likelihood to choose the safe bet in a subsequent trial. The central result from the individual differences analysis was that the activity of the right anterior insula during punishment was modulated by harm avoidance. Participants scoring high on harm avoidance showed a stronger activation during punishment ($r = 0.54$). No correlation was found between harm avoidance and insular activity during risk processing. The study by Paulus and colleagues is the first fMRI study to demonstrate that harm avoidance affects information processing on a neural level. The right anterior insula has also been identified in other work as a hub region in the harm avoidance network.[58] Notably, the relationship between harm avoidance and sensitivity to punishment is in line with the initial claim by Cloninger that harm avoidance relates to the sensitivity to punishment.

The second fMRI study by Most et al.[62] investigated the influence of harm avoidance on attentional modulation. The attentional focus of the participants was manipulated to investigate the influence of attention on the processing of irrelevant but affectively negative information. Specifically, they looked at amygdala activity while participants processed negative or neutral content, which was not of relevance for the instructed task. In total, data of 29 healthy participants (20 women and 9 men) were included in the study. All participants had filled in the harm avoidance scale from the TCI. The participants' task was to identify targets from a stream of briefly flashed pictures of landscapes. Targets were rotated by 90° to the right or the left, and participants were instructed to detect the rotated picture in the stream and to indicate whether it was rotated to the left or to the right. Participants were informed that the target pictures never contained people or animals. On some trials, participants were told to search for the rotated picture of a building (instruction A) or for the rotated picture of a building or a landscape (instruction B). These two different instructions were given to manipulate participants' attentional focus from specific (instruction A) to nonspecific (instruction B). The critical comparison, however, involved the distractors: in the stream of landscape pictures from which the target pictures had to be detected, a set of pictures that either contained negatively valenced scenes involving human and animal violence or affectively neutral scenes as a control condition was randomly interspersed. These negative or neutral pictures were always irrelevant to the task. This deign allowed the authors to compare neural activity related to the processing of negative versus neutral irrelevant information under the two attentional conditions of a specific or a nonspecific attentional focus. Results showed a main effect of distractor valence on the amygdalae, in that negative distractors led to stronger bilateral amygdala activation than neutral distractors. There was also a main effect of attentional focus in that, especially in the ventrolateral prefrontal cortex, activity increased in the specific attentional focus compared to the nonspecific attentional focus condition. For both main effects, no robust correlation with harm avoidance scores was observed. The most critical comparison of the study, however, involved the interaction between the two experimental conditions (distractor valence and attentional focus) and the influence of harm avoidance on that interaction. In contrast to the author's prediction, no interaction between the two experimental factors alone was observed. When, however, harm avoidance was modeled as an additional factor in the analysis, the results yielded a significant three-way interaction: a stronger amygdala activation to negative distractors was observed under the nonspecific attentional set but only in participants scoring high in harm avoidance. This result was interpreted as evidence for the idea that in highly harm avoidant participants, spare attentional resources that were not bound by a specific attentional focus allowed for a stronger processing of negative irrelevant information. This evidence complements the first fMRI study by Paulus et al.[61] by showing that another central region from the harm avoidance network—the amygdala—engages differently in information processing depending on individual harm avoidance levels. Furthermore, the authors describe an effect of harm avoidance on the interplay between bottom-up emotion (reactive processing to negative information) and top-down cognition (maintaining an attentional focus). This illustrates that temperaments may affect various stages of information processing and their interactions during complex goal-oriented behavior.

The third task fMRI study on harm avoidance by Yang et al.[63] used a motor control task to investigate the influence of harm avoidance on error processing in a nonaffective task context. A group of 17 adolescent participants (age range 13–17 years, all female) who had filled in the TCI were included in the study. A stop-signal task was adopted to study motor inhibition. The stop-signal task is a variant of the popular go/no-go task in which the participant are instructed to react to frequent go-stimuli and to withhold their response to infrequent no-go-stimuli. The main difference in the stop-signal task is that all stimuli appear as go-stimuli at first. On trials that require a no-go response, a stop signal occurs with a short delay after stimulus onset. The cognitive task is therefore to stop an already initiated response, while in the classical go/no-go task, the cognitive task is to withhold an automatic and otherwise appropriate response before its initiation. Furthermore, a manipulation of the length of the stop-signal delay (the onset asynchrony of the stimulus and the stop signal) allows for a manipulation of task difficulty because longer stop-signal delays make it harder to stop an already initiated response. In their study, Yang et al. used the letters X and O as target stimuli and asked participants to indicate via button press which of the two letters had appeared on-screen. The stop signal was delivered on 25% of the trials as a tone via headphones. Five different stop-signal delays were used. These delays were customized for each participant to ensure equality in task difficulty. Prior to fMRI scanning, the mean reaction time for the X-O categorization task was assessed, and the stop-signal delay was set as this mean reaction time minus 100, 200, 300, 400, or 500 ms. That is, when the mean reaction time was 750 ms, the stop signal occurred either 650, 550, 450, 350, or 250 ms after the stimulus onset (i.e., the presentation of letters X or O). Yang et al.[63] focused their analysis on the subgenual part of the anterior cingulate. They reported increased activity for easy and medium-difficulty trials compared to hard trials. This main effect was modulated by individual harm avoidance scores: the activation difference between easy and hard trials was larger in participants scoring high on harm avoidance. The anterior cingulate is involved in the self-monitoring of ongoing behavior and error processing. The authors argue that they found an influence of harm avoidance on this component of their task. The study by Yang et al. complements the first two task fMRI studies on harm avoidance by showing that a third important structure from the harm avoidance network, the anterior cingulate, also engages differently in information processing depending on individual temperament levels.

The last two task fMRI studies included in this review are by the same work group and focus both on the influence of harm avoidance on amygdala activity during the processing of emotionally valenced baby faces in female volunteers. In the first study, Baeken et al.[64] studied 40 female volunteers who had all filled in the TCI. In the experiment, participants were asked to passively observe a set of baby pictures. The pictures were either positive (family pictures) or negative (disfigured faces selected from the dermatological literature). Neutral images were indistinguishable noise pictures matched for color and luminosity. Emotional valence of the pictures had previously been assessed in a group of 20 female volunteers with the same background as the study subjects. For the analysis, left and right amygdala masks were created from the automatic anatomical labeling (AAL) atlas. In the entire group, the authors found robust amygdala activation for negative pictures compared to positive pictures. This effect was similar for both hemispheres. When looking at individual differences in harm avoidance, the authors observed that participants scoring high on harm avoidance showed a stronger amygdala activation for negatively valenced stimuli. This effect, however, was restricted to the left amygdala. Results are in line with Most et al.[62] in the way that people scoring high on harm avoidance showed a stronger reactivity of the amygdala to negatively valenced stimulation. Importantly, the results by Baeken et al. indicate that such results can be observed in the absence of attentional manipulations.

The fifth task fMRI study on harm avoidance made use of the same stimuli of baby faces.[65] The only difference was that participants were not just asked to passively observe the stimuli but to indicate their valence (positive or negative) by pressing buttons. In total, 34 female volunteers were included in the study. All participants had filled in the TCI. An analysis of the imaging data with an amygdalar region of interest revealed bilateral amygdala activity in response to negative, but not to positive, pictures. A more detailed analysis revealed that the significant activation of the amygdala was mostly constrained to the superficial nucleus (about 80% of totally activated voxels) and the basolateral nucleus (about 10%). When correlating amygdalar activity with harm avoidance scores and when comparing amygdalar activity between high and low harm avoidant participants, significant associations between harm avoidance and amygdalar reactivity were observed: participants scoring high on harm avoidance showed stronger activity in response to both negative and positive pictures. These effects were most clearly observable in the basolateral nucleus of the amygdala. These results provide a successful replication of the findings by Baeken et al. Furthermore, the authors were able to demonstrate that the influence of harm avoidance on the amygdala seems to be broader than previously reported. Apparently, highly harm avoidant participants also show stronger amygdala engagement in the processing of stimuli with positively valenced content. Another important finding by the authors is that the effect of harm avoidance on amygdalar reactivity is largely confined to the basolateral nucleus. Li et al.[56] have reported gender-specific influences of harm avoidance on the connectivity of different amygdalar subregions. In their work, the basolateral amygdala connectivity was most strongly influenced by harm avoidance in women. Unfortunately, van Schuerbeck et al. included only female participant in their studies. It would be interesting to attempt replication in a male sample, particularly with respect to the involvement of different subnuclei of the amygdala.

5.5.2 NIRS Studies

A total of three studies used NIRS to investigate the role of harm avoidance in modulating neural response to stimulation. Please note that in adult participants, NIRS is mostly restricted to the study of cortical activation. The first study by Ito et al.[66] investigated the influence of various temperaments on neural activity during a finger tapping task. A group of 30 healthy participants (15 men and 15 women) were included in the study, and all participants had filled in the TCI. The task consisted of three blocks of 40 seconds of unilateral finger tapping interspaced by 30 seconds of rest. The order of right and left finger tapping was counterbalanced across participants. Electrodes for NIRS measurement were placed bilaterally with the midpoint of the electrodes between the vertex of the scalp and periauricular points. That is, electrodes were placed above temporal, parietal, and frontal regions even though the authors state "temporal regions" as targets. The signals from all electrodes and task repetitions were averaged and split into different time bins for statistical analysis. While the authors were able to relate neural activity to other temperaments, no significant association with harm avoidance could be detected.

The other two NIRS studies, however, did find associations between harm avoidance and neural activity. Morinaga et al.[67] used NIRS to assess neural activation in anticipation of an electric shock. A total of 56 healthy participants (24 women and 32 men) were included into the analysis. All participants had filled in the TCI prior to scanning. Three different scans were administered. During the first scan, participants were instructed to rest with their eyes closed. During the second scan, electrical shocks were administered to the median nerve of the right wrist. Shock intensities were increased until the maximally tolerable intensity was reached (mean amplitude between 30 and 100 V). Prior to the third scan, participants were informed that they had to expect a further shock at their maximally tolerable intensity at some point during the third scan. This shock was never administered, but for the duration of the third run, participants expected a painful shock. Sensors for NIRS measurement were placed above frontal and anterior temporal regions to allow for measurements of oxygenation and hemodynamics in two regions in the prefrontal cortex. The authors reported a positive relationship between oxyhemoglobin levels in the right prefrontal cortex and harm avoidance during anticipation of a shock. Participants scoring high on harm avoidance showed higher oxyhemoglobin levels. This study provides important additional evidence to the task fMRI studies above, as it shows that not only the anterior insula and the amygdala but also prefrontal areas as further important regions of the harm avoidance network show different behavior during task processing depending on harm avoidance.

The third NIRS study by Takizawa et al.[68] investigated the impact of harm avoidance on neural activity during psychological stress. A group of 70 young and middle aged participants (age range 23–57 years, 25 female and 45 male) who had filled in the TCI were included in the study. The authors used a standardized serial arithmetic test that required the participants to perform serial calculations of random digits to induce psychological stress. The test (Uchida–Kraeplin test) was administered twice. Every minute, a signal was given to terminate a given test block and start with the next one. For NIRS measurement, 24 sensor probes were positioned over frontal areas (covering fronto-polar, ventrolateral, and dorsolateral subdivisions), and the relative concentration of oxyhemoglobin and deoxyhemoglobin was measured while participants performed the two versions of the test. Only data from the 14 anterior channels were analyzed, and the signals were collapsed across time into different data bins: pretask resting block, posttask resting block, the two task blocks, and the break in between task blocks. The whole group analysis revealed an increase in prefrontal oxyhemoglobin and a decrease in prefrontal deoxyhemoglobin during task blocks compared to baseline periods. No difference was observed between the channels. These results confirm an increase in neural activation during task processing. Harm avoidance scores were consistently related to the increase in oxyhemoglobin measured at channel 15, positioned over the left fronto-polar cortex, in both task blocks. In task block two, additional correlations were found between harm avoidance and the increase in oxyhemoglobin at channel 13 (left fronto-polar) and channel 2 (right fronto-polar cortex). In all cases, higher harm avoidance scores related to higher levels of neural activation. The study by Takizawa et al. is the second task study pointing toward an involvement of individual differences in harm avoidance in task-evoked activity of the prefrontal cortex. Most notably, the task used by the authors was not a classic anxiety task (anticipating a painful shock, observing unpleasant pictures, getting punished during gambling) but involved a heavy cognitive and maybe a stressful component (because of the time pressure to perform fast and accurately).

5.5.3 EEG Studies

A total of eight EEG studies have examined the influence of harm avoidance on stimulus-evoked neurophysiological activity. These studies provide valuable insights into temporal properties of the neuromodulatory effects of the harm avoidance temperament. Four EEG studies used simple stimuli (letters or acoustic stimuli of different pitches or different durations) to investigate the biasing of information processing by harm avoidance that extends to neutral stimuli with no apparent affective component. While two studies were successful in linking harm avoidance to different stages of perceptual and attentive information processing,[69,70] two other studies report negative findings.[71,72]

In the first study, Hansenne[69] was successful in linking the P300 component of the stimulus-evoked potential to harm avoidance. The P300 component is a positive deflection peaking approximately 300 ms after stimulus onset and is thought to reflect working memory updating and the allocation of attentional resources.[95] The P300 is a robust component, typically triggered by rare stimuli ("odd balls") in a stream of stimuli. Hansenne's study included 43 healthy participants (18 men and 25 women) who had filled in the TCI prior to the EEG testing session. He used an auditory oddball task with 120 high-pitched tones as frequent stimuli and 30 low-pitched tones as odd balls. Participants were instructed to respond to the rare stimuli by pressing a button. The electroencephalogram was recorded from three electrodes positioned on the midline Pz, Cz, Fz sites from the international 10–20 electrode placing system. P300 amplitude and latency were measured by identifying the peak positive deflection in the time window between 280 and 450 ms after stimulus onset. P300 amplitude at electrode sites Pz (parietal regions) and Cz (central regions) were negatively correlated with harm avoidance. In detail, participants scoring high on harm avoidance showed a reduced peak amplitude of the P300 component. No association was found between harm avoidance and 300 latency or reaction times. Moreover no association could be observed with signals from the frontal Fz electrode. No gender differences were observed. The results show an influence of harm avoidance on post-perceptual information processing on the level of working memory and attention in the absence of temperament relevant information (negatively valenced information or stimulus of punishment or nonreward).

A further study by Hansenne et al.[70] points into the direction that harm avoidance also influences early sensory, reflexive attentive processing. In this study, 32 healthy participants (14 women and 18 men) who had filled in the TCI were also tested with an acoustic oddball task. The authors focused their analysis on an early component of the event-related potential called mismatch-negativity (MMN). The MMN is an index of early sensory processing and automatic attention. It describes a negativation of the signal, starting about 130 ms after stimulus onset that lasts for 120–170 ms. The MMN is elicited by rare deviant stimuli and is thought to arise from the mismatch (hence the name) of sensory input and the expectation on the upcoming event held as a representation in short-term memory. Still, the process seems to be automatic since it can also be observed when attentional resources are allocated elsewhere. In the present study, 150 tones of the same pitch were presented: 120 of these tones had a length of 40ms and the other 30 were twice as long, hence being odd balls. To probe the automatic nature of the MMN, participants were distracted from this task by asking them to solve arithmetic problems while listening to the acoustic stimuli. The electroencephalogram was again recorded from the three midline sites (Pz, Cz, Fz). MMN amplitude and latency were measured as the peak negative deflection from the baseline in the time window between 150 and 260ms after stimulus onset, after calculating the difference wave between frequent and rare tones. Harm avoidance was negatively related to the MMN amplitude at all three electrode sites under scrutiny: Participants scoring high on harm avoidance showed a decreased amplitude of the MMN. Even though no gender differences were obtained for the MMN amplitude, the analyses was repeated separately for both gender groups. The authors report that the same negative relationship between MMN amplitude and harm avoidance was found in women. In men, the association between MMN amplitude and harm avoidance was restricted to the frontal Fz electrode. The stronger effects for women are in line with the notion that, in general, women score higher on harm avoidance than men. The results of the study by Hansenne et al.[70] show that not only late reflexive information processing, but also early attentive processing is modulated by harm avoidance.

The effects found in the two studies by Hansenne et al., however, failed to replicate in two further studies. Vedeniapin et al.[71] studied 58 young and middle-aged volunteers (mean age 42 with a range from 22 to 62 years, 44 women and 14 men) who had filled in the TCI as part of a psychological assessment battery. The authors used a visual oddball task to probe the P300 component of the stimulus-evoked potential. In the present study, the authors administered a stream of 200 stimuli with 150 nontarget stimuli, 25 target stimuli that required a manual response, and 25 novel stimuli. The electroencephalogram was recorded from 19 electrodes placed on the scalp. The main analysis focused on event-related potentials time locked to stimulus onset. P300 amplitudes elicited by rare target stimuli were extracted by an automatic peak detection algorithm and extracted amplitudes were submitted to a principal component analysis. The first unrotated factor was then correlated with the scores from the TCI. No correlation with HA was found, however, the factor correlated significantly with the self-directedness character dimension. Subsequent analyses aimed at the P300 at midline electrode sites (Pz, Cz, Fz in the international 10–20 electrode placing system) revealed no association between harm avoidance and the P300 amplitude.

In a further study, an auditory odd ball paradigm was used to probe the influence of harm avoidance on both early and late information processing. Kim et al.[72] studied 25 healthy volunteers (16 men and 9 women) who had filled in the TCI. The authors presented 200 tones to the participants, 170 nontarget low-pitched and 30 high-pitched target tones, while recording the electroencephalogram from 123 electrode sites. The analysis, however, focused solely on the four midline channels (Fz, Cz, Pz, and Oz). Oz represents the central electrode over the occipital lobe. Amplitudes and latencies of event-related potential (ERP) components were extracted from time windows identified by the visual inspection of grand-averaged waveforms. Besides the P300 component and the MMN (see above), one further prominent component was extracted: the P200 component is a positive deflection about 200ms after stimulus onset and reflects higher-order perceptual processing that is already subject to attentional modulation. By measuring MMN, P200, and P300, the authors were able to look at influences of harm avoidance on different stages of information processing. Again, the stimuli used had no direct relevance to the contents of the harm avoidance construct. Even though the authors report associations between P300 (but not MMN and P200) and other temperaments, no association with harm avoidance was found with either of the three ERP measures.

The two unsuccessful attempts to replicate the initial finding by Hansenne[69] and Hansenne et al.[70] indicate that the relationship between information processing and the harm avoidance temperament is not universally found. On the grounds of these four EEG studies, we should be cautious to assume that harm avoidance influences information processing per se. It would be interesting to see if more consistent results can be obtained if stimuli with direct relevance (e.g., unpleasant stimuli) to harm avoidance were used.

Two studies by Mardaga and Hansenne[73,74] used pictures from the international affective picture system (IAPS[75]) as stimuli to investigate this matter. The IAPS system contains a set of pictures with normative ratings on their valence and arousing potential. In the first study, Mardaga and Hansenne[73] tested 46 healthy participants (23 men and 23 women) with a visual odd ball task. All participants had filled in the TCI. Participants were classified into high and low harm avoidance by median-split. In the experiment, a total of 375 trials were administered with 300 frequent stimulus trials (white and red checkerboards) and 75 IAPS pictures as rare odd balls. The 75 IAPS pictures were either of positive valence (such as playing kids), negative valence (such as the depiction of violent acts), or neutral valence (household items). Twenty-five different pictures were shown in each category. The electroencephalogram was recorded from 12 electrode sites placed on the scalp according to the

10–20 system (frontal, central, parietal, and occipital sites). Five different components were quantified according to peak amplitude after stimulus onset (P100, N100, P200, N200, and P300). Significant associations with harm avoidance were reported for P200 latency and for N200 and P300 amplitude: participants scoring low on harm avoidance had faster P200 after the presentation of pleasant compared to unpleasant pictures at central electrode position C3. Furthermore, N200 amplitudes were smaller in low harm avoidant participants when comparing pleasant to unpleasant or neutral pictures. Lastly, P300 amplitudes to pleasant pictures were larger in low harm avoidant participants compared to neutral or unpleasant pictures. No effects of picture category on ERP components were identified in high harm avoidant participants. These results indicate that people scoring low on harm avoidance bias their information processing for positively valenced stimuli while in high harm avoidant participant, all kinds of stimuli elicit a uniform response.

In the second study, Mardaga and Hansenne[74] used an auditory odd ball task to probe the MMN (see above). Again, a secondary task was applied to direct attention away from the odd ball task: while participants listened to the acoustic stimuli (2070 trials, 80% low-pitched standard and 20% high-pitched rare stimuli), they viewed a selection of IAPS pictures (30 positive, 30 negative, and 30 neutral pictures). Sixty healthy participants (28 men and 32 women) were included in the study. All had filled in the TCI. The EEG was recorded from the three midline electrode sites (Pz, Cz, and Fz) and the MMN was defined as the highest negative deflection in the difference-wave between rare and standard stimuli in a 100–200 ms time window after stimulus onset. Highly harm avoidant participants showed a larger MMN at the frontal electrode site (Fz), but only when they watched negatively valenced pictures simultaneously. A possible explanation for this finding is that negative pictures transiently increase the deployment of attentional resources and thus affect the unrelated odd ball task.

A third EEG study also used IAPS-like stimuli, albeit not in an oddball design. Zhang et al.[76] tested 70 adolescents (aged 12–20, no gender information stated) with a picture perception/evaluation task. Participants viewed pictures taken from the Chinese equivalent of the international affective picture system that were either positively (playing kids), negatively (snakes), or neutrally valenced (household objects). Each picture was presented for 1500 ms, and after each picture, participants were asked to rate the picture's valence. This part of the task was administered to ensure participants payed attention to the pictures. The EEG was recorded from 64 electrodes, but only the signals from 9 central and parietal electrodes were analyzed. Event-related potentials were calculated separately for positive, negative, and neutral stimuli, and the averaged waveforms were examined for the amplitude of the late positive potential (LPP). The LPP is a late positive deflection that starts within a few 100 milliseconds after stimulus onset. It is thought to reflect the appraisal of emotional significance and late selective attentional processes. Harm avoidance showed differential effects on LPP amplitude, depending on the affective content of the picture: LPP amplitude correlated positively with harm avoidance for negative pictures and negatively with harm avoidance for positive pictures. Participants scoring high on harm avoidance had higher LPP in response to negative pictures and lower LPP in response to positive pictures. No effect was observed for neutral pictures. These results show a biasing of emotional information processing by harm avoidance that temporally extends the time window investigated so far. At this late stage, a clear differentiation for differently valenced affective information became apparent. Unfortunately, the authors do not report any results on the early components of the event-related potential.

The last EEG study reviewed here took an entirely different methodological approach. Instead of looking at the time course of the averaged response to stimulation, the authors looked at changes in the frequency band of electrophysiological oscillations recorded from EEG electrodes. Takahashi et al.[77] investigated changes in theta-power during meditation and a control condition. Theta oscillations are slow oscillations in the frequency band, between 4 and 7 Hz, commonly observed during meditative, drowsy, and light sleeping stages. A total of 20 healthy male participants were recruited for the study and the TCI was administered for psychometric assessment. Participants were asked to engage in a meditation technique called "Su-Soku" (i.e., counting one's breaths) or to rest for an equal amount of time (15 min). During mediation and the resting condition, participants' EEG were recorded from six lateralized electrodes placed over frontal, central, and occipital sites according to the 10–20 system. Theta power was increased during meditation compared to rest, an effect that the authors attribute to an increase in mindfulness. This increase in theta power was additionally affected by harm avoidance scores: higher harm avoidance scores went along with enhanced mindfulness, as indexed by increased theta power over frontal electrodes. Enhanced mindfulness has been linked to enhanced cognitive control,[78] which might underlie the finding by Takahashi et al.[77]

5.5.3.1 Genetic Imaging

The studies reviewed so far have investigated direct correlates of harm avoidance. This means that harm avoidance was assessed by a questionnaire, and questionnaire scores were then related to neuroimaging data. A complementary and promising strategy is the more indirect approach of genetic imaging (see Chapter 4).

Temperament dimensions such as harm avoidance have a strong genetic basis, as outlined in the introduction. By means of molecular genetic studies, it has become possible to investigate the genetic underpinnings of personality directly.[79] It is beyond the scope of this present chapter to review all molecular genetic studies, but we would like to give two examples in order to illustrate this approach's potential.

In a genetic association study, attempts are made to relate genetic variation to individual differences in a phenotype (e.g., harm avoidance). This can be exploration driven in a genome-wide association study or hypothesis driven by means of a candidate gene approach. A candidate gene is a gene or variation on a gene that may relate to the construct of interest, given the role of the gene in a distinct biological pathway or findings from previous studies. Two genes that have been identified as candidate genes for harm avoidance and anxiety have been successfully linked to harm avoidance and were then also successfully linked to structural variation in brain circuits with relevance for harm avoidance. The first of these two genes is the gene encoding for the brain derived neurotrophic factor (BDNF). A common variation in the form of a single nucleotide polymorphism (SNP, exchange of one nucleobase) has been identified that leads to an exchange of an amino acid in the BDNF molecule. This SNP is called BDNF val66met, and the minor allele (met-variant) can be found in about 30% of the population (in middle Europeans, prevalence in other populations varies). The polymorphism was artificially created in a mouse model in order to compare the behavior and neural tissue in mice with different genotypes.[80] The knock-in met-variant led to increased anxious behavior and a decrease in hippocampal volume and dendritic complexity. Regarding harm avoidance, two predictions can be made from this data. First, human carriers of the met-variant should be characterized by increased levels of harm avoidance, and second, a decrease in hippocampal tissue should be detectable in human met-carriers as well. The latter prediction can be tested by means of genetic imaging. Both predictions were confirmed in subsequent studies. The rare homozygous met-variant of the BDNF val66met polymorphism has been linked to increased harm avoidance.[81] In detail, homozygous Met carriers showed (compared to Val carriers) elevated scores on the harm avoidance subscales anticipatory worry and fear of uncertainty. For a more detailed review on BDNF Val66Met and personality, please see Montag.[82] Furthermore, Montag et al.[83] used VBM to study the effect of different BDNF variants on gray matter volume: the high-harm avoidance met-variant was associated with reduced volume in the parahippocampal gyrus and the amygdala, in line with the predictions from the mouse model.

The second candidate gene is the CHRNA4 gene that codes for a subunit of the nicotinic acetylcholine receptor. Ross et al.[84] were successful in relating this gene to anxiety in a mouse knock-out model. Taking up on this data, Markett et al.[85] examined CHRNA4 rs1044396, a synonymous SNP (resulting in a synonymous codon) in the coding region of the CHRNA4 gene and its potential role in anxiety. Higher harm avoidance levels were observed in human carriers of the CHRNA4 C/C variant. In a later study, the same SNP was related to gray matter volume in the striatum, a structure rich in nicotinic acetylcholine receptors and crucially implicated in the harm avoidance network.[86]

Genetic imaging is an indirect but promising method of molecular imaging in humans (see Chapter 4). Although beyond the scope of the present chapter, we wanted to exemplify this approach by describing its application to harm avoidance.

5.5.3.2 Synopsis

So what have we learned? We have learned that there is a large array of neuroimaging studies specifically looking at neural correlates of harm avoidance, and these studies apply different methodological approaches to gather complementary information from different imaging modalities. Almost all studies have used the recent version of the Temperament and Character Inventory (TCI) and not the older Tridimensional Personality Questionnaire (TPQ) to assess harm avoidance. The studies were successful in identifying various neural structures with implications for harm avoidance. The application of connectivity mapping techniques has revealed that these structures form a network. This network consists of a temporal complex (hippocampus, amygdala, parahippocampal gyrus, and medial temporal lobe), a prefrontal complex (medial prefrontal cortex), and the insular complex (anterior insula, anterior cingulate, basal ganglia, and thalamus). Please note that this grouping of brain areas into complexes is somewhat arbitrary and requires empirical validation in further imaging studies on harm avoidance. Some additional structures have been identified in single studies, such as the cerebellum or the posterior cingulate. Future studies may want to look at these structures in more detail to establish their role in the harm avoidance network. Task-based imaging studies have confirmed that brain regions implicated in the harm avoidance network respond differently when probed by external stimulation, depending on the individual level of harm avoidance. Furthermore, some studies point toward an even broader effect of harm avoidance on information processing. Variation in the temperament dimension seems not only to affect reactive neural processing, but also interacts with attention to bias perception and general information processing. In a different strain of research, molecular underpinnings of harm avoidance have been addressed, either by means of PET or by genetic imaging approaches.

What do we have to do next? There are some issues that need to be addressed in future work. One of these issues is lateralization. Many studies on harm avoidance have identified structures in both hemispheres, but some studies have described lateralized results, in most cases to the right hemisphere. Further studies are needed to characterize the lateralization of harm avoidance. Sex differences are another issue. In general, women and men differ in their harm avoidance level. First of all, it would be interesting to learn why. And furthermore, it would be interesting to examine if the harm avoidance network differs between men and women. Large samples are required to establish the main effects of gender and interactions with harm avoidance on brain function and structure. While this has been accomplished in some of the structural MRI studies that make it easier to acquire large samples, the majority of functional task-based studies are somewhat underpowered to test for sex differences. This issue should be resolved in future work because functional neuroimaging techniques are becoming more widespread and affordable. Another important individual differences variable is age. In the temperament and character framework, temperaments are conceptualized as largely age invariant over the young and middle adulthood. Character development, however, interacts with the temperaments, and age effects on temperaments are therefore still possible.[11] The human brain matures over the first three decades of ontogenesis, and age effects on brain development have been reported over the entire life span. Even though studies have examined samples of different ages (adolescents, young adults, middle-aged adults, and older adults), no systematic assessment of the relationship between harm avoidance, age, and the brain have been documented. This would be an interesting endeavor for future studies, ideally in a longitudinal design.

Almost all of the studies included into this review have administered the TCI as a whole, rather than the harm avoidance scale alone, sometimes in order to probe neural correlates of other temperaments and sometimes to keep the psychometric scale intact and to standardize assessment across studies. This leaves us with the question whether we should statistically control for other temperaments when assessing neural correlates of harm avoidance. The temperaments are conceptualized as independent, and temperament effects should not be diluted by the influences of other temperaments given sufficiently large data sets. Empirically, however, subtle (and sometimes even strong) intercorrelations have been documented. Future studies should at least report how controlling for other TCI scales affects the results. Another issue is the use of the harm avoidance subscales. Since most studies have administered the TCI as a whole, information on the four subscales "anticipatory worry," "fear of uncertainty," "shyness," and "fatigability" is available in principle. However, surprisingly few studies have characterized their findings in depth by examining the four subscales. As this information is usually available, results should be generally reported in future studies.

A further issue for future work concerns the combination of imaging techniques. So far, the majority of studies have focused on data from a single modality (structural or functional, intraregional activity/morphology or interregional connectivity, etc.). This makes it difficult to relate findings from different studies and integrate them into a single framework. If, for example, a study reports an increase in local gray matter volume, the functional consequence of this alteration remains unclear. Therefore, it would be interesting to relate the influence of harm avoidance on the functional properties of the same structure and vice versa. Similarly, when resting-state functional connectivity studies report an alteration of functional connectivity or its network-wide organization to harm avoidance, it would be interesting to examine how functional connections or network properties change after stimulation. The latter is especially a thriving issue because, to our knowledge, no study on the relationship between harm avoidance and task-evoked connectivity changes has been published so far.

A last issue that needs to be discussed is the use of questionnaire measures. In all studies reviewed here, harm avoidance was assessed via self-report. Self-report data on harm avoidance, however, always come with problems of social desirability when answering items such as "I am an anxious person." Therefore, real human behavior on anxiety might bring clearer results, e.g., when correlating the structure of the human brain with individual differences in experimentally manipulated anxiety. Therefore, future studies should also try to rely on overt behavior rather than on questionnaire measures to circumvent these problems and to draw a clearer picture of individual differences in the harm avoidance temperament.[87]

Finally, all studies reviewed here describe the relationship between harm avoidance and the brain in healthy participants. This is highly valuable in order to understand the normal anxiety response. Furthermore, harm avoidance is a risk factor for anxiety disorders, depression, and personality disorders. Given the dimensionality of such disorders, the characterization of relationships in the normal range are also relevant to understanding these disorders. Therefore, it is of high relevance to evaluate the findings reviewed in the present chapter in patient populations. The costs of psychiatric disorders—both for the inflicted persons and society—are rising dramatically,[88] and we are still far away from establishing biomarkers for these disorders to aid diagnosis, treatment, and prevention. Even though we still have a long way to go to fully understand the neural circuits involved in harm avoidance in particular and anxiety in general, we should always keep in mind that our results need to be translated into clinical practice eventually.

Acknowledgment

Christian Montag is funded by a Heisenberg grant by the German Research Foundation (MO 2363/3-1)

References

1. Cloninger CR. A systematic method for clinical description and classification of personality variants: a proposal. *Arch Gen Psychiatry*. 1987;44(6):573–588.
2. Allport GW. *Personality: A Psychological Interpretation*; 1937.
3. Augustine AA, Larsen RJ. Is a trait really the mean of states? Similarities and differences between traditional and aggregate assessments of personality. *J Indiv Differ*. 2012;33(3):131.
4. Costa Jr PT, McCrae RR. Four ways five factors are basic. *Pers Indiv Differ*. 1992;13(6):653–665.
5. McCrae RR, Costa PT, Del Pilar GH, Rolland J-P, Parker WD. Cross-cultural assessment of the five-factor model: the revised NEO personality inventory. *J Cross Cult Psychol*. 1998;29(1):171–188.
6. Zuckerman M, Kuhlman DM, Joireman J, Teta P, Kraft M. A comparison of three structural models for personality: the Big Three, the Big Five, and the Alternative Five. *J Pers Soc Psychol*. 1993;65(4):757.
7. Ashton MC, Lee K, Perugini M, et al. A six-factor structure of personality-descriptive adjectives: solutions from psycholexical studies in seven languages. *J Pers Soc Psychol*. 2004;86(2):356.
8. Gray JA. *The Neuropsychology of Anxiety: An Enquiry into the Functions of the Septo-hippocampal System*. Oxford: Oxford University Press; 1982.
9. Cloninger CR. A unified biosocial theory of personality and its role in the development of anxiety states. *Psychiatr Dev*. 1986;4:167–226.
10. Cloninger CR, Svrakic DM, Przybeck TR. A psychobiological model of temperament and character. *Arch Gen Psychiatry*. 1993; 50(12):975–990.
11. Cloninger CR. Temperament and personality. *Curr Opin Neurobiol*. 1994;4:166–173.
12. Gillespie NA, Cloninger CR, Heath AC, Martin NG. The genetic and environmental relationship between Cloninger's dimensions of temperament and character. *Pers Indiv Differ*. 2003;35(8): 1931–1946.
13. Gray JA, McNaughton N. *The Neuropsychology of Anxiety*. 2nd ed. Oxford: Oxford University Press; 2000.
14. Eysenck HJ. *The Biological Basis of Personality*. Vol. 689. Transaction Publishers; 1967.
15. Davis KL, Panksepp J. The brain's emotional foundations of human personality and the Affective Neuroscience Personality Scales. *Neurosci Biobehav Rev*. 2011;35(9):1946–1958.
16. Montag C, Eichner M, Markett S, et al. An interaction of a NR3C1 polymorphism and antenatal solar activity impacts both hippocampus volume and neuroticism in adulthood. *Front Human Neurosci*. 2013;7:243.
17. Öngür D, Farabaugh A, Iosifescu DV, Perlis R, Fava M. Tridimensional personality questionnaire factors in major depressive disorder: relationship to anxiety disorder comorbidity and age of onset. *Psychother Psychosom*. 2005;74(3):173–178.
18. Nery FG, Hatch JP, Glahn DC, et al. Temperament and character traits in patients with bipolar disorder and associations with comorbid alcoholism or anxiety disorders. *J Psychiatr Res*. 2008;42(7):569–577.
19. Wittchen HU, Jacobi F, Rehm J, et al. The size and burden of mental disorders and other disorders of the brain in Europe 2010. *Eur Neuropsychopharmacol*. 2011;21(9):655–679.
20. Kapur S, Phillips AG, Insel TR. Why has it taken so long for biological psychiatry to develop clinical tests and what to do about it? *Mol Psychiatry*. 2012;17(12):1174–1179.
21. Lenzenweger MF. Endophenotype, intermediate phenotype, biomarker: definitions, concept comparisons, clarifications. *Depress Anxiety*. 2013;30(3):185–189.
22. Kose S. Psychobiological model of temperament and character: TCI. *Yeni Symp*. 2003;41(2):86–97.
23. LeDoux J. Emotion circuits in the brain. *Ann Reviews Neuroscience*. 2000;23:155–184.
24. Seeley WW, Menon V, Schatzberg AF, et al. Dissociable intrinsic connectivity networks for salience processing and executive control. *J Neurosci*. 2007;27(9):2349–2356.
25. Sylvester CM, Corbetta M, Raichle ME, et al. Functional network dysfunction in anxiety and anxiety disorders. *Trends Neurosci*. 2012;35(9):527–535.
26. Phelps EA, Delgado MR, Nearing KI, LeDoux JE. Extinction learning in humans: role of the amygdala and vmPFC. *Neuron*. 2004;43: 897–905.
27. Lang S, Kroll A, Lipinski SJ, et al. Context conditioning and extinction in humans: differential contribution of the hippocampus, amygdala and prefrontal cortex. *Eur J Neurosci*. 2009; 29:823–832.
28. Andrews-Hanna JR, Reidler JS, Huang C, Buckner RL. Evidence for the default network's role in spontaneous cognition. *J Neurophysiol*. 2010;104(1):322–335.
29. Raichle ME, MacLeod AM, Snyder AZ, Powers WJ, Gusnard DA, Shulman GL. A default mode of brain function. *Proc Natl Acad Sci USA*. 2001;98(2):676–682.
30. Montag C, Reuter M, Jurkiewicz M, Markett S, Panksepp J. Imaging the structure of the human anxious brain: a review of findings from neuroscientific personality psychology. *Rev Neurosci*. 2013;24(2):167–190.
31. Pujol J, López A, Deus J, Cardoner N, Vallejo J. Anatomical variability of the anterior cingulate gyrus and basic dimensions of human personality. *NeuroImage*. 2002;15:847–855.
32. Bush G, Luu P, Posner MI. Cognitive and emotional influences in anterior cingulate cortex. *Trends Cogn Sci*. 2000;4(6):215–222.
33. Kaasinen V, Maguire RP, Kurki T, Brück A, Rinne JO. Mapping brain structure and personality in late adulthood. *NeuroImage*. 2005;24(2):315–322.
34. Iidaka T, Matsumoto A, Ozaki N, et al. Volume of left amygdala subregion predicted temperamental trait of harm avoidance in female young subjects. A voxel-based morphometry study. *Brain Res*. 2006;1125(1):85–93.
35. Yamasue H, Abe O, Suga M, et al. Gender-common and -specific neuroanatomical basis of human anxiety-related personality traits. *Cereb Cortex*. 2008;18:46–52.
36. Sheline YI, Wang PW, Gado MH, Csernansky JG, Vannier MW. Hippocampal atrophy in recurrent major depression. *Proc Natl Acad Sci USA*. 1996;93(9):3908–3913.
37. Gardini S, Cloninger CR, Venneri A. Individual differences in personality traits reflect structural variance in specific brain regions. *Brain Res Bull*. 2009;79(5):265–270.
38. Van Schuerbeek P, Baeken C, De Raedt R, De Mey J, Luypaert R. Individual differences in local gray and white matter volumes reflect differences in temperament and character: a voxel-based morphometry study in healthy young females. *Brain Res*. 2011;1371: 32–42.
39. Laricchiuta D, Petrosini L, Piras F, et al. Linking novelty seeking and harm avoidance personality traits to cerebellar volumes. *Human Brain Mapp*. 2014;35(1):285–296.
40. Buckner RL. The cerebellum and cognitive function: 25 years of insight from anatomy and neuroimaging. *Neuron*. 2013;80(3): 807–815.
41. Westlye LT, Bjørnebekk A, Grydeland H, Fjell AM, Walhovd KB. Linking an anxiety-related personality trait to brain white matter microstructure: diffusion tensor imaging and harm avoidance. *Arch Gen Psychiatry*. 2011;68(4):369–377.

42. Smith SM, Jenkinson M, Johansen-Berg H, et al. Tract-based spatial statistics: voxelwise analysis of multi-subject diffusion data. *NeuroImage*. 2006;31(4):1487–1505.
43. Taddei M, Tettamanti M, Zanoni A, Cappa S, Battaglia M. Brain white matter organisation in adolescence is related to childhood cerebral responses to facial expressions and harm avoidance. *NeuroImage*. 2012;61(4):1394–1401.
44. Montag C, Reuter M, Weber B, Markett S, Schoene-Bake JC. Individual differences in trait anxiety are associated with white matter tract integrity in the left temporal lobe in healthy males but not females. *Neuroscience*. 2012;217(C):77–83.
45. Jones DK, Knosche TR, Turner R. White matter integrity, fiber count and other fallacies: the do's and don'ts of diffusion MRI. *NeuroImage*. 2013;73:239–254.
46. Lei X, Chen C, Xue F, et al. Fiber connectivity between the striatum and cortical and subcortical regions is associated with temperaments in Chinese males. *NeuroImage*. 2014;89:226–234.
47. Laricchiuta D, Petrosini L, Piras F, et al. Linking novelty seeking and harm avoidance personality traits to basal ganglia: volumetry and mean diffusivity. *Brain Struct Funct*. 2014;219(3):793–803.
48. Kaasinen V, Nurmi E, Bergman J, Solin O, Kurki T, Rinne JO. Personality traits and striatal 6-[^{18}F]fluoro-L-dopa uptake in healthy elderly subjects. *Neurosci Lett*. 2002;332(1):61–64.
49. Kim J-H, Son Y-D, Kim H-K, et al. Association of harm avoidance with dopamine D2/3 receptor availability in striatal subdivisions: a high resolution PET study. *Biol Psychol*. 2011;87(1):164–167.
50. Tuominen L, Salo J, Hirvonen J, et al. Temperament trait Harm Avoidance associates with μ-opioid receptor availability in frontal cortex: a PET study using [^{11}C]carfentanil. *NeuroImage*. 2012;61(3):670–676.
51. Fox MD, Raichle ME. Spontaneous fluctuations in brain activity observed with functional magnetic resonance imaging. *Nat Rev Neurosci*. 2007;8(9):700–711. http://dx.doi.org/10.1038/nrn2201.
52. Youn T, Lyoo IK, Kim J-K, et al. Relationship between personality trait and regional cerebral glucose metabolism assessed with positron emission tomography. *Biol Psychol*. 2002;60(2–3):109–120.
53. Hakamata Y, Iwase M, Iwata H, et al. Regional brain cerebral glucose metabolism and temperament: a positron emission tomography study. *Neurosci Lett*. 2006;396(1):33–37.
54. Nishio M, Matsuda H, Ozaki N, Inada T. Gender difference in relationship between anxiety-related personality traits and cerebral brain glucose metabolism. *Psychiatry Res*. 2009;173:206–211.
55. Turner RM, Hudson IL, Butler PH, Joyce PR. Brain function and personality in normal males: a SPECT study using statistical parametric mapping. *NeuroImage*. 2003;19(3):1145–1162.
56. Li Y, Qin W, Jiang T, Zhang Y, Yu C. Sex-dependent correlations between the personality dimension of harm avoidance and the resting-state functional connectivity of amygdala subregions. *PLoS One*. 2012;7(4):e35925.
57. Baeken C, Marinazzo D, Van Schuerbeek P, et al. Left and right amygdala – mediofrontal cortical functional connectivity is differentially modulated by harm avoidance. *PLoS One*. 2014;9(4):e95740.
58. Markett S, Weber B, Voigt G, et al. Intrinsic connectivity networks and personality: the temperament dimension harm avoidance moderates functional connectivity in the resting brain. *Neuroscience*. 2013;240(C):98–105.
59. Kyeong S, Kim E, Park H-J, Hwang D-U. Functional network organizations of two contrasting temperament groups in dimensions of novelty seeking and harm avoidance. *Brain Res*. 2014;1575:33–44.
60. Nakao T, Matsumoto T, Shimizu D, et al. Resting state low-frequency fluctuations in prefrontal cortex reflect degrees of harm avoidance and novelty seeking: an exploratory NIRS study. *Front Syst Neurosci*. 2013;7:115.
61. Paulus MP, Rogalsky C, Simmons A, Feinstein JS, Stein MB. Increased activation in the right insula during risk-taking decision making is related to harm avoidance and neuroticism. *NeuroImage*. 2003;19(4):1439–1448.
62. Most SB, Chun MM, Johnson MR, Kiehl KA. Attentional modulation of the amygdala varies with personality. *NeuroImage*. 2006; 31(2):934–944.
63. Yang TT, Simmons AN, Matthews SC, et al. Adolescent subgenual anterior cingulate activity is related to harm avoidance. *NeuroReport*. 2009;20(1):19–23.
64. Baeken C, De Raedt R, Ramsey N, et al. Amygdala responses to positively and negatively valenced baby faces in healthy female volunteers: influences of individual differences in harm avoidance. *Brain Res*. 2009;1296:94–103.
65. Van Schuerbeek P, Baeken C, Luypaert R, De Raedt R, De Mey J. Does the amygdala response correlate with the personality trait "harm avoidance" while evaluating emotional stimuli explicitly? *Behav Brain Funct*. 2014;10(18):1–13.
66. Ito M, Suto T, Uehara T, Ida I, Fukuda M, Mikuni M. Cerebral blood volume activation pattern as biological substrate of personality: multichannel near-infrared spectroscopy study in healthy subjects. *Int Congr Ser*. 2002;1232:71–75.
67. Morinaga K, Akiyoshi J, Matsushita H. Anticipatory anxiety-induced changes in human lateral prefrontal cortex activity. *Biol Psychol*. 2007;74(1):34–38.
68. Takizawa R, Nishimura Y, Yamasue H, Kasai K. Anxiety and performance: the disparate roles of prefrontal subregions under maintained psychological stress. *Cereb Cortex*. 2013;24(7):1858–1866.
69. Hansenne M. P300 and personality: an investigation with the Cloninger's model. *Biol Psychol*. 1999;50(2):143–155.
70. Hansenne M, Pinto E, Scantamburlo G, et al. Harm avoidance is related to mismatch negativity (MMN) amplitude in healthy subjects. *Pers Indiv Differ*. 2003;34(6):1039–1048.
71. Vedeniapin AB, Anokhin AP, Sirevaag E, Rohrbaugh JW, Cloninger CR. Visual P300 and the self-directedness scale of the Temperament and Character Inventory. *Psychiatry Res*. 2001; 101(2):145–156.
72. Kim M-S, Cho S-S, Kang K-W, Hwang J-L, Kwon JS. Electrophysiological correlates of personality dimensions measured by Temperament and Character Inventory. *Psychiatry Clin Neurosci*. 2002;56(6):631–635.
73. Mardaga S, Hansenne M. Personality modulation of P300 wave recorded within an emotional oddball protocol. *Clin Neurophysiol*. 2009;39(1):41–48.
74. Mardaga S, Hansenne M. Do personality traits modulate the effect of emotional visual stimuli on auditory information processing? *J Indiv Differ*. 2009;30(1):28.
75. Lang PJ, Bradley MM, Cuthbert BN. *International Affective Picture System (IAPS): Technical Manual and Affective Ratings*. Gainesville: National Institute of Mental Health Center for the Study of Emotion and Attention; 1999.
76. Zhang W, Lu J, Ni Z, Liu X, Wang D, Shen J. Harm avoidance in adolescents modulates late positive potentials during affective picture processing. *Int J Dev Neurosci*. 2013;31(5):297–302.
77. Takahashi T, Murata T, Hamada T, et al. Changes in EEG and autonomic nervous activity during meditation and their association with personality traits. *Int J Psychophysiol*. 2005;55(2):199–207.
78. Teper R, Inzlicht M. Meditation, mindfulness and executive control: the importance of emotional acceptance and brain-based performance monitoring. *Soc Cogn Affect Neurosci*. 2013;8(1):85–92.
79. Montag C, Reuter M. Disentangling the molecular genetic basis of personality: from monoamines to neuropeptides. *Neurosci Biobehav Rev*. 2014;43(C):228–239.
80. Chen Z-Y, Jing D, Bath KG, et al. Genetic variant BDNF (Val66Met) polymorphism alters anxiety-related behavior. *Science*. 2006;314(5796):140–143.
81. Montag C, Basten U, Stelzel C, Fiebach CJ, Reuter M. The BDNF Val66Met polymorphism and anxiety: support for animal knock-in studies from a genetic association study in humans. *Psychiatry Res*. 2010;179(1):86–90.

82. Montag C. The brain derived neurotrophic factor and personality. *Adv Biol*. 2014.
83. Montag C, Weber B, Fliessbach K, Elger C, Reuter M. The BDNF Val66Met polymorphism impacts parahippocampal and amygdala volume in healthy humans: incremental support for a genetic risk factor for depression. *Psychol Med*. 2009;39(11):1831.
84. Ross SA, Wong JYF, Clifford JJ, et al. Phenotypic characterization of an α4 neuronal nicotinic acetylcholine receptor subunit knock-out mouse. *J Neurosci*. 2000;20(17):6431–6441.
85. Markett S, Montag C, Reuter M. The nicotinic acetylcholine receptor gene CHRNA4 is associated with negative emotionality. *Emotion*. 2011;11(2):450–455.
86. Markett S, Reuter M, Montag C, Weber B. The dopamine D2 receptor gene DRD2 and the nicotinic acetylcholine receptor gene CHRNA4 interact on striatal gray matter volume: evidence from a genetic imaging study. *NeuroImage*. 2013;64(C):167–172.
87. Markett S, Montag C, Reuter M. In favor of behavior: on the importance of experimental paradigms in testing predictions from Gray's revised reinforcement sensitivity theory. *Front Syst Neurosci*. 2014;8(184):1–3.
88. Simon G, Ormel J. Health care costs associated with depressive and anxiety disorders in primary care. *Am J Psychiatry*. 1995;1:52.
89. Maldjian JA, Laurienti PJ, Kraft RA, Burdette JH. An automated method for neuroanatomic and cytoarchitectonic atlas-based interrogation of fMRI data sets. *NeuroImage*. 2003;19:1233–1239.
90. Xia M, Wang J, He Y. BrainNet Viewer: a network visualization tool for human brain connectomics. *PLoS One*. 2013;8:e68910.
91. Coninger CR, Przybeck, Skravic DM. The Tridimensional Personality Questionnaire: U.S. normative data. *Psychol Rep*. 1991;69(3): 1047–1057.
92. Cloninger CR. *The Temperament and Character Inventory (TCI): A Guide to Its Development and Use*. St. Louis: Washington University; 1994.
93. Beck AT, Ward CH, Mendelson M, Mock J, Erbaugh J. An inventory for measuring depression. *Arch Gen Psychiatry*. 1961;4(6):561–571.
94. Tzourio-Mazoyer LN, Landeau B, Papathanassiou D, et al. Automated Anatomical Labeling of activations in SPM using a Macroscopic Anatomical Parcellation of the MNI MRI single-subject brain. *Neuroimaging*. 2002;15(1):273–289.
95. Polich J. Updating P300: an integrative theory of P3a and P3b. *Clin Neurophysiol*. 2007;118(10):2128–2148.

CHAPTER

6

Impulsiveness and Inhibitory Mechanisms

Andrea Bari[1], *Tanja S. Kellermann*[2], *Bettina Studer*[3]

[1]Department of Brain and Cognitive Sciences, The Picower Institute for Learning and Memory, Massachusetts Institute of Technology, Cambridge, MA, USA; [2]Department of Neurosurgery, Medical University of South Carolina, Charleston, SC, USA; [3]Department of Neurology, St. Mauritius Therapieklinik, Meerbusch, Germany

OUTLINE

1. IMPULSIVE TRAITS AND PERSONALITY

1.1 Impulse Control and Its Failure

Impulsivity is a fundamental construct in several theories of personality and is linked to numerous psychopathological conditions.[1,2] Impulsive traits are moderately to highly heritable[3–7] and seem to be normally distributed in the general population, with some people that regularly plan their future behavior, while others consistently act on the spur of the moment with little or no regard for the consequences of their actions.[8,9] Although some degree of impulsivity is not considered pathological, excessive impulsive behavior can have serious negative repercussions on everyday life and constitutes a risk factor for a large number of life-threatening behaviors, addictions, and mental illnesses.[10–16] The fifth version of the *Diagnostic and Statistical Manual of Mental Disorders (DSM-V)*[17] includes impulsivity as a symptom for several psychopathological conditions, and a new chapter called *"disruptive, impulse-control and conduct disorders"* has been created to group together disorders characterized by impaired emotional and behavioral self-control. Individual differences in the domain of impulse control have long been studied in psychology and psychiatry, but only recently have we been able to correlate these differences with brain activity, morphology, and neurochemistry, thanks to increasingly sophisticated brain imaging tools and preclinical research.[18–23]

Impulsive behavior manifests itself when the fulfillment of strong urges, if inappropriate in a given context, is not inhibited by internal cognitive control mechanisms. Such control mechanisms (often called *executive functions*) rely in part on learned information and past experiences that help in determining what

Neuroimaging Personality, Social Cognition, and Character
http://dx.doi.org/10.1016/B978-0-12-800935-2.00006-3

is appropriate (and what is not) in specific settings in order to selectively inhibit actions that are likely to elicit unwanted consequences.[18] Thus, it is clear that the definition of impulsive behavior heavily depends on environmental and cultural variables. Although impulsive responding can provide some advantages in particular situations when a fast response is desirable,[24] it is by definition maladaptive in the majority of occasions. This is because, apart from morality and social conventions, impulsive acts are often characterized by spontaneous, unplanned (or poorly conceived) reactions to the triggering events at the expense of accuracy. Another class of behaviors that are considered impulsive and that do not require fast responses are represented by the excessive discounting of future rewards, preferring instead immediate but smaller ones, in situations where these two options are likely to be mutually exclusive.[25] The excitement for the immediate fulfillment of strong urges overshadows its negative consequences, as well as the prospective advantages of its inhibition, although both are well-known to the individual.

Since the criteria for the description of impulsive behavior change quite dramatically whether we put emphasis on the environment (internal or external) or on the "response style" of the individual, general consensus on a comprehensive definition of impulsivity (or impulsiveness) has been difficult to achieve. In this regard, it has been recognized for some time that impulsivity is a multifactorial (nonunitary) construct,[26] leading some authors to consider it as an artificial umbrella term.[27] Indeed, studies show that the term impulsivity is used to refer to a number of different traits that are only moderately related.[27–30] However, a first gross distinction can be made between "state" and "trait" impulsivity.[16,31] *State impulsivity* refers to some momentary changes affecting the internal state of the individual that somehow deviate from usual behavior and that influence actions and thoughts for a limited time. These changes can be brought about by the external environment (e.g., stressful or anxiogenic situations, psychotropic drug intoxication, or social interactions) or from within the organism itself (current needs, memories, and thoughts). *Trait impulsivity* is by definition considered a more stable characteristic of the individual that determines a specific style of conduct and which is more or less always present and permeates most of the individual's behavioral repertoire, if counteracting cognitive strategies are inefficient or not exploited at all. Moreover, impulsive traits are considered to be at the core of several psychiatric conditions such as attention deficit/hyperactivity disorder (ADHD),[32] drug addiction,[33] and schizophrenia.[34] Impulsive traits are of course also present to different degrees among the healthy population and have been found to be strong predictors of an unhealthy social life, as well as poor career achievements, problematic behaviors, and subclinical psychopathic features.[2,20,35–37] Such traits are probably phylogenetically conserved because they can be advantageous to the society as a whole by restricting the risks of pursuing potentially dangerous, though yet unexplored, activities to a subgroup of individuals.[38] In higher mammals, the ability to inhibit automatic reflexes, urges, and conditioned responses is likely to have evolved to allow slower cognitive processes to guide the behavior and is thought to involve frontostriatal brain circuitry,[39,40] along with other limbic and paralimbic brain structures.[41,42]

This chapter will focus mainly on trait impulsivity, the strategies that have been used to measure it in the laboratory and in the clinic, and its neurobiological underpinning both in clinical and healthy populations. However, since state impulsivity necessarily interacts with different baseline levels of impulsive traits in different individuals, it will be considered as well. Important research efforts have been directed towards the definition of brain areas, and more recently, brain circuits involved in different subtypes of impulsivity and their dysfunction in psychopathology. This line of research has the potential of bringing enormous advantages to the medical field in terms of prevention, diagnosis, and intervention.[43,44] On the other hand, in a hypothetical future where science will be able to reliably assess people's predispositions towards risky and inappropriate behavior, ethical issues will also have to be considered.

1.2 Impulsivity as a Personality Trait

In the writings of early philosophers and psychiatrists, impulsive individuals were described as dominated by their impulses and incapable of learning from the detrimental consequences of their own actions.[18] Impulsive acts were mostly viewed as morally deviant and destined to social condemnation. Hippocrates (460–377 BC) was perhaps the first to attempt a formal classification of personality traits, two of which (choleric and sanguine temperaments) are intimately associated with impulsivity. Similarly, several modern theorists and students of personality included impulsivity in their classifications of human behavior. One of the most proficient investigation of impulsivity as a personality trait came from Eysenck and Eysenck,[45] who used factor analysis to identify three orthogonal dimensions, all of which are related to impulsivity. According to these authors, individuals presenting elevated scores on extroversion, neuroticism, or psychoticism are more likely to engage in impulsive behaviors. The work of several other personality theorists, such as Ernest Barratt, Arnold Buss, Robert Plomin, Albert Bandura, Walter Mischel, and Jerome Kagan, have importantly contributed to the recognition of the primacy of impulsivity among personality constructs.[46]

The pioneering work of Barratt helped link impulsivity to specific aspects of behavioral performance during laboratory tasks in healthy individuals. He found that impulsive traits are associated with low levels of performance on perceptual-motor tasks and with difficulties in planning ahead.[47–49] Barratt's experiments also revealed that impulsivity and anxiety measures are orthogonal, and that subjects scoring high on impulsivity and low on anxiety have the worst performance on a wide range of laboratory tasks.[49] Other behavioral characteristics that Barratt's work found to be associated with high impulsive traits were fast cognitive tempo and delinquency, especially during adolescence.[50,51] In general, some of the most important contributions of Barratt to the study of impulsivity have been his attempts to systematically link impulsive traits to behavioral measures and the search for their neurobiological correlates.[52,53] Eysenck, in his original theory of extraversion,[54] hypothesized that impulsivity is the consequence of low cortical arousal and poor functioning of the reticular activating system.[55]

Nowadays, it is commonly assumed that personality traits are relatively stable over time and difficult to change. When impulsive behaviors are consistently expressed by individuals, they are likely linked to personality traits. Kagan and colleagues, for instance, have suggested that impulsivity is an enduring behavioral trait that can be readily identified in individuals from a very young age.[56] In fact, several models assume a strong genetic component for personality traits,[57–60] which are present already during childhood and are gradually shaped by the interaction with the environment. However, initial successful approaches aimed at decreasing impulsive and antisocial conduct[61,62] demonstrated that impulsive individuals may benefit from cognitive/behavioral training, especially early in life. Moreover, it has been recognized for some time that personality traits alone are not able to explain all the variance in individual conduct, and situational variables have to be considered as well.[63,64]

Personality research on impulsivity has flourished also due to the repeated observations that impulsive traits often have serious repercussions on health and are closely associated with psychopathology.[1,15] Several lines of research have demonstrated that impulsive traits increase the risk for aggressive behavior,[65] suicidal ideation,[66,67] pathological gambling,[68] and substance use disorders such as smoking[69] and alcohol abuse.[70] The accepted view is that impulsivity includes several different dimensions[26,71] mediated by dissociable, although possibly overlapping, neurobiological substrates.[18,23,72,73] Thus, although impulsivity appears in almost every system of personality and is considered as an important psychological construct,[8] it is not surprising to note the heterogeneity among its conceptualizations.

1.3 Measuring Impulsivity

A multitude of both self-report questionnaires and laboratory measures of impulsivity have been established in the literature. Self-report questionnaires have traditionally been the method of choice to assess individual differences in personality dispositions and continue to be popular in the field of impulsivity research. In such questionnaires, respondents indicate how they typically act in a described situation, and to what extent they endorse statements about their own personality/disposition. Early self-report measures of impulsivity include *the Barratt Impulsiveness Scale* (BIS),[74] *Dickman's Impulsivity Inventory*,[24] and the *Eysenck Impulsiveness Questionnaire*.[75] The BIS, currently in its 11th version,[76] is arguably the most commonly used self-report measure of impulsivity in the literature. The BIS-11 is a 30-item questionnaire and is designed to assess three separable dispositions: (1) *attentional impulsiveness*, defined as the (in-)ability to concentrate or focus attention, (2) *motor impulsiveness*, or the tendency to act without thinking, and (3) *nonplanning impulsiveness*, or the lack of future planning and forethought. Intercorrelations among these subscales range from rho = 0.39 to rho = 0.50.[77] Underlying these subscales are six first-order factors: *attention, motor, self-Control, cognitive complexity, perseverance, and cognitive instability*. The main advantages of the BIS-11 are the availability of normative data and of a large range of translations, the considerable amount of previous research using this scale in a variety of clinical populations, and its good psychometric properties.[77] One drawback is that despite the three-subscale structure of the BIS-11, the majority of studies using this scale have focused on the singular *total impulsiveness score*. This approach ignores the multidimensional nature of impulsivity, and thereby impedes the interpretation of individual differences.[78]

In recent years, the *UPPS*[27,79] and the extended *UPPS-P Impulsive Behavior Scale*[28] have also become established self-report measures of impulsivity. The *UPPS* is a 45-item questionnaire, which was developed based on a factor analysis of eight self-report measures of impulsivity, including the *BIS-11*, *Dickman's Impulsivity Inventory*, and others. The original questionnaire assesses four subdimensions of impulsivity (from which the acronym UPPS is derived): (1) *urgency*, defined by Whiteside and Lynam[27] as "the tendency to commit rash or regrettable actions as a result of intense negative affect" (p. 677), (2) *lack of premeditation*, (3) *lack of perseverance*, and (4) *sensation seeking*. Internal consistencies were high for all subscales, ranging from 0.82 to 0.91. The intercorrelation among the four subscales was 0.22 on average, with the strongest correlation being observed for *lack of premeditation* and *lack of perseverance* (r = 0.45). Cyders and colleagues[28] added a further subscale labeled *positive urgency* to the *UPPS*, defined as the "tendency to act rashly or maladaptively in response to positive mood" (p. 107), and relabeled the original *urgency* subscale to *negative urgency.* The resulting *UPPS-P* entails 59 items. Empirical data on the *UPPS-P* shows that *positive urgency* and *negative urgency* are related with an intercorrelation of 0.37[28] but have distinguishable external correlates. The *UPPS* and *UPPS-P* scales have been particularly well-received by researchers assessing impulsive behavior in clinical conditions such as substance disorders,[80,81] pathological gambling,[82,83] and ADHD.[84]

Self-report measures of impulsivity are widely used and have shown to discriminate well between impulsive behavior in the normal and clinical populations. However, self-report questionnaires do have weaknesses. They rely on the respondent having good insight into their personality, and self-reports are vulnerable to biases induced by social desirability and scale usage. These caveats are particularly problematic when assessing impulsive tendencies in clinical populations. Further, questionnaire measures can only be used in human subjects, but much can be learned about impulsivity and the neurobiological underpinnings of impulsive actions from pharmacological and behavioral research in nonhuman animals. One way to avoid these caveats is to use behavioral measures of impulsivity.

Laboratory measures of impulsivity are often based on cognitive models and aim to assess the cognitive processes underlying impulsive behavior.[85,86] Typically, these tests compare response times or response accuracy between two or more task conditions. Like for questionnaires, many different neurocognitive tests relating impulsivity have been reported in the human literature. Most of these measures can be assigned to one of the six following categories.[80,81,85] The first category comprises measures of *response inhibition*, which assesses subjects' ability to suppress prepotent (motor) responses. For instance, the *Stop signal task* (SST)[87] requires participants to make a fast response to a visual cue on the majority of trials but inhibit/stop this response on trials where the visual cue is unexpectedly followed by a "stop" signal. Other well-established tests in this category are the *go/no-go task*[88] and the *continuous performance task.*[89]

A second category of tasks assesses *resistance to distractor interference*, defined as the ability to suppress interference from task-irrelevant distractors in the external environment,[85] such as in the *Stroop task*[90] or in the *Eriksen flanker task.*[91] In the *Eriksen flanker task*, subjects are required to respond to a target, which is either presented alone or flanked by response-congruent or response-incongruent cues. Participants who have a low resistance to distractors are expected to show a large increase in response times in the incongruent condition.

A third category of tasks assesses *resistance to proactive interference*, defined as the ability to suppress the interference of information that was previously relevant but has become irrelevant.[85] An example of this category is the *Cued recall test*,[92] in which participants are presented with two lists of four words and are later required to recall the cue-associated answer from the second list, ignoring the (interfering) first list.

A fourth category of tests assesses *reflection impulsivity*, defined as the tendency to gather information before making a decision.[93] For instance, the *Information sampling task*[94] presents participants with a matrix of covered squares and asks them to determine whether the majority of these squares are blue or yellow by taking samples. In one condition, participants face a trade-off: taking samples is costly, but collecting sufficient information is necessary to find the correct (and rewarded) answer. Another established laboratory measure of reflection impulsivity is the *matching familiar figures test.*[93]

A fifth category entails *choice impulsivity* as measured by *Delay discounting tasks* (DDT), which estimate the degree to which rewards lose attractiveness with increasing waiting time. Typically, the DDT presents the participant with a choice between a small immediate reward and a larger but delayed reward. A preference for the immediate smaller reward is considered impulsive.[95]

Finally, impulsivity can also be assessed by means of *Time estimation tasks*, which test the judgment of temporal durations. For instance, in the *TIME* paradigm,[86] participants are required to indicate when a predefined time interval has passed, either by pressing a button at the end of the interval or by performing a continuous response for the entire interval. Impulsive subjects show a reduced accuracy in time estimation compared to nonimpulsive individuals.[96]

The heterogeneity between these different categories of tests again underlines the multifaceted nature of impulsivity. Accordingly, previous research has found only a small correlation between self-report and behavioral measures of impulsivity,[97–100] and many studies have failed to find a significant relationship altogether.[77,101,102] These findings suggest that there is little overlap between what is being assessed by behavioral tests versus self-report measures of impulsivity. There can be several explanations for this weak relationship. One possibility is that behavioral tasks offer situation-specific snapshots of behavior that might not necessarily reflect a person's general behavioral tendencies. In other words, behavioral tasks are often thought to assess *state* impulsivity, while self-report measures examine *trait* impulsivity.[80,98,103] Following this rationale, some authors have argued that a strong correlation between these different types of measures should not be expected.[80,98] One notable exception is the performance on the DDT, which has shown long-term stability.[104] Another potential explanation lies in the fact that many studies in the literature have compared total scores of impulsivity questionnaires to often highly specific behavioral tasks.

The increasing use of questionnaires that allow quantification of impulsivity-related subtraits, such as the UPPS-P, might reveal more significant relationships between behavioral and self-report measures in future research. For instance, Dick and colleagues recently hypothesized that self-report measures of *negative* and *positive urgency* should correlate with laboratory measures of *response inhibition*, while self-report scores for *lack of perseverance* should relate to *resistance to proactive interference* and *resistance to distractors interference*.[80] A small but significant relationship between *negative urgency* and *response inhibition* was confirmed by a meta-analysis conducted in 2011,[105,106] while empirical evidence for the other predictions is still warranted.

1.4 The Multidimensional Nature of Impulsivity

As described in the previous section, impulsivity has been conceptualized in multiple ways, and historically, there has been little consensus about the number and exact natures of the dispositions underlying impulsive behavior. Even early accounts describe impulsivity as a multidimensional construct,[107,108] and it is now widely acknowledged that multiple and separable psychological dimensions underlie impulsive behavior. Empirical support for the multidimensional nature of impulsivity has been provided by factor analytical studies.[10,27,78,107,109] These investigations administered multiple existing self-report measures of impulsivity to the same subjects and then analyzed the responses by exploratory factor analysis or principal component analysis (PCA) to identify the shared underlying impulsivity-related dimensions. For instance, Miller and colleagues[10] collected responses from 245 adults on three impulsivity questionnaires: the *Dickman's Impulsivity Inventory*,[24] the *Eysenck Impulsiveness Questionnaire*,[75] and the *BIS*,[76] alongside a self-report measure of reward/punishment sensitivity, the *BIS/BAS* scales,[110] and calculated PCA scores over the subscales of these measures. They found a three-component structure, with the components labeled *nonplanning/dysfunctional impulsive behavior, functional venturesomeness*, and *reward responsiveness and drive*. However, the authors also noted that *reward responsiveness and drive* might reflect a separate, largely unrelated personality construct rather than a subfacet of impulsivity. Another highly influential factor analytical study is the one by Whiteside and Lynam.[27] These researchers administered eight established impulsivity self-report questionnaires, together with the revised *NEO* (Neuroticism-Extraversion-Openness) *Personality Inventory*,[111] and 14 new impulsivity items to 437 undergraduate students. Factor analysis revealed four main underlying factors: *lack of planning, urgency, sensation seeking*, and *lack of perseverance*, which were related to the *Five-Factor Model* of personality.[112] Whiteside and Lynam then proceeded to design the *UPPS* based on these four factors. This four-dimensional construct was later confirmed by empirical data.[78] Smith and colleagues administered the *UPPS* to a sample of nearly 2000 undergraduates and compared the fit of the four-factor model to a model that defined impulsivity as a single, one-dimensional trait. The four-dimensional model fits the empirical data much better than the one-dimensional model supporting the multifaceted nature of the impulsivity construct. Finally, a recent study used PCA to assess the interrelation between behavioral and self-reported measures of impulsivity and found that *BIS-11, choice impulsivity* (DDT), and *response inhibition* (SST) loaded on independent factors.[113]

In addition to the results of factor analyses, the validity of a multidimensional model of impulsivity is also supported by the observation that different impulsivity-related traits have divergent external correlates. For instance, Smith et al. found that in their sample of undergraduates, (negative) *urgency* predicted problem gambling and problem drinking, whereas *sensation seeking* predicted the frequency of drinking and gambling but not whether these behaviors became problematic.[78] Meanwhile, *lack of perseverance* and *lack of planning* predicted poor school performance. Bayard et al.[114] reported that (negative) *urgency* and *sensation seeking*, but not *lack of perseverance* and *lack of planning*, were associated with poorer performance on a laboratory risky decision-making task.

Based on these findings, as well as theoretical considerations, a number of authors have urged researchers to abandon broad measures of impulsivity and instead use specific measures of the aforementioned one-dimensional subtraits.[78,80,115] Similarly, as discussed in the previous section, some of the most widely used laboratory tasks measure related but dissociable forms of response inhibition. Thus, impulsive performance on the DDT is generally linked to affective and reward-related aspects of impulsivity, whereas the SST and the go/no-go task are generally considered to tap more specifically on executive inhibitory mechanisms.[18] Moreover, the subtle distinction between *action restraint* and *action cancellation* (as measured by the SST and go/no-go tasks, respectively) supported a more precise definition of behavioral phenotypes related to impulsivity in both human and nonhuman subjects.[23,116]

2. NEURAL BASIS OF IMPULSIVITY

The neural underpinnings of impulsive behavior and impulsivity-related traits have been a major focus of recent neuroscience research. A multitude of neuropsychological, neuroimaging, and pharmacological studies have investigated the brain areas and networks involved in impulsive control and their dysfunction, allowing for a deeper understanding of the neural underpinnings of impulsiveness in both healthy individuals and clinical populations. In this section of the chapter, we will discuss seminal works and recent findings from research into the neural correlates of cognitive control and impulsivity-related subdimensions.

2.1 Brain Areas Involved in Impulse Control

Neuropsychological studies have provided evidence that several regions within the prefrontal cortex (PFC) play a crucial role in response control. For instance, Berlin and colleagues[117] compared the performance of patients with lesions of the orbitofrontal cortex (OFC), patients with damage to the dorsolateral PFC, and healthy controls on two laboratory measures of impulsivity, the *Matching familiar figures task*[93] and the *Time estimation task*. Patients with OFC lesions showed increased impulsivity compared to healthy control volunteers on both tasks. On the *Matching familiar figures task*, OFC lesion patients took less time to reflect on the correct answer and, as a consequence, made more errors than control participants. On the *Time estimation task*, OFC lesion patients systematically overestimated the passage of time, indicating a faster subjective sense of time. Finally, OFC lesion patients also achieved higher scores on the *BIS*, with significantly elevated scores on the subscales *motor impulsiveness* and *nonplanning impulsiveness*.

Lesions to the OFC were also found to be associated with impulsive responding on temporal discounting tasks, both in humans[118] and nonhuman subjects.[119,120] Sellitto and colleagues administered three temporal discounting tasks to patients with damage to the medial OFC, healthy controls, and a lesion control group with damage outside the frontal lobes. Across the three tasks, the type of reward (primary versus secondary) was varied.[118] Patients with medial OFC damage showed a stronger preference for immediate rewards than healthy volunteers and patients with nonfrontal lesions, and this held for both reward types. Furthermore, a significant correlation between the discounting rate and the lesion volume in the OFC was found. In striking contrast to this behavioral profile, OFC patients did not rate their impulsivity higher than healthy controls on a self-report instrument. Sellitto et al. reasoned that this null finding on the self-report questionnaire is likely to be the result of a lack of self-awareness in OFC lesion patients.[118]

Impulsive and inaccurate responding have sometimes also been observed following damage to the dorsolateral PFC. In the aforementioned study by Berlin et al., the performance of patients with dorsolateral PFC lesions on the Matching familiar figures task and the Time estimation task fell between those of OFC lesion patients and healthy controls. Thus, impulsivity was somewhat elevated in this group but not as strongly as in the patients with OFC damage. A more recent study by Figner and colleagues[121] investigated the role of the lateral PFC in temporal discounting by inducing a temporary disruption of neural activity in this area by means of repetitive transcranial magnetic stimulation (TMS). Disruption of the left, but not the right, lateral PFC increased choices of immediate rewards over delayed rewards. Intriguingly, valuation of delayed rewards per se was not affected. Based on these findings, it has been proposed that in the context of temporal discounting, the medial OFC is critically implicated in the valuation of immediate and delayed rewards,[118,122] while the lateral PFC is crucially involved in exerting self-control[118,121,123] and ensuring that the best available option is chosen.[124]

A third area that lesion studies implicate in impulsivity and response control is the right inferior frontal gyrus (IFG), situated in the ventrolateral PFC. In their seminal study, Aron and colleagues[125] assessed patients with damage to the right frontal lobe on the SST. Compared to healthy controls, these patients showed slower stop signal reaction time (SSRT), which measures the speed of inhibitory processes.[126] Lesion-behavior mapping revealed that this deficit in response inhibition was highly correlated with the lesion volume in the right IFG. These findings were later confirmed by research using TMS or direct electrical stimulation to temporarily disrupt the right IFG: again, IFG disruption resulted in longer SSRTs.[127–129] Furthermore, a recent study by Jacobson and colleagues[130] showed that upregulation of the activity of the right IFG by means of transcranial direct current stimulation significantly and selectively shortened SSRTs, that is to say improved response inhibition. Finally, lesion and TMS studies also indicate that the supplementary motor area (SMA),[131–133] pre-SMA,[134,135] and the subthalamic nucleus[136] are critically involved in the inhibition of motor responses. Transient disruption of neural activity in these areas is associated with longer SSRTs and poorer performance on the go/no-go task. Based on these results and related neuroimaging findings, current neurobiological models propose that the successful inhibition of prepotent motor responses is critically dependent upon the interaction between IFG and pre-SMA and is implemented via the subthalamic nucleus.[137]

2.2 Structural Abnormalities Related to Impulsivity

The neural underpinnings of impulsivity have also been investigated by means of structural neuroimaging studies. This line of research has identified several associations between impulsivity and brain morphology. Using structural magnetic resonance imaging (MRI), recent investigations in healthy individuals demonstrated that trait impulsivity is associated with gray matter (GM) volume in medial[138–142] and lateral[142] prefrontal regions. The structural correlate most often linked to self-reported impulsivity in these studies is the OFC. For instance, a voxel-based morphometry study by Matsuo and colleagues[138] found that total impulsivity scores on the BIS were inversely correlated with GM volume in both the left and right OFC, such that subjects with higher levels of impulsivity had smaller GM volumes. The same study also found differential structural correlates of impulsivity subdimensions as assessed by the BIS. Scores on the nonplanning impulsivity subscale were inversely correlated with GM volume in the right OFC, while scores on the motor impulsivity subscale were inversely correlated with GM volume in the left OFC.[138] Similarly, recent neuroimaging data from a large cohort of healthy adolescents shows a negative correlation between GM volume in the left OFC and scores on the impulsivity subscale of Cloninger's TCI-R.[140] Moreover, a study by Boes and colleagues in healthy boys confirmed the link between morphology of the right OFC and impulsivity. High impulsivity, as assessed by parents and teacher reports, was associated with smaller GM volume in the right OFC.[141] Finally, DeYoung et al. observed that GM volume in the medial OFC was linked to participants' self-reported extraversion, as assessed with the Big Five Model.[143] All of the aforementioned studies found a *negative* association between GM volume in the OFC and impulsivity; however, it should be noted that a *positive* relationship between OFC volume and impulsivity has also been reported.[142,144] In addition to the OFC, GM volumes in the anterior cingulate cortex (ACC) and left dorsolateral PFC have been found to correlate with total, nonplanning, and attentional impulsivity scores on the BIS,[138,142] though again the direction of the relationship between impulsivity scores and GM volume differed between the two studies.

These investigations of the structural correlates of self-reported impulsivity have been complemented by recent studies assessing morphological correlates of performance on behavioral measures assessing impulsivity.[142,145–147] Gray matter volume in the left dorsolateral PFC was positively correlated with participant's inhibition scores on the Stroop task.[145] Meanwhile, another study examined the anatomical correlates of delay discounting and found that GM volume in the OFC and striatum were predictive of temporal discounting in healthy individuals.[142] The connections between these regions are also thought to be of importance. Two recent diffusion tension imaging studies found that stronger delay discounting is associated with reduced integrity of fronto-striatal white matter tracts,[147] weaker connectivity between the striatum and the dorsolateral PFC,[146] and stronger connectivity between the striatum and the amygdala in both hemispheres.[146] In other words, increased striatal connectivity with the dorsolateral PFC was associated with more patient behavior on the DDT, while increased striatal connectivity with the limbic region was associated with higher choice impulsivity.[146]

In summary, structural neuroimaging studies in healthy individuals have consistently found a significant relationship between GM volume in the medial and lateral PFC and both self-reported and behaviorally assessed impulsivity. However, the direction of these effects varied across studies. This divergence in the results might be a consequence of subtle differences in the methodologies used by the individual studies, including the precise parceling of neuroanatomical subdivisions within the PFC. For instance, studies in rodents and primates demonstrate that the lateral and medial aspects of the OFC are involved in dissociable cognitive functions[120,148,149] and have distinct neuroanatomical connections.[150,151] The OFC has strong connections with several subcortical structures that are involved in the regulation of emotions.[152] White matter fiber tracts form reciprocal connections between the OFC and limbic structures such as the amygdala, which together play an important role in the representation of reward value.[153] Thus, a disruption within cortico-limbic circuitry may lead to reduced OFC modulation of subcortical activity, which results in increased impulsivity and reward sensitivity.

A substantial body of research has assessed structural properties and abnormalities related to impulsivity in clinical populations associated with impaired impulse control. In a large part, these studies point to the same anatomical substrates as the aforementioned investigations in healthy individuals, including the OFC and the surrounding ventromedial PFC,[154–156] ACC,[156] and striatum.[157] For instance, two studies assessing GM volume in cocaine users found that impulsivity in this population was linked to the volume of the right OFC[155] and left striatum.[157] Similarly, a recent diffusion tension imaging study in patients with ADHD found that fractional anisotropy—a marker of white matter fiber density and myelination—in the right PFC was negatively correlated with performance on the go/no-go task.[154] More specifically, lower white matter density in the PFC was associated with poorer impulse control.

Specific anatomical characteristics may well represent a vulnerability factor for the development of pathologies characterized by impulse control deficits. A recent line of research has been able to identify genetic polymorphisms associated with impulse control deficits that are also linked to brain morphological and developmental abnormalities found in impulsive individuals.[71,158–160] These important studies have made possible the definition of endophenotypes related to several psychiatric disorders characterized by impulsive behavior, thus facilitating the study, the diagnosis, and the treatment of these conditions.[161,162] Endophenotypes are heritable and measurable traits associated with an increased genetic risk of developing a certain disorder and are supposed to inform on the etiology and diagnosis of complex pathological conditions.[163]

2.3 Functional Correlates of Impulsive Behavior

Functional neuroimaging studies have investigated the patterns of brain activation at the level of single brain areas, as well as at the network level during laboratory tasks of impulsivity. While morphological anomalies may be the result of genetic, developmental, and environmental variables, the functional connectivity between different brain areas involved in impulse control is also modulated by current task demands. For instance, both hypo- and hyperactivation within key brain circuits may subtend impulse control deficits, where a hyperactivation is usually interpreted as an attempt to cope with a compromised system in order to achieve acceptable levels of performance.

Measures of delay discounting have been widely used in functional neuroimaging studies and have helped delineate the circuit involved in this type of impulsive behavior.[31] Generally, in healthy individuals, the choice between an immediate small reward and a delayed larger one correlates with brain activity in the ventral striatum, medial PFC, and posterior cingulate cortex.[164,165] These brain areas show increased activity at the time when subjects receive (or expect) a reward, and the magnitude of the activation is positively correlated to the magnitude, probability, and desirability of the reward.[166–170] However, what studies using the DDT have clearly shown is that neural activity in these areas faithfully matches the behavioral discounting curve as it decreases proportionally to the delay imposed to the reward, thus possibly representing the neural substrate of the subjective value of immediate, as well as delayed, rewards.[164]

The ventral striatum, the posterior cingulate cortex, and the orbital sector of the medial PFC belong to a network involved in computing the subjective value of different options.[171,172] On the other hand, activation in the lateral OFC has been linked to the excitement in pursuing risky activities.[173] This evidence suggests that some forms of impulsive behavior may arise from dysregulation in brain circuitry involved in assigning positive and negative values to actions and their consequences, thus leading to poor decision-making and "myopia for the future."[174] A recent study confirmed this view, but found that after adjusting for subjective value of the reward, dorsal sectors of the PFC were more active in trials when subjects chose the delayed reward option.[175] These results support the distinction between the two main subtypes of impulsive behavior. Namely, a dorsal brain network involved in higher-order executive functions responsible for the inhibition of impulsive responses and a ventral motivational circuitry involved in reward evaluation. Diminished activation of the first and increased activity in the second had been found in a sample of adolescent drug users during DDT performance.[176]

The neural substrates of response inhibition have also been extensively investigated by functional MRI studies that used the go/no-go paradigm and the SST. Brain structures more often linked to "motor" response inhibition have been the premotor cortex,[177] SMA and pre-SMA,[178,179] parietal cortex,[180,181] ventrolateral PFC, and the insula.[182,183] Among these areas, the right inferior frontal cortex, which comprises the (right) IFG, has been shown to exert a prominent role in the inhibition of prepotent responses.[184–186] In keeping with the above results, it has been reported that activation of the right IFG and the adjacent right insula during go/no-go performance is higher in subjects with impulsive traits,[41] possibly accounting for the need of additional cognitive resources in these individuals.

3. INHIBITORY DEFICITS, NEUROPATHOLOGY, AND NEURODEVELOPMENT

3.1 Impulse Control in Clinical Populations

For many psychiatric conditions, inhibitory deficits are likely to derive from more generalized impairments in cognitive control and may not be the most compelling feature of the disease. However, they are considered to be central in certain conditions, such as drug addiction and ADHD.[32,187] Deficits in response inhibition, for example, predict alcohol and drug abuse in humans[188,189] and in animal models of drug addiction.[190,191] Moreover, impulsive traits are expressed in the close relatives of ADHD individuals.[43,192] Research on patients diagnosed with ADHD, obsessive-compulsive disorder (OCD), schizophrenia, drug-addicted individuals, and their unaffected siblings suggests that deficits in response inhibition are potentially inheritable characteristics and, therefore, promising candidate endophenotypes for genetic investigation.[193–196]

3.1.1 *Attention Deficit/Hyperactivity Disorder*

Of the various forms of impulsivity, deficits in inhibitory processes, as assessed by response inhibition tasks, are among the most replicated findings in children with ADHD.[197–201] Impulsivity is so pervasive in the behavior of ADHD children that the father of British pediatrics, Sir George Frederick Still (1868–1941), described them as having little inhibitory volition.[202] Similarly, contemporary theorists recognize the centrality of poor behavioral inhibition in this disorder.[32,203–206] Children with ADHD not only display longer SSRTs,[200,207] but also more variable response times[197,198] compared to the general population. Inconsistent responding from trial to trial is probably the most replicated and stable finding in children with ADHD,[208] but it is not clear whether this deficit is more related to impulsivity or to inattentiveness.[209,210]

ADHD patients show abnormal activation in brain areas involved in response inhibition and cognitive control, such as the inferior frontal and cingulate cortices, as well as the striatal regions.[211–214] A recent meta-analysis has confirmed the association between impaired inhibition in ADHD and functional abnormalities in a right-lateralized fronto-basal ganglia network, plus functional deficits in dorsolateral, parietal, and cerebellar cortices linked to attentional deficits in these patients.[215] In addition, GM volume reductions in fronto-striatal areas, the cerebellum, the bilateral temporo-parietal cortex, and the left cingulate cortex have been reported in patients with ADHD compared to healthy controls.[216] Long-range brain connectivity has been found to be abnormal in ADHD individuals, especially in the default-mode network components.[217–219] In general, the aforementioned deficits are interpreted as a consequence of dysregulated modulation of cortical plasticity during development, which causes abnormal brain connectivity patterns that may persist into adulthood.[220]

Methylphenidate, a first-line ADHD treatment, which increases dopaminergic and noradrenergic modulation of corticostriatal circuitry, has been shown to normalize brain connectivity in ADHD patients both during inhibitory performance and at rest.[221,222] Moreover, both methylphenidate and amphetamine have been shown to modulate fronto-striatal circuitries differentially in ADHD compared to control individuals[223,224] (also during response inhibition).[225] The different modulation of brain areas between normal and ADHD individuals after drug administration is likely to reflect the underlying neurochemical and neuroanatomical abnormalities of the patients, who presumably need different degrees of brain activation when compared to control subjects during the same task.

3.1.2 *Obsessive-Compulsive Disorder*

OCD patients display a profound inability to inhibit intrusive thoughts and compulsive behaviors.[226] Neuropsychological investigations have highlighted these inhibitory deficits during tasks of response inhibition including go/no-go,[227] the anti-saccade task,[228] and the SST.[229] Moreover, an MRI study found a significant correlation between inhibitory deficits in OCD patients and reduced GM density in OFC and ventrolateral PFC; on the other hand, increased GM volume was found in cingulate and parietal cortices and striatum.[230] A more recent study found hyperactivation of the pre-SMA and impaired activation of the right IFC and the inferior parietal cortex in OCD patients engaged in the SST.[231] Similar to ADHD, these neurocognitive features of OCD were present also in unaffected first-degree relatives, thus supporting the potential use of deficient response inhibition as an endophenotype for OCD[232] and OCD-related traits, including impulsivity. In fact, comorbid impulse control disorders are common in OCD[233,234] and are positively linked to symptom severity and treatment outcomes.[235] However, in contrast to ADHD, OCD medications do not seem to improve response inhibition deficits.[236] Since the most commonly used treatment for OCD is represented by serotonergic drugs, these findings are consistent with the reported inefficacy of merely increasing extracellular serotonin (5-HT) levels on inhibitory performance in laboratory tasks.[237–240]

3.1.3 *Drug Abuse*

Impulsivity is often a main characteristic of drug abusers.[33,81,241–243] Impulsive behavior in addicted individuals is thought to arise from both drug use, causing maladaptive plasticity and neurotoxic effects on key brain areas, as well as from preexisting impulsive traits that predispose individuals to drug-taking. In practice, these mechanisms are difficult to disentangle because cognitive and intellectual abilities prior to drug use are not normally known. In their extensive review, Perry and Carroll[242] highlight the possible involvement of a third (presently unknown) factor, which may constitute a link between addiction and impulsivity (e.g., cognitive deficits, hormonal status, or reward sensitivity).

Chronic exposure to stimulant drugs is associated with morphological and functional abnormalities in a number of brain areas involved in impulsivity and reward sensitivity, impairing the ability of vulnerable subjects to refrain from using drugs and discounting negative future consequences of chronic drug abuse.[33,244] Because of the difficulty in delaying gratification, drug users fail to resist the temptation for immediate reward.[25] Moreover, altered reward sensitivity has been hypothesized to be the cause of initial and/or persistent drug use.[245,246] Consistently, both alcohol and tobacco use have been linked to heightened impulsivity on the DDT.[247,248] Specifically, subjects with more severe alcohol abuse show enhanced activation of SMA, insula, OFC, IFG, and precuneus during DDT performance compared to subjects that consume less alcohol.[249] This evidence is supported by findings on alcohol misusers, who have been found to score higher on impulsivity measures, such as novelty seeking and sensation seeking as compared to healthy controls.[250,251] High novelty-seeking traits have been found in adolescent smokers along with decreased activation of the ventral striatum during reward anticipation and impulsive performance on the DDT.[252] A recent study on adolescents with substance abuse disorder has linked impulsive performance on the DDT to enhanced activation of a reward valuation network, including the amygdala, hippocampus, insula, and ventromedial PFC.

On the other hand, the same study revealed a decrease in activity in brain areas related to executive attentional control, including dorsolateral and dorsomedial PFC, parietal and cingulate cortices, and the precuneus.[176]

A recent meta-analysis confirmed the impairments observed during response inhibition tasks in substance abusers and addicted individuals.[187] These inhibitory deficits seem to persist even after protracted periods of abstinence from the addictive substance, thus increasing the risk of relapse.[253] For example, abstinent cocaine users show impaired inhibitory ability in the SST, probably as a consequence of diminished performance monitoring.[254] Several studies have reported impaired response inhibition after amphetamine[255] or cocaine[256] administration in subjects with a drug-dependence history, which is consistent with the hypothesis of impairments in executive functioning being exacerbated by drug use.[257–259] Specifically, stimulant abuse has been consistently linked to abnormal activation of the cingulate cortex during response inhibition tasks.[260–262] The cingulate cortex is thought to subtend error processing, cognitive control, and conflict monitoring,[263] all functions that are impaired in stimulant users, making it difficult to quit drug use.[253,264] A recent study has shown that both inhibitory deficits and abnormal task-related cingulate activity may be remediated by methylphenidate administration in cocaine-dependent individuals.[265] On the other hand, decreased response inhibition in alcohol-dependent subjects has been associated with hyperactivity in the putamen and thalamus, and hypoactivation of the SMA during SST performance.[266] Interestingly, administration of the atypical stimulant modafinil to alcohol-dependent patients has been shown to ameliorate both behavioral and brain activation deficits.[267]

An important role for impulsivity in drug addiction has been recognized due to advancements in animal res earch.[14,16,20,242,244,268,269] It has been found, in both human and nonhuman subjects, that individuals displaying high levels of exploratory behavior, sensation/novelty seeking, and resistance to punishment are more likely to initiate drug use earlier in life and show a higher rate of drug abuse than the general population.[35,190,191,271–278] On the other hand, impulsive laboratory animals, characterized by high levels of premature responses on behavioral laboratory tasks, do not show increased heroin self-administration,[278] although they are more likely to develop cocaine dependence and show elevated novelty seeking, but not anxiety.[279] These preclinical results suggest that different preexisting personality traits may represent risk factors for the development of drug addictions to specific classes of psychotropic substances.

3.1.4 Pathological Gambling

While gambling is a widespread and socially accepted recreational activity, it can turn into a behavioral addiction in approximately 1–3% of the population.[280] Like drug users, regular gamblers and individuals with a gambling disorder indicate elevated impulsivity on self-report measures, including the *BIS*[281–285] and *UPPS*.[82,284,286] Three recent investigations using the *UPPS*[82] or *UPPS-P*[284,286] indicate that gamblers consistently report higher levels of *negative* and *positive urgency* than non-gambling controls, while *lack of perseverance* and *lack of premeditation* were less consistently elevated, and scores for *sensation seeking* were similar to non-gambling controls in all three studies. Studies assessing impulsivity with behavioral measures corroborate these findings: individuals with gambling disorder take longer to inhibit responses on the SST,[82,287] make more errors on the go/no-go task,[282] and show weaker performance on the Stroop task[285,287,288] than healthy non-gambling individuals. Impairments in time estimation[287] and a strong preference for immediate over delayed rewards on DDT (see Ref. 289 for a review) have also been observed. Perhaps unsurprisingly, gamblers also show abnormalities in the form of increased risk-taking on laboratory gambling tasks,[290–293] such as the Iowa gambling task[294] and the game of dice task.[295] In these tasks, selecting highly rewarded options despite increased probability for negative outcomes is considered impulsive, and generally, studies in healthy volunteers found that risk-taking in these tasks does covary with self-reported impulsivity.[114,296–298] Intriguingly, similar to drug addiction, longitudinal studies suggest that impulsivity acts as a vulnerability factor for the development of gambling disorder[299–302] and is associated with a poor prognosis for treatment success.[303]

The neural correlates of gambling disorder have been studied with both functional MRI and positron emission topography (PET). Much of this research has focused on the neuromodulator dopamine and the brain reward circuitry. A consistent finding in the extant functional imaging literature on gambling disorder is that responses in the midbrain, striatum, and OFC during the performance of laboratory gambling tasks differ between pathological/problem gamblers and healthy non-gambling controls.[304–307] However, the direction of this effect measured by blood-oxygen-level dependent (BOLD) responses is inconsistent across studies, with some studies observing increased[306,308] and others decreased[304,306] activations in gamblers. Nonetheless, PET studies found that dopamine release in the ventral striatum during the Iowa gambling task is associated with higher levels of excitement but poorer decision-making performance in pathological gamblers.[308,309]

Some studies have specifically investigated the link between increased impulsivity and brain function in gambling disorder. For example, Potenza and colleagues assessed BOLD responses during performance of the Stroop task and found a reduced activity in the left ventromedial PFC in pathological gamblers compared to healthy controls.[310] Similarly, de Ruiter and colleagues observed reduced activation of the dorsomedial PFC during the SST in problem gamblers compared to healthy controls.[311] In contrast, using a go/no-go task, van Holst et al. found increased activation of the dorsolateral PFC and the ACC during inhibition in gamblers.[312] Van Holst and colleagues argued that this increased activation might reflect compensatory recruitment of control-related brain areas. Finally, a recent PET study[83] found that in patients with gambling disorder, dopamine D2/D3 receptor availability in the striatum was negatively correlated with self-report scores on the *negative* and *positive urgency* subscale of the UPPS-P.

In summary, neuroimaging studies suggest that impulsive decision-making in patients with gambling disorder might arise as a consequence of dopaminergic dysfunction, and poor inhibition in these patients is likely to be related to hypoactivation of prefrontal regions. Further studies are warranted in order to reconcile between inconsistent findings in the extant literature and provide a more in-depth insight into the neurobiological correlates of increased impulsivity in gambling disorders.

3.1.5 *Schizophrenia*

Schizophrenia is characterized by frontal dysfunction.[313] Neurocognitive studies suggest an array of impairments in schizophrenia, in particular concerning frontal/executive abilities.[314,315] Schizophrenia is strongly associated with impulsivity, and patients present abnormal brain activation, as well as impaired performance during DDT and response inhibition tasks.[316–320] Frontal brain areas known to have an important role in response inhibition[128] demonstrate morphological abnormalities in men with schizophrenia.[321,322] Impulsivity in schizophrenia is thought to result from a dysfunction of fronto-temporo-limbic circuitry.[321,323] The same network shows decreased activation when schizophrenics are engaged in response inhibition tasks, and the level of activation is positively correlated with patients' impulsive traits as measured by the BIS.[318,324]

Other brain areas also display dysfunctional activity during response inhibition tasks in patients with schizophrenia,[320,325] including the cingulate cortex,[326,327] the dorsolateral PFC,[328] the caudate nucleus, and the thalamus.[329] Finally, abnormal striatal activation during SST performance was found in both schizophrenic patients and their healthy relatives, suggesting a possible neurobiological and genetic marker for the risk of developing schizophrenia.[194] In fact, individuals with a genetic predisposition to schizophrenia also display signs of inhibitory control failure.[330]

Impulsivity in schizophrenia is generally associated with worse clinical outcomes and constitutes a risk factor for the development of violence, drug addiction, and suicidal behavior.[331,332] Relative to ADHD, fewer studies have investigated inhibitory deficits in schizophrenic patients, with some studies reporting intact[326] and some others impaired[319,333,334] response inhibition. Other studies found more specific impairments in inhibitory capacity assessed by the SST, such as deficient proactive and spared reactive inhibition[335] or impaired conscious but intact unconscious inhibitory processing.[336] Another study found only decreased stop accuracy but normal SSRT,[337] suggesting slower, but still efficient, inhibitory processes. Patients with schizophrenia were also slower than control subjects in inhibiting saccadic responses, but exhibited intact error monitoring.[338] The inconsistency in the behavioral results from different studies is not entirely surprising, given the marked heterogeneity in the behavioral manifestations of this complex pathology. However, a recent meta-analysis partly confirmed the presence of inhibitory deficit in schizophrenia as revealed by longer SSRTs,[339] and the SST has been chosen as part of a battery of tests of executive control in schizophrenic patients in the CNTRICS (Cognitive Neuroscience Treatment Research to Improve Cognition in Schizophrenia) initiative for the study of this condition.[34,340] Poor inhibitory control may also explain the high rate of comorbid substance abuse associated with schizophrenia.[341] Finally, cognitive inhibition, in the form of resistance to distractors, is also impaired and linked to intrusive thoughts in patients with schizophrenia.[342]

3.2 Neurochemistry of Impulsivity

Historically, the neuromodulators 5-HT and dopamine have been linked to impulsive behavior.[343–348] Moreover, the interaction between these two neurotransmitter systems has been noted to determine specific patterns of impulsivity in humans and other animals.[349,350] For instance, genetic polymorphisms regulating the function of both neurotransmitters have additive influence on impulsive traits in humans.[351] In general, genetic markers of 5-HT neurotransmission have been linked to aggressive, violent, and suicidal behaviors,[354–360] whereas genes involved in dopamine neurotransmission have been often related to addictive personality, sensation seeking, and inhibitory executive functions.[72,359–361]

Acute tryptophan depletion in human subjects, which transiently reduces 5-HT brain levels, has been shown to increase impulsivity and aggression,[362,363] whereas pharmacologically increasing 5-HT neurotransmission decreases impulsivity as measured by the DDT.[364] One study investigated the effects of transient 5-HT depletion during a go/no-go task and found a significant modulation of right ventral PFC during response inhibition, although no behavioral effects were detected.[365] Clinical and preclinical neuropharmacology studies confirm the findings above with several reports of 5-HT involvement in the discounting of delayed or probabilistic rewards and go/no-go performance, although with some inconsistent results, especially across DDT studies.[72] In general, it seems that what has been called "waiting" impulsivity is more susceptible to 5-HT manipulation.[23] Subjects carrying the short allele variant of the promoter for the 5-HT transporter (5-HTTLPR), which are expected to have decreased 5-HT transporter expression,[366,367] have low persistency and are more sensitive to immediate reward,[368] suggesting impaired cognitive control, especially for emotional stimuli.[371] On the other hand, there have been mixed results regarding the association between the 5-HT transporter polymorphism and neuroticism (which includes both impulsive and depression traits) as measured by the NEO P-R personality inventory.[370–372]

In line with the results from preclinical models,[191,373] a recent PET study on healthy volunteers found a significant association between impulsive personality traits and dopamine synthesis capacity in the ventral, but not dorsal, striatum.[374] Brain markers of dopaminergic dysfunction have also been consistently associated with ADHD and aggressive behavior.[361,375] Moreover, a host of data has established a link between striatal D2/D3 receptor availability and both drug addiction and impulsivity.[20,21,376] Methylphenidate, which increases extracellular levels of both dopamine and norepinephrine improves measures of response inhibition in children[377–379] and adults with ADHD.[380] On the contrary, drugs specifically affecting the serotonergic system have no effects on stopping performance in ADHD individuals[238] and healthy subjects.[381] However, some anti-ADHD medications improve inhibitory and attentional performance also in normal subjects, although these improvements sometimes depend on task difficulty or sample composition.[381–384] For instance, it was found that in subjects carrying a DRD2 polymorphism associated with a slow SSRT, amphetamine speeded stopping performance, while it had the opposite effect on subjects with a different polymorphism associated with fast SSRTs under placebo conditions.[385]

Atomoxetine administration, which preferentially increases prefrontal extracellular levels of norepinephrine, has been shown to modulate activation of the right IFG during response inhibition in the SST[386] and to speed SSRT.[381] These findings have been replicated in animal models, where both systemically and intra-PFC administered atomoxetine improve stopping performance in a rodent version of the SST.[239,387] Thus, both human and nonhuman studies provide converging evidence that noradrenergic transmission is particularly involved in modulating inhibitory executive functions, consistent with its role in behavioral flexibility and decision-making.[388,389]

Preclinical research has significantly advanced our understanding of the role of dopamine neurotransmission in impulsive behavior. High impulsive rats display lower levels of the D2/3 subtype of dopamine receptors in the ventral striatum and acquire more readily cocaine self-administration, compared to low impulsive subjects.[191] Moreover, recent studies have shown that blocking D2 receptors in the dorsal striatum negatively affects SST performance[390] but normalizes impulsive behavior when infused into the nucleus accumbens.[391] On the other hand, systemic or intra-PFC D2 receptor antagonism does not affect impulsive responding in laboratory animals.[387,392,393] Taken together, these studies highlight the role of dopamine D2 receptors located in the striatum as a critical link between impulsivity and drug addiction,[394] possibly depending on the genetic background of the subjects. It is likely that preclinical research in this field will continue to provide important contributions toward a better definition of the pathological neural substrates in diseases characterized by impulsivity and/or the propensity to develop compulsive behavior and addictions.[20]

3.3 Impulse Control During Development

While impulsivity in the adult individual is usually assessed by self-report questionnaire measurements, rating scales completed by parents, teachers, and caregivers are often used to measure impulsivity in children.[395] However, similar to adults, behavioral tasks are also widely used in children and adolescents. A good correspondence between parent ratings and behavioral performance has been found in two-year-old children,[396] although reliable behavioral measures have been obtained even in younger infants.[397] High levels of impulsivity in preschoolers have been linked to the development of delinquency later in life.[11] Neurodevelopmental origins of impulsive traits have been studied in terms of the interaction between parenting behavior and preexisting temperament features,[398] which are mainly determined by the genetic background. However, impulsive behavior in the developing individual promotes beneficial experimentation with novel adaptive behavioral strategies and may also increase one's popularity among peers.[399,400] Apart from genetic or social determinants of impulsive behavior, prenatal exposure to toxic substances or stress has been linked to abnormalities of the dopaminergic system, the hypothalamic-pituitary-adrenal axis, and to impulsivity during adolescence.[398,401,402] Moreover, a recent study found that preadolescents from families with a history of drug use, which are at heightened risk of developing substance abuse later in life, display elevated impulsivity on the DDT.[403] These results highlight the importance of reliable measures of impulsivity in the young population that may be used for the prediction of future risky behavior.

Adolescence is a critical time in life when behavioral tendencies and personality are established. Particularly, adolescence is characterized by neurobiological transitional episodes and is characterized by impulsive decision-making, sensation seeking, risky behavior and increased emotional reactivity.[404,405] These impulsivity-related behavioral traits are considered to represent risk factors for substance experimentation and other dangerous behaviors in early adolescence.[406] Studies found significant correlations between impulsivity, substance experimentation in adolescence, and later substance abuse during adulthood.[405,407] Drug use during adolescence has a long-lasting neurotoxic impact on brain development, as animal studies have shown, specifically on functions of the frontal lobes,[408] including memory and impulse control.[409–412] Risk-taking behavior and impulsivity in adolescence partly result from an incomplete development of brain regions involved in top-down control and inhibitory processes,[413] in particular the PFC.[414,415] The frontal lobes undergo structural changes over time,[416] a process in which GM decreases as a result of synaptic pruning and myelination, and that results in improved brain connectivity.[417] More precisely, GM volume achieves a peak in early adolescence and declines throughout late adolescence and early adulthood.[413,416] In turn, white matter volume increases as a result of myelination, which leads to increased functional specialization during adulthood.[418,419] It is believed that these neurodevelopmental processes are the reason for the decrease in impulsive behavior throughout adolescence.[420,421] One example of a group of clinical conditions where the brain does not follow normal development is autism spectrum disorders (ASD). ASD are characterized by an early acceleration of brain growth and high levels of impulsive and compulsive behavior.[422,423] Studies assume a possible intermediate phenotype of abnormal timing of brain development[424] or even alternative developmental pathways[425] in ASD. Larger caudate volume in ASD, for instance, predicts increased impulsivity and repetitive behavior,[426,427] whereas increased left OFC and IFG activity has been linked to motor impulsivity in adults with ASD.[428]

The late maturation of PFC networks is thought to underlie the gradual improvement with age in response inhibition performance and risk-taking tendencies, as well as the progressively increasing activation of brain areas involved in these behaviors.[429,430] In particular, frontal-striatal-thalamic and frontal-cerebellar networks display age-related increased activation during response inhibition tasks.[431,432] Neuroimaging studies have found decreased activation of OFC, ventrolateral PFC, and ACC during a reward-related decision-making task in adolescents compared to adults, which correlated with increased risk-taking behavior.[433] OFC function has been reported to be reduced in adolescents with early nicotine or alcohol misuse,[434] which supports the view that abnormal OFC development is implicated in decreased impulse control and drug-seeking behavior during adolescence.[435] Also error-monitoring networks involving the ACC show decreased activation in children during response inhibition tasks.[436] While both children and adults disengage the ACC during correct inhibition, only adults display a robust activation of the ACC during inhibitory errors.[439] Finally, a recent study found that high impulsivity traits in 8–12 year-old typically developing children, as measured by a parent-report questionnaire, is related to lower resting state brain activity in two regions of the default mode network, namely the posterior cingulate cortex and the right angular gyrus.[438]

In summary, studies support the view that during development, the balance between earlier-maturing limbic system networks, important for emotion and reward processing, and later-maturing frontal lobe networks, important for cognitive control, is disturbed in impulsive individuals.[79,439,440] The late development of PFC areas involved in response inhibition and impulse control may leave the individual under the influence of a hypersensitive reward system, leading to excessive sensation seeking.[440–442] Understanding the interplay between limbic and frontal regions is important for advancing our knowledge of decision-making deficits and cognitive/impulse control during brain development.[443]

4. CONCLUDING REMARKS

The research on impulsivity and related personality traits reviewed above clearly links impulsive behavior with a variety of brain pathological states and negative consequences in everyday life. The wide range of behavioral manifestations, brain areas, and neurotransmitters involved in impulsive behavior warrants a better definition of the impulsivity construct and its numerous subdimensions. Although several efforts have been directed toward this goal, a clear taxonomical classification of impulsivity is still lacking.[18] This is particularly important since different subtypes of impulsivity seem to be related to specific pathological states and neurobiological substrates. A shared classification of impulsivity subtypes may greatly enhance collaborative and translational research, leading to better diagnosis, treatment, and prevention of several psychiatric conditions and criminal tendencies.

Advances in the sensitivity of neuroimaging techniques are revealing more detailed subdivisions among brain networks involved in specific aspects of impulse control and reward-related behavior. These more-refined techniques, together with the powerful genetic and optical tools developed for preclinical research,[444–446] will hopefully allow a better understanding of the impulsivity construct in the near future. However, this can happen only through a refinement of the tools currently used to assess trait impulsivity, especially behavioral tasks and taxonomic classifications.[447]

References

1. Moeller FG, Barratt ES, Dougherty DM, Schmitz JM, Swann AC. Psychiatric aspects of impulsivity. *Am J Psychiatry*. November 2001;158(11):1783–1793.
2. Gorenstein EE, Newman JP. Disinhibitory psychopathology: a new perspective and a model for research. *Psychol Rev*. May 1980;87(3):301–315.
3. Hicks BM, Krueger RF, Iacono WG, McGue M, Patrick CJ. Family transmission and heritability of externalizing disorders: a twin-family study. *Arch Gen Psychiatry*. September 2004;61(9):922–928.
4. Rhee SH, Waldman ID. Genetic and environmental influences on antisocial behavior: a meta-analysis of twin and adoption studies. *Psychol Bull*. May 2002;128(3):490–529.
5. Eaves L, Heath A, Martin N, et al. Comparing the biological and cultural inheritance of personality and social attitudes in the Virginia 30,000 study of twins and their relatives. *Twin Res*. June 1999;2(2):62–80.
6. Crosbie J, Arnold P, Paterson A, et al. Response inhibition and ADHD traits: correlates and heritability in a community sample. *J Abnorm Child Psychol*. April 2013;41(3):497–507.
7. Coccaro EF, Bergeman CS, McClearn GE. Heritability of irritable impulsiveness: a study of twins reared together and apart. *Psychiatry Res*. September 1993;48(3):229–242.
8. DeYoung CG. Impulsivity as a personality trait. In: Vohs KD, Baumeister RF, eds. *Handbook of Self-Regulation: Research, Theory, and Applications*. 2nd ed. New York: Guilford Press; 2010:485–502.
9. Spinella M. Normative data and a short form of the Barratt Impulsiveness Scale. *Int J Neurosci*. March 2007;117(3):359–368.
10. Miller E, Joseph S, Tudway J. Assessing the component structure of four self-report measures of impulsivity. *Pers Individ Differ*. July 2004;37(2):349–358.
11. Kagan J, Zentner M. Early childhood predictors of adult psychopathology. *Harv Rev Psychiatry*. March–April 1996;3(6):341–350.
12. Ivanov I, Schulz KP, London ED, Newcorn JH. Inhibitory control deficits in childhood and risk for substance use disorders: a review. *Am J Drug Alcohol Abuse*. 2008;34(3):239–258.
13. Carroll ME, Anker JJ, Perry JL. Modeling risk factors for nicotine and other drug abuse in the preclinical laboratory. *Drug Alcohol Depend*. October 1, 2009;104(suppl 1):S70–S78.
14. Everitt BJ, Belin D, Economidou D, Pelloux Y, Dalley JW, Robbins TW. Review. Neural mechanisms underlying the vulnerability to develop compulsive drug-seeking habits and addiction. *Philos Trans R Soc Lond B Biol Sci*. October 12, 2008;363(1507):3125–3135.
15. Moeller FG, Dougherty DM, Barratt ES, Schmitz JM, Swann AC, Grabowski J. The impact of impulsivity on cocaine use and retention in treatment. *J Subst Abuse Treat*. December 2001;21(4):193–198.
16. Bari A, Robbins TW, Dalley JW. Impulsivity. In: Olmstead MC, ed. *Animal Models of Drug Addiction*. Totawa: Humana Press; 2011.
17. American Psychiatric Association. *Diagnostic and Statistical Manual of Mental Disorders*. 5th ed. Washington: American Psychiatric Association; 2013.
18. Bari A, Robbins TW. Inhibition and impulsivity: behavioral and neural basis of response control. *Prog Neurobiol*. September 2013;108:44–79.
19. Jupp B, Caprioli D, Dalley JW. Highly impulsive rats: modelling an endophenotype to determine the neurobiological, genetic and environmental mechanisms of addiction. *Dis Models Mech*. March 2013;6(2):302–311.
20. Dalley JW, Everitt BJ, Robbins TW. Impulsivity, compulsivity, and top-down cognitive control. *Neuron*. February 24, 2011;69(4):680–694.
21. Dalley JW, Mar AC, Economidou D, Robbins TW. Neurobehavioral mechanisms of impulsivity: fronto-striatal systems and functional neurochemistry. *Pharmacol Biochem Behav*. August 2008;90(2):250–260.
22. Winstanley CA. The utility of rat models of impulsivity in developing pharmacotherapies for impulse control disorders. *Br J Pharmacol*. October 2011;164(4):1301–1321.
23. Eagle DM, Bari A, Robbins TW. The neuropsychopharmacology of action inhibition: cross-species translation of the stop-signal and go/no-go tasks. *Psychopharmacology*. August 2008;199(3):439–456.
24. Dickman SJ. Functional and dysfunctional impulsivity: personality and cognitive correlates. *J Pers Soc Psychol*. January 1990;58(1):95–102.
25. Green L, Myerson J. A discounting framework for choice with delayed and probabilistic rewards. *Psychol Bull*. September 2004;130(5):769–792.
26. Evenden JL. Varieties of impulsivity. *Psychopharmacology*. October 1999;146(4):348–361.
27. Whiteside SP, Lynam DR. The five factor model and impulsivity: using a structural model of personality to understand impulsivity. *Pers Individ Differ*. 2001;30:669–689.
28. Cyders MA, Smith GT, Spillane NS, Fischer S, Annus AM, Peterson C. Integration of impulsivity and positive mood to predict risky behavior: development and validation of a measure of positive urgency. *Psychol Assess*. March 2007;19(1):107–118.
29. Depue RA, Collins PF. Neurobiology of the structure of personality: dopamine, facilitation of incentive motivation, and extraversion. *Behav Brain Sci*. June 1999;22(3):491–517; discussion 518–569.
30. Zuckerman M, Kuhlman D, Joireman J, Teta P, Kraft M. A comparison of the three structural models for personality: the big three, the big five, and the alternative five. *J Pers Soc Psychol*. 1993;65:757–768.
31. Peters J, Buchel C. The neural mechanisms of inter-temporal decision-making: understanding variability. *Trends Cogn Sci*. May 2011;15(5):227–239.
32. Nigg JT. Is ADHD a disinhibitory disorder? *Psychol Bull*. September 2001;127(5):571–598.
33. Jentsch JD, Taylor JR. Impulsivity resulting from frontostriatal dysfunction in drug abuse: implications for the control of behavior by reward-related stimuli. *Psychopharmacology*. October 1999;146(4):373–390.
34. Gilmour G, Arguello A, Bari A, et al. Measuring the construct of executive control in schizophrenia: defining and validating translational animal paradigms for discovery research. *Neurosci Biobehav Rev*. November 2013;37(9 Pt B):2125–2140.
35. Sher KJ, Trull TJ. Personality and disinhibitory psychopathology: alcoholism and antisocial personality disorder. *J Abnorm Psychol*. February 1994;103(1):92–102.
36. Beauchaine TP, Gatzke-Kopp LM. Instantiating the multiple levels of analysis perspective in a program of study on externalizing behavior. *Dev Psychopathol*. August 2012;24(3):1003–1018.
37. Mischel W, Shoda Y, Rodriguez MI. Delay of gratification in children. *Science*. May 26, 1989;244(4907):933–938.
38. Williams J, Taylor E. The evolution of hyperactivity, impulsivity and cognitive diversity. *J R Soc Interface/R Soc*. June 22, 2006;3(8):399–413.
39. Robbins TW. Dissociating executive functions of the prefrontal cortex. *Philosophical Trans R Soc Lond B Biol Sci*. October 29, 1996;351(1346):1463–1470; discussion 1470–1471.
40. Aron AR. The neural basis of inhibition in cognitive control. *Neuroscientist*. June 2007;13(3):214–228.
41. Horn NR, Dolan M, Elliott R, Deakin JF, Woodruff PW. Response inhibition and impulsivity: an fMRI study. *Neuropsychologia*. 2003;41(14):1959–1966.
42. Mirabella G. Should I stay or should I go? Conceptual underpinnings of goal-directed actions. *Front Syst Neurosci*. 2014;8:206.
43. Chamberlain SR, Sahakian BJ. The neuropsychiatry of impulsivity. *Curr Opin Psychiatry*. May 2007;20(3):255–261.

44. Mitchell MR, Potenza MN. Addictions and personality traits: impulsivity and related constructs. *Curr Behav Neurosci Rep.* March 1, 2014;1(1):1–12.
45. Eysenck HJ, Eysenck M. *Personality and Individual Differences.* New York: Plenum Press; 1985.
46. Mccown WG, DeSimone PA. Impulses, impulsivity, and impulsive behaviors: a historical review of a contemporary issue. In: Mccown WG, Johnson JL, Shure MB, eds. *The Impulsive Client.* Washington: American Psychological Association; 1993.
47. Barratt ES. Intra-individual variability of performance: ANS and psychometric correlates. *Tex Rep Biol Med.* 1963;21:496–504.
48. Barratt ES, Patton J, Olsson NG, Zuker G. Impulsivity and paced tapping. *J Mot Behav.* December 1981;13(4):286–300.
49. Barratt ES. Perceptual-motor performance related to impulsiveness and anxiety. *Percept Mot Ski.* October 1967;25(2):485–492.
50. Stanford MS, Barratt ES. Impulsivity and the multi-impulsive personality disorder. *Pers Individ Differ.* 1992;13:831–834.
51. Barratt ES. Time perception, cortical evoked potentials, and impulsiveness among three groups of adolescents. In: Hays JR, Solway KS, eds. *Violence and the Violent Individual.* New York: SP Medical and Scientific Books; 1981:87–96.
52. Creson DL, Barratt ES, Russell GV, Schlagenhauf GK. The effect of lithium chloride on amygdala-frontal cortex evoked potentials. *Tex Rep Biol Med.* Fall 1967;25(3):374–379.
53. Barratt E. Anxiety and impulsiveness: toward a neuropsychological model. In: Speielberger CD, ed. *Anxiety: Current Trends in Theory and Research.* Vol. 1. New York: Academic Press; 1972:195–222.
54. Eysenck HJ, Eysenck SB. On the unitary nature of extraversion. *Acta Psychol.* November 1967;26(4):383–390.
55. Strelau J, Eysenck HJ. *Personality Dimensions and Arousal.* New York: Plenum Press; 1987.
56. Kagan J, Lapidus D, Moore M. Infant antecedents of cognitive functioning: a longitudinal study. *Annu Prog Child Psychiatry Child Dev.* 1979:46–77.
57. Buss A, Plomin RA. *Temperament Theory of Personlaity Development.* London: Wiley-Interscience; 1975.
58. Eaves L, Eysenck HJ, Martin M. *Genes, Culture and Personality: An Empirical Approach.* New York: Academic Press; 1989.
59. Anokhin AP, Golosheykin S, Grant JD, Heath AC. Heritability of delay discounting in adolescence: a longitudinal twin study. *Behav Genet.* March 2011;41(2):175–183.
60. Eisenberg DT, Mackillop J, Modi M, et al. Examining impulsivity as an endophenotype using a behavioral approach: a DRD2 TaqI A and DRD4 48-bp VNTR association study. *Behav Brain Funct.* 2007;3:2.
61. Spivack G, Shure MB. Interpersonal cognitive problem solving (ICPS): a competence-building primary presentation program. *Prev Hum Serv.* 1989;6:151–178.
62. Suedfeld P. *Restricted Environmental Stimulation.* New York: Wiley; 1980.
63. Magnusson D, ed. *Toward a Psychology of Situations: An Interactional Perspective.* Hillsdale: Erlbaum; 1981.
64. Mischel W. *Personality and Assessment.* New York: Wiley; 1968.
65. Nelson RJ, Trainor BC. Neural mechanisms of aggression. *Nat Rev Neurosci.* July 2007;8(7):536–546.
66. Mann JJ, Arango VA, Avenevoli S, et al. Candidate endophenotypes for genetic studies of suicidal behavior. *Biol Psychiatry.* April 1, 2009;65(7):556–563.
67. Iancu I, Bodner E, Roitman S, Piccone Sapir A, Poreh A, Kotler M. Impulsivity, aggression and suicide risk among male schizophrenia patients. *Psychopathology.* 2010;43(4):223–229.
68. Goudriaan AE, Yucel M, van Holst RJ. Getting a grip on problem gambling: what can neuroscience tell us? *Front Behav Neurosci.* 2014;8:141.
69. Krishnan-Sarin S, Reynolds B, Duhig AM, et al. Behavioral impulsivity predicts treatment outcome in a smoking cessation program for adolescent smokers. *Drug Alcohol Depend.* April 17, 2007;88(1):79–82.
70. Stautz K, Cooper A. Impulsivity-related personality traits and adolescent alcohol use: a meta-analytic review. *Clin Psychol Rev.* June 2013;33(4):574–592.
71. Congdon E, Canli T. A neurogenetic approach to impulsivity. *J Pers.* December 2008;76(6):1447–1484.
72. Dalley JW, Roiser JP. Dopamine, serotonin and impulsivity. *Neuroscience.* July 26, 2012;215:42–58.
73. Mitchell MR, Potenza MN. Recent insights into the neurobiology of impulsivity. *Curr Addict Rep.* December 1, 2014;1(4):309–319.
74. Barratt ES. Anxiety and impulsiveness related to psychomotor efficiency. *Percept Mot Ski.* June 1, 1959;9(3):191–198.
75. Eysenck SBG, Pearson PR, Easting G, Allsopp JF. Age norms for impulsiveness, venturesomeness and empathy in adults. *Pers Individ Differ.* 1985;6(5):613–619.
76. Patton JH, Stanford MS, Barratt ES. Factor structure of the barratt impulsiveness scale. *J Clin Psychol.* 1995;51(6):768–774.
77. Stanford MS, Mathias CW, Dougherty DM, Lake SL, Anderson NE, Patton JH. Fifty years of the Barratt Impulsiveness Scale: an update and review. *Pers Individ Differ.* 2009;47(5):385–395.
78. Smith GT, Fischer S, Cyders MA, Annus AM, Spillane NS, McCarthy DM. On the validity and utility of discriminating among impulsivity-like traits. *Assessment.* June 1, 2007;14(2):155–170.
79. Casey BJ, Duhoux S, Malter Cohen M. Adolescence: what do transmission, transition, and translation have to do with it? *Neuron.* September 9, 2010;67(5):749–760.
80. Dick DM, Smith G, Olausson P, et al. Review: understanding the construct of impulsivity and its relationship to alcohol use disorders. *Addict Biol.* April 2010;15(2):217–226.
81. Verdejo-García A, Lawrence AJ, Clark L. Impulsivity as a vulnerability marker for substance-use disorders: review of findings from high-risk research, problem gamblers and genetic association studies. *Neurosci Biobehav Rev.* 2008;32(4):777–810.
82. Billieux J, Lagrange G, Van der Linden M, Lancon C, Adida M, Jeanningros R. Investigation of impulsivity in a sample of treatment-seeking pathological gamblers: a multidimensional perspective. *Psychiatry Res.* July 30, 2012;198(2):291–296.
83. Clark L, Stokes PR, Wu K, et al. Striatal dopamine D(2)/D(3) receptor binding in pathological gambling is correlated with mood-related impulsivity. *NeuroImage.* October 15, 2012;63(1):40–46.
84. Miller DJ, Derefinko KJ, Lynam DR, Milich R, Fillmore MT. Impulsivity and attention deficit-hyperactivity disorder: subtype classification using the UPPS impulsive behavior scale. *J Psychopathol Behav Assess.* September 2010;32(3):323–332.
85. Friedman NP, Miyake A. The relations among inhibition and interference control functions: a latent-variable analysis. *J Exp Psychol General.* March 2004;133(1):101–135.
86. Dougherty DM, Mathias CW, Marsh DM, Jagar AA. Laboratory behavioral measures of impulsivity. *Behav Res Methods.* February 1, 2005;37(1):82–90.
87. Logan GD. On the ability to inhibit simple thoughts and actions: I. Stop-signal studies of decision and memory. *J Exp Psychol Learn Mem Cognition.* 1983;9(4):585–606.
88. Newman JP, Widom CS, Nathan S. Passive avoidance in syndromes of disinhibition: psychopathy and extraversion. *J Pers Soc Psychol.* May 1985;48(5):1316–1327.
89. Rosvold HE, Mirsky AF, Sarason I, Bransome Jr ED, Beck LH. A continuous performance test of brain damage. *J Consult Psychol.* 1956;20(5):343–350.
90. Stroop JR. Studies of interference in serial verbal reactions. *J Exp Psychol.* 1935;18(6):643–662.
91. Eriksen B, Eriksen C. Effects of noise letters upon the identification of a target letter in a nonsearch task. *Percept Psychophys.* January 1, 1974;16(1):143–149.
92. Tolan GA, Tehan G. Determinants of short-term forgetting: decay, retroactive interference, or proactive interference? *Int J Psychol.* 1999;34(5–6):285–292.

93. Kagan J. Reflection–impulsivity: the generality and dynamics of conceptual tempo. *J Abnorm Psychol*. February 1966;71(1):17–24.
94. Clark L, Robbins TW, Ersche KD, Sahakian BJ. Reflection impulsivity in current and former substance users. *Biol Psychiatry*. September 1, 2006;60(5):515–522.
95. Bickel WK, Marsch LA. Toward a behavioral economic understanding of drug dependence: delay discounting processes. *Addiction*. January 2001;96(1):73–86.
96. Reynolds B, Richards JB, Dassinger M, de Wit H. Therapeutic doses of diazepam do not alter impulsive behavior in humans. *Pharmacol Biochem Behav*. September 2004;79(1):17–24.
97. Kirby KN, Petry NM. Heroin and cocaine abusers have higher discount rates for delayed rewards than alcoholics or non-drug-using controls. *Addiction*. 2004;99(4):461–471.
98. Cyders MA, Coskunpinar A. Measurement of constructs using self-report and behavioral lab tasks: is there overlap in nomothetic span and construct representation for impulsivity? *Clin Psychol Rev*. August 2011;31(6):965–982.
99. Aichert DS, Wöstmann NM, Costa A, et al. Associations between trait impulsivity and prepotent response inhibition. *J Clin Exp Neuropsychol*. December 1, 2012;34(10):1016–1032.
100. Gomide Vasconcelos A, Sergeant J, Corrêa H, Mattos P, Malloy-Diniz L. When self-report diverges from performance: the usage of BIS-11 along with neuropsychological tests. *Psychiatry Res*. 2014;218(1):236–243.
101. Coffey SF, Gudleski GD, Saladin ME, Brady KT. Impulsivity and rapid discounting of delayed hypothetical rewards in cocaine-dependent individuals. *Exp Clin Psychopharmacol*. 2003; 11(1):18–25.
102. Verdejo-García A, Lozano Ó, Moya M, Alcázar MÁ, Pérez-García M. Psychometric properties of a Spanish version of the UPPS–P impulsive behavior scale: reliability, validity and association with trait and cognitive impulsivity. *J Pers Assess*. January 1, 2010;92(1): 70–77.
103. Reynolds B, Richards JB, de Wit H. Acute-alcohol effects on the Experiential Discounting Task (EDT) and a question-based measure of delay discounting. *Pharmacol Biochem Behav*. February 2006;83(2):194–202.
104. Kirby KN. One-year temporal stability of delay-discount rates. *Psychon Bull Rev*. June 2009;16(3):457–462.
105. Cyders MA, Zapolski TCB, Combs JL, Settles RF, Fillmore MT, Smith GT. Experimental effect of positive urgency on negative outcomes from risk taking and on increased alcohol consumption. *Psychol Addict Behav*. 2010;24(3):367–375.
106. Wilbertz T, Deserno L, Horstmann A, et al. Response inhibition and its relation to multidimensional impulsivity. *NeuroImage*. December 2014;103(0):241–248.
107. Barratt ES. Factor analysis of some psychometric measures of impulsiveness and anxiety. *Psychol Rep*. April 1, 1965;16(2): 547–554.
108. Twain DC. Factor analysis of a particular aspect of behavioral control: impulsivity. *J Clin Psychol*. 1957;13(2):133–136.
109. Gerbing DW, Ahadi SA, Patton JH. Toward a conceptualization of impulsivity: components across the behavioral and self-report domains. *Multivar Behav Res*. July 1, 1987;22(3):357–379.
110. Carver CS, White TL. Behavioral inhibition, behavioral activation, and affective responses to impending reward and punishment: the BIS/BAS Scales. *J Pers Soc Psychol*. 1994;67(2):319–333.
111. Costa PT, McCrae RR. *NEO PI-R Professional Manual*. Odessa: Psychological Assessment Resources; 1992.
112. McCrae Jr RR, Coasta PT. *Personality in Adulthood*; 1990. New York.
113. Broos N, Schmaal L, Wiskerke J, et al. The relationship between impulsive choice and impulsive action: a cross-species translational study. *PLoS One*. 2012;7(5):e36781.
114. Bayard S, Raffard S, Gely-Nargeot M-C. Do facets of self-reported impulsivity predict decision-making under ambiguity and risk? Evidence from a community sample. *Psychiatry Res*. 2011.
115. Cyders MA, Smith GT. Emotion-based dispositions to rash action: positive and negative urgency. *Psychol Bull*. 2008;134(6):807–828.
116. Schachar R, Logan GD, Robaey P, Chen S, Ickowicz A, Barr C. Restraint and cancellation: multiple inhibition deficits in attention deficit hyperactivity disorder. *J Abnorm Child Psychol*. April 2007;35(2):229–238.
117. Berlin HA, Rolls ET, Kischka U. Impulsivity, time perception, emotion and reinforcement sensitivity in patients with orbitofrontal cortex lesions. *Brain*. May 1, 2004;127(5):1108–1126.
118. Sellitto M, Ciaramelli E, di Pellegrino G. Myopic discounting of future rewards after medial orbitofrontal damage in humans. *J Neurosci*. December 8, 2010;30(49):16429–16436.
119. Rudebeck PH, Walton ME, Smyth AN, Bannerman DM, Rushworth MFS. Separate neural pathways process different decision costs. *Nat Neurosci*. September 2006;9(9):1161–1168.
120. Mar AC, Walker ALJ, Theobald DE, Eagle DM, Robbins TW. Dissociable effects of lesions to orbitofrontal cortex subregions on impulsive choice in the rat. *J Neurosci*. April 27, 2011;31(17):6398–6404.
121. Figner B, Knoch D, Johnson EJ, et al. Lateral prefrontal cortex and self-control in intertemporal choice. *Nat Neurosci*. May 2010;13(5):538–539.
122. Peters J, Buchel C. Neural representations of subjective reward value. *Behav Brain Res*. December 1, 2010;213(2):135–141.
123. Hare TA, Camerer CF, Rangel A. Self-control in decision-making involves modulation of the vmPFC valuation system. *Science*. May 1, 2009;324(5927):646–648.
124. Dixon ML, Christoff K. The lateral prefrontal cortex and complex value-based learning and decision making. *Neurosci Biobehav Rev*. September 2014;45(0):9–18.
125. Aron AR, Fletcher PC, Bullmore ET, Sahakian BJ, Robbins TW. Stop-signal inhibition disrupted by damage to right inferior frontal gyrus in humans. *Nat Neurosci*. February 2003;6(2):115–116.
126. Logan GD, Cowan WB, Davis KA. On the ability to inhibit simple and choice reaction time responses: a model and a method. *J Exp Psychol Hum Percept Perform*. April 1984;10(2):276–291.
127. Chambers CD, Bellgrove MA, Gould IC, et al. *Dissociable Mechanisms of Cognitive Control in Prefrontal and Premotor Cortex*. Vol. 98; 2007.
128. Chambers CD, Bellgrove MA, Stokes MG, et al. Executive "brake failure" following deactivation of human frontal lobe. *J Cogn Neurosci*. March 2006;18(3):444–455.
129. Wessel JR, Conner CR, Aron AR, Tandon N. Chronometric electrical stimulation of right inferior frontal cortex increases motor braking. *J Neurosci*. December 11, 2013;33(50):19611–19619.
130. Jacobson L, Javitt DC, Lavidor M. Activation of inhibition: diminishing impulsive behavior by direct current stimulation over the inferior frontal gyrus. *J Cogn Neurosci*. November 1, 2011;23(11):3380–3387.
131. Floden D, Stuss D. Inhibitory control is slowed in patients with right superior medial frontal damage. *J Cogn Neurosci*. 2006;18(11):1843–1849.
132. Décary A, Richer F. Response selection deficits in frontal excisions. *Neuropsychologia*. October 1995;33(10):1243–1253.
133. Drewe EA. Go - no go learning after frontal lobe lesions in humans. *Cortex*. March 1975;11(1):8–16.
134. Chen C-Y, Muggleton NG, Tzeng OJL, Hung DL, Juan C-H. Control of prepotent responses by the superior medial frontal cortex. *NeuroImage*. January 15, 2009;44(2):537–545.
135. Nachev P, Wydell H, O'Neill K, Husain M, Kennard C. The role of the pre-supplementary motor area in the control of action. *NeuroImage*. 2007;36(suppl 2):T155–T163.
136. Obeso I, Wilkinson L, Casabona E, et al. The subthalamic nucleus and inhibitory control: impact of subthalamotomy in Parkinson's disease. *Brain*. May 2014;137(Pt 5):1470–1480.
137. Aron AR. From reactive to proactive and selective control: developing a richer model for stopping inappropriate responses. *Biol Psychiatry*. June 15, 2011;69(12):e55–68.

138. Matsuo K, Nicoletti M, Nemoto K, et al. A voxel-based morphometry study of frontal gray matter correlates of impulsivity. *Hum Brain Mapp*. April 2009;30(4):1188–1195.
139. Schilling C, Kuhn S, Paus T, et al. Cortical thickness of superior frontal cortex predicts impulsiveness and perceptual reasoning in adolescence. *Mol Psychiatry*. May 2013;18(5):624–630.
140. Schilling C, Kuhn S, Romanowski A, et al. Common structural correlates of trait impulsiveness and perceptual reasoning in adolescence. *Hum Brain Mapp*. February 2013;34(2):374–383.
141. Boes AD, Bechara A, Tranel D, Anderson SW, Richman L, Nopoulos P. Right ventromedial prefrontal cortex: a neuroanatomical correlate of impulse control in boys. *Soc Cogn Affect Neurosci*. March 2009;4(1):1–9.
142. Cho SS, Pellecchia G, Aminian K, et al. Morphometric correlation of impulsivity in medial prefrontal cortex. *Brain Topogr*. July 2013;26(3):479–487.
143. DeYoung CG, Hirsh JB, Shane MS, Papademetris X, Rajeevan N, Gray JR. Testing predictions from personality neuroscience. Brain structure and the big five. *Psychol Sci*. June 2010;21(6):820–828.
144. Lee TY, Kim SN, Jang JH, et al. Neural correlate of impulsivity in subjects at ultra-high risk for psychosis. *Prog Neuropsychopharmacol Biol Psychiatry*. August 1, 2013;45:165–169.
145. Moreno-López L, Soriano-Mas C, Delgado-Rico E, Rio-Valle JS, Verdejo-García A. Brain structural correlates of reward sensitivity and impulsivity in adolescents with normal and excess weight. *PLoS One*. 2012;7(11):e49185.
146. van den Bos W, Rodriguez CA, Schweitzer JB, McClure SM. Connectivity strength of dissociable striatal tracts predict individual differences in temporal discounting. *J Neurosci*. July 30, 2014;34(31):10298–10310.
147. Peper JS, Mandl RCW, Braams BR, et al. Delay discounting and frontostriatal fiber tracts: a combined DTI and MTR study on impulsive choices in healthy young adults. *Cereb Cortex*. July 1, 2013;23(7):1695–1702.
148. Noonan MP, Walton ME, Behrens TE, Sallet J, Buckley MJ, Rushworth MF. Separate value comparison and learning mechanisms in macaque medial and lateral orbitofrontal cortex. *Proc Natl Acad Sci USA*. November 23, 2010;107(47):20547–20552.
149. Kringelbach ML, Rolls ET. The functional neuroanatomy of the human orbitofrontal cortex: evidence from neuroimaging and neuropsychology. *Prog Neurobiol*. April 2004;72(5):341–372.
150. Ferry AT, Ongur D, An X, Price JL. Prefrontal cortical projections to the striatum in macaque monkeys: evidence for an organization related to prefrontal networks. *J Comp Neurol*. September 25, 2000;425(3):447–470.
151. Kondo H, Witter MP. Topographic organization of orbitofrontal projections to the parahippocampal region in rats. *J Comp Neurol*. March 2014;522(4):772–793.
152. Kellermann TS, Caspers S, Fox PT, et al. Task- and resting-state functional connectivity of brain regions related to affection and susceptible to concurrent cognitive demand. *NeuroImage*. May 15, 2013;72:69–82.
153. Rolls ET. *The Brain and Emotion*. Oxford: Oxford University Press; 1999.
154. Casey BJ, Epstein JN, Buhle J, et al. Frontostriatal connectivity and its role in cognitive control in parent-child dyads with ADHD. *Am J Psychiatry*. November 2007;164(11):1729–1736.
155. Crunelle CL, Kaag AM, van Wingen G, et al. Reduced frontal brain volume in non-treatment-seeking cocaine-dependent individuals: exploring the role of impulsivity, depression, and smoking. *Front Hum Neurosci*. 2014;8(7).
156. Hazlett EA, New AS, Newmark R, et al. Reduced anterior and posterior cingulate gray matter in borderline personality disorder. *Biol Psychiatry*. October 15, 2005;58(8):614–623.
157. Ersche KD, Barnes A, Jones PS, Morein-Zamir S, Robbins TW, Bullmore ET. Abnormal structure of frontostriatal brain systems is associated with aspects of impulsivity and compulsivity in cocaine dependence. *Brain*. July 2011;134(Pt 7):2013–2024.
158. Sweitzer MM, Donny EC, Hariri AR. Imaging genetics and the neurobiological basis of individual differences in vulnerability to addiction. *Drug Alcohol Depend*. June 2012;123(suppl 1):S59–S71.
159. Munafo MR, Yalcin B, Willis-Owen SA, Flint J. Association of the dopamine D4 receptor (DRD4) gene and approach-related personality traits: meta-analysis and new data. *Biol Psychiatry*. January 15, 2008;63(2):197–206.
160. Nomura M, Nomura Y. Psychological, neuroimaging, and biochemical studies on functional association between impulsive behavior and the 5-HT2A receptor gene polymorphism in humans. *Ann N Y Acad Sci*. November 2006;1086:134–143.
161. Pezawas L, Meyer-Lindenberg A, Drabant EM, et al. 5-HTTLPR polymorphism impacts human cingulate-amygdala interactions: a genetic susceptibility mechanism for depression. *Nat Neurosci*. June 2005;8(6):828–834.
162. Meyer-Lindenberg A, Buckholtz JW, Kolachana B, et al. Neural mechanisms of genetic risk for impulsivity and violence in humans. *Proc Natl Acad Sci USA*. April 18, 2006;103(16):6269–6274.
163. Gottesman II , Gould TD. The endophenotype concept in psychiatry: etymology and strategic intentions. *Am J Psychiatry*. April 2003;160(4):636–645.
164. Kable JW, Glimcher PW. The neural correlates of subjective value during intertemporal choice. *Nat Neurosci*. December 2007;10(12):1625–1633.
165. McClure SM, Laibson DI, Loewenstein G, Cohen JD. Separate neural systems value immediate and delayed monetary rewards. *Science*. October 15, 2004;306(5695):503–507.
166. Studer B, Clark L. Place your bets: psychophysiological correlates of decision-making under risk. *Cogn Affect Behav Neurosci*. June 2011;11(2):144–158.
167. Abler B, Walter H, Erk S, Kammerer H, Spitzer M. Prediction error as a linear function of reward probability is coded in human nucleus accumbens. *NeuroImage*. June 2006;31(2):790–795.
168. Tanaka SC, Doya K, Okada G, Ueda K, Okamoto Y, Yamawaki S. Prediction of immediate and future rewards differentially recruits cortico-basal ganglia loops. *Nat Neurosci*. August 2004;7(8):887–893.
169. McCoy AN, Platt ML. Risk-sensitive neurons in macaque posterior cingulate cortex. *Nat Neurosci*. September 2005;8(9):1220–1227.
170. Breiter HC, Aharon I, Kahneman D, Dale A, Shizgal P. Functional imaging of neural responses to expectancy and experience of monetary gains and losses. *Neuron*. May 2001;30(2):619–639.
171. Padoa-Schioppa C, Cai X. The orbitofrontal cortex and the computation of subjective value: consolidated concepts and new perspectives. *Ann N Y Acad Sci*. December 2011;1239:130–137.
172. Padoa-Schioppa C, Assad JA. Neurons in the orbitofrontal cortex encode economic value. *Nature*. May 11, 2006;441(7090):223–226.
173. Elliott R, Dolan RJ, Frith CD. Dissociable functions in the medial and lateral orbitofrontal cortex: evidence from human neuroimaging studies. *Cereb Cortex*. March 2000;10(3):308–317.
174. Bechara A, Tranel D, Damasio H. Characterization of the decision-making deficit of patients with ventromedial prefrontal cortex lesions. *Brain*. November 2000;123(Pt 11):2189–2202.
175. Hare TA, Hakimi S, Rangel A. Activity in dlPFC and its effective connectivity to vmPFC are associated with temporal discounting. *Front Neurosci*. 2014;8:50.
176. Stanger C, Elton A, Ryan SR, James GA, Budney AJ, Kilts CD. Neuroeconomics and adolescent substance abuse: individual differences in neural networks and delay discounting. *J Am Acad Child Adolesc Psychiatry*. July 2013;52(7):747–755. e746.
177. Watanabe J, Sugiura M, Sato K, et al. The human prefrontal and parietal association cortices are involved in NO-GO performances: an event-related fMRI study. *NeuroImage*. November 2002;17(3):1207–1216.
178. Simmonds DJ, Pekar JJ, Mostofsky SH. Meta-analysis of Go/No-go tasks demonstrating that fMRI activation associated with response inhibition is task-dependent. *Neuropsychologia*. January 15, 2008;46(1):224–232.

179. Mostofsky SH, Simmonds DJ. Response inhibition and response selection: two sides of the same coin. *J Cogn Neurosci*. May 2008;20(5):751–761.
180. Menon V, Adleman NE, White CD, Glover GH, Reiss AL. Error-related brain activation during a Go/NoGo response inhibition task. *Hum Brain Mapp*. March 2001;12(3):131–143.
181. Rubia K, Russell T, Overmeyer S, et al. Mapping motor inhibition: conjunctive brain activations across different versions of go/no-go and stop tasks. *NeuroImage*. February 2001;13(2):250–261.
182. Swick D, Ashley V, Turken AU. Left inferior frontal gyrus is critical for response inhibition. *BMC Neurosci*. 2008;9(102).
183. Boehler CN, Appelbaum LG, Krebs RM, Hopf JM, Woldorff MG. Pinning down response inhibition in the brain–conjunction analyses of the Stop-signal task. *NeuroImage*. October 1, 2010; 52(4):1621–1632.
184. Aron AR, Robbins TW, Poldrack RA. Inhibition and the right inferior frontal cortex: one decade on. *Trends Cogn Sci*. April 2014;18(4):177–185.
185. Aron AR, Robbins TW, Poldrack RA. Inhibition and the right inferior frontal cortex. *Trends Cogn Sci*. April 2004;8(4):170–177.
186. Garavan H, Hester R, Murphy K, Fassbender C, Kelly C. Individual differences in the functional neuroanatomy of inhibitory control. *Brain Res*. August 11, 2006;1105(1):130–142.
187. Smith JL, Mattick RP, Jamadar SD, Iredale JM. Deficits in behavioural inhibition in substance abuse and addiction: a meta-analysis. *Drug Alcohol Depend*. December 1, 2014;145C:1–33.
188. Nigg JT, Wong MM, Martel MM, et al. Poor response inhibition as a predictor of problem drinking and illicit drug use in adolescents at risk for alcoholism and other substance use disorders. *J Am Acad Child Adolesc Psychiatry*. April 2006;45(4):468–475.
189. Rubio G, Jimenez M, Rodriguez-Jimenez R, et al. The role of behavioral impulsivity in the development of alcohol dependence: a 4-year follow-up study. *Alcohol Clin Exp Res*. September 2008;32(9):1681–1687.
190. Belin D, Mar AC, Dalley JW, Robbins TW, Everitt BJ. High impulsivity predicts the switch to compulsive cocaine-taking. *Science*. June 6, 2008;320(5881):1352–1355.
191. Dalley JW, Fryer TD, Brichard L, et al. Nucleus accumbens D2/3 receptors predict trait impulsivity and cocaine reinforcement. *Science*. March 2, 2007;315(5816):1267–1270.
192. Slaats-Willemse D, Swaab-Barneveld H, de Sonneville L, van der Meulen E, Buitelaar J. Deficient response inhibition as a cognitive endophenotype of ADHD. *J Am Acad Child Adolesc Psychiatry*. October 2003;42(10):1242–1248.
193. Aron AR, Poldrack RA. The cognitive neuroscience of response inhibition: relevance for genetic research in attention-deficit/hyperactivity disorder. *Biol Psychiatry*. June 1, 2005;57(11):1285–1292.
194. Vink M, Ramsey NF, Raemaekers M, Kahn RS. Striatal dysfunction in schizophrenia and unaffected relatives. *Biol Psychiatry*. July 1, 2006;60(1):32–39.
195. Chamberlain SR, Fineberg NA, Menzies LA, et al. Impaired cognitive flexibility and motor inhibition in unaffected first-degree relatives of patients with obsessive-compulsive disorder. *Am J Psychiatry*. February 2007;164(2):335–338.
196. Ersche KD, Jones PS, Williams GB, Turton AJ, Robbins TW, Bullmore ET. Abnormal brain structure implicated in stimulant drug addiction. *Science*. February 3, 2012;335(6068):601–604.
197. Castellanos FX, Sonuga-Barke EJ, Milham MP, Tannock R. Characterizing cognition in ADHD: beyond executive dysfunction. *Trends Cogn Sci*. March 2006;10(3):117–123.
198. Lijffijt M, Kenemans JL, Verbaten MN, van Engeland H. A meta-analytic review of stopping performance in attention-deficit/hyperactivity disorder: deficient inhibitory motor control? *J Abnorm Psychol*. May 2005;114(2):216–222.
199. Alderson RM, Rapport MD, Kofler MJ. Attention-deficit/hyperactivity disorder and behavioral inhibition: a meta-analytic review of the stop-signal paradigm. *J Abnorm Child Psychol*. October 2007;35(5):745–758.
200. Oosterlaan J, Logan GD, Sergeant JA. Response inhibition in AD/HD, CD, comorbid AD/HD + CD, anxious, and control children: a meta-analysis of studies with the stop task. *J Child Psychol Psychiatry*. March 1998;39(3):411–425.
201. Schachar R, Mota VL, Logan GD, Tannock R, Klim P. Confirmation of an inhibitory control deficit in attention-deficit/hyperactivity disorder. *J Abnorm Child Psychol*. June 2000;28(3):227–235.
202. Still GF. Some abnormal physical conditions in children. *Lancet*. 1902;1.
203. Barkley RA. A critique of current diagnostic criteria for attention deficit hyperactivity disorder: clinical and research implications. *J Dev Behav Pediatr*. December 1990;11(6):343–352.
204. Quay HC. Theories of ADDH. *J Am Acad Child Adolesc Psychiatry*. March 1988;27(2):262–263.
205. Schachar R, Tannock R, Marriott M, Logan G. Deficient inhibitory control in attention deficit hyperactivity disorder. *J Abnorm Child Psychol*. August 1995;23(4):411–437.
206. Barkley RA. Behavioral inhibition, sustained attention, and executive functions: constructing a unifying theory of ADHD. *Psychol Bull*. January 1997;121(1):65–94.
207. Schachar R, Logan GD. Impulsivity and inhibitory control in normal development and childhood psychopathology. *Dev Psychol*. 1990;26(5):710–720.
208. Russell VA, Oades RD, Tannock R, et al. Response variability in Attention-Deficit/Hyperactivity Disorder: a neuronal and glial energetics hypothesis. *Behav Brain Funct*. 2006;2:30.
209. Scheres A, Oosterlaan J, Sergeant JA. Response inhibition in children with DSM-IV subtypes of AD/HD and related disruptive disorders: the role of reward. *Child Neuropsychol*. September 2001;7(3):172–189.
210. de Zeeuw P, Aarnoudse-Moens C, Bijlhout J, et al. Inhibitory performance, response speed, intraindividual variability, and response accuracy in ADHD. *J Am Acad Child Adolesc Psychiatry*. July 2008;47(7):808–816.
211. Pliszka SR, Glahn DC, Semrud-Clikeman M, et al. Neuroimaging of inhibitory control areas in children with attention deficit hyperactivity disorder who were treatment naive or in long-term treatment. *Am J Psychiatry*. June 2006;163(6):1052–1060.
212. Dickstein SG, Bannon K, Castellanos FX, Milham MP. The neural correlates of attention deficit hyperactivity disorder: an ALE meta-analysis. *J Child Psychol Psychiatry*. October 2006;47(10): 1051–1062.
213. Rubia K, Halari R, Cubillo A, Mohammad AM, Scott S, Brammer M. Disorder-specific inferior prefrontal hypofunction in boys with pure attention-deficit/hyperactivity disorder compared to boys with pure conduct disorder during cognitive flexibility. *Hum Brain Mapp*. December 2010;31(12):1823–1833.
214. Giedd JN, Blumenthal J, Molloy E, Castellanos FX. Brain imaging of attention deficit/hyperactivity disorder. *Ann N Y Acad Sci*. June 2001;931:33–49.
215. Hart H, Radua J, Nakao T, Mataix-Cols D, Rubia K. Meta-analysis of functional magnetic resonance imaging studies of inhibition and attention in attention-deficit/hyperactivity disorder: exploring task-specific, stimulant medication, and age effects. *JAMA Psychiatry*. February 2013;70(2):185–198.
216. Carmona S, Vilarroya O, Bielsa A, et al. Global and regional gray matter reductions in ADHD: a voxel-based morphometric study. *Neurosci Lett*. December 2, 2005;389(2):88–93.
217. Uddin LQ, Kelly AM, Biswal BB, et al. Network homogeneity reveals decreased integrity of default-mode network in ADHD. *J Neurosci Methods*. March 30, 2008;169(1):249–254.

218. Castellanos FX, Margulies DS, Kelly C, et al. Cingulate-precuneus interactions: a new locus of dysfunction in adult attention-deficit/hyperactivity disorder. *Biol Psychiatry*. February 1, 2008;63(3):332–337.
219. Konrad K, Eickhoff SB. Is the ADHD brain wired differently? A review on structural and functional connectivity in attention deficit hyperactivity disorder. *Hum Brain Mapp*. June 2010;31(6):904–916.
220. Liston C, Malter Cohen M, Teslovich T, Levenson D, Casey BJ. Atypical prefrontal connectivity in attention-deficit/hyperactivity disorder: pathway to disease or pathological end point? *Biol Psychiatry*. June 15, 2011;69(12):1168–1177.
221. Peterson BS, Potenza MN, Wang Z, et al. An FMRI study of the effects of psychostimulants on default-mode processing during Stroop task performance in youths with ADHD. *Am J Psychiatry*. November 2009;166(11):1286–1294.
222. Rubia K, Smith AB, Halari R, et al. Disorder-specific dissociation of orbitofrontal dysfunction in boys with pure conduct disorder during reward and ventrolateral prefrontal dysfunction in boys with pure ADHD during sustained attention. *Am J Psychiatry*. January 2009;166(1):83–94.
223. Moll GH, Heinrich H, Rothenberger A. Methylphenidate and intracortical excitability: opposite effects in healthy subjects and attention-deficit hyperactivity disorder. *Acta Psychiatr Scand*. January 2003;107(1):69–72.
224. Mattay VS, Goldberg TE, Fera F, et al. Catechol O-methyltransferase val158-met genotype and individual variation in the brain response to amphetamine. *Proc Natl Acad Sci USA*. May 13, 2003;100(10):6186–6191.
225. Vaidya CJ, Austin G, Kirkorian G, et al. Selective effects of methylphenidate in attention deficit hyperactivity disorder: a functional magnetic resonance study. *Proc Natl Acad Sci USA*. November 24, 1998;95(24):14494–14499.
226. Chamberlain SR, Blackwell AD, Fineberg NA, Robbins TW, Sahakian BJ. The neuropsychology of obsessive compulsive disorder: the importance of failures in cognitive and behavioural inhibition as candidate endophenotypic markers. *Neurosci Biobehav Rev*. May 2005;29(3):399–419.
227. Malloy P, Rasmussen S, Braden W, Haier RJ. Topographic evoked potential mapping in obsessive-compulsive disorder: evidence of frontal lobe dysfunction. *Psychiatry Res*. April 1989;28(1):63–71.
228. Tien AY, Pearlson GD, Machlin SR, Bylsma FW, Hoehn-Saric R. Oculomotor performance in obsessive-compulsive disorder. *Am J Psychiatry*. May 1992;149(5):641–646.
229. Chamberlain SR, Fineberg NA, Blackwell AD, Robbins TW, Sahakian BJ. Motor inhibition and cognitive flexibility in obsessive-compulsive disorder and trichotillomania. *Am J Psychiatry*. July 2006;163(7):1282–1284.
230. Menzies L, Achard S, Chamberlain SR, et al. Neurocognitive endophenotypes of obsessive-compulsive disorder. *Brain*. December 2007;130(Pt 12):3223–3236.
231. de Wit SJ, de Vries FE, van der Werf YD, et al. Presupplementary motor area hyperactivity during response inhibition: a candidate endophenotype of obsessive-compulsive disorder. *Am J Psychiatry*. October 2012;169(10):1100–1108.
232. Chamberlain SR, Menzies L. Endophenotypes of obsessive-compulsive disorder: rationale, evidence and future potential. *Expert Rev Neurother*. August 2009;9(8):1133–1146.
233. Grant JE, Mancebo MC, Eisen JL, Rasmussen SA. Impulse-control disorders in children and adolescents with obsessive-compulsive disorder. *Psychiatry Res*. January 30, 2010;175(1–2):109–113.
234. Grant JE, Mancebo MC, Pinto A, Eisen JL, Rasmussen SA. Impulse control disorders in adults with obsessive compulsive disorder. *J Psychiatr Res*. September 2006;40(6):494–501.
235. Fontenelle LF, Mendlowicz MV, Versiani M. Impulse control disorders in patients with obsessive-compulsive disorder. *Psychiatry Clin Neurosci*. February 2005;59(1):30–37.
236. Bannon S, Gonsalvez CJ, Croft RJ, Boyce PM. Response inhibition deficits in obsessive-compulsive disorder. *Psychiatry Res*. June 1, 2002;110(2):165–174.
237. Nandam LS, Hester R, Wagner J, et al. Methylphenidate but not atomoxetine or citalopram modulates inhibitory control and response time variability. *Biol Psychiatry*. May 1, 2011;69(9):902–904.
238. Overtoom CC, Bekker EM, van der Molen MW, et al. Methylphenidate restores link between stop-signal sensory impact and successful stopping in adults with attention-deficit/hyperactivity disorder. *Biol Psychiatry*. April 1, 2009;65(7):614–619.
239. Bari A, Eagle DM, Mar AC, Robinson ES, Robbins TW. Dissociable effects of noradrenaline, dopamine, and serotonin uptake blockade on stop task performance in rats. *Psychopharmacology*. August 2009;205(2):273–283.
240. Drueke B, Boecker M, Schlaegel S, et al. Serotonergic modulation of response inhibition and re-engagement? Results of a study in healthy human volunteers. *Hum Psychopharmacol*. August 2010;25(6):472–480.
241. Volkow ND, Fowler JS, Wang GJ. Role of dopamine in drug reinforcement and addiction in humans: results from imaging studies. *Behav Pharmacol*. September 2002;13(5–6):355–366.
242. Perry JL, Carroll ME. The role of impulsive behavior in drug abuse. *Psychopharmacology*. September 2008;200(1):1–26.
243. Olmstead MC. Animal models of drug addiction: where do we go from here? *Q J Exp Psychol*. April 2006;59(4):625–653.
244. de Wit H, Richards JB. Dual determinants of drug use in humans: reward and impulsivity. *Neb Symp Motiv*. 2004;50:19–55.
245. Comings DE, Blum K. Reward deficiency syndrome: genetic aspects of behavioral disorders. *Prog Brain Res*. 2000;126:325–341.
246. Leyton M. Conditioned and sensitized responses to stimulant drugs in humans. *Prog Neuropsychopharmacol Biol Psychiatry*. November 15, 2007;31(8):1601–1613.
247. Bickel WK, Odum AL, Madden GJ. Impulsivity and cigarette smoking: delay discounting in current, never, and ex-smokers. *Psychopharmacology*. October 1999;146(4):447–454.
248. Vuchinich RE, Simpson CA. Hyperbolic temporal discounting in social drinkers and problem drinkers. *Exp Clin Psychopharmacol*. August 1998;6(3):292–305.
249. Claus ED, Kiehl KA, Hutchison KE. Neural and behavioral mechanisms of impulsive choice in alcohol use disorder. *Alcohol Clin Exp Res*. July 2011;35(7):1209–1219.
250. Sher KJ, Bartholow BD, Wood MD. Personality and substance use disorders: a prospective study. *J Consult Clin Psychol*. October 2000;68(5):818–829.
251. Bjork JM, Hommer DW, Grant SJ, Danube C. Impulsivity in abstinent alcohol-dependent patients: relation to control subjects and type 1-/type 2-like traits. *Alcohol*. October–November 2004;34(2–3):133–150.
252. Peters J, Bromberg U, Schneider S, et al. Lower ventral striatal activation during reward anticipation in adolescent smokers. *Am J psychiatry*. May 2011;168(5):540–549.
253. Bell RP, Garavan H, Foxe JJ. Neural correlates of craving and impulsivity in abstinent former cocaine users: towards biomarkers of relapse risk. *Neuropharmacology*. October 2014;85:461–470.
254. Li CS, Milivojevic V, Kemp K, Hong K, Sinha R. Performance monitoring and stop signal inhibition in abstinent patients with cocaine dependence. *Drug Alcohol Depend*. December 1, 2006;85(3):205–212.
255. Fillmore MT, Rush CR, Marczinski CA. Effects of d-amphetamine on behavioral control in stimulant abusers: the role of prepotent response tendencies. *Drug Alcohol Depend*. August 20, 2003;71(2):143–152.

256. Fillmore MT, Rush CR, Hays L. Acute effects of oral cocaine on inhibitory control of behavior in humans. *Drug Alcohol Depend.* July 1, 2002;67(2):157–167.
257. Fillmore MT, Rush CR. Impaired inhibitory control of behavior in chronic cocaine users. *Drug Alcohol Depend.* May 1, 2002;66(3):265–273.
258. Lyvers M. "Loss of control" in alcoholism and drug addiction: a neuroscientific interpretation. *Exp Clin Psychopharmacol.* May 2000;8(2):225–249.
259. Volkow ND, Ding YS, Fowler JS, Wang GJ. Cocaine addiction: hypothesis derived from imaging studies with PET. *J Addict Dis.* 1996;15(4):55–71.
260. Morein-Zamir S, Simon Jones P, Bullmore ET, Robbins TW, Ersche KD. Prefrontal hypoactivity associated with impaired inhibition in stimulant-dependent individuals but evidence for hyperactivation in their unaffected siblings. *Neuropsychopharmacology.* September 2013;38(10):1945–1953.
261. Prisciandaro JJ, Joseph JE, Myrick H, et al. The relationship between years of cocaine use and brain activation to cocaine and response inhibition cues. *Addiction.* December 2014;109(12):2062–2070.
262. Kaufman JN, Ross TJ, Stein EA, Garavan H. Cingulate hypoactivity in cocaine users during a GO-NOGO task as revealed by event-related functional magnetic resonance imaging. *J Neurosci.* August 27, 2003;23(21):7839–7843.
263. Botvinick MM, Braver TS, Barch DM, Carter CS, Cohen JD. Conflict monitoring and cognitive control. *Psychol Rev.* July 2001;108(3):624–652.
264. Castelluccio BC, Meda SA, Muska CE, Stevens MC, Pearlson GD. Error processing in current and former cocaine users. *Brain Imaging Behav.* March 2014;8(1):87–96.
265. Matuskey D, Luo X, Zhang S, et al. Methylphenidate remediates error-preceding activation of the default mode brain regions in cocaine-addicted individuals. *Psychiatry Res.* November 30, 2013;214(2):116–121.
266. Sjoerds Z, van den Brink W, Beekman AT, Penninx BW, Veltman DJ. Response inhibition in alcohol-dependent patients and patients with depression/anxiety: a functional magnetic resonance imaging study. *Psychol Med.* June 2014;44(8):1713–1725.
267. Schmaal L, Joos L, Koeleman M, Veltman DJ, van den Brink W, Goudriaan AE. Effects of modafinil on neural correlates of response inhibition in alcohol-dependent patients. *Biol Psychiatry.* February 1, 2013;73(3):211–218.
268. Dawe S, Loxton NJ. The role of impulsivity in the development of substance use and eating disorders. *Neurosci Biobehav Rev.* May 2004;28(3):343–351.
269. Belin D, Bari A, Besson M, Dalley JW. Multi-disciplinary investigations of impulsivity in animal models of attention-deficit hyperactivity disorder and drug addiction vulnerability. In: Granon S, ed. *Endophenotypes of Psychiatric and Neurodegenerative Disorders in Rodent Models.* London: Transworld Research; 2009.
270. Regier DA, Farmer ME, Rae DS, et al. Comorbidity of mental disorders with alcohol and other drug abuse. Results from the Epidemiologic Catchment Area (ECA) Study. *JAMA.* November 21, 1990;264(19):2511–2518.
271. McGue M, Iacono WG, Legrand LN, Malone S, Elkins I. Origins and consequences of age at first drink. I. Associations with substance-use disorders, disinhibitory behavior and psychopathology, and P3 amplitude. *Alcohol Clin Exp Res.* August 2001;25(8):1156–1165.
272. White HR, Pandina RJ, Chen PH. Developmental trajectories of cigarette use from early adolescence into young adulthood. *Drug Alcohol Depend.* January 1, 2002;65(2):167–178.
273. Hesselbrock VM, Hesselbrock MN, Stabenau JR. Alcoholism in men patients subtyped by family history and antisocial personality. *J Stud Alcohol.* January 1985;46(1):59–64.
274. Piazza PV, Deminiere JM, Le Moal M, Simon H. Factors that predict individual vulnerability to amphetamine self-administration. *Science.* September 29, 1989;245(4925):1511–1513.
275. Cloninger CR. Neurogenetic adaptive mechanisms in alcoholism. *Science.* April 24, 1987;236(4800):410–416.
276. Verdejo-Garcia A, Perez-Garcia M. Ecological assessment of executive functions in substance dependent individuals. *Drug Alcohol Depend.* September 6, 2007;90(1):48–55.
277. Deroche-Gamonet V, Belin D, Piazza PV. Evidence for addiction-like behavior in the rat. *Science.* August 13, 2004;305(5686):1014–1017.
278. McNamara R, Dalley JW, Robbins TW, Everitt BJ, Belin D. Trait-like impulsivity does not predict escalation of heroin self-administration in the rat. *Psychopharmacology.* December 2011;212(4):453–464.
279. Molander AC, Mar A, Norbury A, et al. High impulsivity predicting vulnerability to cocaine addiction in rats: some relationship with novelty preference but not novelty reactivity, anxiety or stress. *Psychopharmacology.* June 2011;215(4):721–731.
280. Ladouceur R, Jacques C, Ferland F, Giroux I. Prevalence of problem gambling: a replication study 7 years later. *Can J Psychiatry.* October 1999;44(8):802–804.
281. Sáez-Abad C, Bertolín-Guillén JM. Personality traits and disorders in pathological gamblers versus normal controls. *J Addict Dis.* February 27, 2008;27(1):33–40.
282. Fuentes D, Tavares H, Artes R, Gorenstein C. Self-reported and neuropsychological measures of impulsivity in pathological gambling. *J Int Neuropsychol Soc.* 2006;12(06):907–912.
283. Petry NM. Substance abuse, pathological gambling, and impulsiveness. *Drug Alcohol Depend.* 2001;63(1):29–38.
284. Lorains FK, Stout JC, Bradshaw JL, Dowling NA, Enticott PG. Self-reported impulsivity and inhibitory control in problem gamblers. *J Clin Exp Neuropsychol.* February 7, 2014;36(2): 144–157.
285. Kräplin A, Bühringer G, Oosterlaan J, van den Brink W, Goschke T, Goudriaan AE. Dimensions and disorder specificity of impulsivity in pathological gambling. *Addict Behav.* November 2014; 39(11):1646–1651.
286. Michalczuk R, Bowden-Jones H, Verdejo-Garcia A, Clark L. Impulsivity and cognitive distortions in pathological gamblers attending the UK National Problem Gambling Clinic: a preliminary report. *Psychol Med.* 2011:1–11. FirstView.
287. Goudriaan A,E, Oosterlaan J, de Beurs E, van den Brink W. Neurocognitive functions in pathological gambling: a comparison with alcohol dependence, Tourette syndrome and normal controls. *Addiction.* 2006;101(4):534–547.
288. Kertzman S, Lowengrub K, Aizer A, Nahum ZB, Kotler M, Dannon PN. Stroop performance in pathological gamblers. *Psychiatry Res.* May 30, 2006;142(1):1–10.
289. Reynolds B. A review of delay-discounting research with humans: relations to drug use and gambling. *Behav Pharmacol.* 2006;17(8):651–667. 610.1097/FBP.1090b1013e3280115f3280199.
290. Brand M, Kalbe E, Labudda K, Fujiwara E, Kessler J, Markowitsch HJ. Decision-making impairments in patients with pathological gambling. *Psychiatry Res.* 2005;133(1):91–99.
291. Brevers D, Bechara A, Cleeremans A, Noel X. Iowa Gambling Task (IGT): twenty years after – gambling disorder and IGT. *Front Psychol.* September 30, 2013;4.
292. Lorains FK, Dowling NA, Enticott PG, Bradshaw JL, Trueblood JS, Stout JC. Strategic and non-strategic problem gamblers differ on decision-making under risk and ambiguity. *Addiction.* July 2014;109(7):1128–1137.
293. Lawrence AJ, Luty J, Bogdan N, Sahakian BJ, Clark L. Problem gamblers share deficits in impulsive decision-making with alcohol-dependent individuals. *Addiction.* 2009;104(6):1006–1015.
294. Bechara A, Damasio AR, Damasio H, Anderson SW. Insensitivity to future consequences following damage to human prefrontal cortex. *Cognition.* 1994;50(1–3):7–15.

295. Brand M, Fujiwara E, Borsutzky S, Kalbe E, Kessler J, Markowitsch HJ. Decision-making deficits of Korsakoff patients in a new gambling task with explicit rules: associations with executive functions. *Neuropsychology*. 2005;19(3):267–277.
296. Franken IHA, van Strien JW, Nijs I, Muris P. Impulsivity is associated with behavioral decision-making deficits. *Psychiatry Res*. 2008;158(2):155–163.
297. Burdick JD, Roy AL, Raver CC. Evaluating the Iowa gambling task as a direct assessment of impulsivity with low-income children. *Pers Individ Dif*. October 2013;55(7):771–776.
298. Sweitzer MM, Allen PA, Kaut KP. Relation of individual differences in impulsivity to nonclinical emotional decision making. *J Int Neuropsychol Soc*. 2008;14(05):878–882.
299. Vitaro F, Arseneault L, Tremblay RE. Impulsivity predicts problem gambling in low SES adolescent males. *Addiction*. 1999;94(4):565–575.
300. Slutske WS, Caspi A, Moffitt TE, Poulton R. Personality and problem gambling: a prospective study of a birth cohort of young adults. *Arch Gen Psychiatry*. July 1, 2005;62(7):769–775.
301. Dussault F, Brendgen M, Vitaro F, Wanner B, Tremblay RE. Longitudinal links between impulsivity, gambling problems and depressive symptoms: a transactional model from adolescence to early adulthood. *J Child Psychol Psychiatry*. 2011;52(2):130–138.
302. Cyders MA, Smith GT. Clarifying the role of personality dispositions in risk for increased gambling behavior. *Pers Individ Differ*. October 2008;45(6):503–508.
303. Maccallum F, Blaszczynski A, Ladouceur R, Nower L. Functional and dysfunctional impulsivity in pathological gambling. *Pers Individ Differ*. 2007;43(7):1829–1838.
304. Power Y, Goodyear B, Crockford D. Neural correlates of pathological gamblers preference for immediate rewards during the Iowa gambling task: an fMRI study. *J Gambl Stud*. December 2012;28(4):623–636.
305. Chase HW, Clark L. Gambling severity predicts midbrain response to near-miss outcomes. *J Neurosci*. May 5, 2010;30(18):6180–6187.
306. Reuter J, Raedler T, Rose M, Hand I, Glascher J, Buchel C. Pathological gambling is linked to reduced activation of the mesolimbic reward system. *Nat Neurosci*. 2005;8(2):147–148.
307. Miedl SF, Fehr T, Meyer G, Herrmann M. Neurobiological correlates of problem gambling in a quasi-realistic blackjack scenario as revealed by fMRI. *Psychiatry Res Neuroimaging*. 2010;181(3):165–173.
308. Linnet J, Moller A, Peterson E, Gjedde A, Doudet D. Dopamine release in ventral striatum during Iowa Gambling Task performance is associated with increased excitement levels in pathological gambling. *Addiction*. February 2011;106(2):383–390.
309. Linnet J, Moller A, Peterson E, Gjedde A, Doudet D. Inverse association between dopaminergic neurotransmission and Iowa Gambling Task performance in pathological gamblers and healthy controls. *Scand J Psychol*. February 2011;52(1):28–34.
310. Potenza MN, Leung HC, Blumberg HP, et al. An FMRI Stroop task study of ventromedial prefrontal cortical function in pathological gamblers. *Am J Psychiatry*. November 2003;160(11):1990–1994.
311. de Ruiter MB, Oosterlaan J, Veltman DJ, van den Brink W, Goudriaan AE. Similar hyporesponsiveness of the dorsomedial prefrontal cortex in problem gamblers and heavy smokers during an inhibitory control task. *Drug Alcohol Depend*. February 1, 2012;121(1–2):81–89.
312. van Holst RJ, van Holstein M, van den Brink W, Veltman DJ, Goudriaan AE. Response inhibition during cue reactivity in problem gamblers: an fMRI study. *PLoS One*. 2012;7(3):e30909.
313. Hill K, Mann L, Laws KR, Stephenson CM, Nimmo-Smith I, McKenna PJ. Hypofrontality in schizophrenia: a meta-analysis of functional imaging studies. *Acta Psychiatr Scand*. October 2004;110(4):243–256.
314. Pantelis C, Barnes TR, Nelson HE, et al. Frontal-striatal cognitive deficits in patients with chronic schizophrenia. *Brain*. October 1997;120(Pt 10):1823–1843.
315. Twamley EW, Palmer BW, Jeste DV, Taylor MJ, Heaton RK. Transient and executive function working memory in schizophrenia. *Schizophr Res*. October 2006;87(1–3):185–190.
316. Kiehl KA, Smith AM, Hare RD, Liddle PF. An event-related potential investigation of response inhibition in schizophrenia and psychopathy. *Biol Psychiatry*. August 1, 2000;48(3):210–221.
317. Heerey EA, Robinson BM, McMahon RP, Gold JM. Delay discounting in schizophrenia. *Cogn Neuropsychiatry*. May 2007;12(3):213–221.
318. Kaladjian A, Jeanningros R, Azorin JM, Anton JL, Mazzola-Pomietto P. Impulsivity and neural correlates of response inhibition in schizophrenia. *Psychol Med*. February 2011;41(2):291–299.
319. Enticott PG, Ogloff JR, Bradshaw JL. Response inhibition and impulsivity in schizophrenia. *Psychiatry Res*. January 15, 2008;157(1–3):251–254.
320. Ford JM, Gray M, Whitfield SL, et al. Acquiring and inhibiting prepotent responses in schizophrenia: event-related brain potentials and functional magnetic resonance imaging. *Arch Gen Psychiatry*. February 2004;61(2):119–129.
321. Hoptman MJ, Volavka J, Johnson G, Weiss E, Bilder RM, Lim KO. Frontal white matter microstructure, aggression, and impulsivity in men with schizophrenia: a preliminary study. *Biol Psychiatry*. July 1, 2002;52(1):9–14.
322. Narayan VM, Narr KL, Kumari V, et al. Regional cortical thinning in subjects with violent antisocial personality disorder or schizophrenia. *Am J Psychiatry*. September 2007;164(9):1418–1427.
323. Hoptman MJ, Ardekani BA, Butler PD, Nierenberg J, Javitt DC, Lim KO. DTI and impulsivity in schizophrenia: a first voxelwise correlational analysis. *Neuroreport*. November 15, 2004;15(16):2467–2470.
324. Kaladjian A, Jeanningros R, Azorin JM, Grimault S, Anton JL, Mazzola-Pomietto P. Blunted activation in right ventrolateral prefrontal cortex during motor response inhibition in schizophrenia. *Schizophr Res*. December 2007;97(1–3):184–193.
325. Joyal CC, Putkonen A, Mancini-Marie A, et al. Violent persons with schizophrenia and comorbid disorders: a functional magnetic resonance imaging study. *Schizophr Res*. March 2007;91(1–3):97–102.
326. Rubia K, Russell T, Bullmore ET, et al. An fMRI study of reduced left prefrontal activation in schizophrenia during normal inhibitory function. *Schizophr Res*. October 1, 2001;52(1–2):47–55.
327. Laurens KR, Ngan ET, Bates AT, Kiehl KA, Liddle PF. Rostral anterior cingulate cortex dysfunction during error processing in schizophrenia. *Brain*. March 2003;126(Pt 3):610–622.
328. Arce E, Leland DS, Miller DA, Simmons AN, Winternheimer KC, Paulus MP. Individuals with schizophrenia present hypo- and hyperactivation during implicit cueing in an inhibitory task. *NeuroImage*. August 15, 2006;32(2):704–713.
329. Barkataki I, Kumari V, Das M, Sumich A, Taylor P, Sharma T. Neural correlates of deficient response inhibition in mentally disordered violent individuals. *Behav Sci Law*. 2008;26(1):51–64.
330. Ross RG, Harris JG, Olincy A, Radant A, Adler LE, Freedman R. Familial transmission of two independent saccadic abnormalities in schizophrenia. *Schizophr Res*. February 27, 1998;30(1):59–70.
331. Dumais A, Potvin S, Joyal C, et al. Schizophrenia and serious violence: a clinical-profile analysis incorporating impulsivity and substance-use disorders. *Schizophr Res*. August 2011;130(1–3):234–237.
332. Gut-Fayand A, Dervaux A, Olie JP, Loo H, Poirier MF, Krebs MO. Substance abuse and suicidality in schizophrenia: a common risk factor linked to impulsivity. *Psychiatry Res*. May 10, 2001;102(1):65–72.

333. Yun DY, Hwang SS, Kim Y, Lee YH, Kim YS, Jung HY. Impairments in executive functioning in patients with remitted and nonremitted schizophrenia. *Prog Neuropsychopharmacol biol Psychiatry*. June 1, 2011;35(4):1148–1154.
334. Nolan KA, D'Angelo D, Hoptman MJ. Self-report and laboratory measures of impulsivity in patients with schizophrenia or schizoaffective disorder and healthy controls. *Psychiatry Res*. May 15, 2011;187(1–2):301–303.
335. Zandbelt BB, van Buuren M, Kahn RS, Vink M. Reduced proactive inhibition in schizophrenia is related to corticostriatal dysfunction and poor working memory. *Biol Psychiatry*. September 6, 2011.
336. Huddy VC, Aron AR, Harrison M, Barnes TR, Robbins TW, Joyce EM. Impaired conscious and preserved unconscious inhibitory processing in recent onset schizophrenia. *Psychol Med*. June 2009;39(6):907–916.
337. Badcock JC, Michie PT, Johnson L, Combrinck J. Acts of control in schizophrenia: dissociating the components of inhibition. *Psychol Med*. February 2002;32(2):287–297.
338. Thakkar KN, Schall JD, Boucher L, Logan GD, Park S. Response inhibition and response monitoring in a saccadic countermanding task in schizophrenia. *Biol Psychiatry*. January 1, 2011;69(1):55–62.
339. Lipszyc J, Schachar R. Inhibitory control and psychopathology: a meta-analysis of studies using the stop signal task. *J Int Neuropsychol Soc*. November 2010;16(6):1064–1076.
340. Barch DM, Braver TS, Carter CS, Poldrack RA, Robbins TW. CNTRICS final task selection: executive control. *Schizophr Bull*. January 2009;35(1):115–135.
341. Chambers RA, Krystal JH, Self DW. A neurobiological basis for substance abuse comorbidity in schizophrenia. *Biol Psychiatry*. July 15, 2001;50(2):71–83.
342. Paulik G, Badcock JC, Maybery MT. Dissociating the components of inhibitory control involved in predisposition to hallucinations. *Cogn Neuropsychiatry*. January 2008;13(1):33–46.
343. Cloninger CR. A systematic method for clinical description and classification of personality variants. A proposal. *Arch Gen Psychiatry*. June 1987;44(6):573–588.
344. Crow TJ, Longden A, Smith A, Wendlandt S. Pontine tegmental lesions, monoamine neurons and varieties of learning. *Behav Biol*. June 1977;20(2):184–196.
345. Soubrie P. Serotonergic neurons and behavior. *Journal de pharmacologie*. April–June 1986;17(2):107–112.
346. Zuckerman M, Ballenger JC, Post RM. The neurobiology of some dimensions of personality. *Int Rev Neurobiol*. 1984;25:391–436.
347. Zuckerman M. *Psychobiology of Personality*. New York: Cambridge University Press; 1991.
348. Gray JA. The neuropsychology of emotion and personality. In: Stahl SM, Iversen SD, Goodman EC, eds. *Cognitive Neurochemistry*. Oxford: Oxford University Press; 1987:171–190.
349. Winstanley CA, Theobald DE, Dalley JW, Cardinal RN, Robbins TW. Double dissociation between serotonergic and dopaminergic modulation of medial prefrontal and orbitofrontal cortex during a test of impulsive choice. *Cereb Cortex*. January 2006;16(1):106–114.
350. Winstanley CA, Theobald DE, Dalley JW, Robbins TW. Interactions between serotonin and dopamine in the control of impulsive choice in rats: therapeutic implications for impulse control disorders. *Neuropsychopharmacology*. April 2005;30(4):669–682.
351. Varga G, Szekely A, Antal P, et al. Additive effects of serotonergic and dopaminergic polymorphisms on trait impulsivity. *Am J Med Genet B Neuropsychiatr Genet*. April 2012;159B(3):281–288.
352. Linnoila M, Virkkunen M, Scheinin M, Nuutila A, Rimon R, Goodwin FK. Low cerebrospinal fluid 5-hydroxyindoleacetic acid concentration differentiates impulsive from nonimpulsive violent behavior. *Life Sci*. December 26, 1983;33(26):2609–2614.
353. Virkkunen M, Linnoila M. Brain serotonin, type II alcoholism and impulsive violence. *J Stud Alcohol Suppl*. September 1993;11:163–169.
354. Li J, Wang Y, Zhou R, et al. Serotonin 5-HT1B receptor gene and attention deficit hyperactivity disorder in Chinese Han subjects. *Am J Med Genet B Neuropsychiatr Genet*. January 5, 2005;132B(1):59–63.
355. Zhou Z, Roy A, Lipsky R, et al. Haplotype-based linkage of tryptophan hydroxylase 2 to suicide attempt, major depression, and cerebrospinal fluid 5-hydroxyindoleacetic acid in 4 populations. *Arch Gen Psychiatry*. October 2005;62(10):1109–1118.
356. Parsey RV, Oquendo MA, Simpson NR, et al. Effects of sex, age, and aggressive traits in man on brain serotonin 5-HT1A receptor binding potential measured by PET using [C-11]WAY-100635. *Brain Res*. November 8, 2002;954(2):173–182.
357. Benko A, Lazary J, Molnar E, et al. Significant association between the C(-1019)G functional polymorphism of the HTR1A gene and impulsivity. *Am J Med Genet B Neuropsychiatr Genet*. March 5, 2010;153B(2):592–599.
358. Bevilacqua L, Goldman D. Genetics of impulsive behaviour. *Proc R Soc Lond Ser B Biol Sci*. 2013;368(1615):20120380.
359. Egan MF, Goldberg TE, Kolachana BS, et al. Effect of COMT Val108/158 Met genotype on frontal lobe function and risk for schizophrenia. *Proc Natl Acad Sci USA*. June 5, 2001;98(12):6917–6922.
360. Paloyelis Y, Asherson P, Mehta MA, Faraone SV, Kuntsi J. DAT1 and COMT effects on delay discounting and trait impulsivity in male adolescents with attention deficit/hyperactivity disorder and healthy controls. *Neuropsychopharmacology*. November 2010;35(12):2414–2426.
361. Faraone SV, Mick E. Molecular genetics of attention deficit hyperactivity disorder. *Psychiatr Clin N Am*. March 2010;33(1):159–180.
362. Cleare AJ, Bond AJ. Experimental evidence that the aggressive effect of tryptophan depletion is mediated via the 5-HT1A receptor. *Psychopharmacology*. January 2000;147(4):439–441.
363. Walderhaug E, Lunde H, Nordvik JE, Landro NI, Refsum H, Magnusson A. Lowering of serotonin by rapid tryptophan depletion increases impulsiveness in normal individuals. *Psychopharmacology*. December 2002;164(4):385–391.
364. Cherek DR, Lane SD. Fenfluramine effects on impulsivity in a sample of adults with and without history of conduct disorder. *Psychopharmacology*. October 2000;152(2):149–156.
365. Rubia K, Lee F, Cleare AJ, et al. Tryptophan depletion reduces right inferior prefrontal activation during response inhibition in fast, event-related fMRI. *Psychopharmacology*. June 2005;179(4):791–803.
366. Reimold M, Smolka MN, Schumann G, et al. Midbrain serotonin transporter binding potential measured with [11C]DASB is affected by serotonin transporter genotype. *J Neural Transm*. 2007;114(5):635–639.
367. Praschak-Rieder N, Kennedy J, Wilson AA, et al. Novel 5-HTTLPR allele associates with higher serotonin transporter binding in putamen: a [(11)C] DASB positron emission tomography study. *Biol Psychiatry*. August 15, 2007;62(4):327–331.
368. Must A, Juhasz A, Rimanoczy A, Szabo Z, Keri S, Janka Z. Major depressive disorder, serotonin transporter, and personality traits: why patients use suboptimal decision-making strategies? *J Affect Disord*. November 2007;103(1–3):273–276.
369. Beevers CG, Pacheco J, Clasen P, McGeary JE, Schnyer D. Prefrontal morphology, 5-HTTLPR polymorphism and biased attention for emotional stimuli. *Genes Brain, Behav*. March 1, 2010;9(2):224–233.
370. Lesch KP, Bengel D, Heils A, et al. Association of anxiety-related traits with a polymorphism in the serotonin transporter gene regulatory region. *Science*. November 29, 1996;274(5292):1527–1531.
371. Greenberg BD, Li Q, Lucas FR, et al. Association between the serotonin transporter promoter polymorphism and personality traits in a primarily female population sample. *Am J Med Genet*. April 3, 2000;96(2):202–216.

372. Terracciano A, Balaci L, Thayer J, et al. Variants of the serotonin transporter gene and NEO-PI-R Neuroticism: no association in the BLSA and SardiNIA samples. *Am J Med Genet B Neuropsychiatr Genet*. December 5, 2009;150B(8):1070–1077.
373. Jupp B, Caprioli D, Saigal N, et al. Dopaminergic and GABA-ergic markers of impulsivity in rats: evidence for anatomical localisation in ventral striatum and prefrontal cortex. *Eur J Neurosci*. February 1, 2013.
374. Lawrence AD, Brooks DJ. Ventral striatal dopamine synthesis capacity is associated with individual differences in behavioral disinhibition. *Front Behav Neurosci*. 2014;8:86.
375. Nemoda Z, Szekely A, Sasvari-Szekely M. Psychopathological aspects of dopaminergic gene polymorphisms in adolescence and young adulthood. *Neurosci Biobehav Rev*. August 2011;35(8): 1665–1686.
376. Volkow ND, Fowler JS, Wang GJ, Swanson JM. Dopamine in drug abuse and addiction: results from imaging studies and treatment implications. *Mol Psychiatry*. June 2004;9(6):557–569.
377. Tannock R, Schachar RJ, Carr RP, Chajczyk D, Logan GD. Effects of methylphenidate on inhibitory control in hyperactive children. *J Abnorm Child Psychol*. October 1989;17(5):473–491.
378. Scheres A, Oosterlaan J, Swanson J, et al. The effect of methylphenidate on three forms of response inhibition in boys with AD/HD. *J Abnorm Child Psychol*. February 2003;31(1):105–120.
379. Tannock R, Schachar R, Logan G. Methylphenidate and cognitive flexibility: dissociated dose effects in hyperactive children. *J Abnorm Child Psychol*. April 1995;23(2):235–266.
380. Aron AR, Dowson JH, Sahakian BJ, Robbins TW. Methylphenidate improves response inhibition in adults with attention-deficit/hyperactivity disorder. *Biol Psychiatry*. December 15, 2003;54(12): 1465–1468.
381. Chamberlain SR, Muller U, Blackwell AD, Clark L, Robbins TW, Sahakian BJ. Neurochemical modulation of response inhibition and probabilistic learning in humans. *Science*. February 10, 2006;311(5762):861–863.
382. Clatworthy PL, Lewis SJ, Brichard L, et al. Dopamine release in dissociable striatal subregions predicts the different effects of oral methylphenidate on reversal learning and spatial working memory. *J Neurosci*. April 15, 2009;29(15):4690–4696.
383. Elliott R, Sahakian BJ, Matthews K, Bannerjea A, Rimmer J, Robbins TW. Effects of methylphenidate on spatial working memory and planning in healthy young adults. *Psychopharmacology*. May 1997;131(2):196–206.
384. Turner DC, Robbins TW, Clark L, Aron AR, Dowson J, Sahakian BJ. Relative lack of cognitive effects of methylphenidate in elderly male volunteers. *Psychopharmacology*. August 2003;168(4):455–464.
385. Hamidovic A, Dlugos A, Skol A, Palmer AA, de Wit H. Evaluation of genetic variability in the dopamine receptor D2 in relation to behavioral inhibition and impulsivity/sensation seeking: an exploratory study with d-amphetamine in healthy participants. *Exp Clin Psychopharmacol*. December 2009;17(6):374–383.
386. Chamberlain SR, Hampshire A, Muller U, et al. Atomoxetine modulates right inferior frontal activation during inhibitory control: a pharmacological functional magnetic resonance imaging study. *Biol Psychiatry*. April 1, 2009;65(7):550–555.
387. Bari A, Mar AC, Theobald DE, et al. Prefrontal and monoaminergic contributions to stop-signal task performance in rats. *J Neurosci*. June 22, 2011;31(25):9254–9263.
388. Bari A, Aston-Jones G. Atomoxetine modulates spontaneous and sensory-evoked discharge of locus coeruleus noradrenergic neurons. *Neuropharmacology*. January 2013;64:53–64.
389. Aston-Jones G, Cohen JD. An integrative theory of locus coeruleus-norepinephrine function: adaptive gain and optimal performance. *Annu Rev Neurosci*. 2005;28:403–450.
390. Eagle DM, Wong JC, Allan ME, Mar AC, Theobald DE, Robbins TW. Contrasting roles for dopamine D1 and D2 receptor subtypes in the dorsomedial striatum but not the nucleus accumbens core during behavioral inhibition in the stop-signal task in rats. *J Neurosci*. May 18, 2011;31(20):7349–7356.
391. Pezze MA, Dalley JW, Robbins TW. Remediation of attentional dysfunction in rats with lesions of the medial prefrontal cortex by intra-accumbens administration of the dopamine D(2/3) receptor antagonist sulpiride. *Psychopharmacology*. January 2009;202(1–3): 307–313.
392. van Gaalen MM, Brueggeman RJ, Bronius PF, Schoffelmeer AN, Vanderschuren LJ. Behavioral disinhibition requires dopamine receptor activation. *Psychopharmacology*. July 2006;187(1):73–85.
393. Granon S, Passetti F, Thomas KL, Dalley JW, Everitt BJ, Robbins TW. Enhanced and impaired attentional performance after infusion of D1 dopaminergic receptor agents into rat prefrontal cortex. *J Neurosci*. February 1, 2000;20(3):1208–1215.
394. Volkow ND, Fowler JS, Wang GJ, Swanson JM, Telang F. Dopamine in drug abuse and addiction: results of imaging studies and treatment implications. *Arch Neurol*. November 2007;64(11):1575–1579.
395. Conners CK. Parent and teacher rating forms for the assessment of hyperkinesis in children. In: Keller PA, Ritt LG, eds. *Innovations in Clinical Practice: A Source Book*. Vol. 1. Sarasota: Professional Resource Exchange; 1982.
396. Silverman IW, Ragusa DM. Child and maternal correlates of impulse control in 24-month-old children. *Genet Soc Gen Psychol Monogr*. November 1990;116(4):435–473.
397. Rothbart MK. Temperament and the development of inhibited approach. *Child Dev*. October 1988;59(5):1241–1250.
398. Liu J, Lester BM. Reconceptualizing in a dual-system model the effects of prenatal cocaine exposure on adolescent development: a short review. *Int J Dev Neurosci*. December 2011;29(8):803–809.
399. Kelley AE, Schochet T, Landry CF. Risk taking and novelty seeking in adolescence: introduction to part I. *Ann N Y Acad Sci*. June 2004;1021:27–32.
400. Dahl RE. Biological, developmental, and neurobehavioral factors relevant to adolescent driving risks. *Am J Prev Med*. September 2008;35(3 suppl):S278–S284.
401. Glover V. Annual Research Review: prenatal stress and the origins of psychopathology: an evolutionary perspective. *J Child Psychol Psychiatry*. April 2011;52(4):356–367.
402. Huizink AC, Mulder EJ. Maternal smoking, drinking or cannabis use during pregnancy and neurobehavioral and cognitive functioning in human offspring. *Neurosci Biobehav Rev*. 2006;30(1):24–41.
403. Dougherty DM, Charles NE, Mathias CW, et al. Delay discounting differentiates pre-adolescents at high and low risk for substance use disorders based on family history. *Drug Alcohol Depend*. October 1, 2014;143:105–111.
404. Spear LP. The adolescent brain and age-related behavioral manifestations. *Neurosci Biobehav Rev*. June 2000;24(4):417–463.
405. Tarter RE, Kirisci L, Mezzich A, et al. Neurobehavioral disinhibition in childhood predicts early age at onset of substance use disorder. *Am J Psychiatry*. June 2003;160(6):1078–1085.
406. Romer D, Betancourt L, Giannetta JM, Brodsky NL, Farah M, Hurt H. Executive cognitive functions and impulsivity as correlates of risk taking and problem behavior in preadolescents. *Neuropsychologia*. November 2009;47(13):2916–2926.
407. Bernheim A, Halfon O, Boutrel B. Controversies about the enhanced vulnerability of the adolescent brain to develop addiction. *Front Pharmacol*. 2013;4:118.
408. Black YD, Maclaren FR, Naydenov AV, Carlezon Jr WA, Baxter MG, Konradi C. Altered attention and prefrontal cortex gene expression in rats after binge-like exposure to cocaine during adolescence. *J Neurosci*. September 20, 2006;26(38):9656–9665.

409. Hankosky ER, Gulley JM. Performance on an impulse control task is altered in adult rats exposed to amphetamine during adolescence. *Dev Psychobiol*. November 2013;55(7):733–744.
410. O'Shea M, Singh ME, McGregor IS, Mallet PE. Chronic cannabinoid exposure produces lasting memory impairment and increased anxiety in adolescent but not adult rats. *J Psychopharmacol*. December 2004;18(4):502–508.
411. Harvey RC, Dembro KA, Rajagopalan K, Mutebi MM, Kantak KM. Effects of self-administered cocaine in adolescent and adult male rats on orbitofrontal cortex-related neurocognitive functioning. *Psychopharmacology*. September 2009;206(1):61–71.
412. Hammerslag LR, Waldman AJ, Gulley JM. Effects of amphetamine exposure in adolescence or young adulthood on inhibitory control in adult male and female rats. *Behav Brain Res*. April 15, 2014;263:22–33.
413. Raznahan A, Lerch JP, Lee N, et al. Patterns of coordinated anatomical change in human cortical development: a longitudinal neuroimaging study of maturational coupling. *Neuron*. December 8, 2011;72(5):873–884.
414. Chambers RA, Potenza MN. Neurodevelopment, impulsivity, and adolescent gambling. *J Gambl Stud*. Spring 2003;19(1):53–84.
415. Crews FT, Boettiger CA. Impulsivity, frontal lobes and risk for addiction. *Pharmacol Biochem Behav*. September 2009;93(3):237–247.
416. Gogtay N, Giedd JN, Lusk L, et al. Dynamic mapping of human cortical development during childhood through early adulthood. *Proc Natl Acad Sci USA*. May 25, 2004;101(21):8174–8179.
417. Giedd JN. The teen brain: insights from neuroimaging. *J Adolesc Health*. April 2008;42(4):335–343.
418. Paus T. Mapping brain maturation and cognitive development during adolescence. *Trends Cogn Sci*. February 2005;9(2):60–68.
419. Paus T, Zijdenbos A, Worsley K, et al. Structural maturation of neural pathways in children and adolescents: in vivo study. *Science*. March 19, 1999;283(5409):1908–1911.
420. Avila C, Cuenca I, Felix V, Parcet MA, Miranda A. Measuring impulsivity in school-aged boys and examining its relationship with ADHD and ODD ratings. *J Abnorm Child Psychol*. June 2004;32(3):295–304.
421. Galvan A, Hare T, Voss H, Glover G, Casey BJ. Risk-taking and the adolescent brain: who is at risk? *Dev Sci*. March 2007;10(2):F8–F14.
422. Courchesne E, Pierce K, Schumann CM, et al. Mapping early brain development in autism. *Neuron*. October 25, 2007;56(2):399–413.
423. Aman MG, Farmer CA, Hollway J, Arnold LE. Treatment of inattention, overactivity, and impulsiveness in autism spectrum disorders. *Child Adolesc Psychiatr Clin N Am*. October 2008;17(4):713–738, vii.
424. Rapoport J, Chavez A, Greenstein D, Addington A, Gogtay N. Autism spectrum disorders and childhood-onset schizophrenia: clinical and biological contributions to a relation revisited. *J Am Acad Child Adolesc Psychiatry*. January 2009;48(1):10–18.
425. Crespi B, Stead P, Elliot M. Evolution in health and medicine Sackler colloquium: comparative genomics of autism and schizophrenia. *Proc Natl Acad Sci USA*. January 26, 2010;107(suppl 1):1736–1741.
426. Voelbel GT, Bates ME, Buckman JF, Pandina G, Hendren RL. Caudate nucleus volume and cognitive performance: are they related in childhood psychopathology? *Biol Psychiatry*. November 1, 2006;60(9):942–950.
427. Hollander E, Anagnostou E, Chaplin W, et al. Striatal volume on magnetic resonance imaging and repetitive behaviors in autism. *Biol Psychiatry*. August 1, 2005;58(3):226–232.
428. Schmitz N, Rubia K, Daly E, Smith A, Williams S, Murphy DG. Neural correlates of executive function in autistic spectrum disorders. *Biol Psychiatry*. January 1, 2006;59(1):7–16.
429. Luna B, Padmanabhan A, O'Hearn K. What has fMRI told us about the development of cognitive control through adolescence? *Brain Cogn*. February 2010;72(1):101–113.
430. Van Leijenhorst L, Gunther Moor B, Op de Macks ZA, Rombouts SA, Westenberg PM, Crone EA. Adolescent risky decision-making: neurocognitive development of reward and control regions. *NeuroImage*. May 15, 2010;51(1):345–355.
431. Stevens MC, Kiehl KA, Pearlson GD, Calhoun VD. Functional neural networks underlying response inhibition in adolescents and adults. *Behav Brain Res*. July 19, 2007;181(1):12–22.
432. Rubia K, Smith AB, Woolley J, et al. Progressive increase of frontostriatal brain activation from childhood to adulthood during event-related tasks of cognitive control. *Hum Brain Mapp*. December 2006;27(12):973–993.
433. Eshel N, Nelson EE, Blair RJ, Pine DS, Ernst M. Neural substrates of choice selection in adults and adolescents: development of the ventrolateral prefrontal and anterior cingulate cortices. *Neuropsychologia*. March 25, 2007;45(6):1270–1279.
434. Whelan R, Conrod PJ, Poline JB, et al. Adolescent impulsivity phenotypes characterized by distinct brain networks. *Nat Neurosci*. June 2012;15(6):920–925.
435. Yurgelun-Todd D. Emotional and cognitive changes during adolescence. *Curr Opin Neurobiol*. April 2007;17(2):251–257.
436. Rubia K, Smith AB, Taylor E, Brammer M. Linear age-correlated functional development of right inferior fronto-striato-cerebellar networks during response inhibition and anterior cingulate during error-related processes. *Hum Brain Mapp*. November 2007;28(11):1163–1177.
437. Velanova K, Wheeler ME, Luna B. Maturational changes in anterior cingulate and frontoparietal recruitment support the development of error processing and inhibitory control. *Cereb Cortex*. November 2008;18(11):2505–2522.
438. Inuggi A, Sanz-Arigita E, Gonzalez-Salinas C, Valero-Garcia AV, Garcia-Santos JM, Fuentes LJ. Brain functional connectivity changes in children that differ in impulsivity temperamental trait. *Front Behav Neurosci*. 2014;8:156.
439. Galvan A, Hare TA, Parra CE, et al. Earlier development of the accumbens relative to orbitofrontal cortex might underlie risk-taking behavior in adolescents. *J Neurosci*. June 21, 2006;26(25):6885–6892.
440. Ernst M, Romeo RD, Andersen SL. Neurobiology of the development of motivated behaviors in adolescence: a window into a neural systems model. *Pharmacol Biochem Behav*. September 2009;93(3):199–211.
441. Blakemore SJ, Robbins TW. Decision-making in the adolescent brain. *Nat Neurosci*. September 2012;15(9):1184–1191.
442. Steinberg L, Albert D, Cauffman E, Banich M, Graham S, Woolard J. Age differences in sensation seeking and impulsivity as indexed by behavior and self-report: evidence for a dual systems model. *Dev Psychol*. November 2008;44(6):1764–1778.
443. Giedd JN, Rapoport JL. Structural MRI of pediatric brain development: what have we learned and where are we going? *Neuron*. September 9, 2010;67(5):728–734.
444. Fenno L, Yizhar O, Deisseroth K. The development and application of optogenetics. *Annu Rev Neurosci*. 2011;34:389–412.
445. Steinberg EE, Christoffel DJ, Deisseroth K, Malenka RC. Illuminating circuitry relevant to psychiatric disorders with optogenetics. *Curr Opin Neurobiol*. September 9, 2014;30C:9–16.
446. Tye KM, Deisseroth K. Optogenetic investigation of neural circuits underlying brain disease in animal models. *Nat Rev Neurosci*. April 2012;13(4):251–266.
447. Burgess MA. Theory and methodology in executive function research. In: Rabbitt P, ed. *Methodology of Frontal and Executive Functions*. Hove: Psychology Press; 1997:81–116.

SECTION III

BRAIN IMAGING PERSPECTIVES ON UNDERSTANDING THE SELF AND OTHERS: FROM PERCEPTION TO SOCIAL COGNITION

CHAPTER

7

The Neuroscience of Social Vision

Ryan M. Stolier, Jonathan B. Freeman
Department of Psychology, New York University, New York, NY, USA

OUTLINE

Humans display impressive fluency in perceiving and understanding one another. Through our senses, we are able to discern the identities, social categories, traits, and minds of our conspecifics. These perceptions are often performed accurately, rapidly, automatically, and simultaneously. This feat is especially impressive considering the complexity of social stimuli. A large body of literature has made great progress in documenting how we infer social information from complex perceptual cues, including both static and dynamic information gleaned from a target's face and body. Nonetheless, it is impressive that this information could be perceived efficiently, especially when it is often ambiguous or buried in noise. Since social stimuli are among the most consequential for perceivers, it is important they still efficiently extract this information.

Increasingly, research has documented the role of top-down forces in assisting and biasing social perceptions.[1,2] Such work argues that our context, culture, prior knowledge, emotional, and motivational states can all have great weight in shaping visual perceptions. While social psychological research has documented the perceptual impact of many of these factors, the emergence of social neuroscience has proven vital in understanding the mechanisms involved and the extent of their influence, guiding and constraining theoretical development. In this chapter, we review the current scope of this contribution and discuss its trajectory moving forward. Given the lion's share of this work has regarded visual social perception, this chapter will focus on discussion of top-down influences in social vision.

The theory that top-down forces influence perception is not new, dating as far back as Helmholtz.[3] The empirical start of this approach was carried out by "New Look" researchers in the mid-twentieth century. The New Look saw perception as shaped by top-down factors such as motivations and expectations.[4] For instance, seminal work by Bruner and Goodman[5] found that children's size estimations of coins were biased by the value of the coins and the wealth background of the children.

Though the New Look eventually tapered off due to criticisms of its methodological and inferential rigor,[6] the ubiquity of twenty-first century top-down perception research in social psychology and neuroscience is

Neuroimaging Personality, Social Cognition, and Character
http://dx.doi.org/10.1016/B978-0-12-800935-2.00007-5

of no surprise. If perception is inferential, the role of top-down factors in *social* perception is even more plausible, given the consequential and ambiguous nature of social stimuli. Perceptions of others determine whom we trust and how we navigate interpersonal interactions. The information gleaned from a face is not always as clear-cut as objects and categories in the nonsocial world. For instance, there is large variation in cues that convey our age, sex, and race. Furthermore, these ostensibly independent categories often overlap in their cues and must all be extracted simultaneously. Cues may also be very ambiguous and fleeting in the decoding of transient aspects of a target, such as their current emotional state, beliefs, and intentions. In addition to the clear, adaptive function top-down factors would have in facilitating accurate perceptions, it may even be adaptive to have slight inaccuracies in perception if those inaccuracies facilitate adaptive behavior, such as erring on the side of caution to rapidly avoid potentially dangerous stimuli (e.g., mistaking a stick for a snake).[7] From the social psychological standpoint, such work has thus taken a functionalist perspective—where top-down influences on social vision streamline or alter visual processing to aid adaptive needs—which has proven productive.[1,2]

Another reason for the ubiquity of this work is due, in part, to current knowledge of pervasive top-down feedback processing in visual perception and the brain.[8–10] Research has identified white-matter feedback projections both within low- and high-level visual processing in the occipito-temporal cortex[11] and between many levels of visual processing and top-down regions, such as afferents from the amygdala, prefrontal, and orbitofrontal cortex to the occipito-temporal cortex.[12–14] Furthermore, functional neuroimaging has documented top-down shifts in visual representation reaching as far upstream as early vision in V1,[15] where a striking proportion of input is from higher-level regions.[16] Integrative work in perception has since benefited from this knowledge and advances in cognitive models, producing productive theories and research.[17,18] Such findings have been pivotal in both galvanizing and constraining theory into how top-down mechanisms may inform perceptual processing.

With its roots in both social perception and cognitive neuroscience, social neuroscience has quickly integrated this knowledge to produce valuable insights into these processes. In the current chapter, we will first review current knowledge of the functional neuroanatomy of social perceptual processes. We will then focus upon literature in social neuroscience regarding how social factors influence the visual perception of other people. We will also discuss research from the vision and cognitive neurosciences that provide valuable insights that may inspire future inquiry and observations from social perception ripe for exploration. Lastly, we will discuss the implications of this research for discourse into the origins and function of top-down influences in social vision.

1. SOCIAL (VISUAL) PERCEPTION

Humans are lay experts in predicting many aspects of one another from mere appearance and behavior. Consider an encounter with any stranger on the street. Visual information alone can make apparent another's current emotions, beliefs, and desires, and it can bring to mind stereotypes and traits belonging to that person, in spite of their personalities and histories as we know them. Often these inferences and recollections are achieved from thin slices of another's nonverbal behavior in extremely brief time frames and are impressively accurate.[19] On the other hand, these perceptions are also prone to processing idiosyncrasies and our biases, which can leave them systematically inaccurate. The study of social perception has a long history, which has unearthed much about these processes. Born and raised in social psychology, social perception research focused predominately on how initial perceptions impact our inferences, evaluations, and behaviors toward others. For example, such work has long shown that particular types of information, such as an individual being recognized as a friend or an individual being recognized as a Black male, guides our evaluation of and behavior toward that individual, often in unconscious and unintended ways.[20]

With the increased integration of social, cognitive, and neural sciences in the twenty-first century, more attention has been placed on understanding the processes that give rise to initial social perceptions. Such work has pored over what happens between the reception of sensory input and the experience of a final social percept, such as how the flood of visual information on the retina is transformed into the happy, familiar face we interpret and act upon. Such work has made much headway, documenting how specific face, body, and voice cues assist in our recognition of others' identity, social category membership, traits, and mental states.[21–23] Such work has also taught us much about the cognitive and neural mechanisms underlying these perceptual processes.[24,25]

In social neuroscience, the bulk of research on top-down influences has explored how they modulate or control evaluative or behavioral responses, such as the regulation of stereotypes or negative attitudes in response to out-group members.[26,27] In this chapter, we focus our attention on how top-down processes impact the initial social perceptions themselves, which in turn trigger those stereotypes and attitudes. We will focus upon how social factors influence the visual processing of faces, including social categorization, trait judgment, emotion recognition, and identity recognition. To lay out a foundation for our discussion of the social cognitive impact on processes on face perception, first we review current knowledge of their functional neuroanatomy.

1.1 Identity Recognition

Any story of face perception typically begins with the recognition of identity, as clearly, the successful recognition of familiar others is fundamental to life quality and survival. Identity recognition is a first necessary step in recalling crucial knowledge necessary to successfully navigate social interaction, and accordingly, face recognition has been a focal topic of research. Although a seemingly straightforward process, the computations required to recognize a face's identity and rapidly integrate configural integration of a complex, dynamic stimulus whose features vary across space and time are quite complex. The same individual face can appear different due to attire, age, facial expression, angle, and lighting. Nevertheless, human face recognition is remarkably accurate and efficient against all odds.

Prominent models of face perception put forth core and extended systems responsible for face recognition. These models carve the face perception process into two paths, one for the processing of dynamic features, such as emotion expression, and one for the processing of static features, such as face identity (as well as social categories, discussed below).[24,25] Convergent evidence from functional neuroimaging[28] and lesion patients with face-recognition deficits[29] (prosopagnosics) suggests key contributions of several regions to successful recognition. The primary system for processing static features, such as face identity, is dominant in the right hemisphere and is located along the hierarchical ventral-visual "what" stream responsible for visual stimulus recognition. In this hierarchical processing stream, stimulus representation begins in its retinotopic visual configuration and becomes increasingly more complex and conceptually constrained along the ventral-visual stream. Specific to faces, early feature processing occurs in regions such as the occipital face area (OFA). This information is then used to form higher-order representations, such as the holistic percept of a face, represented further downstream in the ventral temporal cortex (VTC), including the fusiform gyrus (FG) and fusiform face area (FFA) within it.[30] Such higher-order representations are thought to integrate information about a face into a more visual-independent representation. Interestingly, recent perspectives have argued that visually-independent, abstract representations of a face's identity are housed in the right anterior temporal lobe.[31] Most relevant to our discussion in this chapter, however, is how the extended social brain as a whole is also suggested to play important roles in face recognition.[32]

1.2 Social Categorization

Social categorization is the process through which we group individuals based upon social information. The "Big Three" are sex, race, and age, but numerous other dimensions are categorized as well, such as social status, occupation, and even perceptually ambiguous categories such as sexual orientation.[33,34] Once determined, our social categorizations of others shape downstream evaluation and behavior, often without awareness.[35,36] This can occur largely through stereotypic associations, which can result not only in harmful biases, such as a tendency to accidentally shoot individuals who belong to racial groups stereotyped to be hostile,[37] but also ostensibly trivial biases, such as assumptions about the physical strength of young and elderly adults. Social categorizations also elicit evaluative biases and activate related attitudes (e.g., negative attitudes about Black individuals), which can exert strong impacts on behavior often in unintended ways.[35] Somewhat surprisingly, the neural architecture supporting social categorization has been largely unexplored, in part because much interest in social categorization has focused on the outcomes following the categorization process (e.g.,[38,39]). Indeed, for over half a century, social categorization has been considered a precursor to stereotyping and prejudice.[40] Recent models, however, propose that the reverse may also be true. As a social percept is processed in real time, stereotype and attitude structures may begin to spontaneously activate that in turn shape how that percept is even visually processed, molding it to conform to expectations derived from those stereotypes and attitudes.[36] We explore such reciprocal social categorization processes in this chapter.

The features that give rise to social categorizations are often static, such as the shape of facial features, hair, and skin color. Accordingly, the social categorization of faces is undertaken primarily by the ventral-visual stream, including the OFA and FG/FFA.[24] Corroborating this perspective, research has consistently found unique neural patterns for different races and sexes in these regions,[41–43] and these patterns are highly sensitive to natural gradations in such social category cues.[44] A large network of brain regions respond differentially to different visual social categories, but current knowledge indicates the regions in VTC play a central role in social category representation, consistent with their general role in visual categorization.[45] Given the wide impact of social categorizations, stereotyping, and prejudice on evaluation and behavior, different social categories also elicit unique responses in a number of cortical and subcortical regions, such as those involved in conflict monitoring (anterior cingulate cortex), regulation (dorsolateral prefrontal cortex), and evaluation (amygdala, OFC) (for reviews, see Refs 26,27). While these regions are highly responsive to different social categories, this chapter is concerned more with the visual representation of social categories and how social cognitive processes fundamentally mold that representation, rather than downstream evaluative or regulatory processes.

1.3 Emotion Recognition

Recognition of emotion from facial and body expressions is crucial to adaptive social behavior. Emotion recognition guides response and action toward potential friendly or threatening others. As well, emotion recognition is paramount to successful communication between individuals. In order to identify emotions, we process both static and dynamic cues, such as facial expressions and bodily gestures. To make matters more complex, in naturalistic encounters, these emotional expressions are often quite ambiguous[46] and occur rapidly (even without our awareness) and therefore depend upon one another and context to be accurately identified.

While involving static features, emotion recognition differs from face recognition and social categorization in its dependence upon dynamic cues that are often configural. Dynamic cues processing is considered mostly separate from the static cue system discussed earlier. The independence of two systems involved in decoding static versus dynamic cues is an integral aspect of the functional architecture laid out in the Bruce and Young[25] model of face perception, and current neural models consider dynamic cue processing dependent primarily upon the superior temporal sulcus (STS).[24] Multivoxel pattern analyses have lent support to this model, finding the STS to carry categorical information about multiple emotion expressions (potentially right-lateralized).[47,48] While much work has focused on the role of the STS in emotion recognition, recent studies also suggest that ventral-temporal regions, such as the FG, are involved in carrying information about emotion expression categories as well.[49]

That recognition of and responses to emotion expressions depend on both static and dynamic information evokes questions about if and how this information is integrated. Investigations of face processing white-matter tracts found substantial connectivity between the OFA and FG, but no connections between the OFA/FG and STS.[50,51] Transcranial magnetic stimulation (TMS) delivered to the OFA reduces FFA responses to static and dynamic face stimuli, yet only reduces STS responses to static faces.[52] These results bolster a dissociated network in processing static versus dynamic features, yet refine and complicate its structure. It seems that the STS receives dynamic information from early visual regions outside of static face processing regions (e.g., OFA), whereas it does receive static information from the OFA. Future research will be needed to understand if, where, and how this information is integrated to form emotion expression representations in the STS, as well as how the STS receives static information if not through direct white-matter tracts.[50] Furthermore, a critical question to emotion perception is to what degree static and dynamic emotion computations are integrated, and what contributions do each together or separately make to visual experience and behavioral responses. The susceptibility of emotion expression representations in these regions to top-down factors may come to inform their integrated and separable roles.

1.4 Trait Attribution

Among many other dimensions of social perception, humans naturally infer a wide range of traits from the mere appearance of another's face. These judgments tend to be consensual in that perceivers strongly agree in their evaluations, even with very limited exposure to a face,[53] and in some cases, these judgments can be surprisingly accurate.[19] Whether accurate or not, these perceptions may yield important consequences, such as differential outcomes in court for baby-faced as opposed to mature-faced defendants.[54] Overgeneralization theory accounts for many of these judgments, whereby trait judgments are extracted from facial features that associate with them.[22,55] For instance, neotonous features signal submissiveness and innocence due to their similarity to infants, or features on a neutral face similar to positive emotions signal trustworthiness due to their association with positive interaction.

Research into the brain regions underlying trait representation is still in its infancy, but consistent with the discussion so far of static features, evidence implicates the FG to be generally responsive to various trait attributions. Thus far, studies have found the FG to be responsive to baby-facedness[56] and trustworthiness,[57] though the extent of its involvement is still undetermined. Trustworthiness judgments have received considerable attention in the neuroimaging literature, due to their consequential nature and primacy as a dimension in social perception.[22] While static feature and trait representation may be housed primarily within the ventral-visual stream, the process of overgeneralization requires the involvement of many additional mechanisms. For instance, if trustworthiness judgments are the by-product of subtle emotion perception, they likely involve the interaction of face representation (e.g., FG), emotion expression processing (e.g., STS), and evaluative regions responsive to emotional expression (e.g., the amygdala). A series of studies have substantiated the finding that trustworthiness cues are implicitly tracked by the amygdala,[58] even when faces are presented without subjective awareness.[59] However, while activity in both the fusiform and amygdala may underlie trait attribution processing, recent research has also found that these regions may be responding to the typicality of faces, which covaries with trustworthiness cues.[60] Such findings suggest that some of the curvilinear effects of trustworthiness in the amygdala (i.e., higher responses to faces appearing either more untrustworthy or trustworthy, relative to neutral) may possibly be accounted for by mere typicality effects, where deviations from the "typical" face elicit higher amygdala responses. However, such findings do not account well for the negative-linear effects of trustworthiness (i.e., higher responses to more untrustworthy faces) that also exist in the amygdala in different subregions.[59,61] Nevertheless, such findings suggest that there may be multiple component processes underlying the processing of facial traits, such as emotion overgeneralization and face typicality.[62]

2. SOCIAL INFLUENCES ON VISUAL PERCEPTION

The influence and fundamental role of top-down factors in social perception has received profound attention in social vision research.[1,2] Social perception is susceptible to countless social factors, including familiarity and prior knowledge, stereotypes and attitudes, group motivations and biases, emotion, and social context. Social perception is also malleable to the myriad "nonsocial" top-down influences that impact perceptual processing, such as expectation, processing goals, and attention,[8] in ways that are likely to be quite socially consequential.

2.1 Stereotypes and Attitudes

One abundant source of social information highly likely to guide social perception is stereotypes and attitudes. An impressive literature documents the myriad intergroup stereotypes and attitudes individuals form and their robust influence in social cognition and behavior (e.g., Ref. 38). Stereotypes are trait and behavioral ascriptions generalized to a social category, such as hostile stereotypes of African Americans in the United States. Stereotypes are nuanced, and their precise content guides cognition in specific ways.[38,39,63] We may come to avoid or overcome those we see as dangerous or pity and assist those we positively regard and believe unfortunate and helpless. We also may come to associate perceptual features of a target with stereotypes, such as Afrocentric facial features[64] or trustworthiness cues.[65] These associations make clear the expectations we may accrue about one another based upon stereotypes. Our stereotypes of race, sex, and age, among others, all elicit associations with the many dimensions of social perception, such as facial features and traits, voice, and body language.

Race and sex categories are strongly associated with stereotypes and emotional responses tied to approach-avoidance behaviors, and in the presence of sufficient feedback structures, stereotypes may play a crucial role in driving even the visual perception of those categories. A rapidly growing body of behavioral studies has begun to document an interesting source of stereotype feedback in social category perception. For instance, different social categories that incidentally share stereotypes facilitate recognition of one another.[36] One example of this process is where one category (e.g., male) is perceived more efficiently if it happens to share stereotype contents (e.g., "aggressive") with a presumably unrelated category (e.g., Black), and that "unrelated" category becomes activated. This leads to a number of perceptual effects, such as male categorizations of Black faces being especially facilitated, female Black faces partially activating the male category, and gender-ambiguous Black faces being overwhelmingly categorized as male.[67] Such intercategory relations are even found to drive an array of behaviors from interracial marriage to leadership selection.[66] The initial findings came from social categorization work, where categorization along one social dimension facilitated and inhibited categorizations along other categories. This has been shown to occur for race and sex (Black male, Asian female),[67,68] race and emotion (Black anger),[69,70] and sex and emotion (male anger, female joy),[71,72] among others. For instance, categorization efficiency of Black faces increases when they have an angry expression, a relationship that increases with racial prejudice.[69,70] Recent theoretical work has integrated these and other findings into a computational model, proposing stereotypes as one route through which this intercategory facilitation occurs. By this account, the processing of facial features (e.g., skin tone) begins eliciting category activation (e.g., Black), which in turn begins automatically activating associated stereotypes (e.g., hostile). With stereotypes activated, they become an implicit expectation that then guides the categorization process. The recurrent feedback naturally part of this dynamic system thereby allows activation of stereotypes to return upstream and shape other category activations, including those that did not initially activate the stereotype. Thus, for example, when processing a Black face with a happy expression, race-triggered stereotypes may become activated that then place an immediate top-down constraint on the perception of the face's emotion, leading its perception to be biased toward anger. Overall, this work suggests that the visual perception of social categories is the end-result of a dynamic and malleable process wherein bottom-up facial cues and top-down stereotypes form a "compromise" over time, in some cases biasing perceptions in accord with one's expectations.

We recently conducted an fMRI study to examine the neural mechanisms underlying this dynamic process, specifically in the context of race and emotion.[73] In the scanner, subjects passively viewed faces independently varying along race (from White to Black) and emotion (from happy to angry). Following the scan, they completed a mouse-tracking task that measured individual differences in stereotype associations linking Blacks to anger and Whites to joy. In a mouse-tracking task, mouse trajectories are recorded as participants categorize a stimulus along a particular dimension by clicking on one of two responses in either top corners of the computer screen (e.g., Black versus White, anger versus joy). As participants head toward one response option (e.g., joy), their mouse trajectory may initially curve toward the other response option (e.g., anger) if that response option is stereotypically associated with one of the face's task-irrelevant category memberships (e.g., race is Black). For instance, while categorizing a Black face with a joy expression, participants may initially curve toward the "anger" response before ultimately selecting "joy," due to stereotypes associating Black with hostility and anger. In this particular study, the researchers used this task as an index of individual differences in stereotypes linking race and emotion (Black anger, White joy).

As faces became more stereotypically incongruent, we found that the anterior cingulate cortex (ACC), a region important for conflict monitoring,[74] showed linearly increasing activation. The ACC also showed increased functional connectivity with the FG. They argue that, when viewing a face, the ACC may have been involved in resolving conflicts between the bottom-up cue-driven interpretation (e.g., happy Black face) and the top-down stereotype-driven interpretation (e.g., angry Black face). This, in turn, may have led to greater communication with the FG, either for receiving more perceptual evidence to resolve the conflict, contributing notice of the conflict back to the FG, or both. Furthermore, the dorsolateral PFC (dlPFC), a region implicated in inhibiting prepotent responses,[75] showed heightened responses to stereotype-incongruent targets (e.g., happy Black face) in individuals with stronger stereotypic associations (as assessed using the postscan mouse-tracking task). Thus, one possibility is that the dlPFC, through functional connectivity with the ACC,[76] may have served the function of suppressing a stereotype-driven interpretation to make way for the veridical, cue-based interpretation (also see Ref. 77). It is noteworthy that these results were obtained when subjects were merely viewing faces passively in the scanner. Thus, the findings suggest that conflict monitoring and inhibitory mechanisms may help automatically clear inappropriate, stereotype-driven interpretations from the processing landscape, ultimately allowing us to see faces for what they really are through the veil of stereotypes.[73]

Such results are informative as to the mechanisms underlying our ability to resolve the natural inconsistencies often encountered between our stereotypical expectations and another's actual face. Such stereotypical expectations can become activated by simultaneous category memberships, as with race and emotion above, but also by numerous other sources. Although these results implicate several brain regions in an overall sensitivity to stereotypic incongruences, we were also interested more directly in the representational structure of a face's social categories. Specifically, we were interested in how that structure can become altered by one's stereotypes at multiple levels of cortical processing, in turn reflecting a bias in visual perceptions. In a recent fMRI study, participants viewed faces crossed on gender, race, and emotion categories in the scanner.[78] Participants also completed a postscan mouse-tracking task assessing the degree to which the targets activated similar social categories due to shared stereotypes (e.g., to what degree Black targets are implicitly perceived to be more similar to male than female targets, due to overlapping stereotypes). Such stereotypically biased similarities between categories (e.g., Black and male) in subjective perceptions were reflected in the similarity of the categories' multivoxel representations in the FG and OFC, even while controlling for any possible featural similarities. These results suggest that these regions were involved in representing a face's multiple social categories, and importantly, in a manner systematically biased by stereotype information. It is possible that the OFC may be involved in spontaneously retrieving stereotype knowledge and generating implicit expectations (e.g., Blacks are hostile; men are hostile), congruent with prior lesion work.[79,80] Following stereotype retrieval, the OFC could then provide feedback to the FG to bias social category representations of a face, consistent with top-down feedback models in visual object recognition (Figure 1).[13,17,18] Such findings suggest that visual representations of faces' social categories in the FG may be biased systematically by one's stereotypical expectations, which may be imposed by the OFC.

In addition to stereotypes, individuals also develop strong attitudes and evaluative biases toward others that are typically positive or negative in nature.[81] These attitudes often manifest at an implicit level,[82] especially negative ones, and such implicit negative attitudes often predict less successful intergroup interactions in spite of individuals' explicit goals.[83,84] There are many aspects of visual perception that are likely to be influenced by attitudes. Among them is the perceived similarity of in-group and out-group members, as well as liked and disliked others in general. Out-group members are typically considered more dissimilar than in-group members.[85] One

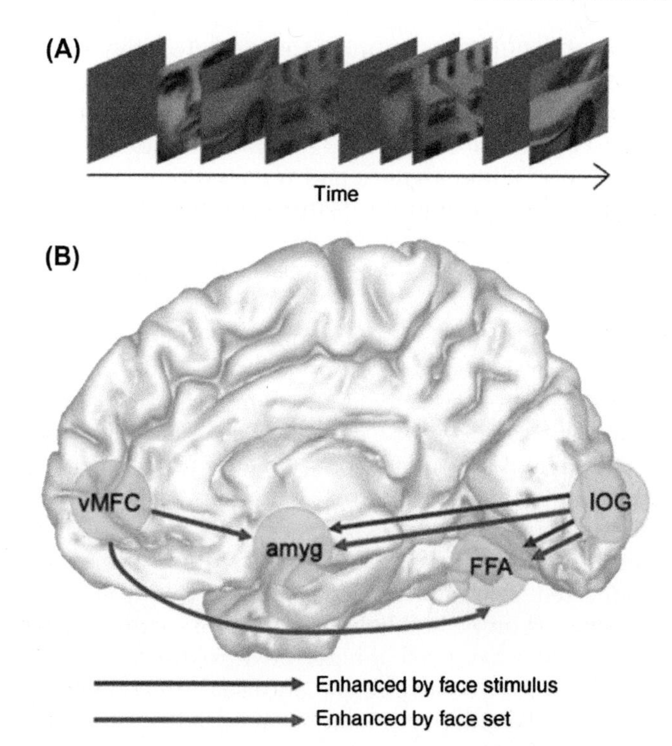

FIGURE 1 Schematic of how top-down feedback regarding expectations is contributed to visual processing of faces.[17] (A) In this study, participants viewed face and house stimuli during sets in which they were to identify one or the other. (B) Connectivity patterns enhanced by face stimuli and face sets. While seeking face stimuli, the OFC (vMFC) showed functional connectivity with the fusiform face area, putatively providing predictions to assist in target recognition. *Adapted from Summerfield and Egner.*[17]

fascinating study explored the neural correlates of this biased tendency, where participants viewed White and Black faces during fMRI, after which they completed an implicit measure of racial attitudes.[86] The authors found that participants with higher pro-White bias held more unique multivoxel representational patterns for White and Black faces in the FFA, suggesting that visual representations of White and Black faces in the FFA were more distinct for those with more biased attitudes. One likely explanation for these effects is the learning and formation of neuronal population codes in the FFA associated with White and Black faces that become more distinct and sharpened over time (or in less biased individuals, they become less distinct and more overlapping over time). In future work, it will be interesting to directly study the tuning of such stereotypically or attitudinally biased face representations over time (whether developmentally or via manipulations), as well as to explore the relevant moderators to better understand these patterns' flexibility and boundary conditions.

Although the effect of social experience and biases on the visual representation of social categories is quite new, there is a rich literature examining their effect on visual representations of a face's identity. The cross-race effect (or own-race bias) is a consistently replicated and highly robust phenomenon whereby own-race faces are better recognized than other-race faces. Its causal mechanisms have been widely documented, extending from perceptual expertise[87] to motivational and cognitive differences in out-group face perception, such as increased individuation toward own-group faces.[88] One such difference in perceptual processing, whether due to perceptual expertise with the out-group or intergroup motivations, is decreased configural face processing.[89] Configural processing of faces is largely dependent on face-specialized regions in the right FG, such as the FFA.[90] To investigate its role in the cross-race effect, a behavioral study presented White and Black faces to the left and right visual fields of participants during a recognition task and found increased cross-race effect for faces presented in the left visual field.[91] Because information presented to one side of the visual field in each eye is routed to the contralateral visual cortex, faces presented to the left visual field were routed to the right visual cortex. The increased cross-race recognition deficit during left field presentations may therefore be explained by right visual cortex dependence on configural processing of faces. It is likely the left visual cortex employed a more similar, featural strategy to face processing and thus processed both races similarly. Thus, configural processing mediated by right visual cortical mechanisms seem to play a particularly important role in the cross-race effect.

To examine the neural correlates of this effect more directly, a seminal study using fMRI found increased bilateral fusiform activity to own-race, compared to other-race, faces during initial encoding, suggesting enhanced processing in face-selective regions may underlie better encoding.[92] Interestingly, as neural responses in the left FG responded more strongly to own- than other-race faces, the cross-race effect increased. That is, larger differences in left FG activity related to better performance for own- compared to other-race faces. However, the lack of right FG correlation with recognition performance is puzzling, due to its primacy in configural face processing. Though inconclusive, findings within the left FG raise interesting questions about its role in face recognition. A related study looking at cross-race deficits in race categorization (better race categorization of own-race faces) have found similar, though opposite, results and suggested featural processing dependence in the left FG to play a part.[93] An event-related potential (ERP) adaptation study has since shed further light on the neural processes underlying the cross-race effect. The N170, a face-selective ERP involved in configural face encoding, showed adaptation (reduced amplitude) to repeated identities in the right hemisphere.[94] These results suggest that other-race faces are perceived as more similar, supporting accounts of the other-race effect as relating to the extent of individuation or categorical perception.[88] These findings may also be due to differences in perceptual expertise, among other factors, but importantly, they show this modulation to occur at core, early stages in perceptual face processing.

We have discussed several ways in which stereotypes, attitudes, and other social biases between groups may influence the visual perception of faces. Stereotypes are a rich source of expectations, and intergroup attitudes may motivate us to see others as dissimilar from ourselves or fail to individuate one another. Although the pronounced impact these factors have on downstream evaluation and behavior has long been recognized, it has only recently been considered that social biases can trickle down to affect lower-level visual processes, shaping our earliest visual perceptions of other people. The nascent study of these topics has left many questions open about the structure and function of these top-down effects on visual perception, which future work in this burgeoning area will need to explore.

2.2 Person Knowledge and Familiarity

Our experience and knowledge of one another are replete with expectations. One expectation is of course appearance, particularly the features and their configuration with which we identify an individual. As we have discussed, a single identity varies in its diagnostic features over space and time. Therefore our familiarity with a face can be crucial to its detection despite shifts in appearance. There is no stimulus with which humans are more adept and experienced than the face, and thus a large volume of research has investigated familiarity in face perception. In one behavioral study, face identities were morphed between two identities, creating a continuum of faces varying between the physical features of both identities.[95] Participants perceived face identity categorically, with an abrupt boundary between morph levels discriminating the two identities from one another. Importantly, participant familiarity with the faces positively predicted the degree of categorical perception. The more familiar a participant was, the more morphed faces were perceived as a specific identity. Another study investigated this during fMRI by looking at stimulus adaptation effects, where neural responses decrease in a region sensitive to properties of a repeated stimulus.[96] Using a morph continuum between Marilyn Monroe and Margaret Thatcher, participants were shown sequential pairs of the morphs, always varying in the same degree of featural changes. They found that the OFA showed similar adaptation to all morph-pairs, implying these regions process featural aspects of the face. However, the FG and FFA showed increased adaptation to morph-pairs recognized as the same identity, implying these regions are sensitive to the identity of a face. Importantly, the degree of these regions' adaptation to identity related positively with participants' familiarity with Monroe and Thatcher before the experiment. These findings suggest that visual FG face representations exist at a higher level than their mere visual features, specific to knowledge about the identities of the targets. While identity and representation of a face depend on visual stimulus features, our familiarity with an individual can constrain these representations and identify faces that do not match perfectly as belonging to the same stimulus. However, an open question remains as to how this learning and modulation occurs.

The neural process underlying familiar face recognition is likely exceedingly more complex. A prominent view now assumes familiarity effects to rely partly upon higher-level knowledge about the target.[32] However, investigation of this is difficult, as familiarity may be due to visual or social experience with the target. Furthermore, the influence of social familiarity is likely manifested through various routes, as social familiarity depends on person knowledge as well as evaluation and attitudes toward that individual. To dissect this process, experiments have often contrasted different forms of familiarity. Adaptation studies have failed to reliably show familiarity effects independent of stimulus features (for a review, see Ref. 97). Therefore it is possible that familiarity effects in the FG are dependent on visual familiarity with the stimulus alone.[96] We may come to better understand exactly how familiarity assists face perception through the study of identity population codes and focus on the extent of the VTC, including anterior regions that may more uniquely represent identity.[31]

A recent study addressed this question by presenting participants with faces of individuals, some of whom were associated with biographical information.[98] These faces varied in orientation to control for visual familiarity. While multivoxel identity patterns were not modulated by the biographical information specifically associated with the target, targets associated with biographical information increased in representational similarity to one another. Similar coding of targets associated with person knowledge suggests social information biases fusiform representations of identities, but the underlying process remains elusive. Nonetheless, this study depended on person knowledge gained through relatively superficial impressions, and the modulatory effects of different forms and degrees of familiarity remain to be seen. If person knowledge does have a unique contribution to visual processing, interesting questions will concern the dynamics of its influence and how it shifts representational structure. For instance, it is possible that familiarity and prior knowledge is fed back to the fusiform online to guide representation.[99] It is also plausible that experience with familiarity of the target structures more permanent population codes in the FG, and familiarity affects the criterion a stimuli must meet to activate this population.

Research outside of neuroimaging has provided convincing evidence that speaks to the role of person knowledge in early face processing. Specifically, individuals associated with negative behaviors are more likely to reach and dominate conscious visual perception.[100] Participants learned about different faces, paired with positive, neutral, or negative information. Participants then completed a binocular rivalry task. In binocular rivalry, different images are presented to both eyes of the participant, and one image comes to dominate conscious visual experience while the other is suppressed. The dominant image in binocular rivalry is largely dependent on competition in early visual processing, while top-down factors, such as attention, modulate this.[101] When different conditions were presented to both eyes, negative-associated targets were found to dominate in binocular rivalry. Impressively, this effect was driven specifically by social information. These results suggest that preconscious processing of the target elicits person knowledge that is fed back to enhance its visual representation. The primacy of negative-associated targets evokes interesting questions about the adaptive role of this feedback, for instance to promote vigilance toward threats (a topic we address later in the chapter).

Neuroimaging work examining the impact of person knowledge on social perception has focused on its impact on trait judgments and impression formation. One study found modulation of both amygdala and fusiform activity when subjects made personality judgments of faces with and without prior knowledge of the target.[102] With no prior knowledge, personality judgments elicited increased amygdala responses for making rapid judgments of faces. However, with prior knowledge of the target, although making the same judgments of faces, amygdala responses were absent with the activation of a cortical mentalizing network in their stead, including the STS and posterior cingulate cortex. These results suggest that prior knowledge modulated the role of the amygdala in personality judgments, supplanting it with mentalizing processes. This provides evidence that, although the amygdala's role in rapid trait inferences is quite spontaneous and can occur even without subjective awareness,[59] such processes may also be sensitive to context and top-down social factors. Future work will be tasked with investigating how these shifts interact with upstream visual processing, as well as possibly adopting neural decoding approaches to better understand the representational content of regional responses to personality traits.[103,104]

Modulation of social perceptual regions through prior knowledge about a target is a largely unexplored topic. The unique contributions of different sources and content of this information is critical to explicating this process. Therefore, future work would do well to consider these distinctions, such as visual and knowledge-based familiarity (for a discussion, see Ref. 97). The contribution of person knowledge to face perception may depend on regions considered key in person knowledge, such as the medial prefrontal cortex (mPFC),[105] which has been shown to make unique contributions to familiarity effects in face perception independent of visual familiarity.[106] Investigation of the coupling between regions involved in person knowledge and face perception will be crucial to understanding the scope of familiarity's influence. Furthermore, the extent and function of this modulation may differ along the ventral visual stream, influencing earlier visual and later higher-level representation differently.

2.3 Motivation and Goals

In contrast to the specific knowledge and expectations provided by social group stereotypes, a number of motivational biases are inherent in merely categorizing ourselves and others as in- or out-group members. Perhaps one of the most consequential of human behaviors, humans are innately coalitional and come to support in-group members while being wary of or hostile toward out-group members. The biases that follow this tendency are comprehensively documented in social psychology and have an equally long history.[107,108] Prominent theories have studied how social identity with a group as well as minimal group categorizations bias cognition and behavior across a variety of domains. In contrast to familiarity, stereotypes, and attitudes learned through culture and intergroup interaction, these biases occur, importantly, in the complete absence of stereotypes and prior experience with both in- and out-group members. The neural substrates guiding in- and out-group categorization and underlying structural schemas that allow these biases to generalize still have yet to be investigated. However, research in social neuroscience has already begun to demonstrate the fascinating ways in which coalitional motivations influence social perception.

A powerful procedure for investigating unadulterated coalitional biases is the minimal group paradigm.[107] This method involves assignment of the participant to a "minimal" group, a group with which the participant has no prior experience or knowledge. Initial application of this in a neuroimaging context assigned participants to one of two arbitrary teams, the Tigers and Leopards.[109] During fMRI, participants learned, then viewed, the faces of both in-group and out-group members, with an equal number of White and Black faces assigned to both teams. In-group member faces elicited stronger bilateral FG responses than out-group members, an effect that was not moderated by target race. Increased in-group responsiveness in the fusiform is consistent with prior research, where such findings have occurred in the context of race.[92,110] Prior work interpreted this bias as potentially reflecting learning and perceptual expertise with the in-group (e.g., cross-race effect, see above).[92] However, these findings support the perspective that increased FG activity corresponds to more effortful processing by the regions toward targets over and above target visual features, perhaps due to individuation processes withheld from out-group faces.[88,94]

The above research raises questions regarding the modulation of race representation in the fusiform. The mitigation of gross response increases toward own-race faces may be interpreted as the fusiform instead focusing on minimal group membership, consistent with behavioral studies finding race to be discounted to a degree when not diagnostic of in- or out-group membership.[111] Regional fMRI signal increases may be indicative of involvement during processing but do not necessarily reflect information represented within any set of voxels. In addition to general increased activity in the fusiform, later research investigated the informational content of the fusiform during a minimal group task. Similar to the study above, participants were presented with novel in- and out-group members, with both White and Black faces assigned to each team.[43] Replicating prior findings, the FG showed increased responding to the faces of in-group members. Despite the absence of race differences in regional activity, multivoxel patterns indicated that race was still represented in the fusiform cortex, even more so than earlier visual cortical regions. Such findings therefore suggest that group membership is unlikely to abate race representation itself, but it may mitigate cross-race perceptual differences through shifts in regional engagement in processing.

Nonetheless, there is much to learn about the impact of coalitional cognition on social perceptual mechanisms. The further exploration of how group membership alters the representation of others is an open area of inquiry, and it will also be important for research to identify how such top-down effects on visual perceptions relate to the robust biases observed in downstream behavior.

2.4 Emotional States

Emotions are internal states, defined by a conglomeration of physiological responses, which serve as powerful catalysts to adaptive cognition and behavior. These states inform us about the threats and affordances in the world around us, and our internal motivations and desires. Their nature is so intuitive that the study of emotion has maintained constructs derived from folk theories of emotion (e.g., fear, anger, and disgust). These internal states have an accordingly long history of functionalist accounts in modern science (e.g., Ref. 112). Traditional theories classified them as emotions with a focus on their functional role in human social and nonsocial behavior (for a review, see Ref. 113). Functionalist accounts of emotion have been valuable to social psychologists interested in how social perception then predicts behavior toward targets, such as how emotion-specific stereotypes and prejudices drive intergroup behavior.[38,63] These theories parsimoniously capture how we avoid others who elicit disgust or fear (due to perceived disease or danger) and approach others who elicit positive emotions (due to perceived benefits) or anger (due to perceived threats that must be overcome).

At the psychophysiological and neural levels of analysis, biomarkers have been identified that reflect accounts of discrete emotional experience. Some physiological research conceptualizes emotions as the psychological categorization and cognitive elaboration of a more fundamental set of underlying physiological states.[114] "Core affect" models the variance in internal states in line with neurophysiological data, reducing them into two dimensions: arousal and valence.[115] For instance, an experience of intensely unpleasant arousal may be construed differently as fear or disgust dependent on the context. Moreover, these internal states show profound impacts on cognition and behavior and may play a crucial role in social perception through the extent of their modulation across the nervous system.[13,116]

Considerable research has focused on the role of the amygdala in processing the affective significance of social stimuli and serving as an important modulator of perception. One such line of work comes from an interesting series of studies exploring the conscious awareness of faces presented subliminally. An initial study presented participants' happy, fearful, and neutral faces outside subjective awareness through a backward masking task.[117] Specifically, participants who were more often aware of the masked fearful faces showed enhanced amygdala activation to subliminal faces. The authors interpreted this as dependence of the amygdala response on conscious awareness of the fearful faces. However, an alternate interpretation was put forth, that only when amygdala responses occurred were participants consciously aware of the masked face.[118] Specifically, it was proposed that known afferents from the amygdala to the ventral-visual stream[12] continuously provide feedback that enhances visual processing to bring the target into awareness. That is, when the amygdala rapidly responds to the momentarily subliminal fearful face, affective information is fed back to enhance and sharpen face representation, increasing awareness.

Providing indirect evidence for this account, a later study had participants complete a binocular rivalry task with happy, angry, and neutral faces.[119] Participants were induced to experience positive, negative, or neutral affect. While negative affect induction exacerbated overall dominance of face stimuli during rivalry, congruency of affect and facial expression valence increased dominance of that facial expression (i.e., angry faces exhibited dominance during negative affect and positive faces during positive affect). The general increase in binocular dominance of face stimuli during negative affect supports one interesting hypothesis, that negative affect states may promote vigilance in perceptual processing toward motivationally relevant stimuli. However, that affective congruence with the stimulus valence increases binocular dominance provokes additional fascinating questions about the role of affect and emotion in perception (e.g., dominance of happy face during positive affect). One possibility is that this effect is due to valence congruency between affect and the stimulus. By this interpretation, positive and negative affect facilitate processing of all congruently valenced stimuli. However, it remains to be seen whether there is a more nuanced functional specificity for individual emotions and motivational states. If emotions do each serve specific adaptive functions, we may expect specific emotions to enhance detection of specific emotional expressions. For instance, it is possible a fearful state shows enhanced facilitation toward threatening stimuli (e.g., anger expressions). Functional specificity of this mechanism would also provoke other interesting questions, such as enhanced processing of different aspects of a target (e.g., disgust enhancing processing of faces with pathogen cues).

Beyond the modulation of perception, emotions may play a more integral role in perception. In fact, research into their role has been the spark and paragon of social perceptual systems inexorable from their top-down contributors. Seminal work demonstrated that bilateral amygdala damage impairs the recognition of fear expressions.[120] From both lesion and fMRI studies, repeated observation of amygdala dependence in fear recognition suggests the affective response to a target plays a causal role in its recognition.[121,122] Lesion deficits may indeed be due to a lack of necessary feedback to visual regions such as the fusiform cortex. In one demonstration, amygdala and control lesion subjects were shown neutral and fear faces during fMRI.[123] Both control and lesion subjects showed face-selective activity in the FG. Intriguingly, while control subjects showed even higher responsiveness in the FG to fearful faces, those with amygdala lesions did not. Furthermore, increases in the extent of amygdala damage parametrically predicted decreases in FG responsiveness to fearful faces. These observations suggest that the amygdala plays a role in the modulation of responses in visual regions, perhaps through increasing responsiveness to certain stimuli. In addition to fearful expression recognition, amygdala damage has been found to relate to many emotion recognition deficits,[121] including an increased deficit toward social emotions (e.g., guilt[124]).

That said, there is still uncertainty concerning the role of emotion in these processes. This has become increasingly clear regarding its role as an essential component in perception. While initial studies hinted at a perceptual system where the lack of emotional feedback to early visual regions precluded emotional expression recognition, later work provided an alternative that is rather compelling. Another potential route through which affective states could impair emotion recognition is through abnormal direction of attention to motivationally relevant stimuli.[121] This evidence comes from studies looking at fear recognition in a patient with complete bilateral amygdala lesions. In a first study, deficits in fear recognition were mediated by lessened attention to the eyes of faces; however, upon directing attention to the eyes, recognition accuracy returned.[125] A later study with the same patient showed normal detection of fearful faces during rapid presentation and masking tasks.[126] These studies therefore suggest the amygdala may only be involved in slow, deliberate, and conscious recognition of the emotional expression (e.g., where attentional search can impact decisions). Whatever the precise role of the amygdala is in modulating perception, Barrett and Bar[13] argue that emotions are predictive in nature and assist and guide perception. This perspective does not require that detriments in emotion preclude perception, but that they reduce its efficiency by preventing predictions of higher-order regions from feeding back to lower-level visual regions. Future research could explore how the lack of such predictions impacts perceptual responses.

2.5 Social Context

In the controlled environment of an experiment, individuals may categorize a cropped face superimposed on a white background. People in the wild are of course seen in meaningful contexts, such as the context of an organic grocery store, or the context of a political gathering. These contexts provide expectations about who we are to perceive within them, informing every facet of social perception. It is natural to anticipate someone's gender in a beauty parlor, emotion at a lively celebration, or identity in the living room of a close friend. In this sense, context is an aspect of the environment that activates expectations that elicit predictions about a target.

A large body of work has demonstrated how expectation influences perception through prediction,[17] and it has done so with the presentation of faces. While activity in the FG, a face-selective region, typically increases to face relative to house stimuli, this increase is also observed in the absence of any face stimuli when participants merely have the expectation that a face is to be presented.[127] FFA activation is also found in response to degraded face images once subjects have learned to detect the face, a process potentially recruiting parietal attention regions.[128] The influence of context as an expectation in perception has been a prominent topic in vision research. As we will return to in our discussion, findings of its profound role in vision pioneered our understanding of prediction in perception and recognition.[18] A predictive account of context allows us to consider a much richer contextual environment (for discussion, see Refs 36,129). Here we discuss context as any aspect of the environment that provides expectations and predictions about the perceptual computation at hand. As we shall see, these contexts include the environment, as well as the body and face, and inferences made from them.

The most patent contextual influence is the scene in which we encounter someone. As we have considered, scenes are ripe with information that predicts who we are to perceive. Yet context is also ripe with information that predicts who we are *not* to see. For instance, one may expect a target to be Asian in the context of a market in the outskirts of Hong Kong. Or, one may expect a target to be White in the context of a corporate supermarket in the Midwestern United States. As comes to mind in consideration of these examples, we may also expect *not* to see a race in the scenic context of another. One line of research has documented how scenic context influences social categorization. In an initial behavioral study,[130] participants categorized faces varying on a White-Asian morph continuum. These faces were presented within scenes associated with each race, as well as a neutral context. Relative to the neutral context, congruent race-context trials facilitated race categorization (e.g., Asian face within Asian scenes), and incongruent contexts interfered with race categorization (e.g., Asian face within American scenes). These findings support a predictive role of context both guiding and limiting potential outcomes.

During fMRI, participants completed the same contextual race categorization task.[131] Both the OFC and retrosplenial cortex (RSC) linearly increased with congruency of face and context. That is, each region increased as facial and context cues became increasingly compatible and decreased as they became increasingly incompatible, all relative to neutral pairs. Prior work on scenic contextual influences in object categorization has documented contributions from the OFC and RSC.[18] The OFC is believed to update knowledge of current context provided by visual regions and play a role in feeding predictions back to early visual regions, such as the FG.[17] The RSC is involved in spatial processing and responsive to full scenes,[132] and increases in activity with contextual-target congruency may underlie contextual associations.[18] Consistent with this role, RSC activity mediated the impact of face-context congruency on reaction times in face categorization. Together with the behavioral findings, these studies show the nuances contextual predictions provide in social perception. Not only do contextually evoked expectations facilitate perceptions, but they can also inhibit perceptions that are incompatible with the context.

Research has extensively documented how context imbues emotional expressions with meaning, especially given the often ambiguous nature of emotional expressions.[133] Scene contexts provide obvious indicators of the emotional state of those around us. When confronted with a long line to the ticket counter for a social event, we have a clear and likely accurate expectation of the frustration on the face of our neighbor. Neuroimaging research into scenes' contextual influences on emotion recognition have shown context to influence processing in early visual regions. In one study, neutral and fearful faces were presented to participants with congruent or incongruent background scenes (e.g., fearful face depicted in front of a burning home).[134] Consistent with an account of contextual prediction influences perceptual processing, activity in the FG increased during congruent face-scene pairs. Also consistent with an account of context inhibiting unlikely outcomes, there was a decrease in FG activity during incongruent face-scene trials. Together, such studies exploring contextual impacts of scenes on social perception implicate a network of regions involved in expectation (OFC), scene perception (RSC), and face perception (FG).

The context in which we perceive a target may also be situational. We would only expect a look of fear on someone in our social event line example if it was known that tickets were running low. One study has investigated how situational context impacts neural responding to emotion expressions, where participants viewed surprise (similar to fearful expressions) expressions in different contexts during fMRI.[135] Faces were preceded with sentences describing a positive or negative context, and faces in negative context evoked stronger amygdala responses. Importantly, these responses were seen in response to faces in context, but not to the contexts independently. Contrasts also found responsivity of the FG between negative and positive conditions. Future research will, however, be needed to fully characterize the mechanisms at play. In addition to exploring the role of situational context, this study provides evidence that context modulates evaluative processing in the amygdala, which as discussed earlier, may play an integral role in emotion perception.[116] Furthermore, in the study of emotion expression perception, increased amygdala responses are typically seen for facial expressions of fear,[136] making it possible that contextual information disambiguated the facial expression systematically due to fear-surprise expression similarity. Multivariate pattern analyses may lend themselves to investigating the neural underpinnings of such shifts more conclusively in the future.

In addition to the scenic and situational context, faces are of course typically atop an entire body. As visual context, the body is especially useful in providing predictions about a face, nonetheless about a target on their own. Akin to the face, body cues convey diagnostic information that reveals a target's intent, experience, traits, and even identity, a process that relies heavily on body-selective regions such as the extrastriate body area (EBA) and fusiform body area (FBA) (for review, see Ref. 137). At the most fundamental level, the mere presence of a body predicts a face situated appropriately upon it. In one study, participants viewed images of faces degraded so as to be unrecognizable as faces and either embedded upon a body appropriately or placed separately elsewhere within the same image.[138] Embedded upon a body, degraded face stimuli elicited FFA activity comparable to nondegraded face stimuli, whereas in other conditions, the degraded face stimuli did not elicit FFA activity. General visual increases in activity to degraded faces upon a body evidences the body as a contextual predictor. Most impressively, this response in the FFA provides convincing evidence of the specificity of this prediction. That the FFA would respond as though a face present is fascinating, as FFA responses are often considered face-specific and are held in higher level regions typically associated with visual experience.[139]

Of course, there are more nuanced predictions the body provides about targets than their possession of a head. Humans, like all animals, behave differently in accordance with our motivations and emotions. The body thus provides a clear window into our internal states. Research into this matter has focused on the congruency of facial and body expressions. For instance, behavioral research has paired fearful and anger face and body stimuli during a face emotion categorization task.[140] Accurate emotion recognition was best in congruent pairs, and when face–body expressions were incongruent, the body expression largely drove emotional categorization of the facial expression. Another study used faces continuously morphed between happy and fearful with happy and fearful body stimuli, finding the body expressions to drive perceptions of the ambiguous facial expressions.[141] Recent research has reiterated the dominance of the body in face–body influences in emotion recognition, focusing even more broadly on valence judgment of emotional expressions.[142] This occurs strikingly to the degree that valence of high intensity natural facial emotional expressions were not discriminated above chance, whereas the valence of body expressions were discriminated accurately and drove valence judgments of facial expressions. Using event-related potentials (ERPs), the congruency of the face and body has been found to modulate neural responding as early as 115 ms.[140] Specifically, Meeren and colleagues found increased occipital P1 amplitude toward incongruent face–body pairs. The P1 is associated largely with attention, and its generator is potentially located in the ventral extrastriate cortex.[143] The response of the P1 to congruency of the face and body implies these separate aspects of the target impact one another early in processing. The rapidity of this modulation leaves open the possibility that their interaction is dynamic and contributes to initial representations that are then fed forward.

Models of expectation and prediction in visual perception propose a predictive role of the OFC in guiding earlier OTC visual representation, especially via context.[18] It is still unclear to what extent these influences modulate visual representation as opposed to visual responding. The introduction of multivariate pattern analyses may serve as a powerful tool in explicating the outcome of these influences in OTC. In combination with research focused upon those percepts most malleable to context (e.g., emotion expression), pattern analysis tools may allow researchers to assess to what degree and how early visual representations are modulated toward those expected. Given its general role in expectation and prediction feedback in vision, the OFC, as we will continue to discuss, may play a general role across contexts.[17] In addition to a general OFC–OTC prediction loop, the nuances of different contextual sources likely depend on different associative cortical regions to exert their influence (e.g., such as scene or body-processing regions). Furthermore, if socially unique effects are continuously encountered in this research (e.g., Ref. 134), an important line of work may pursue the socially domain-specific nature of these effects.

3. MECHANISMS OF SOCIAL VISION

We believe findings in social psychology and neuroscience may provide new avenues for investigation to both extend and elaborate top-down influences in vision. Outside of social psychology and neuroscience, the vision sciences have investigated top-down influences in detail, providing compelling accounts of these processes that speak to their function and origins. At the neural level, top-down influences are likely to be enacted through a series of recurrent feedback connections. For instance, a considerable proportion of input in the visual cortex as early as V1 comes from higher-level regions, and observations of top-down modulation of V1 is accordingly ubiquitous.[8,16,144] Much of this input comes from within the extent of OTC.[11] A sizable amount of modulation also comes from nonperceptual regions of interest to social neuroscientists, such as the amygdala,[12,145] OFC,[13,17] and PFC.[14] While work has documented many direct anatomical sources of afferents in visual regions, much remains to be unveiled. In addition to direct connections, modulation is also likely enacted through successive processing and complex networks extending back to the visual cortex from various extraneous cortical and subcortical regions. Research in cognitive neuroscience has organized the top-down influences of interest to the field predominately under the general processes of attention and expectation, among others less relevant to the current discussion.[8,146] Generally, the current perspective is that top-down influences shift properties of receptive fields as well as the information carried by neuronal populations.

There are findings in social vision and neuroscience that may not fall neatly within these existing frameworks.[8,146] Many of these findings are currently explained by motivational factors. Much recent research has found individual motivations and biases to impact processes such as categorization. One particularly interesting set of questions regards the impact of group membership on perceptual processing. For instance, does heightened FG activity to novel in-group members[109] reflect expectation or attention, or potentially a perceptual processing difference toward this group? Or are these effects merely postperceptual feedback or memory-driven? As well, how do implicitly prejudiced participants show sharpened or more dissimilar representation of in- and out-group races in the FG?[86] Does this reflect the above component processes of expectation or attention? Is this a difference in perceptual processing, or potentially a by-product of learning influenced by the top-down factors? It will be vital for future work in this vein to investigate these processes independently of the processes discussed here, as well as familiarity.

Another interesting avenue for research will be the differential perception of ambiguous stimuli. Much work in social categorization has asked how top-down influences impact category boundaries between stimuli, such as how political orientation, economic scarcity threat, or perceived social status impact the category boundary between in-group White targets and out-group Black targets.[147–149] These effects may reflect perceptual changes in nature or may not elicit differences in perceptual experience. Neuroimaging research can help investigate the extent to which these shifts occur in visual processing regions. An initial glimpse at this process has already been provided by work showing categorical representations of identity in the occipitotemporal cortex related to familiarity with the targets.[96] If there are indeed perceptual processes at play, it will be fascinating to explore how neuronal populations shift their receptive fields and thresholds to represent categories, and whether such shifts comprise a unique process of top-down influence on perception.

4. AN INTEGRATIVE FRAMEWORK

The perceptual and top-down processes we have discussed are numerous. In addition to a more parsimonious account of top-down factors in social perception, recent computational accounts of person perception provide a possible integrative framework for these perceptual and top-down processes and their interactions. The dynamic-interactive theory of person construal provides a connectionist model and dynamical systems approach to understand how the many levels of social perceptual processing functionally interact and develop dynamically.[36] The model is a recurrent connectionist network allowing large-scale interaction to play out over time between parallel and multilevel processes.

There are several key characteristics of the model that dovetail with our current discussion. The model includes multiple levels of processing, such as featural processing (e.g., face, body, voice, and contextual cues), social category representation (e.g., race, sex, emotion, and identity), and top-down factors (e.g., emotion, motivation, context, and prior knowledge). Connections do not only ascend the levels of the model, but feedback connections allow for top-down levels to influence lower-level processing. For instance, emotional states may reach down and activate category nodes relevant to them, facilitating their recognition when featural input activates them (e.g., Ref. 119). Information not only transfers one level at a time in the hierarchy but may also bypass adjacent levels when there is an association with further levels. This characteristic may account for some bottom-up cases where cues directly activate stereotypes[64] and reflects how certain top-down afferents extend directly to early vision.[12] Parallel nodes also interact laterally, allowing them to facilitate and inhibit one another. Therefore multiple inputs, for instance, received in parallel, such as facial features, may influence processing of one another.

These characteristics may sufficiently model expectation effects in social perception. In conjunction with expectation models,[17,18] top-down factors may not only facilitate activation of lower-level representations (such as context facilitating activation of congruent emotional expressions), but lateral interactions allow this activation to cascade and inhibit alternatives (e.g., angry expression inhibits happy expression). Furthermore, this model accounts for complex interactions, such as "bottom-to-top-down" influences, such as when social category interactions occur (for review, see Ref. 150). Cues may activate categories (e.g., Black), which in turn activate stereotypes (e.g., hostile) that wind up feeding back and activating ostensibly unrelated categories (e.g., male). Such category interactions have been observed between various social categories, such as race and sex,[66,67] emotion and sex,[71,72] and race and emotion,[69,70] and these are reflected in multivoxel representations of social categories in the FG and OFC.[78] However, currently the model does not specify how various forms of attention may select and filter information in earlier processing. As we gain insight into how the various forms of attention underlie social perception, future work may come to better integrate this into theoretical models.

5. THE ORIGINS AND FUNCTION OF SOCIAL VISION

As we discuss the myriad ways in which social factors influence perception, it is impossible not to often return to question why such a system would exist. For the social psychologist, functionalist perspectives have been central to theoretical development and have provided a parsimonious account of numerous phenomena. A perspective put forth by Gibson[7] echoes the William James[151] adage that thinking is for doing, in that perception is for doing. Gibson argued that ecological context shifts the affordance provided by any stimulus, and thus vision may guide action by modulating the perception of that stimulus in a way that facilitates appropriate action toward it. For instance, one may mistake a stick for a snake as to guide the less costly action of a false positive than negative. Fiske[152] paraphrased William James to argue that "social thinking is for doing" (p. 877), and in this vein, current work now extends Gibson[7] with the idea that social perception is for doing.

Psychological work in social perception has found supportive evidence in many domains, especially those social.[1,2] For instance, in the face of economic resource scarcity, White participants allocate lesser resources to out-group Black targets while perceiving their skin tone as darker, thereby increasing their out-group perception in a time where in-group favoritism is adaptive.[147] In addition, heterosexual individuals in romantic relationships perceive opposite-sex targets as less attractive, thereby lessening the threat to their relationship satisfaction.[153] Functionalist accounts also include general influences, such as context,[133] emotion,[154] and motivation.[155]

The consistency with which these influences have fit a functionalist framework naturally provokes questions about their adaptiveness and the role of evolution in their emergence. From this perspective, their phylogeny may have been guided by the adaptive benefits they afforded our near or distant ancestors. Mechanisms that allow shifts in perception that facilitate adaptive action, such as wariness of a potential threat or attraction toward a mate, may have been selected for. This evolutionary framing is quite appealing given it outlines one exception to a noteworthy concern with the idea of malleable perceptual experience—that action more often depends on accurate perception than not. An adaptationist account provides a plausible avenue through which a perceptual system that is suboptimal in certain domains (e.g., inaccurate distance perception while tossing an object[156]) may be adaptive in other domains that result in their selection (e.g., vigilance toward a threat by misperceiving its proximity[157]).

Alternatively, an interesting possibility is that social influences are mere by-products of mechanisms adapted for different functions. As has been observed in vision research, many top-down influences in perception serve the purpose of increasing perceptual efficiency.[8,17] Expectation and attention filter and guide information processing in a system flooded with input. Therefore, it is possible that the evolution of these mechanisms put in place the architecture upon which social factors enact their influence as a by-product. By this perspective, allowing factors such as context and emotion to efficiently facilitate accurate perception may have put in place architecture that is subject to prediction errors, such as the misperception of distance or skin pigment. These perceptual errors are often in line with functionalist, adaptive predictions, as the errors facilitate behavior that may have been adaptive in certain contexts (such as exaggerated out-group perception that may cause one to distance themselves from potential threat). Such nonadaptive by-products of traits are discussed in evolutionary biology and are termed "spandrels" after the adorned spaces between indoor arches in architecture, which are not intentionally constructed but a by-product of arch shape.[158] In some cases, they may even provide more plausible accounts of certain phenomena, for instance how social and semantic categories may become entwined due to their overlapping perceptual and top-down features.[36]

There is of course little conclusiveness in the origin of these mechanisms, especially from their mere observation in humans. It is considerable, for instance, that these mechanisms evolved due to the adaptiveness of malleable perception, and their efficiency through prediction was a by-product. As well, they may both have been selected for independently due to adaptive benefits. These theories are also compatible, in that the underlying architecture of this system could have been evolved for perceptual efficiency, then co-opted for adaptive social functions (such as perceiving an object in a way that facilitates adaptive behavior, e.g., a stick for a snake). On the other hand, a malleable perceptual system may have been selected for, then co-opted for general perceptual efficiency purposes. The possibilities are numerous, and all fascinating. Their conclusion will likely rest upon anatomical evolutionary and cross-species research. Regardless, and importantly, these theories each provide fruitful avenues for hypothesis generation, whereby research may investigate where these influences are adaptive, make errors, or arise as by-products of the system's architecture. One benefit of social neuroscience integrating models from cognitive neuroscience is that the complexity and constraints of these models may make apparent by-products that impact social cognition in a manner not intuitive via a functionalist perspective.

Regardless of its origins, the presence of these modulations across the brain provokes core questions in perception and cognition. With the resurgence of research into top-down influences in perception, there has also been a resurgence of the debate as to whether evidence demonstrates top-down factors impact on perceptual *experience* per se.[159,160] The penetrability of conscious perceptual experience to these forces is a captivating question. Yet we must be cautious in interpreting the widespread modulation of activity across visual regions, as there is a lack of conclusiveness regarding neural substrates isomorphic with perceptual experience. For instance, visual activity as early as V1–V4 holds information about a visual stimulus held in working memory without the presence of the stimulus,[161] yet participants do not experience a perceptual experience of the stimulus. Therefore, despite the rich body of work discussed in this chapter, we yet lack conclusive evidence given the perennial challenge of measuring perceptual experience. Nonetheless, the research reviewed here and elsewhere does provide one convincing conclusion: that visual processing, at many levels bar the retina, is affected by the spectrum of top-down social and nonsocial influences. This work thus bolsters the reemerging perspective that fine lines between cognition and perception are worth scrutiny, and we hope its future development may come to bear on this debate.

6. CONCLUSION

Our knowledge of social perception was once limited to its downstream consequences, such as how categorization and perception influence stereotyping and behavior.[40,152] The past decade has seen unprecedented progress in unveiling the processes underlying initial percepts. Such progress has been driven by an equally unprecedented integration between disciplines, including the anticipated wedding of social-cognitive and neural sciences,[162–164] as well as social-cognitive and visual sciences.[1,2,165] Together, these perspectives have engendered productive theoretical accounts of social perception[36] and have galvanized research into its basis across levels of analysis. In this chapter, we have focused on the top-down influences in social perception, and importantly, the interesting and nuanced ways in which they interact with different levels of processing. Specifically, we have reviewed neuroscience research into top-down influences in social vision, a parsimonious account of these influences from the vision and neural sciences, and have discussed how these areas may inform one another and fit within current computational frameworks. We hope the current direction of this research forges a productive collaboration that informs each of the social, cognitive, and neural sciences.

Acknowledgments

This work was supported in part by a National Science Foundation research grant (NSF-BCS-1423708) awarded to J.B.F.

References

1. Balcetis E, Lassiter D. *The Social Psychology of Visual Perception*. New York: Psychology Press; 2010.
2. Adams RB, Ambady N, Nakayama K, Shimojo S. *The Science of Social Vision*. New York: Oxford University Press; 2011.
3. Helmholtz Hv. Concerning the perceptions in general, 1867. In: *Readings in the History of Psychology*. East Norwalk, CT, US: Appleton-Century-Crofts; 1948:214–230.
4. Bruner JS. On perceptual readiness. *Psychol Rev*. 1957;64(2):123–152.
5. Bruner JS, Goodman CC. Value and need as organizing factors in perception. *J Abnorm Soc Psychol*. 1947;42:33–44.
6. Erdelyi MH. A new look at the new look: perceptual defense and vigilance. *Psychol Rev*. 1974;81(1):1–25.
7. Gibson JJ. *The Ecological Approach to Visual Perception*. Boston: Houghton Mifflin; 1979.
8. Gilbert CD, Li W. Top-down influences on visual processing. *Nat Rev Neurosci*. 2013;14(5):350–363.
9. Bar M, Bubic A. Top-down effects in visual perception. In: Ochsner KN, Kosslyn SM, eds. *The Oxford Handbook of Cognitive Neuroscience*. Core Topics; Vol. 1. New York, NY, US: Oxford University Press; 2014:60–73.
10. Engel AK, Fries P, Singer W. Dynamic predictions: oscillations and synchrony in top-down processing. *Nat Rev Neurosci*. 2001;2:704–716.
11. Rockland KS, Van Hoesen GW. Direct temporal-occipital feedback connections to striate cortex (V1) in the macaque monkey. *Cereb Cortex*. 1994;4(3):300–313.
12. Freese JL, Amaral DG. The organization of projections from the amygdala to visual cortical areas TE and V1 in the macaque monkey. *J Comp Neurol*. 2005;486(4):295–317.
13. Barrett LF, Bar M. See it with feeling: affective predictions during object perception. *Philos Trans R Soc B Biol Sci*. 2009;364(1521):1325–1334.
14. Zanto TP, Rubens MT, Thangavel A, Gazzaley A. Causal role of the prefrontal cortex in top-down modulation of visual processing and working memory. *Nat Neurosci*. 2011;14(5):656–661.
15. Kok P, Jehee JF, de Lange FP. Less is more: expectation sharpens representations in the primary visual cortex. *Neuron*. 2012;75(2):265–270.
16. Muckli L, Petro LS. Network interactions: non-geniculate input to V1. *Curr Opin Neurobiol*. 2013;23(2):195–201.
17. Summerfield C, Egner T. Expectation (and attention) in visual cognition. *Trends Cogn Sci*. 2009;13(9):403–409.
18. Bar M. Visual objects in context. *Nat Rev Neurosci*. 2004;5:617–629.
19. Ambady N, Bernieri FJ, Richeson JA. Toward a histology of social behavior: judgmental accuracy from thin slices of the behavioral stream. In: *Advances in Experimental Social Psychology*. Vol. 32. San Diego, CA: Academic Press; 2000:201–271.
20. Fiske ST. Stereotyping, prejudice, and discrimination. In: 4th ed. *The Handbook of Social Psychology*. Vols 1 and 2. New York, NY: McGraw-Hill; 1998:357–411.
21. McAleer P, Todorov A, Belin P. How do you say 'hello'? Personality impressions from brief novel voices. *PLoS One*. 2014;9(3): e90779.
22. Oosterhof NN, Todorov A. The functional basis of face evaluation. *Proc Natl Acad Sci*. 2008;105:11087–11092.
23. Zebrowitz LA. *Reading Faces: Window to the Soul? New Directions in Social Psychology*; 1997.
24. Haxby JV, Hoffman EA, Gobbini MI. The distributed human neural system for face perception. *Trends Cogn Sci*. 2000;4:223–233.

25. Bruce V, Young AW. A theoretical perspective for understanding face recognition. *Br J Psychol*. 1986;77:305–327.
26. Amodio DM. The neuroscience of prejudice and stereotyping. *Nat Rev Neurosci*. 2014;15(10):670–682.
27. Kubota JT, Banaji MR, Phelps EA. The neuroscience of race. *Nat Neurosci*. 2012;15(7):940–948.
28. Haxby JV, Hoffman EA, Gobbini MI. Human neural systems for face recognition and social communication. *Biol Psychiatry*. 2002;51(1):59–67.
29. Gainotti G, Marra C. Differential contribution of right and left temporo-occipital and anterior temporal lesions to face recognition disorders. *Front Hum Neurosci*. 2011;5(55).
30. Kanwisher N, McDermott J, Chun MM. The fusiform face area: a module in human extrastriate cortex specialized for face perception. *J Neurosci*. 1997;17(11):4302–4311.
31. Anzellotti S, Caramazza A. The neural mechanisms for the recognition of face identity in humans. *Front Psychol*. 2014;5:672.
32. Gobbini MI, Haxby JV. Neural systems for recognition of familiar faces. *Neuropsychologia*. 2007;45(1):32–41.
33. Freeman JB, Johnson KL, Ambady N, Rule NO. Sexual orientation perception involves gendered facial cues. *Pers Soc Psychol Bull*. 2010;36:1318–1331.
34. Rule NO, Ambady N. Brief exposures: male sexual orientation is accurately perceived at 50 ms. *J Exp Soc Psychol*. 2008;44(4): 1100–1105.
35. Macrae CN, Bodenhausen GV. Social cognition: thinking categorically about others. *Ann Rev Psychol*. 2000;51:93–120.
36. Freeman JB, Ambady N. A dynamic interactive theory of person construal. *Psychol Rev*. 2011;118:247–279.
37. Correll J, Park B, Judd CM, Wittenbrink B. The police officer's dilemma: using ethnicity to disambiguate potentially threatening individuals. *J Pers Soc Psychol*. 2002;83(6):1314–1329.
38. Cuddy AJ, Fiske ST, Glick P. The BIAS map: behaviors from intergroup affect and stereotypes. *J Pers Soc Psychol*. 2007;92(4):631–648.
39. Fiske ST, Cuddy AJ, Glick P, Xu J. A model of (often mixed) stereotype content: competence and warmth respectively follow from perceived status and competition. *J Pers Soc Psychol*. 2002;82(6):878–902.
40. Allport GW. *The Nature of Prejudice*. Oxford: Addison-Wesley; 1954.
41. Contreras JM, Banaji MR, Mitchell JP. Multivoxel patterns in fusiform face area differentiate faces by sex and race. *PLoS One*. 2013;8(7):e69684.
42. Kaul C, Rees G, Ishai A. The gender of face stimuli is represented in multiple regions in the human brain. *Front Hum Neurosci*. 2011;4(238).
43. Ratner KG, Kaul C, Van Bavel JJ. Is race erased? Decoding race from patterns of neural activity when skin color is not diagnostic of group boundaries. *Soc Cogn Affect Neurosci*. 2012;8(7):750–755.
44. Freeman JB, Rule NO, Adams RB, Ambady N. The neural basis of categorical face perception: graded representations of face gender in fusiform and orbitofrontal cortices. *Cereb Cortex*. 2010;20:1314–1322.
45. Grill-Spector K, Weiner KS. The functional architecture of the ventral temporal cortex and its role in categorization. *Nat Rev Neurosci*. 2014;15(8):536–548.
46. Aviezer H, Hassin RR, Ryan J, et al. Angry, disgusted, or afraid? Studies of the malleability of emotion perception. *Psychol Sci*. 2008;19:724–732.
47. Said CP, Moore CD, Engell AD, Todorov A, Haxby JV. Distributed representations of dynamic facial expressions in the superior temporal sulcus. *J Vis*. 2010;10(5):11.
48. Said CP, Moore CD, Norman KA, Haxby JV, Todorov A. Graded representations of emotional expressions in the left superior temporal sulcus. *Front Syst Neurosci*. 2010;4(6).
49. Harry B, Williams MA, Davis C, Kim J. Emotional expressions evoke a differential response in the fusiform face area. *Front Hum Neurosci*. 2013;7(692):6.
50. Gschwind M, Pourtois G, Schwartz S, Van De Ville D, Vuilleumier P. White-matter connectivity between face-responsive regions in the human brain. *Cereb Cortex*. 2012;22(7):1564–1576.
51. Pyles JA, Verstynen TD, Schneider W, Tarr MJ. Explicating the face perception network with white matter connectivity. *PLoS One*. 2013;8(4):e61611.
52. Pitcher D, Duchaine B, Walsh V. Combined TMS and FMRI reveal dissociable cortical pathways for dynamic and static face perception. *Curr Biol*. 2014;24(17):2066–2070.
53. Willis J, Todorov A. First impressions: making up your mind after a 100-ms exposure to a face. *Psychol Sci*. 2006;17:592–598.
54. Zebrowitz LA, McDonald SM. The impact of litigants' babyfacedness and attractiveness on adjudications in small claims courts. *Law Hum Behav*. 1991;15(6):603.
55. Zebrowitz LA, Fellous J-M, Mignault A, Andreoletti C. Trait impressions as overgeneralized responses to adaptively significant facial qualities: evidence from connectionist modeling. *Pers Soc Psychol Rev*. 2003;7:194–215.
56. Zebrowitz LA, Luevano VX, Bronstad PM, Aharon I. Neural activation to babyfaced men matches activation to babies. *Soc Neurosci*. 2009;4(1):1–10.
57. Todorov A, Engell AD. The role of the amygdala in implicit evaluation of emotionally neutral faces. *Soc Cogn Affect Neurosci*. 2008;3(4):303–312.
58. Engell AD, Haxby JV, Todorov A. Implicit trustworthiness decisions: automatic coding of face properties in the human amygdala. *J Cogn Neurosc*. 2007;19(9):1508–1519.
59. Freeman JB, Stolier RM, Ingbretsen ZA, Hehman EA. Amygdala responsivity to high-level social information from unseen faces. *J Neurosci*. 2014;34(32):10573–10581.
60. Said CP, Dotsch R, Todorov A. The amygdala and FFA track both social and non-social face dimensions. *Neuropsychologia*. 2010;48(12):3596–3605.
61. Mende-Siedlecki P, Said CP, Todorov A. The social evaluation of faces: a meta-analysis of functional neuroimaging studies. *Soc Cogn Affect Neurosci*. 2013;8(3):285–299.
62. Sofer C, Dotsch R, Wigboldus DH, Todorov A. What is typical is good the influence of face typicality on perceived trustworthiness. *Psychol Sci*. 2014. http://dx.doi.org/10.1177/0956797614554955.
63. Cottrell CA, Neuberg SL. Different emotional reactions to different groups: a sociofunctional threat-based approach to "prejudice". *J Pers Soc Psychol*. 2005;88(5):770–789.
64. Blair IV, Judd CM, Sadler MS, Jenkins C. The role of Afrocentric features in person perception: judging by features and categories. *J Pers Soc Psychol*. 2002;83(1):5–25.
65. Dotsch R, Wigboldus DH, Langner O, van Knippenberg A. Ethnic out-group faces are biased in the prejudiced mind. *Psychol Sci*. 2008;19(10):978–980.
66. Galinsky AD, Hall EV, Cuddy AJ. Gendered races: implications for interracial marriage, leadership selection, and athletic participation. *Psychol Sci*. 2013;24(4):498–506.
67. Johnson KL, Freeman JB, Pauker K. Race is gendered: how covarying phenotypes and stereotypes bias sex categorization. *J Pers Soc Psychol*. 2012;102:116–131.
68. Carpinella CM, Chen JM, Hamilton DL, Johnson KL. Gendered facial cues influence race categorizations. *Pers Soc Psychol Bull*. 2015. http://dx.doi.org/10.1177/0146167214567153.
69. Hugenberg K, Bodenhausen GV. Facing prejudice: implicit prejudice and the perception of facial threat. *Psychol Sci*. 2003;14(6):640–643.
70. Hugenberg K, Bodenhausen GV. Ambiguity in social categorization: the role of prejudice and facial affect in race categorization. *Psychol Sci*. 2004;15(5):342–345.

71. Hess U, Adams Jr RB, Kleck RE. Facial appearance, gender, and emotion expression. *Emotion*. 2004;4(4):378–388.
72. Hess U, Senécal S, Kirouac G, Herrera P, Philippot P, Kleck RE. Emotional expressivity in men and women: stereotypes and self-perceptions. *Cogn Emot*. 2000;14(5).
73. Hehman E, Ingbretsen ZA, Freeman JB. The neural basis of stereotypic impact on multiple social categorization. *NeuroImage*. 2014;101:704–711.
74. Botvinick M, Braver T, Barch D, Carter C, Cohen J. Conflict monitoring and cognitive control. *Psychol Rev*. 2001;108:624–652.
75. MacDonald AW, Cohen JD, Stenger VA, Carter CS. Dissociating the role of the dorsolateral prefrontal and anterior cingulate cortex in cognitive control. *Science*. 2000;288(5472):1835–1838.
76. Amodio DM, Frith CD. Meeting of minds: the medial frontal cortex and social cognition. *Nat Rev Neurosci*. 2006;7:268–277.
77. Chee MW, Sriram N, Soon CS, Lee KM. Dorsolateral prefrontal cortex and the implicit association of concepts and attributes. *Neuroreport*. 2000;11(1):135–140.
78. Stolier RM, Freeman JB. Neural pattern similarity reveals the inherent intersection of social categories, invited revision.
79. Knutson KM, Mah L, Manly CF, Grafman J. Neural correlates of automatic beliefs about gender and race. *Hum Brain Mapp*. 2007;28(10):915–930.
80. Milne E, Grafman J. Ventromedial prefrontal cortex lesions in humans eliminate implicit gender stereotyping. *J Neurosci*. 2001;21(12):RC150.
81. Allport G. Attitudes. In: Murchison C, ed. *Handbook of Social Psychology*. Worcester, MA: Clark University Press; 1935.
82. Greenwald AG, McGhee DE, Schwartz JL. Measuring individual differences in implicit cognition: the implicit association test. *J Pers Soc Psychol*. 1998;74(6):1464.
83. McConnell AR, Leibold JM. Relations among the Implicit Association Test, discriminatory behavior, and explicit measures of racial attitudes. *J Exp Soc Psychol*. 2001;37(5):435–442.
84. Dovidio JF, Kawakami K, Gaertner SL. Implicit and explicit prejudice and interracial interaction. *J Pers Soc Psychol*. 2002;82(1):62.
85. Krueger J. On the overestimation of between-group differences. *Eur Rev Soc Psychol*. 1992;3(1):31–56.
86. Brosch T, Bar-David E, Phelps EA. Implicit race bias decreases the similarity of neural representations of black and white faces. *Psychol Sci*. 2013;24(2):160–166.
87. Meissner CA, Brigham JC. Thirty years of investigating the own-race bias in memory for faces: a meta-analytic review. *Psychol Public Policy Law*. 2001;7(1):3.
88. Hugenberg K, Miller J, Claypool HM. Categorization and individuation in the cross-race recognition deficit: toward a solution to an insidious problem. *J Exp Soc Psychol*. 2007;43(2):334–340.
89. Rhodes G, Brake S, Taylor K, Tan S. Expertise and configural coding in face recognition. *Br J Psychol*. 1989;80(3):313–331.
90. Hillger LA, Koenig O. Separable mechanisms in face processing: evidence from hemispheric specialization. *J Cogn Neurosci*. 1991;3(1):42–58.
91. Correll J, Lemoine C, Ma DS. Hemispheric asymmetry in cross-race face recognition. *J Exp Soc Psychol*. 2011;47(6):1162–1166.
92. Golby AJ, Gabrieli JDE, Chiao JY, Eberhardt JL. Differential responses in the fusiform region to same-race and other-race faces. *Nat Neurosci*. 2001;4:845–850.
93. Feng L, Liu J, Wang Z, et al. The other face of the other-race effect: an fMRI investigation of the other-race face categorization advantage. *Neuropsychologia*. 2011;49(13):3739–3749.
94. Vizioli L, Rousselet GA, Caldara R. Neural repetition suppression to identity is abolished by other-race faces. *Proc Natl Acad Sci*. 2010;107(46):20081–20086.
95. Beale JM, Keil CF. Categorical effects in the perception of faces. *Cognition*. 1995;57:217–239.
96. Rotshtein P, Henson RNA, Treves A, Driver J, Dolan RJ. Morphing Marilyn into Maggie dissociates physical and identity face representations in the brain. *Nat Neurosci*. 2005;8:107–113.
97. Natu V, O'Toole AJ. The neural processing of familiar and unfamiliar faces: a review and synopsis. *Br J Psychol*. 2011;102(4):726–747.
98. Verosky SC, Todorov A, Turk-Browne NB. Representations of individuals in ventral temporal cortex defined by faces and biographies. *Neuropsychologia*. 2013;51(11):2100–2108.
99. Summerfield C, Egner T, Greene M, Koechlin E, Mangels J, Hirsch J. Predictive codes for forthcoming perception in the frontal cortex. *Science*. 2006;314(5803):1311–1314.
100. Anderson E, Siegel EH, Bliss-Moreau E, Barrett LF. The visual impact of gossip. *Science*. 2011;332(6036):1446–1448.
101. Tong F, Meng M, Blake R. Neural bases of binocular rivalry. *Trends Cogn Sci*. 2006;10(11):502–511.
102. Freeman JB, Schiller D, Rule NO, Ambady N. The neural origins of superficial and individuated judgments about ingroup and outgroup members. *Hum Brain Mapp*. 2010;31:150–159.
103. Freeman JB, Stolier RM. The medial prefrontal cortex in constructing personality models. *Trends Cogn Sci*. 2014;18(11):571–572.
104. Hassabis D, Spreng RN, Rusu AA, Robbins CA, Mar RA, Schacter DL. Imagine all the people: how the brain creates and uses personality models to predict behavior. *Cereb Cortex*. 2013. bht042.
105. Mitchell JP, Heatherton TF, Macrae CN. Distinct neural systems subserve person and object knowledge. *Proc Natl Acad Sci USA*. 2002;99(23):15238–15243.
106. Cloutier J, Kelley WM, Heatherton TF. The influence of perceptual and knowledge-based familiarity on the neural substrates of face perception. *Soc Neurosci*. 2011;6(1):63–75.
107. Tajfel H. Experiments in intergroup discrimination. *Sci Am*. 1970;223(5):96–102.
108. Tajfel H. Social identity and intergroup behaviour. *Soc Sci Inf*. 1974;13:65–93.
109. Van Bavel JJ, Packer DJ, Cunningham WA. The neural substrates of in-group bias: a functional magnetic resonance imaging investigation. *Psychol Sci*. 2008;19(11):1131–1139.
110. Lieberman MD, Hariri A, Jarcho JM, Eisenberger NI, Bookheimer SY. An fMRI investigation of race-related amygdala activity in African-American and Caucasian-American individuals. *Nat Neurosci*. 2005;8:720–722.
111. Kurzban R, Tooby J, Cosmides L. Can race be erased? Coalitional computation and social categorization. *Proc Natl Acad Sci*. 2001;98(26):15387–15392.
112. Darwin C. *The Expression of the Emotions in Man and Animals*. Harper Perennial; 1872.
113. Keltner D, Gross JJ. Functional accounts of emotions. *Cogn Emot*. 1999;13(5):467–480.
114. Barrett LF. Are emotions natural kinds? *Perspect Psychol Sci*. 2006;1(1):28–58.
115. Russell JA, Barrett LF. Core affect, prototypical emotional episodes, and other things called emotion: dissecting the elephant. *J Pers Soc Psychol*. 1999;76(5):805–819.
116. Adolphs R, Spezio M. Role of the amygdala in processing visual social stimuli. In: Anders S, Ende G, Junghofer M, Kissler J, Wildgruber D, eds. *Progress in Brain Research: Understanding Emotions*. Elsevier Science; 2006:363–377.
117. Pessoa L, Japee S, Sturman D, Ungerleider LG. Target visibility and visual awareness modulate amygdala responses to fearful faces. *Cereb Cortex*. 2006;16(3):366–375.
118. Duncan S, Barrett LF. The role of the amygdala in visual awareness. *Trends Cogn Sci*. 2007;11(5):190–192.
119. Anderson E, Siegel EH, Barrett LF. What you feel influences what you see: the role of affective feelings in resolving binocular rivalry. *J Exp Soc Psychol*. 2011;47(4):856–860.

120. Adolphs R, Tranel D, Damasio H, Damasio A. Impaired recognition of emotion in facial expressions following bilateral damage to the human amygdala. *Nature*. 1994;372(6507):669–672.
121. Adolphs R. Neural systems for recognizing emotion. *Curr Opin Neurobiol*. 2002;12(2):169–177.
122. Adolphs R. Emotional vision. *Nat Neurosci*. 2004;7(11):1167–1168.
123. Vuilleumier P, Richardson MP, Armony JL, Driver J, Dolan RJ. Distant influences of amygdala lesion on visual cortical activation during emotional face processing. *Nat Neurosci*. 2004;7(11):1271–1278.
124. Adolphs R, Baron-Cohen S, Tranel D. Impaired recognition of social emotions following amygdala damage. *J Cogn Neurosci*. 2002;14(12495531):1264–1274.
125. Adolphs R, Gosselin F, Buchanan TW, Tranel D, Schyns P, Damasio AR. A mechanism for impaired fear recognition after amygdala damage. *Nature*. 2005;433(7021):68–72.
126. Tsuchiya N, Moradi F, Felsen C, Yamazaki M, Adolphs R. Intact rapid detection of fearful faces in the absence of the amygdala. *Nat Neurosci*. 2009;12(10):1224–1225.
127. Zhang H, Liu J, Huber DE, Rieth CA, Tian J, Lee K. Detecting faces in pure noise images: a functional MRI study on top-down perception. *Neuroreport*. 2008;19(2):229–233.
128. Dolan RJ, Fink GR, Rolls E, et al. How the brain learns to see objects and faces in an impoverished context. *Nature*. 1997;389(6651):596–599.
129. Johnson KL, Freeman JB. A "New Look" at person construal: Seeing beyond dominance and discreteness. In: Balcetis E, Lassiter D, eds. *The Social Psychology of Visual Perception*. New York: Psychology Press; 2010.
130. Freeman JB, Ma Y, Han S, Ambady N. Influences of culture and visual context on real-time social categorization. *J Exp Soc Psychol*. 2013;49:206–210.
131. Freeman JB, Ma Y, Barth M, Young SG, Han S, Ambady N. The neural basis of contextual influences on face categorization. *Cereb Cortex*. 2015;25:415–422.
132. Henderson JM, Larson CL, Zhu DC. Full scenes produce more activation than close-up scenes and scene-diagnostic objects in parahippocampal and retrosplenial cortex: an fMRI study. *Brain Cogn*. 2008;66(1):40–49.
133. Barrett LF, Mesquita B, Gendron M. Context in emotion perception. *Curr Dir Psychol Sci*. 2011;20(5):286–290.
134. Van den Stock J, Vandenbulcke M, Sinke CBA, Goebel R, de Gelder B. How affective information from faces and scenes interacts in the brain. *Soc Cogn Affect Neurosci*. 2014;9(10):1481–1488.
135. Kim H, Somerville LH, Johnstone T, et al. Contextual modulation of amygdala responsivity to surprised faces. *J Cogn Neurosci*. 2004;16(10):1730–1745.
136. Whalen PJ, Rauch SL, Etcoff NL, McInerney SC, Lee MB, Jenike MA. Masked presentations of emotional facial expressions modulate amygdala activity without explicit knowledge. *J Neurosci*. 1998;18:411–418.
137. Peelen MV, Downing PE. The neural basis of visual body perception. *Nat Rev Neurosci*. 2007;8:636–648.
138. Cox D, Meyers E, Sinha P. Contextually evoked object-specific responses in human visual cortex. *Science*. 2004;304(5667):115–117.
139. Kanwisher N, Yovel G. The fusiform face area: a cortical region specialized for the perception of faces. *Philos Trans R Soc Lond B Biol Sci*. 2006;361:2109–2128.
140. Meeren HKM, van Heijnsbergen CCRJ, de Gelder B. Rapid perceptual integration of facial expression and emotional body language. *Proc Natl Acad Sci USA*. 2005;102:16518–16523.
141. Van den Stock J, Righart R, de Gelder B. Body expressions influence recognition of emotions in the face and voice. *Emotion*. 2007;7(3):487–494.
142. Aviezer H, Trope Y, Todorov A. Body cues, not facial expressions, discriminate between intense positive and negative emotions. *Science*. 2012;338(6111):1225–1229.
143. Martinez A, Anllo-Vento L, Sereno MI, et al. Involvement of striate and extrastriate visual cortical areas in spatial attention. *Nat Neurosci*. 1999;2(4):364–369.
144. Petro LS, Vizioli L, Muckli L. Contributions of cortical feedback to sensory processing in primary visual cortex. *Front Psychol*. 2014;5:1223.
145. Amaral DG, Behniea H, Kelly JL. Topographic organization of projects from the amygdala to the visual cortex in the macaque monkey. *Neuroscience*. 2003;118:1099–1120.
146. Gilbert CD, Sigman M. Brain states: top-down influences in sensory processing. *Neuron*. 2007;54:677–696.
147. Krosch AR, Amodio DM. Economic scarcity alters the perception of race. *Proc Natl Acad Sci USA*. 2014;111(25):9079–9084.
148. Krosch AR, Berntsen L, Amodio DM, Jost JT, Van Bavel JJ. On the ideology of hypodescent: political conservatism predicts categorization of racially ambiguous faces as Black. *J Exp Soc Psychol*. 2013;49(6):1196–1203.
149. Freeman JB, Penner AM, Saperstein A, Scheutz M, Ambady N. Looking the part: social status cues shape race perception. *PLoS One*. 2011;6:e25107.
150. Freeman JB, Johnson KL, Adams Jr RB, Ambady N. The social-sensory interface: category interactions in person perception. *Front Integr Neurosci*. 2012;6.
151. James W. *The Principles of Psychology*. H. Holt; 1890.
152. Fiske ST. Thinking is for doing: portraits of social cognition from Daguerreotype to laserphoto. *J Pers Soc Psychol*. 1992;63(6):877–889.
153. Karremans JC, Dotsch R, Corneille O. Romantic relationship status biases memory of faces of attractive opposite-sex others: evidence from a reverse-correlation paradigm. *Cognition*. 2011;121(3):422–426.
154. Stefanucci JK, Gagnon KT, Lessard DA. Follow your heart: emotion adaptively influences perception. *Social Pers Psychol Comp*. 2011;5(6):296–308.
155. Balcetis E, Dunning D. See what you want to see: motivational influences on visual perception. *J Pers Soc Psychol*. 2006;91:612–625.
156. Balcetis E, Dunning D. Wishful seeing more desired objects are seen as closer. *Psychol Sci*. 2009;21(1):147–152.
157. Cole S, Balcetis E, Dunning D. Affective signals of threat increase perceived proximity. *Psychol Sci*. 2013;24(1):34–40.
158. Gould SJ, Lewontin RC. The spandrels of San Marco and the Panglossian paradigm: a critique of the adaptationist programme. *Proc R Soc Lond B Biol Sci*. 1979;205(1161):581–598.
159. Edelman S. No reconstruction, no impenetrability (at least not much). *Behav Brain Sci*. 1999;22(03):376.
160. Firestone C, Scholl BJ. "Top-down" effects where none should be found: the el greco fallacy in perception research. *Psychol Sci*. 2014;25(1):38–46.
161. Harrison SA, Tong F. Decoding reveals the contents of visual working memory in early visual areas. *Nature*. 2009;458(7238):632–635.
162. Ochsner KN, Lieberman MD. The emergence of social cognitive neuroscience. *Am Psychol*. 2001;56(9):717.
163. Adolphs R. Cognitive neuroscience of human social behavior. *Nat Rev Neurosci*. 2003;4:165–178.
164. Cacioppo JT, Berntson GG. Social psychological contributions to the decade of the brain: doctrine of multilevel analysis. *Am Psychol*. 1992;47(8):1019.
165. Johnson KL, Adams Jr RB. Social vision: an introduction. *Soc Cogn*. 2013;31(6):633–635.

CHAPTER

8

Social Vision: At the Intersection of Vision and Person Perception

Daniel N. Albohn, Reginald B. Adams Jr.

Department of Psychology, The Pennsylvania State University, University Park, PA, USA

OUTLINE

The face is a picture of the mind with the eyes as its interpreter.
Marcus Tullius Cicero

The science of person perception has a long and storied past, much of which is detailed throughout this volume. Philosophers, scientists, and artists have struggled for centuries to capture, as precisely as possible, an accurate perception of the individual, and the experiences that lie therein. Consider, for example, emotion and emotion perception. Substantial thought has been expended on emotions, how they influence the person, and by deductive inference, the perception of the individual. Aristotle spent considerable time defining specific emotions, such as anger, and the consequences that this "distressed desire for conspicuous vengeance" had on the individual.[1] Later, the medieval Christian scholars were preoccupied with sin related to earthly desires, and thus viewed emotions wholly intertwined with these self-serving desires.[2] More recently, contemporary scientists have investigated this issue from multiple angles, including how biological motion,[3–5] voice,[6] olfactory,[7] race,[8–10] and age[11–14] influence the individual, and the perception of the individual.

Neuroimaging Personality, Social Cognition, and Character
http://dx.doi.org/10.1016/B978-0-12-800935-2.00008-7

Due to the increased attention person perception has received in the past century, the face has become perhaps the most studied facet of the field. As Cicero suggests in the quote at the beginning this chapter, the face provides the perceiver with myriad information regarding the intentions, emotional state, mood, attractiveness, age, race, and sex of the actor who is being perceived. Adding to the already complex nature of the face is how these cues interact in perception to inform our impressions of others. Research from our own lab, as well as a plethora of other scientists across diverse fields, are beginning to elucidate the mechanisms by which social cues combine to change visual perception (for a detailed summary see Adams et al.[15]). It is because of the face's rich repository of complex information that it is a central focus of person perception, and it is because of this complex nature that, despite such advances in the field, it requires continued in-depth investigation.

The goal of this chapter is to reduce this complexity by applying an ecological model of vision[16] to guide our understanding of the combinatorial nature of social vision and person perception. Modular approaches to neuroscience and vision have attempted to reduce this complexity by honing in on specific areas of the brain or visual system that act on specific stimuli. Evidence in support of an ecological view can help researchers not only make specific predictions about biological reactions (i.e., brain or vision responses), but behavioral responses as well. While the ecological model of vision is several decades old, and has been applied to the study of face perception,[17] work examining the integration of compound social cues requires further investigation. This point is particularly poignant in the domain of neuroscience, as integrating and studying social cues in person perception has only begun to be studied within a neuroscientific framework.

The remainder of this chapter will summarize recent advancements in the field of face perception before moving on to how an ecological approach to vision can inform future neuroscience practices in the domain of person perception. First, we will outline why the face matters so much in social perception. Then, we will examine several face cues that have received considerable attention. Next, we will explore the neural architecture of social vision. Finally, we examine how these social cues can be studied in an integrated manner, and suggest ways in which neuroscience can add to the burgeoning literature of results to enhance, clarify, and support the complex nature of social vision.

1. WHY IS THE FACE CONSIDERED SPECIAL?

Paul Ekman, a forerunner in the field of emotion expression perception, once said that of all the nonverbal behaviors (body movement, posture, gaze, etc.) the face is "the most commanding and complicated"[18] (p. 45). He went on to say that the face is the "throne" of the five senses, and facial expressions are one of the only, if not *the* only, nonverbal behaviors to occur when we are not in a social context. Consider, for example, conversing with a group of people versus writing a paper alone. When we converse with others, we provide innumerable nonverbal cues to those around us to help them interpret the subtle nuances of our mood, intentions, dominance, and much more. We might gesture with our hands to signal disagreement. We might furrow our brow to signal concentration. We might straighten our posture and move closer to signal dominance and interest. However, when we are writing a paper alone, an activity comparable to a conversation with ourselves, none of the aforementioned nonverbal signals occur save for one—facial expressions. Indeed, watch anyone writing a paper and you will see the "look of concentration" painted across their visage. Perhaps even other expressions will occur, such as when one writes an exceptionally clever or humorous anecdote they might smile to themselves. Field studies of bowling and soccer players during happy and unhappy parts of a game show that smiling occurs outside of just social interactions, albeit at a significantly reduced probability.[19,20] Why, then, are faces so special that they provide information even when there is no socially adaptive need to do so? Moreover, what makes the face the "throne" of information for the perceiver?

Throughout this chapter, we apply the ecological approach to social vision to answer these questions. This approach suggests that the brain evolved to perceive stimuli that convey action affordances that are adaptive to our survival. Gibson summarizes this concept by stating, "A fruit says 'eat me'; water says 'drink me'; thunder says 'fear me'; woman says 'love me'" [16] (p. 138). By Gibson's own logic, a face will tell the perceiver any number of things—mood, intent, sex, age—some of which we outline in this chapter. Recall Ekman's quote about the face; it is commanding because it provides the perceiver with important social information. However, in the second half of Ekman's quote, he states that the face is also complicated. One reason that the face is so complicated is because there is a remarkable amount of information provided by it. At times these cues overlap, supersede, or enhance other information provided by the face—what we call *compound social cues*. Until recently, these compound social cues were largely studied separately. It was believed that the visual system extracts and interprets such information independent of other cues. We now know that this is not true. However, with this advancement also comes a more complicated future for research on face perception. Scientists now need to consider the impact of compound social cues when studying the mechanisms that lie underneath them. Furthermore, we need to develop methods for controlling such cues and paradigms that are sensitive enough to tease apart the influence of each cue on person perception, as well as their interactive influences.

Curiously, faces across species share remarkably similar qualities, consistent with a universal prototype for a face.[21] Most animals have two eyes above a nose, with a mouth that lies underneath. Further, faces tend to be more or less bilaterally symmetrical. This is thought to be the product of nonpreferential lateral forces from the environment (compared to an anterior preference in animals causing the differentiation between front and back). However, despite marked similarity, humans do possess facial qualities that make them distinct from other animals. Bruce and Young[21] outline a number of these features, including facial hair, eyebrows, and speech production. Hair isolated to the top of the human head is drastically different compared to other animals that are covered in a thick coat of facial hair similar to the rest of their bodies. Further, in light of the relative baldness of the human face, compared to other primates, the remainder of hair above the eye—eyebrows—is a curiosity. From an evolutionary perspective, the eyebrows are believed to provide an adaptive mechanism for keeping sweat, rain, dust, and insects out of the eyes.[21] Convincing evidence from psychology also suggests that the eyebrows play an important role in expressivity and communication,[18] as well as face recognition.[22,23] Subtle changes to the height and position of the eyebrows can change the perceived expression on a face and increase efficiency in gender recognition.[24] In one illuminating study, participants viewed faces of famous actors with eyebrows or eye regions erased. When the eyebrows were erased, the participants had more difficulty in accurately identifying the actor than when the entire eye regions were erased.[23]

Even though humans share a set of distinct facial features, faces are highly recognizable and distinguishable. Given the enormous amount of information present on the face, multiplied by the hundreds of faces we see on a daily basis, humans are surprisingly adept at distinguishing one face from another. This adaptive ability to distinguish a face from a non-face, and one face from another, suggests that the brain may have evolved to fit our species' unique social demands. Indeed, perceiving faces is so salient that it is easy to see faces in inanimate objects, such as in a mound of moon dirt or in a stick of margarine, a phenomenon called pareidolia. Further, artistic representations that violate the face template are still recognizable as faces, despite atypical and sometimes gruesome alterations. The brain appears to become more specialized for face perception as it matures. Before nine months of age, infants are able to identify individual faces of both humans and *Barbary macaques*. However, after nine months, infants lose the ability to discriminate between *macaques* and human faces unless continued training occurs.[25,26]

2. SOCIAL FACE PERCEPTION

Social vision is a complicated beast. Scientists and practitioners have long been trying to decipher the mechanisms by which an individual can "read" accurate information about one's personality from the face. Perhaps the most infamous attempt to do so was with a practice called physiognomy. This practice, dating back as early as Aristotle, attempts to associate facial landmarks with certain personality characteristics. For example, the early physiognomist Giambattista della Porta[27] attempted to infer personality of an individual by equating similarity of human faces with animal faces (e.g., certain similarities to bovine faces in humans was equated with stubbornness). Physiognomy probably reached the height of its popularity in the eighteenth century and is now largely discredited as pseudoscience.

However, it is a well-documented phenomenon that individuals extract similar information from a face with high reliability across raters and is formed in what appears to be an effortless manner regardless of the accuracy of such extractions.[28–33] There must exist, then, a reason for these consistent judgments. Secord[34] recognized that although physiognomy had fallen out of favor in the scientific community, there might be some merit to its claims. More recently, Diane Berry and her colleagues[35–38] showed that while the impressions gleaned from the face alone may be inaccurate at times, there is a "kernel of truth"[a] to the information gathered. In one study,[37] individuals were photographed and then rated themselves on a number of personality characteristics, including baby-facedness, attractiveness, dominance, and physical strength. Following this, a separate group of participants rated the photographs on the same personality characteristics. The researchers found a positive correlation between ratings based solely on facial appearance and the self-reported ratings of the models themselves. In another study,[36] college students were asked to rate the other students in their class on personality variables similar to those mentioned in the previous study at the beginning, the fifth, and the ninth week of the semester. They were then photographed and

[a] For a similar construct in person perception, the reader is directed to Ambady and colleagues' *thin slice vision*.[39–41] In this phenomenon, individuals can accurately identify relevant personality traits, mental states, and sexuality from short (<30 s) video clips of nonverbal behavior. Critically, participants are usually more accurate in their identification of these cues when the clip is silenced than when verbal behavior is maintained, suggesting the spontaneous and important inferences that we draw from nonverbal behavior.

separate participants rated each individual on the personality traits. As hypothesized, impressions formed by raters who viewed only the photographs predicted classmates' impressions for all nine weeks. Lastly, in a study[38] directly examining how perceived facial qualities may affect behavior, researchers asked participants to agree or disagree to take part in a number of studies involving deception. Participants knew that the studies they were agreeing or disagreeing to participate in involved deception, and their willingness or unwillingness to participate in these studies provided the researchers with a "scale of willingness to take part in deceptive behavior." Subsequently, a second group of participants rated a photograph of the participants who agreed or disagreed to take part in the deceptive studies. Perhaps unsurprisingly at this point, the more honest a face was rated from a static image alone, the less likely they were to engage in the deceptive studies.

The results discussed so far in support of the "kernel of truth" involve only personality traits extracted from the face. However, there exists evidence that it extends to other dimensions as well. Research has revealed evidence that perceivers can accurately identify political-race winners,[42] political affiliation,[43] business leaders' salaries,[44] sexual orientation,[31,45] and religious affiliation[46] just by reading information available from the face alone. For example, Rule and Ambady[44] showed pictures of managing partners from America's top 100 law firms to participants, who rated each picture on dominance and affiliation. Their results showed that dominance ratings correlated significantly with profit margins for each firm, whereas warmth ratings showed no effect. Importantly, these effects were still apparent after controlling for extraneous variables. In another study, participants were able to correctly identify Mormons from non-Mormons at above-chance levels, with perceived health (related to skin tone) apparently driving this effect.[46] Critically, participants' perceived accuracy for categorization judgments did not match their actual accuracy, suggesting that these judgments happen implicitly.

Taken together, the aforementioned results provide some support for the "kernel of truth" hypothesis. That is, inferences drawn from the face alone are accurate, to a degree, of the individual's true personality and social role. However, what is still uncertain is whether or not these inferences are truly diagnostic, or if the perceived is merely subservient to a self-fulfilling prophecy. Indeed, Bruce and Young[21] raise a point similar to this when discussing an observation involving children who either had or had not suffered physical abuse.[47] When features of the abused and nonabused children were compared, the abused children were shown to have more adult-like features. Bruce and Young point out that there are several potential explanations to elucidate these results. For one, it may be that children who appear older than their chronological age are unjustly punished for behavior that may be misconstrued as juvenile. It may also be that baby-facedness triggers caring and maternal instincts in individuals, thus reducing the potential for physical abuse among more baby-faced children and increased susceptibility to corporeal punishment in children who lack the baby-face qualities. It is also just as likely that baby-facedness covaries with a third variable not measured.

The same logic can be applied to the "kernel of truth" hypothesis. That is, facial features may be diagnostic of an individual's personality, an individual may be victim to a self-fulfilling prophecy, or a third variable may be at play. For example, the complex nature of identity cannot be captured in a single photograph. This idea was detailed nicely in a study where United Kingdom college students were shown different pictures of the same individual, in this case Dutch celebrities, and asked to sort them into categories of the same individual. Because United Kingdom students do not recognize Dutch celebrities, and view them as any other stranger, they often mistook pictures of the same individual as completely different identities.[48] Critically, when the experiment was repeated with Dutch participants, they performed at near flawless levels.

Todorov and Porter[49] attempted to test a related idea by having participants rate multiple static images of the same individuals (taken at different times), and varying in expression, on a number of personality dimensions. Critically, these individuals were instructed to be expressive but were not given specific instructions on what to express, so their poses from time one and time two varied sometimes drastically in expression. The most interesting result from their findings was that "variability in impressions of the same individual was comparable to variability in impressions of different individuals"[49] (p. 1414). Todorov and Porter[49] use these results to support the notion that despite some evidence for the contrary, information extricated from the face is generally an unrevealing snapshot of a dynamic canvas. This study seems to be at odds with the kernel of truth hypothesis, but it need not be. Another interpretation of this study is that it demonstrates how powerfully expression informs person perception. Where and when individuals choose to pose particular expressions, therefore, may be very revealing of their true natures.

Next, we attempt to briefly detail decades of face perception research in two primary areas: identity and facial expression. Then, we will discuss their intersection. We start by summarizing dynamic facial cues, specifically emotional expressivity and gaze behavior, before moving on to relevant research related to static, morphological cues that form the basis for our social identities.

2.1 Facial Expressivity

The human face is capable of complex and varied expressivity through both facial muscle patterning and the behavior of the eyes. Next, we review both in turn, starting with emotional expression.

2.1.1 Emotional Expression

The study of facial expressions as an adaptive social signal owes a debt of gratitude to Charles Darwin. His seminal work, *The Expression of the Emotions in Man and Animals*,[50] is still influential to this day, and it has helped shaped contemporary emotion expression research. However, like all good scientists, Darwin built upon the ideas of others for his work, and it is easy to see connections from his theory of expressions to other work being conducted in the nineteenth century and even prior. For example, Duchenne de Boulogne,[51] a French neurologist in the nineteenth century, used electric shock to stimulate specific facial muscles on participants' faces in an attempt to reproduce precise emotional expressions. Additionally, Sir Charles Bell,[52] a leading scientist on expression in the early 1800s, believed that God had endowed humans with certain facial muscles to allow one to express the emotion that he or she experienced. Darwin himself noted that these two individuals significantly influenced his decision to write down his thoughts on expressions, but it is apparent from his writing that he wrote *Expressions* to add support to his grand theory of evolution. In this section, we start with a brief synopsis of Darwin's work. Next, we summarize two contemporary theories that have drawn heavily from Darwin's *Expressions*. Finally, we end with what we know about how facial expression signals are extracted from the face and processed.

The leading view on emotions and emotional expression when Darwin was contemplating writing *Expressions* was that they served no evolutionary purpose, and thus, had been divinely gifted to humans to express their feelings.[53] Darwin meticulously picked apart this notion, drawing from his decades of research and observation on expression. Darwin's basic premise posited that emotional expressions had evolved and are, or were (at least at some point in the past), adaptive to our survival.[53] His theory was threefold, culminating in the implication that emotional expressions were universally recognized.[b] Darwin's theory is often cited in support of the idea that at least some (if not all) emotional expressions can be recognized by all individuals regardless of culture or geographical location.

This idea, later termed the Universality hypothesis, was put forth by Paul Ekman and his colleagues.[54–56] Universality posits that humans can recognize certain facial displays of emotion regardless of their culture background.[57] In its simplest form, then, these communicative expressions are thought to be easily recognizable in terms of discrete emotion categories. Evidence for the universality of emotional expression has been back and forth. Ekman et al.[54] as well as Ekman and Friesen,[55] provided some of the earliest scientific evidence for the universality of emotion. In their experiments, investigators showed both literate and illiterate populations specific pictures of emotional expression and asked them to categorize each emotion. Participants between and within each culture, as expected, had high agreement on categorization. This evidence was interpreted to suggest that there are pan-cultural elements in the recognition of human affect.[54]

Since his initial findings, Ekman and his colleagues have revisited, revised, and replicated these original findings.[56,58] However, some evidence has surfaced that refutes the universality claim in favor of an alternative approach.[57,59–61] An often-cited criticism of Ekman's original work was the forced-choice type format that he utilized in gathering data.[57,62] Haidt and Kelther[62] found that when utilizing a free-choice format, most of the findings presented by Ekman and colleagues[54,55] were supported, but also found that several other factors influence pan-culture recognition (e.g., non-Western ideals of emotions, and word-based versus situation-based emotion comparisons). Still, some evidence suggests that universality is overrepresented in the scientific literature. For example, in their recent review of 57 studies, Nelson and Russell[63] found that the only emotion consistently recognized across literate, illiterate, Western, and Eastern cultures was happiness. There was little agreement for surprise, sadness, anger, fear, and disgust. Interestingly, Haidt and Kelther[62] found that in their sample of Orissa (a small providence in Eastern India) participants, the most easily recognizable emotion was anger, followed by fear, and *then* happiness.

Needless to say, there have been many criticisms to Darwin's *Expressions* and the Universality hypothesis. But perhaps one of the most notable criticisms was William James' rebuttal to Darwin's theory. James' theory posits that emotions are not the result of mental states that cause changes in the body, but are in fact just the opposite. James indicated, "my thesis on the contrary is that the bodily changes follow directly the perception of the exciting fact, and that our feeling of the same changes

[b] Interestingly, although Darwin maintained that expressions of emotion are universal, he believed gestures to be culture specific, and thereby learned. He did, however, point out a few caveats. For example, Darwin noted that shrugging, while not entirely universal, is almost so global that it is hard to not think of it as an evolutionary byproduct. Similarly, he also pointed out the near-universality of kissing as a sign of affection in many cultures.

as they occur is the emotion"[64] (pp. 189–190). James was ahead of his time, and his theory was met with criticism as well. Even he noted that upon initial inspection his theory seemed quite contrary to a common sense understanding of emotion and emotional expression.

James Russell[65] later put forth an affect model in opposition to Ekman's theory, which was later adapted into the broadly defined Conceptual Act Model of emotion.[61,66–68] We can broadly define these theories as psychological constructivist theories, because they posit that emotions are *not* universal and are *not* biologically endowed. Instead, these theories state that emotions are made up of even more basic psychological "ingredients." While constructionists suggest that emotions consist of many of these ingredients (see Lindquist[69] for a full review), two have received considerable attention and research: "core affect," or internal bodily states, and "conceptualization," or an individual's interpretive process.[60,69] Each of these aspects of emotion construction is thought to be perpetually present, and as a result, do not activate in response to any external stimuli[60,70].

Core affect is different from other theories' emotion elicitors because of its primitive nature; that is, it is indivisible, one of the basic building blocks of emotion.[71] For example, in basic emotion theory, discrete emotions can be separated into the emotion feeling and what that emotion feeling is being directed at; core affect does not need to be directed at anything in order for it to exist (e.g., one can feel an internal state of anxiety without its cause being known[71]). According to some, core affect is some combination of valance and arousal.[67] Specifically, it refers to the representation of bodily changes or symptoms and need not necessarily be directed at anything.[60,68,69] Examples of core affect include cardiovascular rhythms, the release of certain neuropeptides, or a change in gastrointestinal fluid.

Growing evidence from affective neuroscience suggests that specific brain regions are involved across an array of discrete human emotions and play a substantial role in the generation of core affective states.[60,69,72] Recent evidence suggests that discrete emotions are not localized in the brain, but instead, a set of brain regions are consistently involved in subjective feelings of valance and arousal.[72,73] Wilson–Mendenhall et al.[72] reported functional magnetic resonance imaging (fMRI) evidence that suggests the subjective experience of several emotions (fear, sadness, and happiness) activate the same neural structures. Across emotions, valence was significantly correlated with medial orbitofrontal cortex activity, and arousal correlated significantly with amygdala activity. It is hypothesized that these brain areas help to integrate and analyze the continuous influx of external and internal sensory information so that the individual can make meaning of these stimuli.[60,72] Indeed, emotion is not just core affect alone. An individual's experiences are used to interpret their core affective states in the context of their prior social and personal experiences.

Many constructionist theories have an aspect that takes core affective properties and translates them into meaningful experiences through conceptualization. Conceptualization includes past experiences, emotional knowledge, and memories, among others,[60,66,69] and helps to transform core affective feelings into what is thought of as discrete emotional experiences. Conceptualizations occur quickly and automatically, and are situation specific.[60,66] Some evidence supports this notion. When participants feel an unpleasant affective state, those who have access to fear knowledge are more likely to experience "fear"-related experiences (e.g., experiencing the world as full of danger and unpleasantries), than those who do not have access to knowledge about "fear."[66] These results are similar to those purported by Schachter and Singer's[74] classic study involving the labeling of an ambiguous internal physiological state as context dependent.

Another emerging area of conceptualization is studying how language forms emotion. Interestingly, the language processing center of the brain is in the left hemisphere, encompassing the same neural network as face processing (discussed later) in the right hemisphere. Structurally, these two areas are similar and contain similar neural architecture. It makes sense, then, that language and emotional face processing are so interrelated. The interconnected relationship between language and face processing is shown through studies involving semantic satiation and semantic priming. Semantic satiation involves having participants repeat a word over and over again until that word loses meaning. You can try this yourself by repeating a word approximately 30 times. Researchers found that response latency and accuracy for identifying emotional expressions was longer when emotion words were satiated than for when they were primed with congruent emotion words.[75] In a similar study, participants were slower and less accurate at identifying which of two faces (e.g., an angry face versus a similar face) was displaying a certain emotion when that emotion was congruent with a word that had been previously semantically satiated.[76] Additionally, work with patients who have semantic dementia (a neurodegenerative disease that causes words and categories to lose meaning) has proven insightful to understanding language-forming emotional experiences. One would speculate that if emotion were biologically ingrained, and did not require language to be recognized, that even persons without the ability to use language would be able to classify emotion. However, this is not the case. When asked to place faces with specific discrete emotions displayed into matching categories, individuals with semantic dementia had difficulty sorting emotions based solely on facial configurations.[60]

2.1.2 *Eye Gaze Behavior*

Having singled out the face as a particularly salient channel for the transmission of nonverbal expression, researchers have further singled out the eyes as a particularly salient region of the face for conveying social information. Stern[77] provided evidence for the significance of eye gaze behavior very early in social development. It is a way for infants to approach or withdraw from others in an effort to regulate an ideal level of arousal. Indeed, direct gaze has been shown to increase arousal whereas averted eye gaze reduces it. In fact, before becoming mobile, gaze behavior is the primary mechanism by which an infant can approach or avoid others. This ability to approach or avoid using gazing behavior has been extensively documented in adults as well.[78] According to Mehrabian's Immediacy Model of social intimacy,[79] eye gaze is an important cue for enhancing psychological closeness, and research supports the claim that eye gaze can be used to regulate psychological distance (see Argyle and Cook[78]; Rutter[80] for reviews).

In spite of the subtlety of movement in eye gaze (mere millimeters), adults are more sensitive to direct eye gaze than postural behaviors, and they tend to become frustrated when others do not respond to their looks.[81] Notwithstanding this frustration, people will almost always wait until gaze is returned before engaging another verbally. When direct gaze is made, it both indicates that another is attending to us and elicits our attention toward them. Without such visual attention, conversation between two participants seldom occurs in Western societies.[82] Considering the fact that the movements of the eye balls (i.e., "looking behavior") is by far the most studied behavior of the eyes, it is especially surprising that this particular aspect of the eyes has been, until recently, overlooked in the emotion and face literature. Although Ekman and Friesen[54] regard eye contact as one of the primary regulators of human social interaction, they later neglect to mention its potential contribution in the expression of emotion.[83] Notably, Ekman and Oster[84] later expressed surprise over the dearth of such research.

It has been argued that an attraction to the eye region of the face is innately prepared.[78] Compelling evidence for such an innately prepared predilection is the fact that eye-spot configurations are prominently branded on lower-order animals, such as birds, reptiles, primates, fish, and insects.[78] Such eye-spot configurations, in turn, tend to lead to avoidance behavior on the part of other animals, especially when such eye-spots represent "staring" eyes relative to averted eyes.[78] The evolution of such patterns therefore has been described in terms of "mimicry" (in that the "eyes" are suggestive of a larger animal that is prepared to aggress) and is suggestive of an independent evolution across species by similar environmental constraints.[78]

In humans, Hess and Petrovich[85] showed greater pupil dilation in responses to the visual presentation of two concentric circles (eye-like shapes), as opposed to one or three concentric circles, evidencing the possibility of an innately prepared tendency for responding to eye-like stimuli. The eye region of the face does appear to draw significantly more attention from observers than other areas of the face.[86] The eyes have also been reported to be a minimal visual stimulus required to generate a smiling response in infants.[87] In fact, Ahrens[87] found that before the first two months of life, the presentation of two dots (or "eye spots") generate a smile response, whereas the presentation of other features of the face do not until later in development; furthermore, a whole face presented with eyes concealed also fails to produce smiling behavior. Hains and Muir[88] add to this by demonstrating that infants aged 14–26 weeks smile more in response to direct than averted eye gaze. Further supporting the contention that there is an innate predilection for responding to eye-like visual stimuli, particularly eye gaze direction, areas of the brain have been found that appear specialized for processing eye gaze direction and movement, such as the superior temporal sulcus[89,90] (discussed later).

2.2 Social Identity

It should be clear by now that facial expressivity, such as emotional expression and gaze, are powerful cues that need to be integrated by the visual system in order to make an accurate and appropriate judgment about the stimulus being perceived. However, faces differ on dimensions other than just expression and gaze. Indeed, while expression and gaze are commonly studied dynamic cues, there are static morphological cues that are (relatively) constant in face perception.

2.2.1 *Gender*

When individuals view a face, they immediately make inferences about that face, and the person behind the face. Individuals make these inferences incredibly quickly and without much effort, sometimes even unconsciously. Some of the cues that individuals use are obvious, while others are not so obvious. That being said, some of these cues *appear* obvious, but upon closer inspection, are more complex then they appear. A prime example of an apparently obvious, but complex, social cue is perceived gender. For example, we think we can readily distinguish between male and female faces, yet when asked what exactly separates one gender's face from the other, we might stumble in coming up with a precise answer. One might say that typically only men have facial hair; they have more angular facial features, thicker eyebrows, and shorter hair. Conversely, one might conclude that typically women are clean-shaven, have more round facial features, thinner eyebrows, and longer hair. Perhaps this conclusion is correct for some individuals, but not every male or female fits nicely into these prototypes. Some men have longer hair and thinner eyebrows, while some women have angular facial features and thin lips. Despite the wide range of facial appearances in men and women, we are still remarkably adept at distinguishing between the faces of the two genders.

Perceived gender is so apparently obvious, in fact, that individuals are remarkably accurate (~95%) at deciphering between male and female faces.[91,92] This high accuracy remains consistent across cultures,[93] and when attention is divided among tasks.[94] Curiously, even when only parts of a face are presented to individuals, they can still correctly classify these face parts as either being distinctly "male" or "female." Brown and Perrett[95] showed participants isolated facial features from both male and female faces (e.g., brows, noses, or lips). Although accuracy drops, participants were still able to correctly classify images into their correct gender when presented with these isolated features. Interestingly, every part used (brows, eyes, nose, mouth, chin) carried gender information except the nose. When phenotypic differences between female faces and male faces are examined, sex-specific characteristics emerge. For example, it has been noted that male faces tend to be longer than female faces,[96] have broader noses, and lower, thicker brows. In contrast, female faces tend to show more square/oval faces, thinner noses, and higher, thinner brows. When 3D models of faces are evaluated, male faces tend to differ from female faces by the greater protrusion of the nose and brows, as well as a more pronounced chin.[97] To obtain an objective difference in gender identification of faces, Burton et al.[92] measured simple distances between critical landmarks on male and female faces (e.g., distance between nose point to nose corner). They showed that classification of gender based on between 12 and 16 of these sets of measurements reached about 85% accuracy. Burton et al.[92] note that while this accuracy is still above chance, it is significantly lower than human observers. Similarly, Fellous[98] used discriminate analysis to show that "femaleness" was largely related to the distance between external eye corners, and large distances between eyes and eyebrows. In contrast, "maleness" was related to large nostril-to-nostril width, long faces, small distance between eyebrows and eyes, and wide cheekbones.

More recently, the importance of specific facial features for gender classification has been shown using a technique termed *Bubbles*.[99] Bubbles involve graphically manipulating photographs by superimposing noise, or "bubbles," over the image so that only parts of the image are visible. The part of the image that is obscured by the bubbles is randomized by trials. Participants are asked to make judgments about these pictures on a dimension that is pertinent to the experimenter (e.g., "What gender is this face?"). These judgments are aggregated over several hundred trials, and what remains on the aggregate image should be clearer parts that are used more often by the perceiver to make the requested judgment. For example, if individuals use a particular eye or ear to make gender classification judgments, these areas on the aggregate image should be clearer relative to the rest of the face that remains distorted by the bubbles. Several studies using this technique show that the eyes and eye region, and to a lesser extent the mouth region, are important for gender classification.[99,100]

Bubbles is the predecessor to a similar technique termed *reverse correlation*.[29] Reverse correlation is similar to Bubbles, but instead of using bubbles to block out parts of a target stimulus, it uses random noise. The procedure starts with a neutral base image, say an averaged neutral face, and applies random noise over the image. Two images are then made from the random noise image: one with the random noise pattern and another with the inverse of the random noise pattern. Participants then select between these two images which one appears more like a target, say a typical male or female face. Participants complete hundreds of trials in this fashion (>300), and because the superimposed noise is random, each trial's images appear different. After many trials, the selected images' noise patterns are averaged across all trials and then superimposed over the base image. The resultant image, called a *classification image* (CI), is a meaningful mental representation of the target. Another way of thinking about this task is that over many hundreds of trials, meaningful information "emerges" from the noise. One study using a variant of reverse correlation found that color and skin pigmentation was an important cue to gender classification.[101]

Up until now, we have only mentioned physical differences between male and female faces. There exists some evidence that male and female faces differ in terms of relative skin luminance as well.[102–104] Further, while there exists subtle phenotypical differences between male and female faces, these differences can become exaggerated by the use of cosmetics. Russell[104] measured the luminance of photographs of males and females. He showed that females typically have greater luminance differences between the eyes and lips than males. Interestingly, changing only the luminosity of androgynous faces also changed the ratings of "masculinity" and "femininity" for that face (darker luminance was rated as more masculine, while lighter luminance was rated as more feminine).

Another big indicator of gender is hairstyle. In an early study examining how subtle cues affect unconscious learning, Lewicki[105] took yearbook photos of long- and short-haired females matched for head position, gaze, race, eye color, relative attraction, among a host of other attributes. He showed participants three faces from either the long- or short-haired group, along with a vignette emphasizing how "kind" or "capable" the individual represented in the face appeared. Later, participants were shown a novel set of long- or short-haired women and asked to make "kind" and "capable" judgments about them. Lewicki found that with as little as three trials of learning, individuals' response latency between these conditions was drastically different. Further, when participants were asked if they noticed any difference between the photographs or groups, they were unable to articulate a difference! Lewicki's evidence suggests that we pick up social cues with minimal effort and retain them as we navigate our social world. Extending this concept, when androgynous expressive faces are manipulated only by differing gender-typical hairstyles, apparent women are rated as more angry and men as more happy.[106]

Surprisingly, the neuroscientific evidence for the neural substrates related to gender classification is relatively limited in comparison to its behavioral evidence. Furthermore, the small amount of literature that exists attempting to localize specific brain areas related to gender classification is mixed at best. For example, an early positron emission tomography (PET) study indicated that there was a difference in cerebral blood flow for a gender–face identification task compared to resting state. Specifically, greater activation in the right extrastriate cortex was observed.[107] In an encoding and retrieval paradigm, fMRI analyses revealed a gender of the actor by gender of the participant interaction for certain brain regions,[108] including greater activation in the left parahippicampal region, right insula, and left amygdala. Interestingly, the greatest negative activation was observed for participants who viewed faces of the opposite gender than themselves, and greatest positive activation was observed when participants viewed same gender faces. Lastly, a study replicated previous fMRI work showing greater activation for faces in the core (inferior occipital gyrus, fusiform gyrus, and superior temporal sulcus) and extended (amygdala, inferior frontal gyrus, insula, and orbitofrontal cortex) face neural network (discussed later),[109,110] and average BOLD patterns across viewing male and female faces were indistinguishable.[111] However, by using multivariate pattern analysis, it was shown that patterns of BOLD signals in the fusiform gyrus, inferior occipital gyrus, superior temporal sulcus, insula, inferior frontal gyrus, and medial orbitofrontal cortex showed activation patterns that were significantly above chance for decoding accuracy.[111] However, no subject gender by face gender interaction was observed.

2.2.2 *Age*

Other than gender and race (which we will discuss next), age is the other major relatively static facial cue that perceivers pick up on quickly and automatically when viewing faces. Surprisingly, despite its pronounced appearance, relatively few studies have examined the effect of age on the perception of faces. Research dating back several decades has shown that individuals use information about craniofacial growth and the amount of wrinkles to make judgments about the age of a face.[112] Later, Burt and Perrett[113] extended these results using a more sophisticated method involving averaging a number of young and elderly faces to make a composite prototypical young or elderly face. The researchers then applied shape and color cues from the averaged elderly face to young faces and showed that this increased the perceived age of the young faces by approximately 12 years. However, the averaging process used for this research does not capture adequately the complexities that wrinkles provide on elderly faces. One study addressing this issue suggests that age-related wrinkles reduces the signal clarity of emotional expressions due to neutral faces taking on emotionally expressive characteristics because of the wrinkles.[11]

The aforementioned studies suggest that the elderly are victim to an overall negativity bias or age stereotype. Indeed, research suggests that this may be the case. Kite and Johnson[114] performed a meta-analysis of 43 studies, showing that 30 of them reported effect sizes indicating an overall negative perception of the elderly. More recently, Ebner[12] showed the same effect for the perception of elderly faces. She had young and elderly participants rate young and elderly faces. Overall, elderly participants rated elderly faces as more positive, but overall elderly faces were rated as more negative and less attractive than younger faces.

Evidence related to the neural mechanisms that underlie age perception is sparse. Some research has shown that there is a left visual field (left face) bias for age perception, which is in line with previous research showing this bias for expression and gender identity, as well as attractiveness ratings.[115] These results suggest that age is processed in a similar manner to other static cues, such as gender and attractiveness in the right hemisphere (the optic nerve from the left and right visual fields run to contralateral sides of the brain). Other neuroimaging research suggests that the elderly are less efficient at processing faces (especially other-age faces), but this does not affect processing of face age.

2.2.3 *Race*

One way that individuals socially stratify the world is through race. Although genetics has shown that there are no clear divisions between races, race cues are visually salient, and thus socially powerful. For example, skin tone and hairstyle provide an acute visible signal to cue the perceiver into the race of the individual. Typical race cues for African Americans versus European Americans include darker skin, broader noses, fuller lips, and coarser hair.[21] Blair and Judd, through a series of studies, showed that skin luminance, lip fullness, nose shape, hair texture, and hair quantity significantly predicted ratings of African American versus European American males.[116] Facial features lead to social group categorization, from which stereotypes are often activated. Blair et al.'s work has shown that even when directed to abstain from stereotyping, it still occurs, even when social categories are not activated explicitly. For example, when prison inmates matched for severity of crime were evaluated, there appears to be no racial difference in length of incarceration or severity of punishment. Judges, of all people, should show unbiased behavior when evaluating evidence of crimes. However, within racial groups, those with more stereotypical Afrocentric features were given harsher punishment.[117] Presumably, this discrepancy occurs because individuals with greater Afrocentric (or any other race category) features are believed to possess more stereotypical behaviors associated with that race. The fact that this occurs, even when explicitly told to abstain, suggests that stereotyping happens outside of conscious categorization.

Racial differences do not only exist at the level of the stimulus (i.e., face), but also at the perceiver level. Said another way, some individuals are prejudiced against certain groups or individuals. Some individuals hold these prejudices outside of conscious awareness.[118–120] In a particularly informative study,[121] researchers measured participants' prejudice toward a particularly salient out-group in the area that they conducted the research. Following this, individuals performed a reverse correlation task indicating which face looked more like the target out-group. When high-, moderate-, and low-prejudice individuals' classification images were averaged together, pronounced differences in facial structure became apparent. Compared with low-prejudice individuals', high prejudice individuals' mental representation of the out-group face was rated as appearing more criminal-like, and less trustworthy. It is important to remember that each participant, whether they were of high or low prejudice, saw the same base face image (a racially-ambiguous average of many faces) superimposed with random noise. Thus, it appears that the mental representations that we hold about particular groups of individuals also affect the way in which we perceive them.

3. THE NEUROANATOMY OF SOCIAL VISION

In the first section of this chapter, we outlined facial cues that have received considerable attention over the past few decades. However, we have yet to discuss the neuroscience of face perception in depth. The modern understanding of neuroscience as a distinct and separate field is still young. Recent advancements in neuroscientific practices, such as higher resolution imaging and more cost-effective methods, have allowed the field to grow and evolve at an exponential rate. One benefit of this exponential growth is the adoption of typically neuroscientific-only practices into other related fields, such as social psychology and cognitive psychology. Further, neuroscience techniques are becoming standard in fields as diverse as animal cognition[122] and to behavioral economics.[123]

Of particular interest to this chapter, as well as to our understanding of the social relevance of face perception, are the subfields of social neuroscience and social visual perception. Darwin's seminal work on facial expressions[50] suggested that facial expressions are a mechanism by which internal states and intentions become externally signaled. Despite wide disagreement over whether emotional displays are universal or culturally determined, cognitive or biological in origin, most recent researchers nonetheless seem able to agree that facial expressions act to forecast an organism's behavioral tendencies.[124–129] In Darwin's original theory, he did not suggest that facial expressions evolved initially to communicate behavioral intent, but rather that they were a behaviorally adaptive response to recurrent survival obstacles. As the result of the repetitive occurrence of these behavioral responses to such obstacles, animals that learned to take advantage of the predictive value of their conspecifics' facial expressions likely gained a survival advantage, which in turn conceivably led to the evolution of facial expressions as a signaling system.

If the visual system directly evolved to fit our world and provide an adaptive communicative advantage, then we should see a similar evolution in other primates. This hypothesis maintains that in environments where viewing facial expressions is less important or hindered (e.g., by dense foliage), then facial visual cues should be less evident. In fact, this appears to be the case. Old World monkeys, which are terrestrial, possess a greater variety of facial expressions than New World monkeys, which are arboreal,[130] suggesting that in environments with less obstructed views of primate faces, expressivity plays a more important role in social navigation. Applying this same concept to humans, it is easy to see that visual cues from the face play an important role in the perception of others. One might want to signal approach to a potential mate, or dominance and stability to a particularly hardheaded in-law. Humans as a species compete with themselves for resources—both social and environmental—thus necessitating the continued evolution of our social perceptual abilities.

Vision scientists and social psychologists have, until recently, contended that the ability to engage in social exchange is biologically endowed, and by extension, studied face perception in a highly modular fashion. That is, a plethora of previous research has been conducted under the assumption that specific areas of the brain are utilized predominately for singular purposes. Brain modularity has been intensely debated, dating back to the work of Hubel and Weisel[131] up and through the contemporary advent of fMRI. Advancements in neuroscience in the 1990s and 2000s led to an influx of brain imaging studies that culminated in aptly named modules for higher level visual processing. For example, some of the more reliably studied "expert" modules of the brain include the fusiform face area for faces,[132,133] the extrastriate body area for bodies,[134] the parahippicampal place area for places,[135] and the visual word form area for words.[136]

While it certainly appears that at least *some* higher level processing is localized in specific parts of the brain, there is a much clearer case for low-level visual processing. Indeed, for certain low-level visual stimuli, such as lines, contours, and holes, evidence suggests that they are processed in a highly modular fashion. However, a growing body of literature suggests that social vision is influenced by any number of parameters, ranging from the perceiver's internal state and prior knowledge to the stimulus itself. For example, individuals who are induced to be in a sad mood report the steepness of a hill as being much steeper than it actually is in reality.[137] Similarly, the facial cues outlined earlier (e.g., age, gender, emotion) are certainly not low-level stimuli, especially when they are combined. Social visual information should not be studied as if it were the same as perceiving holes, shapes, and contours. From an ecological point of view, these cues provide social affordances that our visual system is equipped to perceive. We have already alluded to some of these social affordances earlier and will expound on a few of them shortly, but first let us examine the intersection of vision and social perception. In this section, we will discuss ways in which neuroscientific evidence has influenced and enlightened the field of social vision, with a particular focus on face perception.

3.1 Neuroanatomy of Basic Vision

If social vision is a complicated beast, then the sense of vision is its perplexing brute of a parent. It is not surprising, then, to see that introductory biological psychology and human biology textbooks devote whole chapters to this one sense, all the while leaving the other four senses lumped into a separate singular chapter. Consider for a moment the intricate process of simply watching the classroom door for a professor to enter. First, light has to enter the pupil and be focused by the lens and cornea so that it can pass through the clear vitreous humor between the lens and the retina. Next, the retina needs to interpret this light through a series of dedicated cells. Light activates specialized receptor cells (i.e., rods and cones), which in turn activate bipolar cells that send signals to ganglion cells.

In the primate brain, ganglion cells come in three flavors, each with a specific purpose: *parvocellular*, *magnocellular*, and *koniocellular*. The first type of cell, parvocellular, has small receptive fields and is adept at detecting visual detail, so it may pick up on the pattern that the professor is wearing, or the fine color detail of his or her briefcase. In contrast, magnocellular ganglion cells have large receptive fields and are colorblind; they respond largely to movement and therefore might be attuned to the gait of the entering professor or the sudden motionlessness of the class as he or she enters. Lastly, koniocellular ganglion cells appear to have several functions and exist throughout the retina. Of the three specialized types of cells, koniocellular are the least studied, and their exact purpose is just now becoming elucidated (see Hendry and Reid[138] for further review).

Advancing past the retina, the ganglion cell axons bundle together to form the optic nerve, which exits at the back of the eye. As the optic nerve travels out of the eye further into the brain, half of the axons from the nasal side of each eye cross at the optic chiasm and proceed to the contralateral side of the brain. The other half of the axon bundle (i.e., the temporal side) travel to the ipsilateral side of the brain. Most of the ganglion bundle goes to the lateral geniculate nucleus, and from there, other axons are sent to the thalamus and the occipital cortex, known as the primary vision center of the brain. From the occipital cortex, visual information may be sent to a number of places, such as the prefrontal cortex, to be further systematically analyzed, or the fusiform face area for specialized processing. Other visual information skips some of the downstream visual cortices entirely and proceeds on a more direct route to the amygdala for quicker survival-related processing and action. Perhaps most fascinating is that this whole process takes place continually while we navigate our social world and occurs effectively at speeds within hundreds of milliseconds of viewing a stimulus.

3.2 Neuroanatomy of Face Processing

3.2.1 The Fusiform Face Area

Perhaps the most famous brain area related to face perception is the fusiform face area (FFA) in the fusiform gyrus. Kanwisher and colleagues'[132] paramount fMRI study found that the FFA had greater activation when subjects viewed faces compared to when they viewed objects such as houses. The implication of these results is that there is a brain region in the cortex that is preferentially attuned to process faces. Follow-up studies showed that the FFA is most activated for human faces, and to a much lesser extent, activated for similar stimuli, such as human heads, bodies, and animal faces, suggesting a specialized area attuned to human face processing.[139] Grill-Spector et al.[140] measured activity in the FFA while participants completed a behavioral face and object identification paradigm. The researchers showed that FFA activity correlated with face identification on a trial-by-trial basis, but the same could not be said for objects. Interestingly, FFA activation in "car experts" did correlate with behavioral data on identifying cars. The researchers suggest that the FFA is critically involved in face processing and identification but has little to do with non-face object recognition and processing. Similarly, a recent review[141] found support for domain-specific face processing and suggests that this face-selective region evolved for adaptive purposes.

The impact of this research launched a frenzy in the neuroscientific community, culminating in an intense debate as to the precise function of the FFA. There is some evidence that the FFA is recruited for familiarity and expert recognition of things other than, but including, faces.[142–145] For example, researchers have found that training participants to be experts at recognizing novel shapes/creatures called "greebles" activates the right FFA.[142] When this concept is extended to other experts, such as experts in the domains of birds and cars, experts' FFA were preferentially activated for their expertise group, compared to their nonexpertise group.[143] The FFA appears to be activated in nonexperts even when they simply classify objects into subordinate versus superordinate categories, both visually[143] and auditorially.[146] In a similar study, when the brains of expert and novice chess players were examined while they viewed chess boards, expert chess players showed increased FFA activation compared to novice players.[147] Critically, experts still show greater activation in the FFA for faces than other stimuli.[143]

However, these results do not undermine previous research suggesting that the FFA is recruited in the processing of faces. Borrowing again from Gibson's ecological model of visual perception, we can think of the face as a visual object that is processed in an expert way. The information derived from faces is much more ecologically relevant to the viewer than chess pieces or greebles. Much like the meaning derived from birds or chess pieces in experts of said areas, every human without impairment can be considered a "face expert" who derives special meaning from faces. Indeed, even if one is an expert at birds, cars, or chess, they have much more practice at being a "face expert" just by living and participating in a social world. Moreover, even if faces are not "special" in terms of the neural machinery required to process them, the information we glean from faces is special to perception and social interaction.

3.2.2 The Occipital Face Area

While the FFA has received considerable attention for being face-sensitive, there are many other areas that are also involved in the visual recognition of faces. Moreover, increasing evidence suggests that the FFA is not sufficient in and of itself for the recognition of faces.[148] One example comes from a patient with neurological damage. Researchers[148] compared patient D.F., who had prosopagnosia (inability to recognize faces) and damage to the lateral occipital cortex and the medial occipito-parietal region, to age-matched controls. Behaviorally, patient D.F. was able to perform some lower-level face tasks, such as discriminating faces from other non-face objects, but could not perform higher-level face tasks, including expression and identity recognition, or face gender classification. Surprisingly, D.F. showed near-normal activation in the FFA compared to controls. However, D.F. had no activation in the area of the inferior occipital gyrus, termed the occipital face area (OFA), due to her damage in that area. All controls showed activation in the OFA, presumably allowing them to complete higher-level face tasks that D.F. could not complete. These results suggest that normal FFA activity does not equate to normal face identification performance, and that other areas must be involved to function normally.

The OFA has not received the extensive analysis that the FFA has been treated to over the last couple of decades, and as such, not as much is known about the OFA's specific function in the face processing network. The OFA is more often than not located in the right hemisphere, consistent with other face-sensitive areas.[149] While the OFA region has been observed since the advent of the fMRI, it was Gauthier et al.[143] who coined the term "occipital face area" for the region in the occipital gyrus. Later, the OFA was studied more extensively, and some posit that it is an early visual detection agent in the neural network of face processing.[150,151] Indeed, the OFA appears to receive information early from the visual cortex before being analyzed by higher cortical areas.[150,151] Along these same lines, the OFA has been implicated to be involved in processing fluid physical face changes, rather than more static identity cues.[150] More recently, the OFA has been shown to be selectively responsive to face parts, such as the eyes and mouth.[149,152–154] In one study using transcranial magnetic stimulation (TMS), a device that delivers strong magnetic pulses to a selective areas of the brain, researchers artificially and temporarily produced the equivalent of lesions in the OFA.[154] As suspected, temporarily disabling the OFA resulted in an inability to discriminate face parts, but an intact ability to discriminate houses. Interestingly, the researchers conducted another study where they delivered TMS pulses at varying times after stimulus onset. They showed that pulses delivered at 60ms and 100ms disrupted face part processing, but delivering pulses at other times did not, suggesting that the OFA is associated with early face detection processes.

3.2.3 *The Superior Temporal Sulcus*

The last face processing area we will single out is the superior temporal sulcus (STS). A vast majority of the early neuropsychological work done involving the STS was conducted with nonhuman primates and single-cell recordings. These studies revealed a "face-sensitive neuron" located in the temporal sulcus. With further investigation, it was shown that the STS appeared to be responsive more to facial movement, eye-gaze, and expression rather than identity cues.[155–157] Functional MRI studies with human subjects have been divided as to the exact purpose of the STS; some studies show an increased activation to speech and language processing, while others have shown its importance in audio-visual processing and theory of mind.[158] More recently, it has become quite clear that the STS region is not only involved in interpreting cues such as facial movement, expression, and eye-gaze direction, but also social perception in general.[159] A plethora of studies have shown that the STS is responsive to cues, such as intentions, derived from biological motion,[160] social gaze,[161] audio perception,[162] and general directional information,[163] in addition to facial expression.[164] While the exact function of the STS appears to be more complicated than the other two face areas previously discussed, the STS is certainly involved in some aspect of facial processing, with a specific emphasis on social cues.

3.2.4 *Extended Areas Utilized in Face Processing*

While we only discussed three areas involved in face processing, there is an extended network of brain regions involved in face processing. Based on high interconnectivity with areas involved in cognitive, perceptual, and behavioral responses, Adolphs[165] argues for a disproportionately important role of the amygdala in the processing of emotion and social information. The amygdala, which is uniquely placed to receive projections from many areas of the brain, is a densely interconnected hub with regions where exteroceptive and interoceptive information can be integrated.[166–168] Thus, Adolphs suggests that the amygdala is involved in orchestrating an integrated response within the brain and body. Similarly, the amygdala is of primary importance in Theory of Mind development, going so far as to generate an "amygdala theory of autism" on its functioning. The amygdala is reciprocally connected to (1) the fusiform gyrus, which is specialized for identity processing;[132,139] (2) the orbital frontal cortex (OFC), which is known to be involved in adaptive behavioral responding, conceptual knowledge retrieval, and decision-making during the processing of emotion information;[169–171] and (3) the STS.[109] Thus, the amygdala is equipped to act as a sort of relay station and integrative command center for the adaptive processing of and response to socially relevant facial information. Notably, these brain regions have been found to be structurally abnormal in autistic populations[172] and is thought to play a critical role in their profound deficits in social perception.

The OFC, amygdala, and STS together comprise a densely interconnected network that arguably sets the stage for basic person perception. Collectively referred to as the "Social Brain" by Brothers,[173] this three-node pathway has been implicated in human social perception, underlying the ability to "mind read" (i.e., infer mental states based on nonverbal information). Numerous studies specifically implicate these three structures in the recognition of emotion displays (see Adolphs[166]), and in gaze direction detection/shared attention processes (see Emery[174]). Since the amygdala is densely and reciprocally interconnected with the STS and OFC, which are in turn directly connected to one another, these three structures are ideal for the transfer and integration of social information. In this way, the brain appears to have the functional architecture necessary for integrating and pooling visual information from facial cues, assessing their combined social significance, and acting on the product accordingly.

3.2.5 *Models of Face Perception*

Until recently, the majority of behavioral and neuroscientific work conducted on face processing examined various facial cues independent of one another, often resulting in entirely separate fields of study (e.g., gaze direction, facial identity, and facial expression). This approach is consistent with models of face processing that suggest that it is adaptive for functionally distinct types of facial information to be processed via separate and thus independent processing routes. Bruce and Young,[175] for instance, proposed these processes were noninteracting, thereby avoiding perceptual interference. Indeed, extensive research suggests that there is a neural separation with regard to how the brain processes facial identity and the recognition of facial expressivity. This contention has been supported via behavioral data (e.g., Bruce and Young[175]), clinical cases (e.g., Tranel and Damasio[176]), single cell recordings done using primates (e.g., Hasselmo et al.[177]), and brain imaging studies done using humans (e.g., Hoffman and Haxby[90]). Although the fusiform gyrus and the STS have both been shown to have cells preferentially responsive to facial stimuli,[132,139] the former appears to be more involved in processing static appearance, such as race and gender and the latter expressive aspects of the face, including both emotional expression and eye gaze (Hasselmo et al.[177]; Phillips et al.[178]; Puce et al.[179]).

Haxby et al.[109] proposed a neurological model that took into account previous behavioral studies, what appeared to be double-dissociable deficits, and neuroimaging research. Their model outlines a neurological difference between the processing of morphological and expressive features of the face. Features that are invariant, which are critical in determining identity, sex, and race, are primarily processed in the fusiform gyrus. However, processing of expressive features of the face, such as expression, speech perception, or eye gaze, activates the fusiform gyrus to a lesser degree,[109] whereas expressive cues activate the STS. In this model, projections from the STS influence a variety of systems. Emotional information is processed in the amygdala and the limbic system, while gaze direction is processed in the intraparietal sulcus. Information about mouth movements involved in speech projects to the auditory cortex, where it facilitates speech comprehension. These regions also project back to the face processing areas guiding continued processing.

Where this model corresponds with the Bruce and Young[175] model is the notion for functionally independent routes for the processing of morphological and expressive cues of the face. Recent evidence suggests that processing of the invariant features and those features that change dynamically within a face may be more linked than previously thought.[180] Given reciprocal connections among many regions included in this model, however, the neuromechanics of compound cue processing would allow for complex interactions.

4. SOCIAL VISION AND COMPOUND CUE INTEGRATION

The primate visual system has evolved to be highly efficient for things such as detection of subtle changes in our environment, as well as being able to distinguish between the highly similar hundreds of faces we see on a daily basis. But this complexity also comes at a cost. Consider the case of threat perception. An early reflexive response to quickly presented threatening faces is thought to bypass some of the downstream visual cortices and instead proceed on a subcortical route from the retina directly to the superior colliculus, then the pulvinar nucleus, and finally onto the amygdala.[167] Theoretically, then, a female fearful face with averted eye gaze (congruent pairing of compound cues) would be responded to faster than a different pairing of similar cues (e.g., a male face with direct-gaze fear expression). Indeed, this appears to be the case, at least at the behavioral level (see Adams and Kveraga[181]). Thus, even before information reaches the retina, social perception is arguably occurring at the level of the stimulus itself. For example, a male face may phenotypically resemble the stereotypical emotion cues related to an anger expression, which when perceived by the visual system, gets overgeneralized in the brain's interpretation of that particular face, influencing impression formation and emotion perception (see Adams et al.[182]).

Take for example work done on facial neotony.[17] The visual system arguably evolved to respond to babyish cues in the face by triggering a caretaking response. This action-based affordance fits with ecological views of vision, which stipulate that when viewed, objects offer opportunities to the observer to act on or be acted upon by the stimulus. That fear expressions physically resemble neotenous cues (large round eyes, high brows) could be the result of fear taking on the form of a baby face in order to exploit their social affordances.[183] When in distress then, we may contort our face into a babyish visage in order to elicit caretaking responses from those around us. Thus, in this sense, facial expressions may have evolved to exploit preexisting neural responses of more stable facial cues.

Further, there are strong cultural stereotypes driving expectations for women to display more happiness, fear, and sadness than men, and for men to display more anger, despite little evidence of gender differences in how much men and women *feel* these emotions.[184,185] Moreover, male and female faces differ structurally, with stereotypical female faces consisting of more round and babyish features that are typically perceived as more affiliative. It makes sense, then, that female faces that combine these characteristics in any pattern are perceived differently. Friedman and Zebrowitz[186] show that when these cues combine to form in a stereotypically congruent pattern (e.g., a female face with babyish facial features expressing a smile) it augments the stereotype (e.g., "women are more expressive," and "expressivity in women is normal"). However, when the cues combine in a way that is incongruent with the stereotype (e.g., a mature appearing female face expressing a smile), it overrides the stereotype. In this particular example, a mature-faced woman will not be viewed as less dominant than a comparable mature-faced male but in some cases as even more dominant.

Relatively new evidence has elucidated several compound social cues that may be influencing the perception of the individual, especially when the only information that is available to the perceiver is the (relatively complex and ambiguous) face. In the remainder of this chapter, we examine several compound cues that have received some behavioral support and are beginning to receive attention from a neural perspective.

4.1 The Nonneutrality of Neutral

Before we discuss the compound influence of social cues that share signal value, a brief review of attempting to understand the neutral face may help elucidate the importance of the current research. This review is brief, not because it is outside the scope of this chapter, but because understanding an emotional, or otherwise, neutral face has received little attention from a social perspective until relatively recently. Most studies examining trait inferences from facial appearance alone have focused on isolating a single trait (e.g., attractiveness) represented by prototypical faces for that trait based off of previously gathered ratings aggregated from a large number of participants. Todorov et al.[187] instead suggest utilizing emotionally neutral faces without any defined anchor constraints to study personality inferences. In an initial investigation of this procedure, they subjected unconstrained descriptors of neutral faces from participants to principle components analysis to reveal a 2D space defined by valence/trust and power/dominance. These two dimensions accounted for nearly 80% of the variance in the judgments, and in the extreme, resulted in what appeared to be emotionally expressive faces (e.g., high trustworthy faces appear to smile).

While this work provides a suitable launching point for future endeavors researching social affordances perceived from a face, it nonetheless fails to account for compound and contextual cues. For example, from a purely affective domain, evidence suggests that a prototypical neutral face may not be as emotionally neutral as one might expect. Carroll and Russell[188] show that the same facial expression shown in two different contexts may be perceived as qualitatively different. Indeed, Russell and Fehr[189] concluded that viewing one facial expression "anchor" shifts the judged perception of a following expression. Specifically, they found that an objectively rated neutral expression was viewed as sad when preceded by a group of happy expressions. Similarly, Pixton[190] (Experiment 2) found that neutral faces were rated more toward "sad" on an 11-point scale ranging from "sad" through "happy," and more toward "angry" on an 11-point scale ranging from "friendly" through "angry." Finally, Lee et al.[191] reported that although participants explicitly rated neutral faces as "neutral," when they completed the Extrinsic Affective Simon Task (a task similar to the Implicit Associations Test), their responses revealed that neutral faces were more similar to the responses made for negative faces than those of positive faces. Thus, it appears that what is thought to be a prototypical neutral face may not always be considered emotionally neutral, at least on an implicit level.

Further, when considering dimensions other than affective, neutral faces also appear to have some intrinsic emotional value. In one study, neutral female faces were rated as more submissive, naïve, cooperative, fearful, happy, and less angry than neutral male faces.[192] In a related study, Adams et al.[192] then "warped" neutral faces over corresponding anger and fear expression, thereby keeping their neutral visage while resembling these emotions. Fear-warped faces, both male and female, were rated as more submissive, naïve, cooperative, fearful, happy, and less angry than neutral male faces, suggesting that in the first study, emotion resembling cues in male and female neutral faces were a primary mechanism of person perception. The apparent nonneutrality of prototypical neutral faces, therefore, poses a considerable obstacle for further understanding face perception unless we can better understand the underlying behavioral and neural mechanisms that drive these differences.

4.2 Emotion Overgeneralization

Overgeneralization effects in face perception assume a biologically hard-wired response to certain cues that indicate traits such as affiliation or physical fitness; these traits are then generalized to faces that incidentally resemble such cues. This approach fits nicely with an ecological approach to visual perception, as it maintains that errors produced by overgeneralizations are less maladaptive than those that might occur had the overgeneralization not occurred.[33] Zebrowitz and her colleagues[8,17,33,193–195] have done considerable amounts of behavioral work and computational modeling to suggest that there are at least four separate overgeneralization effects that influence face perception: (1) attractiveness, (2) baby-facedness, (3) specific emotion, and (4) familiarity. In addition, some neuroscientific evidence supports overgeneralization effects, yet much has yet to be done to identify the specific neural substrates of these neural mechanisms. Because gender-specific emotion overgeneralization effects will be mentioned in the next section, here we discuss these principles in more general circumstances by reviewing overgeneralization effects for attraction and specific emotions.

One robust effect in the face literature is the "attractiveness halo." Multiple studies have shown that objectively rated attractive individuals are rated as more outgoing, competent, intelligent, and healthier than unattractive faces.[194,196,197] These effects occur cross-culturally, regardless of perceiver or face.[198] It is thought that there is an adaptive advantage to recognizing the stereotypically attractive qualities in individuals because it is a sign of health, good genes, and ultimately, a way to pass on our genes. The overgeneralization effect occurs when our responses are generalized to individuals whose faces resemble these unfit or fit characteristics. Kalick et al.[196] showed through a longitudinal analysis of attractiveness and health that the two were unrelated, despite relatively attractive individuals being rated as healthier than their peers. However, more recent evidence has suggested the opposite. At least for some age groups, attractive individuals are perceived and actually are healthier and more intelligent than unattractive individuals.[194,199] Despite mixed results, facial attractiveness and abnormalities do provide some information regarding intelligence, and consequently, fitness as a mate. From a neuroscientific standpoint, evidence shows that attractive faces elicit greater activation in regions of the brain that respond to positive stimuli (specifically the reward network), and unattractive faces activate regions in the brain that respond to negative stimuli.[200,201]

Specific emotion overgeneralization effects have also received substantial attention in the literature. Secord[34] introduced emotion-related overgeneralization effects by suggesting that perceivers mistake temporary expressions for more enduring personality attributes. For example, someone who is perceived as angry may be mistaken to possess a low affiliative nature. Emotion overgeneralization even occurs in individuals whose phenotypical facial appearance naturally resembles emotional expressions.[8,17,202] For example, consider an individual whose mouth corners naturally turn upward. This particular individual may more often be mistaken to be happy when they are not. Similarly, consider an individual whose mouth corners naturally turn down when in a resting baseline state. This individual may more often be mistaken as displeased or angry. This emotion overgeneralization effect has been shown for a number of groups. For example, baby faces tend to resemble fear and surprise more than non-baby faces, both by human rater and by connectionist modeling.[203] Further, human raters tend to rate highly dominant neutral faces as angrier and more male appearing, and highly affiliative faces as more happy/fearful/surprised and female appearing.[187,192,204,205] Similarly, connectionist neural networks trained on a variety of faces rated male faces as appearing angrier than female faces, and female faces as more surprised than male faces.[8] The emotion overgeneralization effect extends to impression formation as well. Said et al.[206] had individuals rate numerous emotionally neutral faces on a number of traits. They then subjected the faces to a Bayesian network classifier trained to detect emotional expressions. Overall, they found that faces that were rated on positive personality traits structurally resembled happiness, and faces rated on personality traits related to dominance structurally resembled angry expressions.

4.3 Gender and Emotion Interactions

One social cue that has received considerable attention over the past few decades is gender. Research examining explicit gender-stereotypic expectations has shown that women are expected to show and express more emotion. This effect is robust and is consistent regardless of gender or age.[185,207,208] Specifically, expressions of happiness, fear, and sadness are thought to be more characteristic of women than men.[184,185] Men, on the other hand, are expected to be less emotional, more stoic, and express more anger.[184] Gender-emotion stereotypes appear early and are consistent. Children as young as three show this same pattern.[209] Research indicating a link between the compound nature of gender-related appearance, facial expression, and maturity has led to the proposition that gender stereotypes may be driven by typical gender facial appearances being similar to specific facial expressions. In essence, a gender-specific emotion overgeneralization effect is occurring.

As mentioned briefly in the introduction to this section, certain facial features are associated with specific traits. Dominance is related to a square jaw, thin lips, and low-set eyebrows, all features more typically found in male faces,[210,211] whereas affiliation/baby-facedness is associated with large, round eyes, and high eyebrows—facial features more typical of the female visage. Evidence supports that when facial appearance is gender stereotypical, it augments the stereotype. However, when facial appearance is not gender stereotype-typical, the perceivers look to facial cues to derive dominance and affiliation.[186]

Expounding on this idea are observations showing that some facial configurations of expressions resemble the structural appearance of stereotypical gender features. For example, wide eyes and raised eyebrows—features more typical of female faces—characterize fear expressions. Similarly, low, tight-knitted eyebrows, thin lips, and narrow eyes—features more typical of male faces—characterize anger expressions.[83] Just as furrowing the brows, as in a typical anger expression, increases perceptions of dominance,[212,213] so too do high versus low brows on otherwise neutral faces affect dominant/submissive ratings, as well as anger/fear attributions, respectively.[211,214] The combination of happy mouths and fearful eyes bias classification of androgynous faces toward female classification, whereas both angry eyes and mouth bias perception of androgynous faces toward male classification.[205] Particularly compelling is a connectionist model that found a correspondence between gender appearance and emotion, based purely on facial metric data, and thereby free from any cultural learning or stereotypic associations. Once trained to categorize emotion, male neutral faces were found to activate the anger nodes and the female faces the surprise and happy nodes.[8]

4.4 Age and Emotion Interactions

Another social identity that the face displays is apparent age. Researchers and laypersons alike have known for millennia, through simple observation, that age affects facial appearance. Growth of the head from childhood to adulthood results in smaller head-to-body ratio, as well as more angular and defined features. In infants the eyes are relatively larger and lower within the face space than adults, resulting in the typical baby-faced "cute" look.[21] Indeed, perceptual judgments of skull images manipulated to replicate human growth show reliable accuracy for apparent age.[215] Interestingly, reliable judgments of age based on computational models approximating skull growth are so profound that this work has been applied to nonhuman animals, and even cars, with high accuracy.[216] Further, as the face ages, skin loses elasticity, resulting in wrinkles and sagging, especially around the eyes. Despite knowing that age transforms faces, until recently very few studies have examined the perceptual influences of age on face perception.

Perhaps the most recognizable features of an aged face are wrinkles and folds. Malatesta et al.[13] found that affective expressions were harder to decode in older adult faces, presumably due to these age-related face structures. Malatesta et al.[217] found that there were more misattributions of emotion to objectively rated neutral faces of older adults. More recently, Hess et al.[11] examined whether wrinkles degrade the signal clarity of emotional expression in elderly faces. Computer generated avatars of young and elderly faces were shown to participants expressing the exact same expression. When expressions were displayed on elderly faces, the expression was rated as *less* intense with regard to the actual expression being displayed, while being rated as *more* intense with regard to every other emotion. This overall pattern suggests that while individuals (in this case young perceivers) can accurately detect facial expressions on elderly faces, the clarity with which what they are intending to forecast with said expression is degraded.

Aside from degrading and changing signal clarity of expressive cues on the face, there also appears to be distinct age-related stereotypes. We already discussed the general negativity bias for the elderly,[12,114] but let us expound on it briefly from an ecological viewpoint. Neuberg and Sng[218] found an age by sex interaction on stereotypes. Replicating previous work, male faces were found to be more agentic than female faces. Interestingly, this was largest for males aged 18–28. Older males (age 60 and over) were perceived as more communal than agentic. Young females were perceived to be equally agentic and communal. While previous research has found an explicit negative stereotype for the elderly,[114] Hummert et al.[219] found an implicit negative stereotype for the elderly, despite conflicting explicit ratings in the opposite direction. From an ecological point of view, perhaps elderly faces signal reduced health that consequently should be avoided—a hypothesis that certainly parallels the face attraction literature. Some preliminary results from our own lab confirm this negative emotion age bias for mental representations of the elderly (in prep). We employed a reverse correlation task with FaceGen avatars of prototypical male and female elderly faces portraying a neutral face as the base image. A second group of participants rated the first participant group's CIs and a number of random foils on how angry, disgusted, fearful, happy, sad, surprised, and trustworthy each face appeared. Results across both elderly male and female faces revealed that elderly neutral faces looked more angry, disgusted, fearful, sad, and surprised than comparable foils. Additionally, elderly faces were rated as less happy and less trustworthy than the foils. Taken together, effects of age on the signal clarity of expressions, coupled with established age stereotypes, compound together to complicate the perception of faces, similar to gender stereotypes.

4.5 Race and Emotion Interactions

Perhaps unsurprisingly, race-associated features, such as face width and skin luminance, combine with other social cues to change face and person perception. Evidence for the compound nature of race and other cues (e.g., gender[220,221]) is becoming prolific, so for our brief overview we will focus on how race and emotion interact to change perception.

Like all perception, the compound effect of race and emotion is influenced by both the stimulus (i.e., face) and by the internally held beliefs (implicit or explicit prejudices) of the perceiver. Starting at the level of the stimulus itself, connectionist modeling has shown that neutral Caucasian faces are counterstereotypically more structurally similar to anger expressions than African American faces, while neutral African American faces structurally resembled happy and surprised expressions more than Caucasian faces.[8] Further, these structural resemblances seem to affect impression formation, such that neutral faces that physically resemble anger expressions were rated as less trustworthy and less likable and more hostile than faces that did not physically resemble anger, while the opposite held true for faces that were more physically similar to happy expressions. When face race was examined as a moderator variable, individuals of different races responded to structurally different faces in interesting ways. For example, when Caucasian raters viewed Caucasian, Korean, and African American neutral faces structurally similar to anger expressions, they rated the Caucasian and Korean faces as less likable and more dangerous than other comparable faces, but there was no differences in ratings for African American neutral faces that structurally resembled anger expressions. The authors conclude that perceiver race may effect face race judgments on the basis of already-held beliefs about that race.[8]

Just as stimulus-driven effects compound the nature of race and emotion, internally held beliefs of the perceiver can influence the compound nature of race and emotion in a top-down process. Implicit and explicit stereotypes about specific races are prolific and certainly change the way a face is perceived, including how the brain interprets such cues.[222,223] For example, a number of studies showed that Caucasian participants higher in implicit prejudice toward African Americans identified anger expressions on African American faces quicker than anger expressions on comparable Caucasian faces.[224,225] This same stereotype-consistent decoding advantage extends to other races as well. Moroccan- and Dutch-stereotypic emotional displays (i.e., anger and sadness, respectively) are recognized faster than the opposite pairing.[226] Moreover, individual implicit stereotype–congruent associations predicted the strength of decoding advantage for stereotypic emotional displays. Similarly, anger expressions (threat) on both African American and Caucasian faces facilitated implicit stereotyping, such that individuals were quicker to identify a gun as opposed to a tool.[227] However, responses were more accurate and quicker for associations of Caucasian faces with tools than for African American faces with tools, suggesting an even deeper stereotype.

4.6 Eye Gaze and Emotion Interactions

Eye gaze is unique as compound social cue in the face, because it can occur in a "nonoverlapping" fashion. What we mean by nonoverlapping is that eye gaze can change without physically influencing other cues in the face. For example, eye gaze can combine with structural maturity or baby-facedness, or facial expressions.

In the last 10 years, significant interdisciplinary attention has been devoted to the combined processing of eye gaze and emotion, demonstrating both the complexities and insights that can result by examining the combinatorial nature of face processing. Much remains to be learned, but what is clear is that eye gaze meaningfully influences the processing of emotional expression and vice versa. Although both of these cues are expressive, and thus thought to be predominantly processed along the same neural pathways, they nevertheless are believed to be represented by two distinct neural systems.[164] Further because these cues occupy nonoverlapping space within the face, interdependencies found cannot be attributable to visually confounded visual properties, as is the case with at least some of the effects reported above for appearance-based (gender, race, age) and expression interactions. Mounting evidence supports the interdependency of processing of eye gaze and emotion, using psychophysical, self-report ratings and neuroimaging paradigms, as well as demonstrated interactivity (for reviews see Adams and Kveraga[181]; Graham and LaBar[228]).

In an initial examination of the role of eye gaze on emotion perception, Adams and colleagues examined the "shared signal hypothesis."[229–231] This hypothesis suggests that since both emotion and eye gaze behaviors are associated with underlying motivational orientations to approach or avoid, when combined in a congruent manner (i.e., both signaling approach such as direct gaze anger and averted gaze fear), processing should be facilitated compared to when they are combined in an incongruent manner (i.e., direct gaze fear and averted gaze anger).

Using speeded reaction time tasks and self-reported perception of emotional intensity, Adams and Kleck[229,230] found that direct gaze reduced processing speed, increased accuracy, and increased the perceived intensity of facially communicated approach-oriented emotions (e.g., anger and joy), whereas averted gaze facilitated processing efficiency, accuracy, and perceived intensity of facially communicated avoidance-oriented emotions (e.g., fear and sadness). Similar effects were also found by Sander et al.[232] using dynamic threat displays, and by Hess et al.[233] who found that direct relative to averted anger expressions and averted relative to direct fear expressions elicited more negative responses in observers. Perhaps the most compelling demonstration of gaze and emotion interactions comes from an attentional blink paradigm, which found that direct gaze anger and averted gaze fear were detected more frequently in the attentional blink window than averted anger or direct fear, suggesting integration occurring at a preattentive level and then capturing attention very early in visual processing.[234]

Facial emotion also influences how eye gaze is perceived. Direct eye gaze is recognized faster when paired with angry faces and averted eye gaze is recognized faster when paired with fearful faces.[235] In addition, perceivers tend to judge eye gaze as more often looking at them when presented on happy and angry faces,[236] though this effect was particularly pronounced for happy faces. Fear expressions, however, did not yield greater perceived averted gaze.

As already noted, gaze itself has been found to have a significant impact on allocation of attention (for a review, see Frischen et al.[237]). This work reveals that eye gaze triggers what appears to be attention cuing in the direction of eye gaze. Thus, when viewing averted gaze, participants are faster to correctly indicate the presence of visual stimuli presented in the same direction cued by eye gaze. One study demonstrated this effect even when gaze direction served as an erroneous cue on 80% of the trials, underscoring the reflexive nature of this effect.[238] These effects have also been examined within the context of facial expression. From a shared-signals hypothesis point of view, we might expect that congruent cues should cause faster and more robust gaze shifting than neutral faces or approach-oriented emotions, such as anger or joy.

Mathews et al.[239] found a faster cueing effect for fear faces than neutral faces for those with high anxiety but not low anxiety (see also Holmes et al.[240]; Putman et al.[241]). Anxious participants also show greater capturing of attention by direct eye gaze when presented on angry faces rather than neutral faces, suggesting that emotion influences how participants process eye gaze on threatening faces with a congruent approach motivation as well.[242] When eye gaze was presented as a dynamic cue, fearful faces were found to induce higher levels of cueing for all participants regardless of anxiety level.[241,243] More recently, Fox et al.[242] report interactive effects of gaze and emotion using a visually mediated attention paradigm, where emotional expression was found to differentially influence attention *capture* of direct eye gaze and attention *shift* of averted eye gaze. Specifically, they found that fear expressions with averted gaze yielded greater reflexive orienting compared to either anger or neutral expressions, whereas anger expressions with direct gaze yielded greater attention capture than either fear or neutral expressions, effects that were moderated by trait anxiety. Notably, the pattern of these effects fits with the shared signal hypothesis.

4.7 Other Compound Cues

4.7.1 Emotion Residue

Recently, our lab has begun preliminary investigations to examine the effects of emotional content remaining on the face after an expresser has completed an expression, a term we call *emotion residue*. Anatomically, over 40 muscles control the face. These muscles are never fully relaxed, save for paralysis due to stroke or a similar disorder. Indeed, observations of patients with facial nerve paralysis show the distinctive "facial droop" associated with such disorders, an approximation of what it would look like if our facial muscles were completely relaxed. Because the face is never fully relaxed, and because individuals do not always display a specific prototypical emotion, it makes ecological sense to investigate (1) whether there are residual effects of emotional expressions, and, if so, (2) how emotion residue influences face perception. We propose that studying emotion residue on faces is often more ecologically relevant than studying prototypical emotions or even microexpressions. Moreover, and important to the field of face perception, it is important to understand how the residual effects of emotion interact with already-known compound cues to change the perception of emotion, trait inferences, memory, and orienting.

Initial results[244] suggest that emotion residue is discernable on postexpressive displays and has some influence on face perception. In one study, we examined whether participants were able to identify emotion residue on faces. In order to study emotion residue, images were extracted from videos of actors beginning with a neutral expression, moving to displaying a target emotion, and then ending with another neutral expression.[245] The result was two neutral faces from the same individual, a preaffective neutral (PRN) and a postaffective neutral (PTN). Participants were shown both faces in a random order and tasked with indicating which face they believed came after the expression. Results revealed that participants were able to identify PTN correctly at above chance levels. When participants were asked to identify the preceding emotion of PTN faces as positive or negative, they again categorized the faces correctly. Interestingly, when specific emotions were looked at individually, all were correctly identified at above-chance levels except for disgust and surprise. In a follow-up study, we employed a paradigm similar to Todorov et al.[49] to examine how emotion residue may influence perceiver behavior. We showed participants the same PRN and PTN faces as before and asked them to identify which image should be chosen for (1) a social media account profile picture, (2) a political campaign ad, (3) a dating profile, and (4) an actor's headshot. For all conditions, participants preferred the PRN when the emotion was of negative valence. For the dating profile and social media conditions, participants preferentially selected the PTN when the expression was of positive valance.

Taken together, these results suggest that there are explicit and identifiable differences between PRN and PTN faces. More importantly, these identifiable differences appear to affect perceiver behavior in some manner. While further research is necessary to identify the exact cause and mechanisms by which these differences are occurring, the preliminary results are nonetheless important to consider for future research on face perception. Indeed, as Gibson notes in his ecological view of vision perception, studying an environment (stimulus) may help elucidate the underlying neural mechanisms that evolved to perceive it.

4.7.2 Faces, Bodies, Voices, and Scenes

While the face provides a plethora of information to the perceiver, the primate visual system integrates cues from a variety of sources at once. From an ecological vision stance, bodies, voices, and scenes, among other facets related to person perception, all hold adaptive value and should follow similar patterns to that of faces. This has been shown by de Gelder and colleagues, who have amply demonstrated that body language, visual scenes, and vocal cues all influence the processing facial displays,[246] and at the earliest stages of face processing, even along purportedly "nonconscious" processing routes.[247] Accuracy and reaction time for deciphering emotion on faces and bodies expressing congruent and incongruent pairings is biased in the direction of the body, especially when the facial expression is ambiguous.[246,248] Indeed, when threat is congruent across faces, bodies, and scenes evoked greater threat responses than did incongruent compounds.[249] Further, when the emotional expression is intense (and thus more ambiguous), participants can only decipher the expression accurately when the body is present.[250,251] In one study, researchers placed disgust expressions on disgust, angry, sad, or fearful body postures (context).[252] Results revealed that both the accuracy and valance and arousal ratings of the facial expression depended on the context in which the expression appeared. Further, eye-tracking fixation analysis showed that context in which the face appeared changed the pattern of fixations for the same facial expressions.

4.8 Neuroscience of Compound Social Cue Processing

The assumption that facial cues, especially identity and emotion, are processed by functional and neurologically independent systems is a strong assumption. When defined in the cognitive processing literature, independence refers to two cues having no influence on the processing of one another.[253] Essentially, if two cues are processed independently, varying one cue should have no influence on the perception of another cue. Yet, a growing body of evidence suggests that social cues communicated by the face convey information about internal states such as wishes, desires, feelings, and intentions (see Baron-Cohen[254]). Underlying such social motives are the core dimensions of social meaning. Dominance and affiliation emerge as key factors in person perception based on facial appearance and expression, such as perceived facial maturity,[17] gender perception,[186] gaze,[78] and emotional displays.[212,213] Psychophysiological studies in nonhuman primates further reveal distinct neural mechanisms for the perception of dominance and affiliation.[255] Underlying these basic social motives are even more fundamental tendencies driving biological movement: those of approach and avoidance,[254,256] which also appear to have a distinct neural basis in nonhuman primates.[257] Davidson and Hugdahl state that "approach and withdrawal are fundamental motivational dimensions that are present at any level of phylogeny where behavior itself is present"[256] (p. 362). As such, these behavioral motivations are likely inherent in all forms of social communication. Given shared signal values, it arguably makes adaptive sense that combinations of such cues would be visually integrated in an efficient manner. Such integrative processing would help ensure a timely assessment of an adaptive response to social cues conveyed by others. Although early studies have using the Garner paradigm to examine perceptual interference concluded that expression and appearance are processed independently,[258,259] several newer studies have revealed influences of compound cues. Asymmetric interference (only one cue influences the other) has been found where identity interferes with expression processing during classification tasks, but not vice versa,[260–262] while symmetric interference has been found in a matching task.[263] Regardless of whether asymmetrical or symmetrical interference is found, one can make the assumption that the perception of these two stimulus dimensions are not entirely independent. Several studies have also found that variation in gender interferes with the processing of emotion, both asymmetrically[264] or symmetrically.[265–267]

Additionally, symmetric interference is found between eye gaze and emotion for upright but not inverted faces,[265,268] suggesting that interdependencies are not necessarily due to overlapping neural substrates. Compelling evidence for nonoverlapping neural networks is revealed using transcranial magnetic stimulation to disrupt neural responses, which showed that left STS disruption influenced eye gaze but not emotion processing, and right somatosensory cortex disruptions influenced emotion but not eye gaze processing.[269] These findings offer compelling evidence that early visual input of eye gaze and emotion may well involve doubly dissociable neural pathways, but this does not undermine the existence or functional value of interactive processes. Top-down modulation of even low-level visual processing occurs rapidly. Magnocellular projections are known to activate early occipital visual cortex responses that project to the orbitofrontal cortex, which projects back to guide simple visual processing in the fusiform gyrus.[270] This top modulation can occur as early as 150ms after visual input.[271] Thus, top-down processing can organize low-level visual processing in functionally meaningful ways. Such a litany of evidence supporting interference in processing various facial cues rebuts evidence of a model where different facial cues are processed in parallel and independently of one another (e.g., Bruce and Young[175];Calder and Young[180]).

Other likely candidates involved in the neural integration of compound social cues include the amygdala, OFC, and STS. The amygdala has been implicated in the integration of complex emotional, social, and behavioral information.[272] Furthermore, pathways leading from the STS to the amygdala[273] offer a means for passing along and integrating information regarding the behavioral tendencies of others and to whom and where that intent is directed so that an adaptive response can be made.[257] Finally, the amygdala is richly interconnected with structures in the OFC, which have been shown to be involved in adaptive behavioral responding and decision-making.[171] The OFC has also been reported to be involved in the processing of emotion information during decision-making.[169] Thus, the amygdala, which potentially acts as a relay station for social signals given off by the face, is likely key to generating the type of adaptive responses that would necessarily occur as part of an early warning system.

5. CONCLUSION

We stated in the beginning that social-visual face perception is a complicated beast. Throw into that mix social factors influencing face perception, and the complexity may seem insurmountable. Yet, what we hope we have convinced the reader of here is that social influences on face perception are so intertwined with visual perception that studying the two separately may be a fool's errand. Our goal throughout this chapter was to provide an alternative approach for viewing this evidence, especially evidence from social face perception that has been considered contradictory or studied independent of each other. We have examined how social cues, visual processes, and neuroanatomy align with an ecological approach to social-visual perception. While previous models often attempted to control for compound social cues focusing on differentiating the "source" of information (e.g., expression versus appearance), central to this approach is the underlying meaning conveyed by these cues and their combined ecological relevance to the observer.

Vision is complicated, and continuing to study individual, isolated features will not shed new light on this complexity. Without integrating these cues, we fail to fully explain how the social-visual system operates in humans and how it contributes to person perception. In this chapter, we have outlined several compound cues that have substantial or burgeoning research to suggest that they influence person perception in a combinatorial manner. Therefore, moving forward, scientists will likely benefit from examining how the visual system extracts social meaning, such as dominance and affiliation, through an integrative process that combines context and compound social cues. Even if focusing on a particular social cue, we aim to underscore the intricacies and potential confounding factors that certain stimuli can have on the social-visual system when attempting to examine the neural mechanics of person perception.

The neuroscience of person and social visual perception is a budding field of study, and one that has yet to identify all of the intricacies that influence person perception. While we focused on face perception, the ecological approach to person perception theoretically should apply cross-modally and specifically to cross-model integration as well. As future research takes into account the combinatorial nature of person perception, perhaps we may eventually tame the social visual beast.

References

1. Aristotle. De Interpretatione. In: McKeon R, ed. *The Basic Works of Aristotle.* ; 1941.
2. Solomon RC. The philosophy of emotions. In: *Handbook of Emotions.* 3rd ed. Guilford Press; 2008.
3. Lick DJ, Johnson KL, Gill SV. Deliberate changes to gendered body motion influence basic social perceptions. *Soc Cogn.* 2013;31(6): 656–671. http://dx.doi.org/10.1521/soco.2013.31.6.656.
4. Pollick FE, McKay LS, Johnson KL. Social constraints on the visual perception of biological motion. In: Adams RB, Ambady N, Nakayama K, Shimojo S, eds. *The Science of Social Vision.* Oxford University Press; 2011:264–277.
5. Shiffrar M, Kaiser MD, Chouchourelou A. Seeing human movement as inherently social. In: Adams RB, Ambady N, Nakayama K, Shimojo S, eds. *The Science of Social Vision.* Oxford University Press; 2011:248–263.
6. Bachorowski J-A, Owren MJ. Vocal expression of emotion: acoustic properties of speech are associated with emotional intensity and context. *Psychol Sci.* 1995;6(4):219–224. http://dx.doi.org/10.2307/40063021.
7. Chen D, Haviland-Jones J. Human olfactory communication of emotion. *Percept Mot Skills.* 2000;91(3):771–781. http://dx.doi.org/10.2466/pms.2000.91.3.771.
8. Zebrowitz LA, Kikuchi M, Fellous J-M. Facial resemblance to emotions: group differences, impression effects, and race stereotypes. *J Pers Soc Psychol.* 2010;98(2):175–189. http://dx.doi.org/10.1037/a0017990.
9. Axt JR, Ebersole CR, Nosek BA. The rules of implicit evaluation by race, religion, and age. *Psychol Sci.* 2014. http://dx.doi.org/10.1177/0956797614543801. 0956797614543801.
10. Humphreys G, Hodsoll J, Campbell C. Attending but not seeing: the "other race" effect in face and person perception studied through change blindness. *Vis Cogn.* 2005;12(1):249–262. http://dx.doi.org/10.1080/13506280444000148.
11. Hess U, Adams RB, Simard A, Stevenson MT, Kleck R. Smiling and sad wrinkles: age-related changes in the face and the perception of emotions and intentions. *J Exp Soc Psychol.* 2012;48(6): 1377–1380. http://dx.doi.org/10.1016/j.jesp.2012.05.018.
12. Ebner NC. Age of face matters: age-group differences in ratings of young and old faces. *Behav Res.* 2008;40(1):130–136. http://dx.doi.org/10.3758/BRM.40.1.130.
13. Malatesta CZ, Izard CE, Culver C, Nicolich M. Emotion communication skills in young, middle-aged, and older women. *Psychol Aging.* 1987;2(2):193–203.
14. Freudenberg M, Adams RB, Kleck RE, Hess U. Through a glass darkly: facial wrinkles affect our processing of emotion in the elderly. *Front Psychol.* 2015;6:1476. http://dx.doi.org/10.3389/fpsyg.2015.01476.
15. Adams RB, Ambady N, Nakayama K, Shimojo S, eds. *The Science of Social Vision.* New York, NY: Oxford University Press; 2011.
16. Gibson JJ. *The Ecological Approach to Visual Perception.* Psychology Press; 1986.
17. Zebrowitz LA. *Reading Faces: Window to the Soul?* Boulder, CO: Westview Press; 1997.
18. Ekman P. Methods for measuring facial action. In: Scherer KR, Ekman P, eds. *Handbook of Methods in Nonverbal Behavior Research.* 1982:45–135.
19. Ruiz-Belda MA, Fernandez-Dols J-M, Carrera P. Spontaneous facial expressions of happy bowlers and soccer fans. *Cogn Emot.* 2003. http://dx.doi.org/10.1080/02699930244000327.

20. Kraut RE, Johnston RE. Social and emotional messages of smiling: an ethological approach. *J Pers Soc Psychol*. 1979;37(9):1539–1553. http://dx.doi.org/10.1037/0022-3514.37.9.1539.
21. Bruce V, Young AW. *Face Perception*. 1st ed. New York, NY: Psychology Press; 2012.
22. Saavedra C, Smith P, Peissig J. The relative role of eyes, eyebrows, and eye region in face recognition. *J Vis*. 2013;13(9):410. http://dx.doi.org/10.1167/13.9.410.
23. Sadr J, Jarudi I, Sinha P. The role of eyebrows in face recognition. *Perception*. 2003;32(3):285–293. http://dx.doi.org/10.1068/p5027.
24. Campbell R, Benson PJ, Wallace SB, Doesbergh S, Coleman M. More about brows: how poses that change brow position affect perceptions of gender. *Perception*. 1999;28(4):489–504. http://dx.doi.org/10.1068/p2784.
25. Di Giorgio E, Leo I, Pascalis O, Simion F. Is the face-perception system human-specific at birth? *Dev Psychol*. 2012;48(4): 1083–1090. http://dx.doi.org/10.1037/a0026521.
26. Pascalis O, Scott LS, Kelly DJ, et al. Plasticity of face processing in infancy. *Proc Natl Acad Sci*. 2005;102(14):5297–5300. http://dx.doi.org/10.1073/pnas.0406627102.
27. Porta Della G. *De Humane Physiognomonia*; 1586.
28. Bar M, Neta M, Linz H. Very first impressions. *Emotion*. 2006;6(2):269–278. http://dx.doi.org/10.1037/1528-3542.6.2.269.
29. Dotsch R, Todorov A. Reverse correlating social face perception. *Soc Psychol Pers Sci*. 2012;3(5):562–571. http://dx.doi.org/10.1177/1948550611430272.
30. Hassin RR, Trope Y. Facing faces: studies on the cognitive aspects of physiognomy. *J Pers Soc Psychol*. 2000;78(5):837–852.
31. Rule NO, Ambady N. Brief exposures: male sexual orientation is accurately perceived at 50ms. *J Exp Soc Psychol*. 2008;44(4): 1100–1105. http://dx.doi.org/10.1016/j.jesp.2007.12.001.
32. Willis J, Todorov A. First impressions: making up your mind after a 100-ms exposure to a face. *Psychol Sci*. 2006;17(7):592–598. http://dx.doi.org/10.1111/j.1467-9280.2006.01750.x.
33. Zebrowitz LA, Montepare JM. Social psychological face perception: why appearance matters. *Soc Pers Psychol Compass*. 2008;2(3): 1497–1517. http://dx.doi.org/10.1111/j.1751-9004.2008.00109.x.
34. Secord PF. Facial features and inferential processes on interpersonal perception. In: *Person Perception and Interpersonal Behavior*. Standford, CA: Standford University Press; 1958: 300–316.
35. Berry DS. Accuracy in social perception: contributions of facial and vocal information. *J Pers Soc Psychol*. 1991;61(2):298–307.
36. Berry DS. Taking people at face value: evidence for the kernel of truth hypothesis. *Soc Cogn*. 1990;8(4):343–361. http://dx.doi.org/10.1521/soco.1990.8.4.343.
37. Berry DS, Brownlow S. Were the physiognomists right? Personality correlates of facial babyishness. *Pers Soc Psychol Bull*. 1989;15(2): 266–279. http://dx.doi.org/10.1177/0146167289152013.
38. Bond CF, Berry DS, Omar A. The kernel of truth in judgments of deceptiveness. *Basic Appl Soc Psychol*. 1994;15(4):523–534. http://dx.doi.org/10.1207/s15324834basp1504_8.
39. Weisbuch M, Ambady N. *Thin-Slice Vision*. Oxford University Press; 2010. http://dx.doi.org/10.1093/acprof:oso/9780195333176.003.0014. 228–247.
40. Ambady N, Rosenthal R. Half a minute: predicting teacher evaluations from thin slices of nonverbal behavior and physical attractiveness. *J Pers Soc Psychol*. 1993;64(3):431–441. http://dx.doi.org/10.1037/0022-3514.64.3.431.
41. Ambady N, Rosenthal R. Thin slices of expressive behavior as predictors of interpersonal consequences: a meta-analysis. *Psychol Bull*. 1992;111(2):256–274. http://dx.doi.org/10.1037/0033-2909.111.2.256.
42. Todorov A. Inferences of competence from faces predict election outcomes. *Science*. 2005;308(5728):1623–1626. http://dx.doi.org/10.1126/science.1110589.
43. Rule NO, Ambady N. Democrats and republicans can be differentiated from their faces. In: Hendricks M, ed. PLoS One; 2010. http://dx.doi.org/10.1371/journal.pone.0008733. 5(1):e8733.
44. Rule NO, Ambady N. Face and fortune: inferences of personality from Managing Partners "faces predict their law firms" financial success. *Leadership Quart*. 2011;22(4):690–696. http://dx.doi.org/10.1016/j.leaqua.2011.05.009.
45. Rule NO, Ambady N, Adams RB, Macrae CN. Accuracy and awareness in the perception and categorization of male sexual orientation. *J Pers Soc Psychol*. 2008;95(5):1019–1028. http://dx.doi.org/10.1037/a0013194.
46. Rule NO, Garrett JV, Ambady N. On the perception of religious group membership from faces. *PLoS One*. 2010;5(12):e14241. http://dx.doi.org/10.1371/journal.pone.0014241.
47. McCabe V. Abstract perceptual information for age level: a risk factor for maltreatment? *Child Dev*. 1984;55(1):267. http://dx.doi.org/10.2307/1129851.
48. Jenkins R, White D, Van Montfort X, Burton AM. Variability in photos of the same face. *Cognition*. 2011:1–11. http://dx.doi.org/10.1016/j.cognition.2011.08.001.
49. Todorov A, Porter JM. Misleading first impressions: different for different facial images of the same person. *Psychol Sci*. 2014;25(7): 1404–1417. http://dx.doi.org/10.1177/0956797614532474.
50. Darwin C, Ekman P, Prodger P. *The Expression of the Emotions in Man and Animals*. Oxford; New York: Oxford University Press; 1998.
51. Duchenne GB, Cuthbertson RA. *The Mechanism of Human Facial Expression*. New York, NY: Cambridge University Press; 1990.
52. Bell SC. *Essays on the Anatomy and Philosophy of Expression*; 1824.
53. Hess U, Thibault P. Darwin and emotion expression. *Am Psychol*. 2009;64(2):120–128. http://dx.doi.org/10.1037/a0013386.
54. Ekman P, Sorenson ER, Friesen W. Pan-cultural elements in facial displays of emotion. *Science*. 1969.
55. Ekman P, Friesen W. Constants across cultures in the face and emotion. *J Pers Soc Psychol*. 1971;17(2):124–129.
56. Ekman P, Friesen W, O'Sullivan M, et al. Universals and cultural differences in the judgments of facial expressions of emotion. *J Pers Soc Psychol*. 1987;53(4):712–717.
57. Nelson NL, Russell JA. Universality revisited. *Emot Rev*. 2013;5(1):8–15. http://dx.doi.org/10.1177/1754073912457227.
58. Biehl M, Matsumoto D, Ekman P, Hearn V. Matsumoto and Ekman's Japanese and Caucasian Facial Expressions of Emotion (JACFEE): reliability data and cross-national differences. *J Nonverbal Behav*. 1997;21(1):3–21. http://dx.doi.org/10.1023/A:1024902500935.
59. Russell JA. Facial expressions of emotion: what lies beyond minimal universality? *Psychol Bull*. 1995;118(3):379–391.
60. Lindquist KA. Emotions emerge from more basic psychological ingredients: a modern psychological constructionist model. *Emot Rev*. 2013;5(4):356–368. http://dx.doi.org/10.1177/1754073913489750.
61. Barrett LF. Are emotions natural kinds? *Perspect Psychol Sci*. 2006;1(1):28–58. http://dx.doi.org/10.1111/j.1745-6916.2006.00003.x.
62. Haidt J, Keltner D. Culture and facial expression: open-ended methods find more expressions and a gradient of recognition. *Cognit Emot*. 1999;13(3):225–266. http://dx.doi.org/10.1080/026999399379267.
63. Nelson NL, Russell JA. Dynamic facial expressions allow differentiation of displays intended to convey positive and hubristic pride. *Emotion*. 2014;14(5):857–864. http://dx.doi.org/10.1037/a0036789.
64. James W. What is an Emotion? *Mind*. 1884;9(34):188–205.
65. Russell JA. A circumplex model of affect. *J Pers Soc Psychol*. 1980;39(6):1161–1178. http://dx.doi.org/10.1037/h0077714.
66. Lindquist KA, Barrett LF. Constructing emotion: the experience of fear as a conceptual act. *Psychol Sci*. 2008;19(9):898–903. http://dx.doi.org/10.1111/j.1467-9280.2008.02174.x.

67. Barrett LF, Bliss-Moreau E. Affect as a psychological primitive. *Adv Exp Soc Psychol*. 2009;41:167–218. http://dx.doi.org/10.1016/S0065-2601(08)00404-8.
68. Russell JA, Barrett LF. Core affect, prototypical emotional episodes, and other things called emotion: dissecting the elephant. *J Pers Soc Psychol*. 1999;76(5):805–819.
69. Lindquist KA, Barrett LF. A functional architecture of the human brain: emerging insights from the science of emotion. *Trends Cogn Neurosci*. 2012;16(11):533–540. http://dx.doi.org/10.1016/j.tics.2012.09.005.
70. Gendron M, Barrett LF. Reconstructing the past: a century of ideas about emotion in psychology. *Emot Rev*. 2009;1(4):316–339. http://dx.doi.org/10.1177/1754073909338877.
71. Fernandez-Dols J-M, Russell JA. Emotion, affect, and mood in social judgments. In: Weiner IB, Tennen HA, Suls JM, eds. *Handbook of Psychology, Personality and Social Psychology*. Hoboken, NJ, USA: John Wiley & Sons, Inc.; 2003:283–298. http://dx.doi.org/10.1002/0471264385.wei0512.
72. Wilson-Mendenhall CD, Barrett LF, Barsalou LW. Neural evidence that human emotions share core affective properties. *Psychol Sci*. 2013;24(6):947–956. http://dx.doi.org/10.1177/0956797612464242.
73. Lindquist KA, Wager TD, Kober H, Bliss-Moreau E, Barrett LF. The brain basis of emotion: a meta-analytic review. *Behav Brain Sci*. 2012;35(03):121–143. http://dx.doi.org/10.1017/S0140525X11000446.
74. Schachter S, Singer JE. Cognitive, social, and physiological determinants of emotional state. *Psychol Rev*. 1962;69(5):379–399. http://dx.doi.org/10.1037/h0046234.
75. Gendron M, Lindquist KA, Barsalou LW, Barrett LF. Emotion words shape emotion percepts. *Emotion*. 2012;12(2):314–325. http://dx.doi.org/10.1037/a0026007.
76. Lindquist KA, Barrett LF, Bliss-Moreau E, Russell JA. Language and the perception of emotion. *Emotion*. 2006;6(1):125–138. http://dx.doi.org/10.1037/1528-3542.6.1.125.
77. Stern D. *The First Relationship: Infant and Mother*. Cambridge: Harvard University Press; 1977.
78. Argyle M, Cook M. *Gaze and Mutual Gaze*. Cambridge, England; New York: Cambridge University Press; 1976.
79. Mehrabian A. Attitudes inferred from non-immediacy of verbal communications. *J Verbal Learning Verbal Behav*. 1967;6(2):294–295. http://dx.doi.org/10.1016/S0022-5371(67)80113-0.
80. Rutter DR. *Looking and Seeing: the Role of Visual Communication in Social Interaction*. Chichester: Wiley; 1984.
81. Ellsworth PC. Direct gaze as a social stimulus: the example of aggression. In: *Nonverbal Communication of Aggression*. Boston, MA: Springer US; 1975:53–75. http://dx.doi.org/10.1007/978-1-4684-2835-3_3.
82. Cary MS. The role of gaze in the initiation of conversation. *Soc Psychol*. 1978;41(3):269. http://dx.doi.org/10.2307/3033565.
83. Ekman P, Friesen W. *Unmasking the Face: A Guide to Recognising Emotions from Facial Expressions*. CA: Consulting Psychologists Press; 1975. http://dx.doi.org/10.3389/neuro.09.023.2009/full.
84. Ekman P, Oster H. Facial expressions of emotion. *Ann Rev Psychol*. 1979;30:527–554.
85. Hess U, Petrovich SB. Pupillary behavior in communication. In: *Nonverbal Behavior and Communication*. 2nd ed. Hobokon: Psychology Press; 1987.
86. Janik SW, Wellens AR, Goldberg ML, Dell'Osso LF. Eyes as the center of focus in the visual examination of human faces. *Percept Mot Skills*. 1978;47(3 Pt 1):857–858. http://dx.doi.org/10.2466/pms.1978.47.3.857.
87. Benthall J, Argyle M, Cook M. Gaze and mutual gaze. *Rain*. 1976;12(7). http://dx.doi.org/10.2307/3032267.
88. Hains SM, Muir DW. Infant sensitivity to adult eye direction. *Child Dev*. 1996;67(5):1940–1951.
89. Puce A, Smith A, Allison T. Dealing with a poker face?: ERPs elicited to changes in gaze direction and mouth movement. *NeuroImage*. 1998;4(7):S347.
90. Hoffman EA, Haxby JV. Distinct representations of eye gaze and identity in the distributed human neural system for face perception. *Nat Neurosci*. 2000;3(1):80–84. http://dx.doi.org/10.1038/71152.
91. O'Toole AJ, Deffenbacher KA, Valentin D, McKee K, Huff D, Abdi H. The perception of face gender: the role of stimulus structure in recognition and classification. *Mem Cognit*. 1998;26(1):146–160.
92. Burton AM, Bruce V, Dench N. What's the difference between men and women? Evidence from facial measurement. *Perception*. 1993;22(2):153–176.
93. Zhao M, Hayward WG. Holistic processing underlies gender judgments of faces. *Atten Percept Psychophys*. 2010;72(3):591–596. http://dx.doi.org/10.3758/APP.72.3.591.
94. Reddy L, Wilken P, Koch C. Face-gender discrimination is possible in the near-absence of attention. *J Vis*. 2004;4(2):4. http://dx.doi.org/10.1167/4.2.4.
95. Brown E, Perrett DI. What gives a face its gender? *Perception*. 1993;22(7):829–840.
96. Ferrario VF, Sforza C, Pizzini G, Vogel G, Miani A. Sexual dimorphism in the human face assessed by euclidean distance matrix analysis. *J Anat*. 1993;183(Pt 3):593–600.
97. Bruce V, Burton AM, Hanna E, et al. Sex discrimination: how do we tell the difference between male and female faces? *Perception*. 1993;22(2):131–152.
98. Fellous JM. Gender discrimination and prediction on the basis of facial metric information. *Vision Res*. 1997;37(14):1961–1973. http://dx.doi.org/10.1016/S0042-6989(97)00010-2.
99. Gosselin F, Schyns PG. Bubbles: a technique to reveal the use of information in recognition tasks. *Vision Res*. 2001;41(17):2261–2271.
100. Dupuis-Roy N, Fortin I, Fiset D, Gosselin F. Uncovering gender discrimination cues in a realistic setting. *J Vis*. 2009;9(2):10. http://dx.doi.org/10.1167/9.2.10.
101. Nestor A, Tarr MJ. Gender recognition of human faces using color. *Psychol Sci*. 2008;19(12):1242–1246. http://dx.doi.org/10.1111/j.1467-9280.2008.02232.x.
102. Russell R, Sinha P, Biederman I, Nederhouser M. Is pigmentation important for face recognition? Evidence from contrast negation. *Perception*. 2006;35(6):749–759. http://dx.doi.org/10.1068/p5490.
103. Russell R. Sex, beauty, and the relative luminance of facial features. *Perception*. 2003;32(9):1093–1107. http://dx.doi.org/10.1068/p5101.
104. Russell R. A sex difference in facial contrast and its exaggeration by cosmetics. *Perception*. 2009;38(8):1211–1219. http://dx.doi.org/10.1068/p6331.
105. Lewicki P. Processing information about covariations that cannot be articulated. *J Exp Psychol Learn Mem Cogn*. 1986;12(1):135–146. http://dx.doi.org/10.1037//0278-7393.12.1.135.
106. Hess U, Adams RB, Kleck R. Facial appearance, gender, and emotion expression. *Emotion*. 2004;4(4):378–388. http://dx.doi.org/10.1037/1528-3542.4.4.378.
107. Sergent J, Ohta S, MacDonald B. Functional neuroanatomy of face and object processing. A positron emission tomography study. *Brain*. 1992;115(Pt 1):15–36.
108. Ino T, Nakai R, Azuma T, Kimura T, Fukuyama H. Gender differences in brain activation during encoding and recognition of male and female faces. *Brain Imaging Behav*. 2010;4(1):55–67. http://dx.doi.org/10.1007/s11682-009-9085-0.
109. Haxby JV, Hoffman E, Gobbini M. The distributed human neural system for face perception. *Trends Cogn Sci*. 2000;4(6):223–233.
110. Ishai AA. *Cortical Network for Face Perception*; 2011. pp. 1–13.
111. Kaul C, Rees G, Ishai A. The gender of face stimuli is represented in multiple regions in the human brain. *Front Hum Neurosci*. 2011;4:1–12. http://dx.doi.org/10.3389/fnhum.2010.00238.

112. Mark LS, Pittenger JB, Hines H, Carello C, Shaw RE, Todd JT. Wrinkling and head shape as coordinated sources of age-level information. *Percept Psychophys.* 1980;27(2):117–124. http://dx.doi.org/10.3758/BF03204298.
113. Burt DM, Perrett DI. Perception of age in adult caucasian male faces: computer graphic manipulation of shape and colour information. *Proc R Soc B Biol Sci.* 1995;259(1355):137–143. http://dx.doi.org/10.1098/rspb.1995.0021.
114. Kite ME, Johnson BT. Attitudes toward older and younger adults: a meta-analysis. *Psychol Aging.* 1988;3(3):233–244. http://dx.doi.org/10.1037//0882-7974.3.3.233.
115. Burt DM, Perrett DI. Perceptual asymmetries in judgements of facial attractiveness, age, gender, speech and expression. *Neuropsychologia.* 1997;35(5):685–693. http://dx.doi.org/10.1016/s0028-3932(96)00111-x.
116. Blair IV, Judd CM. Afrocentric facial features and stereotyping. In: Adams RB, Ambady N, Nakayama K, Shimojo S, eds. *The Science of Social Vision.* Oxford University Press; 2011:347–362. http://dx.doi.org/10.1093/acprof:oso/9780195333176.003.0019.
117. Blair IV, Judd CM, Chapleau KM. The influence of Afrocentric facial features in criminal sentencing. *Psychol Sci.* 2004;15(10): 674–679. http://dx.doi.org/10.1111/j.0956-7976.2004.00739.x.
118. Greenwald AG, McGhee DE, Schwartz JL. Measuring individual differences in implicit cognition: the implicit association test. *J Pers Soc Psychol.* 1998;74(6):1464.
119. Nosek BA, Banaji MR, Greenwald AG. Harvesting implicit group attitudes and beliefs from a demonstration web site. *Group Dyn Theor Res Pract.* 2002;6(1):101–115. http://dx.doi.org/10.1037//1089-2699.6.1.101.
120. McLallen AS, Johnson BT, Dovidio JF. Black and white: the role of color bias in implicit race bias. *Soc Cogn.* 2006;24(1):46–73. http://dx.doi.org/10.1521/soco.2006.24.1.46.
121. Dotsch R, Wigboldus DHJ, Langner O, van Knippenberg A. Ethnic out-group faces are biased in the prejudiced mind. *Psychol Sci.* 2008;19(10):978–980. http://dx.doi.org/10.1111/j.1467-9280.2008.02186.x.
122. Berns GS, Brooks AM, Spivak M. Scent of the familiar: an fMRI study of canine brain responses to familiar and unfamiliar human and dog odors. *Behav Processes.* 2015;110:37–46. http://dx.doi.org/10.1016/j.beproc.2014.02.011.
123. Sanfey AG, Rilling JK, Aronson JA, Nystrom LE, Cohen JD. The neural basis of economic decision-making in the ultimatum game. *Science.* 2003;300(5626):1755–1758. http://dx.doi.org/10.1126/science.1082976.
124. Ekman P. Universals and cultural differences in facial expressions of emotion. *J Pers Soc Psychol.* 1971;53.
125. Plutchik R. A psychoevolutionary theory of emotions. *Soc Sci Inf.* 1982;21(4–5):529–553. http://dx.doi.org/10.1177/053901882021004003.
126. Ekman P. Universal facial expressions in emotion. *Studia Psychol.* 1973.
127. Fridlund AJ. *Human Facial Expression: An Evolutionary Perspective;* 1994.
128. Izard CE. *The Faces of Emotion.* New York: Appleton-Century-Crofts; 1971.
129. Plutchik RA. General psychoevolutionary theory of emotion. In: *Emotion, Theory, Research, and Experience.* Elsevier; 1980:3–33. http://dx.doi.org/10.1016/B978-0-12-558701-3.50007-7.
130. Redican WK. An evolutionary perspective on human facial displays. In: Ekman P, ed. *Emotion in the Human Face: Guidelines and Practice for Research.* 2nd ed. New York, NY: Cambridge University Press; 1982:212–280.
131. Hubel DH, Wiesel TN. Receptive fields of single neurones in the cat's striate cortex. *J Physiol.* 1959;148(3):574–591. http://dx.doi.org/10.1113/jphysiol.1959.sp006308.
132. Kanwisher N, McDermott J, Chun MM. The fusiform face area: a module in human extrastriate cortex specialized for face perception. *J Neurosci.* 1997;17(11):4302–4311.
133. Kanwisher N. Domain specificity in face perception. *Nat Neurosci.* 2000;3(8):759–763. http://dx.doi.org/10.1038/77664.
134. Downing PE, Jiang Y, Shuman M, Kanwisher N. A cortical area selective for visual processing of the human body. *Science.* 2001;293(5539): 2470–2473. http://dx.doi.org/10.1126/science.1063414.
135. Epstein R, Kanwisher N. A cortical representation of the local visual environment. *Nature.* 1998;392(6676):598–601. http://dx.doi.org/10.1038/33402.
136. Cohen L, Dehaene S, Naccache L, Lehéricy S. The visual word form area Spatial and temporal characterization of an initial stage of reading in normal subjects and posterior split-brain patients. *Brain.* 2000.
137. Riener CR, Stefanucci JK, Proffitt DR, Clore G. An effect of mood on the perception of geographical slant. *Cogn Emot.* 2011;25(1): 174–182. http://dx.doi.org/10.1080/02699931003738026.
138. Hendry SHC, Reid RC. The koniocellular pathway in primate vision. *Ann Rev Neurosci.* 2000;23(1):127–153. http://dx.doi.org/10.1146/annurev.neuro.23.1.127.
139. Kanwisher N, Stanley D, Harris A. The fusiform face area is selective for faces not animals. *Neuroreport.* 1999;10(1):183–187.
140. Grill-Spector K, Knouf N, Kanwisher N. The fusiform face area subserves face perception, not generic within-category identification. *Nat Neurosci.* 2004;7(5):555–562. http://dx.doi.org/10.1038/nn1224.
141. McKone E, Kanwisher N, Duchaine BC. Can generic expertise explain special processing for faces? *Trends Cogn Sci.* 2007;11(1): 8–15. http://dx.doi.org/10.1016/j.tics.2006.11.002.
142. Gauthier I, Tarr MJ, Anderson AW, Skudlarski P, Gore JC. Activation of the middle fusiform "face area" increases with expertise in recognizing novel objects. *Nat Neurosci.* 1999;2(6):568–573. http://dx.doi.org/10.1038/9224.
143. Gauthier I, Skudlarski P, Gore JC, Anderson AW. Expertise for cars and birds recruits brain areas involved in face recognition. *Nat Neurosci.* 2000;3(2):191–197. http://dx.doi.org/10.1038/72140.
144. McGugin RW, Newton AT, Gore JC, Gauthier I. Robust expertise effects in right FFA. *Neuropsychologia.* 2014;63(C):135–144. http://dx.doi.org/10.1016/j.neuropsychologia.2014.08.029.
145. Mason MF, Macrae CN. Categorizing and individuating others: the neural substrates of person perception. *J Cogn Neurosci.* 2004;16(10):1785–1795. http://dx.doi.org/10.1162/0898929042947801.
146. Adams RB, Janata P. A comparison of neural circuits underlying auditory and visual object categorization. *NeuroImage.* 2002;16(2):361–377. http://dx.doi.org/10.1006/nimg.2002.1088.
147. Bilalic M, Langner R, Ulrich R, Grodd W. Many faces of expertise: fusiform face area in chess experts and novices. *J Neurosci.* 2011;31(28):10206–10214. http://dx.doi.org/10.1523/JNEUROSCI.5727-10.2011.
148. Steeves JKE, Culham JC, Duchaine BC, et al. The fusiform face area is not sufficient for face recognition: evidence from a patient with dense prosopagnosia and no occipital face area. *Neuropsychologia.* 2006;44(4):594–609. http://dx.doi.org/10.1016/j.neuropsychologia.2005.06.013.
149. Pitcher D, Walsh V, Duchaine BC. The role of the occipital face area in the cortical face perception network. *Exp Brain Res.* 2011;209(4): 481–493. http://dx.doi.org/10.1007/s00221-011-2579-1.
150. Rotshtein P, Henson RNA, Treves A, Driver J, Dolan RJ. Morphing Marilyn into Maggie dissociates physical and identity face representations in the brain. *Nat Neurosci.* 2004;8(1):107–113. http://dx.doi.org/10.1038/nn1370.
151. Hemond CC, Kanwisher NG, Op de Beeck HP. Ferrari P, ed. A preference for contralateral stimuli in human object- and face-selective cortex. *PLoS One.* 2007;2(6):e574–e575. http://dx.doi.org/10.1371/journal.pone.0000574.

152. Nichols DF, Betts LR, Wilson HR. Decoding of faces and face components in face-sensitive human visual cortex. *Front Psychol.* 2010:1–13. http://dx.doi.org/10.3389/fpsyg.2010.00028.
153. Pitcher D, Duchaine BC, Walsh V. Combined TMS and fMRI reveal dissociable cortical pathways for dynamic and static face perception. *Curr Biol.* 2014;24(17):2066–2070. http://dx.doi.org/10.1016/j.cub.2014.07.060.
154. Pitcher D, Walsh V, Yovel G, Duchaine BC. TMS evidence for the involvement of the right occipital face area in early face processing. *Curr Biol.* 2007;17(18):1568–1573. http://dx.doi.org/10.1016/j.cub.2007.07.063.
155. Heywood CA, Cowey A, Rolls ET. The role of the 'face-cell' area in the discrimination and recognition of faces by monkeys [and discussion]. *Philos Trans R Soc B Biol Sci.* 1992;335(1273):31–38. http://dx.doi.org/10.1098/rstb.1992.0004.
156. Campbell R, Heywood CA, Cowey A, Regard M, Landis T. Sensitivity to eye gaze in prosopagnosic patients and monkeys with superior temporal sulcus ablation. *Neuropsychologia.* 1990;28(11):1123–1142. http://dx.doi.org/10.1016/0028-3932(90)90050-X.
157. Desimone R. Face-selective cells in the temporal cortex of monkeys. *J Cogn Neurosci.* 1991;3(1):1–8. http://dx.doi.org/10.1162/jocn.1991.3.1.1.
158. Hein G, Knight RT. Superior temporal sulcus–it's my area: or is it? *J Cogn Neurosci.* 2008;20(12):2125–2136. http://dx.doi.org/10.1162/jocn.2008.20148.
159. Allison T, Puce A, McCarthy G. Social perception from visual cues: role of the STS region. *Trends Cogn Sci.* 2000.
160. Saxe R, Xiao DK, Kovacs G, Perrett DI, Kanwisher N. A region of right posterior superior temporal sulcus responds to observed intentional actions. *Neuropsychologia.* 2004;42(11):1435–1446. http://dx.doi.org/10.1016/j.neuropsychologia.2004.04.015.
161. Pelphrey KA, Viola RJ, McCarthy G. When strangers pass: processing of mutual and averted social gaze in the superior temporal sulcus. *Psychol Sci.* 2004;15(9):598–603. http://dx.doi.org/10.1111/j.0956-7976.2004.00726.x.
162. Redcay E. The superior temporal sulcus performs a common function for social and speech perception: implications for the emergence of autism. *Neurosci Biobehav Rev.* 2008;32(1):123–142. http://dx.doi.org/10.1016/j.neubiorev.2007.06.004.
163. Materna S, Dicke PW, Thier P. The posterior superior temporal sulcus is involved in social communication not specific for the eyes. *Neuropsychologia.* 2008;46(11):2759–2765. http://dx.doi.org/10.1016/j.neuropsychologia.2008.05.016.
164. Engell AD, Haxby JV. Facial expression and gaze-direction in human superior temporal sulcus. *Neuropsychologia.* 2007;45(14):3234–3241. http://dx.doi.org/10.1016/j.neuropsychologia.2007.06.022.
165. Adolphs R. Is the human amygdala specialized for processing social information? *Ann N Y Acad Sci.* 2003;985(1):326–340. http://dx.doi.org/10.1111/j.1749-6632.2003.tb07091.x.
166. Adolphs R. Neural systems for recognizing emotion. *Curr Opin Neurobiol.* 2002;12(2):169–177. http://dx.doi.org/10.1016/S0959-4388(02)00301-X.
167. Morris JS, Ohman A, Dolan RJ. A subcortical pathway to the right amygdala mediating "unseen" fear. *Proc Natl Acad Sci.* 1999;96(4):1680–1685. http://dx.doi.org/10.2307/47262.
168. Adolphs R, Tranel D. Emotion recognition and the human amygdala. In: Aggleton JP, ed. *The Amygdala. A Functional Analysis.* London: Oxford University Press; 2000:587–630.
169. Bechara A, Damasio H, Damasio AR. Emotion, decision making and the orbitofrontal cortex. *Cereb Cortex.* 2000;10(3):295–307. http://dx.doi.org/10.1093/cercor/10.3.295.
170. Eslinger PJ. Orbital frontal cortex: behavioral and physiological significance: an introduction to special topic papers: part II. *Neurocase.* 1999;5(4):299–300. http://dx.doi.org/10.1080/13554799908411983.
171. Schoenbaum G, Chiba AA, Gallagher M. Neural encoding in orbitofrontal cortex and basolateral amygdala during olfactory discrimination learning. *J Neurosci.* 1999;19(5):1876–1884.
172. Salmond CH, de Haan M, Friston KJ, Gadian DG, Vargha-Khadem F. Investigating individual differences in brain abnormalities in autism. *Philos Trans R Soc B Biol Sci.* 2003;358(1430):405–413. http://dx.doi.org/10.1098/rstb.2002.1210.
173. Brothers L. The neural basis of primate social communication. *Motiv Emot.* 1990;14(2):81–91. http://dx.doi.org/10.1007/BF00991637.
174. Emery NJ. The eyes have it: the neuroethology, function and evolution of social gaze. *Neurosci Biobehav Rev.* 2000;24(6):581–604. http://dx.doi.org/10.1016/S0149-7634(00)00025-7.
175. Bruce V, Young A. Understanding face recognition. *Br J Psychol.* 1986;77(Pt 3):305–327.
176. Tranel D. Dissociated verbal and nonverbal retrieval and learning following left anterior temporal damage. *Brain Cogn.* 1991;15(2):187–200. http://dx.doi.org/10.1016/0278-2626(91)90025-4.
177. Hasselmo ME, Rolls ET, Baylis GC. The role of expression and identity in the face-selective responses of neurons in the temporal visual cortex of the monkey. *Behav Brain Res.* 1989;32(3):203–218. http://dx.doi.org/10.1016/S0166-4328(89)80054-3.
178. Phillips ML, Bullmore ET, Howard R, et al. Investigation of facial recognition memory and happy and sad facial expression perception: an fMRI study. *Psychiatry Res.* 1998;83(3):127–138. http://dx.doi.org/10.1016/S0925-4927(98)00036-5.
179. Puce A, Allison T, Bentin S, Gore JC, McCarthy G. An fMRI study of changes in gaze direction and mouth position. *NeuroImage.* 1997.
180. Calder AJ, Young AW. Understanding the recognition of facial identity and facial expression. *Nat Rev Neurosci.* 2005;6(8):641–651. http://dx.doi.org/10.1038/nrn1724.
181. Adams RB, Kveraga K. Social vision: functional forecasting and the integration of compound social cues. *Rev Philos Psychol.* 2015:1–20. http://dx.doi.org/10.1007/s13164-015-0256-1.
182. Adams RB, Hess U, Kleck R. The intersection of gender-related facial appearance and facial displays of emotion. *Emot Rev.* 2015;7(1):5–13. http://dx.doi.org/10.1177/1754073914544407.
183. Marsh AA, Adams RB, Kleck R. Why do fear and anger look the way they do? Form and social function in facial expressions. *Pers Soc Psychol Bull.* 2005;31(1):73–86. http://dx.doi.org/10.1177/0146167204271306.
184. Fabes RA, Martin CL. Gender and age stereotypes of emotionality. *Pers Soc Psychol Bull.* 1991;17(5):532–540. http://dx.doi.org/10.1177/0146167291175008.
185. Briton NJ, Hall JA. Gender-based expectancies and observer judgments of smiling. *J Nonverbal Behav.* 1995;19(1):49–65. http://dx.doi.org/10.1007/BF02173412.
186. Friedman H, Zebrowitz LA. The contribution of typical sex differences in facial maturity to sex role stereotypes. *Pers Soc Psychol Bull.* 1992;18(4):430–438. http://dx.doi.org/10.1177/0146167292184006.
187. Todorov A, Baron SG, Oosterhof NN. Evaluating face trustworthiness: a model based approach. *Soc Cogn Affect Neurosci.* 2008;3(2):119–127. http://dx.doi.org/10.1093/scan/nsn009.
188. Carroll JM, Russell JA. Do facial expressions signal specific emotions? Judging emotion from the face in context. *J Pers Soc Psychol.* 1996;70(2):205–218.
189. Russell JA, Fehr B. Relativity in the perception of emotion in facial expressions. *J Exp Psychol.* 1987;116(3):223–237.
190. Pixton TS. *Expecting Happy Women, Not Detecting the Angry Ones: Detection and Perceived Intensity of Facial Anger, Happiness, and Emotionality* [Unpublished Dissertation]. Stockholm University; 2011.
191. Lee E, Kang JI, Park IH, Kim J-J, An SK. Is a neutral face really evaluated as being emotionally neutral? *Psychiatry Research.* 2008;157(1–3):77–85. http://dx.doi.org/10.1016/j.psychres.2007.02.005.

192. Adams RB, Nelson AJ, Soto JA, Hess U, Kleck R. Emotion in the neutral face: a mechanism for impression formation? *Cogn Emot*. 2012;26(3):431–441. http://dx.doi.org/10.1080/02699931.2012.666502.
193. Zebrowitz LA, Andreoletti C, Collins MA. Bright, bad, babyfaced boys: appearance stereotypes do not always yield self-fulfilling prophecy effects. *J Pers Soc Psychol*. 1998;75(5):1300–1320. http://dx.doi.org/10.1037//0022-3514.75.5.1300.
194. Zebrowitz LA, Hall JA, Murphy NA, Rhodes G. Looking smart and looking good: facial cues to intelligence and their origins. *Pers Soc Psychol Bull*. 2002;28(2):238–249. http://dx.doi.org/10.1177/0146167202282009.
195. Zebrowitz LA, Fellous J-M, Mignault A, Andreoletti C. Trait impressions as overgeneralized responses to adaptively significant facial qualities: evidence from connectionist modeling. *Pers Soc Psychol Rev*. 2003;7(3):194–215.
196. Kalick SM, Zebrowitz LA, Langlois JH, Johnson RM. Does human facial attractiveness honestly advertise health? longitudinal data on an evolutionary question. *Psychol Sci*. 1998;9(1):8–13. http://dx.doi.org/10.1111/1467-9280.00002.
197. Zebrowitz LA, Franklin RG. The attractiveness halo effect and the babyface stereotype in older and younger adults: similarities, own-age accentuation, and older adult positivity effects. *Exp Aging Res*. 2014;40(3):375–393. http://dx.doi.org/10.1080/0361073X.2014.897151.
198. Coetzee V, Greeff JM, Stephen ID, Perrett DI. Cross-cultural agreement in facial attractiveness preferences: the role of ethnicity and gender. *PLoS One*. 2014;9(7):e99629. http://dx.doi.org/10.1371/journal.pone.0099629.
199. Gray AW, Boothroyd LG. Female facial appearance and health. *Evol Psychol*. 2012;10(1):66–77.
200. Liang X, Zebrowitz LA, Zhang Y. Neural activation in the "reward circuit" shows a nonlinear response to facial attractiveness. *Soc Neurosci*. 2010;5(3):320–334. http://dx.doi.org/10.1080/17470911003619916.
201. Winston JS, Strange BA, O'Doherty J, Dolan RJ. Automatic and intentional brain responses during evaluation of trustworthiness of faces. *Nat Neurosci*. 2002;5(3):277–283. http://dx.doi.org/10.1038/nn816.
202. Zebrowitz LA, Voinescu L, Collins MA. "Wide-eyed" and "crooked-faced": determinants of perceived and real honesty across the life span. *Pers Soc Psychol Bull*. 1996;22(12):1258–1269. http://dx.doi.org/10.1177/01461672962212006.
203. Zebrowitz LA, Kikuchi M, Fellous J-M. Are effects of emotion expression on trait impressions mediated by babyfaceness? Evidence from connectionist modeling. *Pers Soc Psychol Bull*. 2007;33(5):648–662. http://dx.doi.org/10.1177/0146167206297399.
204. Hess U, Adams RB, Kleck R. The categorical perception of emotions and traits. *Soc Cogn*. 2009;27(2):320–326. http://dx.doi.org/10.1521/soco.2009.27.2.320.
205. Hess U, Adams RB, Grammer K, Kleck R. Face gender and emotion expression: are angry women more like men? *J Vis*. 2009;9(12):19. http://dx.doi.org/10.1167/9.12.19.
206. Said CP, Sebe N, Todorov A. Structural resemblance to emotional expressions predicts evaluation of emotionally neutral faces. *Emotion*. 2009;9(2):260–264. http://dx.doi.org/10.1037/a0014681.
207. Fischer AH. Sex differences in emotionality: fact or stereotype? *Fem Psychol*. 1993;3(3):303–318. http://dx.doi.org/10.1177/0959353593033002.
208. Shields SA. Thinking about gender, thinking about theory: gender and emotional experience. In: Fischer AH, ed. *Gender and Emotion: Social Psychological Perspectives*. Cambridge: Cambridge University Press; 2000:3–23.
209. Birnbaum DW. Preschoolers' stereotypes about sex differences in emotionality: a reaffirmation. *J Gen Psychol*. 1983;143(1):139–140. http://dx.doi.org/10.1080/00221325.1983.10533542.
210. Keating CF. Gender and the physiognomy of dominance and attractiveness. *Soc Psychol Quart*. 1985;48(1):61. http://dx.doi.org/10.2307/3033782.
211. Keating CF, Mazur A, Segall MH. Facial gestures which influence the perception of status. *Sociometry*. 1977;40(4):374. http://dx.doi.org/10.2307/3033487.
212. Hess U, Blairy S, Kleck R. The influence of facial emotion displays, gender, and ethnicity on judgments of dominance and affiliation. *J Nonverbal Behav*. 2000;24(4):265–283. http://dx.doi.org/10.1023/A:1006623213355.
213. Knutson B. Facial expressions of emotion influence interpersonal trait inferences. *J Nonverbal Behav*. 1996;20(3):165–182. http://dx.doi.org/10.1007/BF02281954.
214. Laser PS, Mathie VA. Face facts: an unbidden role for features in communication. *J Nonverbal Behav*. 1982;7(1):3–19. http://dx.doi.org/10.1007/BF01001774.
215. Pittenger JB, Shaw RE. Aging faces as viscal-elastic events: implications for a theory of nonrigid shape perception. *J Exp Psychol Hum Percept Perform*. 1975;1(4):374–382.
216. Pittenger JB, Shaw RE, Mark LS. Perceptual information for the age level of faces as a higher order invariant of growth. *J Exp Psychol Hum Percept Perform*. 1979;5(3):478–493.
217. Malatesta CZ, Fiore MJ, Messina JJ. Affect, personality, and facial expressive characteristics of older people. *Psychol Aging*. 1987;2(1):64–69. http://dx.doi.org/10.1037/0882-7974.2.1.64.
218. Neuberg SL, Sng O. A life history theory of social perception: stereotyping at the intersections of age, sex, ecology (and race). *Soc Cogn*. 2013;31(6):696–711. http://dx.doi.org/10.1521/soco.2013.31.6.696.
219. Hummert ML, Garstka TA, O'Brien LT, Greenwald AG, Mellott DS. Using the implicit association test to measure age differences in implicit social cognitions. *Psychol Aging*. 2002;17(3):482–495. http://dx.doi.org/10.1037//0882-7974.17.3.482.
220. Carpinella CM, Chen JM, Hamilton DL, Johnson KL. Gendered facial cues influence race categorizations. *Pers Soc Psychol Bull*. 2015. http://dx.doi.org/10.1177/0146167214567153.
221. Johnson KL, Freeman JB, Pauker K. Race is gendered: how covarying phenotypes and stereotypes bias sex categorization. *JPSP*. 2012;102(1):116–131. http://dx.doi.org/10.1037/a0025335.
222. Phelps EA, O'Connor KJ, Cunningham WA, et al. Performance on indirect measures of race evaluation predicts amygdala activation. *J Cogn Neurosci*. 2000;12(5):729–738. http://dx.doi.org/10.1162/089892900562552.
223. Phelps EA, Thomas LA. Race, behavior, and the brain: the role of neuroimaging in understanding complex social behaviors. *Polit Psychol*. 2003;24(4):747–758. http://dx.doi.org/10.1046/j.1467-9221.2003.00350.x.
224. Hugenberg K, Bodenhausen GV. Facing prejudice: implicit prejudice and the perception of facial threat. *Psychol Sci*. 2003;14(6):640–643. http://dx.doi.org/10.1046/j.0956-7976.2003.psci_1478.x.
225. Hugenberg K. Social categorization and the perception of facial affect: target race moderates the response latency advantage for happy faces. *Emotion*. 2005;5(3):267–276. http://dx.doi.org/10.1037/1528-3542.5.3.267.
226. Bijlstra G, Holland RW, Dotsch R, Hugenberg K, Wigboldus DHJ. Stereotype associations and emotion recognition. *Pers Soc Psychol Bull*. 2014;40(5):567–577. http://dx.doi.org/10.1177/0146167213520458.
227. Kubota JT, Ito TA. The role of expression and race in weapons identification. *Emotion*. 2014;14(6):1115–1124. http://dx.doi.org/10.1037/a0038214.
228. Graham R, LaBar KS. Neurocognitive mechanisms of gaze-expression interactions in face processing and social attention. *Neuropsychologia*. 2012;50(5):553–566. http://dx.doi.org/10.1016/j.neuropsychologia.2012.01.019.
229. Adams RB, Kleck R. Perceived gaze direction and the processing of facial displays of emotion. *Psychol Sci*. 2003;14(6):644–647.

230. Adams RB, Kleck R. Effects of direct and averted gaze on the perception of facially communicated emotion. *Emotion*. 2005;5(1):3–11. http://dx.doi.org/10.1037/1528-3542.5.1.3.
231. Adams RB, Ambady N, Macrae CN, Kleck R. Emotional expressions forecast approach-avoidance behavior. *Motiv Emot*. 2006;30(2):177–186. http://dx.doi.org/10.1007/s11031-006-9020-2.
232. Sander D, Grandjean D, Kaiser S, Wehrle T, Scherer KR. Interaction effects of perceived gaze direction and dynamic facial expression: evidence for appraisal theories of emotion. *Eur J Cogn Psychol*. 2007;19(3):470–480. http://dx.doi.org/10.1080/09541440600757426.
233. Hess U, Adams R, Kleck R. Looking at you or looking elsewhere: the influence of head orientation on the signal value of emotional facial expressions. *Motiv Emot*. 2007;31(2):137–144. http://dx.doi.org/10.1007/s11031-007-9057-x.
234. Milders M, Hietanen JK, Leppänen JM, Braun M. Detection of emotional faces is modulated by the direction of eye gaze. *Emotion*. 2011;11(6):1456–1461. http://dx.doi.org/10.1037/a0022901.
235. Adams RB, Franklin RG. Influence of emotional expression on the processing of gaze direction. *Motiv Emot*. 2009;33(2):106–112. http://dx.doi.org/10.1007/s11031-009-9121-9.
236. Lobmaier JS, Tiddeman BP, Perrett DI. Emotional expression modulates perceived gaze direction. *Emotion*. 2008;8(4):573–577. http://dx.doi.org/10.1037/1528-3542.8.4.573.
237. Frischen A, Bayliss AP, Tipper SP. Gaze cueing of attention: visual attention, social cognition, and individual differences. *Psychol Bull*. 2007;133(4):694–724. http://dx.doi.org/10.1037/0033-2909.133.4.694.
238. Driver J, Davis G, Ricciardelli P, Kidd P, Maxwell E, Baron-Cohen S. Gaze perception triggers reflexive visuospatial orienting. *Vis Cogn*. 1999;6(5):509–540. http://dx.doi.org/10.1080/135062899394920.
239. Mathews A, Fox E, Yiend J, Calder A. The face of fear: effects of eye gaze and emotion on visual attention. *Vis Cogn*. 2003;10(7):823–835. http://dx.doi.org/10.1080/13506280344000095.
240. Holmes A, Richards A, Green S. Anxiety and sensitivity to eye gaze in emotional faces. *Brain Cogn*. 2006;60(3):282–294. http://dx.doi.org/10.1016/j.bandc.2005.05.002.
241. Putman P, Hermans E, van Honk J. Anxiety meets fear in perception of dynamic expressive gaze. *Emotion*. 2006;6(1):94–102. http://dx.doi.org/10.1037/1528-3542.6.1.94.
242. Fox E, Mathews A, Calder AJ, Yiend J. Anxiety and sensitivity to gaze direction in emotionally expressive faces. *Emotion*. 2007;7(3):478–486. http://dx.doi.org/10.1037/1528-3542.7.3.478.
243. Tipples J. Fear and fearfulness potentiate automatic orienting to eye gaze. *Cogn Emot*. 2006;20(2):309–320. http://dx.doi.org/10.1080/02699930500405550.
244. Albohn DN, Adams RB. The perception of emotion residue and its influences on impression formation. In: Albohn DN, Adams RB, eds. *Is a Neutral Face Really Neutral?* Philadelphia: PA; 2015.
245. Kaulard K, Cunningham DW, Bülthoff HH, Wallraven C. The MPI facial expression database – a validated database of emotional and conversational facial expressions. *PLoS One*. 2012;7(3):e32321. http://dx.doi.org/10.1371/journal.pone.0032321.
246. Meeren HKM, van Heijnsbergen CCRJ, de Gelder B. Rapid perceptual integration of facial expression and emotional body language. *Proc Natl Acad Sci*. 2005;102(45):16518–16523. http://dx.doi.org/10.1073/pnas.0507650102.
247. de Gelder B, Morris JS, Dolan RJ. Unconscious fear influences emotional awareness of faces and voices. *Proc Natl Acad Sci*. 2005;102(51):18682–18687. http://dx.doi.org/10.1073/pnas.0509179102.
248. Van den Stock J, Righart R, de Gelder B. Body expressions influence recognition of emotions in the face and voice. *Emotion*. 2007;7(3):487–494. http://dx.doi.org/10.1037/1528-3542.7.3.487.
249. Kret ME, Roelofs K, Stekelenburg JJ, de Gelder B. Emotional signals from faces, bodies and scenes influence observers' face expressions, fixations and pupil-size. *Front Hum Neurosci*. 2013;7:810. http://dx.doi.org/10.3389/fnhum.2013.00810.
250. Aviezer H, Trope Y, Todorov A. Body cues, not facial expressions, discriminate between intense positive and negative emotions. *Science*. 2012;338(6111):1225–1229. http://dx.doi.org/10.1126/science.1224313.
251. Hassin RR, Aviezer H, Bentin S. Inherently ambiguous: facial expressions of emotions, in context. *Emot Rev*. 2013;5(1):60–65. http://dx.doi.org/10.1177/1754073912451331.
252. Aviezer H, Hassin RR, Ryan J, et al. Angry, disgusted, or afraid? Studies on the malleability of emotion perception. *Psychol Sci*. 2008;19(7):724–732. http://dx.doi.org/10.1111/j.1467-9280.2008.02148.x.
253. Ashby FG, Townsend JT. Varieties of perceptual independence. *Psychol Rev*. 1986;93(2):154–179.
254. Baron-Cohen S. Are children with autism blind to the mentalistic significance of the eyes? *Br J Dev Psychol*. 1995;13:379–398.
255. Pineda JA, Sebestyen G, Nava C. Face recognition as a function of social attention in non-human primates: an ERP study. *Cogn Brain Res*. 1994;2(1):1–12.
256. Davidson RJ, Hugdahl K, eds. *Brain Asymmetry*. Cambridge, MA: Bradford Books; 1997.
257. Brothers L, Ring B. Mesial temporal neurons in the macaque monkey with responses selective for aspects of social stimuli. *Behav Brain Res*. 1993;57(1):53–61. http://dx.doi.org/10.1016/0166-4328(93)90061-T.
258. Etcoff NL. Selective attention to facial identity and facial emotion. *Neuropsychologia*. 1984;22(3):281–295. http://dx.doi.org/10.1016/0028-3932(84)90075-7.
259. Le Gal PM, Bruce V. Evaluating the independence of sex and expression in judgments of faces. *Percept Psychophys*. 2002;64(2):230–243. http://dx.doi.org/10.3758/BF03195789.
260. Schweinberger SR, Soukup GR. Asymmetric relationships among perceptions of facial identity, emotion, and facial speech. *J Exp Psychol Hum Percept Perform*. 1998;24(6):1748–1765.
261. Schweinberger SR, Burton AM, Kelly SW. Asymmetric dependencies in perceiving identity and emotion: experiments with morphed faces. *Percep Psychophys*. 1999;61(6):1102–1115. http://dx.doi.org/10.3758/BF03207617.
262. Baudouin JY, Tiberghien G. Gender is a dimension of face recognition. *J Exp Psychol Learn Mem Cogn*. 2002;28(2):362–365. http://dx.doi.org/10.1037//0278-7393.28.2.362.
263. Levy Y, Bentin S. Interactive processes in matching identity and expressions of unfamiliar faces: evidence for mutual facilitation effects. *Perception*. 2008;37(6):915–930.
264. Atkinson AP, Tipples J, Burt DM, Young AW. Asymmetric interference between sex and emotion in face perception. *Percept Psychophys*. 2005;67(7):1199–1213. http://dx.doi.org/10.3758/BF03193553.
265. Ganel T, Goshen-Gottstein Y, Goodale MA. Interactions between the processing of gaze direction and facial expression. *Vision Res*. 2005;45(9):1191–1200. http://dx.doi.org/10.1016/j.visres.2004.06.025.
266. Ganel T, Goshen-Gottstein Y. Perceptual integrality of sex and identity of faces: further evidence for the single-route hypothesis. *J Exp Psychol Hum Percept Perform*. 2002;28(4):854–867.
267. Ganel T, Valyear KF, Goshen-Gottstein Y, Goodale MA. The involvement of the "fusiform face area" in processing facial expression. *Neuropsychologia*. 2005;43(11):1645–1654. http://dx.doi.org/10.1016/j.neuropsychologia.2005.01.012.
268. Graham R, LaBar KS. Garner interference reveals dependencies between emotional expression and gaze in face perception. *Emotion*. 2007;7(2):296–313. http://dx.doi.org/10.1037/1528-3542.7.2.296.
269. Pourtois G, Sander D, Andres M, et al. Dissociable roles of the human somatosensory and superior temporal cortices for processing social face signals. *Eur J Neurosci*. 2004;20(12):3507–3515. http://dx.doi.org/10.1111/j.1460-9568.2004.03794.x.

270. Kveraga K, Boshyan J, Bar M. Magnocellular projections as the trigger of top-down facilitation in recognition. *J Neurosci*. 2007;27(48):13232–13240. http://dx.doi.org/10.1523/JNEUROSCI.3481-07.2007.
271. Kveraga K, Ghuman AS, Bar M. Top-down predictions in the cognitive brain. *Brain Cogn*. 2007;65(2):145–168. http://dx.doi.org/10.1016/j.bandc.2007.06.007.
272. Emery NJ, Amaral DG. *The Role of the Amygdala in Primate Social Cognition*. New York, NY: Oxford University Press; 2000.
273. Aggleton JP, Burton MJ, Passingham RE. Cortical and subcortical afferents to the amygdala of the rhesus monkey (*Macaca mulatta*). *Brain Res*. 1980;190(2):347–368. http://dx.doi.org/10.1016/0006-8993(80)90279-6.

CHAPTER

9

Neuroimaging Investigations of Social Status and Social Hierarchies

Jasmin Cloutier, Carlos Cardenas-Iniguez, Ivo Gyurovski, Anam Barakzai, Tianyi Li

Department of Psychology, University of Chicago, Chicago, IL, USA

OUTLINE

1. INTRODUCTION

Social hierarchies are omnipresent in the lives of many species. The ability to successfully navigate complex social environments with consideration of the relative rank of conspecifics is an essential skill not only for humans, but also for numerous other social beings. From maintaining rank and reducing conflict, to communication and reproduction, organisms of varying complexity rely on social hierarchies to support social interactions. Hierarchical social structures can provide order and clarify the roles of individual group members, thus facilitating social coordination. Furthermore, in some instances, status-based hierarchies can incentivize those lower in relative rank to progress and achieve higher standing among their peers, thus providing motivation to perform a variety of behaviors.

Social hierarchies have been identified across a broad range of organisms, from simpler model systems such as insects, to nonhuman and human primates. For example, reliance on status cues to organize important social behavior is identified in ants[1] and other insects, such as bees, who infer higher ranking in the social hierarchy based on physical body size.[2] Many species of fish are also known to rely on social hierarchies. For example, in *Cichlasoma dimerus* males (South American cichlid fish known to have stable and linear hierarchies), a greater social position within the hierarchy has been linked to lower relative stress levels and increased reproductive success.[3]

Neuroimaging Personality, Social Cognition, and Character
http://dx.doi.org/10.1016/B978-0-12-800935-2.00009-9

Greater complexity can be found in the social hierarchies of mammals, such as rats and primates, with increased research demonstrating the impact of relative social status on communication, reproductive behavior, and access to resources.[4,5] Among nonhuman primates, a diverse array of social structures and organizations support, sometimes simultaneously, the establishment and maintenance of social hierarchies. Positions within these more complex hierarchies often determine central aspects of their members' social interactions and life outcomes.[6–13] In line with the important benefits often associated with the possession of greater status (i.e., better health and reproductive success[14,15]), primates appear to value group members with greater social status. Whereas primates often find conspecifics rewarding compared to other stimuli,[16–19] research has found that macaques are willing to sacrifice primary rewards in order to see the faces of high-status others.[20] High-status primates often utilize overt displays of dominance, such as "chest-beating," to assert their position in the hierarchy; their status is often suggested to correlate with power, resource control, and preferential mate selection.[21–23] However, whereas such rigid status hierarchies may be prevalent among macaques and baboons, it appears not to be the case in other primate species such as the gibbon and howler monkey.[24,25]

Whereas the omnipresence of status-based hierarchical social organization among animals and humans alike is generally agreed upon, it is a challenge to provide a precise and inclusive definition of social status. When referring to humans, historians, sociologists, psychologists, and economists utilize a variety of generalizable definitions of social status. Perhaps one of the most common terms referring to human social status is socioeconomic status (SES). SES is a multidimensional construct usually based on objectively assessed factors related to education, occupation, and income.[26] Depending on the ages of the individuals, measures of SES will take into account their own education, occupation, and income but also those of their parents. Other measures of SES consider an individual's neighborhood of residence[27] or the subjective assessment of his or her perceived social standing relative to others.[28] Even if sizeable correlations between the factors comprising SES have been reported, it is important to note that distinct factors of SES, such as income and education, often reflect discrete past experiences and may often not be interchangeable or appropriately used as proxy variables for one another.[29,30] More fundamentally, a single and generalizable measure of social status is difficult to formulate when considering that social hierarchies can be based on various social dimensions and that the meaning of "being at the top" differs across individuals. For instance, some may perceive high status as referring to the possession of vast amounts of disposable income (i.e., high financial status), while others may place greater emphasis on physical characteristics, such as attractiveness or fitness, as symbols of high status. To others, high status may be conferred by prestigious occupations demonstrating intellectual accomplishments rather than financial wealth or by the enactment of prosocial behavior and the possession of well-developed moral principles.

Whereas research suggests that young children tend to ascribe higher status to individuals perceived to be more dominant,[31–33] adults base their judgments of others' social status on a wide range of socially valued dimensions that may or may not be perceptually available. For example, individuals believed to be immoral or bad tend to be assigned lower status.[34–37] Although what conveys social status may not always generalize across individuals and social groups, perceived differences in standing typically appear to be based on social dimensions valued by members of a given group.

Focusing henceforth solely on humans, the goal of the current chapter is to provide a review of recent insights into how differences in social status may impact brain structure and function. Although, considering the centrality of social status as a construct guiding social interactions, relatively little research has been done on the topic. Recent brain imaging investigations have begun to explore how our own social status shapes us and how the social status of others shapes our responses to them. In the subsequent sections, we will first begin by briefly examining available evidence of the influence of social status on brain structure and cognitive development. Subsequently, we will focus on recent functional Magnetic Resonance Imaging (fMRI) investigations on the impact of social status on how we construe others. Finally, we will present fMRI studies suggesting that individual differences in social status shape how we respond to others. Throughout the chapter, we also intend to highlight some of the behavioral data complementing these early brain imaging investigations of the impact that social hierarchies have on ourselves and on how we construe others.

2. SOCIAL STATUS AND BRAIN STRUCTURE: STATUS AS AN ENVIRONMENTAL FACTOR

Recent research indicates that social status, defined here as a social and environmental factor, has a sizeable impact on the development of neural structures and their functions. Advances in neuroimaging methodologies provides opportunities to investigate how various environmental factors associated with social status, particularly SES, impact brain functioning. In this section, we first review the somewhat disparate literature suggesting that different measures of SES may be associated with different structural brain differences. We will subsequently attempt to provide an overview of the growing body of literature exploring how differences in SES during development impact cognition and brain functions.

2.1 Measuring Social Status

As mentioned in the introduction of this chapter, social status is often measured in terms of one's socioeconomic status, or SES. Socioeconomic status typically refers to an individual's access to economic and social resources and is therefore considered a multidimensional construct.[26,29] As such, researchers investigating the link between SES and brain structure must make a number of careful decisions when defining SES. Research on SES often relies on a number of "proxies" to operationalize complex environmental factors. The most common factors selected by researchers to measure SES include income, educational attainment, occupational prestige, and information regarding an individual's neighborhood SES. In some circumstances, for instance, depending on the age of the individual, these factors may be assessed in terms of both the individual and his/her parents. Although many of these factors may be correlated, they should not be thought of as interchangeable since, for example, they may differentially impact developmental outcomes.[29]

2.2 Income

Income may be calculated using household, familial, or parental income in studies with child populations or solely by the income of the individual when studying adult populations.[30,38–45] While income has been used widely as a marker for SES, and therefore social status in general, in recent years it has fallen out of favor as a reliable measure due to the unreliability of self-report data from participants and the marked fluctuations of income over time at both an individual and familial level.[29] More recent studies have instead begun to use the Income-to-Needs (ITN) ratio. This ratio divides total family income by the official federal poverty threshold for a family of that size.[46] The ITN ratio now allows researchers to assess family income while also taking into account other important factors, such as national norms, family size, and cost of living, thereby providing a clearer measure of a family's financial standing.

2.3 Educational Attainment

Educational attainment, defined simply as the highest level of education completed by either the parents or the individual, is another component of SES that is often used to assess social status. Commonly used as a proxy for a number of factors related to cognitive stimulation in one's home environment, educational attainment is thought to measure the qualitative aspects of the relationship between the caregiver and child, such as exposure to complex language, parent–child interactions, and the quality of guardian caregiving practices.[47–53] The results of a number of studies focusing on maternal educational attainment, which is believed to be associated with better cognitive stimulation in the home environment, suggest that education may be the best predictor of a number of developmental outcomes.[54–56]

2.4 Additional Ways to Measure SES

Other ways of assessing SES include occupation (of either the parent(s) or the individual), average neighborhood SES, and subjective social status. Occupation is typically correlated with education and income, as particular occupations are normally associated with distinct levels of education and earnings.[57] Neighborhood SES, measured as the average SES of the individual's immediate neighborhood, is often found to be associated with exposure to environmental stressors, such as greater police presence, poverty, and higher prevalence of physical and social disorder, as well as limited access to institutional resources, such as libraries, medical care facilities, and overall employment opportunities.[27] Another measure commonly used by researchers is subjective social status: a self-report index that refers to an individual's perception of his or her own social rank relative to others within a defined group. Subjective social status is typically measured using the MacArthur Scale of Subjective Social Status. This scale requires individuals to indicate their place on a ten-rung ladder said to represent their larger community and has been found to predict a number of physical and mental health outcomes, above and beyond other, possibly more objective, measures of SES.[28,58]

Given the wide variety of measures used to define social status, many researchers prefer to use composite measures of SES, including a combination of two or more of the previously mentioned factors. Composite measures of SES commonly used in brain imaging research include the Hollingshead scale,[59] which combines occupation and education (Two-Factor Index), or education, occupation, marital status, and employment status (Four-Factor Index), as well as the Barratt Simplified Measure of Social Status,[60] which combines educational attainment, occupational prestige, and income.

2.5 Investigating the Impact of SES on Adult and Child Populations

Studies investigating the role of SES on brain development may choose to focus on adult populations, adolescents, or young children, each of which brings its own challenges. Since difficulties exist in assessing the SES of children and adolescents, studies examining these populations often rely on information from the child's guardian(s). Studies involving adult populations, however, can assess an individual's SES by taking into account the various measures discussed above. While this may provide a more accurate measure of current social standing, researchers do acknowledge that it may not accurately characterize the environmental and social factors that occurred during the individual's childhood (since biases in retroactive self-report data can affect the factors being assessed). In addition, researchers must also consider unforeseen circumstances that may cause changes in SES (in both adult and children), such as sudden unemployment or moving to a different neighborhood. In an attempt to address these issues, some researchers are conducting longitudinal studies following participants over the course of a given time period. Such studies allow researchers to examine the stability of an individual's status over time and the patterns of fluctuations in aspects of social status, which have been suggested to represent sources of stress.[45,61,62]

2.6 Challenges in the Study of SES and Brain Functions

Unfortunately, a number of challenges and difficulties face researchers investigating the role of SES on brain structure and development. As it may now be apparent, SES can be a difficult construct to define and measure; although SES has been shown to impact behavioral, educational, and life outcomes,[26,63–73] several studies have failed to establish a relationship between SES and brain function or structure.[41,42,44] Challenges arising when measuring brain morphometry, such as a lack of consensus on whether to measure brain volume or surface area of the cortex, are believed to contribute to these difficulties.[29] To further complicate matters, indices used to measure SES can be highly correlated with important mediators being considered, such as stress, nutrition, and family environment, making it difficult to isolate the effect of SES.[70,74] In spite of these various challenges, dedicated cognitive and developmental neuroscientists have begun to uncover how variations in social status lead to differences in neural structure and cognitive development.

2.7 Impact of SES on Brain and Cognitive Development

Typically demonstrating that low-SES individuals do not perform as well as their higher-SES counterparts, a number of studies found that SES is associated with the development of cognitive functions, such as language, executive function, memory, and visual cognition.[26,47,68,73–76] These findings tend to parallel results from behavioral studies, indicating that deprivation of resources (e.g., less cognitive stimulation from caregivers or home environments) leads to various impairments in cognitive performance.[47,77,78] In light of these findings, establishing how SES impacts the neural processes supporting these cognitive functions is believed to be a vital component of the development of interventions to improve the educational and life outcomes of lower SES individuals.

2.8 Executive Function

The term executive function is often used to describe cognitive processes involved in planning, execution, reasoning, and problem-solving.[79–82] Researchers often break down executive functions into a number of subprocesses often suggested to rely on prefrontal brain regions. Some examples of these subprocesses include working memory (the ability to hold information in the mind and use it to complete a task), inhibitory control (the ability to stop oneself from performing an action), and mental flexibility (the ability to sustain or flexibly switch between sets of behaviors, tasks, rules, or mental states).[79–82]

Decreased performance in executive function among lower-SES children has been reported in the literature. For example, using a delayed-response paradigm, in which infants had to search for an object hidden in one of two hiding places after a delay period, Lipina et al.[83] reported that low-SES infants made more errors when asked to inhibit an incorrect response and had worse memory for the spatial location of objects. In addition, low-SES infants had a greater rate of A-not-B errors—not correcting for the new hidden location of an object in a consecutive trial—which has been widely associated with an immaturity of object permanence.[84] Similarly, a study using a flanker task to assess cognitive operations related to alerting, orienting, and executive attention reported reduced speed and accuracy among low-SES children, indicating difficulty in inhibiting distracting information.[85]

Recent neuroimaging evidence suggests that brain regions associated with executive functions may develop differently for individuals with lower-SES backgrounds. In a study of children and adolescents aged 4–18 years old, Lawson et al.[43] reported that higher parental education predicts increased cortical thickness in the left superior frontal gyrus and right anterior cingulate gyrus, frontal brain areas linked to the ability to suppress or override competing responses while appropriately adjusting the effort required to do so.[86,87] These results are consistent with a study by Noble et al.[88] using diffusion tensor imaging (DTI) to investigate the structure of white matter in the brain of participants that varied on educational attainment. Their results suggest that lower educational attainment correlates with alteration of white matter tracts believed to be important for aspects of cognitive control when compared to participants with higher educational attainment.

Recent studies explored the potential relationship between gray matter volume and maternal occupation and education. In one such study, gray matter volume in a number of brain regions was found to be larger for children with mothers who have higher levels of educational attainment and greater job prestige (using the Hollingshead 2-factor index).[89] In another study, Raizada et al.[90] report a marginal association (possibly due to a low sample size) in five-year-old children between greater gray and white matter volumes, particularly in inferior frontal regions, and greater maternal education levels and occupational prestige. However, possibly the most convincing piece of evidence to support this relationship comes from a study by Noble et al.[91] investigating the relationship between SES and brain morphometry in a large sample of 1099 typically-developing individuals aged between 3 and 20 years. Results from this study also indicate a positive relationship between surface area in regions related to executive function and spatial skills and SES. Interestingly, the increase in surface area was found to be logarithmic, such that a subtle increase in income for lower-SES individuals is associated with a relatively large increase in surface area, suggesting that extremely disadvantaged children are the most negatively impacted.

Dysfunction in attention and executive function in low-SES children has also been demonstrated with the use of electrophysiology experiments. Using a technique measuring event-related potentials (ERPs), researchers have been able to investigate differences in attention allocation in children from different SES environments. ERPs allow researchers to measure electrical brain activity at the scalp and provide superior temporal resolution but poorer spatial resolution than fMRI. In one particular set of studies, where children were asked to listen to a story presented in one ear (the attended story) while ignoring a story playing in the other ear (unattended story), researchers found differences in the amplitude of the P1 component, a waveform associated with attention allocation occurring about 100 milliseconds following the presentation of the stimuli. Whereas higher-SES children displayed a greater P1 response corresponding to the hemisphere in which the attended story was presented, indicating correct discrimination of the distractor (or unattended) story playing in the other ear, lower-SES children did not.[92,93]

Although further research is needed to isolate the components of social status shaping brain structure and functions, convergent evidence from behavioral, fMRI, and ERP studies support the claim that low-SES environments are associated with impairments in attention and executive function.

2.9 Language

Studies exploring the relationship between brain morphometry and SES have also led to the suggestion that low-SES individuals have impairments in brain structures associated with language and reading. Focusing on 10- to 12-year-old children differing in SES (based on parental income), Eckert et al.[94] investigated the relationship between phonological skill (the ability to break down spoken and written words as individual units of sound) and cortical surface area of the left and right planum temporale (regions in the temporal lobe that are highly involved with language processing). Eckert et al.[94] found that, when taking into account SES, planar asymmetry (the difference of left minus right planum temporale surface area, which has been associated with greater phonological awareness[95]) was positively correlated with phonological skill, with low-SES participants showing lower phonological skills and lower asymmetry in the planum temporale. Similarly, evidence from Raizada et al.[90] suggest that possessing higher SES, as determined by a composite of parental income and educational attainment, is associated with a greater degree of hemispheric specialization in Broca's area, or left inferior frontal gyrus, during a rhyming task.

In a study considering the neighborhood SES of participants 35–64 years of age, Krishnadas et al.[96] found that even after controlling for age and alcohol use, participants living in the most deprived neighborhoods have significant cortical thinning in bilateral perisylvian cortices. These brain areas, which divide the frontal and parietal lobes from the temporal lobe below, are believed to be involved in language processing as it relates to verbal short-term memory.[97]

Research by Noble et al.[98] has shown that SES in children, measured as a composite of parental education, occupation, and income-to-needs ratio, can moderate the relationship between certain brain behavior relationships important for reading. Phonological processing ability is positively correlated with activity in the left fusiform and perisylvian cortices, brain areas believed to support learning to read.[99,100] However, Noble et al.[98] have found that this relationship differs among children of high versus low SES. In low-SES children, greater scores in phonological processing strongly predict greater activity in left fusiform and perisylvian regions during reading, whereas in high-SES children, this relationship is less strong. These findings provide further evidence that SES is intertwined with the development of reading and language abilities. Together, these studies indicate that possessing a lower level of SES may result in impaired brain functions related to language and reading, while having a higher level of SES may safeguard individuals from language and reading impairments.

2.10 Stress

A number of studies have suggested that greater exposure to stressors in individuals from low-SES environments can result in changes in neural structures regulating stress response, particularly the hippocampus and amygdala.[74,86,101–103] Impairments in function and development of both the hippocampus and amygdala are important, as they can lead to dysregulation of the hypothalamus-pituitary-adrenal (HPA) axis, one of the systems regulating responses to stressful stimuli.[104]

In a study investigating gray and white matter volume in children 6–12 years old, Luby et al.[39] found that hippocampal and amygdala volumes increased as the income-to-needs ratio of an individual increased. Interestingly, the authors also reported that individuals with greater number of stressful life events could be expected to display greater reduction in hippocampal volume, suggesting that the relationship between the hippocampus and SES may be due to experience- or environment-related stress.

A number of studies have shown reduced hippocampal and amygdala volume in low-SES individuals when assessed based on income, economic hardship, or educational attainment.[40,105–108] These findings are often suggested to be the result of a strong correlation between higher educational attainment and parenting practices promoting socioemotional development, which may have a protective effect on responses to environmental stressors.[47,48,53] Consistent with the idea that educational attainment may be a "proxy" for the quality of home environment, smaller volumes in areas suggested to be related to stress regulation (i.e., hippocampus and amygdala) have been found in individuals from homes with lower educational attainment.[39,40,89,105–109] These data suggest that greater educational attainment may be related not only to greater access to financial resources, but also with a greater ability to deal with stress.[64,109]

Taken together, this research suggests the importance of early factors related to social status on the developmental trajectory of the hippocampus and amygdala. Furthermore, it highlights the possibility that the impaired development of these brain areas in low-SES individuals may result in an impaired ability to respond to later environmental stressors.

2.11 Conclusion

Environmental factors found to associate with social status have important effects on the development, morphology, and function of various brain networks supporting language, attention, executive function, and stress responses. Whereas results may vary based on how researchers characterized SES, these findings indicate that future research integrating neuroimaging methodologies and rigorous examinations of socioeconomic factors should lead to a better understanding of the relationship between social status and brain function. To accomplish this goal, further research may also benefit from investigating the impact of mechanisms related to prenatal and genetic influences. Furthermore, better understanding the impact of social status on social interactions may provide insights into its pervasive role in multiple facets of our lives. Indeed, recent functional neuroimaging investigations suggest that one's own social status and the social status of others shape fundamental social cognitive processes.

3. HOW THE SOCIAL STATUS OF CONSPECIFICS SHAPES PERSON PERCEPTION AND PERSON EVALUATION

As previously mentioned, social status not only impacts cognitive and brain development, but also guides many facets of social behavior.[21,22,35,110–112] A small, but growing, body of neuroimaging research has explored how social status shapes social cognition and provides preliminary insights into how social hierarchies guide how we respond to conspecifics. Once again, because it is such a multifaceted construct, investigations into the neural substrates of social status processing produced a complex pattern of results spanning a number of brain regions. Nonetheless, some of these findings converge to provide insight into how status is identified and impacts evaluations during person perception.

Broadly speaking, areas of the prefrontal, parietal, hippocampal complex, and amygdala appear responsive to variations in the social status of the individuals we perceive. For example, within the prefrontal cortex, the ventral medial prefrontal cortex (VMPFC), ventrolateral prefrontal cortex (VLPFC), dorsolateral prefrontal cortex (DLPFC), and dorsomedial prefrontal cortex (DMPFC) have all been suggested to be recruited when processing the social status of conspecifics.[113–121] Whereas a number of social cognitive processes may be impacted by the availability of social status information, we will focus our review on brain regions believed to support status identification and status-based evaluation (see Figure 1).[114,116]

3.1 Perceived Social Status from Perceptual Cues of Dominance

As previously stated, research to date has construed social status in a number of ways. In some instances, often building upon a rich body of research focused on

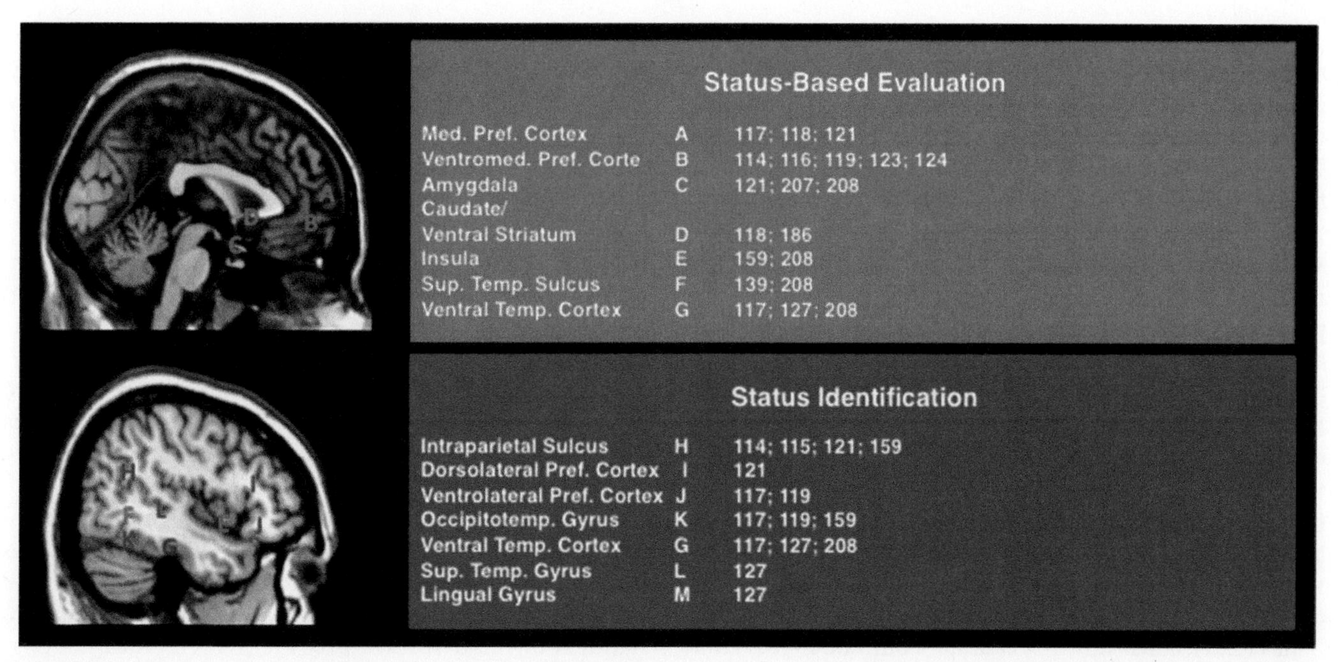

FIGURE 1 Brain regions believed to index status-based evaluation or status identification. Brain regions designated in red have been hypothesized to be involved status-based evaluations. Brain regions designated in blue have been hypothesized to be involved in social status identification.

nonhuman social hierarchies, status is suggested to be identifiable from available perceptual cues (i.e., social dominance).[4,20,119,122–125]

Dominance impacts social organization across a number of species, from ants to nonhuman primates.[2,4,14,20,31,119,122–129] Dominant individuals tend to enjoy greater health, fertility, access to resources, and reproductive success.[14,31,126,128,129] Its impact on human social organization may be less ubiquitous, and the endorsement of dominance as a guiding principle varies considerably across individuals and social groups.[125] Nonetheless, human perceivers readily infer dominance from cues when encountering unknown others.[33,118,119,127,130–136] Dominance can be inferred from static facial cues (i.e., direct eye gaze and upward head tilt), dynamic facial cues (i.e., overt facial anger expressions, speaking quickly and confidently), bodily cues (i.e., crossed arms, open chest, broad shoulders), and perceived nonverbal cues during social interactions (i.e., increased eye contact while speaking compared to listening).[35,118,130–132,137–140] Facial expressions of emotions have also been shown to convey dominance, with expressions of anger being perceived as highly dominant and fearful expressions as highly submissive.[131,132] Given the extent to which social status shapes how we construe others, the efficiency with which we perceive dominance cues in our social environment may not be surprising. Accordingly, recent brain imaging research has begun to explore how the human brain may support social cognitive mechanisms sensitive to variations in the perceived dominance of conspecifics.

In studying the effects of social status on how we perceive others, a number of fMRI studies have focused on the neural underpinnings of perceived dominance from facial expression and body posture.[118,119,127,139] In one such fMRI experiment, Marsh et al.[119] presented photographs of actors varying their gaze orientations, body postures, and gestures to convey low, average, and high dominance to participants who were asked to perform gender judgments. High status was depicted by actors holding their hands behind their back, harboring a dominant facial expression, and generally portraying an "open" body posture. Low status was depicted by actors holding their chin, harboring a submissive facial expression, crossing their legs under their seat, and generally portraying a "closed" body posture. Greater activity in the VLPFC in response to individuals displaying high-status cues, relative to neutral and low-status cues, was observed. The authors suggest that activity in this region may mediate changes in behavior dependent on the social context following status identification. This explanation is consistent with previous work suggesting that the VLPFC is recruited when socioemotional cues, including cues relevant to social hierarchy, are available to influence the selection of appropriate behavioral responses.[117,119,127,141–145] Interestingly, the authors also suggest that the VLPFC/lateral orbitofrontal cortex (OFC) could also be involved in processing the reward value of the targets. Considering that the lateral, in contrast to the medial, OFC has previously been associated with negative evaluations of stimuli,[146,147] it is possible that the dominance portrayed by the high-status individuals may have led to negative evaluations.

Another study examined brain responses to dominance cues in the context of the exploration of cultural differences between individualistic and collectivistic individuals during person perception.[118] Indeed, it has been suggested that American culture tends to place higher value on dominance, whereas Japanese culture tends to reinforce subordinate behavior.[148] In this study, social status was communicated via silhouettes or "figural outlines" that displayed either a dominant body posture (i.e., crossing their arms and looking straight ahead) or a subordinate body posture (i.e., looking down and placing their feet together).[118] The results revealed that American perceivers had greater activation in the caudate and MPFC in response to dominant postures, whereas Japanese perceivers had greater activation in the same brain areas when presented with subordinate postures. The authors interpret these findings to suggest that these brain regions can be flexibly shaped by a cultural tendency to value more dominant or subordinate behavior. Accordingly, the caudate could be involved in learning both the negative and positive value of stimuli during complex social interactions.[149–151] Whereas distinct regions of the MPFC have been found to be involved in mentalizing, self-referential processing, and social evaluations,[152–154] the authors suggest that in the context of their findings, the MPFC may reflect the processing of culturally-specific secondary reinforcers (i.e., dominance for Americans participants and subordination for Japanese participants).[155,156] These findings suggest that response to dominance cues can be flexible and shaped by one's culture and the values it tends to reinforce.

Whereas little research has directly addressed how perceived dominance conveys status in human social hierarchies, the fact that perceptual dominance cues efficiently shape neural responses to others suggest its potential importance in guiding social behavior. Departing from this literature on the perception of dominance, the following sections will review neuroimaging research examining how knowledge of the social standing of others shapes person perception.

3.2 Status Identification and Attention to Social Status

In light of the prevalence of social hierarchies in our lives, it is not surprising that various cognitive and attentional processes have been suggested to be particularly sensitive to the detection of social status. For instance, how we allocate attention toward others is believed to vary as a function of their social status. Although sensitivity to difference in social rank has repeatedly been demonstrated in nonhuman animals,[157,158] it has only recently been systematically investigated in humans. To do so, Foulsham et al.[158] tracked the eye gaze (i.e., frequency and duration of eye movements) of perceivers while they observed individuals of varying social status engaged in a decision-making task (in this study, status was conveyed through peer ratings based on the task performance and leadership abilities of each individual). They found that participants gazed longer and more often at the eyes and faces of high, compared to low, status individuals. Interestingly, recent research suggests that not only are we more likely to orient attention towards high-status conspecifics, but that we also tend to orient our attention in the same direction as them. Indeed, using a gaze-cuing task, Dalmaso et al.[157] found that perceivers shift their attention in response to the averted eye gaze of high-status individuals (in this study status was conveyed through an occupation title presented in a CV).

These findings have been interpreted to suggest that allocating and orienting attention based on social status may be a spontaneous social cognitive skill essential in navigating our social environment. However, only a handful of fMRI studies provide insights into the cognitive processes involved in social status identification. Whereas specific areas of the parietal and prefrontal cortices are typically hypothesized to support status identification, other areas, such as the amygdala, have been shown to be responsive to variations in the social status of others.

Existing research suggests that regions within the parietal cortex are involved in differentiating others based on their social status.[113–115,159,160] More specifically, the intraparietal sulcus (IPS) has been suggested to represent social status. Although this region of the parietal cortex has previously been suggested to support the representation of numerical magnitudes,[161–165] a number of recent studies suggest it may also index social distances, including differences in social status. In one such study, participants were either asked to perform numerical comparisons (by comparing the magnitude of numbers) and social status comparison (by comparing the rank of military positions).[159] The results revealed bilateral IPS activation in response to both numerical and social status comparison. Specifically, greater IPS activity was found when comparing targets closer in distance than those further apart. However, the processes involved in hierarchy differentiation may not be specific to social status comparisons but rather be indicative of the assessments of social distances. Indeed, in another study, participants were asked to either estimate physical distances by indicating which one of two objects is closer to them or social distances by indicating which one of two faces is closer to them.[160] Interestingly, a nearby area of the parietal cortex, the superior parietal lobule (SPL), is also believed to be involved in egocentric representations of spatial information relating external objects to the self.[166,167] Accordingly, it has been proposed that the mental representation of numbers, physical space, and social distance may all share common neural operations supported by the IPS.[159,160,168]

In addition to its involvement in explicit status comparisons, recent studies reveal that areas of the parietal cortex are spontaneously sensitive to social status information when perceiving or interacting with others.[114,121] In one such study, greater IPS activity was found in response to faces paired with low financial status information (e.g., "earns $25,000") compared to those paired with high financial status information (e.g., "earns $350,000").[114] In this study, the participants were never asked to evaluate the social status of the targets and were instead required to form a holistic impression of the individuals.

We replicated and extended these findings in a subsequent study revealing, once again, greater IPS activity in response to faces paired with lower financial status compared to equal and higher financial status, but also greater activity to faces paired with equal and higher moral status compared to lower moral status.[115] Importantly, social status was not conveyed through salary information but instead inferred based on colored backgrounds paired with the faces. For instance, during a prescan training period, darker red backgrounds may have been associated with "lower financial status" (than the participants themselves), whereas lighter red backgrounds may have been associated with "higher financial status." In this case, a darker blue background would have been associated with "lower moral status," while a lighter blue background would have been associated with "higher moral status." These cues were employed to convey levels and types of social status in order to minimize potential confounds associated with the presentation of statements indicating professional occupations or yearly earnings. In contrast to previous research where the status-related information was easily quantifiable, for instance, status based on salary[114] or military rank,[159] this study extends earlier efforts by using status labels that may not as easily lend perceivers to number-processing operations believed to be supported by the IPS.[162–164,169] One possibility is that IPS activity may be indexing the spontaneous assessment of the status of others in comparison to our own.[114,115] The ability to efficiently assess the relative status of conspecifics and their status in relation to our own, and therefore "map-out" a given social hierarchy, may be fundamental to our ability to successfully navigate the social environment. Although this hypothesis seems plausible based on previous research,[114,159,160] future research using larger samples are needed to directly test this potential mechanism.

Assuming the existence of cognitive processes supporting the spontaneous identification of the social status of conspecifics, our response to the social status of others may also depend on the nature of social hierarchical structures.[14,170] Testing this hypothesis, a study by Zink et al.[121] examined how social attention may vary within stable versus unstable social hierarchies. In this study, social status was conveyed by the skill level (e.g., a "one star" for low status versus a "three stars" for high status) of fictitious opponents, depicted via a photograph, with whom the participants were ostensibly playing a game. In reality, the game was simulated and the outcomes were predetermined regardless of participants' actual performance. The results revealed greater activity in the right inferior parietal cortex when participants viewed higher status players relative to lower status ones in a stable hierarchy condition. In contrast, during an unstable hierarchy condition, additional regions, specifically the amygdala and the MPFC (regions suggested to be implicated in social cognition, behavioral readiness, and emotion processing[152,171,172]), were preferentially responsive to viewing high status, compared to low status, players. These findings suggest that perceivers are not only sensitive to others' social rank relative to their own, but that they are also influenced by the stability of the hierarchy. When in an unstable hierarchy, individuals may be more likely to experience emotional arousal stemming, for example, from perceived threats to their relative rank.

3.3 Status-Based Evaluation

Possibly because of the benefits conferred from possessing higher SES in our society, greater social status is often suggested to lead to prestige and positive evaluations from others. Accordingly, individuals with higher status are suggested to be perceived as more competent, valuable to the group, prominent, generous, and reputable, compared to individuals with lower social standing.[35,173–175] However, although social status may influence who we interact with, how we perceive them, and our behavior toward them, we may often be inaccurate when inferring others' personal characteristics based on social status information.[176] Whereas high status may be highly valued by others and confers prestige, it is still unclear whether individuals possessing high social status are consistently evaluated in a positive light. For instance, differences in status-based evaluations may depend on the context in which we encounter others or on the social dimension conferring their status. For example, individuals working on Wall Street may be more likely to be positively evaluated by their peers based on their wealth compared to individuals working for nonprofit charitable organizations.

The characteristics conferring greater status, and possibly positive evaluations, can differ as a function of the nature of social hierarchies. In addition to increased access to material resources, financial status may have such prominence in our society because of the positive life outcomes it confers.[177–181] Nonetheless, whereas high-status individuals may often be evaluated positively, being rich may not necessarily lead to positive evaluations.[182] For instance, rich people tend to be seen as lower in warmth than other social groups.[183]

In contrast, status conferred by relative moral standing, which is suggested to be integral to the maintenance of human social hierarchies,[34,184] may typically lead to positive evaluations and may be detected at an early age.[36,37] In fact, it is conceivable that greater financial wealth may also, in some instances, evoke lower moral status and lead to negative evaluations.[182,183] In contrast, high moral status may confer the respect required to maintain one's standing within hierarchies.[34,175,185]

Recent fMRI investigations suggest that the VMPFC may index status-based evaluations.[114,115,124,186] Lesion studies have involved the VMPFC in processes tied to mentalizing (thinking about the thoughts and feelings of others), emotion processing, decision-making, and person evaluation.[187–190] For example, individuals with damage to the VMPFC show impairments in moral judgment[191] and deficits during facial expressions of emotion recognition tasks.[192,193] Although this region is believed to support numerous social cognitive functions, there is increased evidence suggesting it may index social evaluations in the context of person perception. In the context of social status judgments, evidence derived from human lesion studies suggest that patients with VMPFC damage maintain the ability to recognize social status but exhibit deficits in moral judgment and social norm comprehension.[124] This suggests that the VMPFC may not support cognitive processes related to status identification or differentiation but may instead support the status-based evaluation of others. The VMPFC has been shown to be involved during the evaluation of a wide variety of stimuli.[154,187,194–201] In the context of person perception, fMRI studies suggest that the VMPFC is recruited when individuals are asked to evaluate others.[114,202,203] Interestingly, preferential VMPFC responses are observed not only when evaluating others, but also when reporting one's own affective state.[204,205] The region may integrate affective and social information not only when forming impressions of others, but also when reflecting about the self.

Given its putative involvement in social evaluative processes, it is not surprising that activity in the VMPFC is responsive to the social status of others. In a recent study, greater VMPFC activation was observed in response to individuals paired with high, compared to low, moral status, as indicated by their professional occupation.[114] In this study, participants were shown photographs of faces preceded by information denoting their low or high moral status (e.g., "is a tobacco executive" or "does cancer research"). Once again, participants were asked to form an impression of the targets, not to evaluate their social status. Results revealed that activation of the VMPFC was sensitive to the moral status of individuals, such that higher activation was observed in response to targets paired with person knowledge denoting high moral status, compared to those paired with person knowledge denoting low moral status.

In a subsequent study, we again examined VMPFC response to targets varying in these status dimensions. As previously described, rather than using descriptive knowledge to convey social status, such as pay level or professional occupation, participants were presented with individual faces paired with colored background indicative of a given social status level (high, equal, or low) and dimension (financial or moral) learned during a prescan training session. Analyses revealed an interaction in VMPFC activity between status dimension (financial vs moral) and status level, such that greater activity was not only observed in response to targets with higher compared to lower moral status, as previously shown,[114] but also in response to targets with lower compared to higher financial status. The behavioral evidence collected also revealed that participants judged high moral status individuals to be more likeable. Taken together, these findings suggest that, at least in some contexts, higher status may not always lead to positive evaluations. In fact, supporting previous evidence,[182,183] lower financial status individuals may indeed be evaluated more positively than higher financial status targets.

In addition to the VMPFC, which supports person evaluation and, more generally, the generation of affective meaning,[124,154,202] there are a number of other brain regions that appear to be involved in the perception and evaluation of individuals varying in social status. Although future research will be required to better specify their functions, studies to date suggest that the amygdala, superior temporal sulcus, insula, fusiform gyrus, and lingual gyrus may be components of networks recruited during the processing of information related to social hierarchies.[127,139,206–209]

Whereas brain imaging research covered thus far provides evidence of the impact of social status on brain regions involved in the perception and evaluation of others, a number of studies have also begun to explore how individual differences in the social status of perceivers impact brain responses during social cognitive tasks. The final section in this chapter takes a brief look at this research.

3.4 Individual Differences in Social Status

Recent research has shown different ways in which one's own perceived social status impacts brain activity in response to others.[44,120,186] Muscatell and colleagues[120] explored how subjective social status shapes brain activity when thinking and feeling about others. In the first of two fMRI experiments, participants varying in subjective social status engaged in tasks requiring them to either think about people or objects. The results revealed greater activity in regions previously implicated in the mentalizing network, such as MPFC, posterior Superior Temporal Sulcus (pSTS) and precuneus/posterior Cingulate Cortex (PCC),[210,211] in response to trials in which participants were thinking about people. In addition, a negative correlation was observed between participants' subjective social status and activation in MPFC and precuneus/PCC, such that participants with lower subjective social status exhibited greater activity in these regions. Based on previous work implicating these brain areas in mentalizing processes (thinking about the thoughts and feelings of others)[152] and evidence suggesting that, at least in some contexts, lower status individuals display greater social engagement, empathic accuracy, and perspective taking skills,[212–214] the authors concluded that low-status individuals, relative to their high-status counterparts, may be more likely to focus on others' mental states and infer how they might think and feel.

In their second experiment, Muscatell et al.[120] extended the previous findings by exploring the relationship between perceivers' SES and threat perception. Participants in this experiment were adolescents; therefore, parental income and educational attainment were used to assess their level of SES. When in the scanner, participants were asked to view a series of threatening faces (i.e., facial expressions of anger). In line with previous behavioral findings,[215,216] results revealed a negative correlation between participants' social status and brain activation in regions of interest, such that perceivers with lower SES tended to have increased activity in the MPFC and left amygdala in response to threatening images relative to a fixation baseline. The authors interpret these results as evidence of a relationship between social status and the recruitment of brain regions involved both in mentalizing (i.e., MPFC) and threat detection (i.e., amygdala). Similar to the first study, they suggest that those lower in social status may have greater abilities to recognize others' mental and emotional states.

Finally, a recent study explored how the subjective social status of perceivers may impact reward responses when processing information about others varying in social status. In this study, Ly et al.[186] found that viewing individuals of the same rank as oneself elicited greater activity in the ventral striatum, an area widely associated with reward processing. During an fMRI session, participants were shown two faces serially along with a caption indicating whether the individual had a higher or lower social status than them. Following the serial presentation of these faces, the two faces were presented one above the other along with a statement (i.e., which person has been fired from more than one job?). Participants were then instructed to indicate to which of the two individuals the statement was more likely to pertain. The results revealed that participants' subjective social status was associated with differences in ventral striatal activity in response to information paired with either high or low-status targets. More specifically, perceivers who reported higher subjective social status demonstrated greater ventral striatal responses to targets of higher status, whereas participants with low subjective social status displayed greater ventral striatal activity when viewing targets of lower status. In contrast to work with nonhuman primates suggesting that higher status conspecifics may typically elicit greater reward responses,[16–20] these findings suggest that individuals with similar social statuses may be more rewarding. Taken together, the results of these three experiments suggest that our own social status shapes how we respond to others.

4. TOWARDS AN INTEGRATION OF BRAIN IMAGING INVESTIGATIONS OF SOCIAL STATUS

Initial brain imaging efforts suggest that social hierarchies and social status impact cognitive functioning, shape brain structures, and impact the neural substrates of person perception and evaluation. Although the previously described findings are indicative of progress toward our understanding of how this important construct shapes us and our response to others, much remains to be done to better understand the neural processes at play and to integrate these findings across many relevant research areas.

For example, according to a status characteristics theory,[217] higher social status is associated with certain social groups, such as being white, male, middle-aged, having higher educational attainment, and having greater occupational prestige. Although, as discussed, income, education, and occupation are commonly used as objective markers of social status, race and gender may also serve as status cues (in a similar way dominance cues are suggested to indicate social status) and interact with other status indicators during person perception and evaluation.[123] Indeed, in the context of contemporary American culture, social status and race are often intertwined. Members of racial minority groups, such as African Americans, are often assumed to be socially disadvantaged, whereas white individuals are often assumed to possess high status.[218] Future studies may benefit from exploring the interaction of race and gender with various indicators of social status to better understand the variables shaping the distributed network of brain regions involved in person perception. When doing so, important individual differences, such as contact to racial outgroup members[219] and endorsement of status-legitimizing beliefs, should be considered.[220] For example, research suggests that whites endorsing status-legitimizing beliefs view rises in social status of black individuals as threatening.[218] Similarly, males who endorse social dominance orientation to a greater extent demonstrate greater gender differences in issues relevant to gender equity.[221] Accordingly, group membership (i.e., gender and race) may often impact how social status guides person perception and, therefore, research considering both the social status of targets and their gender and race is needed.

In recent years, a large body of literature has uncovered the detrimental effects of subjective social status on health, even after controlling for objective measures, such as socioeconomic status.[14,170,222,223] A meta-analysis conducted by Thayer et al.[224] identified that changes in the autonomic nervous system, suggested by some to index potential detrimental effects of possessing lower SES, are associated with brain regions involved in emotional processing, such as the amygdala and the MPFC. Indeed, greater exposure to stressors associated with possessing a lower social status can cause structural and functional changes in the brain and predisposes the brain to a disrupted stress response.[102,225–227] Stress-related responses in the neuroendocrine system, the autonomic nervous system, and the immune system occur in order to protect an organism against these adversities in life, a process often called "allostasis."[102,228] Although these physiological adaptations are beneficial in the short run, allostatic load in the face of chronic stress may lead to worse health over an extended period of time.

In addition to investigating how environmental factors associated with possessing lower SES impact stress-related health outcomes, social cognitive investigations integrating behavioral, psychophysiological, and brain imaging methodologies may be beneficial to our understanding of the dynamic mechanism by which social status impacts function.[229] Indeed, one possibility is that VMPFC activity indexing status-based evaluations may also provide important indications of psychophysiological responses to others varying in social status.[116,230,231]

Brain imaging investigation of social hierarchies and social status has made great strides in recent years, but it remains highly underrepresented in the literature. We suspect that the great variability in definitions used when investigating both the impact of SES on brain functions and the impact of status on social cognition greatly contributes to this state of affairs. Nonetheless, with its immense relevance to central aspects of stress, coping, and health, as well as to social cognition, it seems essential to tackle these difficulties and move forward in our understanding of the pervasive impact that social status has on human life.

References

1. Casacci LP, Thomas JA, Sala M, et al. Ant pupae employ acoustics to communicate social status in their colony's hierarchy. *Curr Biol*. 2013;23(4):323–327.
2. Wilson EO. *Sociobiology*. Cambridge: Harvard University Press; 2000.
3. Morandini L, Honji RM, Ramallo MR, Moreira RG, Pandolfi M. The interrenal gland in males of the cichlid fish *Cichlasoma dimerus*: relationship with stress and the establishment of social hierarchies. *General Comp Endocrinol*. 2014;195:88–98.
4. Cheney D, Seyfarth R. Attending to behaviour versus attending to knowledge: examining monkeys' attribution of mental states. *Anim Behav*. 1990;40(4):742–753.
5. Wesson DW. Sniffing behavior communicates social hierarchy. *Curr Biol*. 2013;23(7):575–580.
6. Alberts SC, Watts HE, Altmann J. Queuing and queue-jumping: long-term patterns of reproductive skew in male savannah baboons, *Papio cynocephalus*. *Anim Behav*. 2003;65(4):821–840.
7. Clutton–Brock TH, Harvey PH. Primate ecology and social organization. *J Zool*. 1977;183(1):1–39.
8. Dunbar R. *Primate Social Systems*. London: Chapman & Hall; 1988.
9. Eisenberg JF, Muckenhirn NA, Rundran R. The relation between ecology a social structure in primates. *Science*. 1972;176(4037): 863–874.
10. Janson CH. Primate socio-ecology: the end of a golden age. *Evol Anthropol*. 2000;9(2):73–86.
11. Silk JB, Alberts SC, Altmann J. Social bonds of female baboons enhance infant survival. *Science*. 2003;302(5648):1231–1234.
12. Strier KB. *Primate Behavioral Ecology*. Boston: Allyn and Bacon; 2007.
13. Widdig A, Bercovitch FB, Streich WJ, Sauermann U, Nürnberg P, Krawczak M. A longitudinal analysis of reproductive skew in male rhesus macaques. *Proc R Soc Lond Ser B Biol Sci*. 2004;271 (1541):819–826.
14. Sapolsky RM. The influence of social hierarchy on primate health. *Science*. 2005;308(5722):648–652.
15. Hoffman CL, Ayala JE, Mas–Rivera A, Maestripieri D. Effects of reproductive condition and dominance rank on cortisol responsiveness to stress in free-ranging female rhesus macaques. *Am J Primatol*. 2010;72(7):559–565.
16. Anderson JR. Social stimuli and social rewards in primate learning and cognition. *Behav Process*. 1998;42(2):159–175.
17. Andrews MW, Bhat MC, Rosenblum LA. Acquisition and long-term patterning of joystick selection of food-pellet vs social-video reward by bonnet macaques. *Learn Motiv*. 1995;26(4):370–379.
18. Haude RH, Graber JG, Farres AG. Visual observing by rhesus monkeys: some relationships with social dominance rank. *Anim Learn Behav*. 1976;4(2):163–166.
19. Sackett GP. Monkeys reared in isolation with pictures as visual input: evidence for an innate releasing mechanism. *Science*. 1966;154(3755):1468–1473.
20. Deaner RO, Khera AV, Platt ML. Monkeys pay per view: adaptive valuation of social images by rhesus macaques. *Curr Biol*. 2005;15(6):543–548.
21. Cheney DL, Seyfarth RM. *Baboon Metaphysics: The Evolution of a Social Mind*. Chicago: University of Chicago Press; 2008.
22. Hare B, Tomasello M. Chimpanzees are more skilful in competitive than in cooperative cognitive tasks. *Anim Behav*. 2004;68(3): 571–581.
23. Maestripieri D. Maternal encouragement of infant locomotion in pigtail macaques, *Macaca nemestrina*. *Anim Behav*. 1996;51(3): 603–610.
24. Carpenter CA. A field-study in Siam of the behavior and social relations of the gibbon (Hylobales lar). *Comp Psychol Monog*. 1940; 16:1–212.
25. Carpenter CR. The howlers of Barro Colorado Island. In: Devore I, ed. *Primate Behavior: Field Studies of Monkeys and Apes*. New York: Holt, Rinehart & Winston; 1965:250–291.
26. McLoyd VC. Socioeconomic disadvantage and child development. *Am Psychol*. 1998;53(2):185–204.

27. Leventhal T, Brooks-Gunn J. The neighborhoods they live in: the effects of neighborhood residence on child and adolescent outcomes. *Psychol Bull.* 2000;126(2):309–337.
28. Adler NE, Epel ES, Castellazzo G, Ickovics JR. Relationship of subjective and objective social status with psychological and physiological functioning: preliminary data in healthy, white women. *Health Psychol.* 2000;19(6):586–592.
29. Brito NH, Noble KG. Socioeconomic status and structural brain development. *Front Neurosci.* 2014;8:276.
30. Duncan GJ, Magnuson K. Socioeconomic status and cognitive functioning: moving from correlation to causation. *Wiley Interdiscip Rev Cogn Sci.* 2012;3(3):377–386.
31. Hawley PH. The ontogenesis of social dominance: a strategy-based evolutionary perspective. *Dev Rev.* 1999;19(1):97–132.
32. Strayer F, Strayer J. An ethological analysis of social agonism and dominance relations among preschool children. *Child Dev.* 1976:980–989.
33. Thomsen L, Frankenhuis WE, Ingold-Smith M, Carey S. Big and mighty: preverbal infants mentally represent social dominance. *Science.* 2011;331(6016):477–480.
34. Boehm C. Ancestral hierarchy and conflict. *Science.* 2012;336(6083): 844–847.
35. Fiske ST. Interpersonal stratification: status, power, and subordination. In: Fiske S, Gilbert D, Lindzey G, eds. *Handbook of Social Psychology*. Vol. 2. Hoboken: Wiley; 2010:941–982.
36. Hamlin JK, Wynn K. Young infants prefer prosocial to antisocial others. *Cogn Dev.* 2011;26(1):30–39.
37. Hamlin JK, Wynn K, Bloom P. Three-month-olds show a negativity bias in their social evaluations. *Dev Sci.* 2010;13(6): 923–929.
38. Hanson JL, Hair N, Shen DG, et al. Family poverty affects the rate of human infant brain growth. *PLoS One.* 2013;8(2):e80954.
39. Luby J, Belden A, Botteron K, et al. The effects of poverty on childhood brain development: the mediating effect of caregiving and stressful life events. *JAMA Pediatr.* 2013;167(12):1135–1142.
40. Hanson JL, Chandra A, Wolfe BL, Pollak SD. Association between income and the hippocampus. *PLoS One.* 2011;6(5):e18712.
41. Lange N, Froimowitz MP, Bigler ED, Lainhart JE, Brian Development Cooperative Group. Associations between IQ, total and regional brain volumes, and demography in a large normative sample of healthy children and adolescents. *Dev Neuropsychol.* 2010;35(3):296–317.
42. Brain Development Cooperative Group. Total and regional brain volumes in a population-based normative sample from 4 to 18 years: the NIH MRI study of normal brain development. *Cereb Cortex.* 2012;22(1):1–12.
43. Lawson GM, Duda JT, Avants BB, Wu J, Farah MJ. Associations between children's socioeconomic status and prefrontal cortical thickness. *Dev Sci.* 2013;16(5):641–652.
44. Gianaros PJ, Horenstein JA, Cohen S, et al. Perigenual anterior cingulate morphology covaries with perceived social standing. *Soc Cogn Affect Neurosci.* 2007;2(3):161–173.
45. Hackman DA, Gallop R, Evans GW, Farah MJ. Socioeconomic status and executive function: developmental trajectories and mediation. *Dev Sci.* 2015;18.
46. U.S. Census Bureau. *Current Population Survey: Definitions and Explanations*. Population Division, Fertility & Family Statistics Branch; 2004.
47. Bradley RH, Corwyn RF. Socioeconomic status and child development. *Annu Rev Psychol.* 2002;53(1):371–399.
48. Duncan GJ, Brooks–Gunn J, Klebanov PK. Economic deprivation and early childhood development. *Child Dev.* 1994;65(2): 296–318.
49. Evans GW, English K. The environment of poverty: multiple stressor exposure, psychophysiological stress, and socioemotional adjustment. *Child Dev.* 2002;73(4):1238–1248.
50. Hart B, Risley TR. *Meaningful Differences in the Everyday Experience of Young American Children*. Baltimore: Paul Brookes Publishing Company; 1995.
51. Hoff-Ginsberg E, Tardif T. Socioeconomic status and parenting. In: Bornstein MH, ed. *Handbook of Parenting: Biology and Ecology of Parenting*. Biology and Ecology of Parenting; Vol. 2. Hillsdale: Lawrence Erlbaum Associates, Inc; 1995:161–188.
52. Hoff E. The specificity of environmental influence: socioeconomic status affects early vocabulary development via maternal speech. *Child Dev.* 2003;74(5):1368–1378.
53. McLoyd VC. The impact of poverty and low socioeconomic status on the socioemotional functioning of African-American children and adolescents: mediating effects. In: Taylor RD, Wang MC, eds. *Social and Emotional Adjustment and Family Relations in Ethnic Minority Families*. Mahwah: Lawrence Erlbaum Associates, Inc; 1997:7–34.
54. Aubret F. Prediction of elementary school success. *Enfance.* 1977:2–4.
55. Haveman R, Wolfe B. The determinants of children's attainments: a review of methods and findings. *J Econ Literature.* 1995; 33(4):1829–1878.
56. Smith JR, Brooks-Gunn J, Klebanov PK. Consequences of living in poverty for young children's cognitive and verbal ability and early school achievement. In: Duncan G, Brooks-Gunn J, eds. *Consequences of Growing Up Poor*. New York: Russell Sage Foundation; 1997:132–189.
57. Jencks C, Perman L, Rainwater L. What is a good job? A new measure of labor-market success. *Am J Sociol.* 1988;93(6):1322–1357.
58. Demakakos P, Nazroo J, Breeze E, Marmot M. Socioeconomic status and health: the role of subjective social status. *Soc Sci Med.* 2008;67(2):330–340.
59. Hollingshead AB. *Four Factor Index of Social Status*. New Haven: Yale University Press; 1975.
60. Barratt W. *The Barratt Simplified Measure of Social Status (BSMSS): Measuring SES*. Terre Haute: Indiana State University; 2006.
61. Ackerman BP, Izard CE, Schoff K, Youngstrom EA, Kogos J. Contextual risk, caregiver emotionality, and the problem behaviors of six-and seven-year-old children from economically disadvantaged families. *Child Dev.* 1999:1415–1427.
62. Marcynyszyn LA, Evans GW, Eckenrode J. Family instability during early and middle adolescence. *J Appl Dev Psychol.* 2008; 29(5):380–392.
63. Braveman PA, Cubbin C, Egerter S, et al. Socioeconomic status in health research: one size does not fit all. *JAMA.* 2005;294 (22):2879–2888.
64. Duncan GJ, Magnuson KA. Off with Hollingshead: socioeconomic resources, parenting, and child development. In: Bornstein M, Bradley R, eds. *Socioeconomic Status, Parenting, and Child Development*. Mahwah: Lawrence Erlbaum; 2003:83–106.
65. Adler NE, Boyce T, Chesney MA, et al. Socioeconomic status and health: the challenge of the gradient. *Am Psychol.* 1994;49(1): 15–24.
66. Goodman E, Adler NE, Kawachi I, Frazier AL, Huang B, Colditz GA. Adolescents' perceptions of social status: development and evaluation of a new indicator. *Pediatrics.* 2001;108(2):e31.
67. Krieger N, Williams DR, Moss NE. Measuring social class in US public health research: concepts, methodologies, and guidelines. *Annu Rev Public Health.* 1997;18(1):341–378.
68. Brooks-Gunn J, Duncan GJ. The effects of poverty on children. *Future Child.* 1997;7(2):55–71.
69. Duncan GJ, Yeung WJ, Brooks-Gunn J, Smith JR. How much does childhood poverty affect the life chances of children? *Am Sociol Rev.* 1998;63(3):406–423.
70. Evans GW. The environment of childhood poverty. *Am Psychol.* 2004;59(2):77–92.

71. Guo G, Harris KM. The mechanisms mediating the effects of poverty on children's intellectual development. *Demography*. 2000;37(4):431–447.
72. Heckman JJ. Skill formation and the economics of investing in disadvantaged children. *Science*. 2006;312(5782):1900–1902.
73. Sirin SR. Socioeconomic status and academic achievement: a meta-analytic review of research. *Rev Educ Res*. 2005;75(3):417–453.
74. Hackman DA, Farah MJ. Socioeconomic status and the developing brain. *Trends Cogn Sci*. 2009;13(2):65–73.
75. Miller P, Votruba-Drzal E, Setodji CM. Family income and early achievement across the urban–rural continuum. *Dev Psychol*. 2013;49(8):1452–1465.
76. Lipina SJ, Posner MI. The impact of poverty on the development of brain networks. *Front Hum Neurosci*. 2012;6:238.
77. Bradley RH, Corwyn RF, Burchinal M, McAdoo HP, Coll CG. The home environments of children in the United States Part II: Relations with behavioral development through age thirteen. *Child Dev*. 2001;72(6):1868–1886.
78. Rowe ML, Goldin–Meadow S. Early gesture selectively predicts later language learning. *Dev Sci*. 2009;12(1):182–187.
79. Miyake A, Friedman NP, Emerson MJ, Witzki AH, Howerter A, Wager TD. The unity and diversity of executive functions and their contributions to complex "frontal lobe" tasks: a latent variable analysis. *Cogn Psychol*. 2000;41(1):49–100.
80. Baddeley A. Working memory: looking back and looking forward. *Nat Rev Neurosci*. 2003;4(10):829–839.
81. Best JR, Miller PH. A developmental perspective on executive function. *Child Dev*. 2010;81(6):1641–1660.
82. Huizinga M, Dolan CV, van der Molen MW. Age-related change in executive function: developmental trends and a latent variable analysis. *Neuropsychologia*. 2006;44(11):2017–2036.
83. Lipina SJ, Martelli MI, Colombo J. Performance on the A-not-B task of Argentinean infants from unsatisfied and satisfied basic needs homes. *Interam J Psychol*. 2005;39(1):49–60.
84. Piaget J. *The Construction of Reality in the Child*. New York: Basic Books; 1954.
85. Mezzacappa E. Alerting, orienting, and executive attention: developmental properties and sociodemographic correlates in an epidemiological sample of young, urban children. *Child Dev*. 2004;75(5):1373–1386.
86. Noble KG, McCandliss BD, Farah MJ. Socioeconomic gradients predict individual differences in neurocognitive abilities. *Dev Sci*. 2007;10(4):464–480.
87. Casey B, Tottenham N, Fossella J. Clinical, imaging, lesion, and genetic approaches toward a model of cognitive control. *Dev Psychobiol*. 2002;40(3):237–254.
88. Noble KG, Korgaonkar MS, Grieve SM, Brickman AM. Higher education is an age-independent predictor of white matter integrity and cognitive control in late adolescence. *Dev Sci*. 2013;16(5):653–664.
89. Jednoróg K, Altarelli I, Monzalvo K, et al. The influence of socioeconomic status on children's brain structure. *PLoS One*. 2012;7(8):e42486.
90. Raizada RD, Richards TL, Meltzoff A, Kuhl PK. Socioeconomic status predicts hemispheric specialisation of the left inferior frontal gyrus in young children. *NeuroImage*. 2008;40(3):1392–1401.
91. Noble KG, Houston SM, Brito NH, et al. Family income, parental education and brain structure in children and adolescents. *Nat Neurosci*. 2015;18(5):773–778.
92. Stevens C, Lauinger B, Neville H. Differences in the neural mechanisms of selective attention in children from different socioeconomic backgrounds: an event-related brain potential study. *Dev Sci*. 2009;12(4):634–646.
93. Neville H, Stevens C, Klein S, et al. Improving behavior, cognition and neural mechanisms of attention in lower SES children. In: *Paper Presented at: The Society for Neuroscience*; 2011. Washington, DC.
94. Eckert MA, Lombardino LJ, Leonard CM. Planar asymmetry tips the phonological playground and environment raises the bar. *Child Dev*. 2001;72(4):988–1002.
95. Leonard CM, Lombardino LJ, Mercado LR, Browd SR, Breier JI, Agee OF. Cerebral asymmetry and cognitive development in children: a magnetic resonance imaging study. *Psychol Sci*. 1996:89–95.
96. Krishnadas R, Kim J, McLean J, et al. The envirome and the connectome: exploring the structural noise in the human brain associated with socioeconomic deprivation. *Front Hum Neurosci*. 2013;7:722.
97. Koenigs M, Acheson DJ, Barbey AK, Solomon J, Postle BR, Grafman J. Areas of left perisylvian cortex mediate auditory–verbal short-term memory. *Neuropsychologia*. 2011;49(13):3612–3619.
98. Noble KG, Wolmetz ME, Ochs LG, Farah MJ, McCandliss BD. Brain–behavior relationships in reading acquisition are modulated by socioeconomic factors. *Dev Sci*. 2006;9(6):642–654.
99. Shaywitz BA, Shaywitz SE, Pugh KR, et al. Disruption of posterior brain systems for reading in children with developmental dyslexia. *Biol Psychiatry*. 2002;52(2):101–110.
100. Turkeltaub PE, Gareau L, Flowers DL, Zeffiro TA, Eden GF. Development of neural mechanisms for reading. *Nat Neurosci*. 2003;6(7):767–773.
101. Buss C, Lord C, Wadiwalla M, et al. Maternal care modulates the relationship between prenatal risk and hippocampal volume in women but not in men. *J Neurosci*. 2007;27(10):2592–2595.
102. McEwen BS, Gianaros PJ. Central role of the brain in stress and adaptation: links to socioeconomic status, health, and disease. *Ann N Y Acad Sci*. 2010;1186(1):190–222.
103. Tottenham N, Sheridan MA. A review of adversity, the amygdala and the hippocampus: a consideration of developmental timing. *Front Hum Neurosci*. 2009;3:68.
104. Dowd JB, Simanek AM, Aiello AE. Socio-economic status, cortisol and allostatic load: a review of the literature. *Int J Epidemiol*. 2009:dyp277.
105. Butterworth P, Cherbuin N, Sachdev P, Anstey KJ. The association between financial hardship and amygdala and hippocampal volumes: results from the PATH through life project. *Soc Cogn Affect Neurosci*. 2012;7(5):548–556.
106. Noble KG, Houston SM, Kan E, Sowell ER. Neural correlates of socioeconomic status in the developing human brain. *Dev Sci*. 2012;15(4):516–527.
107. Piras F, Cherubini A, Caltagirone C, Spalletta G. Education mediates microstructural changes in bilateral hippocampus. *Hum Brain Mapp*. 2011;32(2):282–289.
108. Staff RT, Murray AD, Ahearn TS, Mustafa N, Fox HC, Whalley LJ. Childhood socioeconomic status and adult brain size: childhood socioeconomic status influences adult hippocampal size. *Ann Neurol*. 2012;71(5):653–660.
109. Noble KG, Grieve SM, Korgaonkar MS, et al. Hippocampal volume varies with educational attainment across the life-span. *Front Hum Neurosci*. 2012;6:307.
110. Magee JC, Galinsky AD. Social hierarchy: the self–reinforcing nature of power and status. *Acad Manag Ann*. 2008;2(1):351–398.
111. Sapolsky RM. *Why Zebras Don't Get Ulcers: The Acclaimed Guide to Stress, Stress-Related Diseases, and Coping-Now Revised and Updated*. New York: Henry Holt and Company, LLC; 2004.
112. Stephens NM, Markus HR, Townsend SS. Choice as an act of meaning: the case of social class. *J Pers Soc Psychol*. 2007;93(5):814–830.
113. Chiao JY. Neural basis of social status hierarchy across species. *Curr Opin Neurobiol*. 2010;20(6):803–809.
114. Cloutier J, Ambady N, Meagher T, Gabrieli J. The neural substrates of person perception: spontaneous use of financial and moral status knowledge. *Neuropsychologia*. 2012;50(9):2371–2376.
115. Cloutier J, Gyurovski I. Intraparietal sulcus activity during explicit self-referential social status judgments about others. *Int J Psychol Res*. 2013;6(SPE):68–79.

116. Cloutier J, Gyurovski I. Ventral medial prefrontal cortex and person evaluation: forming impressions of others varying in financial and moral status. *NeuroImage*. 2014;100:535–543.
117. Farrow TF, Jones SC, Kaylor-Hughes CJ, et al. Higher or lower? the functional anatomy of perceived allocentric social hierarchies. *NeuroImage*. 2011;57(4):1552–1560.
118. Freeman JB, Rule NO, Adams Jr RB, Ambady N. Culture shapes a mesolimbic response to signals of dominance and subordination that associates with behavior. *NeuroImage*. 2009;47(1):353–359.
119. Marsh AA, Blair KS, Jones MM, Soliman N, Blair RJR. Dominance and submission: the ventrolateral prefrontal cortex and responses to status cues. *J Cogn Neurosci*. 2009;21(4):713–724.
120. Muscatell KA, Morelli SA, Falk EB, et al. Social status modulates neural activity in the mentalizing network. *NeuroImage*. 2012;60(3):1771–1777.
121. Zink CF, Tong Y, Chen Q, Bassett DS, Stein JL, Meyer-Lindenberg A. Know your place: neural processing of social hierarchy in humans. *Neuron*. 2008;58(2):273–283.
122. Fragale AR, Overbeck JR, Neale MA. Resources versus respect: social judgments based on targets' power and status positions. *J Exp Soc Psychol*. 2011;47(4):767–775.
123. Freeman JB, Penner AM, Saperstein A, Scheutz M, Ambady N. Looking the part: social status cues shape race perception. *PLoS One*. 2011;6(9):e25107.
124. Karafin MS, Tranel D, Adolphs R. Dominance attributions following damage to the ventromedial prefrontal cortex. *J Cogn Neurosci*. 2004;16(10):1796–1804.
125. Sidanius J, Pratto F. *Social Dominance: An Intergroup Theory of Social Hierarchy and Oppression*. New York: Cambridge University Press; 1999.
126. Pusey A, Williams J, Goodall J. The influence of dominance rank on the reproductive success of female chimpanzees. *Science*. 1997;277(5327):828–831.
127. Chiao JY, Adams RB, Peter UT, Lowenthal WT, Richeson JA, Ambady N. Knowing who's boss: fMRI and ERP investigations of social dominance perception. *Group Process Intergr Relat*. 2008;11(2):201–214.
128. Fedigan LM. Dominance and reproductive success in primates. *Am J Phys Anthropol*. 1983;26(S1):91–129.
129. Gangestad SW, Simpson JA, Cousins AJ, Garver-Apgar CE, Christensen PN. Women's preferences for male behavioral displays change across the menstrual cycle. *Psychol Sci*. 2004;15(3):203–207.
130. Ellyson SL, Dovidio JF. *Power, Dominance, and Nonverbal Behavior: Basic Concepts and Issues. Power, Dominance, and Nonverbal Behavior*. New York: Springer; 1985: 1–27.
131. Hess U, Blairy S, Kleck RE. The influence of facial emotion displays, gender, and ethnicity on judgments of dominance and affiliation. *J Nonverbal Behav*. 2000;24(4):265–283.
132. Knutson B. Facial expressions of emotion influence interpersonal trait inferences. *J Nonverbal Behav*. 1996;20(3):165–182.
133. Said CP, Sebe N, Todorov A. Structural resemblance to emotional expressions predicts evaluation of emotionally neutral faces. *Emotion*. 2009;9(2):260.
134. Todorov A. Evaluating faces on social dimensions. In: *Social Neuroscience: Toward Understanding the Underpinnings of the Social Mind*; 2011:54–76.
135. Todorov A, Said CP, Engell AD, Oosterhof NN. Understanding evaluation of faces on social dimensions. *Trends Cogn Sci*. 2008;12(12):455–460.
136. Oosterhof NN, Todorov A. The functional basis of face evaluation. *Proc Natl Acad Sci*. 2008;105(32):11087–11092.
137. Dovidio JF, Ellyson SL. Decoding visual dominance: attributions of power based on relative percentages of looking while speaking and looking while listening. *Soc Psychol Q*. 1982:106–113.
138. Hall JA, Coats EJ, LeBeau LS. Nonverbal behavior and the vertical dimension of social relations: a meta-analysis. *Psychol Bull*. 2005;131(6):898–924.
139. Mason M, Magee JC, Fiske ST. Neural substrates of social status inference: roles of medial prefrontal cortex and superior temporal sulcus. *J Cogn Neurosci*. 2014;26(5):1131–1140.
140. Mignault A, Chaudhuri A. The many faces of a neutral face: head tilt and perception of dominance and emotion. *J Nonverbal Behav*. 2003;27(2):111–132.
141. Aron AR, Robbins TW, Poldrack RA. Inhibition and the right inferior frontal cortex. *Trends Cogn Sci*. 2004;8(4):170–177.
142. Blair R. The roles of orbital frontal cortex in the modulation of antisocial behavior. *Brain Cogn*. 2004;55(1):198–208.
143. Cools R, Clark L, Owen AM, Robbins TW. Defining the neural mechanisms of probabilistic reversal learning using event-related functional magnetic resonance imaging. *J Neurosci*. 2002;22(11):4563–4567.
144. Elliott R, Friston KJ, Dolan RJ. Dissociable neural responses in human reward systems. *J Neurosci*. 2000;20(16):6159–6165.
145. Zald DH, Curtis C, Chernitsky LA, Pardo JV. Frontal lobe activation during object alternation acquisition. *Neuropsychology*. 2005;19(1):97.
146. Kringelbach ML. The human orbitofrontal cortex: linking reward to hedonic experience. *Nat Rev Neurosci*. 2005;6(9):691–702.
147. O'Doherty J, Kringelbach ML, Rolls ET, Hornak J, Andrews C. Abstract reward and punishment representations in the human orbitofrontal cortex. *Nat Neurosci*. 2001;4(1):95–102.
148. Triandis HC, Gelfand MJ. Converging measurement of horizontal and vertical individualism and collectivism. *J Pers Soc Psychol*. 1998;74(1):118–128.
149. Graybiel AM. The basal ganglia: learning new tricks and loving it. *Curr Opin Neurobiol*. 2005;15(6):638–644.
150. King-Casas B, Tomlin D, Anen C, Camerer CF, Quartz SR, Montague PR. Getting to know you: reputation and trust in a two-person economic exchange. *Science*. 2005;308(5718):78–83.
151. Rilling JK, Gutman DA, Zeh TR, Pagnoni G, Berns GS, Kilts CD. A neural basis for social cooperation. *Neuron*. 2002;35(2):395–405.
152. Amodio DM, Frith CD. Meeting of minds: the medial frontal cortex and social cognition. *Nat Rev Neurosci*. 2006;7(4):268–277.
153. Kelley WM, Macrae CN, Wyland CL, Caglar S, Inati S, Heatherton TF. Finding the self? An event-related fMRI study. *J Cogn Neurosci*. 2002;14(5):785–794.
154. Roy M, Shohamy D, Wager TD. Ventromedial prefrontal-subcortical systems and the generation of affective meaning. *Trends Cogn Sci*. 2012;16(3):147–156.
155. Grüsser SM, Wrase J, Klein S, et al. Cue-induced activation of the striatum and medial prefrontal cortex is associated with subsequent relapse in abstinent alcoholics. *Psychopharmacology*. 2004;175(3):296–302.
156. Tzschentke T. The medial prefrontal cortex as a part of the brain reward system. *Amino Acids*. 2000;19(1):211–219.
157. Dalmaso M, Pavan G, Castelli L, Galfano G. Social status gates social attention in humans. *Biol Lett*. 2011;8(3):450–452.
158. Foulsham T, Cheng JT, Tracy JL, Henrich J, Kingstone A. Gaze allocation in a dynamic situation: effects of social status and speaking. *Cognition*. 2010;117(3):319–331.
159. Chiao JY, Harada T, Oby ER, Li Z, Parrish T, Bridge DJ. Neural representations of social status hierarchy in human inferior parietal cortex. *Neuropsychologia*. 2009;47(2):354–363.
160. Yamakawa Y, Kanai R, Matsumura M, Naito E. Social distance evaluation in human parietal cortex. *PLoS One*. 2009;4(2):e4360.
161. Kadosh RC, Henik A, Rubinsten O, et al. Are numbers special? The comparison systems of the human brain investigated by fMRI. *Neuropsychologia*. 2005;43(9):1238–1248.
162. Dehaene S, Piazza M, Pinel P, Cohen L. Three parietal circuits for number processing. *Cogn Neuropsychol*. 2003;20(3–6):487–506.

163. Pinel P, Dehaene S, Riviere D, LeBihan D. Modulation of parietal activation by semantic distance in a number comparison task. *NeuroImage*. 2001;14(5):1013–1026.
164. Pinel P, Piazza M, Le Bihan D, Dehaene S. Distributed and overlapping cerebral representations of number, size, and luminance during comparative judgments. *Neuron*. 2004;41(6):983–993.
165. Shuman M, Kanwisher N. Numerical magnitude in the human parietal lobe: tests of representational generality and domain specificity. *Neuron*. 2004;44(3):557–569.
166. Naito E, Scheperjans F, Eickhoff SB, et al. Human superior parietal lobule is involved in somatic perception of bimanual interaction with an external object. *J Neurophysiol*. 2008;99(2):695–703.
167. Neggers S, Van der Lubbe R, Ramsey N, Postma A. Interactions between ego-and allocentric neuronal representations of space. *NeuroImage*. 2006;31(1):320–331.
168. Ramachandran VS, Hubbard EM. Synaesthesia–a window into perception, thought and language. *J Conscious Stud*. 2001; 8(12):3–34.
169. Kadosh RC, Kadosh KC, Kaas A, Henik A, Goebel R. Notation-dependent and-independent representations of numbers in the parietal lobes. *Neuron*. 2007;53(2):307–314.
170. Sapolsky RM. Social status and health in humans and other animals. *Annu Rev Anthropol*. 2004:393–418.
171. Adolphs R. Is the human amygdala specialized for processing social information? *Ann N Y Acad Sci*. 2003;985(1):326–340.
172. Mitchell JP, Macrae CN, Banaji MR. Forming impressions of people versus inanimate objects: social-cognitive processing in the medial prefrontal cortex. *NeuroImage*. 2005;26(1):251–257.
173. Anderson C, Kilduff GJ. The pursuit of status in social groups. *Curr Dir Psychol Sci*. 2009;18(5):295–298.
174. Flynn FJ, Reagans RE, Amanatullah ET, Ames DR. Helping one's way to the top: self-monitors achieve status by helping others and knowing who helps whom. *J Pers Soc Psychol*. 2006;91(6):1123–1137.
175. Ridgeway CL, Walker HA. Status structures. In: Cook K, Fine G, House J, eds. *Sociological Perspectives on Social Psychology*. Needham Heights: Allyn & Bacon; 1995:281–310.
176. Varnum ME. What are lay theories of social class? *PLoS One*. 2013;8(7):e70589.
177. Marmot MG. *The Status Syndrome: How Social Standing Affects Our Health and Longevity*. London: Bloomsbury Publishing Plc; 2004.
178. Werner S, Malaspina D, Rabinowitz J. Socioeconomic status at birth is associated with risk of schizophrenia: population-based multilevel study. *Schizophr Bull*. 2007;33(6):1373–1378.
179. Boushey H, Weller CE. Has growing inequality contributed to rising household economic distress? *Rev Political Econ*. 2008; 20(1):1–22.
180. Ellis LE. A Comparative Biosocial Analysis. *Social Stratification and Socioeconomic Inequality*. Vol. 1. Westport: Praeger Publishers/ Greenwood Publishing Group; 1993.
181. Singh D. Female judgment of male attractiveness and desirability for relationships: role of waist-to-hip ratio and financial status. *J Pers Soc Psychol*. 1995;69(6):1089–1101.
182. Ribstein LE. How movies created the financial crisis. *Mich St L Rev*. 2009:1171.
183. Fiske ST, Cuddy AJ, Glick P, Xu J. A model of (often mixed) stereotype content: competence and warmth respectively follow from perceived status and competition. *J Pers Soc Psychol*. 2002;82(6):878–902.
184. Rai TS, Fiske AP. Moral psychology is relationship regulation: moral motives for unity, hierarchy, equality, and proportionality. *Psychol Rev*. 2011;118(1):57–75.
185. Yzerbyt V, Demoulin S. Intergroup relations. In: Fiske S, Gilbert D, Lindzey G, eds. *Handbook of Social Psychology*. Vol. 2. Hoboken: Wiley; 2010:1024–1083.
186. Ly M, Haynes MR, Barter JW, Weinberger DR, Zink CF. Subjective socioeconomic status predicts human ventral striatal responses to social status information. *Curr Biol*. 2011;21(9):794–797.
187. Adolphs R. The social brain: neural basis of social knowledge. *Annu Rev Psychol*. 2009;60:693–716.
188. Gläscher J, Adolphs R, Damasio H, et al. Lesion mapping of cognitive control and value-based decision making in the prefrontal cortex. *Proc Natl Acad Sci*. 2012;109(36):14681–14686.
189. Leopold A, Krueger F, dal Monte O, et al. Damage to the left ventromedial prefrontal cortex impacts affective theory of mind. *Soc Cogn Affect Neurosci*. 2012;7(8):871–880.
190. Shamay-Tsoory SG, Tomer R, Berger B, Aharon-Peretz J. Characterization of empathy deficits following prefrontal brain damage: the role of the right ventromedial prefrontal cortex. *J Cogn Neurosci*. 2003;15(3):324–337.
191. Croft KE, Duff MC, Kovach CK, Anderson SW, Adolphs R, Tranel D. Detestable or marvelous? Neuroanatomical correlates of character judgments. *Neuropsychologia*. 2010;48(6):1789–1801.
192. Heberlein AS, Padon AA, Gillihan SJ, Farah MJ, Fellows LK. Ventromedial frontal lobe plays a critical role in facial emotion recognition. *J Cogn Neurosci*. 2008;20(4):721–733.
193. Hornak J, Rolls E, Wade D. Face and voice expression identification in patients with emotional and behavioural changes following ventral frontal lobe damage. *Neuropsychologia*. 1996; 34(4):247–261.
194. Berridge KC, Kringelbach ML. Affective neuroscience of pleasure: reward in humans and animals. *Psychopharmacology*. 2008;199(3): 457–480.
195. Bouret S, Richmond BJ. Ventromedial and orbital prefrontal neurons differentially encode internally and externally driven motivational values in monkeys. *J Neurosci*. 2010;30(25):8591–8601.
196. Chib VS, Rangel A, Shimojo S, O'Doherty JP. Evidence for a common representation of decision values for dissimilar goods in human ventromedial prefrontal cortex. *J Neurosci*. 2009;29(39): 12315–12320.
197. Fellows LK. Advances in understanding ventromedial prefrontal function the accountant joins the executive. *Neurology*. 2007;68(13):991–995.
198. Fellows LK, Farah MJ. The role of ventromedial prefrontal cortex in decision making: judgment under uncertainty or judgment per se? *Cereb Cortex*. 2007;17(11):2669–2674.
199. Frith CD, Frith U. Mechanisms of social cognition. *Annu Rev Psychol*. 2012;63:287–313.
200. Henri-Bhargava A, Simioni A, Fellows LK. Ventromedial frontal lobe damage disrupts the accuracy, but not the speed, of value-based preference judgments. *Neuropsychologia*. 2012;50(7): 1536–1542.
201. Valentin VV, Dickinson A, O'Doherty JP. Determining the neural substrates of goal-directed learning in the human brain. *J Neurosci*. 2007;27(15):4019–4026.
202. Mende-Siedlecki P, Said CP, Todorov A. The social evaluation of faces: a meta-analysis of functional neuroimaging studies. *Soc Cogn Affect Neurosci*. 2013;8(3):285–299.
203. Bzdok D, Langner R, Hoffstaedter F, Turetsky BI, Zilles K, Eickhoff SB. The modular neuroarchitecture of social judgments on faces. *Cereb Cortex*. 2012;22(4):951–961.
204. Gusnard DA, Akbudak E, Shulman GL, Raichle ME. Medial prefrontal cortex and self-referential mental activity: relation to a default mode of brain function. *Proc Natl Acad Sci*. 2001;98(7):4259–4264.
205. Moran JM, Macrae CN, Heatherton TF, Wyland C, Kelley WM. Neuroanatomical evidence for distinct cognitive and affective components of self. *J Cogn Neurosci*. 2006;18(9):1586–1594.
206. Klein JT, Shepherd SV, Platt ML. Social attention and the brain. *Curr Biol*. 2009;19(20):R958–R962.
207. Kumaran D, Melo HL, Duzel E. The emergence and representation of knowledge about social and nonsocial hierarchies. *Neuron*. 2012;76(3):653–666.
208. Singer T, Kiebel SJ, Winston JS, Dolan RJ, Frith CD. Brain responses to the acquired moral status of faces. *Neuron*. 2004;41(4):653–662.

209. Swencionis JK, Fiske ST. How social neuroscience can inform theories of social comparison. *Neuropsychologia*. 2014;56:140–146.
210. Frith CD, Frith U. The neural basis of mentalizing. *Neuron*. 2006;50(4):531–534.
211. Gallagher HL, Frith CD. Functional imaging of 'theory of mind'. *Trends Cogn Sci*. 2003;7(2):77–83.
212. Kraus MW, Keltner D. Signs of socioeconomic status a thin-slicing approach. *Psychol Sci*. 2009;20(1):99–106.
213. Kraus MW, Côté S, Keltner D. Social class, contextualism, and empathic accuracy. *Psychol Sci*. 2010;21(11):1716–1723.
214. Lammers J, Gordijn EH, Otten S. Looking through the eyes of the powerful. *J Exp Soc Psychol*. 2008;44(5):1229–1238.
215. Allan S, Gilbert P. Anger and anger expression in relation to perceptions of social rank, entrapment and depressive symptoms. *Pers Individ Differ*. 2002;32(3):551–565.
216. Wilkinson RG. Health, hierarchy, and social anxiety. *Ann N Y Acad Sci*. 1999;896(1):48–63.
217. Berger J, Cohen BP, Zelditch Jr M. Status characteristics and social interaction. *Am Sociol Rev*. 1972;37(3):241–255.
218. Wilkins CL, Kaiser CR. Racial progress as threat to the status hierarchy implications for perceptions of anti-white bias. *Psychol Sci*. 2014;25(2):439–446.
219. Cloutier J, Li T, Correll J. The impact of childhood experience on amygdala response to perceptually familiar black and white faces. *J Cogn Neurosci*. 2014;26(9):1992–2004.
220. Wilkins CL, Wellman JD, Babbitt LG, Toosi NR, Schad KD. You can win but I can't lose: bias against high-status groups increases their zero-sum beliefs about discrimination. *J Exp Soc Psychol*. 2015;57:1–14.
221. Pratto F, Stallworth LM, Sidanius J. The gender gap: differences in political attitudes and social dominance orientation. *Br J Soc Psychol*. 1997;36(1):49–68.
222. Singh-Manoux A, Marmot MG, Adler NE. Does subjective social status predict health and change in health status better than objective status? *Psychosom Med*. 2005;67(6):855–861.
223. Derry HM, Fagundes CP, Andridge R, Glaser R, Malarkey WB, Kiecolt-Glaser JK. Lower subjective social status exaggerates interleukin-6 responses to a laboratory stressor. *Psychoneuroendocrinology*. 2013;38(11):2676–2685.
224. Thayer JF, Åhs F, Fredrikson M, Sollers JJ, Wager TD. A meta-analysis of heart rate variability and neuroimaging studies: implications for heart rate variability as a marker of stress and health. *Neurosci Biobehav Rev*. 2012;36(2):747–756.
225. Hackman DA, Farah MJ, Meaney MJ. Socioeconomic status and the brain: mechanistic insights from human and animal research. *Nat Rev Neurosci*. 2010;11(9):651–659.
226. Taylor SE, Eisenberger NI, Saxbe D, Lehman BJ, Lieberman MD. Neural responses to emotional stimuli are associated with childhood family stress. *Biol Psychiatry*. 2006;60(3):296–301.
227. Kim P, Evans GW, Angstadt M, et al. Effects of childhood poverty and chronic stress on emotion regulatory brain function in adulthood. *Proc Natl Acad Sci*. 2013;110(46):18442–18447.
228. McEwen BS. Stress, adaptation, and disease: allostasis and allostatic load. *Ann N Y Acad Sci*. 1998;840(1):33–44.
229. Cloutier J, Norman G, Li T, Berntson G. Person perception and autonomic nervous system response: the costs and benefits of possessing a high social status. *Biol Psychol*. 2013;92(2):301–305.
230. Eisenberger NI, Master SL, Inagaki TK, et al. Attachment figures activate a safety signal-related neural region and reduce pain experience. *Proc Natl Acad Sci*. 2011;108(28):11721–11726.
231. Phelps EA, Delgado MR, Nearing KI, LeDoux JE. Extinction learning in humans: role of the amygdala and vmPFC. *Neuron*. 2004;43(6):897–905.

CHAPTER

10

Cognitive Neuroscience of Self-Reflection

Joseph M. Moran[1,2]

[1]Center for Brain Science, Harvard University, Cambridge, MA, USA; [2]US Army Natick Soldier Research, Development, and Engineering Center, Natick, MA, USA

OUTLINE

The self has been a serious topic of inquiry since the emergence of reflexive consciousness, which one may estimate commenced around the period when humans began to use language in their thought processes and began to represent their perceptual milieu in cave art. "Know thyself" was considered important enough by the ancient Greeks to warrant inscription on the Temple of Delphi, perhaps the sole maxim given by Apollo to make the cut.[1] Famously brief, a more modern translation implies that it meant "pay no attention to the opinion of the multitude."[2] If we take this to mean that we should aim to understand ourselves independently from how others view us, then it becomes clear that questions about the *social psychology* of the self have been of great importance to individuals for at least three millennia. Over the last 100 years or so, there has been deep interest in the field of psychology in understanding the self: how we characterize it, how it develops, how it changes in injury or disease, what cognitive processes it includes and does not include, what perceptual objects are considered part of the self, whether our social partners are considered part of the self, and indeed whether our sense of self in the wild is colored by us paying attention to the multitudes. Over the last 50 years or so, that work has been complemented by the insistent charge of neuroscience, first through the technique of mapping the sequelae of accidental lesions in humans, then through intentionally modifying animal brains, and finally through noninvasive brain imaging using techniques such as positron emission tomography (PET), functional magnetic resonance imaging (fMRI), and electroencephalography.

Given the broad scope and variety of research themes and questions concerning the self, it is no surprise to learn that a quick Google Scholar search ("self psychology") unearths over 2.5 million hits, should we wish to gather for ourselves a complete understanding of, well, ourselves. Since the domain is so vast, this chapter

Neuroimaging Personality, Social Cognition, and Character
http://dx.doi.org/10.1016/B978-0-12-800935-2.00010-5

will be unable to cover all of these developments and instead will focus on a particular aspect of the self that has received a great deal of attention from psychologists and cognitive neuroscientists alike: self-reflection. Self-reflection is the cognitive process we use when someone asks us "What kind of music do you like?" It is the mechanism we use to know that we are introverted but not shy, and it is the means by which we transport ourselves to the hypothetical to answer the question of whether we would prefer a pet hippopotamus or a tiger.

From this inquiry, several principles have emerged. First, it appears that the self is a superordinate cognitive schema, such that information about the self is encoded in a separate cognitive structure than information about other people, animals, or nonliving things. Second, there is a brain network at least partially dedicated to processing information about ourselves. Third, this brain network overlaps heavily with a network that is active when we rest, or when our mind wanders. Fourth, such overlaps suggest that much of our mental machinery is in fact given over to social processes, a conclusion which would have been unthinkable prior to the onset of functional neuroimaging techniques that allow us to observe the living brain in action. Finally, disorders of the self, or lesions to specific brain regions associated with self-reflection, leave individuals with difficulties in recalling the stable nature of their selves, which has implications for our philosophy of self-knowledge. Future directions for how we may investigate self-reflection at both the cognitive and neural levels of analysis are discussed at the end of the chapter. The rest of the chapter is devoted to examining the evidence for each of these claims in turn, but we begin by expanding upon how the concept of the self is characterized within psychology.

1. WHAT IS THE SELF?

No concept within psychology is as ephemeral as the self, save for perhaps consciousness, a process and experience that looms large among the cognitive, emotional, and social ingredients that combine to form the self. A useful place to begin when trying to define what we mean by "the self" is with William James, a forebear whose writings serve as the bedrock of so many areas of psychological inquiry. According to James, the empirical self "of each of us is all that he is tempted to call by the name of me. " [3] Further, James discussed how the line between *me* and *mine* was barely distinct, foreshadowing a large portion of current work in cognitive neuroscience. While noting the correspondence of *me* and *mine*, James does subdivide the self into "its constituents, the feelings and emotions they arouse, and the actions to which they prompt."[3] In modern parlance, we can see that these concepts map on well to the cognition, emotion, and motivation of the self, useful, if fictional, distinctions that certainly hold sway in cognitive neuroscience.

Damasio has distinguished between the "core" and "extended" selves.[4] The core self consists in that collection of processes that make up the ongoing functions that differentiate the self from others, while the extended self describes a more narrative structure, embodied in the here and now, but also in possession of a future and a past and mental access to both through processes of memory and autonoetic consciousness.[5] These distinctions adhere closely to the Jamesian *me* and *mine* distinction, developing the concept further to include abstract consideration of future and past. Partitioning the self along similar lines, LeDoux describes differences between implicit and explicit selves, where the explicit self contains those processes that are present in consciousness, such as ongoing self-awareness, and implicit self refers to aspects of the self that are not accessible to conscious awareness.[6] This conceptualization has its roots in the psychodynamic psychology of Freud, who described conscious, preconscious, and unconscious aspects of the self, but whose investigations of the self remained at odds with the prevailing culture in experimental psychology.

Prior to Freud's time, traditional studies of the self in the introspectionist school of psychology focused squarely on explicit or conscious aspects of the self. These aspects of the self are wholly more amenable to psychological inquiry than their implicit counterparts, given that the implicit self remains ineffable and thus a poor candidate for study. Indeed, for a significant period of time, the prevailing view was that even explicit aspects of the self were inaccessible by scientific methods and should therefore be ignored (methodological behaviorism) or disavowed entirely (radical behaviorism).[7] Rejecting the strict adherence to the stimulus–response psychology of behaviorism, the architects of the cognitive revolution in the middle part of the last century allowed for the return of processes between the input and output stages of behavior. They reasoned that the workings of the mind resembled those of the computers that were beginning to exhibit complex behaviors, and hence began an emphasis on the *processes* occurring in cognition that continues in present day cognitive psychology, cognitive neuroscience, and social cognition research. Cognitive neuroscience has expressly positioned itself as the discipline capable of further investigating such *processes*, enabling us to go beyond boxes-and-arrows diagrams to the internal workings underlying what happens when we open our eyes and the light from a stimulus object in the environment falls upon our retinas. Whether the gifts of cognitive neuroscience justify their expensive wrapping is a debate taken up by many commentators in the literature,[8–13] but it is safe to say that such methods are here to stay, and at the least, have provided us with some insights into cognition that were unavailable prior to their inception.[14]

1.1 Psychology and the Self

The self can be thought of as both a physical and psychological entity: our conceptions of ourselves contain both physical and psychological aspects. Aspects of the physical self include the body proper, the boundary between ourselves and everything else, and things or objects that we might consider belonging to us. For example, our cars, clothes, and houses may be considered parts of our self-concept, as items that are fundamentally self rather than other. The physical self can also be either an agent or an object of actions, and in many circumstances, both simultaneously. Thus, when considering the physical self, the possibility for first, second, third, and even higher-order representations must not be overlooked. Similarly, these higher-order representations are also possible when considering the psychological self. Investigations of the psychological self have been divided into those researching knowledge about the self and those researching the unique first-person perspective.[15] Investigations of agency combine information about both the physical and psychological selves—"I see myself throwing the ball and see myself in the process of watching myself throw the ball," and so on, ad infinitum. This chapter focuses on questions about the psychological self, and specifically, those concerning knowledge about the self. The research discussed here concerns how we represent and access knowledge about our own self, such as trait information or other types of personal semantic knowledge, and how these representations interact with the emotional experience of the self. These types of representations all feed into what is most often considered to be a unitary sense of self that exists across space and time; a sense of self that includes expectations, beliefs, memories, autobiographical experiences, personal identity, and somatosensory experience among other things. Behavioral psychological research on the self has focused on delineating the nature of self and characterizing how the self and other aspects of psychology converge and diverge.

1.2 The Self as a Superordinate Cognitive Schema

Much of this research has played a role in igniting a major debate within the psychological literature that hinges upon whether or not the self enjoys privileged status as a "special" constituent of our cognitive architecture.[16–26] Evidence for the self as a special mechanism in our mental lives comes from work by Rogers et al.[16] who first demonstrated the self-reference effect in memory. The self-reference effect in memory refers to the observation that information encoded with reference to the self tends to be later recalled with greater accuracy than information encoded with reference to another person or encoded simply in a general semantic manner (e.g., "is this word abstract or concrete?"). This effect has been shown to be highly replicable,[27] but its origins have been ambiguous. One prevailing viewpoint has argued that the self is in some way special, a superordinate cognitive structure that possesses extraordinary mnemonic abilities. An alternative explanation is that the knowledge already contained in memory about the self encourages more elaborative encoding of material that is processed in reference to the self, and hence leads to better memory. While both possibilities are compelling, resolution of the two proved difficult in the absence of converging experimental methods. Neuroimaging[28,29] and neuropsychological[30,31] evidence that will be detailed in later sections suggests a privileged position for the self, at least in the domain of memory.

A related strand of research within the literature has investigated the existence and nature of a privileged role for self-relevant information in the biasing of attention.[32,33] After Miller's[34] groundbreaking work on short-term memory, we know that the capacity of our short-term store is very limited, and thus the problem of which information to focus our limited cognitive/executive/conscious resources upon becomes most pertinent. Research in the domain of attention has indicated that information that impinges upon the self-concept (self-relevant information) receives preferential selection with respect to further processing than information that is not distinctive in this manner.[32,35] Indeed, early research on the selectivity of attention revealed a prescient awareness of the issues: Postman, Bruner, and McGinnies[36] wrote that what one "selects from a near infinitude of potential percepts... [is] a servant of one's interests, needs, and values," while Kelly[37] suggested that our system of social constructs operates as a "scanning pattern" that "picks up blips of meaning" from the environment.[35]

This early research demonstrated significant benefits in reaction time for self-relevant information and is complemented well by research showing that self-relevant information is able to divert precious attentional resources towards itself in the context of focusing attention elsewhere. In the first demonstration of the famous "cocktail party effect," Moray[38] revealed that presenting the subject's own name to them on the unattended channel during a dichotic listening task reliably resulted in subjects perceiving the name, whereas other information presented on the unattended channel remained obscure. Building on this early work, Bargh[32] demonstrated facilitation of a probe response-time task when self-relevant information was presented on the attended channel and inhibition of the probe task when self-relevant information was presented on a to-be-ignored channel. An important difference between the studies is that in Bargh's work, subjects remained entirely unaware of the contents of the unattended channel, even when that information was self-relevant (adjectives describing a trait for which subjects were schematic). This raises an important point. While self-relevant information is able to grab attention and hence speed or inhibit performance on tasks requiring attention, this attentional mechanism operates typically outside the realm of conscious awareness, that is, automatically. Thus, in some circumstances, those items of information that remain chronically accessible in the self-concept are able to bias ongoing processing in favor of themselves, all without the subject's being consciously aware of this biasing. Here we can see an interaction between the implicit and explicit selves in navigating the world and the putative existence of a mechanism that causes us to focus on that which is important to the self without our ever becoming aware of its operation.

These attentional mechanisms, in concert with the memorial mechanisms described above, appear to argue for a privileged ("special") interpretation of self with respect to the rest of the cognitive architecture. How we might confirm our suspicions with respect to the self's "specialness" will be addressed in the next section, which details some of the important findings from cognitive neuroscience regarding self-referential processing, perspective taking, experiencing agency, and inferring intentionality.

2. COGNITIVE NEUROSCIENCE OF THE SELF

In review, Gillihan and Farah[15] delineate four possible criteria for "specialness." These are anatomical, functional uniqueness, functional independence, and species specificity. Anatomically, "engaging distinct brain areas, as demonstrated by functional neuroimaging studies, or requiring distinct brain areas, as demonstrated by lesion studies, is one sense in which a system can be said to be special."[15] Functional uniqueness refers to specificity in the *way* in which information is processed by a given network of regions. Functional independence consists in one system's operating entirely independently of another's, while species specificity refers to the existence of a given cognitive domain in one species but not others. As we have already seen, many researchers have argued for species specificity at least for the explicit, consciously aware aspects of self, such as autonoetic consciousness.[5] The studies referenced in this section will provide evidence for the first of these criteria, anatomical specificity. The particular characteristics of functional neuroimaging make it currently ill-specified to be brought to bear on questions of functional uniqueness, although analytical developments in multivoxel pattern analysis are beginning to chip away at this particular fortress,[39] highlighting, for instance, that the fusiform face area identified as being uniquely sensitive to faces[40] may, in addition, represent other visual object categories, such as houses.[41] The correlational nature of neuroimaging makes it difficult to answer questions of functional independence, but here again, newer design techniques, such as those that use real-time fMRI,[42] are beginning to demonstrate that particular brain states have particular effects on the processing of subsequently presented experimental stimuli.[43] In addition, findings from neuroimaging may be informed by those from single-unit recording and lesion studies to make it possible to garner arguments in support of both functional uniqueness and functional independence for the self.

Neuroimaging studies of the self and self-referential processing began obliquely, with studies designed more specifically to investigate episodic memory[44] or attention to internal versus external aspects of the environment.[45] Andreasen and colleagues[44] investigated common activations during focused and "random" episodic recall. Focused and random recall conditions were chosen to mirror respectively the "taking a history" and "free association" methods used in psychiatric evaluation. Common activations occurred in regions of medial prefrontal (MPFC) and retrosplenial/posterior cingulate cortices (PCC). The retrosplenial region was also activated in a study requiring the retrieval of affectively laden autobiographical memories.[46] In the study of Lane et al.[45] subjects were imaged while viewing emotional pictures. In one condition, they were required to respond about spatial aspects of the pictures, whereas in the other condition, they were required to attend to how the pictures made them feel. Contrasting internal (feelings) versus external (spatial) attention revealed activity in the rostral anterior cingulate/MPFCs. Thus, early neuroimaging studies that fell broadly under the remit of self-referential processing tended to reveal activity in cortical midline regions, such as medial prefrontal/paracingulate cortices and in retrosplenial/PCCs. Corroborating these findings with paradigms that are more explicitly self-related are two studies by McGuire and colleagues.[47,48] In the first study, subjects were required to read out loud visually presented words.[47] Crucially, in one condition, subjects' verbal responses were overshadowed by that of an experimenter, such that subjects did not hear their own voice, but instead an "alien" voice. Contrasting this condition with a condition in which they heard their own voice produced activity in MPFC. At first glance, this might appear counterintuitive with respect to self-referential processing in this region, until we read that "subjects reported that they spent much of [this condition] wondering why they were hearing someone else's voice instead of their own."[47]

This observation offers a brief glimpse into the sorts of stimulus-independent thoughts (SIT) that can elicit MPFC activity, and indeed, the second study from this group[48] revealed that the amount of self-reported SIT during various cognitive tasks (across two experiments) correlated with activity during those tasks solely in MPFC. In two follow-up studies, McKiernan and colleagues,[49,50] using event-related fMRI, have shown that the amount of SIT increases linearly with increasing activity in these regions. We shall return to this point later when discussing the broad literature on task-induced deactivations in these regions and their strong functional and anatomical overlap with regions of the brain's default mode network (DMN).[51]

Building on these early studies, Zysset and colleagues describe two experiments in which subjects were required to perform evaluative and semantic judgments[52] and episodic retrieval.[53] Semantic judgments took the form "George Bush is President" (yes/no); evaluative judgments took the form "George Bush is a good president" (yes/no), while episodic retrieval took the form "I voted for George Bush" (yes/no). Contrasting activity between conditions revealed greater anterior medial prefrontal and precuneal involvement during evaluative judgments relative to semantic judgments, and during episodic retrieval relative to semantic judgments. The regions were differentiated such that anterior MPFC was more active during evaluative judgments relative to episodic retrieval, while the reverse pattern occurred in precuneus. Zysset and colleagues interpreted these results to suggest that the critical difference between evaluative and semantic judgments is a reference to the narrative self:[54] "[E]valuative judgments ... are a special type of judgment, in which the internal scale is related to the person's value system (preferences, norms, aesthetic values, etc.)."[52] Thus, controlling for processes associated with memory retrieval and response selection, MPFC and PCC are engaged specifically when task demands encourage reference to some internal evaluative scale. Contributing to the notion that internal evaluative judgments are sufficient to drive MPFC activity, recent research from Johnson and colleagues contrasted activity during an internal subjective decision (ISD) versus activity during an external subjective decision (ESD).[55] During both conditions, subjects were asked to differentiate between two colors that were equidistant in terms of their RGB values from a third target color. The crucial difference between conditions was induced via task instructions; specifically, in the ISD, subjects were asked which color they *preferred* with the target color, while in the ESD, subjects were asked which color was more *similar* to the target color. Activity in the internal versus the external decision was apparent in MPFC and PCC. This result strengthens the idea that one critical factor in eliciting activity in these cortical midline structures (CMS) is evaluation with reference to some internal dimension.

The preceding paragraphs demonstrate a role for what have been termed CMS[56] in instantiating the sorts of mental operations that give rise to a sense of self, such as internal evaluations, episodic memory retrieval, and SIT. While these studies all contribute to the canon of knowledge about the self and the brain structures that are involved in processing some aspects of the self, these aspects of self are tangential to the topic at hand. Our discussion will now turn to those studies that have investigated directly the contribution of self-referential mental processes to brain activity.

Furthering understanding of the cortical structures involved in instantiating the self and bringing the discussion squarely into the realm of directly self-related cognitions, Craik et al.[28] using PET, and Kelley et al.[29] using fMRI, investigated the neural processes associated with encoding trait adjectives with reference to the self and a famous other. In the first study,[28] Craik and colleagues had subjects judge trait adjectives with reference to the self ("How well does this word describe you?"), with reference to a famous other ("How well does this word describe [former Canadian prime minister] Brian Mulroney?"), semantically ("How socially desirable is the trait described by the adjective?"), or phonologically ("How many syllables does this word contain?"). Contrasting activity among all conditions revealed a general trend for greater activity in MPFC in each condition relative to phonological processing. Using a partial least-squares approach, the authors further revealed that activity appeared to be greater in MPFC during self-referential processing than during all other conditions combined. Adapting Craik et al.'s paradigm for event-related fMRI, Kelley and colleagues[29] revealed greater activity in MPFC during self-referential processing relative to both other-referential processing (former President George W. Bush) and a shallow processing control task (case processing). In a further demonstration of medial prefrontal and posterior cingulate activity during self-referential processing, Johnson et al.[57] contrasted activity during the evaluation of potentially self-relevant statements such as "I get angry easily" with activity during evaluation of general semantic statements, such as "10 s is more than a minute." They found significant activity in both regions in this contrast in each of 11 subjects. These studies together formed the bedrock of cognitive neuroscience investigations of self-reflection. Since these early studies, numerous investigations have looked at the neural correlates of self-reflection, to the point where large meta-analyses are possible. One technology for this work can be found at www.neurosynth.org,[58] where researchers can run automated meta-analyses. One such analysis, of 102 neuroimaging studies that use the term

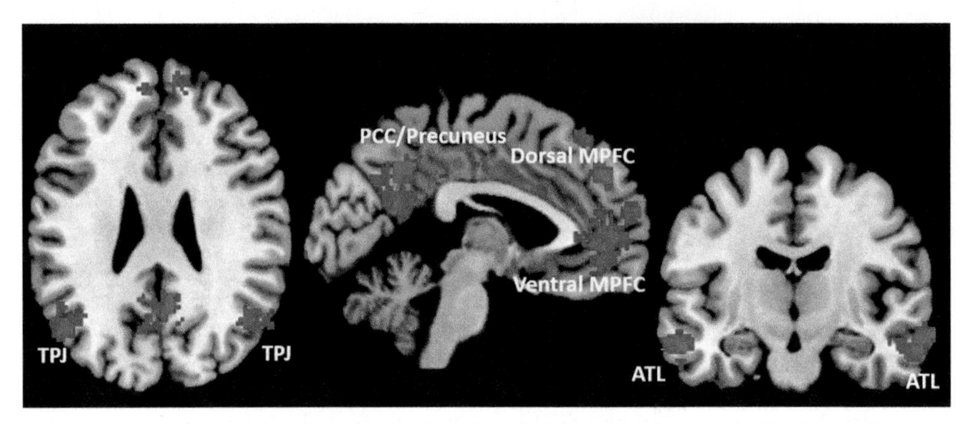

FIGURE 1 Automated meta-analysis of 102 neuroimaging studies differentially using the term "self-referential" from www.neurosynth.org_ENREF_58. Regions more active when studies used the term self-referential than when they did not included the dorsal and ventral medial prefrontal cortex (MPFC), posterior cingulate cortex (PCC) and precuneus, bilateral temporoparietal junction (TPJ), and bilateral anterior temporal lobes (ATL).

"self-referential," reveals the consistent involvement ($p<0.001$, corrected for false discovery rate) of MPFC, PCC, and bilateral anterior temporal lobes and temporoparietal junction (TPJ) (Figure 1). This "reverse inference" map implies that the 102 studies that use the term self-referential are more likely to find activation in those regions than the >9000 studies available in the database that do not use the term "self-referential."

Fossati and colleagues[59] extended these early findings by comparing activity during self-referencing of negative and positive emotional words. Replicating the extant literature, they found greater activity during self versus general semantic-referential processing. However, results did not reveal differences in activity between negative and positive emotional words.

In two complementary studies, Macrae et al.[60] and Fossati et al.[61] investigated the neural correlates of successful encoding and retrieval, respectively, of words processed in relation to the self. In the study of Macrae and colleagues, later successful recall of trait adjectives encoded with reference to the self was associated with greater hippocampal and medial prefrontal activity during incidental memory encoding. The implication is that the hippocampus serves to bind together information in long-term memory, and the regions engaged by a given task are important for memory formation during said task (i.e., greater left inferior prefrontal activity during successful memory encoding of words processed semantically, see Wagner et al.[62]). More simply, greater MPFC activity during encoding reflects greater self-referential processing, and this is reflected in later memory performance. Fossati and colleagues[61] compared brain activity during the memory retrieval of words encoded with reference to the self. Activity in the MPFC during memory retrieval predicted retrieval success; that is, subjects were more likely to correctly remember having seen a given word in the context of greater activity in this region. These results suggest behavioral consequences of medial prefrontal activity such that this region is implicated in passing verbal information into and out of memory, but only when that information is processed in a way that impinges upon the self-concept, rather than in a general semantic manner.

Two further studies controlled for the important potential confound of valence information in interpreting prior results. Although the previous work suggested that MPFC and PCC were specifically related to processing, encoding, and retrieving information about the self, almost all studies had included trait adjectives as stimuli. And since people are notoriously self-serving when evaluating their own personalities ("I'm definitely sincere and not at all lazy"; e.g., Taylor and Brown[63]), it is important to determine whether these activations are in fact related to self-relevance, or rather simply reflect the asymmetric proportions of positively- and negatively-valenced items in the self-relevant and not-self-relevant bins respectively. Moran and colleagues[64] used fMRI and had people rate the self-descriptiveness of 540 trait adjectives that had previously been normed for valence. The authors found that MPFC and PCC were more active when participants viewed high versus low self-relevant words, and this activation was not modulated as a function of valence: both MPFC and PCC's responses were wholly independent of the valence of the words. An independent region in the ventral anterior cingulate cortex (ACC) was more responsive to positive versus negative words, but only for those that were high in self-relevance. This suggests that the CMS encode and retrieve self-relevant information but are not differentially concerned with the valence of that information. In contrast, ventral ACC may be encoding valence information, but only in the context of information that is salient to the self. Phan et al.[65] used pictures instead of words, but found an almost identical pattern of results:

activation in the MPFC was higher when participants explicitly judged self-relatedness rather than pleasantness (valence) and increased linearly with increasing self-relatedness but had no relationship with pleasantness (either positive or negative). These two findings suggest together that activity in these regions occurs as a function of both *process* and *outcome*: when we think about whether information is self-relevant, we activate MPFC and PCC, and over and above this process, we activate these CMS more when the information *is* self-relevant. This correspondence between process and outcome will be developed in more detail as we discuss the overlapping nature of the regions involved in self-reflection and in the DMN.

Further investigations of self-reflection have sought to uncover whether the CMS are engaged when participants are not required to self-reflect. That is, do these regions perform self-relevance computations even when ongoing processing goals are orthogonal to such calculation? Two studies had participants make non-self-relevant judgments about items that had been preselected to be self-relevant to the participants. In the study of Moran and colleagues,[66] greater activity was observed in the CMS when participants passively viewed highly self-relevant information (e.g., their current zip code or social security number). Rameson et al.[67] observed activity in both the CMS and in extended regions associated with salience processing (such as the amygdala) when participants made judgments about whether images contained people, but only when those images depicted scenes that were schematic for each participant (e.g., an athletic scene for a participant that was a member of the track team). Further, Moran et al.[66] did not observe an effect of implicit self-relevance for trait adjectives—items whose self-relevance is arguably less than personal semantic information, such as the name of one's spouse. Taken together with the positive findings from both studies, this observation suggests that the CMS may encode self-relevance only if such information is self-relevant enough to grab attention in the context of an orthogonal processing task. This result can be compared to the results of binaural listening experiments[38] that demonstrate that participants are unable to report any information from an unattended second auditory channel aside from when their own name is presented. Gray et al.[35] expand on this finding by showing that self-relevant information (such as home address) produces a P300 event-related potential indexing attentional orienting when it is presented on a nonattended channel. While most investigations note coactivation of the MPFC and PCC in the service of self-reflection, Moran, Lee, and Gabrieli[68] were able to dissociate activity in these regions by having participants consider physical (e.g., "do I have a beard?") versus psychological aspects of the self (e.g., "am I careful?"). When participants considered psychological aspects of the self (and a close other), the ventral MPFC was more active, which contrasted with PCC that was more active when participants considered physical aspects of themselves or others. The take-home point from these studies is that the CMS likely form part of a network of brain regions that consistently scans the environment for information that is relevant to the self. Of particular note is that these regions become active when encountering self-relevant information that is presented visually, verbally, or auditorially, suggesting a heteromodal representation of self-relevance that in turn argues that these regions are quite downstream of the primary sensory cortices.

Activity in the CMS is observed not just in tasks that encourage explicit cognitive self-referential processing. A major factor eliciting CMS activity has been first-person perspective taking.[69,70] Vogeley and colleagues[71] demonstrated in a virtual reality environment that taking a first-person perspective (counting the number of balls visible to the subject) versus taking a third-person perspective (counting the number of balls visible to a virtual reality character seen in the same stimulus field) yielded increased activity in both CMS. Indeed, Schilbach and colleagues[72] compared neural activity in a virtual reality simulation of a character either looking at the subject or looking at a third character in the environment. This comparison revealed greater MPFC activity when the subject was the object of attention. Thus, not only is the MPFC important for instantiating a sense of first-person perspective taking, but also for recognizing when the self is the target of someone else's attention. Further exploring the relationship between perspective taking and self-reflection, a clever experiment revealed that distinct regions of the MPFC are related to self-reflection and perspective taking.[73] Both dorsal and ventral aspects of *anterior* MPFC were more active when participants reflected on themselves versus another person, whereas a distinct region of *posterior* dorsal MPFC was more active when participants adopted the perspective of a third person relative to their own perspective.[73] What was most exciting about this approach was that the decision and stimuli were identical across first and third person perspectives: participants always decided whether (e.g.,) "kind" described themselves but had to say whether they believed it of themselves (first-person) of whether a predefined close friend believed it of the participant (third-person). Simply changing the simulated decision-maker demonstrated activation in a different aspect of the MPFC. Here we see that self-reflection per se is enough to drive the anterior MPFC regions described previously, but that adopting a different perspective engages slightly different cortical real estate.

Complementary to these findings is a broad literature on Theory of Mind (ToM) processing. ToM refers to the "meta-representational capacity to attribute opinions, perceptions or attitudes to others."[70] Several extensive reviews and meta-analyses have detailed numerous studies that elicit CMS activation when subjects are required to "mind read."[74–78] Most paradigms typically involve subjects inferring the intentions of another individual and answering questions based on short stories detailing the individual's behavior. Chief among these paradigms has been the false belief task,[79,80] which has participants read stories about another person's beliefs that are in conflict with the true state of the world. The participant must inhibit their own perspective to successfully answer a question that probes their knowledge of the story protagonist's belief about the world. In this task, activity is observed in the MPFC, PCC, and in a region on the outer part of the cortex known as the TPJ, among others.[80,81] Activity in the regions of MPFC most commonly seen during self-referential processing has also been associated with the playing of economic games,[82,83] thinking about communicative intentions (deception),[84,85] viewing real-time interactions between third-party individuals,[86] and taking part in a live interaction with another person.[87] Gobbini et al.[88] revealed that this MPFC region is similarly active when participants answer questions about others' false beliefs and infer intentions in the movements of geometric shapes. In our work, we note an overlap in MPFC and PCC when participants are required to answer questions about false beliefs and when they are required to self-reflect.[89] Further investigations have shown a role for both CMS in processing others' intentions when making moral judgments about those others,[90] and in a complementary finding, ventral MPFC damage is associated with reduced attention to others' intentions.[91] In a thorough review, Amodio and Frith[92] detail several differing paradigms that elicit MPFC activity and argue for a putative general-purpose role in thinking about social attributes for MPFC. These authors suggest that MPFC activation in general is most closely associated with social consideration of any kind, rather than self-reflection per se. While the question of whether self-reflection, social reflection, ToM, and other related social phenomena definitively share neural substrates is an interesting one and the center of much current debate, it is outside the main focus of this chapter.

The studies summarized in this section point to the conclusion that the CMS are at least partially dedicated to representing the self. Moran, Kelley, and Heatherton[93] suggest three possible models for the function of the CMS. The first argues that the CMS are specialized for representing social information, with the self being the canonically most powerful social stimulus, and hence activating these regions most consistently. The second argues that these regions (in particular, MPFC) serve as a hub to integrate internal and external information. Such a role would be consistent with high activity during self-reflection, as self-reflection is the quintessential task that requires the integration of what we know with what we are currently observing. The final model suggests that MPFC is responsible for directing conscious thought processes, a position consistent again with its canonical role in self-reflection. An inescapable observation in the literature on self-reflection has been that the CMS are invariably activated both when we reflect on ourselves and when we are resting quietly. The implications of this observation for our understanding of the neural basis of self-reflection are discussed in the next section.

3. THE BRAIN NETWORK FOR SELF-REFLECTION OVERLAPS STRONGLY WITH THE BRAIN NETWORK OBSERVED IN THE RESTING BRAIN

Putting this point another way, almost all fMRI and PET studies observe that activity in the CMS is negative-going with respect to baseline fixation. That is, when participants are engaged in a goal-directed task provided by the experimenter, activity in the MPFC and PCC is typically decreased relative to the comparison condition of resting quietly with eyes open or closed. Several meta-analyses of experiments involving disparate cognitive activation paradigms (e.g., working memory, visual object processing, language tasks, etc.) have revealed a network of brain regions that consistently deactivates in the presence of an external task—the DMN.[94–96] The CMS are both members of the DMN,[51,97] and the functional significance of their high activity during rest is a topic of some controversy. In review, Gusnard[97] argues strongly for the notion of a "physiologic baseline"[98] during rest that reflects ongoing information processing.[99,100] This processing is "spontaneous and virtually continuous, being attenuated only under certain behavioral conditions when such functionality is either less salient or might actually interfere with the implementation of other more salient processing demands."[97]

The fact that when we reflect upon ourselves we do not reduce activity relative to baseline in the CMS stands in stark counterpoint to the findings from almost every other domain examined.[29,60,66] This observation suggests that resting in the MRI or PET scanner might naturally engage just the sorts of processes encouraged when we ask participants "Are you hard-working?"[101–103] Putting this point more forcefully, Singer came to the conclusion that "when man is denied sources of stimulation from the outside, he either produces more inner stimulation or perforce attends more actively to the ever-present stream of imagery or fantasy."[104] Thus, intuitive linkages exist between the physiologic baseline, the DMN, self-referential processing, and daydreaming.

Going beyond the intuitive, several research groups have demonstrated empirical evidence of direct overlap between the brain networks engaged during rest and during self-reflection. An early meta-analysis demonstrated overlap between self-reflection and rest in the MPFC in a small number of studies.[96] D'Argembeau and colleagues[105] used PET to look at the overlap between self-reflection and rest in the brain. By asking participants about their thoughts during each scan, the authors were able to regress the amount of self-referential thoughts against activity during the scan. A region of ventral MPFC during scans in which participants thought about themselves relative to others was more active during the conjunction of self and rest scans and also increased its activity linearly with the amount of thoughts about the self subjects reported having during all scan types. These results provide nice evidence that self-reflection and rest share a neural substrate in the MPFC. Research from fMRI corroborated these findings,[106] showing that DMN activity during the resting period immediately following (implicit) presentation of self-relevant items was increased, suggesting that the self-relevant items prompted self-reflection during the postpresentation rest periods. A study directly contrasted self-reflection and rest against a goal-directed semantic task[107] and showed an association between self and rest in two regions: ventral MPFC and PCC. Further, there were dissociations such that dorsal MPFC was more active when participants self-reflected, and a region in the precuneus was more active during rest than both self-reflection and semantic processing. These findings show that there is a core set of regions (ventral MPFC and PCC) that appears to engage in similar processing, both when we rest and when we self-reflect, but there are indeed distinct regions that are more active respectively during rest and self-reflection. The final piece of evidence regarding rest and self-reflection comes from Mason et al.[108] who trained participants on simple working-memory tasks to a point of automation. Once participants performed the tasks relatively automatically, the authors used thought-sampling to verify that participants were generating SIT during blocks of tasks that were well-practiced. Practiced tasks led to more SIT than novel tasks, and activity in the DMN during practiced tasks was much higher than during novel tasks, even when task performance metrics were identical. Further, an individual participant's tendency to let the mind wander, as measured by the imaginal processes inventory,[104] correlated significantly with activation in the DMN during practiced, relative to novel, tasks. People who are more likely to let their mind wander showed increased activity in the DMN during a simple task in which they were free to let their mind wander.

4. THIS OVERLAP IMPLIES THAT A LARGE PORTION OF OUR BRAIN'S METABOLIC BUDGET DURING REST IS GIVEN OVER TO CONSIDERATION OF OURSELVES AND OTHERS

But what is the cognitive significance of such findings? Given that cognitive neuroscientists aim to use brain-level data to inform theorizing at the level of psychology (or cognition), what can we infer from findings that show that our psychological resting state consists mostly of thoughts about the self? In this section, I will summarize evidence that differences in default mode activity can be observed in at least two disorders, and that these differences map onto psychological differences observed in those conditions.

4.1 Default Mode Network Differences in Disorder

Most cognitive neuroscience studies, along with most psychology studies, use young, healthy participants. These studies have arrived at a general conception that the CMS form the core of the DMN, an observation that is among the most replicable in all of cognitive neuroscience. Here I will detail populations and areas where results diverge from this standard model and discuss the implications of these findings for a neuroscience of self-reflection.

4.1.1 Schizophrenia

One area that has received a great deal of attention is schizophrenia. Schizophrenia is a common disorder affecting social processing and self-related cognition, among many other functions. Zhou et al.[109] found evidence for increased resting state connectivity in the DMN in a schizophrenic group, along with increased anticorrelation between the DMN and another brain network known as the task-positive network (TPN). Such anticorrelations are found in healthy individuals[110] and are thought to reflect the ongoing switching of attention between the external and internal words, mediated respectively by the TPN and DMN. Zhou et al.[109] suggest that their findings reflect greater competition than is usual between the DMN and TPN, which may serve as a neural substrate for the increased mentalizing and overgeneralization of self-relevance seen in schizophrenia. Whitfield-Gabrieli et al.[111] report that schizophrenia patients and their first-degree relatives do not show the typical DMN deactivations in working-memory tasks as seen in controls, and that the degree to which participants did deactivate the DMN in service of a goal-directed task was predictive of task performance. Further, patients showed abnormally larger functional connectivity within the DMN during task performance (recall that the DMN is typically decoupled during external task performance), and this greater connectivity was associated with the degree of psychopathology in the patients. Whitfield-Gabrieli et al.[111] also corroborated the findings of Zhou and colleagues, in that there were reduced anticorrelations between MPFC, a part of the DMN, and the lateral prefrontal cortex, a part of the TPN, in patients relative to controls. Pomarol-Clotet et al.[112] report a similar finding of a failure to deactivate a region of ventral MPFC in schizophrenics in a goal-directed task. Such findings converge on the notion that unusually high activity and connectivity in the DMN in schizophrenia may underlie both cognitive (poor working memory) and social (excessive self-relevance and self-focus) aspects of the disorder. Liemburg and colleagues[113] probed this question further by comparing DMN connectivity between two groups of participants with schizophrenia who differed on their level of insight into their illness. Patients with low insight showed reduced functional connectivity within and between MPFC and PCC relative to patients with high insight. This result can be interpreted to mean that reduced illness insight is associated with greater impairment in self-reflection than is usually seen in schizophrenia.

Our own work using fMRI[114] showed that patients with schizophrenia activated the MPFC region less than did controls during self-reflection compared with a semantic control task. This finding fits well with the prior work showing less deactivation for the MPFC in schizophrenia, implying that the sort of ongoing self-reflective processes engaged during rest and self-reflection may not be suppressed in typical goal-directed tasks, or in the general semantic processing task used in our work.[114] Supporting this view, other research demonstrated that patients with schizophrenia did not show the classic self-reference effect in memory, arguing that this result might have been obtained due to impaired MPFC functioning in the patient group.[115]

4.1.2 Autism

In the domain of autism spectrum disorders (ASDs), researchers report almost diametrically opposite findings regarding activity and connectivity within the DMN. ASDs are common neurodevelopmental disorders characterized by impairments in communication and social functioning. In one such paper, participants completed a simple number task designed to suppress default mode activity, and this activation was compared with rest.[116] Control participants showed the expected deactivation, reducing activity in the CMS in response to the goal-directed math task relative to rest, whereas individuals with ASD failed to show such a deactivation. Further investigation of the ventral MPFC showed that this region's activity during number processing versus rest correlated positively with autism symptom severity in the ASD group; individuals with more severe autism symptoms actually had less activity during rest than during the goal-directed task. While at first blush, ASD and schizophrenia both present with a "failure to deactivate' in the DMN in response to goal-directed tasks, the difference is in the details: people with schizophrenia appear to have higher activity and connectivity at rest than controls,[111] whereas people with ASDs tend to have lower activity in the DMN regions at rest.[116] Further, individuals with ASD present with altered DMN functional connectivity (reduced MPFC-PCC connectivity), but not altered TPN connectivity,[117] suggesting a neural substrate for the pattern of impaired social functioning but intact general cognition observed in ASD. These observations together argue that ASD may be characterized by a reduced self-focus, in opposition to the excessive self-focus and hyperactivity in schizophrenia. A couple of papers tackled this question directly[118,119] and found indeed that the standard self-reference effect observed in control participants was largely absent in individuals with ASD. That is, people with ASD did not remember more items that were encoded in reference to the self than items that were encoded through general semantic processing. Further, control participants showed a greater increase in memory for self-related items than for those related to a well-known other (Harry Potter) than did participants with ASD.[118] Lombardo and colleagues[119] followed up on their own work and showed that the neural correlates of self-reflection in ASD were also different, in ways that we might have expected based on our reading of the literature on default mode activity and connectivity in autism. In particular, ASD individuals showed equal activation of MPFC in response to both self and other (British Queen) trials. Follow-up analyses showed that the amount of activation difference in the MPFC between self and other was negatively predicted by autism symptom severity: people with higher severity differentiated between self and other less in this region than people with lower sensitivity. This region generally shows greater activation for self-reflection versus reflection about a famous but not personally known other (three extensive meta-analyses support this point).[120–122] These behavioral and neural data support the idea that ASD is associated with a reduced self-focus and suggest the possibility that other cognitive functions that involve the DMN (such as prospective memory and navigation)[123] may also be compromised in ASDs.

These findings in schizophrenia and ASD have been driven by discoveries in cognitive neuroscience that illuminated a path for researchers looking to better understand the nature of disturbances in both disorders. Drawing on the robust replicability of the DMN, researchers have been able to demonstrate that activity and connectivity differences in nodes and networks associated with self and other reflection correspond to psychological differences in autism and schizophrenia and have brought our understanding of these disorders squarely into the realm of the cognitive processes that are implicated. In turn, investigating these disorders has enriched our basic science understanding, confirming many of the suspicions about how *typical* self-related and social cognition are organized at the level of the neural circuits implementing them. In sum, the advent of functional neuroimaging has allowed for the discovery that the association areas largely associate social information, that much of what we do when resting is associate social information with ourselves, that ASD and schizophrenia may represent two extremes of a spectrum of self-focus (with reduced self-focus in autism and heightened self-focus in schizophrenia), and finally, that there exists an always-on brain network dedicated to representing, simulating, and understanding ourselves and others.

In the final Section, I will discuss findings from neuropsychology, the study of how cognitive processes are affected by disease or injury to the brain, that impinge upon our understanding of how the brain represents ourselves.

5. LESIONS TO THE BRAIN NETWORK ASSOCIATED WITH SELF-REFLECTION CAUSE DIFFICULTIES IN UNDERSTANDING OURSELVES

In the previous sections, I covered a significant amount of research that related activity in the CMS to the process and outcome of self-reflection. Since fMRI and PET data provide evidence that is correlational, it is important to consider data that can go beyond such observations to determine whether activity in these regions is *necessary* for reflecting on ourselves. One such example that used neuroimaging evidence as its foundation investigated the effects of damage to a region of the MPFC that had been specifically linked to self-reflection in prior research. Carefully selecting their patient group to have damage overlapping in the MPFC region of interest in Kelley et al.'s work,[29] Philippi and her colleagues[124] demonstrated that this patient group did not show the self-reference effect in memory. That is, these patients did not have a memorial advantage for words they had encoded in reference to the self. This result provides clear evidence that the MPFC is necessary for self-reflection but is mute regarding other aspects of self. These researchers followed up with a case study of an individual who suffered extensive anterior cingulate, insula, and MPFC damage due to herpes simplex encephalitis[125] and demonstrated that he did not suffer from any problems of self-awareness (as measured by self-face recognition and consistency of recollection of his own personality). Such findings converge to suggest that the MPFC's role in self-reflection is more high-level than basic self-awareness and may involve the construction of episodic information about ourselves based on memory.[93]

Other neuropsychological evidence corroborates the data from neuroimaging cited earlier that supported a partial distinction between self and other at the level of semantic memory.[30,31] Over the course of a number of investigations into the intact and compromised memory systems of several patient groups, researchers have demonstrated that episodic and semantic memory are dissociable with respect to the self. In a comprehensive review of their findings, Klein, Rozendal, and Cosmides[31] detailed several disparate cases of individuals who show strikingly dissociable patterns of deficit. The first case concerned a woman (WJ) who had suffered a traumatic brain injury (TBI) in college.[126] Because of her TBI, WJ was unable to recall any recent episodic autobiographical memories when cued with familiar nouns. Over a period of 4 weeks following the injury, her amnesia remitted, and she made a full recovery of her episodic memory faculties. Interestingly, the authors also had WJ rate herself on a series of trait adjectives immediately following the accident and again after the amnesia had remitted. Her ratings of herself did not differ between her amnestic and recovered periods. Since she was unable to produce any episodic memories of her traits during her period of amnesia, she must have been relying on a semantic memory system to assess her traits. Further differentiating self from other, there is evidence of several cases demonstrating dissociations within the domain of semantic memory.[31] It appears that self and other forms of general semantic memory are dissociable in certain patient groups, and individuals can retain their ability to know what they are like in the presence of severely compromised general semantic knowledge.

The evidence from neuropsychology indicates that within the broad domain of self, the ability to self-reflect upon one's own traits and the ability to perform this task accurately are skills dissociable from other forms of memory and are not dependent upon episodic retrieval processes. Other conclusions warranted from these studies are that the MPFC is necessary for self-reflection, but not self-awareness, suggesting a processing role in the MPFC that is (a) downstream from basic encoding processes, (b) relatively far removed from sensory perception, and (c) critical for the accurate encoding and retrieval of information about the self.

6. FUTURE DIRECTIONS

The work covered in this chapter has combined insights from cognitive psychology, neuropsychology, and cognitive neuroscience to arrive at the conclusion that there are dedicated neural circuits for the processing of information about the self, that these neural circuits are often engaged when we process information about others (at least intimately known others), and that damage to these neural circuits results in deficits that largely hew to the self-reflective cognitive processes that researchers have inferred to be subserved by such neural circuits. Several open questions remain, however, that future research efforts might aim to shed light upon. First, what therapeutic agents (be they cognitive or pharmacological) might be designed to improve people's ability to reflect upon themselves in the distinct and divergent disorders of ASD and schizophrenia? Both disorders present very differently, yet with a similar neural substrate that causes a reduction in the ability to encode information relative to the self. Such deficits in the self-reference effect in memory point to a larger memorial encoding problem that may lead to failures in insight (and hence socialization) in both disorders. Second, what is the evolutionary significance of a network dedicated to self-reflection that consumes a large portion of the brain's energy budget? Why is it that humans have evolved a powerful system for self-reflection that seemingly sets us apart from other animals? Third, what educational interventions might be designed to take advantage of the natural tendency to mind-wander and self-reflect to improve educational outcomes across the board?

7. CONCLUSIONS

This chapter has argued for the notion that the MPFC and PCC together form a core network dedicated to reflecting upon ourselves, encoding and retrieving information about ourselves, and which is at least partially associated with similar decisions about other social targets. The preponderance of cognitive neuroscience research in this area has proceeded through exploring the boundaries of these cognitive processes and has erected an impressive edifice of knowledge about how humans are able to reflect upon ourselves, and the ensuing effects of such self-reflection. Through this knowledge, scientists have been able to create a language that explores the deficits in self-reflection associated with autism, schizophrenia, and neurological insult, and in turn, suggests ways in which we might mitigate the debilitating effects of such disorders. Many critics of neuroimaging research[10] have suggested that the data derived from functional imaging methods are unable to adjudicate between competing psychological theories (or even inform psychological theory) and thus form an expensive epiphenomenological curiosity that excludes real cognitive research that is better placed to uncover regularities about cognition. While neuroimaging research is undoubtedly expensive, I argue that the insights summarized in this chapter may never have been gained from behavioral methods alone (certainly in the case of the "self-is-special" debate). New knowledge about how autism and schizophrenia differ in their neural organization of self-reflection may well lead to better therapeutic targets for these disorders that may ultimately improve the quality of life of millions of individuals affected by them.

References

1. *Pausanius, Descriptions of Greece*. London: Heinemann; 1918.
2. Philebus RC. *Suda on Line*; 2002. http://www.stoa.org/sol-entries/gamma/334. Accessed 12.01.14.
3. James W. *The Principles of Psychology*. New York, NY: Dover Publications; 1890.
4. Damasio AR. *The Feeling of What Happens: Body and Emotion in the Making of Consciousness*. New York, NY: Harcourt Brace; 1999.
5. Tulving E. Memory and consciousness. *Can Psychol*. 1985; 26(1):1–12.
6. LeDoux J. *Synaptic Self: How Our Brains Become Who We Are*. New York, NY: Viking Press; 2002.
7. Watson JB. Psychology as the behaviorist views it. *Psychol Rev*. May 1913;20(2):158–177.
8. Henson R. What can functional neuroimaging tell the experimental psychologist? *Q J Exp Psychol A*. February 2005;58(2):193–233.
9. Poldrack RA. Can cognitive processes be inferred from neuroimaging data? *Trends Cogn Sci*. February 2006;10(2):59–63.
10. Coltheart M. What has functional neuroimaging told us about the mind (so far)? *Cortex*. April 2006;42(3):323–331.
11. Gul F, Pesendorfer W. The case for mindless economics. In: Caplin A, Shotter A, eds. *The Foundations of Positive and Normative Economics*. Oxford, England: Oxford University Press; 2008.
12. McCabe DP, Castel AD. Seeing is believing: the effect of brain images on judgments of scientific reasoning. *Cognition*. April 2008;107(1):343–352.
13. Weisberg DS, Keil FC, Goodstein J, Rawson E, Gray JR. The seductive allure of neuroscience explanations. *J Cogn Neurosci*. March 2008;20(3):470–477.
14. Moran JM, Zaki J. Functional neuroimaging and psychology: what have you done for me lately? *J Cogn Neurosci*. June 2013;25(6):834–842.
15. Gillihan SJ, Farah MJ. Is self special? A critical review of evidence from experimental psychology and cognitive neuroscience. *Psychol Bull*. January 2005;131(1):76–97.
16. Rogers TB, Kuiper NA, Kirker WS. Self-reference and encoding of personal information. *J Pers Soc Psychol*. 1977;35(9):677–688.
17. Markus H. Self-schemata and processing information about self. *J Pers Soc Psychol*. 1977;35(2):63–78.
18. Bower GH, Gilligan SG. Remembering information related to ones-self. *J Res Pers*. 1979;13(4):420–432.
19. Greenwald AG, Banaji MR. The self as a memory system - powerful, but ordinary. *J Pers Soc Psychol*. July 1989;57(1):41–54.
20. Kihlstrom JF, Cantor N. Mental representations of the self. In: Berkowitz L, ed. *Advances in Experimental Social Psychology*. Vol. 17. New York: Academic Press; 1984.
21. Maki RH, Mccaul KD. The effects of self-reference versus other reference on the recall of traits and nouns. *Bull Psychon Soc*. 1985;23(3):169–172.
22. Klein SB, Kihlstrom JF. Elaboration, organization, and the self-reference effect in memory. *J Exp Psychol Gen*. March 1986;115(1):26–38.
23. Klein SB, Loftus J. The nature of self-referent encoding - the contributions of elaborative and organizational processes. *J Pers Soc Psychol*. July 1988;55(1):5–11.

24. Kihlstrom JFA, Klein SB. The self as a knowledge structure. In: 2nd ed. Wyer RS, Srull TK, eds. *Handbook of Social Cognition*. Vol. 1. Hillsdale, NJ: Erlbaum; 1994:153–208.
25. Linville P, Carlston DE. Social cognition of the self. In: Devine PG, Hamilton DL, Ostrom TM, eds. *Social Cognition: Impact on Social Psychology*. San Diego: Academic Press; 1994:143–193.
26. Greenwald AG, Pratkanis AR. The self. In: Wyer RS, Srull TK, eds. *Handbook of Social Cognition*. Hillsdale, NJ: Erlbaum; 1984:129–178.
27. Symons CS, Johnson BT. The self-reference effect in memory: a meta-analysis. *Psychol Bull*. May 1997;121(3):371–394.
28. Craik FIM, Moroz TM, Moscovitch M, et al. In search of the self: a positron emission tomography study. *Psychol Sci*. January 1999;10(1):26–34.
29. Kelley WM, Macrae CN, Wyland CL, Caglar S, Inati S, Heatherton TF. Finding the self? An event-related fMRI study. *J Cogn Neurosci*. July 1, 2002;14(5):785–794.
30. Klein SB, Chan RL, Loftus J. Independence of episodic and semantic self-knowledge: the case from autism. *Soc Cogn*. 1999; 17(4):413–436.
31. Klein SB, Cosmides L, Costabile KA, Mei L. Is there something special about the self? A neuropsychological case study. *J Res Pers*. October 2002;36(5):490–506.
32. Bargh JA. Attention and automaticity in the processing of self-relevant information. *J Pers Soc Psychol*. 1982;43(3):425–436.
33. Bargh JA, Chartrand TL. The unbearable automaticity of being. *Am Psychol*. July 1999;54(7):462–479.
34. Miller GA. The magical number 7, plus or minus 2-some limits on our capacity for processing information. *Psychol Rev*. 1956;63(2):81–97.
35. Gray HM, Ambady N, Lowenthal WT, Deldin P. P300 as an index of attention to self-relevant stimuli. *J Exp Soc Psychol*. March 2004;40(2):216–224.
36. Postman L, Bruner JS, Mcginnies E. Personal values as selective factors in perception. *J Abnorm Soc Psych*. 1948;43(2):142–154.
37. Kelly G. *The Psychology of Personal Constructs*. New York: Norton; 1955.
38. Moray N. Attention in dichotic-listening—affective cues and the influence of instructions. *Q J Exp Psychol*. 1959;11(1):56–60.
39. Norman KA, Polyn SM, Detre GJ, Haxby JV. Beyond mind-reading: multi-voxel pattern analysis of fMRI data. *Trends Cogn Sci*. September 2006;10(9):424–430.
40. Kanwisher N, McDermott J, Chun MM. The fusiform face area: a module in human extrastriate cortex specialized for face perception. *J Neurosci*. June 1, 1997;17(11):4302–4311.
41. Hanson SJ, Halchenko YO. Brain reading using full brain support vector machines for object recognition: there is no "face" identification area. *Neural Comput*. February 2008;20(2):486–503.
42. deCharms RC, Christoff K, Glover GH, Pauly JM, Whitfield S, Gabrieli JD. Learned regulation of spatially localized brain activation using real-time fMRI. *Neuroimage*. January 2004; 21(1):436–443.
43. Hinds O, Thompson TW, Ghosh S, et al. Roles of default-mode network and supplementary motor area in human vigilance performance: evidence from real-time fMRI. *J Neurophysiol*. March 2013;109(5):1250–1258.
44. Andreasen NC, O'Leary DS, Cizadlo T, Arndt S, Rezai K, Watkins GL, et al. Remembering the past: two facets of episodic memory explored with positron emission tomography. *Am J Psychiatry*. November 1995;152(11):1576–1585.
45. Lane RD, Fink GR, Chau PML, Dolan RJ. Neural activation during selective attention to subjective emotional responses. *Neuroreport*. December 22, 1997;8(18):3969–3972.
46. Fink GR, Markowitsch HJ, Reinkemeier M, Bruckbauer T, Kessler J, Heiss WD. Cerebral representation of one's own past: neural networks involved in autobiographical memory. *J Neurosci*. July 1, 1996;16(13):4275–4282.
47. McGuire PK, Silbersweig DA, Frith CD. Functional neuroanatomy of verbal self-monitoring. *Brain*. June 1996;119:907–917.
48. McGuire PK, Paulesu E, Frackowiak RSJ, Frith CD. Brain activity during stimulus independent thought. *Neuroreport*. September 2, 1996;7(13):2095–2099.
49. McKiernan KA, D'Angelo BR, Kaufman JN, Binder JR. Interrupting the "stream of consciousness": an fMRI investigation. *Neuroimage*. February 15, 2006;29(4):1185–1191.
50. McKiernan KA, Kaufman JN, Kucera-Thompson J, Binder JR. A parametric manipulation of factors affecting task-induced deactivation in functional neuroimaging. *J Cogn Neurosci*. April 1, 2003;15(3):394–408.
51. Raichle ME, MacLeod AM, Snyder AZ, Powers WJ, Gusnard DA, Shulman GL. A default mode of brain function. *Proc Natl Acad Sci USA*. January 16, 2001;98(2):676–682.
52. Zysset S, Huber O, Ferstl E, von Cramon DY. The anterior frontomedian cortex and evaluative judgment: an fMRI study. *Neuroimage*. April 2002;15(4):983–991.
53. Zysset S, Huber O, Samson A, Ferstl EC, von Cramon DY. Functional specialization within the anterior medial prefrontal cortex: a functional magnetic resonance imaging study with human subjects. *Neurosci Lett*. January 2, 2003;335(3):183–186.
54. Gallagher HL, Happe F, Brunswick N, Fletcher PC, Frith U, Frith CD. Reading the mind in cartoons and stories: an fMRI study of 'theory of mind' in verbal and nonverbal tasks. *Neuropsychologia*. 2000;38(1):11–21.
55. Johnson SC, Schmitz TW, Kawahara-Baccus TN, et al. The cerebral response during subjective choice with and without self-reference. *J Cogn Neurosci*. 2005;17(12):1897–1906.
56. Northoff G, Bermpohl F. Cortical midline structures and the self. *Trends Cogn Sci*. 2004;8(3):102–107.
57. Johnson SC, Baxter LC, Wilder LS, Pipe JG, Heiserman JE, Prigatano GP. Neural correlates of self-reflection. *Brain*. August 2002;125(Pt 8):1808–1814.
58. Yarkoni T, Poldrack RA, Nichols TE, Van Essen DC, Wager TD. Large-scale automated synthesis of human functional neuroimaging data. *Nat Methods*. August 2011;8(8):665–U695.
59. Fossati P, Hevenor SJ, Graham SJ, et al. In search of the emotional self: an fMRI study using positive and negative emotional words. *Am J Psychiat*. November 2003;160(11):1938–1945.
60. Macrae CN, Moran JM, Heatherton TF, Banfield JF, Kelley WM. Medial prefrontal activity predicts memory for self. *Cereb Cortex*. June 2004;14(6):647–654.
61. Fossati P, Hevenor SJ, Lepage M, et al. Distributed self in episodic memory: neural correlates of successful retrieval of self-encoded positive and negative personality traits. *Neuroimage*. 2004;22(4):1596–1604.
62. Wagner AD, Schacter DL, Rotte M, et al. Building memories: remembering and forgetting of verbal experiences as predicted by brain activity. *Science*. August 21, 1998;281(5380):1188–1191.
63. Taylor SE, Brown JD. Illusion and well-being—a social psychological perspective on mental-health. *Psychol Bull*. March 1988;103(2):193–210.
64. Moran JM, Macrae CN, Heatherton TF, Wyland CL, Kelley WM. Neuroanatomical evidence for distinct cognitive and affective components of self. *J Cogn Neurosci*. September 2006;18(9):1586–1594.
65. Phan KL, Taylor SF, Welsh RC, Ho SH, Britton JC, Liberzon I. Neural correlates of individual ratings of emotional salience: a trial-related fMRI study. *Neuroimage*. 2004;21(2):768–780.
66. Moran JM, Heatherton TF, Kelley WM. Modulation of cortical midline structures by implicit and explicit self-relevance evaluation. *Soc Neurosci UK*. 2009;4(3):197–211.
67. Rameson LT, Morelli SA, Lieberman MD. The neural correlates of empathy: experience, automaticity, and prosocial behavior. *J Cogn Neurosci*. January 2012;24(1):235–245.

68. Moran JM, Lee SM, Gabrieli JD. Dissociable neural systems supporting knowledge about human character and appearance in ourselves and others. *J Cogn Neurosci*. September 2011;23(9): 2222–2230.
69. Newen A, Vogeley K. Self-representation: searching for a neural signature of self-consciousness. *Conscious Cogn*. December 2003;12(4):529–543.
70. Vogeley K, Fink GR. Neural correlates of the first-person-perspective. *Trends Cogn Sci*. January 2003;7(1):38–42.
71. Vogeley K, May M, Ritzl A, Falkai P, Zilles K, Fink GR. Neural correlates of first-person perspective as one constituent of human self-consciousness. *J Cogn Neurosci*. June 2004;16(5):817–827.
72. Schilbach L, Wohlschlaeger AM, Kraemer NC, et al. Being with virtual others: neural correlates of social interaction. *Neuropsychologia*. 2006;44(5):718–730.
73. D'Argembeau A, Ruby P, Collette F, et al. Distinct regions of the medial prefrontal cortex are associated with self-referential processing and perspective taking. *J Cogn Neurosci*. June 2007; 19(6):935–944.
74. Frith CD, Frith U. Interacting minds—a biological basis. *Science*. November 26, 1999;286(5445):1692–1695.
75. Frith U, Frith C. The biological basis of social interaction. *Curr Dir Psychol Sci*. October 2001;10(5):151–155.
76. Gallagher HL, Frith CD. Functional imaging of 'theory of mind'. *Trends Cogn Sci*. February 2003;7(2):77–83.
77. Stuss DT, Gallup GG, Alexander MP. The frontal lobes are necessary for 'theory of mind'. *Brain*. February 2001;124:279–286.
78. Van Overwalle F, Baetens K. Understanding others' actions and goals by mirror and mentalizing systems: a meta-analysis. *Neuroimage*. November 15, 2009;48(3):564–584.
79. Zaitchik D. When representations conflict with reality—the preschoolers problem with false beliefs and false photographs. *Cognition*. April 1990;35(1):41–68.
80. Saxe R, Kanwisher N. People thinking about thinking people. The role of the temporo-parietal junction in "theory of mind". *Neuroimage*. August 2003;19(4):1835–1842.
81. Fletcher PC, Happé F, Frith U, et al. Other minds in the brain: a functional imaging study of "theory of mind" in story comprehension. *Cognition*. November 1995;57(2):109–128.
82. McCabe K, Houser D, Ryan L, Smith V, Trouard T. A functional imaging study of cooperation in two-person reciprocal exchange. *Proc Natl Acad Sci USA*. September 25, 2001;98(20): 11832–11835.
83. Rilling JK, Sanfey AG, Aronson JA, Nystrom LE, Cohen JD. The neural correlates of theory of mind within interpersonal interactions. *Neuroimage*. August 2004;22(4):1694–1703.
84. Grezes J, Frith C, Passingham RE. Brain mechanisms for inferring deceit in the actions of others. *J Neurosci*. June 16, 2004;24(24):5500–5505.
85. Walter H, Adenzato M, Ciaramidaro A, Enrici I, Pia L, Bara BG. Understanding intentions in social interaction: the role of the anterior paracingulate cortex. *J Cogn Neurosci*. December 2004;16(10):1854–1863.
86. Saxe R, Carey S, Kanwisher N. Understanding other minds: linking developmental psychology and functional neuroimaging. *Annu Rev Psychol*. 2004;55:87–124.
87. Redcay E, Dodell-Feder D, Pearrow MJ, et al. Live face-to-face interaction during fMRI: a new tool for social cognitive neuroscience. *Neuroimage*. May 1, 2010;50(4):1639–1647.
88. Gobbini MI, Koralek AC, Bryan RE, Montgomery KJ, Haxby JV. Two takes on the social brain: a comparison of theory of mind tasks. *J Cogn Neurosci*. November 2007;19(11):1803–1814.
89. Saxe R, Moran JM, Scholz J, Gabrieli J. Overlapping and non-overlapping brain regions for theory of mind and self reflection in individual subjects. *Soc Cogn Affect Neurosci*. December 2006;1(3):229–234.
90. Young L, Cushman F, Hauser M, Saxe R. The neural basis of the interaction between theory of mind and moral judgment. *Proc Natl Acad Sci USA*. May 15, 2007;104(20):8235–8240.
91. Young L, Bechara A, Tranel D, Damasio H, Hauser M, Damasio A. Damage to ventromedial prefrontal cortex impairs judgment of harmful intent. *Neuron*. March 25, 2010;65(6):845–851.
92. Amodio DM, Frith CD. Meeting of minds: the medial frontal cortex and social cognition. *Nat Rev Neurosci*. April 2006;7(4):268–277.
93. Moran JM, Kelley WM, Heatherton TF. What can the organization of the brain's default mode network tell us about self-knowledge? *Front Hum Neurosci*. 2013;7:391.
94. Mazoyer B, Zago L, Mellet E, et al. Cortical networks for working memory and executive functions sustain the conscious resting state in man. *Brain Res Bull*. February 2001;54(3):287–298.
95. Shulman GL, Fiez JA, Corbetta M, Buckner RL, Miezin FM, Raichle ME, et al. Common blood flow changes across visual tasks. 2. Decreases in cerebral cortex. *J Cogn Neurosci*. September 1997;9(5):648–663.
96. Wicker B, Ruby P, Royet JP, Fonlupt P. A relation between rest and the self in the brain? *Brain Res Brain Res Rev*. 2003;43(2):224–230.
97. Gusnard DA. Being a self: considerations from functional imaging. *Conscious Cogn*. December 2005;14(4):679–697.
98. Raichle ME, Mintun MA. Brain work and brain imaging. *Annu Rev Neurosci*. 2006;29:449–476.
99. Tononi GE, Edelman GM. Consciousness and the integration of information in the brain. In: Jasper HH, Descarries L, Castelluci VF, Rossignol S, eds. *Consciousness: At the Frontiers of Neuroscience*. Philadelphia: Lippincott-Raven; 1998.
100. Gusnard DA, Raichle ME. Searching for a baseline: functional imaging and the resting human brain. *Nat Rev Neurosci*. October 2001;2(10):685–694.
101. Antrobus JS, Singer JL, Greenber S. Studies in stream of consciousness—experimental enhancement and suppression of spontaneous cognitive processes. *Percept Mot Ski*. 1966;23(2):399.
102. Singer JL, Mccraven VG. Some characteristics of adult daydreaming. *J Psychol*. 1961;51(1):151–164.
103. Singer JL, Antrobus JS. Daydreaming, imaginal processes, and personality: a normative study. In: Sheehan PW, ed. *The Nature and Function of Imagery*. New York: Academic Press; 1972:175–202.
104. Singer JL. *Daydreaming: An Introduction to the Experimental Study of Inner Experience*. New York: Random House; 1966.
105. D'Argembeau A, Collette F, Van der Linden M, et al. Self-referential reflective activity and its relationship with rest: a PET study. *Neuroimage*. April 1, 2005;25(2):616–624.
106. Schneider FC, Royer A, Grosselin A, et al. Modulation of the default mode network is task-dependant in chronic schizophrenia patients. *Schizophr Res*. February 2011;125(2–3):110–117.
107. Whitfield-Gabrieli S, Moran JM, Nieto-Castanon A, Triantafyllou C, Saxe R, Gabrieli JD. Associations and dissociations between default and self-reference networks in the human brain. *Neuroimage*. March 1, 2011;55(1):225–232.
108. Mason MF, Norton MI, Van Horn JD, Wegner DM, Grafton ST, Macrae CN. Wandering minds: the default network and stimulus-independent thought. *Science*. January 19, 2007;315(5810): 393–395.
109. Zhou Y, Liang M, Tian L, et al. Functional disintegration in paranoid schizophrenia using resting-state fMRI. *Schizophr Res*. December 2007;97(1–3):194–205.
110. Fox MD, Snyder AZ, Vincent JL, Corbetta M, Van Essen DC, Raichle ME. The human brain is intrinsically organized into dynamic, anticorrelated functional networks. *Proc Natl Acad Sci USA*. July 5, 2005;102(27):9673–9678.
111. Whitfield-Gabrieli S, Thermenos HW, Milanovic S, et al. Hyperactivity and hyperconnectivity of the default network in schizophrenia and in first-degree relatives of persons with schizophrenia. *Proc Natl Acad Sci USA*. January 27, 2009;106(4):1279–1284.

112. Pomarol-Clotet E, Salvador R, Sarró S, et al. Failure to deactivate in the prefrontal cortex in schizophrenia: dysfunction of the default mode network? *Psychol Med*. August 2008;38(8):1185–1193.
113. Liemburg EJ, van der Meer L, Swart M, et al. Reduced connectivity in the self-processing network of schizophrenia patients with poor insight. *PLoS One*. August 9, 2012;7(8).
114. Holt DJ, Cassidy BS, Andrews-Hanna JR, et al. An anterior-to-posterior shift in midline cortical activity in schizophrenia during self-reflection. *Biol Psychiatry*. March 1, 2011;69(5):415–423.
115. Harvey PO, Lee J, Horan WP, Ochsner K, Green MF. Do patients with schizophrenia benefit from a self-referential memory bias? *Schizophr Res*. April 2011;127(1–3):171–177.
116. Kennedy DP, Redcay E, Courchesne E. Failing to deactivate: resting functional abnormalities in autism. *Proc Natl Acad Sci USA*. May 23, 2006;103(21):8275–8280.
117. Kennedy DP, Courchesne E. The intrinsic functional organization of the brain is altered in autism. *Neuroimage*. February 15, 2008;39(4):1877–1885.
118. Lombardo MV, Barnes JL, Wheelwright SJ, Baron-Cohen S. Self-referential cognition and empathy in autism. *PLoS One*. 2007;2(9):e883.
119. Lombardo MV, Chakrabarti B, Bullmore ET, et al. Atypical neural self-representation in autism. *Brain*. December 14, 2009:611–624.
120. Denny BT, Kober H, Wager TD, Ochsner KN. A meta-analysis of functional neuroimaging studies of self- and other judgments reveals a spatial gradient for mentalizing in medial prefrontal cortex. *J Cogn Neurosci*. August 2012;24(8):1742–1752.
121. Murray RJ, Debbane M, Fox PT, Bzdok D, Eickhoff SB. Functional connectivity mapping of regions associated with self- and other-processing. *Hum Brain Mapp*. 2015;36(4):1304–1324.
122. Murray RJ, Schaer M, Debbane M. Degrees of separation: a quantitative neuroimaging meta-analysis investigating self-specificity and shared neural activation between self- and other-reflection. *Neurosci Biobehav R*. March 2012;36(3):1043–1059.
123. Spreng RN, Mar RA, Kim AS. The common neural basis of autobiographical memory, prospection, navigation, theory of mind, and the default mode: a quantitative meta-analysis. *J Cogn Neurosci*. March 2009;21(3):489–510.
124. Philippi CL, Duff MC, Denburg NL, Tranel D, Rudrauf D. Medial PFC damage abolishes the self-reference effect. *J Cogn Neurosci*. 2012;24(2):475–481.
125. Philippi CL, Feinstein JS, Khalsa SS, Damasio A, Tranel D, Landini G, et al. Preserved self-awareness following extensive bilateral brain damage to the insula, anterior cingulate, and medial prefrontal cortices. *PLoS One*. August 22, 2012;7(8).
126. Klein SB, Loftus J, Kihlstrom JF. Self-knowledge of an amnesic patient: toward a neuropsychology of personality and social psychology. *J Exp Psychol Gen*. September 1996;125(3):250–260.

SECTION IV

BRAIN IMAGING PERSPECTIVES ON AFFECT AND EMOTION REGULATION

CHAPTER

11

The Neural Basis of Frustration State

Rongjun Yu

Department of Psychology, National University of Singapore, Singapore

OUTLINE

When we fail to reinforce a response that has previously been reinforced...we set up an emotional response—perhaps what is often meant by frustration.[1]

1. FROM MOTIVATION TO FRUSTRATION TO AGGRESSION

Feeling frustrated is undeniably one of the most frequent experiences we have in response to many different events in our daily lives, such as a computer program crashing and taking the last two hours of work with it, finding that your flight is canceled due to a volcanic eruption in Iceland, or learning your papers have been rejected by editors. Frustration, like many other psychological concepts originating in everyday language, is susceptible to many different meanings. It has been frequently described as an "ambiguous negative state"[2] or "a hypothetical (defined) state or condition of an organism,"[3] partially because there are so many different antecedents to frustration, such as physical barriers, delay, an incompatible response tendency, effort, unfairness, or the blocking of reward. Thus, frustration is a broadly defined concept and it has not been well studied.

Reward omission or reward blockage is one major elicitor of frustration. The experimental studies of frustration can be dated back to Pavlov's experiments on reward omission and classical conditioning. Pavlov found "rapid and more or less smoothly progressive weakening of the reflex to a conditioned stimulus which is repeated a number of times without reinforcement..." and referred to this operation as "experimental extinction."[4] The observation of reduced response after nonreinforcement in a classical conditioning context can also be extended to the operant conditioning context. Indeed, Thorndike proposed the famous "Law of Effect," saying that "responses that produce a satisfying effect in a particular situation become more likely to occur again in that situation, and responses that produce a discomforting effect become less likely to occur again in that situation."[5] A huge amount of evidence support the law of effect theory.[6] However, discouraging responses that produce negative outcomes are not the only behavioral consequences of reward omission. Whereas the reduced frequency of the learned response following nonreward has been studied extensively, other consequences of frustration are less appreciated. A main topic of this thesis is to look at these other consequences of frustration.

Neuroimaging Personality, Social Cognition, and Character
http://dx.doi.org/10.1016/B978-0-12-800935-2.00011-7

Since frustration or reward omission is so common and so frequent in the path of goal attainment, one puzzle is how we can still continue our reward pursuit activity in the face of obstacles. For example, Thomas Edison tried hundreds of materials without success before he eventually invented the electric light bulb. Without even knowing that the electric light bulb is possible, how could Thomas Edison persist in searching for proper materials when repeatedly being frustrated? The law of effect would predict that Thomas Edison would be less likely to search for materials for a light bulb because such behaviors had produced negative outcomes. Following the same logic, many human achievements are impossible because they usually come after repeated failure. Given the prevalence of frustration, it is intuitively unlikely that individuals would simply terminate their learned response or give up when reward is frustrated. There must be mechanisms that allow individuals to maintain or even increase motivation in order to overcome the obstacles between them and the final goals. Consistent with this idea, several lines of research suggest that frustration does not always lead to response reduction, as shown in some extinction studies, but rather can produce other behavioral changes, such as enhanced response vigor.

One example of increased vigor in response to frustration is aggression. The famous frustration aggression theory posits that frustration has an important role in producing aggressive behaviors.[7] Amsel further emphasized the role of frustration in behavior learning.[8] Nevertheless, many questions about frustration remain unanswered. Clearly, not all reward omissions are equally frustrating, and not all frustrations generate the same amount of aggression. What causes the frustrated feelings? What determines the intensity of frustration, from a weak unpleasant feeling from not being able to open an attachment in an email to an intensive frustration at being denied a promotion? What is the consequence of being frustrated? In this chapter, I explore the psychological and neural mechanisms of frustration using functional magnetic resonance imaging (fMRI).

1.1 Motivation-Frustration Hypothesis

When hopes or desires are suddenly thwarted, we feel frustrated. Amsel[8] defined frustration as an emotional state that results from the nonreinforcement of an instrumental response that previously was consistently reinforced. Intuitively, it should have a close link with reward expectation or motivation. Indeed, it has been clearly pointed out that frustration is possible only when there is an "intent to gratify the primary drive."[9] More precisely, it has been proposed that the variation in the amount of frustration is a function of the strength of the desire to obtain the goal.[7] This motivation-frustration hypothesis is supported by evidence from several experiments in infants. It has been shown that the amount of crying in infants following the omission of milk was significantly correlated with the amount of milk received before the feeding interruption.[10,11] These studies indicate that "the strength of a frustration-reaction varies directly with the strength of instigation to the frustrated goal-response."[10,11] By training infants to pull with the right arm in order to receive a reward (e.g., slides and music), it was found that when the expected reward was missed after pulling, infants displayed increased sad and angry expressions, heart rates,[12] and pulling responses.[13–16] These studies demonstrate that reward expectancy violation produces frustration in infants.

Other experiments on frustration in adults also explicitly manipulated subjects' expectations. In one study, participants selected a particular prize as a reward but received another relatively unattractive reward instead. These participants expressed the strongest aggression toward the person who was responsible for providing the reward.[17] In another study, the experimenter deliberately cut into a line in front of participants at various stores, banks, restaurants, and ticket windows. Participants displayed more aggression if they were closer to the front of the queue than far away from it.[18] Presumably, the person closer to the front had a stronger motivation than the one at the rear, and therefore felt more frustrated if someone cut in the line in front of him/her. In summary, in both infants and adults, frustration is influenced by reward motivation or expectation.

Understanding what increases motivation is a complex issue. Several factors, such as the reward magnitude, the proximity to the reward, the amount of effort expended, and the social expectation based on fairness norms, may influence our reward expectation or motivation. It is noted that the magnitude of a reward exerts a profound effect on performance in different tasks. Rats responded with faster reaction times as the size of an expected reward increased.[19] Larger reward leads to improved efficiency in orienting and reorienting of exogenous spatial attention.[20] In a memory task, recognition accuracy was greater for high-reward than low-reward scenes.[21] Higher rewards motivate individuals to work harder (e.g., squeeze a handgrip harder).[22–24] Although the detrimental influences of reward on performance have also been found in some studies,[25,26] the majority of studies found a facilitative effect of reward.

Accumulating evidence using delay discounting paradigms has demonstrated that people prefer immediate reward (e.g., 10 pounds today) over distal reward (e.g., 10 pounds next month).[27] The preference to immediate reward is even stronger in other animals. For example, pigeons, rats, and monkeys usually cannot tolerate temporal delays of a few seconds.[28–32] Hull's goal-gradient hypothesis states that animals' motivation to reach a goal increases with proximity (i.e., spatial and temporal) to the goal.[33] This has been demonstrated across species such that rats run progressively faster as they come closer to the food[34] and monkeys' error rates and reaction times decrease significantly as the final trial approaches.[35–38] Similar effects have also been observed in humans.[39] However, individual differences also play an important role. As a distant future achievement task approached in time, individuals high in motivation to approach success (approach-oriented) increased their performance, whereas avoidance-oriented individuals decreased their performance.[40] Other work has shown that low anxiety individuals increased their performance as the final test approached, whereas high anxious individuals show a decrease in performance.[41] Taken together, these results suggest that reward proximity can increase individuals' motivation to obtain a reward, although other factors, such as personality, can also modulate this effect.

The influence of previous investments on motivational state is debated. Rational decision theories would predict that previous investment should not influence decision-making because the investment cannot be recovered no matter what current choices are made.[42] For this reason, the previous investment is called a sunk cost. However, such a view is challenged by observations suggesting that work expended in trying to reach a goal enhances the motivation to obtain it, known as the "sunk-costs" effects in economics,[43] or "work ethic" in ecology.[44–46] People usually invest more money simply because "so much has already been invested." This phenomenon has strong parallels with striking observations in ecology, whereby animals seem to prefer rewards that have previously taken more effort to accrue to those that were acquired at little cost. The conventional wisdom regarding the prevalent tendency to honor sunk costs focuses on social factors, such as self-justification and the overgeneralization of a social norm akin to "do not waste." It states that people have a psychological need to justify past behavior and have a preference that past efforts not be perceived as being wasteful by others.[44] No matter what drives sunk costs bias, the effect of sunk costs on building up motivation seems to be quite common and robust.

To test the motivation-frustration hypothesis, it is important to examine the effects of different psychological factors that can influence motivation, as well as frustration. Here, I tested a variety of psychological factors that can potentially influence individuals' motivation and experienced frustration.

1.2 Emotional and Physiological Responses to Frustration

There is a large amount of evidence to suggest that subjects not only reduce their response rates when they are no longer rewarded as they used to be but also exhibit emotional responses to the surprising termination of a reward. Some early studies have investigated the variety of emotional responses following surprising reward omission (i.e., frustrative nonreward) or blocked reward. After training rats in a runway situation with food reward in a final goal location, they found that during the extinction trials in which the food reward was missing, agitated behavior, such as sniffing, head movement, and grooming, occurred more frequently.[47] Skinner also observed that "a pigeon which has failed to receive reinforcement turns away from the key, cooing, flapping its wings and engaging in other emotional behaviour."[48] Another well-controlled experiment showed how animals would respond aggressively when reward expectation is unmet, such as when a more attractive expected reward is replaced by a much less attractive reward. In that experiment, monkeys first observed a piece of banana being placed under a cup. Then, after a while, monkeys found that under the same cup there was only a piece of lettuce. Monkeys "shrieked at them in apparent anger and would not accept the lettuce."[49] These animal studies suggest that frustrative nonreward induces anger-like emotion.

Darwin proposed that the angry facial expression and other bodily cues to emotion are the behavior patterns that occur when access to a goal is thwarted.[50,51] Infants showed wary facial expressions when they were restricted to their seats compared to when they could freely move about the room.[52] It has also been shown that abrupt inaccessibility of milk leads to increases in crying, clearly indicating the aversiveness of the reward withholding.[10] However, from studies in animals and infants, we cannot infer for sure whether they actually felt frustrated, or at least not simply from their behavior. Physiological measurement, such as corticosteroid levels, skin conductance response, heart rate, and blood pressure, may allow us to examine individuals' emotional responses more directly.

Corticosteroids are known to increase after individuals experience stress induced by pain, aggression, and other aversive stimuli.[53] During extinction, rats trained to lever-press for water showed increased corticosterone levels, relative to a control group in which water was delivered on a response-independent manner.[54] The corticosterone level changes cannot be simply due to the novelty of the new situation, because surprisingly, increased reward ratios actually led to a decrease in corticosterone levels.[55]

Aversive stimuli, such as an electric shock, have been shown to influence electrodermal activity, measured by skin conductance response, and heart rate.[56] In a response-feedback study with human participants, the experimenter first gave participants accurate feedback about their choice and then always gave them negative feedback regardless of their choices. It was shown that human participants showed an increase in skin conductance when repeatedly receiving negative feedback.[57] Similar findings were found with other reward-response extinction experiments.[58]

Failures also can result in increases in blood pressure. In one study, the experimenter interrupted participants' work in an intelligence test and eventually told them that they had failed the test. The interrupted and failed group showed significant increases in blood pressure relative to the control group that completed the test successfully without interruption.[59] Similarly, participants' blood pressure was raised significantly after being told by the experimenter that they were not trying hard in the experiment and could not get the monetary reward promised.[60] All these changes associated with frustration indicate that frustration may activate the approach system and will make individuals more active and aroused rather than producing withdrawal and a passive or depressed state.

1.3 Frustration-Aggression Theory

> The occurrence of aggression presupposes frustration... Frustration produces instigations to a number of different types of responses, one of which is an instigation to some form of aggression.[7]

In both animals and humans, aggression is a prevalent behavior. It may have evolved in the context of defending or obtaining valued resources, such as food, reproductive partners, territory, and social status.[61] Given its survival role in securing reward, a strong link between aggression and frustration is not surprising.

One of the most popular and influential theories in social psychology is the frustration-aggression hypothesis. Dollard et al. highlighted the strong link between frustration and aggression.[7] They stated firmly that "the occurrence of aggressive behavior always presupposes the existence of frustration" and "the existence of frustration always leads to some form of aggression."[7] This hypothesis, although it sounds too strong, is indeed supported by a massive amount of daily life experiences and experimental data. For example, people behave aggressively if they have just had a bad day. Football crowds can become aggressive when their team loses, fighting or throwing objects at fans and players from the opposing team. The discontinuation of monetary reward produces increased aggression-like punching behavior.[62] A famous experiment with children shows how frustration induced by reward delay can induce aggression.[63] In that experiment, the experimenter put some toys behind a wire screen where children could see them but not play with them. After a while, children were allowed to play with the toys. The study shows that children then became quite destructive because they had been frustrated by not being able to get to them sooner.[63] In another experiment, an experimenter took a toy that the child played with and placed it behind a Plexiglass barrier, causing the children to show frustration and distress.[64] These studies with young children suggest that people tend to respond to frustration with aggression at a very early stage. In a field study, it was found that when the workers' payoff was temporarily cut by 15%, workers theft rates increased significantly in the companies.[65] Furthermore, even socially justified frustration can generate anger and hostile behavior. In an early study on community residents and college students, 11% of the sample reported that they were angered by another's actions even though the instigator had behaved legitimately, and another 7% said that they were provoked by unavoidable accidents or events.[66]

In animals, aggression after frustrative reward omission is even more common. Researchers trained pigeons to peck keys with grain as reinforcement.[166] When the schedule shifted from reinforcement to extinction, the pigeons attacked the other pigeon in the same box more frequently. Pigs also exhibited increased aggressive fighting during the extinction of a panel-pressing response that had been reinforced with food.[67] Likewise, rats also exhibited displaced aggression during an extinction period (e.g., biting a plastic target).[68] Taken together, all of this evidence suggests that aggression might be a prepotent response to frustration in both animals and humans.

Researchers have also noticed that there are many circumstances in which frustration does not evoke obvious aggressive behavior. Unlike the classic frustration-aggression theory which states "aggression is always a consequence of frustration," Berkowitz's frustration-aggression hypothesis admits that frustration is not the only antecedent of aggression, and frustrations do not always lead to overt aggression.[69] The link between frustration and aggression is modulated by several factors, like personality, social context, and so on. For example, in Germany, it was found that participants who expected to be unemployed or who were currently unemployed reported stronger aggressive inclinations than controls, but this aggression-eliciting effect of unemployment only occurred for participants with low self-awareness.[70]

Berkowitz[69] revised Dollard's hypothesis[7] by arguing that it is the intensity of the negative affect elicited by frustration that determines the likelihood of aggression.[69] Here, aggression is a more general example of the relationship between aversive stimuli and negative emotions. Put simply, frustration produces aggression because frustration is unpleasant. It is the aversiveness that leads to aggression. Negative emotions, such as frustration, anger, annoyance, or pain, can trigger either "fight" or "flight," as well as other associated thoughts and behaviors, determined by how individuals evaluate the situation and control their emotions.[69] However, it is worth noting that the revised frustration-aggression theory is so broad that it is therefore almost not falsifiable.[71] Berkowitz also made no distinction among frustration and other negative emotions, such as anger.[69] Anger is the emotion people experience when they feel offended. So the intention behind the action rather than the consequence itself plays an important role in determining anger. Anger can be experienced even when the objective outcome is good. For example, imagine someone fixed a bug in your script but did it with an arrogant attitude; you may still feel being offended and thus angry because it hurt your self-esteem. On the other hand, imagine someone tried to fix a bug in your script, but unfortunately he failed; you may feel frustrated (the bug is not fixed) but not angry (you still appreciate his help).

Amsel extended the frustration aggression hypothesis by proposing that general response vigor is increased after frustration. Amsel describes frustration as having "a transient energizing effect on responses with which it coincides, increasing particularly the intensity with which these responses are performed."[8] The main idea of Amsel's frustration theory is the frustration effect, which refers to an observable increase in response vigor immediately following nonreward of previously rewarded events.[8] Although frustration is a negative emotion, it is a motivational condition. Like hunger, pain, or fear, it has motivational properties, increasing the strength of some ongoing behavior. By using the double-runway apparatus, research with rats has provided evidence to support the motivational properties of frustration.[8,72–74] The increase in responding at the outset of the removal of a reward or goal is also known as an "extinction burst."[75] It is also proposed that blockage in goal-approach situations leads to increased efforts to overcome the blockage.

1.4 Fear = Frustration Hypothesis

Frustration has been theorized to be functionally and physiologically very similar to fear, known as the "fear = frustration" hypothesis.[76] Fear is the emotion associated with punishment. The defining characteristic of a punishment is that it is aversive, and the organism will work to terminate or avoid it. Frustrative nonreward shares the most obvious properties of punishment. Thus, it was hypothesized that fear equals frustration.[76]

One demonstration of the similarity between fear and frustration is in aversive odor experiments. Rats can use smell to detect places in which other rats have experienced an electric shock and avoid them. Rat's frustration odor is similar to its fear odor. In one study, rats were trained to find food reward in a box. Then, the reward was withheld, thus generating frustration. Consequently, other rats avoided the "frustrating box," possibly because of the frustration odor left in that box.[77]

Frustrative nonreward is as effective as punishment in shaping individuals' ability to learn to avoid aversion.[76] In the usual extinction period, rats are placed in a start box and the number of trials they continue to run to a goal box is measured as an index of "resistance to extinction." However, a new manipulation was introduced by Adelman and Maatsch.[78] The rats were allowed to jump up to a ledge around the goal box and were then rewarded. Previously frustrated rats learned to jump out more quickly than food-rewarded controls.[78] Frustrative nonreward was able to potentiate the startle reflex, in a similar way as a threatening stimulus (e.g., a shock) can.[3] Drugs (sodium amylobarbitone and alcohol), which can reduce fear, also can effectively reduce frustration.[76] Amylobarbitone, also known as amobarbital, is a drug that has sedative-hypnotic and analgesic properties.

Animals are good at associating frustration with punishment. Using a partial reinforcement schedule in which animals are only rewarded on a proportion of the trials, when not rewarded, a punishment (i.e., noise) was delivered and animals could jump across the barrier to turn the stimulus off. It was found that the animals on the frustrative trials jumped more often than controls that experienced the punishment only.[79] Similar improved learning to escape was found in a reduced reward paradigm.[80]

Taken together, these studies suggest that frustration is similar to fear functionally and physiologically. However, the "fear = frustration" hypothesis is not well accepted generally,[81] because of the lack of concrete evidence to support such similarity. Fear and frustration are distinct emotions. The ultimate source of fear is danger, which is a negative stimulus that individuals want to keep away from. Frustration, on the other hand, is caused by a blocked reward. The reward is still a positive stimulus that individuals want to obtain. While frustration is more associated with an approach tendency, fear is more associated with an avoidance tendency.

In summary, frustration is a negative emotion elicited by being blocked from obtaining a desired reward. Here, the desired reward is context-dependent. A bottle of water may be extremely desired for a thirsty person but not for one who is not thirsty. A reward can be either material or social. For example, watching someone cut into someone else's queue can also make other people feel frustrated because the social norm one values is violated, although no material reward is blocked. Frustration is associated with an approach, rather than withdrawal, tendency and enhanced skin conductance responses and blood pressure.[57] In terms of appraisal, when people feel frustrated, they might think that if they try harder, they can overcome the obstacle and still reach the goal. These characteristics distinguish frustration from other negative emotions such as fear and anger. In a fearful condition, people usually withdraw from the dangerous source rather than approach it. Anger is usually against a social target. The angry person usually finds the cause of their anger in an intentional person. If someone hits your car intentionally, you will feel angry. But if the damage is caused by situational forces (e.g., a hailstorm), you will mainly feel frustrated instead. Nevertheless, given the complexity in real-life situations, people may feel more than one emotion at a time.

From the findings and theories I have introduced, a link from motivation to frustration and aggression emerges. Motivation can be influenced by several factors, such as reward magnitude, reward proximity, and expended effort.[21,33,43] The degree of motivation can modulate the amount of frustration,[7] which further determines the levels of aggression.[7,8] Although relationships between motivation and frustration and between frustration and aggression have been studied separately, to my knowledge, no studies have systematically examined these associations together. Moreover, the neural mechanisms underlying these associations are not yet clear.

2. THE SEEKING, AVERSION, AND RAGE CIRCUITS

Reward pursuit is one of the primary functions of a mobile creature. Reward is fundamental to emotion, motivation, learning, and goal-directed behavior. It is not surprising that our brain has specific circuits devoted to processing reward in an efficient manner, providing a critical evolutionary advantage for survival. Understanding the neural basis of reward processing is important for investigating the neural basis of frustration, which is mainly elicited by reward blockage. Electrophysiological studies in nonhuman primates have established that single-cell firing rates are modulated by reward within the basal ganglia, ventral tegmental area, striatum, orbitofrontal and other areas of the prefrontal cortex.[82–85] Brain stimulation reward (BSR) opened the window to study the relationship between the reward and the brain. BSR is a phenomenon originally described by Olds and Milner.[86–89] In these studies, a lever was placed in a cage and connected to a stimulator so that each press would result in the delivery of a pulse. It was found that rats with the electrode implanted in the lateral hypothalamus rapidly learned to press the lever and would press it up to 7000 times per hour for several hours if allowed, ignoring other basic needs such as water and food.[89] This method was then used to determine the location of reward-relevant neurons and map neural pathways that are directly affected by brain stimulation. The mesolimbic pathway, the ventral tegmental area (VTA), striatum, and the medial prefrontal cortex are associated with the strongest reward effects of stimulation.[85,90–93] Neuroimaging studies in humans have corroborated the electrophysiology data in studies using a variety of methods and different types of rewards. The basal ganglia have been found to play a key role in reward processing. The basal ganglia are comprised of the striatum, the internal (medial) and external (lateral) segments of the globus pallidus, the subthalamic nucleus (STN), the VTA and the substantia nigra pars compacta (SNc) and pars reticulata (SNr). The striatum consists of the nucleus accumbens/ventral striatum, the caudate, and the putamen. Brain responses in the reward-sensitive regions (e.g., striatum) have been elicited by primary rewards, such as juice and water, and by pleasant tastes and smells.[94–97] A number of other "rewarding" stimuli also engage these regions, including secondary rewards such as money, beautiful faces, humor, and other social rewards, suggesting that the brain may process rewards along a single common pathway.[98] A common pathway is important since it allows widely different rewards, from different modalities and units, to be directly compared on a same scale for the purpose of choosing between possible courses of action.

Living in a world with both potential reward and potential punishment, avoiding punishment, either physical pains or monetary losses, is as important as, if not more important than, chasing reward. Thus, it is also important to encode negative events rapidly. Previous studies have focused on brain responses to unpleasant stimuli, like pain, disgusting images, threat, and so on, and have identified several neural substrates associated with aversion. Frustration is one of the aversive emotions and may share the aversion circuits implicated in processing pain and anger.

The insula is a portion of the cerebral cortex folded deep within the lateral sulcus between the temporal lobe and the frontal lobe. It can be divided into anterior and posterior parts. The anterior insula receives direct projections from the thalamus and has reciprocal connections with the central nucleus of the amygdala.[99,100] The posterior insula connects reciprocally with the primary sensory cortex and receives inputs from thalamic nuclei.[99] The insula is highly specialized to convey homeostatic information, e.g., the physiological condition of the body.[101,102] Inhaled bad-smelling odorants can activate the insula, as well as video clips showing the emotional facial expression of disgust in others.[103] Moreover, anterior insula has been implicated in aversive conditioning.[104] More specifically, it seems that the insula is more dedicated to recognizing disgust than other negative emotions. Viewing facial signals of disgust or disgusting scenes activated the insula.[105,106] Taken together, this evidence strongly indicates the crucial role the insula plays in the registration of disgust.[107] Recent studies suggest that the insula is also involved in processing negative social moral emotions, like social disgust and anger.

Using the classic Ultimatum game, it has been found that unfairness activated the anterior cingulate cortex (ACC) as well as the insula. Moreover, activity in the insula predicted the probability that individuals rejected the unfair offers.[108,109] Social disgust has been proposed to have evolved from the phylogenetically more primitive sensation of disgust.[110] These studies suggest that moral disgust and primitive disgust may share the same neural substrate, namely the insula.[111]

The amygdala is located in the medial temporal cortex and is densely interconnected with the hippocampus, situated medially to the lateral ventricles, forming the amygdala–hippocampal complex. The amygdala receives projections from the frontal cortex, temporal cortex, olfactory system, and other parts of the limbic system.[112] In return, it sends its afferents to the frontal and prefrontal cortex, orbitofrontal cortex, hypothalamus, and hippocampus.[113–117] There is a wealth of evidence to suggest that the amygdala (as well as the hippocampus) is one of the most important brain regions underpinning the processing of negative emotion such as fear and threat. In a fMRI study attempting to examine the brain response to real-life fear, Mobbs et al. found that activity in the amygdala increased as a spider approached participants' feet (as they believed).[118] The amygdala response to threat can be attenuated by courageously choosing to overcome the fear, as shown by a recent fMRI study using an approaching snake to represent a real-life threat.[119] Threatening pictures intended to induce a fearful response activated the bilateral amygdala.[120] The amygdala is also involved in the subconscious detection of threat.[121,122] The amygdala plays a crucial role in processing social threat, as well as nonsocial threat. Participants with complete bilateral amygdala damage rated unfamiliar individuals to be more approachable and more trustworthy than did normal and brain damaged controls.[123]

Periaqueductal gray (PAG) is the gray matter located around the cerebral aqueduct within the tegmentum of the midbrain. Its ascending projections go to the thalamus, hypothalamus, amygdala (through the central nucleus), and its descending projections go to the nucleus of the solitary tract (NTS) and other brain stem areas.[124–126] Based on its functional significance, the PAG can be divided into the ventral and dorsal parts. The dorsal PAG (also called lateral PAG) receives projections from the hypothalamus, the central and basal nuclei of the amygdala, and the septal nuclei.[127,128] The ventral PAG receives projections from the central nucleus of the amygdala.[128] The midbrain PAG coordinates distinct patterns of behavior and physiological reactions critical for survival.[129,130] Early studies reported that defensive or "aversive" reactions, hypertension, and reduced sensitivity to pain are elicited by electrical stimulation at sites in the PAG.[131–134] It has been proposed that an immediate threat will activate the dorsal PAG, resulting in escape and defensive aggression, as well as nonopioid analgesia. By contrast, less immediate threats activate the amygdala, which in turn commands the ventral PAG to produce freezing and opioid analgesia.[135,136] The dorsal PAG may inhibit the ventral PAG to suppress freezing when a vigorous defense is required.[135] Consistent with this idea, it has been found that dorsolateral PAG ablation disrupts the unconditioned activity burst that accompanies footshock, but not the freezing response, whereas more ventrally placed PAG lesions disrupt the freezing response but not the activity burst.[136] Further, chemical stimulation of the dorsal PAG increases fight and flight responses in rats,[137] whereas chemical stimulation of more ventral PAG produces immobility and behavior quiescence.[138] Stimulation of the ventral PAG evoked more passive behaviors, e.g., quiescence and escape.[129,139,140] Stimulation of the dorsal PAG in rats elicited fear reactions[141] and displays of threatening behavior, e.g., vocalization, flattening of the ears, and arching of the back.[142] Using fear-potentiated startle effect, it has been found that the amygdala mediates the potentiation of startle by moderate fear and PAG mediates potentiated startle at higher levels of fear.[143] It seems that the transition from passive to active behaviors, as threat levels increase, is mediated neurally by a shift in activity from ventral to dorsal PAG elements.[135] Using a Pacman-like game combined with fMRI, Mobbs et al., found that as the threat of an electric shock drew closer, brain activation shifted from the prefrontal cortex to the PAG.[144,145] Activity in the PAG was also significantly correlated with self-reported dread intensity.[145]

Given the close link between frustration and aggression,[7] the neural basis of aggression may give some hint of the neural basis of frustration. Indeed, it has recently been proposed that frustration should activate the aggression-related regions.[146] Some researchers have also proposed that frustration is mild aggression.[81] Electrical stimulation studies conducted on animals have identified that the hypothalamus, amygdala, and PAG are associated with aggression. In cats, two general types of aggression have been elicited by electrical stimulation of the hypothalamus: affective attack, which is emotional and reactive and can be elicited by medial hypothalamic stimulation,[147,148] and goal-directed attack, also known as quiet-biting in cats, which can be elicited by lateral hypothalamic stimulation.[148] Affective attack and rage responses are also obtained by stimulation of the dorsomedial amygdala.[149] In rhesus monkeys, bilateral amygdala lesions in the brain of the most dominant monkey led to this monkey falling to the bottom of the hierarchy. This study suggests that the amygdala is important for the aggression normally needed in maintaining a position in the social hierarchy.[150] Comparative research also suggests that stimulation of the midbrain PAG elicits aggressive behavior,[151–153] whereas lesions result in a pronounced reduction in aggression.[154,155]

In recent years, researchers have begun to use neuroimaging methods to explore the neural basis of human aggression. Different paradigms have been used: (1) the traditional Taylor Aggression Paradigm (TAP), which examines aggression directly; (2) the social exclusion paradigm, which specifically tests social rejection based aggression; (3) the interpersonal insult paradigm; (4) the moral decision paradigm, which pits aggression (e.g., harming a person) against a utilitarian goal (e.g., save many others); and (5) basic emotional and cognitive tasks (e.g., viewing negative images), which investigate basic valence evaluation and cognitive control functions that are essential for aggression.

In one study, the modified TAP was used.[156] During the decision phase, high provocation (high voltage electric shock delivered by the opponent), relative to low provocation, activated the bilateral insula and the ACC.[157] The selection of high punishment, contrasted with the selection of low punishment, activated the dorsal striatum, suggesting that aggression is rewarding (the "sweetness of revenge").[157] Using a similar paradigm, another fMRI study found that dorsal medial PFC activity was associated with the selected punishment strength, whereas ventral medial PFC activation was correlated with the skin conductance response during observation of the opponent suffering.[158] Not surprisingly, psychopathic individuals responded differently on such a task. For example, when psychopaths punished others with a high amount of shock they showed increased activation in the hypothalamus and other regions (e.g., the lateral prefrontal cortex, the posterior cingulate cortex, and the amygdala).[159] Moreover, activity in hypothalamus was positively correlated with trait "physical aggression." However, when psychopaths observed the paired opponent being punished, they showed increased activation in the dorsal and ventral medial prefrontal cortex, which was positively associated by measured impulsivity and antisocial traits.[159]

Aggression can be elicited by social rejection, known as the rejection-hostility link in romantic relationships.[160] Using a social exclusion paradigm, combined with genotype analysis,[161] it has been found that individuals with the Monoamine oxidase A gene (MAOA)-L gene reported higher trait aggression and showed greater dorsal ACC activity to social exclusion, compared with MAOA-H individuals.[162]

Interpersonal insult can easily provoke anger and aggression. In an fMRI study examining the brain response, an experimenter spoke in a rude and upset tone of voice to participants being scanned. It was found that increased activation in the ACC and insula following the provocation predicted subsequent angry rumination.[163] When choosing to cooperate, psychopathic individuals had increased activity in the dorsolateral prefrontal cortex (DLPFC),[164] implicated in high-level cognitive control. The authors concluded that these results suggest that it might be more difficult for psychopathic individuals to cooperate.

Taken together, these fMRI studies indicate that abnormal activity in emotional regions (e.g., amygdala) and cognitive control regions (e.g., DLPFC) might underlie aggression. In contrast to the findings in animal studies, which predominantly found subcortical regions such as the amygdala and PAG in aggression, human fMRI studies seem to emphasize the role of the fronto-temporal cortex in aggression. Given the consistent findings in functional neuroimaging studies showing that aggression is characterized by abnormal activity in certain brain regions, it is reasonable to expect that aggression can be explained by certain brain structural abnormities. Indeed, neuroanatomical correlates of aggression have been found in a number of recent studies.[165]

Three different neuroanatomical brain circuits were described. One circuit, mainly including the striatum and VTA, is involved in reward processing. This circuit is also called the SEEKING system.[81] In both human and animal studies, the striatum is consistently activated in the majority of reward studies. The VTA is less frequently found in human neuroimaging studies. One reason is that brain regions, such as the striatum, are larger and therefore more easily imaged with fMRI. The ACC-insula-amygdala-PAG circuit is involved in processing aversive stimuli. Frustrating events are generally aversive. It is still unclear whether there is a general aversion circuitry that processes all aversive stimuli or different types of aversion are processed in different brain regions. The available evidence seems to support the latter. For example, disgust and fear have dissociable neural underpinnings.[107] Another circuit, including the amygdala, hypothalamus, and PAG, is implicated in aggression. This circuit is also called the RAGE system.[81] The RAGE system overlaps with regions implicated in processing aversive stimuli.

3. REWARD BLOCKAGE INDUCES FRUSTRATION

From missing the bus to helplessly observing someone procure our taxi, daily life throws numerous obstacles in the path of our desired goals. Frustration is a natural response to such circumstances. It is widely held that frustration instigates aggression,[7] where one might use profanity or displaced aggression (e.g., "kick the cat") to ease this negative emotion. Studies on rodents demonstrate that blocking a scheduled reinforcement reliably elicits bursts of aggression (i.e., "frustrative nonreward").[6,72,166,167] Understanding the laws that engender frustration is a crucial step to understanding the mechanisms that instigate aggression and possibly criminal violence.

As stated previously, the variation in the amount of frustration is a function of the strength of the desire to obtain the goal.[7] Hull's goal-gradient hypothesis states that the effort animals expend to reach a goal increases with proximity (i.e., spatial and temporal proximity) to the goal.[33] This has been demonstrated across species where rats run progressively faster as they come closer to food[168] and monkeys' error rates and reaction times decrease significantly as the final trial is approached.[35–37,169] For example, using a visually cued multitrial reward schedule paradigm in which monkeys were required to successfully complete a schedule of trials (one, two, three, or four trials) in order to obtain a reward, it has been shown that monkeys made progressively fewer errors as the final rewarded trial approached. Given the hypothesized link between motivation and frustration,[7] one might theorize from the goal-gradient hypothesis that the closer one is to a goal, the greater the strength of the motivation to approach the goal, and the stronger the experienced frustration when the goal/reward is blocked.

One manifestation of frustration is displaced aggression, which can take the form of thumping a desk or slamming a door. In fact, surveys report that one-third of aggressive responses to provocation were against a nonhuman object.[170] In addition, empirical studies have shown that induced frustration is vented out through displacing aggression towards objects such as a computer keyboard, telephone, or plunger (used to turn a buzzer off in one experiment, explained below).[171–175] For example, Haner and Brown examined the effect of goal proximity on response pressure, an index of frustration. In that experiment, participants were asked to place marbles in 36 holes during the experiment.[171] They were interrupted by the experimenters after they had filled 25%, 50%, 75%, 89%, or 100% of the holes. Interruption was signaled by a buzzer, followed by the marbles falling down. Participants could then push a plunger to turn the buzzer off and restart the next trial. The pressure the participants exerted when pushing the plunger was used as an index of frustration. The results revealed that the intensity of the push response was modulated by the goal proximity at the time of blocking. However, no self-reported frustration was measured in this study. It is unclear whether participants actually experienced frustration or not.

To my knowledge, no study has systematically examined response force and subjective accounts of frustration. Here, I conducted experiments to investigate if the motivation to obtain the final reward and the frustration elicited by the failure to obtain the reward are modulated by reward proximity. To systematically manipulate reward proximity, I developed a reward schedule task in which participants were required to successfully complete a series of consecutive trials in order to obtain the reward. Hence, as the number of remaining trials decreased, the proximity to the final reward increased. I predicted that the experienced frustration following blocking would be modulated by reward proximity, such that missing a proximal reward produces the maximal frustration. Behavior theorists propose that frustrative nonreward appears to have an invigorating effect on rodents' subsequent behaviors.[8,73] Based on this theory, I hypothesized a significant relationship between the amount of frustration and the response force levels when reward is blocked.

In conjunction with goal gradients, frustration and the motivational drive to acquire a reward are presumably influenced by other contingencies, such as reward magnitude and expended effort. Anticipation of a more valued reward leads to stronger motivation, as evidenced by measures of arousal, attention, and intensity of motor output.[22,25] Experienced effort could also influence motivation. The commitment to a course of action is more likely to be escalated as more effort is invested, known as the "sunk cost" effect.[43] To examine the possible interactions between reward proximity and other contingencies, I dissociated the effects of reward proximity and the effects of invested effort. Further, to achieve an objective measurement of frustration, I recorded forces applied to buttons when participants responded to confirm the outcomes. Previous studies have found that computer users applied significantly more pressure to a mouse when the task is frustrating (e.g., a delay in the internet network) than in the control condition (no network delay).[176] These mouse pressure data, together with other channels of affect-related information, can detect a computer user's response to frustration at a rate better than chance.[176,177] Thus, I predicted that the force applied to buttons after being blocked from obtaining a reward would correlate with the degree of experienced frustration.

To disentangle the effects of reward proximity and invested effort on frustration, I used a multitrial reward schedule task and tested whether proximity effects of frustration are still robust when invested efforts are controlled. In this experiment, participants had to complete four (1/4, 2/4, 3/4, 4/4), three (1/3, 2/3, 3/3), two (1/2, 2/2) and one (1/1) trials successfully to obtain a reward (two pounds). At the beginning of each schedule, participants were presented with a cue image indicating how many trials were left (the number of white squares) and how many trials are given as a bonus (the number of green squares with a stripe). For example, when participants started on the third trial, trial one and two were given as bonuses. The rest of the experiment was similar to the Experiments one and two. For the first trial of each schedule (1/4, 1/3, 1/2, and

1/1), the proximity to reward was different, but the invested effort (finishing one trial) was constant. The same logic applied to the second trials (2/4, 2/3, and 2/2) and third trials (3/4 and 3/3) of each schedule. On the other hand, for the last trials of each schedule (4/4, 3/3, 2/2, and 1/1), the invested effort is different, but the distance to reward (0 trials left) was constant. The same logic applied to the penultimate trials (3/4, 2/3, 1/2) and the antepenultimate trials (2/4 and 1/3). I predicted that both motivation (reaction time (RT) and self-reported motivation) and frustration (response force and self-reported frustration) would be independently modulated by proximity and expended effort.

3.1 Paradigm: Multitrial Reward Schedule Task

Participants were required to successfully perform a multitrial reward schedule comprising four (1/4, 2/4, 3/4, 4/4), three (1/3, 2/3, 3/3), two (1/2, 2/2) and one (1/1) trials in order to obtain a reward of two pounds (see Figures 1 and 2). The arrow cue on each trial was preceded by a two second presentation of a schedule cue indicating how many trials were left (i.e., two blank boxes represent one trial to complete). The two blank boxes were used instead of one blank box to give the impression of increased spatial distance to the reward. Progress was also indicated by proportions of a complete image of two pounds. Two squares would be filled in green when the corresponding trial was performed successfully.

After each schedule cue, participants were presented with an array of three arrows for one second and were required to indicate the direction of the arrows as quickly and accurately as possible to advance through the schedule. Participants were told that the response criterion for each trial was set by the computer in an unpredictable fashion. If their reaction time was slower than the criterion or they responded incorrectly, the schedule was terminated and they would fail to win the reward and the appropriate feedback ("Blocked") was presented for two seconds. Only after successfully finishing four consecutive trials would they win the reward, and the feedback "Win" would be presented. As for Experiments one and two, the criterions were predetermined so that participants were blocked about 14 times at each schedule state. Participants won about 33% trials within each schedule. After the feedback, participants were presented with words "Win? Blocked?" (or "Blocked? Win?") on the screen and were required to press the corresponding key to confirm the outcomes within four seconds. The stimulus presentation and response recording was controlled by E-Prime 2.0 software (Psychology Software Tools, Pittsburgh, PA).[178]

After the second and fourth blocks, participants complete a postscan questionnaire, which asked them to

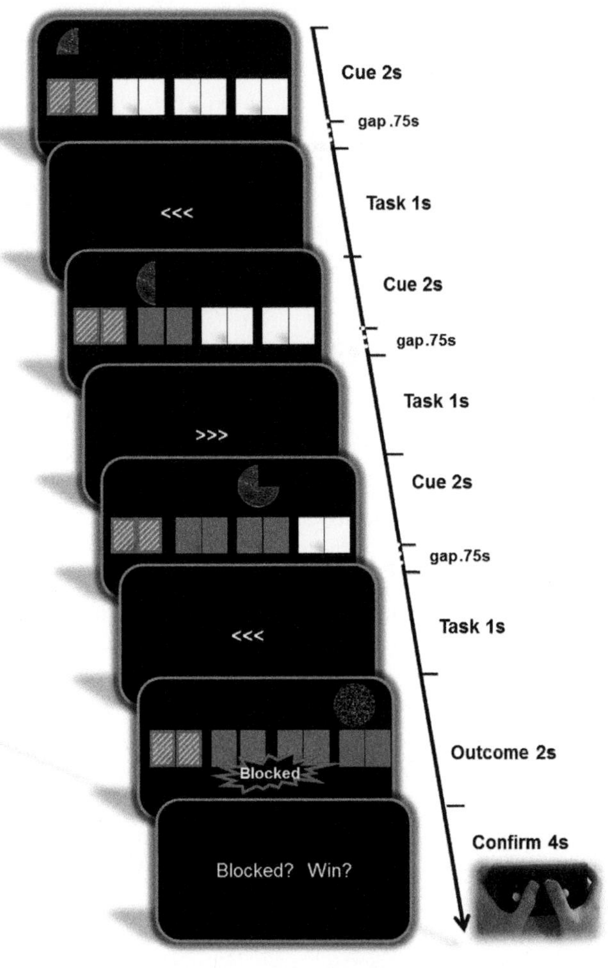

FIGURE 1 The multitrial reward schedule task. Illustration of a three-trial schedule in which the participant has been blocked on the last trial.

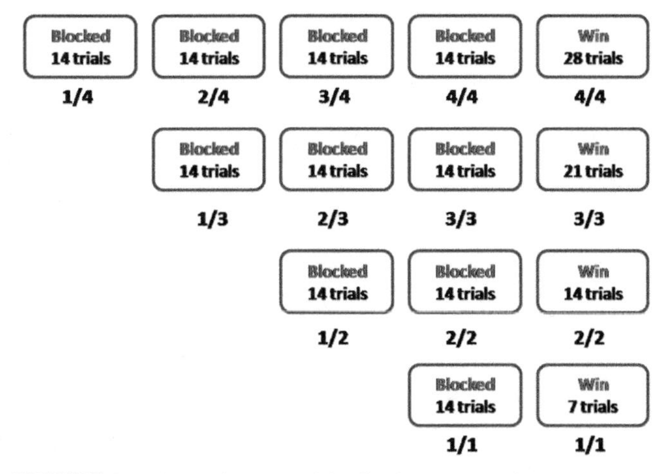

FIGURE 2 The predetermined feedback structure of Experiment 1a. The number of trials for blocked and win outcomes in each schedule state. The outcomes were predetermined such that the percentage of schedules rewarded was 33%, and participants were blocked the same number of times at each schedule state.

indicate on 10-point analogue Likert scales (1 = not at all, 10 = very intensely) how frustrated and surprised they felt about the blocked outcomes, as well as how motivated they felt as they progressed successfully through the different stages of the reward schedule. The whole experiment included 210 trials and lasted about 52 min.

To provide a more objective trial-by-trial measure of frustration, I used the response force after reward blockage as an additional measure of frustration. A specially designed pressure button box was used to record the force participants applied when responding (Magconcept® Sunnyvale, CA). The digitized force signal was recorded with a resolution of ~0.3 Newton (N). The sampling rate was 500 Hz. This allowed RTs to be measured to the nearest 2 ms. RTs were computed as the time at which the force first exceeded 2 N. This value is well within the range used by standard all-or-none response keys recording RTs.

The results (Figure 3) demonstrate that as the reward becomes close, self-reported motivation to obtain the reward and frustration after being blocked from obtaining the goal increase. Increased motivation as a function of decreased goal distance was evident in self-reported motivation to gain the reward and faster RTs to arrow

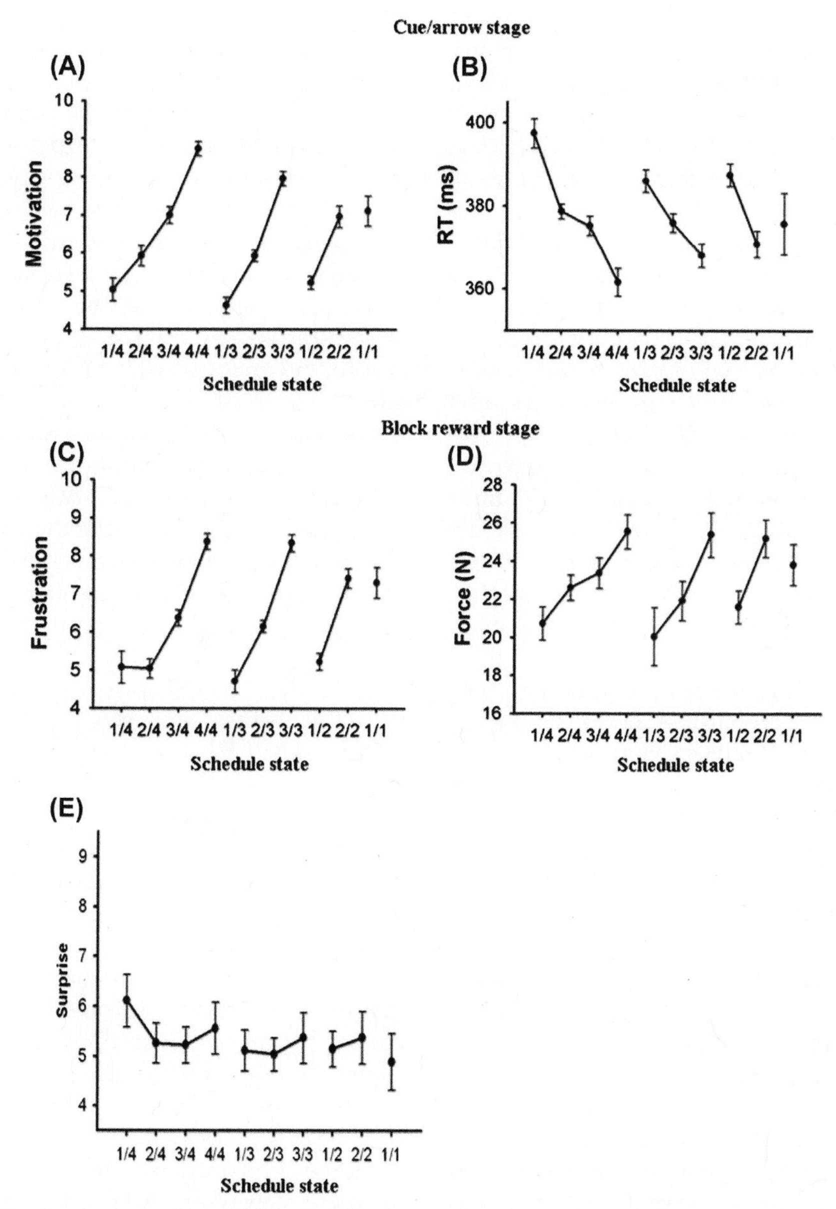

FIGURE 3 Behavioral data from Experiment 1a (behavioral study). Self-reported motivation (A) to obtain the reward and RTs (B) to arrows at each of the different schedule states; Self-reported frustration (C) applied response force to confirm the blocked outcome (D), and self-reported surprise (E) at being blocked. Error bars indicate standard error of the mean, after between-subject variability has been removed, which is appropriate for repeated-measures comparisons. This applies to all error bars shown in this figure.

cues. Participants reported stronger frustration after being blocked from obtaining proximal rather than distal rewards, and the self-reported frustration was significantly correlated with the motivation to obtain the reward and the force applied to buttons after blockage. The effect of reward proximity on motivation and frustration were still robust after controlling for the effects of invested effort. The findings therefore support the hypothesis that frustration transfers unfulfilled motivation into response vigor.

The present work is grounded in Hull's goal-gradient hypothesis, which states that the organism will increase its effort when a goal is proximal[33] and the notion that the amount of frustration is modulated by the strength of the desire to obtain the goal.[7] While the goal-gradient hypothesis remains largely unexplored in humans, these results support the animal literature showing that increasing proximity to reward over successive trials speeds RTs. One influence of motivation on behavior is the "energizing" effect, which determines the vigor of actions.[179] The ability to adjust motivational level as a function of reward proximity is crucial to yield adaptive goal-directed behavior. Exerting more strength at the final steps presumably increases the chance of obtaining a reward that is within reach.

Such augmented motivation energy seems to be transferred into the frustration-induced response vigor after goal blockage. I propose that the adaptive value of frustration may reside in its ability to motivate or energize subsequent behaviors, allowing individuals to continue pursuing goal-directed activities despite frustrative nonrewards and obstacles. My proposal can be related back to Amsel's proposal based on studies in animals that frustrative nonreward appears to have an invigorating effect on behaviors that immediately follow it.[8,72,73] Along these lines, early scientific accounts found that rats restrained near a food goal pulled on a harness harder than those restrained farther away.[180] Similarly, children push on a plunger more vigorously when they are blocked close to a goal.[171–173] The present results extend these findings by showing that after being blocked from a proximal reward, self-reported frustration and force applied to buttons were highly correlated in adult human participants. The association between applied force and the level of frustration found in human subjects provides evidence that task-irrelevant actions are also energized by frustration. Following a frustrating event, the energy associated with frustration may then be applied to an immediately following behavior, no matter whether it is task relevant or not. The finding that frustration was stronger for missing the higher magnitude reward than the lower magnitude reward also supports this notion. As a consequence of such "energizing" effects of frustration, frustrated individuals are more likely to engage in direct or displaced aggressive behaviors.

The proximity effect on frustration is independent from invested effort. It has been shown that people have a greater tendency to continue an endeavor once an investment in money, effort, or time has been made.[43,181] An extension of this effect would be that the larger the invested effort is, the stronger the motivation to get the reward and the stronger the frustration experienced when the reward is blocked. Indeed, I found that frustration at being blocked in the final trials of each schedule increased as the number of previous completed trials increased, suggesting the expenditure of effort enhances frustration after failure to get the reward. While I focused on the reward proximity effect in this work, it will be interesting to investigate more closely the effect of effort on motivation and frustration in future studies.

In conclusion, the results demonstrate that proximity to reward enhanced the motivation to obtain the reward and heightened the experienced frustration when blocked. Enhanced frustration was associated with harder button pressing after reward blockage. I have attempted to advance the goal-gradient theory by proposing that the enhanced motivation toward a proximal goal is transferred to enhanced frustration after goal blockage. I believe that the evolutionary significance of frustration is to transfer the unfulfilled motivation aroused by proximal reward into subsequent behavioral vigor.

Although the behavioral study shows that blocked reward induces frustration and aggression, the underlying neural mechanism of this process is still unclear. Understanding how frustration is transferred into aggression at the neural level can identify the critical neural substrates involved in aggression and consequently the manner in which they are dysfunctional.

4. THE NEURAL BASIS OF THE FRUSTRATION STATE

Previous neuroimaging studies have focused on the neural basis of physically aversive states, such as fear, pain, and disgust.[107,182,183] In real-life situations, frustration is an equally, if not more common, aversive event than receiving a physical pain or viewing/smelling disgusting stimuli. However, the neural basis of frustration in humans is largely unexplored. Here, I sought to examine the neural correlates of increasing frustration in the context of reward proximity and expended effort.

Studies on rodents and nonhuman primates demonstrate that surprising omission of a scheduled reinforcement (i.e., "frustrative nonreward") reliably elicits bursts of aggression. Animal research suggests that reactive aggression is mediated by the RAGE circuitry, which runs from the amygdala through the hypothalamus and down to the PAG of the midbrain.[81] Comparative research suggests that stimulations of the amygdala or midbrain PAG elicit aggressive behavior, whereas lesions to these areas result in a pronounced reduction in aggression. Human fMRI studies also found that the amygdala was activated by angry prosody[184] and angry faces.[185–187] Given that frustrating events are the main elicitors of aggression, Blair recently proposed that frustration should activate the RAGE circuitry.[146] Thus, my first hypothesis is that frustration should activate the amygdala-hypothalamus-PAG circuitry.

The fear = frustration theory proposes that frustration, in many aspects, functions similarly to the fear of physical threat. Following this link, similar to fear, frustration has also been integrated into a neuropsychological model of a hierarchical defense system. In the hierarchical defense theory, also known as the Reinforcement Sensitivity Theory (RST), there are two fundamental dimensions.[188,189] One dimension is categorical, making a distinction between the Behavior Approach System (BAS) and the Behavior Inhibition/Avoidance System (BIS). Another dimension is continuous, applying to both the approach and avoidance systems, appropriate to the defense distance. It is known as the "Fight-Flight-Freeze System" (FFFS). FFFS controls the types of defensive behavior—to fight, to flight, or to freeze. Animal studies indicate that the neural substrates associated with FFFS respond to threat progressively.[188] Low levels of threat from a distant threat, e.g., a tiger two kilometres away, induce sophisticated planning, cognitive control, and may activate the frontal cortex. Higher levels of threat produce freezing and attempts to escape the dangerous environment. As the threat is very close and escape is unlikely, the highest danger will instigate fight and reactive aggression. Accordingly, the basic threat system that runs from the amygdala, via the hypothalamus, to the periaqueductal gray will be more activated for high intense threat. However, higher-level processing in subcortical regions does not imply less involvement in fundamental features of defense in cortical regions.[188] The hierarchical responses to different levels of threat have recently been documented in humans. Studies using fMRI have shown that the increasing proximity to a threat activated the amygdala and PAG.[118,144,145]

If fear = frustration, as theorized,[76] I would predict that as the frustration level increases, the amygdala-hypothalamus-PAG circuit would be more activated. Similar to the FFFS hierarchical threat system, I propose a hypothetical hierarchical model of the frustration system. For example, in the context of reward proximity, if participants are blocked when a reward is far away, the negative feedback may encourage cognitive problem solving. The negative emotion at this stage is very weak. If participants are blocked when they are more close to obtaining a reward, mild aversive emotion is experienced and the ACC and the insula regions may be activated. If participants are blocked when a reward is proximal, the frustration level continues to increase progressively and the amygdala-hypothalamus-PAG circuitry is more activated. The highest level of frustration will activate the terminal of this circuitry (the PAG) and elicit affective aggression. This frustration system controls the types of behavioral responses to frustration, from cognitive problem solving to affective aggression. Thus, based on both the fear = frustration hypothesis and the hierarchical defense theory,[188] I predict that the intensity of frustration should be associated with activity in basic threat regions, including the amygdala, hypothalamus, and PAG.

In addition, activation in the insula and ACC is consistently seen in response to anger and social provocation.[157,163,190–193] Using an interpersonal insult paradigm (i.e., an experimenter spoke in a rude and upset tone of voice to participants being scanned), a recent study found that increased activation in the ACC and insula following the provocation predicted subsequent angry rumination.[163] Another fMRI study also found increased ACC and insula activation in response to high versus low social provocation, using the TAP, in which the winner can choose to punish the loser after a reaction time competition task.[157] Since anger and frustration are intimately connected, I predicted that the ACC and insula might also be involved in processing frustration in the current experimental context.

Using the multitrial schedule task combined with fMRI, I explored the neural basis of frustration elicited by frustrative nonreward. I therefore reasoned that blocked reward should activate this basic circuitry, including the ACC, insula, amygdala, and PAG, and that activation in these regions would be modulated by proximity and invested effort. In addition, my previous behavior studies demonstrated that the motivation to obtain the reward increases as a function of reward proximity and expended effort. Recent fMRI studies have also identified the striatum as a reward region using different types of rewarding stimuli, from monetary reward to pretty faces.[194] Based on previous research on reward processing, I also predicted that the activity in brain reward regions (e.g., the striatum), would be modulated by both reward proximity and effort.

To investigate how reward motivation and frustration are generally encoded in the brain, I examined the brain responses to the schedule cues during each trial (excluding the actual reward delivery), and compared it to responses to blocked reward feedback that indicated reward blockage. I found that cue minus blocked reward activated bilateral ventral striatum and other regions. The striatum has been implicated in reward processing.[194] Blocked reward versus cue activated the anterior cingulate cortex, left insula, bilateral amygdala, and left periaqueductal gray. These regions have been

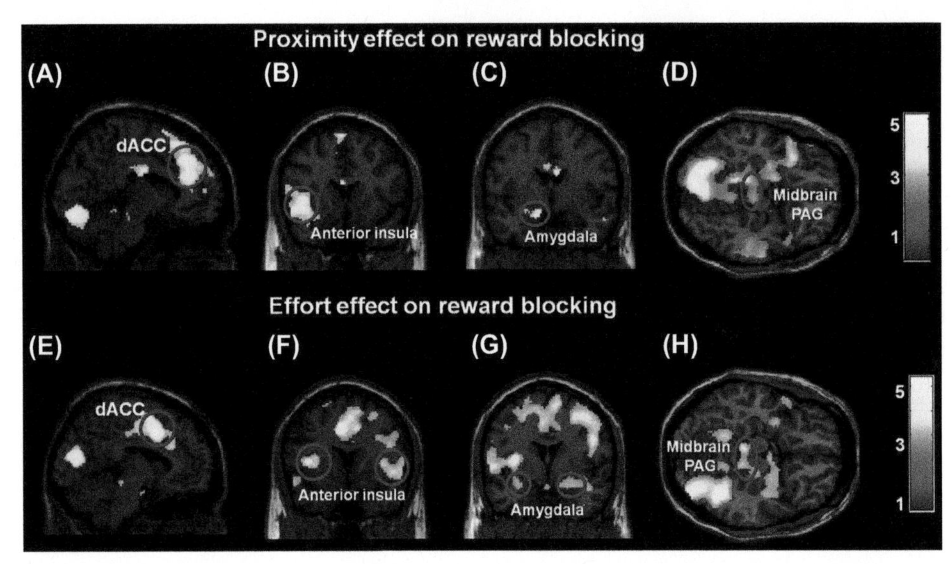

FIGURE 4 Parametric increases in Blood-oxygen-level dependent (BOLD) response to block outcomes associated with increasing proximity and increasing expended effort. Positive associations with increasing proximity were found in (A) left dorsal anterior cingulate cortex (dACC), (B) left anterior insula, (C) left amygdala, and (D) left midbrain PAG. Positive associations with increasing expended effort were found in (E) left dACC, (F) bilateral anterior insula, (G) bilateral amygdala, and (H) bilateral midbrain PAG. All activations are significant at $p<0.05$, small volume corrected. For display purposes, maps are thresholded at $p<0.005$, uncorrected.

frequently implicated in coding aversive stimuli and aggression.[81,101,107,195]

To investigate the specific effect of reward proximity and effort on motivation, I examined the parametric effects of proximity and effort on schedule cues, using the reaction times as a surrogate measure of reward motivation.[196] I found that the proximity effect was associated with activity in the bilateral caudate, bilateral putamen, bilateral ventral striatum, and right insula. Effort effect was associated with activity in the bilateral caudate and other regions.

To investigate the conjunction effect of proximity and effort, I did a conjunction analysis using a liberal threshold (for each contrast, $P<0.05$ uncorrected). This analysis revealed significant activations in the left caudate, left putamen, and right caudate, suggesting a similar neural network contributing to both the proximity and effort effects during the cue stage.

To investigate the specific effects of proximity and effort on frustration, I then examined how reward proximity and prior effort modulate the brain responses to blocked reward feedback that induce frustration. This revealed a correlation with the proximity effect in the dorsal ACC, left anterior insula, left amygdala, and left PAG. The specific effect of effort was associated with activity in the dACC, bilateral anterior insula, right amygdala, and bilateral PAG.

To examine the similarity between the two effects, I did a conjunction analysis between the proximity effect and the effort effect using a liberal threshold (for each contrast, $P<0.05$ uncorrected). The conjunction analysis revealed activations in the ACC, left anterior insula, left amygdala, and left PAG. These results suggest that similar neural networks underlie the effects of proximity and effort (although at a liberal threshold, see Figure 4, results reported in a research paper[197]).

Hull's goal-gradient hypothesis states that animals' motivation to reach a goal increases with proximity to the goal.[33] It has been supported by evidence from studies in animals.[34–38] In a series of experiments, I replicate these findings and extend them to human adults by showing that both adults' motivation to obtain a reward and frustration after reward blockage increased with increasing reward proximity. Moreover, the enhanced frustration was associated with stronger keyboard strikes, an aggression-like behavior. Here, using fMRI, I found that when the enhanced motivation is unmet, the brain regions implicated in aggression are engaged in response to reward blockage.

The effect of threat proximity on brain activity has been understudied. Only recently, it has been shown that when a threat becomes close, the amygdala and PAG are more activated.[118,144,145] The similarity between the threat circuitry identified in these studies and the frustration/aggression circuitry identified in current study may support the fear = frustration hypothesis and the Reinforcement Sensitivity Theory.[188,189] The association between increasing activation in aggression circuitry and increasing reward proximity also support the animal model of the FFFS. FFFS coordinates response patterns to different levels of threat, such that low levels of threat (e.g., distant threat) induce sophisticated planning, and the highest danger instigates fight and reactive aggression. Blocking a proximal reward might resemble a threat for animals. Intuitively, a proximal reward (e.g., food) is a reward that an individual is most likely to obtain. If a predator cannot even capture a prey that is proximal, it would be even more unlikely for the predator to capture a distal prey. Thus, if a proximal reward is blocked, the animal may end up with no resource to survive (e.g., starve to death). In this sense, it is not surprising that blocking a proximal reward may be encoded as the highest level of threat, and it engages aggression regions that are similarly activated by the highest threat. Our results support the hypothetical neuropsychological model of the hierarchical frustration system.

In addition to the proximity effect, I showed that expended effort also influences human motivation and frustration, in that increased effort produced heightened motivation and heightened frustration when blocked. This provides neural support for the existence of work-ethic or "sunk-cost" effects and illustrates that retrospective discounting occurs alongside the better-studied prospective discounting. According to social psychological theories, reduction of an expected reward should induce a stronger effort–reward imbalance when effort has been expended. High effort and low reward signals that reciprocity is violated and induces possible stress and anger.[198,199] Moreover, from an ecology perspective, the more effort one has expended to obtain a reward, the less energy one has to obtain other rewards. For example, a predator that spent hours trying to capture a prey will have little energy to capture another prey. Thus, blocking a high-effort reward is also a high threat to animals' survival. Our findings suggest that blocking high-effort reward may be encoded as a high threat and thus induces frustration and aggression.

Although it has been posited that FEAR and RAGE circuits are intimately related,[81] it by no means suggests that the two systems cannot be separated. For example, it has been found that the amygdala in the FEAR circuit is more lateral and more medial in the RAGE circuit.[81] However, the functional territories of the human amygdala or PAG are difficult to dissociate due to the limited spatial resolution of fMRI. Given that the amygdala and PAG are implicated in both the FEAR and RAGE systems, it would be interesting to compare functional and anatomical specificities of these two regions in the fear and rage context.

One influence of motivation on behavior is the "energizing" effect, which determines the vigor of actions. Here I showed that the augmented motivation energy may be transferred into displaced aggression when the goal is blocked. This is consistent with the proposal that a rapid suppression of activity within the reward system should promote the arousal of RAGE circuitry.[81] Rats tend to bite when rewarding brain stimulation is terminated.[200] I propose that the adaptive value of this process may reside in its ability to motivate or energize subsequent behaviors, allowing individuals to continue pursuing goal-directed activities in the face of obstacles. This is consistent with the notion that the primary evolved function of anger is to motivate individuals to compete effectively for environment resources.[50,81]

Although the insula and ACC are not part of the threat circuitry that are hypothesized to respond to high threat, these two regions have also been found in previous studies on anger.[157,163,190–193] The ACC is associated with a number of negative emotions, including the intensity of social distress following social rejection[161,201,202] and envy.[203] Both social rejection and envy are common antecedents of aggression. The ACC and insula have also been implicated in pain induced distress[204] and the perception of other people in pain.[205] As being blocked from obtaining a reward is generally aversive, the ACC and insula may encode the aversive component associated with frustration/aggression in our task.

Frustration/aggression is complex and poorly understood. Although the present study highlights a key role of the amygdala and PAG in frustration and aggression, it does not exclude the possibility that there are other regions that also play an important role in aggression. For example, electrical stimulation of the cerebellum has been used to control violent pathological aggression, a treatment known as "cerebellar pacemaker."[206,207] Complex emotions and behaviors, such as frustration and aggression, are usually not the result of single localized brain areas but of complex interactions within the brain regions. More studies are needed in order to fully understand the frustration/aggression neural systems.

Some limitations in the present study are worth mentioning. All the behavioral ratings were conducted after the experiment. They may not accurately represent the true emotional responses during the task. It is also unknown whether participants' experienced frustration changed during the course of the experiment or not. Future studies should use both online ratings and postscan ratings and examine the dynamic fluctuations of frustration during the experiment. Another pitfall in this study is that I did not record the response force due to technical issues. Future studies may examine the relationships between response force and brain activity. Finally, I only examined male participants, and these findings may not be generalized to the female population. Gender difference on frustration may be an interesting topic for future studies.

In conclusion, these data provide both behavioral and neurophysiological evidence that human motivation and frustration/aggression are constructed by both prospective reward proximity and retrospective investment of effort. Enhanced frustration was accompanied by increased displaced aggression, manifested as a harder button press after reward blockage. Blocking goal-directed behavior activated brain regions associated with basic aversive processes and aggression. I advance the goal-gradient theory by proposing that the enhanced motivation toward a proximal goal transfers to escalated frustration/aggression after goal blockage. The evolutionary significance of this process might be to transfer the unfulfilled motivation into subsequent behavior vigor to overcome obstacles. Furthermore, understanding the laws and neural substrates that govern the frustration-aggression association has implications for patient populations with pathological aggression.

5. FUTURE DIRECTIONS

The work presented here could lead to a number of further investigations.

An interesting extension of this research would be to examine the neural basis of frustration and aggression in individuals with psychiatric disorders.

An increased risk for aggression is common in many psychiatric disorders, such as posttraumatic stress disorder (PTSD), bipolar disorder, borderline personality disorder, intermittent explosive disorder, psychopathy, or antisocial personality disorder (APD) and conduct disorder (CD). According to a large-scale meta-analytic review of mental health in worldwide prison systems, 47% of male prisoners (i.e., 5113 of 10,797) and 21% of female prisoners (i.e., 631 of 3047) were diagnosed with antisocial personality disorder, whereas only roughly 3.7% male and 4.0% female prisoners were diagnosed with a psychotic illness.[208] Many of the regions reported in my study (e.g., the amygdala and anterior insula), have already been linked to these psychiatric disorders.[209,210] Future studies could use the paradigms developed in this thesis in psychiatric studies to test whether frustration activates the same or distinct brain regions in individuals with psychiatric disorders compared with healthy controls and whether or not patients exhibit enhanced response to frustration in some regions. Here, I use conduct disorder as an example to show how my paradigms can be used to investigate frustration/aggression in clinical populations, such as adolescents with CD.

One of the psychiatric disorders that may be most relevant to frustration is CD. Conduct disorder is a serious antisocial disorder, occurring in childhood and adolescence that is characterized by a longstanding pattern of violations of the rights of others or the current social norms. Symptoms used to classify CD fall into the four main categories: (1) aggression toward people and animals; (2) destruction of property without aggression toward people or animals; (3) deceitfulness, lying, and theft; and (4) serious violations of rules.[211] Based on these symptoms, I predict that, compared to healthy controls, participants with CDs are more sensitive to the experience of frustration; more risk prone in decision-making since they have a deficit in anticipating potential frustration; lacking in concern for others' welfare in self-advantage conditions; and more likely to overreact to social provocation in self-disadvantage conditions.

Recent neuroimaging studies on individuals with CD revealed an abnormal brain response to affective information. CD adolescents displayed reduced response to angry faces in regions associated with aggression (e.g., the amygdala and insula).[210] CD patients showed enhanced left-lateralized amygdala activation in response to negative pictures.[212] Individuals with CD, who are also characterized by high levels of aggression, usually attack without empathy for their victims. Atypical brain responses to others' pain has been found in CD subjects, for example, they exhibit less amygdala-prefrontal coupling when watching images showing pain in someone harmed by another person.[213] Consistent with fMRI findings, structural MRI studies also found abnormalities in the insula and amygdala in CD patients.[214,215] It has also been reported that children with conduct problems showed increased gray matter (GM) concentration in the medial orbital frontal cortex and anterior cingulate cortex, as well as increased GM volume and concentration in the temporal lobes.[216,217] Taken together, these findings suggest that the amygdala, insula, and ACC, regions involved in frustration, also play an important role in the pathology of CD.

Since individuals with CD seem to overreact to social provocation, I would expect them to display exaggerated emotional response to frustration. Counterintuitively, in one study, using a psychosocial stress procedure designed to elicit frustration and provocation, it was found that adolescents with CD showed blunted cortisol and cardiovascular responses to frustration.[218] These findings are consistent with a previous study showing that boys with CD exhibited reduced autonomic responses to all categories of pictures (positive, neutral, and negative), although they reported lower levels of emotional response only to aversive pictures.[219] Future empirical studies have the promise to better our understanding of the biological mechanisms underpinning CD and thus lead to new approaches to its treatment.

6. CONCLUSIONS

The present research extends previous frustrative nonreward studies in animals and children into human adults and demonstrates similarities in the mechanisms underpinning frustration and aggression. I demonstrated that frustrative nonreward activated the amygdala and PAG, regions found to be involved in rage and aggression in animal studies. These findings also provide preliminary support that the aversive signal in the insula, which reflects anticipatory frustration, plays a key role in mediating decision biases. These findings merit further investigation of frustration in clinical populations.

Acknowledgments

Funding support for writing this chapter was provided by a grant from the National University of Singapore (R-581-000-166-133).

References

1. Skinner BF. *The Behavior of Organisms*. New York: Appleton-Century-Crofts; 1938.
2. Lazarus RS. *Emotion and Adaptation*. New York: Oxford University Press; 1991.
3. Brown JS, Farber IE. Emotions conceptualized as intervening variables - with suggestions toward a theory of frustration. *Psychol Bull*. 1951;48:465–495.
4. Pavlov IP. *Conditioned Reflexes. An Inverstigation of the Physiological Activity of the Cerebral Cortex*. Oxford: Oxford University Press; 1927.
5. Thorndike EL. Animal intelligence: an experimental study of the associative processes in animals. *Psychol Rev Monogr Suppl*. 1898;2(4):i.
6. Skinner BF. Are theories of learning necessary? *Psychol Rev*. 1950:193–216.
7. Dollard J, Doob LW, Miller NE, Mowrer OH, Sears RR. *Frustration and Aggression*. New Haven: Yale University Press; 1939.
8. Amsel A. *Frustration Theory: An Analysis of Dispositional Learning and Memory*. Cambridge: Cambridge University Press; 1992.
9. Yates AJ. *Frustration and Conflict*. London: Methuen; 1962.
10. Sears RR, Sears PS. Minor studies of aggression: V. Strength of frustration-reaction as a function of strength of drive. *J Psychol*. 1940;9:297–300.
11. Marquis DP. A study of frustration in newborn infants. *J Exp Psychol*. 1943;32:123–138.
12. Lewis M, Hitchcock DF, Sullivan MW. Physiological and emotional reactivity to learning and frustration. *Infancy*. 2004;6(1):121–143.
13. Sullivan MW, Lewis M, Alessandri SM. Cross-age stability in emotional expressions during learning and extinction. *Dev Psychol*. 1992;28:58–63.
14. Sullivan MW, Lewis M. Contextual determinants of anger and other negative expressions in young infants. *Dev Psychol*. 2003;39(4):693–705.
15. Alessandri SM, Sullivan MW, Lewis W. Violation of expectancy and frustration in early infancy. *Dev Psychol*. 1990;26:738–744.
16. Lewis M, Sullivan MW, Ramsay DS. Individual differences in anger and sad expressions during extinction: antecedents and consequences. *Infant Behav Dev*. 1992;15:443–452.
17. Worchel S. The effect of three types of arbitrary thwarting on the instigation to aggression. *J Pers*. 1974;42:300–318.
18. Harris MB. Mediators between frustration and aggression in a field experiment. *J Exp Soc Psychol*. 1974;10:561–571.
19. Brown VJ, Bowman EM. Discriminative cues indicating reward magnitude continue to determine reaction time of rats following lesions of the nucleus accumbens. *Eur J Neurosci*. 1995;7(12):2479–2485.
20. Engelmann JB, Pessoa L. Motivation sharpens exogenous spatial attention. *Emotion*. 2007;7(3):668–674.
21. Adcock RA, Thangavel A, Whitfield-Gabrieli S, Knutson B, Gabrieli JD. Reward-motivated learning: mesolimbic activation precedes memory formation. *Neuron*. 2006;50(3):507–517.
22. Pessiglione M, Schmidt L, Draganski B, et al. How the brain translates money into force: a neuroimaging study of subliminal motivation. *Science*. 2007;316(5826):904–906.
23. Schmidt L, d'Arc BF, Lafargue G, et al. Disconnecting force from money: effects of basal ganglia damage on incentive motivation. *Brain*. 2008;131(Pt 5):1303–1310.
24. Schmidt L, Palminteri S, Lafargue G, Pessiglione M. Splitting motivation: unilateral effects of subliminal incentives. *Psychol Sci*. 2010;21(7):977–983.
25. Mobbs D, Hassabis D, Seymour B, et al. Choking on the money: reward-based performance decrements are associated with midbrain activity. *Psychol Sci*. 2009;20(8):955–962.
26. Beilock SL, Decaro MS. From poor performance to success under stress: working memory, strategy selection, and mathematical problem solving under pressure. *J Exp Psychol Learn Mem Cogn*. 2007;33(6):983–998.
27. Weatherly JN, Terrell HK, Derenne A. Delay discounting of different commodities. *J Gen Psychol*. 2010;137(3):273–286.
28. Simon NW, LaSarge CL, Montgomery KS, et al. Good things come to those who wait: attenuated discounting of delayed rewards in aged Fischer 344 rats. *Neurobiol Aging*. 2010;31(5):853–862.
29. Wilhelm CJ, Mitchell SH. Strain differences in delay discounting using inbred rats. *Genes Brain Behav*. 2009;8(4):426–434.
30. Mazur JE, Biondi DR. Delay-amount tradeoffs in choices by pigeons and rats: hyperbolic versus exponential discounting. *J Exp Anal Behav*. 2009;91(2):197–211.
31. Freeman KB, Green L, Myerson J, Woolverton WL. Delay discounting of saccharin in rhesus monkeys. *Behav Process*. 2009;82(2):214–218.
32. Hwang J, Kim S, Lee D. Temporal discounting and inter-temporal choice in rhesus monkeys. *Front Behav Neurosci*. 2009;3:9.
33. Hull CL. The goal gradient hypothesis and maze learning. *Psychol Rev*. 1932;39:25–43.
34. Stewart DS. A new conjunctival-fold fixation forceps. *Br J Ophthalmol*. 1934;18(5):274–275.
35. Bowman EM, Aigner TG, Richmond BJ. Neural signals in the monkey ventral striatum related to motivation for juice and cocaine rewards. *J Neurophysiol*. 1996;75(3):1061–1073.
36. Shidara M, Aigner TG, Richmond BJ. Neuronal signals in the monkey ventral striatum related to progress through a predictable series of trials. *J Neurosci*. 1998;18(7):2613–2625.
37. Shidara M, Richmond BJ. Anterior cingulate: single neuronal signals related to degree of reward expectancy. *Science*. 2002;296(5573):1709–1711.
38. Liu Z, Murray EA, Richmond BJ. Learning motivational significance of visual cues for reward schedules requires rhinal cortex. *Nat Neurosci*. 2000;3(12):1307–1315.
39. Rowe JB, Eckstein D, Braver T, Owen AM. How does reward expectation influence cognition in the human brain? *J Cogn Neurosci*. 2008;20(11):1980–1992.
40. Gjesme T. Slope of gradients for performance as a function of achievement motive, goal distance in time, and future time orientation. *J Psychol*. 1975;91(1st Half):143–160.
41. Gjesme T. Future-time gradients for performance in test anxious individuals. *Percept Mot Ski*. 1976;42(1):235–242.
42. Friedman M. *Essays in Positive Economics*. Chicago: University of Chicago Press; 1953.
43. Staw B. Knee deep in the big muddy. *Organ Behav Hum Decis Process*. 1976;35:124–140.
44. Arkes HR, Ayton P. The sunk cost and concorde effects: are humans less rational than lower animals? *Psychol Bull*. 1999;125(5):591–600.
45. Clement TS, Feltus JR, Kaiser DH, Zentall TR. "Work ethic" in pigeons: reward value is directly related to the effort or time required to obtain the reward. *Psychon Bull Rev*. 2000;7(1):100–106.

46. Pompilio L, Kacelnik A, Behmer ST. State-dependent learned valuation drives choice in an invertebrate. *Science*. 2006;311(5767):1613–1615.
47. Miller NE, Stevenson SS. Agitated behavior of rats during experimental extinction and a curve of spontaneous recovery. *J Comp Psychol*. 1936;21:205–231.
48. Skinner BF. *Science and Human Behavior*. New York: Macmillan; 1953.
49. Tinklepaugh OL. An experimental study of representative factors in monkeys. *J Comp Psychol*. 1928;8:197–236.
50. Darwin C. *The Expression of the Emotions in Man and Animals*. Chicago: University of Illinois Press; 1965.
51. Darwin C, Britannica E. *The Origin of Species by Means of Natural Selection*; 1872.
52. Lewis M, Sullivan MW, Brooks-Gunn J. Emotional behavior during the learning of a contingency in early infancy. *Br J Dev Psychol*. 1985;3:307–316.
53. Panksepp J. Neurochemical control of moods and emotions: amino acids to neuropeptides. In: Lewis M, Haviland JM, eds. *Handbook of Emotions*. New York: Guilford Press; 1993.
54. Coe CL, Stanton ME, Levine S. Adrenal responses to reinforcement and extinction: role of expectancy versus instrumental responding. *Behav Neurosci*. 1983;97:654–657.
55. Goldman L, Coover GD, Levine S. Bidirectional effects of reinforcement shifts on pituitary adrenal activity. *Physiol Behav*. 1973;10:209–214.
56. Andreassi JL. Psychophysiology. In: Hillsdale, ed. *Human Behavior and Physiological Response*. New Jesey: Erlbaum; 1989.
57. Germana JJ, Pavlik WB. Autonomic correlates of acquisition and extinction. *Psychon Sci*. 1964;1:109–110.
58. Pittenger DJ, Pavlik WB. An investigation of the partial reinforcement extinction effect in humans and corresponding changes in physiological variables. *Bull Psychon Soc*. 1989;27:253–256.
59. Gentry WD. Sex differences in the effects of frustration and attack on emotion and vascular processes. *Psychol Rep*. 1970;27:383–390.
60. Doob AN, Kirshenbaum HM. The effects of arousal of frustration and aggressive films. *J Exp Soc Psychol*. 1973;9:57–64.
61. Blanchard DC, Blanchard RJ. What can animal aggression research tell us about human aggression? *Horm Behav*. 2003;44(3):171–177.
62. Kelly JF, Hake DF. An extinction-induced increase in an aggressive response with humans. *J Exp Anal Behav*. 1970;14(2):153–164.
63. Barker RG, Dembo T, Lewin K. *Frustration and Regression: An Experiment with Young Children*. New York and London: McGraw-Hill; 1941.
64. Calkins SD, Johnson MC. Toddler regulation of distress to frustrating events: temperamental and maternal correlates. *Infant Behav Dev*. 1998;21:379–395.
65. Greenberg J. Organizational justice: yesterday, today, and tomorrow. *J Manag*. 1990;16(2):399–432.
66. Averill JR. Studies on anger and aggression: implications for theories of emotion. *Am Psychol*. 1983;38:1145–1160.
67. Dantzer R, Arnone M, Mormede P. Effects of frustration on behaviour and plasma corticosteroid levels in pigs. *Physiol Behav*. 1980;24:1–4.
68. Tomie A, Carelli R, Wagner GC. Negative correlation between tone (S−) and water increases target biting during S− in rats. *Anim Learn Behav*. 1993;21:355–359.
69. Berkowitz L. Frustration-aggression hypothesis: examination and reformulation. *Psychol Bull*. 1989;106(1):59–73.
70. Fischer P, Greitemeyer T, Frey D. Unemployment and aggression: the moderating role of self-awareness on the effect of unemployment on aggression. *Aggress Behav*. 2008;34(1):34–45.
71. Buss AH. Anger, frustration, and aversiveness. *Emotion*. 2004; 4(2):131–132. discussion 151–155.
72. Amsel A, Roussel J. Motivational properties of frustration: I. Effect on a running response of the addition of frustration to the motivational complex. *J Exp Psychol*. 1952;43(5):363–368.
73. Amsel A, Hancock W. Motivational properties of frustration: relation of frustration effect to antedating goal factors. *J Exp Psychol*. 1957;53(2):126–132.
74. Johnson RN, Lobdell P, Levy RS. Intracranial self-stimulation and the rapid decline of frustrative nonreward. *Science*. 1969; 164(882):971–972.
75. Miller LK. *Principles of Everyday Behavior Analysis*. Pacific Grove: Brooks/Cole; 1996.
76. Gray JA. *The Psychology of Fear and Stress*. Cambridge: Cambridge University Press; 1987.
77. Collerain I. Frustration odor of rats receiving small numbers of prior rewarded running trials. *J Exp Psychol Anim Behav Process*. 1978;4(2):120–130.
78. Adelman HM, Maatsch JL. Resistance to extinction as a function of the type of response elicited by frustration. *J Exp Psychol*. 1955;50(1):61–65.
79. Wagner AR. Conditioned frustration as a learned drive. *J Exp Psychol*. 1963;66:142–148.
80. Daly HB. Learning of a hurdle-jump response to escape cues paired with reduced reward or frustrative nonreward. *J Exp Psychol*. 1969;79(1):146–157.
81. Panksepp J. *Affective Neuroscience*. Oxford: Oxford University Press; 2005.
82. Apicella P. Tonically active neurons in the primate striatum and their role in the processing of information about motivationally relevant events. *Eur J Neurosci*. 2002;16(11):2017–2026.
83. Schultz W. Dopamine signals for reward value and risk: basic and recent data. *Behav Brain Funct*. 2010;6:24.
84. Hikosaka O, Sakamoto M, Usui S. Functional properties of monkey caudate neurons. I. Activities related to saccadic eye movements. *J Neurophysiol*. 1989;61(4):780–798.
85. Wise RA, Rompre PP. Brain dopamine and reward. *Annu Rev Psychol*. 1989;40:191–225.
86. Olds J. Self-stimulation of the brain; its use to study local effects of hunger, sex, and drugs. *Science*. 1958;127(3294):315–324.
87. Olds J. Self-stimulation experiments. *Science*. 1963;140:218–220.
88. Olds J, Killam KF, Bach-Y-Rita P. Self-stimulation of the brain used as a screening method for tranquilizing drugs. *Science*. 1956;124(3215):265–266.
89. Olds J, Milner P. Positive reinforcement produced by electrical stimulation of septal area and other regions of rat brain. *J Comp Physiol Psychol*. 1954;47(6):419–427.
90. Wise RA. Addictive drugs and brain stimulation reward. *Annu Rev Neurosci*. 1996;19:319–340.
91. Wise RA, Bauco P, Carlezon Jr WA, Trojniar W. Self-stimulation and drug reward mechanisms. *Ann N Y Acad Sci*. 1992;654:192–198.
92. Wise RA. Brain reward circuitry: insights from unsensed incentives. *Neuron*. 2002;36(2):229–240.
93. Wise RA. Opiate reward: sites and substrates. *Neurosci Biobehav Rev*. 1989;13(2–3):129–133.
94. Berns GS, McClure SM, Pagnoni G, Montague PR. Predictability modulates human brain response to reward. *J Neurosci*. 2001;21(8):2793–2798.
95. O'Doherty JP, Deichmann R, Critchley HD, Dolan RJ. Neural responses during anticipation of a primary taste reward. *Neuron*. 2002;33(5):815–826.
96. Pagnoni G, Zink CF, Montague PR, Berns GS. Activity in human ventral striatum locked to errors of reward prediction. *Nat Neurosci*. 2002;5(2):97–98.
97. Small DM. Toward an understanding of the brain substrates of reward in humans. *Neuron*. 2002;33(5):668–671.
98. Yu R, Zhou X. Neuroeconomics: opening the "black box" of economic behavior. *Chin Sci Bull*. 2007;52(9):1153–1161.
99. Augustine JR. Circuitry and functional aspects of the insular lobe in primates including humans. *Brain Res Rev*. 1996;22(3): 229–244.

100. Yasui Y, Breder CD, Saper CB, Cechetto DF. Autonomic responses and efferent pathways from the insular cortex in the rat. *J Comp Neurol*. 1991;303(3):355–374.
101. Craig AD. How do you feel? Interoception: the sense of the physiological condition of the body. *Nat Rev Neurosci*. 2002;3(8):655–666.
102. Craig AD. How do you feel–now? The anterior insula and human awareness. *Nat Rev Neurosci*. 2009;10(1):59–70.
103. Wicker B, Keysers C, Plailly J, Royet JP, Gallese V, Rizzolatti G. Both of us disgusted in My insula: the common neural basis of seeing and feeling disgust. *Neuron*. 2003;40(3):655–664.
104. Seymour B, O'Doherty JP, Koltzenburg M, et al. Opponent appetitive-aversive neural processes underlie predictive learning of pain relief. *Nat Neurosci*. 2005;8(9):1234–1240.
105. Phillips ML, Young AW, Scott SK, et al. Neural responses to facial and vocal expressions of fear and disgust. *Proc Biol Sci*. 1998;265(1408):1809–1817.
106. Phillips ML, Marks IM, Senior C, et al. A differential neural response in obsessive-compulsive disorder patients with washing compared with checking symptoms to disgust. *Psychol Med*. 2000;30(5):1037–1050.
107. Calder AJ, Lawrence AD, Young AW. Neuropsychology of fear and loathing. *Nat Rev Neurosci*. 2001;2(5):352–363.
108. Tabibnia G, Satpute AB, Lieberman MD. The sunny side of fairness: preference for fairness activates reward circuitry (and disregarding unfairness activates self-control circuitry). *Psychol Sci*. 2008;19(4):339–347.
109. Sanfey AG, Rilling JK, Aronson JA, Nystrom LE, Cohen JD. The neural basis of economic decision-making in the Ultimatum Game. *Science*. 2003;300(5626):1755–1758.
110. Rozin P, Fallon AE. A perspective on disgust. *Psychol Rev*. 1987;94(1):23–41.
111. Chapman HA, Kim DA, Susskind JM, Anderson AK. In bad taste: evidence for the oral origins of moral disgust. *Science*. 2009;323(5918):1222–1226.
112. Herzog AG, Van Hoesen GW. Temporal neocortical afferent connections to the amygdala in the rhesus monkey. *Brain Res*. 1976;115(1):57–69.
113. Bacon SJ, Headlam AJ, Gabbott PL, Smith AD. Amygdala input to medial prefrontal cortex (mPFC) in the rat: a light and electron microscope study. *Brain Res*. 1996;720(1–2):211–219.
114. Garcia R, Vouimba RM, Baudry M, Thompson RF. The amygdala modulates prefrontal cortex activity relative to conditioned fear. *Nature*. 1999;402(6759):294–296.
115. Llamas A, Avendano C, Reinoso-Suarez F. Amygdaloid projections to prefrontal and motor cortex. *Science*. 1977;195(4280):794–796.
116. Brutus M, Shaikh MB, Edinger H, Siegel A. Effects of experimental temporal lobe seizures upon hypothalamically elicited aggressive behavior in the cat. *Brain Res*. 1986;366(1–2):53–63.
117. Krettek JE, Price JL. A direct input from the amygdala to the thalamus and the cerebral cortex. *Brain Res*. 1974;67(1):169–174.
118. Mobbs D, Yu R, Rowe JB, Eich H, FeldmanHall O, Dalgleish T. Neural activity associated with monitoring the oscillating threat value of a tarantula. *Proc Natl Acad Sci USA*. 2010;107(47):20582–20586.
119. Nili U, Goldberg H, Weizman A, Dudai Y. Fear thou not: activity of frontal and temporal circuits in moments of real-life courage. *Neuron*. 2010;66(6):949–962.
120. Hariri AR, Tessitore A, Mattay VS, Fera F, Weinberger DR. The amygdala response to emotional stimuli: a comparison of faces and scenes. *Neuroimage*. 2002;17(1):317–323.
121. Morris JS, Ohman A, Dolan RJ. Conscious and unconscious emotional learning in the human amygdala. *Nature*. 1998;393(6684):467–470.
122. Whalen PJ, Rauch SL, Etcoff NL, McInerney SC, Lee MB, Jenike MA. Masked presentations of emotional facial expressions modulate amygdala activity without explicit knowledge. *J Neurosci*. 1998;18(1):411–418.
123. Adolphs R, Tranel D, Damasio AR. The human amygdala in social judgment. *Nature*. 1998;393(6684):470–474.
124. Mantyh PW. Connections of midbrain periaqueductal gray in the monkey. II. Descending efferent projections. *J Neurophysiol*. 1983;49(3):582–594.
125. Mantyh PW. Connections of midbrain periaqueductal gray in the monkey. I. Ascending efferent projections. *J Neurophysiol*. 1983;49(3):567–581.
126. Bandler R, Tork I. Midbrain periaqueductal grey region in the cat has afferent and efferent connections with solitary tract nuclei. *Neurosci Lett*. 1987;74(1):1–6.
127. Vianna DM, Brandao ML. Anatomical connections of the periaqueductal gray: specific neural substrates for different kinds of fear. *Braz J Med Biol Res*. 2003;36(5):557–566.
128. Mattson MP. *Neurobiology of Aggression: Understanding and Preventing Violence*. New Jersey: Humana Press; 2003.
129. Bandler R, Shipley MT. Columnar organization in the midbrain periaqueductal gray: modules for emotional expression? *Trends Neurosci*. 1994;17(9):379–389.
130. Siegel A. *The Neurobiology of Aggression and Rage*. Boca Roton: CRC Press; 2005.
131. Carrive P, Bandler R, Dampney RA. Anatomical evidence that hypertension associated with the defence reaction in the cat is mediated by a direct projection from a restricted portion of the midbrain periaqueductal grey to the subretrofacial nucleus of the medulla. *Brain Res*. 1988;460(2):339–345.
132. Carrive P, Bandler R, Dampney RA. Somatic and autonomic integration in the midbrain of the unanesthetized decerebrate cat: a distinctive pattern evoked by excitation of neurones in the subtentorial portion of the midbrain periaqueductal grey. *Brain Res*. 1989;483(2):251–258.
133. Keay KA, Clement CI, Depaulis A, Bandler R. Different representations of inescapable noxious stimuli in the periaqueductal gray and upper cervical spinal cord of freely moving rats. *Neurosci Lett*. 2001;313(1–2):17–20.
134. Shaikh MB, Siegel A. GABA-mediated regulation of feline aggression elicited from midbrain periaqueductal gray. *Brain Res*. 1990;507(1):51–56.
135. Fanselow MS. The midbrain periaqueductal gray as a coordinator of action in response to fear and anxiety. In: depaulis A, Bandler RJ, eds. *The Midbrain Periaqueductal Gray Matter: Functional, Anatomical, and Neurochemical Organization*. New York: Plenum Press; 1991:151–174.
136. Fanselow MS. Neural organization of the defensive behavior system responsible for fear. *Psychon Bull Rev*. 1994;1:429–438.
137. Depaulis A, Keay KA, Bandler R. Longitudinal neuronal organization of defensive reactions in the midbrain periaqueductal gray region of the rat. *Exp Brain Res*. 1992;90(2):307–318.
138. Depaulis A, Keay KA, Bandler R. Quiescence and hyporeactivity evoked by activation of cell bodies in the ventrolateral midbrain periaqueductal gray of the rat. *Exp Brain Res*. 1994; 99(1):75–83.
139. Zhang SP, Bandler R, Carrive P. Flight and immobility evoked by excitatory amino acid microinjection within distinct parts of the subtentorial midbrain periaqueductal gray of the cat. *Brain Res*. 1990;520(1–2):73–82.
140. Bandler R, Keay KA. Columnar organization in the midbrain periaqueductal gray and the integration of emotional expression. *Prog Brain Res*. 1996;107:285–300.
141. Jenck F, Moreau JL, Martin JR. Dorsal periaqueductal gray-induced aversion as a simulation of panic anxiety: elements of face and predictive validity. *Psychiatry Res*. 1995;57(2):181–191.
142. Bandler R, Carrive P. Integrated defence reaction elicited by excitatory amino acid microinjection in the midbrain periaqueductal grey region of the unrestrained cat. *Brain Res*. 1988; 439(1–2):95–106.

143. Walker DL, Cassella JV, Lee Y, De Lima TC, Davis M. Opposing roles of the amygdala and dorsolateral periaqueductal gray in fear-potentiated startle. *Neurosci Biobehav Rev*. 1997;21(6):743–753.
144. Mobbs D, Marchant JL, Hassabis D, et al. From threat to fear: the neural organization of defensive fear systems in humans. *J Neurosci*. 2009;29(39):12236–12243.
145. Mobbs D, Petrovic P, Marchant JL, et al. When fear is near: threat imminence elicits prefrontal-periaqueductal gray shifts in humans. *Science*. 2007;317(5841):1079–1083.
146. Blair RJ. Psychopathy, frustration, and reactive aggression: the role of ventromedial prefrontal cortex. *Br J Psychol*. 2010;101:383–399.
147. Karli P. The Norway rat's killing response to the white mouse. *Behavior*. 1956;10:81–103.
148. Wasman M, Flynn JP. Directed attack elicited from hypothalamus. *Arch Neurol*. 1962;6:220–227.
149. MacLean P,D, Delgado JMR. Electrical and chemical stimulation of fronto-temporal portions of limbic system in the waking animal. *Electroenceph Clin Neurophysiol*. 1953;5:91–100.
150. Rosvold HE, Mirsky AF, Pribram KH. Influence of amygdalectomy on social behavior in monkeys. *J Comp Physiol Psychol*. 1954;47(3):173–178.
151. Potegal M, Ferris CF, Hebert M, Meyerhoff J, Skaredoff L. Attack priming in female Syrian golden hamsters is associated with a c-fos-coupled process within the corticomedial amygdala. *Neuroscience*. 1996;75(3):869–880.
152. Potegal M, Hebert M, DeCoster M, Meyerhoff JL. Brief, high-frequency stimulation of the corticomedial amygdala induces a delayed and prolonged increase of aggressiveness in male Syrian golden hamsters. *Behav Neurosci*. 1996;110(2):401–412.
153. Siegel A, Roeling TA, Gregg TR, Kruk MR. Neuropharmacology of brain-stimulation-evoked aggression. *Neurosci Biobehav Rev*. 1999;23(3):359–389.
154. Siegel A, Victoroff J. Understanding human aggression: new insights from neuroscience. *Int J Law Psychiatry*. 2009;32(4):209–215.
155. Greenberg N, Scott M, Crews D. Role of the amygdala in the reproductive and aggressive behavior of the lizard, *Anolis carolinensis*. *Physiol Behav*. 1984;32(1):147–151.
156. Taylor SP. Aggressive behavior and physiological arousal as a function of provocation and the tendency to inhibit aggression. *J Pers*. 1967;35(2):297–310.
157. Kramer UM, Jansma H, Tempelmann C, Munte TF. Tit-for-tat: the neural basis of reactive aggression. *Neuroimage*. 2007;38(1):203–211.
158. Lotze M, Veit R, Anders S, Birbaumer N. Evidence for a different role of the ventral and dorsal medial prefrontal cortex for social reactive aggression: an interactive fMRI study. *Neuroimage*. 2007;34(1):470–478.
159. Veit R, Lotze M, Sewing S, Missenhardt H, Gaber T, Birbaumer N. Aberrant social and cerebral responding in a competitive reaction time paradigm in criminal psychopaths. *Neuroimage*. 2010;49(4):3365–3372.
160. Romero-Canyas R, Downey G, Berenson K, Ayduk O, Kang NJ. Rejection sensitivity and the rejection-hostility link in romantic relationships. *J Pers*. 2010;78(1):119–148.
161. Eisenberger NI, Lieberman MD, Williams KD. Does rejection hurt? An FMRI study of social exclusion. *Science*. 2003;302(5643):290–292.
162. Eisenberger NI, Way BM, Taylor SE, Welch WT, Lieberman MD. Understanding genetic risk for aggression: clues from the brain's response to social exclusion. *Biol Psychiatry*. 2007;61(9):1100–1108.
163. Denson TF, Pedersen WC, Ronquillo J, Nandy AS. The angry brain: neural correlates of anger, angry rumination, and aggressive personality. *J Cogn Neurosci*. 2009;21(4):734–744.
164. Rilling JK, Glenn AL, Jairam MR, et al. Neural correlates of social cooperation and non-cooperation as a function of psychopathy. *Biol Psychiatry*. 2007;61(11):1260–1271.
165. Yang Y, Glenn AL, Raine A. Brain abnormalities in antisocial individuals: implications for the law. *Behav Sci Law*. 2008;26(1):65–83.
166. Azrin NH, Hutchinson RR, Hake DF. Extinction-induced aggression. *J Exp Anal Behav*. 1966;9:191–204.
167. Thompson T, Bloom W. Aggressive behavior and extinction-induced response rate increase. *Psychon Sci*. 1966;5:335–336.
168. Hull CL. The Rats' speed of locomotion gradient in the approach to Food. *J Comp Psychol*. 1934;17(3):393–422.
169. Liu Z, Richmond BJ. Response differences in monkey TE and perirhinal cortex: stimulus association related to reward schedules. *J Neurophysiol*. 2000;83(3):1677–1692.
170. Vasquez EA, Denson TF, Pedersen WC, Stenstrom DM, Miller N. The moderating effect of trigger intensity on triggered displaced aggression. *J Exp Soc Psychol*. 2005;41:61–67.
171. Haner CF, Brown PA. Clarification of the instigation to action concept in the frustration-aggression hypothesis. *J Abnorm Psychol*. 1955;51(2):204–206.
172. Holton RB. Amplitude of an instrumental response following the cessation of reward. *Child Dev*. 1961;32:107–116.
173. Longstreth LE. Frustration and secondary reinforcement concepts as applied to human instrumental conditioning and extinction. *Psychol Monogr*. 1966;80(11):1–29.
174. Benderlioglu Z, Nelson RJ. Digit length ratios predict reactive aggression in women, but not in men. *Horm Behav*. 2004;46(5):558–564.
175. Benderlioglu Z, Sciulli PW, Nelson RJ. Fluctuating asymmetry predicts human reactive aggression. *Am J Hum Biol*. 2004;16(4):458–469.
176. Reynolds C. *The Sensing and Measurement of Frustration with Computers*. Massachusetts: Massachusetts Institute of Technology; 2001.
177. Kapoor A, Burleson W, Picard RW. Automatic prediction of frustration. *Int J Hum Comput Stud*. 2007;65:724–736.
178. Inc PST. E-Prime (Version 2.0) (computer software). Psychology Software Tools, Pittsburgh, PA; 2002.
179. Niv Y, Joel D, Dayan P. A normative perspective on motivation. *Trends Cogn Sci*. 2006;10(8):375–381.
180. Brown JS. The generalization of approach responses as a function of stimulus intensity and strength of motivation. *J Comp Psychol*. 1942;33:209–226.
181. Lewis M. Psychological effect of effort. *Psychol Bull*. 1965;64: 183–190.
182. Tracey I. Neuroimaging of pain mechanisms. *Curr Opin Support Palliat Care*. 2007;1(2):109–116.
183. Ploghaus A, Tracey I, Gati JS, et al. Dissociating pain from its anticipation in the human brain. *Science*. 1999;284(5422):1979–1981.
184. Sander D, Grandjean D, Pourtois G, et al. Emotion and attention interactions in social cognition: brain regions involved in processing anger prosody. *Neuroimage*. 2005;28(4):848–858.
185. Cremers HR, Demenescu LR, Aleman A, et al. Neuroticism modulates amygdala-prefrontal connectivity in response to negative emotional facial expressions. *Neuroimage*. 2010;49(1):963–970.
186. Stanton SJ, Wirth MM, Waugh CE, Schultheiss OC. Endogenous testosterone levels are associated with amygdala and ventromedial prefrontal cortex responses to anger faces in men but not women. *Biol Psychol*. 2009;81(2):118–122.
187. Ewbank MP, Lawrence AD, Passamonti L, Keane J, Peers PV, Calder AJ. Anxiety predicts a differential neural response to attended and unattended facial signals of anger and fear. *Neuroimage*. 2009;44(3):1144–1151.
188. McNaughton N, Corr PJ. A two-dimensional neuropsychology of defense: fear/anxiety and defensive distance. *Neurosci Biobehav Rev*. 2004;28(3):285–305.
189. Gray JA, McNaughton N. *The Neuropsychology of Anxiety*. New York: Oxford University Press; 2000.
190. Dougherty DD, Shin LM, Alpert NM, et al. Anger in healthy men: a PET study using script-driven imagery. *Biol Psychiatry*. 1999;46(4):466–472.

191. Damasio AR, Grabowski TJ, Bechara A, Ponto LL, Parvizi J, Hichwa RD. Subcortical and cortical brain activity during the feeling of self-generated emotions. *Nat Neurosci*. 2000;3(10):1049–1056.
192. Pietrini P, Guazzelli M, Basso G, Jaffe K, Grafman J. Neural correlates of imaginal aggressive behavior assessed by positron emission tomography in healthy subjects. *Am J Psychiatry*. 2000;157(11):1772–1781.
193. Strauss MM, Makris N, Aharon I, et al. fMRI of sensitization to angry faces. *Neuroimage*. 2005;26(2):389–413.
194. Knutson B, Cooper JC. Functional magnetic resonance imaging of reward prediction. *Curr Opin Neurol*. 2005;18(4):411–417.
195. Nelson RJ, Trainor BC. Neural mechanisms of aggression. *Nat Rev Neurosci*. 2007;8(7):536–546.
196. Minamimoto T, La Camera G, Richmond BJ. Measuring and modeling the interaction among reward size, delay to reward, and satiation level on motivation in monkeys. *J Neurophysiol*. 2009;101(1):437–447.
197. Yu R, Mobbs D, Seymour B, Rowe JB, Calder AJ. The neural signature of escalating frustration in humans. *Cortex*. 2014;54:165–178.
198. Niedhammer I, Tek ML, Starke D, Siegrist J. Effort-reward imbalance model and self-reported health: cross-sectional and prospective findings from the GAZEL cohort. *Soc Sci Med*. 2004;58(8):1531–1541.
199. van Wolkenten M, Brosnan SF, de Waal FB. Inequity responses of monkeys modified by effort. *Proc Natl Acad Sci USA*. 2007;104(47):18854–18859.
200. Hutchinson RR, Renfrew JW. Functional parallels between the neural and environmental antecedents of aggression. *Neurosci Biobehav Rev*. 1977;2:33–58.
201. Onoda K, Okamoto Y, Nakashima K, Nittono H, Ura M, Yamawaki S. Decreased ventral anterior cingulate cortex activity is associated with reduced social pain during emotional support. *Soc Neurosci*. 2009;4(5):443–454.
202. Masten CL, Eisenberger NI, Borofsky LA, et al. Neural correlates of social exclusion during adolescence: understanding the distress of peer rejection. *Soc Cogn Affect Neurosci*. 2009;4(2):143–157.
203. Takahashi H, Kato M, Matsuura M, Mobbs D, Suhara T, Okubo Y. When your gain is my pain and your pain is my gain: neural correlates of envy and schadenfreude. *Science*. 2009;323(5916):937–939.
204. Xu X, Zuo X, Wang X, Han S. Do you feel my pain? Racial group membership modulates empathic neural responses. *J Neurosci*. 2009;29(26):8525–8529.
205. Singer T, Seymour B, O'Doherty J, Kaube H, Dolan RJ, Frith CD. Empathy for pain involves the affective but not sensory components of pain. *Science*. 2004;303(5661):1157–1162.
206. Heath RG, Rouchell AM, Llewellyn RC, Walker CF. Cerebellar pacemaker patients: an update. *Biol Psychiatry*. 1981;16(10):953–962.
207. Heath RG, Llewellyn RC, Rouchell AM. The cerebellar pacemaker for intractable behavioral disorders and epilepsy: follow-up report. *Biol Psychiatry*. 1980;15(2):243–256.
208. Fazel S, Danesh J. Serious mental disorder in 23000 prisoners: a systematic review of 62 surveys. *Lancet*. 2002;359(9306):545–550.
209. Etkin A, Prater KE, Schatzberg AF, Menon V, Greicius MD. Disrupted amygdalar subregion functional connectivity and evidence of a compensatory network in generalized anxiety disorder. *Arch Gen Psychiatry*. 2009;66(12):1361–1372.
210. Passamonti L, Fairchild G, Goodyer IM, et al. Neural abnormalities in early-onset and adolescence-onset conduct disorder. *Arch Gen Psychiatry*. 2010;67(7):729–738.
211. Lahey BB, Loeber R, Quay HC, et al. Validity of DSM-IV subtypes of conduct disorder based on age of onset. *J Am Acad Child Adolesc Psychiatry*. 1998;37(4):435–442.
212. Herpertz SC, Huebner T, Marx I, et al. Emotional processing in male adolescents with childhood-onset conduct disorder. *J Child Psychol Psychiatry*. 2008;49(7):781–791.
213. Decety J, Michalska KJ, Akitsuki Y, Lahey BB. Atypical empathic responses in adolescents with aggressive conduct disorder: a functional MRI investigation. *Biol Psychol*. 2009;80(2):203–211.
214. Fairchild G, Passamonti L, Hurfor G, et al. Brain structure abnormalities in early-onset and adolescence-onset conduct disorder. *Am J Psychiatry*. 2011;168(6):624–633.
215. Sterzer P, Stadler C, Poustka F, Kleinschmidt A. A structural neural deficit in adolescents with conduct disorder and its association with lack of empathy. *Neuroimage*. 2007;37(1):335–342.
216. Blair RJ. Too much of a good thing: increased grey matter in boys with conduct problems and callous-unemotional traits. *Brain*. 2009;132(Pt 4):831–832.
217. De Brito SA, Mechelli A, Wilke M, et al. Size matters: increased grey matter in boys with conduct problems and callous-unemotional traits. *Brain*. 2009;132(Pt 4):843–852.
218. Fairchild G, van Goozen SH, Stollery SJ, et al. Cortisol diurnal rhythm and stress reactivity in male adolescents with early-onset or adolescence-onset conduct disorder. *Biol Psychiatry*. 2008;64(7):599–606.
219. Herpertz SC, Mueller B, Qunaibi M, Lichterfeld C, Konrad K, Herpertz-Dahlmann B. Response to emotional stimuli in boys with conduct disorder. *Am J Psychiatry*. 2005;162(6):1100–1107.

CHAPTER

12

Emotional Learning and Regulation in Social Situations

Andreas Olsson, Irem Undeger, Jonathan Yi

Emotion Lab at Karolinska Institutet, Department of Clinical Neuroscience, Division of Psychology, Stockholm, Sweden

OUTLINE

1. INTRODUCTION

Emotional learning and regulation are fundamental abilities needed to adaptively navigate in our environment. These processes are present at all walks of human life and come in many different forms. For example, you might learn to dislike and avoid your new neighbor because of a personally aversive experience—maybe a heated quarrel or just an unfriendly gaze. The same aversion might be acquired indirectly by watching, or hearing about the neighbor harassing someone else. Similarly, watching a trusted person cheerfully interacting with the neighbor might help you to attenuate your aversions and regulate your future behavioral responses toward the neighbor. These are examples of how other individuals might serve as both targets and sources of emotional learning. Understanding the mechanisms of emotional learning and regulation in social situations is central to how we view the emergence, maintenance, and change of social behavior with relevance for a myriad of life events, ranging from neighborly quarrels and bullying to cultural taboos and clinical disorders.

Recent research in the fields of cognitive and affective neuroscience has made tremendous gains in terms of understanding the neurobiological bases of social cognitions, traits, and mental states, and how such psychological

Neuroimaging Personality, Social Cognition, and Character
http://dx.doi.org/10.1016/B978-0-12-800935-2.00012-9

constructs affect behavior.[1,2] Yet, relatively little effort has been placed on examining the associated learning processes.[3] This is surprising because learning is ubiquitous; humans and other species constantly encode and use information to predict emotionally relevant events in their environment and to regulate their behavior to attain their goals.[4] Indeed, the lion share of emotional learning and regulation in our species might be social in nature. For example, many of our fears and anxieties might be transmitted *from* others and might in fact be *about* others.

This chapter focuses on recent research on the biological bases of emotional learning and regulation. We describe emerging lines of research aiming to capture the dynamic situations in which social learning usually occurs. In particular, we will survey research using functional magnetic resonance imaging (fMRI), but we will also take efforts to describe the behavioral paradigms that are at the core of every successful imaging study, because brain activity is interpretable only in light of meaningful behavioral responses. Throughout, we will discuss social learning and regulation with a focus on fears and aversions. These are the emotional domains currently most explored in terms of social-emotional learning.

We begin by surveying direct emotional learning (Pavlovian conditioning) and regulation (extinction): how we come to learn new, and modify existing, emotional responses in the light of direct, personal experiences of threat and safety. We then introduce social learning and regulation, here illustrated mainly by research on the transmission of information through observation. Shared and unique features of social and direct forms of learning and regulation are discussed and related to a recently emerging literature on the impact of social perception and cognition on observational emotional fear learning. To further the understanding of the brain basis of learning from other individuals' emotional expressions, we review relevant research on empathy and appraisals of others' mental states in humans. This line of research is highly related to, but has so far been largely unconnected with, the study of observational learning.[3] Finally, we describe a simple neurobiological model of observational fear learning that may serve as a basic building block in the understanding of far more complex instances of emotional learning and regulation in social situations. We conclude by outlining a few challenges and unanswered questions for future research in the field.

2. EMOTIONAL LEARNING AND REGULATION THROUGH FIRSTHAND EXPERIENCES

The last two decades have seen an explosion of knowledge about the biological foundation of emotional learning through direct experiences using Pavlovian or "classical" fear conditioning, which has served as a model for understanding human emotional learning at large.[5] In a traditional fear-conditioning procedure, a neutral conditioned stimulus (CS), such as a tone or a colored square, is paired with the direct experience of a naturally aversive stimulus (unconditioned stimulus, or US), such as an electric shock (see Figure 1(A)). The US results in a collection of defensive unlearned fear responses (unconditioned responses, or UR). Importantly, these responses are transferred to the predictive CS and persist in the absence of the US when the CS is later presented alone (conditioned response, or CR).

2.1 The Amygdala and the Emotional Learning Network

Research in nonhuman animals has provided a detailed understanding of the brain mechanisms and pathways underlying fear conditioning. Across species, the primary brain region involved in the learning, storage, and expression of conditioned fear is the amygdala, a nut-shaped subcortical structure deep within the bilateral medial temporal lobes. The amygdala serves as a hub in fear learning and is functionally interconnected with both other subcortical and more distant cortical regions that critically shape learning and regulation (see Figure 2(A)). Although the amygdala processes a wide range of emotionally relevant information,[6,7] its anatomy and functional role has been especially well studied using fear conditioning paradigms.[5,8]

The amygdala is a conglomerate of several subnuclei with specific roles in fear conditioning. In short, research in rodents shows that sensory information arrives in the lateral nucleus of the amygdala from regions in the thalamus and sensory cortices.[9,10] The lateral nucleus also receives nociceptive information specific for the US, providing a biological basis for the coupling of CS-US information. In fact, research suggests that this is the location of synaptic plasticity, shaping associations between representations of the CS and US.[11–13] The lateral nucleus subsequently channels information to the basal and central nuclei that mediate the output to other brain regions that regulate the expression of fear and anxiety.[14] For example, projections from the central nucleus to the hypothalamus are important for the mediation of autonomic responses,[15] which in humans can be measured through the skin conductance response.[16] Other areas of projection, such as the ventral tegmental area[17] and the periaqueductal gray,[18] serve roles in regulation of basic defensive behaviors. Another fear-related behavior, avoidance, is mediated by input to the basal ganglia from the basal nucleus.[19] The striatum, which serves as the input region of the basal ganglia, signals prediction errors during reward learning,[20] and its role in aversive

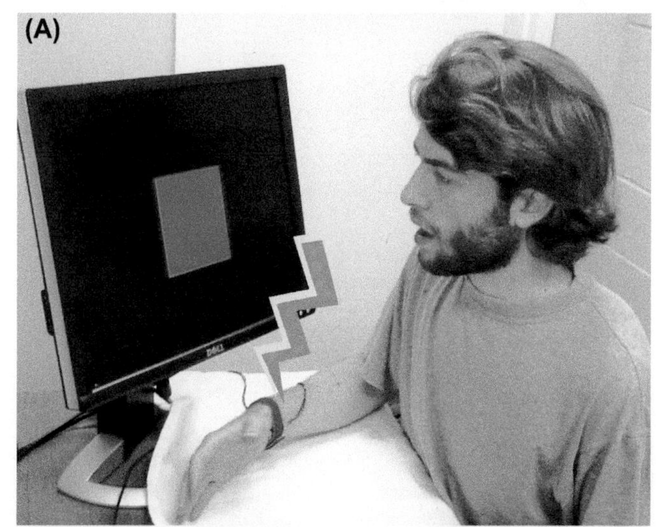

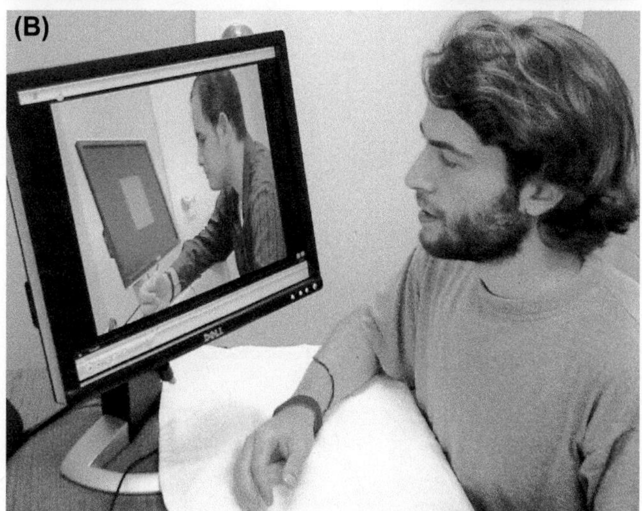

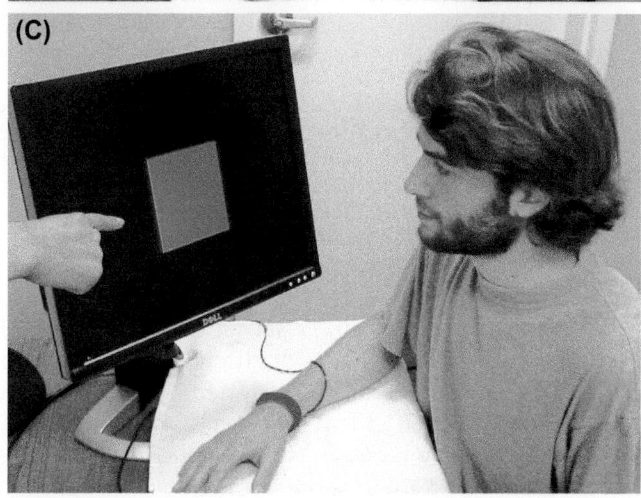

FIGURE 1 Nonsocial and social fear learning in humans. An individual learns to fear a conditioned stimulus (CS) through its pairing with (A) an electric shock to the wrist (Pavlovian conditioning), (B) a learning model's ("demonstrator's") expression of distress (observational or vicarious fear learning), and (C) verbal information about its aversive properties (instructed fear learning). *From Ref. 34.*

learning has recently become more clear.[21,22] In support of a cross-species view of these mechanisms, research using fMRI in humans has largely reported similar functional roles of the amygdala and its connectivity with other regions.[5,23] However, the amygdala is not only playing a key role in aversive learning. Recent evidence shows that this region is important also in the processing of appetitive information.[24,25]

As will be discussed in more detail later, the interconnectivity both within and between the amygdala and other subcortical regions, such as the striatum, appears to have important implications for social-emotional learning in humans. The same applies to the functional connectivity between the amygdala and cortical regions involved in the following: (1) the processing and regulation and of affective information, such as the anterior insula (AI), anterior cingulate cortex (ACC), and ventral medial prefrontal cortex (vmPFC); (2) contextual and mnemonic aspects, such as the hippocampus; and (3) the perception of, and thinking about, other people, such as the fusiform gyrus (FG), superior temporal sulcus (STS), temporal-parietal regions belonging to the "mirror neuron system" (MNS), and dorsal medial prefrontal cortex (dmPFC). The functional roles of these distributed regions and their functional connections with the amygdala provide the bases for emotional (especially aversive) learning in social situations. Figure 2(A) displays a selection of functional connections between the amygdala and other brain regions involved in Pavlovian fear conditioning.

2.2 The Prefrontal Cortex and Emotional Regulation

Across species, the prefrontal cortex (PFC) is critical to the regulation of learned and nonlearned affective responses through its connectivity with subcortical regions, such as the amygdala.[26,27] In particular, the ventral (infralimbic) region of the medial prefrontal cortex (mPFC) is necessary for the retention of extinction (i.e., the downregulation of learned fear responses through repeated exposure to nonreinforced presentations of the CS) of conditioned fear in rats.[28] Supporting the view that basic regulation of learned aversions also draws on evolutionarily conserved mechanisms, the human homolog of this region is involved in extinction learning in humans.[29] Active regulation of affective expressions by appraisal of the emotional meaning of a given situation involves more dorsal and lateral regions of the PFC. For example, the dorsolateral prefrontal cortex (dlPFC) has been assigned a key role in the "upregulation" and "downregulation" of affective responses to images through deliberate appraisal strategies by their impact on amygdala activity.[30] In accordance with these findings, a study by Delgado

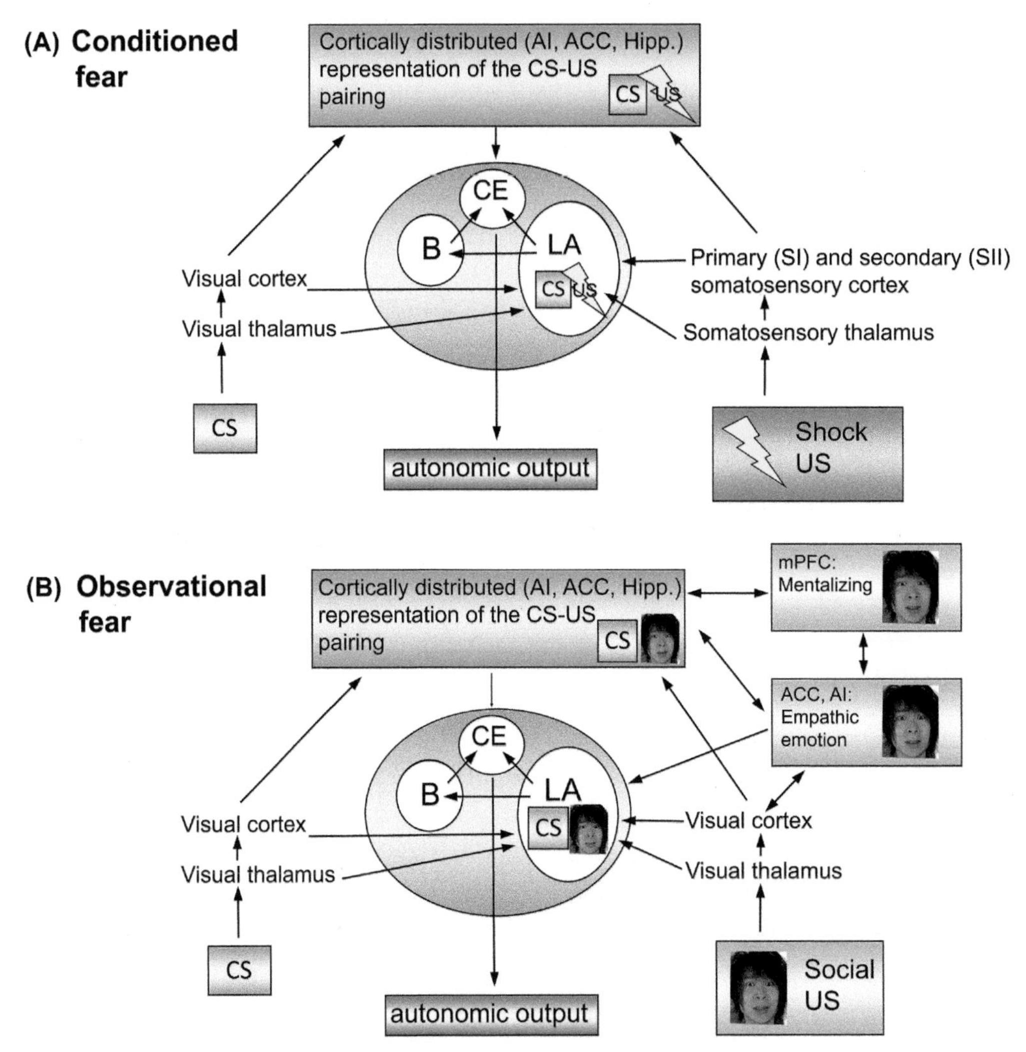

FIGURE 2 A simple neural model of direct, Pavlovian, and observational fear learning in humans. The arrows describe the flow of information between different functional brain regions. Although the arrows point only in one direction, the functional connectivity is bidirectional in most cases. (A) Pavlovian fear conditioning occurs by associating the visual representation of the CS with the somatosensory representation of the conditioned stimulus (CS), such as a blue square with the somatosensory representation of the aversive unconditioned stimulus (US—such as a shock). The lateral nucleus (LA), in which sensory information of the CS and US converge, is believed to be the site of learning. The amygdala also receives input from the hippocampal memory system (hipp.), anterior insula (AI), and anterior cingulate cortex (ACC) containing secondary representations of the CS and US, information about the learning context and the internal state of the organism. (B) In observational fear learning, the visual representation of the distressed model serves as the US. As in Pavlovian conditioning, it is hypothesized that representations of the CS and "social" US converge in the LA. The strength of the social US may be modified by the medial prefrontal cortex (mPFC) input related to the interpretation of the model's mental state, as well as cortical representations of empathic pain through the AI and ACC.

and colleagues[21] reported that subjects using appraisal strategies recruited the dlPFC to downregulate their conditioned fear responses, and related amygdala activity, to the CS. Notably, this study showed activation in the mPFC overlapped with regions previously implicated in studies of extinction of conditioned fear in humans, suggesting both similarities (mPFC) and differences (dlPFC) during passive (extinction) and active (reflective) regulation of learned emotional responses. Consistent with this, a recent study[31] demonstrated that spontaneous modulation of aversive emotions involved a combination of two factors: (1) moment-to-moment modulations depending on dlPFC and dmPFC regions that support reappraisal and (2) an impact of individual differences in modulation depending on a region in the vmPFC that overlapped with the region commonly identified in fear extinction.

The connectivity between the amygdala and the prefrontal cortex is likely to serve a functionally important role in socially mediated emotional regulation. During such situations, information about the social present and memories of the past need to be integrated with a valuation process. Indeed, vmPFC is believed to integrate information from distributed brain regions involved in signaling affective value, episodic memory, and social cognition.[32]

3. SOCIAL LEARNING AND REGULATION OF FEAR

Whereas research on the biological foundation of direct fear conditioning using simple CS and US has made enormous leaps forward, far less is known about the mechanisms supporting social transmission of fear and other emotions. This is surprising because social channels of information transmission might be more representative of the natural environment of our own species. Indeed, humans routinely acquire emotionally significant information at different levels: through abstract language, social observation, and interaction[33,34] (Figure 1(B) and (C)). This section will keep with the focus on fear, and to illustrate the social nature of learning and regulation, we will primarily draw on studies using observational manipulations.

3.1 Learning by Observing Others: Basic Behavioral Paradigms

3.1.1 Shared Processes in Observational and Pavlovian Learning

In a social species like our own, expressions of fear and distress are extremely salient cues that have a rapid and strong impact on observers.[6] This profound impact stems from the important survival value of detecting, and learning about, potential threats in the environment. Research on social fear learning in humans and other species has shown that the mere sight of stress or pain in a conspecific (a "model" or "demonstrator") can serve as a potent proxy for the individual's own direct experience, thereby offering a more safe and efficient route to learning, as compared to individual trial and error.[34–40] The acquired learning can be expressed at a later point in time in the absence of the model, thus distinguishing observational learning from related social phenomena that are expressed in the presence of the model, such as in "imitation" and "social facilitation."[34]

The behavioral impact of observational fear learning on an observer individual has been studied in many species, including birds[41], rodents,[24,42–45] cats,[46] cows,[47] and primates.[48–54] Taken together, these lines of research suggest that observational and direct Pavlovian fear learning operates through partly overlapping associative mechanisms.[34,50] This claim was explicitly tested in a seminal series of behavioral studies by Mineka and colleagues. In these studies, cage-reared monkeys were shown model monkeys reacting fearfully to snakes (toy or real) or to non-fear-relevant objects.[50,51] When fear-relevant objects (snakes) were used, the relationships between the strength of a learning model's expressed distress, the observer's immediate emotional response to the model's distress, and the learned fear in the observer as measured at a later time were similar to the known relationships between the US, UR, and CR in Pavlovian fear conditioning.[50,51] This single social learning episode with a fearful fellow monkey produced a strong and robust fear response that was measured several months after the encounter.[50] These findings strongly indicated that observational fear learning draws on the same underlying processes as direct fear conditioning.

Behavioral and psychophysiology research in humans has broadly demonstrated consistent results.[36,49,52,54] Providing direct support of similarities between these kinds of fear learning, a study by Olsson and Phelps[52] compared the effects of observational and instructed fear learning with Pavlovian fear conditioning within the same subjects (see Figure 1). The results demonstrated fear learning, measured as skin conductance, of equal magnitude following the three ways of learning when the CSs were presented supraliminally (seen and explicitly reported by the subjects). Interestingly, when CSs were presented subliminally (and reportedly not seen), the conditioned learned fear responses were comparable in the groups that had learned either by watching a model's aversive responses to the CS or through direct, Pavlovian conditioning. This lent further support to the idea that similar learning processes and representations underlie observational and direct, conditioned fear. In other words, although the learning *procedures* are different, the underlying *processes* might be partly overlapping. This proposal has later been strengthened by neurobiological evidence from nonhuman animals and imaging research on humans[42,53,55] (see Figure 2). Before we examine these lines of research, we will discuss features distinguishing observational from direct emotional learning.

3.1.2 Processes Distinguishing Observational from Pavlovian Learning

In spite of the similarities as discussed in previous sections, observational learning involves the processing of social information, which is per definition absent in traditional fear conditioning procedures. The learning effect of the emotional expression of the other individual is likely to be mediated by the observer's perception of the other individual and their interpersonal learning history, as well as the observer's stable social cognitive abilities. Indeed, mice observing a familiar, but not an unfamiliar, mouse experiencing pain displayed enhanced sensitization to pain on a later test.[56,57] Research that more directly examined learning showed that the social transmission of emotionally relevant information was enhanced by relatedness,[45,58] familiarity,[59] and social status.[58] In primates, the intrinsic aversiveness of observing a conspecific in pain was evidenced by the willingness of monkeys to starve themselves as long as a shock was administered to a fellow monkey every time the observer attempts to eat,[60] but again, this altruistic behavior was influenced by familiarity and past experience of the conspecific.[60,61]

If social information profoundly impacts fear learning through social channels in rodents, as discussed in the previous section, this impact is likely to be at least as potent in our own species. In an early study of observational fear learning in humans, Berber[48] showed that another person's arm movement in response to a shock served as a US, but only when the observer believed that the movement was caused by a shock, and not when the belief was that the model's arm moved without a shock or when a shock was delivered without the arm moving. These findings supported the conclusion that perceptual properties of the learning model interact with the observer's understanding of the model's (the demonstrator's) mental states to instigate an unconditioned response. Similarly, information about a model's alleged spider phobia induced an aversive response to a spider that was presented to the "phobic" model, even without any physical cues of distress,[49] which is consistent with later findings that the affective response in an observer can be modified by contextual information transmitted through instructions about the intentions of a target person who expresses distress and pain.[62,63] In a recent study, Olsson et al.[64] directly supports the conjecture that empathic processes, including appraisals about the model's mental states, causally affects observational learning. This study manipulated empathy appraisals using standard instructions[65,66] to "turn up" and "turn down" empathy with a learning model, who received mild electric shocks to the wrist. The results confirmed that both active appraisal, as well as the observer's measured trait empathy, affected the expression of learned autonomic fear (skin conductance) at a later point in time, when the model was no longer present. More specifically, actively taking the model's emotional perspective increased the observer's expression of learned fear relative to not taking the model's perspective or responding naturally to the model's pain and distress. Interestingly, these effects were stronger in subjects that scored high in trait empathy.

Learning from others not only affects the observer's responses to a CS during passive viewing during a later test phase. Historically, the study of observational learning has often been about the effects on instrumental behavior (e.g., skills) of observing a model's behavior,[67] and it has been argued that learning behaviors from others is especially adaptive in a dangerous environment.[68] Such social instrumental learning should, however, be applied critically, implying that an evaluation of the model's competence should be advantageous.[69] In accordance with this assumption, research using an instrumental observational learning procedure has shown that learning from a skilled versus an unskilled model is more likely to optimize the observer's behavior in order to avoid aversive punishments.[70] Interestingly, although unrelated to the observed models actual skill, receiving verbal information about skill level also impacted the ensuing learning.[71] In fact, models believed to be skilled versus unskilled promoted better learning (i.e., optimized instrumental avoidance), an effect that was likely to be mediated by attention.

Taken together, the studies reviewed in this section show that social cognition plays an important role in shaping behavior based on observational experiences. This realization uncovers conceptual links to the field of research on empathy and mental state attributions, as well as their brain bases, which we will turn to next.

4. PROCESSING SOCIAL INFORMATION: EMPATHY AND MENTAL STATE ATTRIBUTIONS

The demonstration that social information, through processes such as empathy and mental state attributions (e.g., about distress, safety, or competence), causally can affect social learning and regulation is consistent with accumulating evidence in related, but unconnected, lines of research in the social and affective neurosciences. These lines of research describe the effects of observing other peoples' emotional expressions and how these effects depend on various situational and contextual factors.

Empathic responses are supported partly by the same mechanisms as self-experienced distress and pain, including the AI and the ACC,[72,73] together with attributions of mental states to the other, involving the mPFC.[74–76] Importantly, empathic responses depend on attributions of mental states and traits to the target persons in predicted ways. For example, observers who believe that social targets are competitive or untrustworthy exhibit inhibited spontaneous empathic responses,[62] and attenuated empathic responses to cheaters are related to decreased activity in empathy-related brain regions.[63,77] In contrast, perceiving the target as a cooperator[62] or attending to his or her painful experiences[66] enhances empathic responses. In fact, many perspective-taking manipulations strongly alter individuals' social behavior and their brain bases across a variety of contexts.[65,78,79]

The link between empathic and learning processes in the brain was substantiated by an imaging study[53] showing that activity in the AI and the ACC during observation of a demonstrator's pain predicted the strength of the learning when later tested. The impact of empathy on learning might be possible because the demonstrator's expression of fear and pain serves as a "social" US, similar to a personally experienced US (Figure 2(B)). Accordingly, the quality of the social US might determine the outcome of vicarious learning, similar to how the quality of a directly experienced, tactile US, such as a mild electric shock, determines the outcome of Pavlovian conditioning (see Figure 2(A) and (B)). In light of this associative model of social fear learning, which will be discussed further in Section 7 below, deliberate attempts to alter ("turning up" and "turning down") empathic appraisals to a demonstrator's emotional responses should produce differences in vicarious fear learning along the corresponding gradients. Indeed, this was exactly what was shown in the previously discussed study by Olsson and colleagues.[64]

Not only aversive emotions are vicariously transmitted. Research on human[80–82] and nonhuman[83] primates on the neural and behavioral bases of watching others' reward expressions show that these processes are supported by partly the same neural processes as self-experienced reward. In analogy with observational fear learning, vicarious reward learning might therefore partly involve the same mechanisms as direct reward learning together with appraisals of the model's emotions and mental states. This remains to be explored in future research.

In addition to situational factors, such as appraisals, individuals vary strongly in their empathic ability. Past research has documented considerable interindividual difference in empathic ability as measured by self-reports,[84,85] physiological concurrence over time,[86] accurate understanding of others' emotions,[87,88] and activity in brain regions implicated in empathic processes, such as the medial prefrontal, insular, and temporal cortices.[72,76] Finally, individual differences and contextual factors often interact to produce empathy. For instance, individuals high, versus low, in empathy may differ from each other with respect to prosociality or the accuracy of their social inferences, but only when targets' group membership or expressivity provides an opportunity for individual differences to manifest themselves.[87,89]

5. LINKING OBSERVATIONAL FEAR LEARNING AND THE PROCESSING OF SOCIAL INFORMATION IN THE BRAIN

Taken together, the research on observational fear learning and empathic processes, as surveyed in previous sections, suggests that the brain bases of appraising other peoples' emotions and thoughts are closely linked to the mechanisms of emotional learning. This assertion is substantiated by work in both rodents and primates that will be reviewed next. In a study by Knapska and colleagues,[24] the rodent amygdala was shown to be engaged simply by interacting with a conspecific that has recently undergone fear conditioning, and brief interactions with conditioned cage mates facilitated avoidance learning and enhanced later conditioned freezing without affecting pain sensitivity. These results showed that socially transferred arousal can motivate aversive learning by priming the neural circuitry that is known to encode and express fear. In a later study on observational fear learning in rodents, Jeon and colleagues[45] directly investigated the brain structures critically involved in connecting pain perception and social learning. In this study, mice watched a conspecific receiving painful foot shocks in an adjacent chamber separated by a transparent divider. Following selective inactivation of the ACC, which is involved in the affective aspects of pain perception (see also Section 4 for a discussion of the human homologue in empathic processing), observational fear learning was impaired. Inactivation of thalamic nuclei involved in the sensory component of pain left learning intact, suggesting that it is the perceived, and not direct experience of, pain that enables the observational learning of fear. Interestingly, the functional connectivity between the amygdala and the ACC was augmented and more synchronized during and after learning relative to before. These results emphasize the critical role of the ACC in the learning of vicarious aversions in rodents. It also demonstrates the importance of the functional links between the amygdala and the ACC, which might play a modulatory role in determining learning outcome. Recent research in rodents has suggested that the critical role of the ACC during observational learning in rodents might be limited to the right hemisphere.[44]

The limited research on the brain bases of observational fear learning in humans is largely consistent with both the findings from the rodent research on observational fear learning that emphasizes the roles of the ACC and the amygdala and the existing imaging work on emotional sharing and mental state attributions in humans. Olsson and colleagues[53] asked subjects to watch a movie of another person expressing distress when receiving electric shocks paired with a CS. Later, subjects expected to receive shocks along with the same stimulus that was paired with the model's distress in the movie they just had watched. Importantly, no shocks were administered to the subjects during the test stage to ensure that their representation of the US-CS pairing was based solely on indirect, social experiences. The results showed that, similar to previous studies on direct fear conditioning, the bilateral amygdala was involved during both the learning (observation) and the subsequent expression (test) of learned fear, again supporting the assumption that similar associative computations and their underlying neural processes support both conditioned and observational fear learning. Similar conclusions can be drawn from an fMRI study by Hooker and colleagues.[55] In this study, subjects watched images of fearful and happy faces alone or together with arbitrary figures (learning objects). The results showed that the amygdala was more responsive to faces when their emotional expressions were directed toward a learning object than presentations of the emotional expressions or objects alone, pointing toward the importance of facial expressions as a source of emotional learning.

The lines of evidence reviewed here and in previous sections, as well as the rich connectivity known to exist between the amygdala and visual and ventral parts of the mPFC indicates that, at least in primates, representation of fear learning through observation and classical conditioning may be rather similar within the amygdala. However, as stressed previously, the procedural differences between nonsocial and social forms of learning also differ in fundamental ways, implying the involvement of partially dissociable neural networks outside the amygdala. For example, a conspecific's expression of distress may signal an imminent threat that serves as a US and elicits an immediate, unconditioned response in the observer that becomes associated with a CS. However, as supported by behavioral evidence (see Section 3.1), this response is also mediated by the observer's perception of the model. Accordingly, it was shown that activity in brain regions implicated in empathic processes and mental state attributions during the observation of another in distress predicted the strength of the expression of learned fear at a later time in the absence of the model.[53] This study was recently followed up by an fMRI experiment directly contrasting observational and Pavlovian fear learning in a within-subjects design. Accordingly, each subject was submitted to both an observational and a Pavlovian fear conditioning procedure in a counterbalanced order. To optimize the paradigm for examining the computational properties of these forms of learning, subjects were asked to rate their expectancy to receive shocks on a trial-by-trial basis. Computational reinforcement learning (RL) modeling was used to characterize the learning mechanisms and showed that both the direct and observational learning was best explained by a Hybrid Rescorla-Wagner/Pearce-Hall model.[22,90] In this model, the learning rate was gated trial-by-trial by associability, which broadly corresponds to an unsigned "surprise signal." The AI and amygdala showed overlapping associability-related activity for both observational and direct fear learning, including responses to the US. These findings show that the extension of a computational learning approach from activity in the striatum to the amygdala[22] also applies to socially transmitted threat information. In terms of divergences between the two kinds of learning, the observational condition was characterized by an enhanced cue-related activity in the dmPFC. Taken together, these results suggest overlapping cue, outcome (US), and learning-related representations between observational and direct learning. Moreover, the study shows that observational learning was distinguished by its involvement of the dmPFC, known to code for mental state attributions.[3,91] These findings dovetail closely with previous demonstrations of the importance of the mPFC for the strength of observational fear learning.[53]

6. BASIC SOCIAL REGULATION: OBSERVATIONAL EXTINCTION OF LEARNED FEAR

The influence of the presence of others and their behavior on learning is not limited to the effects of vicarious aversions. In fact, other individuals may serve to facilitate the regulation of fears and stress.[92–94] The attenuating effect of a conspecific in a threatening situation has long been observed in rodents, as shown by an early study by Anderson.[95] More recent studies have demonstrated that "social buffering" can mitigate conditioned fear in rodents.[96] In particular, it was found that the attenuating (buffering) effect was realized through the olfactory peduncle linking the main olfactory blub to the amygdala. Research in primates has shown that prior exposure to a nonfearful conspecific approaching a fear-relevant stimulus (a snake) can serve to "immunize" the observer from subsequent observational conditioning to the snake.[97] This effect is likely to depend on latent inhibition, meaning that a stimulus previously associated with safety has a retarded function as a CS in subsequent fear conditioning.

Similar to the effect of social information in the acquisition of fears through observation, socially transmitted safety information is dependent on the perception and appraisal of the safety model. In an attempt to study precisely this, Lungwitz and colleagues[98] used a habituation paradigm to examine the effects of social familiarity of a cage mate on anxiety-like behavior in rats. It was shown that social familiarity selectively reduced anxiety-like behaviors. The anxiolytic effect of social familiarity could be elicited over multiple training sessions and was specific to both the presence of the anxiogenic stimulus and the familiar social partner. The role of the familiar conspecifics as a safety signal was confirmed by the fact that anxiety-like responses returned in the absence of the familiar partner. Moreover, inactivation of the PFC selectively blocked the expression of social familiarity-induced anxiolysis, while having no effect on a novel partner. This PFC-dependent effect is likely to involve the integration of social cues of familiarity with contextual and emotional information to regulate anxiety-like behavior.

In humans, observational safety learning has long been used as a part of exposure treatment of phobias. In such treatment, the phobic individual watches the therapist—acting as a learning model—interact with the feared stimulus before the phobic individual is directly exposed to it.[99] Earlier behavioral research[36] has described the underlying processes of this form of vicarious treatment, but the conclusions drawn from these studies have been limited due to methodological constraints. These older, as well as the few existing recent, studies[100] have focused on phobic participants, thus preventing the generalization of the results to healthy individuals. In a more recent experimental study on healthy adults[94] (Figure 3), the additive effect of observational information beyond the traditional, direct extinction was examined. The results showed that subjects extinguished their learned fear more readily during observational extinction as compared to direct extinction learning, including only exposure to the CS. Intriguingly, and in contrast to traditional extinction, observationally attenuated learning did not recover when later tested. Similar to the research reviewed earlier, this study demonstrates that safety learning depends on social cognition: on perceiving the learning model to be safe (and not only the presence of a model per se). A follow-up[101] of the 2013 study aimed to examine the neural bases of observational extinction. In brief, the results replicated the finding that vicariously extinguished conditioned fear responses did not recover when later tested. In other words, although the subject did not receive shocks to any of the CS during extinction, only the CS that was safely approached by the model blocked the return of learned fear when provoked in a so-called "reinstatement test" (measuring recovery of fear after the presentation of unsignaled US presentations). Moreover, this social safety enhancement was accompanied by changes in connectivity between the vmPFC and other regions. In short, during extinction, the CS that the model expressed safety toward triggered less activity in the vmPFC and less vmPFC-amygdala connectivity as compared to the CS that the model expressed distress toward. Conversely, during the later reinstatement test, the "safe" CS enhanced vmPFC activity and enhanced functional coupling between the vmPFC and the hippocampus, which is consistent with the previously suggested role of this network in successful recall of context-dependent extinction memory.[102,103]

The attribution of safety is not the only kind of social information processing that influences the success of social regulation. In a related study by Golkar, Castro, and Olsson,[64] it was found that both observational fear and safety (extinction) learning from a racial in-group member was superior to learning from a racial out-group member. Critically, whereas fear was successfully acquired from both an in- and out-group demonstrator, safety learning was only transmitted between in-group members.

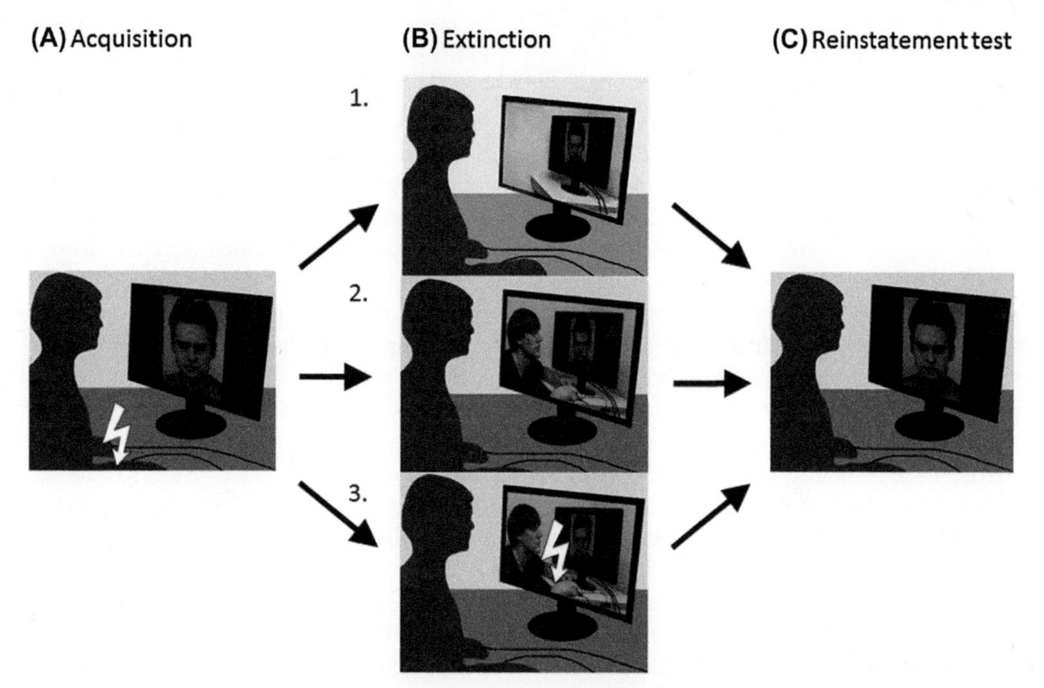

FIGURE 3 Observational/vicarious extinction paradigm. During the acquisition stage (A), all participants saw two faces, presented separately. One of the faces (shown here) was paired with a mildly aversive shock (US) to the participant's wrist, whereas the other face was never paired to the US. During the subsequent extinction stage (B), participants were divided up into three different groups. (1) In the "Direct (traditional) extinction" group, participants watched a movie of nonreinforced presentations of the CS (no model present). (2) In the "Observational extinction" group, participants watched a movie in which a model was placed in front of the screen where nonreinforced presentations of the CS were presented. The model appeared calm and safe. (3) To control for the presence of the model, participants in the "Observational reinforcement" group watched a movie in which a model was shown that reinforced presentations of the CS. The model responded with distress when receiving the shocks. The last experimental stage (C) assessed the effects of vicarious extinction on the recovery of learned fear. To this end, all participants received three unsignaled shocks, followed by a reinstatement-test stage, in which each CS was presented nonreinforced. *Modified from Ref. 94.*

These results suggest that social information about social group can determine whether social safety learning is possible. This dramatic impact might be mediated by regions in both the ventral and dorsal mPFC believed to integrate information from distributed brain regions involved in computing and integrating affective valuations of stimuli, episodic memory, and social cognition,[32,104] as well as the use of this information to provide a selective safety signal that indicates which stimuli are safe to ignore.[105]

7. A NEURAL MODEL OF EMOTIONAL LEARNING AND REGULATION

The research reviewed so far shows both important similarities and differences between direct and social forms of emotional learning and regulation. In short, the involvement of the amygdala and the AI is shared between direct and observational learning, whereas the activity of the vmPFC seems to be shared between direct and observational regulation (extinction). In contrast, observational forms of learning and extinction appear to be distinguished from their direct counterparts by the involvement of brain regions known to process social information, such as empathy and mental state attributions in the dorsal parts of the mPFC. These patterns of activity could be integrated in a neurobiological model of social learning and regulation.

Figure 2 describes a simple neurobiological model of social learning and extinction. The model proposes that the functional mechanisms underlying learning through social observation may be similar, but that the representations underlying the formation of associations are, at least partially, different. For example, the US in observational fear learning may be the perceived fear expression of a conspecific giving rise to an emotional response (a "social" UR). Similarly to how information about the naturally aversive shock is processed in Pavlovian conditioning, the representation of the fearful face is conveyed to the lateral nucleus through the sensory cortices and possibly the sensory thalamus. Importantly, the strength of the US in the lateral nucleus may be modified by cortical processes of empathic pain through input from the ACC and insular cortex and the perception and interpretation of the learning model's mental state as supported by the mPFC. The model suggests that, similar to classical fear conditioning, the lateral nucleus is a site of plasticity underlying memory for the CS-US association, in addition to a distributed cortical representation of the CS-US association acquired through the hippocampal memory system. The hippocampus and prefrontal regions might contribute to the meaning of the "social" US by linking the stimuli to contextual information. For example, is the fearful other a friend or a foe, authentic or a pretender, experienced or a novice? Although learning may occur regardless, such appraisals will affect the responding. The output mechanism for observational fear learning does not differ from that for fear conditioning.

The presented model remains speculative and should be viewed in light of two important caveats. First, the model highlights unidirectional projections between brain regions, but most of the regions we have discussed have bidirectional connections with the amygdala. Second, the striatum is not highlighted in the model. Yet, human brain-imaging studies on both conditioned[23,106] and observational[53] fear learning report activation of the striatum. This brain region is known to update the learner about the predictive value of both social and nonsocial cues and has been shown to be involved in learning during social observation and interaction.[22,107,108]

8. FUTURE CHALLENGES FOR THE STUDY OF SOCIAL LEARNING AND REGULATION

The majority of the surveyed studies have examined normally developing and healthy subjects reflecting the important role of social transmission of information in our everyday life. Indeed, social learning is likely to not only shape our evaluations and behavior toward particular events or people, but may in fact be at the core of human culture more generally,[67] including social phenomena (such as cultural avoidance traditions),[108] conformity, and norm compliance.[109,110] An important challenge for future research, along these lines, is to link learning in the brain with emerging properties in larger social networks. To this end, experimental methods in social and affective neuroscience should be combined with computational modeling, which has the advantage that it can be implemented in simulated agents in, for example, agent-based modeling.[108,111] The joint use of these methods might help us to characterize how social learning mechanisms at an individual level (e.g., via observation) can give rise to emergent properties on larger scales: networks of individuals, groups, and societies. Recent work has just begun to explore this research route.[108] It is our hope that the neurobiological model for social learning outlined above might serve to facilitate research efforts to link the brain to the societal level.

In many instances, socially transmitted information is adaptive. Recently, however, there has been a growing understanding of the role of social information in the development of a variety of psychological disorders, most prominently anxiety disorders,[35] which is reflected by the inclusion of social origins of traumas and phobias in the fifth edition of the *Diagnostic and Statistical Manual* (DSM V).[112] This means that social information is considered equally important as direct experiences in the etiology of, for example, posttraumatic stress disorders

(PTSD) and specific phobias. A related question for future research is to specify the effects of interacting sources of emotional information. For example, how are different combinations of verbally communicated, observationally gleaned, and self-experienced threatening experiences affecting the ensuing learning? Social learning of avoidance might also play a role in many different forms of dysfunctional behavior that do not necessarily fulfill diagnostic criteria. For example, socially learned avoidance behaviors might involve elements of magical thinking or superstition that might have been adaptive to our ancestors (e.g., food taboos)[108] but can be detrimental to the individual in our modern society.[113]

Another important methodological challenge for the study of emotional learning and regulation in social situations concerns ecological validity. For a better understanding of social learning as we know it from our everyday life (our human cultural ecology), we need to continue to enhance the naturalism of the learning situations that are investigated without compromising the experimental control. To this end, experimental, virtual social environments should be better used and integrated with both behavioral and neural measures of learning. More naturalistic environments will also increase the focus on different forms of learning from social feedback. To this end, computational approaches implementing reinforcement learning modeling has already begun to bear fruit, but much remains to be done to better capture the dynamic learning environment in social situations.

Although the importance of socially transmitted information about emotionally relevant events and stimuli was described long ago,[33,36] the specific processes underlying social transmission have only recently begun to be understood.[34,35,113] In this chapter, we have used research on observationally transmitted information related to fear and safety information to describe the processes underlying social learning and regulation. Future research should continue to map out the processes underlying observational learning and regulation of reward information.

Acknowledgments

This work was supported by a grant from the Riksbankens Jubileumsfond P11-1017 and an Independent Starting Grant (284366; Emotional Learning in Social Interaction) from the European Research Council to Andreas Olsson.

References

1. Lieberman MD. Social cognitive neuroscience: a review of core processes. *Annu Rev Psychol*. 2007;58:259–289. http://www.ncbi.nlm.nih.gov/entrez/query.fcgi?cmd=Retrieve&db=PubMed&dopt=Citation&list_uids=17002553.
2. Frith CD, Frith U. Mechanisms of social cognition. *Annu Rev Psychol*. 2012;63:287–313. http://dx.doi.org/10.1146/annurev-psych-120710-100449.
3. Olsson A, Ochsner KN. The role of social cognition in emotion. *Trends Cogn Sci*. 2008;12(2):65–71. http://dx.doi.org/10.1016/j.tics.2007.11.010. pii: S1364-6613(07)00337-3.
4. LeDoux JE. Coming to terms with fear. *Proc Natl Acad Sci USA*. 2014;111(8):2871–2878.http://www.pubmedcentral.nih.gov/articlerender.fcgi?artid=3939902&tool=pmcentrez&rendertype=abstract.
5. Phelps EA, LeDoux JE. Contributions of the amygdala to emotion processing: from animal models to human behavior. *Neuron*. 2005;48(2):175–187.http://www.ncbi.nlm.nih.gov/entrez/query.fcgi?cmd=Retrieve&db=PubMed&dopt=Citation&list_uids=16242399.
6. Adolphs R. The biology of fear. *Curr Biol*. 2013;23(2). http://dx.doi.org/10.1016/j.cub.2012.11.055.
7. Janak PH, Tye KM. From circuits to behaviour in the amygdala. *Nature*.2015;517(7534):284–292.http://dx.doi.org/10.1038/nature14188.
8. Johansen JP, Cain CK, Ostroff LE, Ledoux JE. Molecular mechanisms of fear learning and memory. *Cell*. 2011;147(3):509–524.
9. Amaral DG. Amygdalohippocampal and amygdalocortical projections in the primate brain. *Adv Exp Med Biol*. 1986;203:3–17.
10. LeDoux JE, Farb C, Ruggiero DA. Topographic organization of neurons in the acoustic thalamus that project to the amygdala. *J Neurosci*. 1990;10(4):1043–1054.
11. Schafe GE, Nader K, Blair HT, LeDoux JE. Memory consolidation of Pavlovian fear conditioning: a cellular and molecular perspective. *Trends Neurosci*. 2001;24(9):540–546. http://dx.doi.org/10.1016/S0166-2236(00)01969-X.
12. Romanski LM, Clugnet MC, Bordi F, LeDoux JE. Somatosensory and auditory convergence in the lateral nucleus of the amygdala. *Behav Neurosci*. 1993;107(3):444–450. http://dx.doi.org/10.1037/0735-7044.107.3.444.
13. Quirk GJ, Armony JL, LeDoux JE. Fear conditioning enhances different temporal components of tone-evoked spike trains in auditory cortex and lateral amygdala. *Neuron*. 1997;19(3):613–624. http://www.ncbi.nlm.nih.gov/entrez/query.fcgi?cmd=Retrieve&db=PubMed&dopt=Citation&list_uids=9331352.
14. Ledoux JE, Gorman JM. A call to action: overcoming anxiety through active coping. *Am J Psychiatry*. 2001;158(12):1953–1955. http://dx.doi.org/10.1176/appi.ajp.158.12.1953.
15. Price JL, Amaral DG. An autoradiographic study of the projections of the central nucleus of the monkey amygdala. *J Neurosci*. 1981;1(11):1242–1259.
16. Davis M, Whalen PJ. The amygdala: vigilance and emotion. *Mol Psychiatry*. 2001;6(1):13–34. http://www.ncbi.nlm.nih.gov/entrez/query.fcgi?cmd=Retrieve&db=PubMed&dopt=Citation&list_uids=11244481.
17. Simon H, Le Moal M, Calas A. Efferents and afferents of the ventral tegmental-A10 region studied after local injection of [^{3}H] leucine and horseradish peroxidase. *Brain Res*. 1979;178(1):17–40. http://dx.doi.org/10.1016/0006-8993(79)90085-4.
18. Hopkins DA, Holstege G. Amygdaloid projections to the mesencephalon, pons and medulla oblongata in the cat. *Exp Brain Res*. 1978;32(4):529–547. http://dx.doi.org/10.1007/BF00239551.
19. Everitt BJ, Robbins TW. Amygdala-ventral striatal interactions and reward-related processes. *Amygdala Neurobiol Asp Emot Mem Ment*. 1992:401–429.
20. Schultz W, Dayan P, Montague PR. A neural substrate of prediction and reward. *Science*. 1997;275(5306):1593–1599. http://dx.doi.org/10.1126/science.275.5306.1593.
21. Delgado MR, Gillis MM, Phelps EA. Regulating the expectation of reward via cognitive strategies. *Nat Neurosci*. 2008;11(8):880–881. http://dx.doi.org/10.1038/nn.2141.
22. Li J, Schiller D, Schoenbaum G, Phelps EA, Daw ND. Differential roles of human striatum and amygdala in associative learning. *Nat Neurosci*. 2011;14(10):1250–1252. http://dx.doi.org/10.1038/nn.2904.

23. Buchel C, Morris J, Dolan RJ, Friston KJ. Brain systems mediating aversive conditioning: an event-related fMRI study. *Neuron*. 1998;20(5):947–957. http://www.ncbi.nlm.nih.gov/entrez/query.fcgi?cmd=Retrieve&db=PubMed&dopt=Citation&list_uids=9620699.
24. Knapska E, Walasek G, Nikolaev E, et al. Differential involvement of the central amygdala in appetitive versus aversive learning. *Learn Mem*. 2006;13(2):192–200. http://dx.doi.org/10.1101/lm.54706.
25. Everitt BJ, Cardinal RN, Parkinson JA, Robbins TW. Appetitive behavior: impact of amygdala-dependent mechanisms of emotional learning. *Ann N Y Acad Sci*. 2003;985:233–250. http://dx.doi.org/10.1111/j.1749-6632.2003.tb07085.x.
26. Ochsner KN, Gross JJ. The cognitive control of emotion. *Trends Cogn Sci*. 2005;9(5):242–249. http://www.ncbi.nlm.nih.gov/entrez/query.fcgi?cmd=Retrieve&db=PubMed&dopt=Citation&list_uids=15866151.
27. Robbins TW. Chemistry of the mind: neurochemical modulation of prefrontal cortical function. *J Comp Neurol*. 2005;493:140–146. http://dx.doi.org/10.1002/cne.20717.
28. Quirk GJ, Garcia R, González-Lima F. Prefrontal mechanisms in extinction of conditioned fear. *Biol Psychiatry*. 2006;60(4):337–343. http://dx.doi.org/10.1016/j.biopsych.2006.03.010.
29. Phelps EA, Delgado MR, Nearing KI, LeDoux JE. Extinction learning in humans: role of the amygdala and vmPFC. *Neuron*. 2004;43(6):897–905. http://www.ncbi.nlm.nih.gov/entrez/query.fcgi?cmd=Retrieve&db=PubMed&dopt=Citation&list_uids=15363399.
30. Ochsner K, Gross J. Cognitive emotion regulation insights from social cognitive and affective neuroscience. *Curr Dir Psychol*. 2008;17(2):153–158. http://dx.doi.org/10.1111/j.1467-8721.2008.00566.x.
31. Silvers Ja, Wager TD, Weber J, Ochsner KN. The neural bases of uninstructed negative emotion modulation. *Soc Cogn Affect Neurosci*. 2014;10. http://www.ncbi.nlm.nih.gov/pubmed/24493847.
32. Roy M, Shohamy D, Wager TD. Ventromedial prefrontal-subcortical systems and the generation of affective meaning. *Trends Cogn Sci*. 2012;16(3):147–156. http://dx.doi.org/10.1016/j.tics.2012.01.005.
33. Rachman S. The conditioning theory of fear-acquisition: a critical examination. *Behav Res Ther*. 1977;15(5):375–387. http://www.ncbi.nlm.nih.gov/entrez/query.fcgi?cmd=Retrieve&db=PubMed&dopt=Citation&list_uids=612338.
34. Olsson A, Phelps EA. Social learning of fear. *Nat Neurosci*. 2007;10(9):1095–1102. http://dx.doi.org/10.1038/nn1968. pii: nn1968.
35. Askew C, Field AP. The vicarious learning pathway to fear 40 years on. *Clin Psychol Rev*. 2008;28(7):1249–1265. http://dx.doi.org/10.1016/j.cpr.2008.05.003.
36. Bandura A. *Social Learning Theory*. New York: General Learning Press; 1977. 1–46.
37. Mommaerts J, Goubert L, Devroey D. Empathy beyond the conceptual level. *Perspect Biol Med*. 2012;55(2):176–182.
38. Cook M, Mineka S. Observational conditioning of fear to fear-relevant versus fear-irrelevant stimuli in rhesus monkeys. *J Abnorm Psychol*. 1989;98(4):448–459. http://www.ncbi.nlm.nih.gov/entrez/query.fcgi?cmd=Retrieve&db=PubMed&dopt=Citation&list_uids=2592680.
39. Rachman S. The return of fear. *Behav Res Ther*. 1979;17(2):164–166. http://dx.doi.org/10.1016/0005-7967(79)90028-7.
40. Goubert L, Vlaeyen JWS, Crombez G, Craig KD. Learning about pain from others: an observational learning account. *J Pain*. 2011;12(2):167–174.
41. Zimmermann U, Curio E. Two conflicting needs affecting predator mobbing by great tits, *Parus major*. *Anim Behav*. 1988;36(3):926–932. http://dx.doi.org/10.1016/S0003-3472(88)80175-1.
42. Debiec J, Sullivan RM. Intergenerational transmission of emotional trauma through amygdala-dependent mother-to-infant transfer of specific fear. *Proc Natl Acad Sci*. 2014:1–6. http://dx.doi.org/10.1073/pnas.1316740111.
43. Kavaliers M, Choleris E. Antipredator responses and defensive behavior: ecological and ethological approaches for the neurosciences. *Neurosci Biobehav Rev*. 2001;25(7–8):577–586. http://dx.doi.org/10.1016/S0149-7634(01)00042-2.
44. Kim S, Matyas F, Lee S, Acsady L, Shin H-S. Lateralization of observational fear learning at the cortical but not thalamic level in mice. *Proc Natl Acad Sci*. 2012;109(38):15497–15501. http://dx.doi.org/10.1073/pnas.1213903109.
45. Jeon D, Kim S, Chetana M, et al. Observational fear learning involves affective pain system and $Ca_v1.2$ Ca^{2+} channels in ACC. *Nat Neurosci*. 2010;13(4):482–488. http://dx.doi.org/10.1038/nn.2504.
46. John ER, Chesler P, Bartlett F, Victor I. Observation learning in cats. *Science*. 1968;159(3822):1489–1491. http://dx.doi.org/10.1126/science.159.3822.1489.
47. Rushen J, Munksgaard L, Marnet PG, DePassillé AM. Human contact and the effects of acute stress on cows at milking. *Appl Anim Behav Sci*. 2001;73(1):1–14. http://dx.doi.org/10.1016/S0168-1591(01)00105-8.
48. Berber SM. Conditioning through vicarious instigation. *Psychol Rev*. 1962;69:450–466. http://dx.doi.org/10.1037/h0046466.
49. Hygge S, Ohman A. Modeling processes in the acquisition of fears: vicarious electrodermal conditioning to fear-relevant stimuli. *J Pers Soc Psychol*. 1978;36(3):271–279. http://dx.doi.org/10.1037/0022-3514.36.3.271.
50. Mineka S, Cook M. Mechanisms involved in the observational conditioning of fear. *J Exp Psychol Gen*. 1993;122(1):23–38. http://www.ncbi.nlm.nih.gov/entrez/query.fcgi?cmd=Retrieve&db=PubMed&dopt=Citation&list_uids=8440976.
51. Mineka S, Davidson M, Cook M, Keir R. Observational conditioning of snake fear in rhesus monkeys. *J Abnorm Psychol*. 1984;93(4):355–372. http://www.ncbi.nlm.nih.gov/entrez/query.fcgi?cmd=Retrieve&db=PubMed&dopt=Citation&list_uids=6542574.
52. Olsson A, Phelps EA. Learned fear of "unseen" faces after Pavlovian, observational, and instructed fear. *Psychol Sci*. 2004;15(12):822–828. http://www.ncbi.nlm.nih.gov/entrez/query.fcgi?cmd=Retrieve&db=PubMed&dopt=Citation&list_uids=15563327.
53. Olsson A, Nearing KI, Phelps EA. Learning fears by observing others: the neural systems of social fear transmission. *Soc Cogn Affect Neurosci*. 2007;2(1):3–11. http://dx.doi.org/10.1093/scan/nsm005.
54. Vaughan KB, Lanzetta JT. Vicarious instigation and conditioning of facial expressive and autonomic responses to a model's expressive display of pain. *J Pers Soc Psychol*. 1980;38(6):909–923. http://www.ncbi.nlm.nih.gov/entrez/query.fcgi?cmd=Rerieve&db=PubMed&dopt=Citation&list_uids=7391931.
55. Hooker CI, Germine LT, Knight RT, D'Esposito M. Amygdala response to facial expressions reflects emotional learning. *J Neurosci*. 2006;26(35):8915–8922. http://www.ncbi.nlm.nih.gov/entrez/query.fcgi?cmd=Retrieve&db=PubMed&dopt=Citation&list_uids=16943547.
56. Langford DJ, Crager SE, Shehzad Z, et al. Social modulation of pain as evidence for empathy in mice. *Science*. 2006;312(5782):1967–1970. http://dx.doi.org/10.1126/science.1128322.
57. Ben-Ami Bartal I, Decety J, Mason P. Empathy and prosocial behavior in rats. *Science*. 2011;334(6061):1427 1430. http://dx.doi.org/10.1126/science.1210789.
58. Kavaliers M, Colwell DD, Choleris E. Kinship, familiarity and social status modulate social learning about "micropredators" (biting flies) in deer mice. *Behav Ecol Sociobiol*. 2005;58(1):60–71. http://dx.doi.org/10.1007/s00265-004-0896-0.
59. Jones CE, Riha PD, Gore AC, Monfils M-H. Social transmission of Pavlovian fear: fear-conditioning by-proxy in related female rats. *Anim Cogn*. 2014;17(3):827–834. http://dx.doi.org/10.1007/s10071-013-0711-2.
60. Masserman JH, Wechkin S, Terris W. "Altruistic" behavior in rhesus monkeys. *Am J Psychiatry*. 1964;121:584–585. http://dx.doi.org/10.1176/appi.ajp.121.6.584.

61. Hauser MD, Chen MK, Chen F, Chuang E. Give unto others: genetically unrelated cotton-top tamarin monkeys preferentially give food to those who altruistically give food back. *Proc Biol Sci*. 2003; 270(1531):2363–2370. http://dx.doi.org/10.1098/rspb.2003.2509.
62. Lanzetta JT, Englis BG. Expectations of cooperation and competition and their effects on observers' vicarious emotional responses. *J Pers Soc Psychol*. 1989;56:543–554.
63. Singer T, Seymour B, O'Doherty JP, Stephan KE, Dolan RJ, Frith CD. Empathic neural responses are modulated by the perceived fairness of others. *Nature*. 2006;439(7075):466–469. http://www.ncbi.nlm.nih.gov/entrez/query.fcgi?cmd=Retrieve&db=PubMed&dopt=Citation&list_uids=16421576.
64. Olsson A, McMahon K, Papenberg G, Bolger N, Zaki J, Ochsner KN. Vicarious fear learning depends on empathic appraisals and trait empathy. *Psychol Sci*. In press.
65. Batson CD, Lishner DA, Carpenter A, et al. "...As you would have them do unto you": Does imagining yourself in the other's place stimulate moral action? *Pers Soc Psychol Bull*. 2003;29(9):1190–1201. http://www.ncbi.nlm.nih.gov/entrez/query.fcgi?cmd=Retrieve&db=PubMed&dopt=Citation&list_uids=15189613.
66. Lamm C, Batson CD, Decety J. The neural substrate of human empathy: effects of perspective-taking and cognitive appraisal. *J Cogn Neurosci*. 2007;19(1):42–58. http://www.ncbi.nlm.nih.gov/entrez/query.fcgi?cmd=Retrieve&db=PubMed&dopt=Citation&list_uids=17214562.
67. Tomasello M, Kruger AC, Ratner HH. Cultural learning. *Behav Brain Sci*. 2010;16(3):495. http://www.journals.cambridge.org/abstract_S0140525X0003123X.
68. Kendal RL, Coolen I, van Bergen Y, Laland KN. Trade-offs in the adaptive use of social and asocial learning. *Adv Study Behav*. 2005;35:333–379.
69. Enquist M, Eriksson K, Ghirlanda S. Critical social learning: a solution to Rogers' s paradox of nonadaptive culture. *Am Anthropol*. 2007;109(4):727–734. http://dx.doi.org/10.1525/AA.109.4.727.728.
70. Selbing I, Lindström B, Olsson A. Demonstrator skill modulates observational aversive learning. *Cognition*. 2014;133(1):128–139.
71. Selbing I, Olsson A. Interaction of actual and described demonstrator ability during observational avoidance learning. In: *Poster Presented at: Society of Neuroeconomics Annual Meeting, Miami*; 2015.
72. Singer T, Seymour B, O'Doherty J, Kaube H, Dolan RJ, Frith CD. Empathy for pain involves the affective but not sensory components of pain. *Science*. 2004;303(5661):1157–1162. http://www.ncbi.nlm.nih.gov/entrez/query.fcgi?cmd=Retrieve&db=PubMed&dopt=Citation&list_uids=14976305.
73. Lamm C, Decety J, Singer T. Meta-analytic evidence for common and distinct neural networks associated with directly experienced pain and empathy for pain. *NeuroImage*. 2011;54(3):2492–2502. http://dx.doi.org/10.1016/j.neuroimage.2010.10.014.
74. Shamay-Tsoory SG, Aharon-Peretz J, Perry D. Two systems for empathy: a double dissociation between emotional and cognitive empathy in inferior frontal gyrus versus ventromedial prefrontal lesions. *Brain*. 2009;132(3):617–627. http://dx.doi.org/10.1093/brain/awn279.
75. Wilhelm I, Wagner U, Born J. Opposite effects of cortisol on consolidation of temporal sequence memory during waking and sleep. *J Cogn Neurosci*. 2011;23(12):3703–3712. http://dx.doi.org/10.1162/jocn_a_00093.
76. Zaki J, Ochsner K. The neuroscience of empathy: progress, pitfalls and promise. *Nat Neurosci*. 2012;15(5):675–680. http://dx.doi.org/10.1038/nn.3085.
77. Cikara M, Bruneau EG, Saxe RR. Us and them: intergroup failures of empathy. *Curr Dir Psychol Sci*. 2011;20(3):149–153. http://dx.doi.org/10.1177/0963721411408713.
78. Galinsky AD, Moskowitz GB. Perspective-taking: decreasing stereotype expression, stereotype accessibility, and in-group favoritism. *J Pers Soc Psychol*. 2000;78(4):708–724. http://dx.doi.org/10.1037/0022-3514.78.4.708.
79. Bernhardt BC, Singer T. The neural basis of empathy. *Annu Rev Neurosci*. 2012;35(1):1–23.
80. Kätsyri J, Hari R, Ravaja N, Nummenmaa L. Just watching the game ain't enough: striatal fMRI reward responses to successes and failures in a video game during active and vicarious playing. *Front Hum Neurosci*. June 2013;7(278). http://www.pubmedcentral.nih.gov/articlerender.fcgi?artid=3680713&tool=pmcentrez&rendertype=abstract.
81. Mobbs D, Yu R, Meyer M, et al. A key role for similarity in vicarious reward. *Science*. 2009;324(5929):900.
82. Burke CJ, Tobler PN, Baddeley M, Schultz W. Neural mechanisms of observational learning. *Proc Natl Acad Sci USA*. 2010;107(32): 14431–14436.
83. Chang SWC, Winecoff AA, Platt ML. Vicarious reinforcement in rhesus macaques (*Macaca mulatta*). *Front Neurosci*. March 2011;5.
84. Davis MH. Measuring individual differences in empathy: evidence for a multidimensional approach. *J Pers Soc Psychol*. 1983;44(1): 113–126. http://dx.doi.org/10.1037/0022-3514.44.1.113.
85. Mehrabian A. Pleasure-Arousal. Dominance: a general framework for describing and measuring individual differences in temperament. *Curr Psychol*. 1996;14:261–292. http://dx.doi.org/10.1007/BF02686918.
86. Levenson RW, Ruef AM. Empathy: a physiological substrate. *J Pers Soc Psychol*. 1992;63(2):234–246. http://dx.doi.org/10.1037/0022-3514.63.2.234.
87. Zaki J, Bolger N, Ochsner K. It takes two: the interpersonal nature of empathic accuracy: Research article. *Psychol Sci*. 2008;19(4): 399–404. http://dx.doi.org/10.1111/j.1467-9280.2008.02099.x.
88. Mayer JD, Salovey P, Caruso DR. Emotional intelligence: new ability or eclectic traits? *Am Psychol*. 2008;63(6):503–517. http://dx.doi.org/10.1037/0003-066X.63.6.503.
89. Stürmer S, Snyder M, Kropp A, Siem B. Empathy-motivated helping: the moderating role of group membership. *Pers Soc Psychol Bull*. 2006;32(7):943–956. http://dx.doi.org/10.1177/0146167206287363.
90. Roesch MR, Esber GR, Li J, Daw ND, Schoenbaum G. Surprise! Neural correlates of Pearce-Hall and Rescorla-Wagner coexist within the brain. *Eur J Neurosci*. 2012;35(7):1190–1200. http://dx.doi.org/10.1111/j.1460-9568.2011.07986.x.
91. Mitchell JP, Heatherton TF, Macrae CN. Distinct neural systems subserve person and object knowledge. *Proc Natl Acad Sci USA*. 2002;99(23):15238–15243. http://www.ncbi.nlm.nih.gov/entrez/query.fcgi?cmd=Retrieve&db=PubMed&dopt=Citation&list_uids=12417766.
92. Kikusui T, Winslow JT, Mori Y. Social buffering: relief from stress and anxiety. *Philos Trans R Soc Lond B Biol Sci*. 2006;361(1476):2215–2228.
93. Beckes L, Coan JA. Social baseline theory: the role of social proximity in emotion and economy of action. *Soc Personal Psychol Compass*. 2011;5(12):976–988.
94. Golkar A, Selbing I, Flygare O, Ohman A, Olsson A. Other people as means to a safe end: vicarious extinction blocks the return of learned fear. *Psychol Sci*. 2013;24(11):2182–2190. http://www.ncbi.nlm.nih.gov/pubmed/24022651.
95. Anderson EE. The Effect of the presence of a second animal upon emotional behavior in the male albino rat. *J Soc Psychol*. 1939;10(2): 265–268. http://dx.doi.org/10.1080/00224545.1939.9713365.
96. Kiyokawa Y, Wakabayashi Y, Takeuchi Y, Mori Y. The neural pathway underlying social buffering of conditioned fear responses in male rats. *Eur J Neurosci*. 2012;36(10):3429–3437. http://dx.doi.org/10.1111/j.1460-9568.2012.08257.x.
97. Mineka S, Cook M. Immunization against the observational conditioning of snake fear in rhesus monkeys. *J Abnorm Psychol*. 1986;95(4):307–318. http://www.ncbi.nlm.nih.gov/entrez/query.fcgi?cmd=Retrieve&db=PubMed&dopt=Citation&list_uids=3805492.

98. Lungwitz EA, Stuber GD, Johnson PL, et al. The role of the medial prefrontal cortex in regulating social familiarity-induced anxiolysis. *Neuropsychopharmacology*. 2014;39(4):1009–1019. http://dx.doi.org/10.1038/npp.2013.302.
99. Öst LG, Reuterskiöld L. Specific phobias. In: *CBT for Anxiety Disorders: A Practitioner Book*; 2013:107–133. http://dx.doi.org/10.1002/9781118330043.ch5.
100. Gilroy L, Kirkby K, Daniels B, Menzies R, Montgomery I. Controlled comparison of computer-aided vicarious exposure versus live exposure in the treatment of spider phobia. *Behav Ther*. 2000;31(4):733–744. http://dx.doi.org/10.1016/S0005-7894(00)80041-6.
101. Haaker J, Golkar A, Selbing I, Olsson A. Observational extinction blocks learned fear by recruitment of the vmPFC in humans. In: *Poster Presented at: Society of Neuroeconomics Annual Meeting 2015;* Miami Society for Neuroscience, November 2014; Washington, D.C. ; 2014.
102. Kalisch R, Korenfeld E, Stephan KE, Weiskopf N, Seymour B, Dolan RJ. Context-dependent human extinction memory is mediated by a ventromedial prefrontal and hippocampal network. *J Neurosci*. 2006;26(37):9503–9511. http://dx.doi.org/10.1523/JNEUROSCI.2021-06.2006.
103. Milad MR, Wright CI, Orr SP, Pitman RK, Quirk GJ, Rauch SL. Recall of fear extinction in humans activates the ventromedial prefrontal cortex and hippocampus in concert. *Biol Psychiatry*. 2007;62(5):446–454. http://dx.doi.org/10.1016/j.biopsych.2006.10.011.
104. Zaki J, López G, Mitchell JP. Activity in ventromedial prefrontal cortex co-varies with revealed social preferences: evidence for person-invariant value. *Soc Cogn Affect Neurosci*. 2014;9(4):464–469.
105. Schiller D, Levy I, Niv Y, LeDoux JE, Phelps EA. From fear to safety and back: reversal of fear in the human brain. *J Neurosci*. 2008;28(45):11517–11525.http://dx.doi.org/10.1523/JNEUROSCI.2265-08.2008.
106. LaBar KS, Gatenby JC, Gore JC, LeDoux JE, Phelps EA. Human amygdala activation during conditioned fear acquisition and extinction: a mixed-trial fMRI study. *Neuron*. 1998;20(5):937–945. http://www.ncbi.nlm.nih.gov/entrez/query.fcgi?cmd=Retrieve&db=PubMed&dopt=Citation&list_uids=9620698.
107. Hampton AN, Bossaerts P, O'Doherty JP. Neural correlates of mentalizing-related computations during strategic interactions in humans. *Proc Natl Acad Sci USA*. 2008;105(18):6741–6746. http://dx.doi.org/10.1073/pnas.0711099105. pii: 0711099105.
108. Lindström B, Olsson A. Mechanisms of social avoidance learning can explain the emergence of adaptive and arbitrary behavioral traditions in humans. *J Exp Psychol Gen*. 2015;144(3):688–703. http://dx.doi.org/10.1037/xge0000071.
109. Fehr E, Fischbacher U. Social norms and human cooperation. *Trends Cogn Sci*. 2004;8(4):185–190. http://www.ncbi.nlm.nih.gov/entrez/query.fcgi?cmd=Retrieve&db=PubMed&dopt=Citation&list_uids=15050515.
110. Seymour B, Singer T, Dolan R. The neurobiology of punishment. *Nat Rev Neurosci*. 2007;8(4):300–311. http://dx.doi.org/10.1038/nrn2119. pii: nrn2119.
111. Gray K, Rand DG, Ert E, Lewis K, Hershman S, Norton MI. The emergence of "us and them" in 80 lines of code: modeling group genesis in homogeneous populations. *Psychol Sci*. 2014;25(4):982–990. http://www.ncbi.nlm.nih.gov/pubmed/24590382.
112. American Psychiatric Association. *Diagnostic and Statistical Manual of Mental Disorders*. 5th ed. 2013. http://dx.doi.org/10.1176/appi.books.9780890425596.744053.
113. Mineka S, Zinbarg R. A contemporary learning theory perspective on the etiology of anxiety disorders: it's not what you thought it was. *Am Psychol*. 2006;61(1):10–26.

C H A P T E R

13

Emotion and Aging: The Impact of Emotion on Attention, Memory, and Face Recognition in Late Adulthood

Maryam Ziaei[1], *Håkan Fischer*[2]

[1]School of Psychology, University of Queensland, St. Lucia, QLD, Australia; [2]Department of Psychology, Stockholm University, Stockholm, Sweden

O U T L I N E

It has been estimated that by 2050, 1.5billion people will be aged 65 or older, representing 16% of the world's population. Hence, understanding the full picture of aging could provide a new lens to think about long-term planning for health, work policies, and opportunities for engaging and collaborating with older adults. Although aging is associated with cognitive deficits and these are associated with some functional costs this does not provide a complete picture of psychological changes that occur with aging. In particular increasing, evidence from the emotional aging literature offers a different perspective on how we age. Recent discoveries in functional neuroimaging also provide important insights into how the brain functions during various cognitive and emotional tasks change as we age, granting a more comprehensive view of the aging brain. Therefore, the primary focus of this chapter is to provide an overview of multidisciplinary evidence from both behavioral and neuroimaging studies in the emotional aging literature. The chapter is organized based on the impact of age-related changes in emotional processing on three main categories of cognitive function: *attention, memory,* and *face recognition.* Before discussing the main findings from each of these three categories, some of the major discoveries and dominant models in the cognitive aging domain will be discussed briefly. Throughout this chapter, several questions are addressed: what are the underlying cognitive and neural mechanisms of the attentional biases toward

Neuroimaging Personality, Social Cognition, and Character
http://dx.doi.org/10.1016/B978-0-12-800935-2.00013-0

positive items in aging? Do older adults have difficulties in processing negative emotions, or do they process positive emotions differently than younger adults? What factors influence the processing of emotional facial expressions in late adulthood? Do the temporal features of stimuli help older adults overcome difficulties in recognizing emotions? Are there any age differences in processing the six main emotions expressed by the face?

1. COGNITIVE AGING

Emotional and cognitive functions are interrelated, and overlooking either of these two critical aspects of aging leaves us with an incomplete picture of aging. Although the fine details of the cognitive aging literature are beyond the scope of this chapter, we briefly acknowledge and discuss the central existing theories of the cognitive aspects of aging before discussing the emotion and aging studies. Understanding the existing theories in cognitive domains provides a deeper understanding of the underlying mechanisms for emotional processes and subsequently will help reconcile emotional and cognitive aging theories/domains.

A substantial body of evidence now shows cognitive deficits among older adults in relation to speed of processing, memory, and attention (for review, see[1,2]). Multiple cognitive aging theories have been proposed and have provided a foundation for understanding the underlying mechanisms associated with cognitive deficit as we age. In the following, we briefly present four particularly influential discoveries, and then we describe some of the dominant theories that have been proposed to explain such findings in the cognitive aging literature.

1.1 Major Discoveries about the Aging Brain

Overactivation: older adults activate some areas of the brain to a greater extent than younger adults during the performance of cognitive tasks. These patterns of activity have been thought of as the neural correlates of cognitive decline.[1] However, overactivation has been observed in high-performing older individuals.[3,4] Therefore, overactivation could also reflect the compensatory mechanisms utilized by "successful" older adults, who have greater cognitive ability and might recruit different neural networks relative to younger adults. However, it is still unclear whether older adults are using these additional brain regions to implement different strategies, or whether they are using the same cognitive strategies as young adults, but are relying on different brain areas. In other words, overactivation might have a function, or it might be a by-product of aging influencing the brain. The former view has been supported by the compensatory viewpoint, which suggests that the increased bilateral recruitment of both hemispheres among older adults is associated with enhanced cognitive performance and has a compensatory role.[5]

The hemispheric asymmetry reduction in older adults (HAROLD)[6] model is consistent with the second interpretation of overactivation as a by-product of aging. Evidence from episodic memory, semantic retrieval, working memory (WM), perception, and inhibitory control provide converging evidence that older adults show increased activity in both hemispheres relative to younger adults, indicating that the overactivation can be considered a by-product of aging.

Dedifferentiation: dedifferentiation refers to reduced regional specialization or specificity in a particular area.[7] Overactivity of the prefrontal regions and dedifferentiation of the ventral visual system are typically construed as being consistent with a compensatory mechanism.

Frontal compensation: overrecruitment of prefrontal regions in a wide range of tasks is often seen in studies of older adults. Importantly, this overrecruitment of alternative brain networks is related to improvement in cognitive performances.[5] Recruiting more anterior prefrontal cortex (PFC) regions is accompanied by deactivation of posterior regions, such as the medial temporal lobe and ventral visual cortex. This pattern also relates to the posterior-anterior shift in aging (PASA), which has been thought to reflect compensatory mechanisms with advancing age.[8]

Default mode network (DMN): this network includes the medial PFC, posterior cingulate, medial and lateral parietal regions, and seems to be involved "during mental explorations referenced to oneself including remembering, considering hypothetical social interactions, and thinking about one's own future."[9] In healthy young adults, the DMN is deactivated with increasing task demands. Several studies, however, reported that older adults failed to "deactivate" DMN during many cognitive tasks.[10,11] One reason for such a pattern might be that older adults have a reduced ability to suspend DMN activity when other cognitive tasks require their attention.[11] Results from longitudinal studies also indicate that the task-induced deactivation in DMN remains stable over time among older adults.[12]

1.2 Theories in Cognitive Aging

Three influential models, including the *inhibitory deficit hypothesis, the Compensation-Related Utilization of Neural Circuits Hypothesis* (CRUNCH), and the *Scaffolding Theory of Aging and Cognition* (STAC model) have been proposed to explain the underlying mechanisms of the cognitive deficit that accompanies aging.

One of the dominant hypotheses is the *inhibitory deficit hypothesis,* which explains a wide range of age-related cognitive difficulties in WM,[13,14] decision-making, and social functioning.[15,16] Consistent with the notion that reduced inhibitory control underlies many age-related losses (Hasher and Zacks[17]), this hypothesis focuses on the effect of inhibitory control on WM by suppressing irrelevant information. A number of studies have also shown that specific deficits in inhibitory control can lead to WM impairment in aging. It has also been shown that age-related deficits in inhibitory control over distractors disrupt WM performance.[13,18] Recently, a top-down modulation model in aging has been proposed.[19] According to this model, the "top-down modulation, defined as the modulation of neural activity in neurons of sensory or motor areas based on an individual's goals, involves the enhancement of task-relevant representations and/or suppression for task-irrelevant representations."[20] Two forms of top-down modulations have been examined: external (e.g., environmental stimuli) and internal (e.g., goals). Several studies have suggested that changes in any form of top-down modulations can lead to the WM deficit observed with aging.[21,22]

The *Compensation-Related Utilization of Neural Circuits Hypothesis (CRUNCH)* accounts for age-related changes as a function of task difficulty. According to this hypothesis, older adults are likely to engage more neural circuits than younger adults to meet task demands. When task demands are low, older adults show more overactivation (frontal and bilateral), whereas younger adults show more focal activation. As task load increases, younger adults may show more bilateral recruitment or overactivation of the anterior regions. However, because of the restricted dynamic range of neural responses among older adults, they may reach their resources' limit and may show underactivation and cognitive decline. Thus, neural compensation is more effective when task demands are low. However, by increasing task difficulty, cognitive resources' limits are reached, and decline in performance becomes more apparent.[1,23]

Despite the changes in brain structure that come with age, such as dopamine receptor depletion, white matter/gray matter deteriorations, and decline in capacity for neurogenesis, synaptogenesis, and angiogenesis, all systems remain functional and provide a mean for building alternative neural circuitry. These structural challenges are accompanied by functional changes (i.e., overactivation, dedifferentiation, compensation, and DMN changes), and recruiting alternative neural circuitry (scaffolds) helps maintain a high level of cognitive functioning—though it might work less efficiently. The *Scaffolding Theory of Aging and Cognition (STAC model)* describes how the scaffolding process starts during childhood when individuals face different challenges and their brains must adapt. Scaffolding is thus influenced by experiences such as learning new skills, engaging in new challenging activities, and cognitive training, all of which have the potential to enhance the brain's ability to maintain high-level function by creating new scaffolding.

2. EMOTION AND AGING

Despite cognitive decline, there are small to moderate gains in some emotional domains, including emotion regulation and biases toward positive emotions in memory and attention. A considerable number of empirical studies support the relative preference of positive (over negative) materials in cognitive processing—known as the "positivity effect" (PE).[24] Despite the converging evidence supporting the PE, the underlying mechanisms for such an effect remain unclear. Several theories, such as Socioemotional Selectivity Theory (SST),[25] the Aging Brain Model (ABM),[26] and the selective optimization with compensation model,[27] have received considerable attention over the past decades in order to explain the PE.

2.1 Theories in Emotional Aging

2.1.1 Socioemotional Selectivity Theory

SST is the dominant theory in the field of emotional and social aging.[25] This theory posits that as people grow older and their time starts to be perceived as limited, their motivational orientation begins to change. A limited time perception results in chronic activation of the goals related to emotional meaning and influences motivational preferences, which changes goal hierarchies—that is, goals will be more person-focused (such as seeking emotion and meaning) rather than future oriented (such as gaining new knowledge or establishing new social contacts).[28] Goals, preferences, and cognitive processes change systematically as a subjective sense of remaining time becomes limited.[29] In one study testing this model, older and younger adults were asked to imagine that they were moving to the other side of the country. Findings showed that the effect of age disappeared, and both groups preferred to spend more time with their familiar social partners. This study and similar studies showed how perceptions of time could influence the way people prioritize their social contacts (see[30] for more details).

Well-being is also affected by such a goal shift. People are attuned to the relevance of incoming information to their goals. They experience negative or positive emotions if something obstructs their goals or if their goals have been attained. Given the limitation of WM and attention, for instance, it seems to be an adaptive strategy to prioritize features of events that facilitate or obstruct goals.[31] As a result of this motivational shift, positive emotions will be prioritized, and older adults reallocate their resources to attain positive emotions and reduce negative emotions. In line with this possibility, longitudinal studies have provided evidence of improvement in overall well-being with advancing age.[32] The positivity preference in attention and memory will be discussed in more detail in Section 3 (Emotion and Attention in Aging) and Section 4 (Emotional Memory and Aging).

2.1.2 *Aging Brain Model*

An assumption underlying social neuroscience is that all humans' social behaviors are implemented biologically. The social neuroscience perspective focuses on fundamental changes in brain functions and how these changes in cognition and decision-making are associated with subjective well-being. The ABM, as a derivative from social neuroscience, attempts to explain a link between affective processing and age-related changes in brain functions.[26] For a long time, cognitive and emotional aging were considered two separate constructs; ABM provides an important connection between these two fields of study.

In order to better understand the ABM, we provide one study as an example and interpret the results from both SST and ABM perspectives. In a 2004 functional Magnetic Resonance Imaging (fMRI) study, Mather et al.[33] presented negative, positive, and neutral pictures to older and younger participants, who were then asked to rate how excited or calm they felt when viewing each picture. Enhanced amygdala activation in response to positive (relative to negative) stimuli was found among older adults, whereas for younger participants, amygdala activation to positive and negative stimuli did not significantly differ. According to SST, this result would be interpreted as the preference of focusing on positive emotional goals and recruiting regulatory strategies among older adults, which led to the reduced cognitive focus on negative stimuli. On the other hand, ABM would interpret the enhanced activity for positive over negative stimuli among older adults as a reduced arousal response to negative items due to attenuated amygdala function.

To further test the ABM, patients with a lesion in the amygdala/medial temporal regions were asked to rate their response toward emotional pictures.[34] It was found that patients with amygdala lesions rated negative pictures as lower in arousal, whereas the emotion categorization for these pictures remained intact. Thus, according to this model, age-related changes in memory for negative items among older adults could reflect changes in the function of the amygdala. Changes in amygdala function have an impact on how the arousal level of emotions is perceived (especially for negative emotions) and therefore reduce the impact of arousal on the memorability of emotional items. The reduction in arousal level to negative emotions can also be associated with enhanced well-being among older adults. Therefore, both SST and ABM models predict that amygdala activation will be smaller for negative stimuli than for positive stimuli among older adults; however, SST predicts that an increased focus on goals and emotional regulatory strategies may be the underlying mechanism leading to the reduced attention toward negative items among older adults.[35] Alternatively, ABM suggests that "these amygdala changes are the cause of the reduced impact of negative stimuli and, consequently, diminished depressive symptomatology and improved subjective well-being."[26] Such changes in amygdala responses could partially explain the age-related PE (see Section 3.2).

2.1.3 *Selective Optimization with Compensation Model*

Another proposal to reconcile the emotional and social aspects of aging is the Selective Optimization with Compensation Model. According to this model, people become aware of their losses and gains across adulthood, and due to the naturally diminished resources that come with aging, they select goals that are important or can be realistically obtained in their lifetimes. Therefore, less important goals will be sacrificed at the cost of obtaining other more important goals. If some of the goals cannot be achieved, people will engage in compensatory activities. According to this model, older adults engage in any task that is important for them despite their physical or biological constraints. In each case, there is individualization of selection, optimization, and compensation. To make it more tangible, consider a person who desires to run a marathon. If the runner wants to reach this level of running performance, he should give up other activities (selection) and should increase or optimize his conditions, such as daily diet (optimization), in order to become an expert in the activity of marathon running. This will then reduce the impact of losing in other activities (compensation). Therefore, the combination of these three elements can contribute to successful completion of the aging process, which requires adaptation and concentration on domains that are high priority for older adults.[27,28]

Taken together, various models and hypotheses have been developed in order to explain the underlying mechanisms of age-related changes in emotional processing, including the SST, The ABM, and the Selective Optimization with Compensation Model. Emotions can influence various cognitive functions, such as attention, memory, and face recognition. Hence, in the following sections, we will discuss the existing literature on the impact of emotion on these three critical domains in regard to the changes that occur in each of these domains with advancing age. In each section, relevant behavioral and neurological findings will also be discussed.

3. EMOTION AND ATTENTION IN AGING

In order to better understand how emotions have an impact on attentional processes in advancing age, it is crucial to first understand how emotion and attention interact. Interest in this area of research is rapidly growing empirically, but it is still in its infancy theoretically. Theories of selective attention have been borrowed and adapted to explain the processing of emotional materials by considering these stimuli as highly salient information. Therefore, these theories indicate that highly salient information, such as emotions, is often prioritized for attentional processing.

The feature integration theory of visual attention,[36] for instance, describes how certain perceptual characteristics, such as orientation or color, can be processed automatically prior to any attentional selection. "Attention is conceived as the process by which representations of more complex stimuli are formed through the combination of individual features ('conjunctions of features')".[36] In visual search paradigms, participants are presented with an array of stimuli and asked to identify or locate the different item as quickly as possible. Using this paradigm, the question of whether emotional information has a feature that acts as a "pop out" effect, or whether emotional information instead requires complex processing in order to evaluate its emotional significance and capture attention, has been investigated extensively. Empirical studies support the former account that emotional information is a highly salient conjunction of features and pops out from the rest of the visual environment. For similar models pertinent to this theory, please see Ref. 37.

Another influential theory in selective attention literature is the "biased competition model."[38] The central tenet of this model is that limited capacity in our information processing systems leads to competition for attention between represented information. Emotions can enhance perceptual processing due to their greater perceptual distinctiveness, leading to the bottom-up prioritization of emotional stimuli. Moreover, top-down factors (such as past experience or environmental context) can similarly influence the competitive bias to enhance the prioritization of one stimulus over the rest. The main implication is that selective attention occurs when stimuli presentation allows for direct competition between stimuli (Figure 1).

In order to understand the interaction of emotion and attention at the neural level, researchers have looked at how particular brain regions, such as the amygdala,

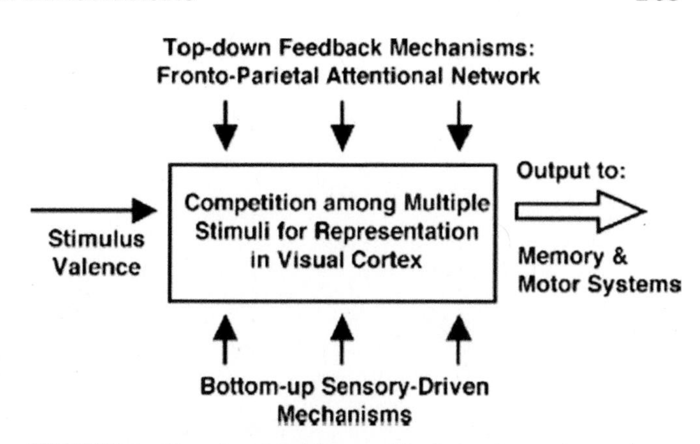

FIGURE 1 Biased competition model of visual attention and process of emotional information presented by Pessoa and Ungerleider (2004). *Courtesy of Dr. Pessoa.*

process emotional information when attentional resources are available and when they are limited. Previous brain imaging studies have reported mixed findings regarding the engagement of the amygdala under attention-control conditions. While some studies found amygdala activation in response to threatening stimuli without engaging attentional resources, others have provided evidence that amygdala responses may be modulated by selective attention.[39,40] In a study by Vuilleumier et al.,[41] emotion and attention were independently manipulated in a paradigm where participants were instructed to complete a matching task. Participants were presented with two faces and two houses arranged in horizontal or vertical pairs and asked to match two faces in either the horizontal pairs or on the vertical pairs. The faces were sometimes emotional (fearful) and sometimes neutral in expressions (Figure 2). Findings showed that while activation of the fusiform face area (a primary region involved in face processing) was modulated by spatial attention manipulation, the amygdala showed consistent responses to the emotional expression of the faces, irrespective of the attentional manipulation.

Studies such as Vuilleumier et al.[41] suggest that processing of facial expressions can occur automatically and support the possibility that the amygdala detects emotionally relevant stimuli quickly even without conscious awareness. Because the amygdala was unresponsive to attentional manipulation in this study, it was suggested that emotional processing does not require attentional resources. In order to test this further, Pessoa et al.[42] examined how emotional responses are modulated by attentional resources. Their design used the same images of faces but with two conditions: an attended face condition and an unattended face condition. In the attended face condition, subjects were required to indicate whether the face was male or female, while in the unattended face condition, subjects indicated whether the bars were of similar orientation

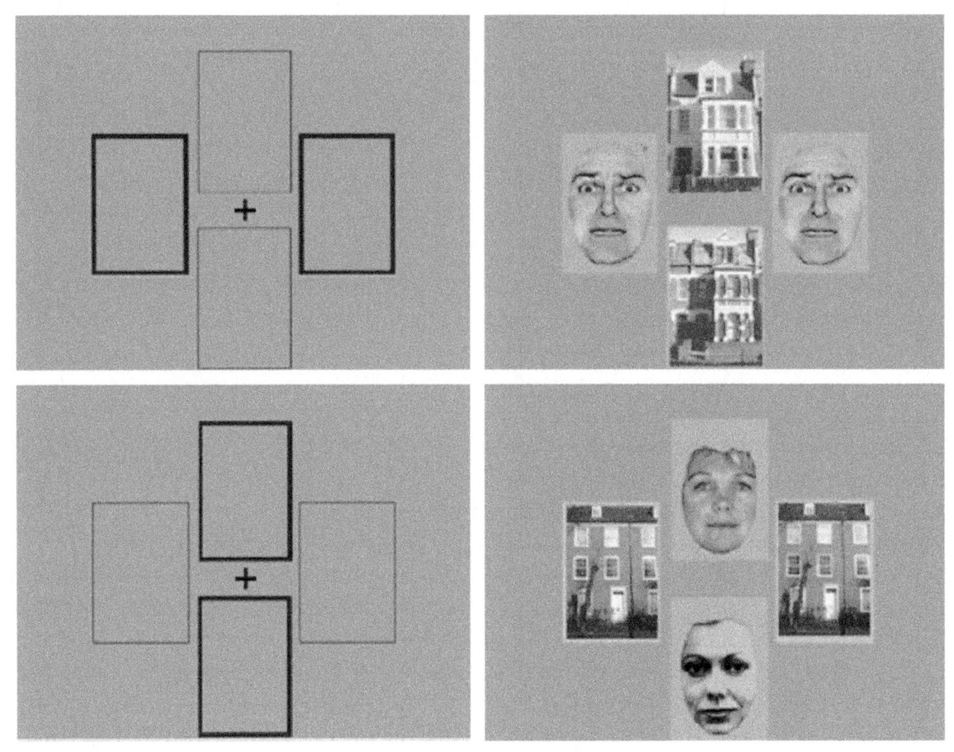

FIGURE 2 Example of experimental design presented by Vuilleumier et al. (2001) to examine the emotion and attention relationship. *Courtesy of Dr. Vuilleumier.*

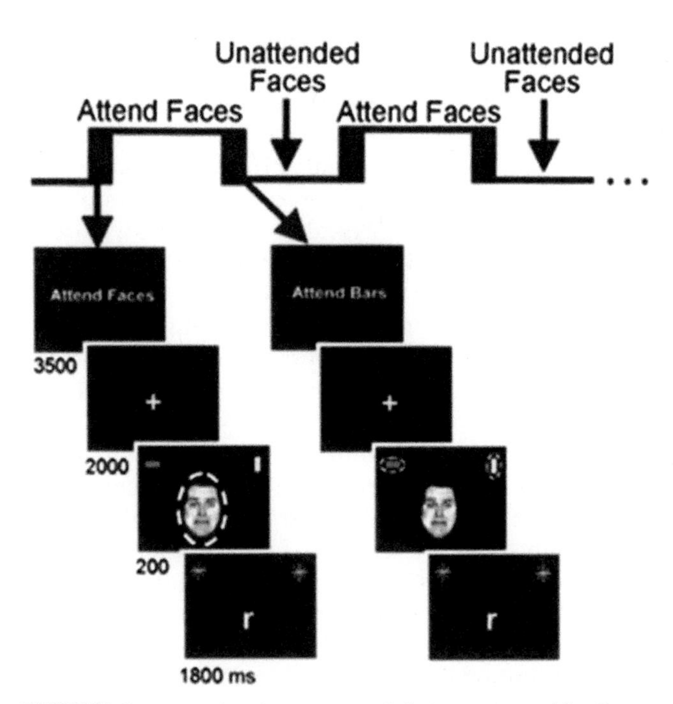

FIGURE 3 Example of experimental design presented by Pessoa et al. (2002) to examine the emotion and attention relationship. *Copyright (2002) National Academy of Sciences, USA. Courtesy of Dr. Pessoa.*

(Figure 3). Contrary to previous findings such as Vuilleumier et al.,[41] it was found that when attentional resources were available, all brain regions responded to the emotional faces differently than the neutral ones. However, when attentional resources were depleted by another task (judging the orientation of the bars), the differential activation in responses to emotional versus neutral faces was diminished. Unlike the Vuilleumier study, which showed no evidence for modulation of activity in the amygdala in response to attentional demand, Pessoa and colleagues concluded that attention is necessary for the processing of emotional items. It is worth mentioning that Pessoa and colleagues used a task manipulation that was more effective in depleting attentional resources than the Vuilleumier's study.

In addition to the studies mentioned previously, there is now a considerable empirical literature focused on better understanding the relation between emotion and attention using various methods, such as filtering (Stroop task, dichotic listening), searching (visual search), cueing (dot-probe, spatial cueing), and multiple-task paradigms (attentional blink). Using these methods, behavioral studies shed light on some interesting issues, such as whether attention to emotional items occurs with/without awareness, if it requires attentional control, or what is the sufficient intensity level for threatening stimuli in order to influence our attentional biases. The basic conclusion that can be drawn from these studies is that emotional information, particularly negative information, can elicit prioritization of attentional resource allocation relative to nonemotional information. However, the main challenge facing researchers is to define how the classification of valence occurs and to what extent different types of classification of emotional information may bias attentional systems—for instance, perceptual (perceptual features of faces such as eye region) versus semantic (subjective rating) classification. Details of these studies are beyond the scope of this chapter, but for further information we refer our readers to Yiend's paper.[43] Now that we have provided a brief background on how emotional information can influence attentional processes in general, we discuss studies that examined the age-related changes in attentional biases toward emotional items.

3.1 Behavioral Findings

One of the early pieces of experimental evidence of attentional positivity preference in aging comes from works by Mather and Carstensen[44] using the dot probe task. In the dot probe paradigm, participants were shown a pair of faces: one emotional (either positive or negative) and one neutral. After the faces were removed, a dot appeared in the location of one of the faces. The participants were then required to respond by detecting the dot as quickly as possible. Older adults exhibited faster responses in detecting the dot when it was in the same location as neutral faces, compared to when the dot was presented in the location of negative faces. Based on these findings, it was argued that during the initial attentional bias, older adults were avoiding negative information and therefore detecting the dots in the location of negative faces took longer than detecting the dots in the location of neutral faces.

Isaacowitz et al.[45] investigated attentional biases among older adults using the dot probe task in conjunction with an eye-tracking device. Using an eye-tracker in this context has several advantages: first, it enables the researcher to track the gaze of participants, and second, it can provide an indirect measure of participants' attentional biases. The results from Isaacowitz et al. study showed that older adults spent more time looking at happy faces than sad and neutral faces, but younger adults spent equal amounts of time looking at both happy and sad faces. Moreover, among older adults, faster reaction times in detecting the dots were found when the dots were positioned with the happy faces. In addition to supporting the attentional bias toward positive items, these studies suggest that the bias toward positive items occurs at a relatively early stage of attentional processing.

To clarify more precisely how early the positivity bias emerges during attentional processing, Isaacowitz et al.[46] examined the time course of the PE and demonstrated that the positivity preference among older adults emerged 500 ms following stimuli onset and increased over time. Thus, the question becomes whether we can conclude that positivity bias among older adults requires cognitive resources as it happens at a later stage of attention, or if it occurs without cognitive control resources. According to the cognitive control hypothesis, the PE relies on the availability of cognitive resources,[47] but some studies failed to support this account, suggesting that the PE might not be fully dependent on cognitive control resources.[48,49] Thus, it remains unclear whether the PE relies on cognitive control mechanisms or if it is an automatic.

In order to investigate the underlying cognitive resource dependency of the PE, some studies have used a dual task paradigm. In such paradigms, participants are asked to perform one task, while their attentional resources are divided and deployed in another secondary task. Knight et al.[50] asked young and older adults to passively view emotional–emotional and emotional–neutral pairs of faces with angry, happy, and neutral valences while a secondary task was presented (auditory music). Both fixation and proportion of the fixations toward emotional items relative to the neutral items were measured using an eye-tracking device. The results suggested that when attentional resources are limited (divided attention condition or doing the dual task), attention is more likely to be drawn toward negative stimuli. When full cognitive resources were available (full attention condition), older adults exhibited positivity preference, thus suggesting that the PE requires cognitive resources. Following this study, Allard and Isaacowitz[51] conducted a similar study using the dual task paradigm but with a within-subjects rather than between-subjects design. The results indicated that regardless of conditions (divided or full attention), older adults showed preference in their fixation for positive and neutral pictures relative to negative pictures. Hence, contrary to Knight et al. findings,[50,52] this study suggests that the PE effect might not be fully dependent on cognitive resources.

It has been argued that positive gaze preference toward positive items in attention reflects the regulatory mechanism that older adults benefit from downregulating their negative emotions and moods. In the context of SST, the PE in information processing can help older adults regulate their mood and optimize their affect. In order to test this possibility, several studies have examined gaze preference as a tool when older adults are in a negative mood. In line with hypotheses from SST, older adults demonstrated enhanced gaze preference to positive items when they were in a negative mood.[53] This idea has prompted the question of whether the positive gaze preference functions to improve mood over time.

For instance, Isaacowitz et al.[54] investigated the link between fixation patterns and mood as a function of age. Older adults who started a task in a positive mood retained their positive mood throughout the experiment. Interestingly, those older adults who resisted a decline in their mood over time were those with good executive control ability. They also displayed positive gaze preferences by looking at happy faces and away from angry faces. Therefore, it seems that older adults do not only use their positive gaze preferences as a strategy for emotion regulation but they are also capable of using positive gaze preferences to regulate their mood in real time. In support of this finding, Larcom and Isaacowitz[55] found that older adults were more likely to rapidly regulate their emotions than younger adults. Moreover, older individuals who rapidly regulated their negative mood had lower trait anxiety, neuroticism, and depressive symptoms, and a higher level of optimism, relative to those who did not regulate their negative mood. Thus, it seems that positive gaze preference can function as a regulatory mechanism tool among older adults, which they might use over an extended period of time.

There is also the possibility that the positive gaze preference among older adults reflects their motivation or desire to change their bad mood, and consequently requires a substantial level of cognitive control. Therefore, a number of studies questioned whether full cognitive effort is required to display positive emotional preference among older adults. If positive gaze preference is a regulatory mechanism, then it requires cognitive resources to operate. Allard et al.[56] conducted an experiment to test whether the gaze is used among older adults as a regulatory strategy in the absence of explicit instruction. They investigated this effect by using mood induction (happy, negative, and neutral) in addition to recording pupillary responses while participants were viewing emotional pictures. The results suggested that lower cognitive effort (as reflected in pupillary response) was expended when older adults were engaged in positive gaze preferences while experiencing a negative mood. Their findings suggest that gaze acts as an effortless regulatory tool used by older adults in order to display positivity preference while experiencing a negative mood.

In this section, we first discussed two influential studies by Vuilleumier et al.[41] and Pessoa et al.[42] regarding how emotion and attention interact and whether attentional resources are needed to process emotional items. However, there are still debates about how emotion influences attentional control and whether processing emotional information requires full attentional resources. Then, we discussed the key studies investigating the impact of emotion on attentional biases in late adulthood. It seems that older adults' attentional bias toward positive emotions is perhaps an adaptive strategy to regulate their negative moods or emotions, which is consistent with the motivational account of the PE (e.g., SST). The cognitive mechanisms underlying such an effect are still an open for question, though. It is unclear whether the attentional bias toward positive items is effortless or requires cognitive control resources. Neuroimaging studies have attempted to address similar questions and can provide further understanding of the underlying cognitive mechanisms of the PE.

3.2 Neurological Findings

Although a number of studies have investigated the role of various brain regions such as the hypothalamus, amygdala, and prefrontal regions during emotional processes, it has been shown that these regions do not work in isolation, but instead form a highly connected network.[40,57] The role of regions such as the hypothalamus, thalamus, basal ganglia, cingulate cortex, anterior insula, orbitofrontal cortex, and cerebellum in processing emotional items has been documented in a number of studies.[58,59] However, it has recently been shown that each emotional category is uniquely associated with a pattern of activity across multiple brain regions and is not constrained with any one region or system.[59]

The amygdala has been considered a particularly important region in emotional processes, and it is a highly connected node, integrating sensory and higher cognitive information (Figure 4). Also in the aging literature, amygdala atrophy has been considered a major contributor to the PE in attention and memory, for instance in the ABM. However, there are still unresolved arguments regarding the underlying mechanism of the PE. In the following paragraphs, we briefly explain neuroimaging studies in this field and provide a summary of the current knowledge on the role of the amygdala in processing emotional items in the aging brain.

Using event-related potential (ERP) and fMRI modalities, Williams et al.[60] tested teenagers, young adults, and older adults in order to understand the brain mechanisms that underlie emotional processing with advancing age. Their findings suggest a significant decline of negative emotion recognition and an increase of positive emotion recognition by age. Additionally, during the processing of fearful faces, older adults showed increased activation in the medial prefrontal cortex (mPFC) and decreased amygdala activation. However, unlike with happy faces (within 150 ms), fearful faces elicited increased neural activity during the later phase (180–450 ms) of processing stimuli (measured by ERP). Williams et al.[60] have argued that this shift of resource allocation from early to later phases might support the selective control over negative stimuli by advancing age. Their results suggest that a shift toward increased control over negative emotions in the later stage and less control over positive emotions in an early stage might be predictive of better emotional stability with advancing age.

It seems that older adults might engage cognitive control or the emotional network differently than do younger adults. Older adults might recruit more prefrontal regions

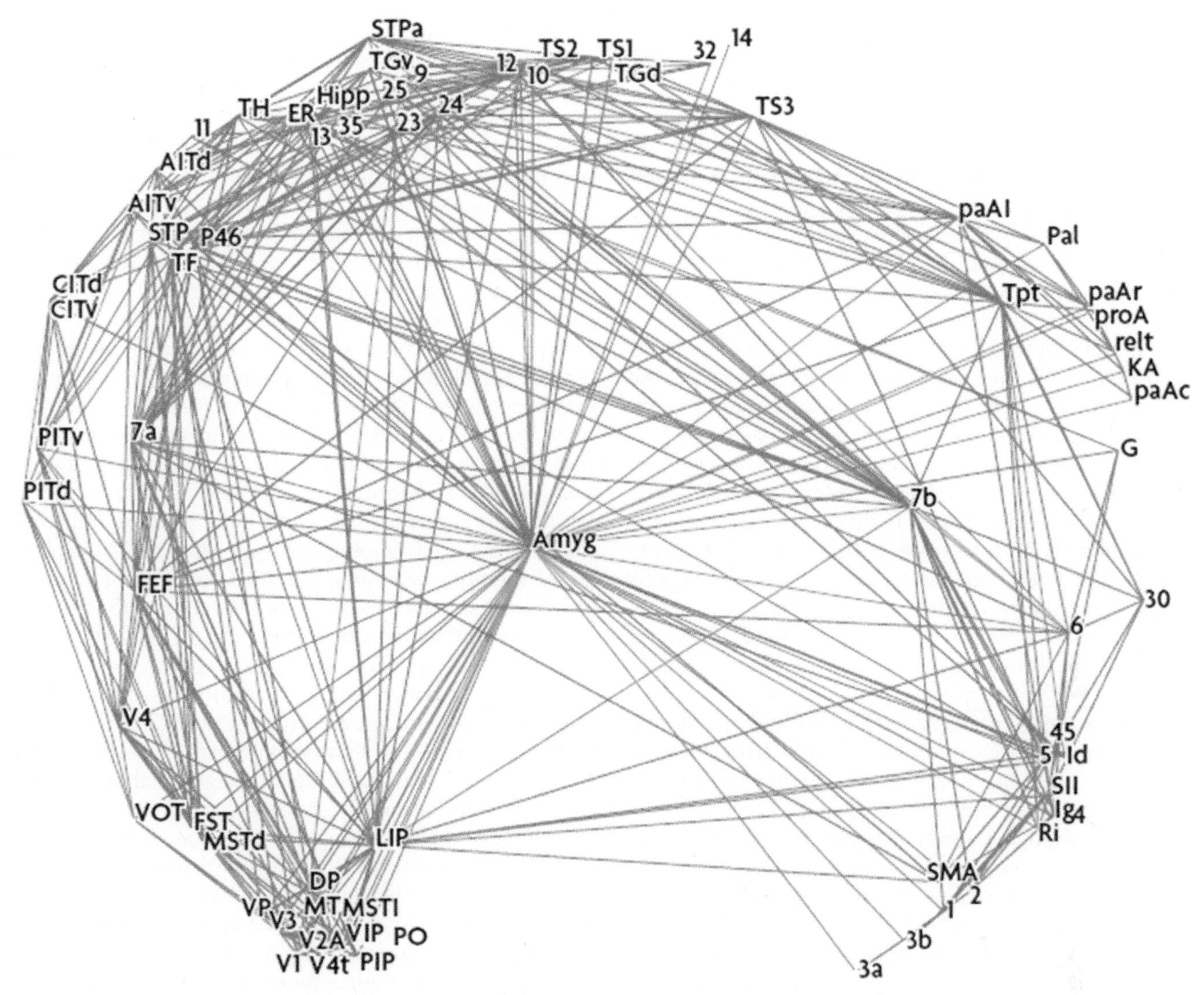

FIGURE 4 Brain connectivity graph presented by Pessoa (2008)[40]. This graph shows that the amygdala is a strong region integrating emotional and cognitive information as it links multiple hubs from separate functional clusters. *Courtesy of Dr. Pessoa.*

for emotional processing relative to younger adults (for a review see[61]), which might reflect older adults' motivational tendency to regulate their negative affect. These neural patterns are generally consistent with the PASA hypothesis[8] (see Section 1.1). Recall from our earlier discussion of cognitive aging (Section 1) that healthy aging is associated with increased engagement of the anterior brain regions and decreased activity in the posterior regions during the cognitive task. Enhanced activity of the frontal cortices and reduced activity of the amygdala and parieto-occipital regions has been reported during processing emotional items among older adults[62] as well, which resembles the PASA pattern. St Jacques et al.[63] found support for the PASA pattern of activity during the processing of emotional stimuli in a study in which participants were asked to view three sets of emotional pictures and rate each picture. Their findings suggest that the right amygdala was functionally connected to the ventral anterior cingulate cortex among older adults and was connected to the posterior regions—including the parahippocampus and visual cortices—among younger adults. Interestingly, this study also found functional activity preservation in the amygdala among older adults, which is consistent with previous findings of Wright et al.[64] but not in line with some other studies, such as Iidaka et al.[62] and Gunning-Dixon et al.[65] These inconsistent patterns in amygdala activity might be related to the different paradigms used. For instance, in the study by Wright et al.,[64] novel fearful faces were compared to familiarized neutral faces, which revealed greater amygdala activity. In the studies by Gunning-Dixon et al.,[65] an emotion discrimination task was used, and participants were asked to rate the intensity of each emotional expression. Besides the differences in paradigm, preserved amygdala activity among older adults in these studies suggests that the aging positivity bias might not be due to impaired processing of negative emotions, as has been suggested with the ABM (for details, please refer to Section 2.1.2).

Given the changes in amygdala and prefrontal activity observed during emotional processes in aging, the pattern of Fronto-Amygdala Differences in Emotion (FADE) has been suggested by St Jacques et al.[66] According to the FADE hypothesis, amygdala function remains intact throughout aging. This FADE pattern might reflect three potential mechanisms in age-related emotional processing. First, the increased activity in frontal regions could be related to the PASA, which has been observed in cognitive domains as well. This shift could reflect the compensatory mechanism, as increased frontal activity was found to be predictive of subsequent memory performance for negative stimuli. Second, the age-related increased activity in the medial PFC might reflect the recruitment of self-referential processes, which have been shown to be associated with this region. In line with this interpretation, Leclerc and Kensinger[67] argued that older adults recruit frontal region more for positive items and less for negative, suggesting that older adults might interpret positive stimuli in a self-relevant manner. Third, FADE could reflect enhanced emotion regulation strategies by older adults for processing negative stimuli. This is consistent with the results from emotion regulation literature indicating that emotion regulation imposes substantial demands on cognitive control operations,[68,69] although some studies have failed to provide support for this model (see[62,70,71]).

It seems that neurological evidence supports the enhanced activity of prefrontal regions in response to emotionally valenced items. There are a number of possible explanations for this change, as proposed by St Jacques et al.[66] Such enhanced PFC activity might be related to the emotion regulation strategies older adults are utilizing for downregulating negative emotions, or it may be due to the general pattern observed in aging (PASA pattern). However, it is still unclear whether older (relative to younger) adults recruit different networks for processing emotionally valenced items or whether the strength of connectivity between brain regions involved in processing emotional items might change as we age. Furthermore, there is a need to explore how the attentional processes during encoding have an impact on subsequent outcome measures, such as memory for emotional items.

4. EMOTIONAL MEMORY AND AGING

The effect of emotion is not only limited to attentional functions. There is now a considerable literature on how emotion can influence memories, including working, autobiographical, and long-term memories. Some studies emphasize how emotion can enhance or disrupt memory in different circumstances,[31] others have examined the distinctive role of valence and arousal features of emotional stimuli on memory functions,[72] and some researchers have investigated how emotional information is likely to be forgotten. A number of studies have also investigated the role of different brain regions during different stages of memory, such as encoding, storage, and retrieval.

Emotional information can constrain memory, a concept usually referred to as "memory narrowing." This effect reflects the phenomenon whereby memory is better for central items, which in turn disrupts memory for unrelated or peripheral items. However, not all studies have reported memory narrowing for emotional items. Important emotional items are sometimes forgotten and peripheral information is preserved. According to Levine and Edelstein,[31] it seems that, "memory narrowing as a result of emotion, and a number of violations of the memory narrowing pattern, can be explained by the view that emotion enhances memory for information relevant to currently active goals."

Recall that according to the biased competition model,[38] information is prioritized for further processing based on the competition between stimuli to reach our attentional focus. Furthermore, the "Arousal-Based Competition" (ABC) model[73] suggests that arousal modulates the strength of mental representation and determines which item will be dominated for selective attention. This competitive process begins during perception and continues into long-term memory. Evidence indicates that arousal can lead to memory narrowing, enhance the memory for gist (rather than details), and enhance consolidation, although there are still somewhat mixed findings regarding these effects (for a review see[73]). Previous studies have also reported a distinction in recruitment of brain networks as a function of arousal and valence.[72] Therefore, both the valence and arousal dimension of the emotional items have the potential to modulate the strength of the memory for that item.

4.1 Behavioral Findings

In a recent meta-analysis,[74] it has been shown that the aging PE in memory occurs in a wide range of paradigms, including long-term memory, WM, and autobiographical memory, as well as decision-making. A variety of stimuli, such as faces, words, and pictures have also been used. Detailed descriptions of these studies are beyond the scope of this chapter, but in the following paragraphs, we briefly review some of the key behavioral findings in this field before discussing the underpinning neurological mechanisms. For further details we refer our readers to a paper by Reed and Carstensen[24] and a recent meta-analysis by Reed et al.[74]

One of the earliest studies to provide important insights into age-related changes in emotional memory was conducted by Charles et al.,[75] in which they assessed both recognition and recall for emotional items. Overall, they found greater age differences for negative items compared to positive items as a function of age. In both recognition and recall memory tests, the memory advantage for negative items decreased with increasing age. These findings have been replicated over the last decade, and a number of studies examined this effect by using different materials, such as words and pictures (for a review see[74]), and different memory indices, such as response bias.

For instance, in one study participants had to decide whether they truly remembered each of the items or if they knew them more vaguely. The response biases of both age groups were examined, as well as memory accuracy. Findings showed that younger adults tended to say "old" or "remember" to negative items more frequently than positive and neutral items, whereas older adults responded "old" and "remember" to both positive and negative items equally frequently.[76] These findings suggest that the PE in late adulthood needs to be construed more broadly to include not only memory enhancement but also more receptiveness to positive stimuli indicated by response biases.

Although several studies showed age-related positivity bias in memory, some studies failed to support such changes.[77,78] The impact of instruction on the PE has been suggested to be the main factor explaining some of these discrepancies in the literature. In the Charles et al.[75] study, participants were instructed to passively view pictures as if they were watching TV. Across two experiments, participants were also instructed to emotionally evaluate the items (e.g., rating the picture based on how it makes them feel while viewing it) or perceptually evaluate them (e.g., rating the visual complexity of the picture). Perceptual evaluation decreased older adults' ability to recall emotional items, whereas the emotional evaluation did not have any impact on older adults' memory. Hence, it seems that older adults' memory for emotional items is at an optimum level when they are able to encode them using their own strategies like passive viewing, or when they can focus on the emotional ratings of the pictures, such as when they emotionally evaluated the pictures (see[79]).

As mentioned, according to the Arousal Biased Competition (ABC) model,[73] the arousal dimension of emotions can influence memory storage. In line with this model, two separate neural networks processing arousal and valence have also been proposed previously.[79] In the aging literature, the question has become whether the PE occurs equally for arousing and nonarousing words.[80] To date, the PE has been found for nonarousing words but not for arousing words. Given that nonarousing words are thought to be remembered as a result of engaging controlled processes, Kensinger[80] concluded that the PE relies on "controlled processing of emotional information."

In this section, we first provided a brief overview of how emotional items can constrain memory processing, referring to memory narrowing. Arousal features of emotional information have been shown to influence memory narrowing, as well as enhance memory for gist. Then, we discussed evidence suggesting that the PE in aging might occur in different stages of the memory process. However, it should be noted that several studies failed to find the PE (for examples see[77,81,82]). One of the potential reasons for the lack of PE in some of the previous studies might be due to the use of instructions. Using instructions is thought to constrain the PE and consequently does not allow for the positive chronic goals to emerge naturally. Moreover, the impact of arousal and valence on memory of emotional items needs to be considered in interpreting emotional memory results in aging (see next section).

4.2 Neurological Findings

The amygdala is a major node in processing emotional items, but the role of this region is not limited to the attentional stage. The amygdala can modulate both the encoding and storage of information by interacting with the hippocampal region.[83] Buchanan and Adolphs[84] described the role of the amygdala in emotional memory by providing converging evidence from lesion, pharmacological, and neuroimaging studies. One pathway by which the amygdala is argued to influence memory is via neurotransmitter and hormonal output. Specifically, the connection between the amygdala and hypothalamus mediates the release of both epinephrine and cortisol when encountering emotionally arousing information, which consequently can influence the central nervous system and amygdala. The other pathway by which the amygdala may influence memory is via noradrenergic neurotransmission in the lateral/basolateral nuclei of the amygdala. Animal studies have also provided important insights into how the manipulation of activity within these subnuclei of the amygdala could influence hippocampal and cortical function. Additionally, in a review paper, LaBar and Cabeza formulated how the amygdala mediates the role of arousal of emotional stimuli on memory, depicted in Figure 5 (for more details, see[85]).

In addition to examining the role of the amygdala, neuroimaging studies have shown that so-called "emotional regions" (e.g., the amygdala) and medial temporal regions (e.g., the hippocampus) can influence higher cognitive areas (e.g., the prefrontal regions) in order to

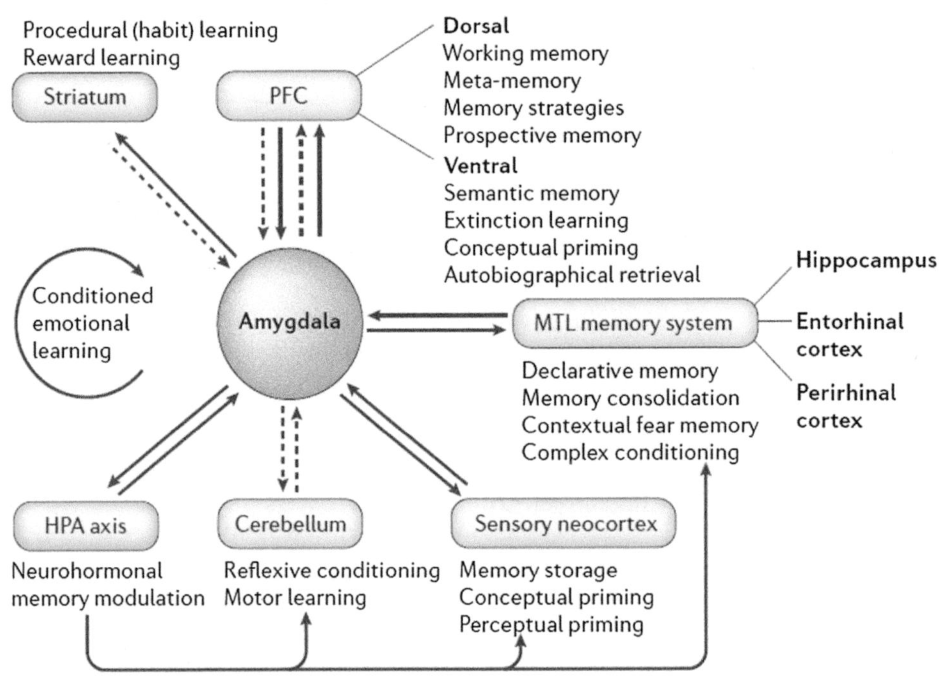

FIGURE 5 The mechanism by which the amygdala mediated the influence of emotion on memory proposed by LaBar and Cabeza (2006). *Courtesy of Dr. LaBar.*

encode and store emotional information. The influence of emotional regions during each memory stage serves to modulate mnemonic function and increase the likelihood of retaining the details of attended information (see[86] for more details). This conclusion has been supported by a review paper using the Activation Likelihood Estimate (ALE) method[87] to examine a wide range of brain regions involved during the encoding of emotional items, including the bilateral amygdalae, anterior hippocampus, anterior and posterior parahippocampal gyri, and the ventral visual stream (including the middle temporal, middle occipital, fusiform gyri, left lateral PFC, and the right ventral parietal cortex). It has been concluded that the amygdala interacts with the prefrontal and parietal cortices in order to enhance perceptual processing, semantic elaboration, and attention, which in turn influence subsequent emotional memory performance.

Interactions between emotional and cognitive processes occur at multiple levels of processing and across many different types of memory and cognitive tasks. However, relatively little is known about the relation between emotion and WM, specifically. Recent investigations have found that emotional distraction during a WM task reduces performance.[88] This behavioral effect has been shown to be related to enhanced activation in brain regions involved in emotional processing, along with decreased activity in brain regions associated with active maintenance of relevant information in WM. Furthermore, it has been shown that top-down attentional regions were involved while attending to the relevant items and ignoring the irrelevant items during the encoding stage of the WM task. The important consequence of goal-directed attention during encoding is to enhance the impact of behaviorally relevant stimuli at the expense of ignoring irrelevant stimuli. Thus, emotional stimuli may affect goal-directed behavior by either enhancing or disrupting task-relevant processes during both the encoding and retrieval stages of WM.

Underlying neural mechanisms that might explain behavioral asymmetry in memory with aging have been examined in a number of studies. In one study, Kensinger and Schacter[89] examined the emotional memory network by using a subsequent memory paradigm. In this paradigm, participants viewed a series of pictures during encoding and were then asked to judge whether each picture would fit inside a cabinet drawer. Outside the scanner, participants had to indicate whether each picture was the same, similar, or new. This paradigm previously showed enhanced memory for negative items in younger adults, and overall emotional memory bias relative to neutral items in older adults.[78,90] Kensinger and Schacter[89] found an age-related effect emerged during the encoding of positive items, where older adults showed stronger activity than younger adults in the medial PFC. As this region has been shown to be activated during self-referential tasks,[71] it was argued that older adults might process positive information in a more self-referential manner, which in turn will result in better memory for positive items.[79] The activity of the frontal region was replicated in another study during the processing of positive (relative to negative) words but not pictures.[91]

Effective connectivity has been used to examine age-related changes in connectivity between regions involved in emotional memories.[92] Age-related changes were observed only for successful encoding of positive items, and not negative. The results showed that older adults' hippocampal activity was modulated by the amygdala and ventromedial regions, and no age differences were observed during encoding of negative items. These results indicate that there might be age-related changes in how positive items are being processed, rather than changes in the processing of negative items.[92]

The functional relationship between subcortical regions (such as the amygdala and hippocampus) and cortical regions during both memory encoding and retrieval has been examined by Murty et al.[93] They investigated age-related changes in declarative memory using neutral and aversive stimuli in both younger and older adults. During encoding, older adults showed significantly greater dorsolateral prefrontal cortex (dlPFC) activity for aversive stimuli relative to younger adults. During retrieval, older adults showed greater coupling between the dlPFC and amygdala, whereas younger adults showed increased valence-related activity (aversive vs. neutral) in the amygdala and hippocampus. These findings are in line with the compensatory hypothesis (see Section 1.1), as older adults recruited more prefrontal regions during both recognition and encoding. Moreover, these results also might suggest that older adults recruited PFC regions to a greater extent than younger adults in order to downregulate their negative emotional reactions.

There is a mounting evidence regarding age-related changes in the way that older adults process and remember emotional expressions on faces (see Section 5 for more details). To extend previous studies on emotion recognition, Fischer et al.[94] investigated the memory of facial expressions of fearful and neutral faces among younger and older adults. No memory recognition advantage for emotional items (relative to neutral items) was found in either age group, but older adults recruited more PFC and insular regions while processing negative faces. This finding is consistent with the general pattern of age-related shift in cortical-subcortical recruitment, as discussed in Section 1 (Cognitive Aging) and Section 3 (Emotion and Attention in Aging). It is also important to consider the link between different stages of cognitive processes, such as attention, recognition, and memory, in order to form a more comprehensive picture of emotional processes in aging.

In this section, we first provided evidence of how the brain processes emotional items and how different brain structures interact to form memories of emotional items. Then, we presented results from behavioral and neuroimaging studies supporting the possibility that older adults might process positive and negative items differentially. However, there is need for further investigation in this area to provide direct evidence on the age-related changes in recruiting the brain networks involved during the encoding of different emotionally valenced items. It seems that there might be different networks engaged in processing emotionally valenced items as a function of age, but the evidence is disparate. Although some studies failed to observe positivity bias in memory (for instance see[77,82]), this discrepancy might be influenced by the conditions, the task instructions, and the paradigms that have been used across different studies.

5. EMOTION RECOGNITION AND AGING

Recognizing emotions on the face is one of the most fundamental aspects of our social communications. There is now a considerable literature focused on how different facial emotions are processed in late adulthood. In this section we introduce behavioral evidence in relation to the most studied emotions, happy and angry emotions, and then describe some of the neuroimaging findings on the topic. There is an increasing interest in understanding the link between the way older adults process emotional items and their ability to detect and respond to social cues in everyday life interaction. However, the social cognitive aspect of aging is beyond the scope of this chapter. For more information on this topic, please refer to Phillips et al.[95] and Von Hippel and Henry.[96]

Empirical evidence suggests that, relative to younger adults, older adults have difficulty recognizing negative emotions such as fear, anger, and sadness—findings which have been reviewed in a meta-analysis by Ruffman et al.[97] However, critical issues have been raised regarding the studies examining emotion recognition in aging. First, the people who model various emotional facial expressions in the photos used in these studies were either young or middle-aged. Secondly, presenting static images lowered the ecological validity of these studies. Such methodological issues make the conditions for facial emotion recognition dissimilar from everyday life experiences and have been argued to affect our understanding of emotion recognition with advancing age. Therefore, it is important to address some of these methodological issues in more detail before discussing the relevant findings.

One of the methodological caveats that has been addressed in several studies is the age of the models used for facial expression photos. Some of the earliest evidence regarding the potential importance of model age was presented in a study using the Multi-Source Interference Task (MSIT) by Ebner and Johnson.[98] It was shown that both younger and older adults were more distracted by faces similar in age to themselves. They extended this initial study by investigating the visual scan patterns of different ages displayed with different facial expressions.[99] The results suggested that the age of the model influences visual inspection of the face (measured by total gaze times). That is, both age groups spent more time looking at faces similar to their own age, relative to other-age faces. Moreover, they found that spending a longer time looking at the own-age faces predicted better own-age emotion identifications. According to a review paper by Folster et al.,[100] "[l]ower expressivity, age-related changes in the face, less elaborated emotion schemas for older faces, negative attitudes toward older adults, and different visual scan patterns may lower decoding accuracy for older faces." Therefore, it seems that using older models might add valuable insight into our understanding of emotion recognition, a view that previously had not been considered widely.

Also, in order to enhance the ecological validity of stimuli used in emotion recognition studies, more dynamic stimuli sets have been used to examine age differences in emotion expression identifications. Sze et al.,[101] for instance, found that older adults had difficulty judging emotional expressions from static images (e.g., angry and disgusted), while they outperformed younger adults in continuous emotion recognition using a dyadic interaction. Therefore, it seems that using dynamic stimuli that resemble everyday situations will influence the accuracy of emotion recognition and will enhance judgment of the authenticity of the expressions. Using dynamic emotional expressions will not only increase ecological validity but will also enhance activity in areas involved in emotion and processing socially relevant information.[102] For instance, Schultz and Pilz[103] reported more activation in response to dynamic faces in face-sensitive areas, including the bilateral fusiform gyrus, the left inferior occipital gyrus (IOG), and the right superior temporal sulcus (STS).

It is worth noting that there are mixed findings regarding dynamic stimuli such as video clips.[104] For instance, using video stimuli for recognition of basic emotions in the face and body, as well as incorporating body, facial, and situational cues, some studies have shown that older adults still perform worse than younger adults in recognizing some emotions[105] (also see[106,107]). Indeed, Ruffman[108] has argued that using static images for emotion recognition has reasonable ecological validity and, more importantly, explains age differences in various tasks, such as social understanding (using a faux pas task[109]), verbosity,[110] and deception.[111] Therefore, it seems that although using dynamic stimuli is an important methodological consideration and a number of studies provide support for the use of dynamic stimuli, static emotional stimuli are also informative.

Given that the most commonly assessed of the six basic emotions are the emotions of anger and happiness, in the following section, empirical evidence regarding these two specific emotions will be discussed.

5.1 Behavioral Findings

5.1.1 Anger

There are evolutionary reasons why threat-related stimuli should be detected more rapidly from a survival point of view. There is also empirical support for this perspective, with younger adults able to detect angry faces more quickly than nonthreatening faces.[112] Mather and Knight[113] demonstrated a similar effect in older adults by using a visual search task in which nine schematic faces were presented in a search array. Both age groups tended to identify angry faces more quickly than sad and happy faces, known as the pop-out effect. In another study, real faces were used in addition to the schematic faces.[114] In the first experiment, participants were asked to identify whether all faces were identical or if one of the faces was different in the array of faces, and in the second experiment, they were asked to label the emotion of the face that was different from the rest of the faces in the search array. The results indicated that although older adults were less accurate in explicitly labeling the angry schematic and real faces compared to younger adults, they responded as quickly as younger adults to the angry faces when they had to identify the different face from the array. On the basis of such findings, it has been argued that at some level (probably more an implicit level), older adults are able to differentiate angry faces equally as fast as younger adults.

Older adults' responses to angry facial expressions have also been studied using electromyography (EMG).[115] EMG measures changes in activity of the facial muscles, so can be used to index implicit mimicry responses toward stimuli. No age differences in *corrugator supercilii* (which pulls the eyebrow muscles into frowning) responses to angry expressions were found in the early stage (200–500 ms poststimulus onset). However, older adults' *corrugator supercilii* responses to anger in the later stage (500–800 ms poststimulus onset) were associated with difficulties in labeling the emotions. Although older adults demonstrate intact implicit mimicry responses to angry facial expressions, mimicry in the poststimulus-onset stage might have been confounded by their difficulties in labeling the expressions. Future studies examining dynamic stimuli and unfolding emotional responses over time by assessing mimicry responses are necessary in order to better understand when age-related differences in identifying angry emotions emerge, and whether such differences have consequences in everyday life interactions.

5.1.2 Happiness

While converging evidence supports the idea that younger adults outperform older adults in recognizing negative emotions, the evidence for age-related changes in identifying positive emotions is less clear. In a review paper by Isaacowitz et al.,[116] 11 out of 13 studies reviewed showed no age effect for recognition of happiness, possibly due to the ceiling effect. However, Ruffman et al.[97] reported moderate sized age deficits recognizing happiness from voices (mean effect size (M) = 0.37) and from matching voices to faces (M = 0.42). That is, dynamic stimuli provide additional temporal cues and resemble more naturalistic occurrences of daily social interaction, they may allow for more accurate emotion discrimination. Using dynamic stimuli, rather than static images, may allow older adults to use their lifetime experience in distinguishing different emotional expressions, such as happiness.[117]

Given older adults exhibit a preference toward positive stimuli, it has been assessed whether this preference is seen in relation to different types of smiles, and specifically posed versus spontaneous smiles. In order to investigate older adults' ability in discriminating smile types, Murphy et al.[117] conducted a study in which actors were videotaped while they gave expressions in response to three situations: receiving news that they won a prize, being invited on a cruise with friends, or being excluded from a group. The actors posed and spontaneous smiles were identified by coders using the Facial Action Coding System (FACS).[118] Results from two experiments using these video clips suggested that older adults were better than younger adults in distinguishing posed versus spontaneous smiles. Given that dynamic stimuli contain temporal information about emotional expressions and are an approximation of real-world experiences, older adults may have used this information to judge and discriminate between smile types.

Age-related changes in the emotion recognition of faces has been brought forward to investigate whether sensitivity in discriminating different types of smiles can impact the social meaning underlying these smiles. The ability to distinguish smiles has been shown to play a crucial role in interpersonal functioning.[119] For instance, increased difficulty distinguishing between smiles is associated with greater social functioning impairments among autistic individuals.[120] The question is how these changes in emotion recognition are linked to older adults' social functioning. In one study,[121] older adults were asked to indicate whether they would ask a favor from a target displaying either a spontaneous enjoyable smile, a posed enjoyable smile, or a deliberately-posed nonenjoyment smile. Although there were no age differences in discriminating between spontaneous and posed smiles, contrary to the Murphy et al.[117] study, older adults were more likely to ask for a favor from a person displaying a nonenjoyment smile. Hence, it seems that older adults are less likely to base their judgment on these social cues (enjoyment smiles/genuine), which can offer cooperation, approachability, or reciprocation in social interactions. Whether this age difference in approaching the nonenjoyment smile could influence older adults' social functioning, and in turn, could make them more prone to fraud as a result of approaching deceptive smiles, is a topic for further investigation in the field of emotion recognition and aging.

Additionally, dynamic stimuli were used to investigate the age-related changes in identifying emotional experiences accompanying three different smile expressions.[122] In this study, smile expressions were shown in different emotional contexts, including when people feel happy or amused. Thus, rather than using two types of smiles (posed/nonenjoyment versus spontaneous/enjoyment), actors were filmed in three different conditions where they were asked to display a smile: first, as if they were being accused of something unfairly; second, as if they were watching an amusing film or cartoon; and third, in an affectively neutral situation. Older and younger adults were asked to identify the emotional experience of each of these models. Overall, they found that younger adults were more accurate than older adults in identifying the emotional experiences accompanying each smile. Older adults were better at correctly ascribing positive affective experiences to the older models than the younger ones. Although these results are in contrast with some previous findings,[117,121] it seems that the context of displayed emotions associated with different types of smiles is an important factor in understanding the age-related changes in recognition of happy facial expressions.

Slessor et al.[123] have looked at the time course of mimicry responses to enjoyment and nonenjoyment smiles. The activity of the *Zygomaticus major* muscle region (which pulls the corner of the lips into a smile) and the *Orbicularis oculi* muscle region (which forms wrinkles around the outer corners of the eyes) were recorded. No age differences were found in the pattern and timing of *Zygomaticus major* activity in response to either type of smile. However, older adults showed extended *Orbicularis oculi* contraction in response to the nonenjoyment smiles. Moreover, older adults rated both types of smile as happy, independent of the *Orbicularis oculi* responses. These results suggest that emotion recognition and mimicry responses might rely on different processes. Further investigation is warranted to replicate these findings and examine the timings of these two processes more closely by using high temporal resolution techniques such as Electroencephalography (EEG).

5.2 Neurological Findings

In addition to the role of the amygdala in attention and memory functions, this region plays a crucial role in processing emotional expressions from faces and decoding other socially relevant cues, such as eye gaze (for a review see[124,125]). One of the earliest insights into the key role of the amygdala in processing facial expression was provided by a review on brain lesion and neuroimaging studies by Adolphs.[126] Most of the neuroimaging studies of emotion at the time confounded the encoding, retrieval, experience, and reaction to emotions, which makes it difficult to specifically attribute activation of the amygdala to any one of these psychological functions. However, our understanding of the neural mechanisms involved in processing facial expression has advanced from brain lesions and neuroimaging studies.

Two visual streams have been identified in nonhuman studies: one for object recognition leading to the temporal lobe, called the ventral stream, and one for localizing objects in space leading to the parietal lobe, called the dorsal stream.[127] Although a number of studies have addressed the dissociation between the two streams in processing concrete information, this dissociation has not been addressed in relation to emotion. Adolphs et al.[128] investigated the naming and recognition of emotion in a patient known as patient B., who presented with deficits retrieving conceptual knowledge. They hypothesized that this patient might be able to recognize emotion from dynamic stimuli, but not from static ones. The results suggested that patient B. was able to recognize emotions at the basic level of categorization if the stimuli contained temporal information. The fact that this patient was impaired in recognizing and verbally labeling emotions from static pictures suggests that recognizing emotions and labeling emotions rely on distinct neural systems. The second interesting finding from this study was that the patient B. had difficulties recognizing fear and anger, but not happiness and sadness. This finding could suggest that "conceptual knowledge of emotions might be organized in subordinate and superordinate levels."[128] Primary categorization of emotions falls into superordinate categories, such as "happy" and "unhappy," distinctions based on the valence. Patient B. can recognize happiness and sadness in most cases, perhaps because these sadness and happiness categorizations are examples of superordinate categories, but anger and fear fall in between and require more basic level recognition (subordinate). Therefore, it seems that such dissociation might be related to the impairment in subordinate categorization level rather than superordinate. The other explanation might be that this dissociation reflects that he fails to retrieve knowledge about arousal but is able to recognize information regarding valence.

Another reported case is patient Roger by Feinstein,[129] who has bilateral damage to the insula, ACC, medial PFC, OFC, basal forebrain, hippocampus, amygdala, parahippocampal gyrus, and temporal poles. One of the interesting findings from comparing patient Roger and B. is that emotion relies on a highly distributed network and is not attributable to a specific brain region. Although these two case studies are informative and provide valuable knowledge about the causal role of these brain regions in emotion recognition, neuroimaging techniques can provide alternative perspectives in order to understand the underlying neural networks and the relation between brain regions during emotion recognition.

One of the early studies in emotion recognition in aging using an fMRI was carried out by Iidaka et al.,[62] in which participants were shown pictures of the six basic facial emotions (happy, angry, disgusted, fearful, sad, and surprised). Participants were asked to label the gender of the actor modeling the emotional expressions, as well as the intensity of each emotion. Relative to younger participants, older adults showed reduced activity in the medial temporal lobe in response to both negative (e.g., left amygdala) and positive emotions (e.g., right parahippocampal gyrus). None of these regions showed increased activity among older adults. This dissociation in decreased activity of the medial temporal lobe may suggest that aging affects the neural networks involved in processing emotionally valenced facial expressions differently.

In another study, Gunning-Dixon et al.[65] investigated facial emotion processing using a task in which they asked participants to discriminate the age of the models during an emotion discrimination task. Older adults recruited more frontal regions relative to the temporo-limbic regions, which were more activated among younger adults. The authors argued that the decreased activity of the amygdala in the emotion discrimination task relative to younger adults was consistent with postmortem studies of age-related loss of neurons in the limbic regions. However, these data need to be interpreted cautiously, as the sample size of this study was very small (eight participants), and consequently, low statistical power could affect the generalizability of the findings.

It has been argued that the emotion recognition difficulties observed with aging might be associated with the cognitive declines associated with advanced age. In order to minimize the role of cognitive mechanisms involved in processing emotional items, a passive viewing paradigm was administered by Fischer et al.[70] They examined age-related changes in emotional face processing by using a relatively larger sample size than previous studies (24 younger and 22 older adults). Pictures of angry and neutral faces were presented to participants. The results showed reduced amygdala activity for negative faces among older adults relative to younger adults. Moreover, increased activity in the anterior-ventral insula cortex was found among older adults in response to negative faces. Again, the age-related increases in the engagement of cortical regions for processing emotional faces could reflect "compensatory mechanisms" older adults utilize for processing emotional faces.[70] The inconsistencies of the results of amygdala activity in Fischer's study and previous studies mentioned previously[62,65] could simply be related to methodological differences. Specifically, Fischer's[70] study used a passive viewing paradigm, whereas the other two studies included labeling and emotion discrimination tasks, which could potentially involve more top-down modulation of sensory information contributing to the discrepancy between the results.

As discussed, the ecological validity of stimuli, such as using own-age faces, is a critical methodological factor in emotion recognition studies in aging. In a neuroimaging study, Ebner et al.[130] addressed the role of identified regions of interests (ROIs), such as the ventromedial prefrontal cortex (vmPFC), dorsomedial prefrontal cortex (dmPFC), and amygdala, in identifying emotional expressions as a function of age of the faces presented to participants. First, the results showed that greater vmPFC activity emerged for identifying happy (relative to angry) faces. Second, both age groups showed enhanced activity of the amygdala in identifying happy (relative to negative) emotions, suggesting the saliency of positive affective processing. Greater activity of dmPFC was found in response to angry faces among older adults, indicating older adults' difficulty in recognizing angry faces. Thus, Ebner et al.'s results[130] highlight the underlying neural mechanisms involved in processing own-age faces during emotion recognition tasks.

Ebner et al.[131] study further demonstrates the importance of the age of faces used as stimuli in aging studies. In this study, the ability to identify emotion was examined in younger and older adults as a function of own-age versus other-age biases and emotional expressions of the faces. In both age groups, greater activity in the ventromedial PFC and insula for own-age faces was found, particularly for happy and neutral faces, but not for angry faces. These findings provide additional support for the differential neural mechanisms involved in processing own-age faces relative to other-age faces, which need to be considered in facial emotion recognition studies when comparing different age groups.

6. CLOSING REMARKS

The main focus of this chapter was to introduce recent findings in the field of emotional aging. We focused on three main domains (attention, memory, and face recognition systems), and for each, explored the influence of emotion. To gain a better understanding of age-related changes in attention and memory domains, the underlying mechanisms of how emotions impact various cognitive functions needs to be fully understood. As has been discussed, the distinction between whether attention influences emotions or vice versa has not been fully clarified in recent studies. Instead, widespread networks linking emotional and cognitive domains suggest that these areas are working in concert, and this perspective needs to be considered in life-span developmental research as well. Hence, rather than looking at changes with age in isolated areas (such as the amygdala), changes in the networks and interregional connections needs to be a primary focus.

To reach a more comprehensive understanding of emotional changes in aging, there is a need for research that integrates emotion and cognitive domains. Existing hypotheses and discoveries from the cognitive literatures can be adapted to obtain a better understanding of emotional aging. Several studies have explored the link between cognitive and emotional aging literature over the past few years. For instance, the proposed FADE hypothesis in emotional aging has been influenced by the PASA hypothesis from cognitive aging studies. However, further research is required to reach a more comprehensive and integrative model in aging.

There are still several questions in the emotional aging literature that need further investigation. Can different biases toward emotional items be beneficial for older adults and, if so, through what mechanisms? Can emotional enhancement or emotion regulation be considered a compensatory mechanism for cognitive deficits that occur with advancing age? Can changes in emotional processing have consequences/implications for social aspects of aging? Future research is required to provide a more comprehensive and integrative model of existing findings in the cognitive, emotional, and social domains of aging.

In the face recognition literature specifically, happy and angry expressions have been investigated extensively. Further studies are now required to examine age-related changes in other domains (such as trustworthiness, pride, and shame) by considering the ecological validity of the stimuli set. Given that some behavioral studies provided support for the importance of temporal cues and the age of face models used to measure emotion recognition ability with advancing age, using more ecologically valid stimuli in combination with fMRI, Magnetoencephalography (MEG), and EEG methods would open new avenues for better understanding the underlying neural mechanism of age-related changes in emotional processes in general.

Acknowledgment

Authors would like to thank Caitlin B. Hawley for her help in editing the chapter. Håkan Fischer was supported by the Swedish Research Council (grant number 421-2013-854). Authors would like to thank Bill von Hippel and Julie D. Henry for their helpful comments on this chapter

References

1. Reuter-Lorenz PA, Park DC. Human neuroscience and the aging mind: a new look at old problems. *J Gerontol Ser B Psychol Sci Soc Sci*. 2010;65B:405–415.
2. Grady CL. Cognitive neuroscience of aging. *Ann NY Acad Sci*. 2008;1124:127–144.
3. Bergerbest D, Gabrieli JDE, Whitfield-Gabrieli S, et al. Age-associated reduction of asymmetry in prefrontal function and preservation of conceptual repetition priming. *NeuroImage*. 2009;45(1):237–246.
4. Cabeza R, Anderson ND, Locantore JK, McIntosh AR. Aging gracefully: compensatory brain activity in high-performing older adults. *NeuroImage*. 2002;17(3):1394–1402.
5. Grady C. The cognitive neuroscience of ageing. *Nat Rev Neurosci*. 2012;13(7):491–505.
6. Cabeza R. Hemispheric asymmetry reduction in older adults: the Harold model. *Psychol Aging*. 2002;17(1):85–100.
7. Li S-C, Lindenberger U, Sikström S. Aging cognition: from neuromodulation to representation. *Trends Cognit Sci*. 2001;5(11):479–486.
8. Davis SW, Dennis NA, Daselaar SM, Fleck MS, Cabeza R. Que PASA? The posterior-anterior shift in aging. *Cereb Cortex*. 2008;18(5):1201–1209.
9. Buckner RL, Andrews-Hanna JR, Schacter DL. The brain's default network: anatomy, function, and relevance to disease. *Ann NY Acad Sci*. 2008;1124:1–38.
10. Sambataro F, Murty VP, Callicott JH, et al. Age-related alterations in default mode network: impact on working memory performance. *Neurobiol Aging*. 2010;31(5):839–852.
11. Persson J, Lustig C, Nelson JK, Reuter-Lorenz PA. Age differences in deactivation: a link to cognitive control? *J Cogn Neurosci*. 2007;19(6):1021–1032.
12. Persson J, Pudas S, Nilsson L-G, Nyberg L. Longitudinal assessment of default-mode brain function in aging. *Neurobiol Aging*. 2014;35(9):2107–2117.
13. Gazzaley A, Cooney JW, Rissman J, D'Esposito M. Top-down suppression deficit underlies working memory impairment in normal aging. *Nat Neurosci*. 2005;8:1298–1300.
14. Gazzaley A, Clapp W, Kelley J, McEvoy K, Knight RT, D'Esposito M. Age-related top-down suppression deficit in the early stages of cortical visual memory processing. *Proc Natl Acad Sci USA*. 2008;105(35):13122–13126.
15. Von Hippel W, Dunlop SM. Aging, inhibition, and social inappropriateness. *Psychol Aging*. 2005;20:519.
16. Henry JD, Von Hippel W, Baynes K. Social inappropriateness, executive control, and aging. *Psychol Aging*. 2009;24:239.
17. Hasher L, Zacks RT. Working memory, comprehension, and aging: a review and a new view. In: Bower GH, ed. *The Psychology of Learning and Motivation*. vol. 22. New York: Academic Press; 1988:193–225.
18. Gazzaley A, Cooney GW, McEvoy K, Knight RT, D'Esposito M. Top-down enhancement and suppression of the magnitude and speed of neural activity. *J Cogn Neurosci*. 2005;17:507–517.
19. Gazzaley A. Top-down modulation deficit in the aging brain: an emerging theory of cognitive aging. In: Stuss DT, Knight RT, eds. *Principles of Frontal Lobe Function*. 3rd ed. Oxford, New York: Oxford University Press; 2012.
20. Zanto TP, Gazzaley A. Attention and ageing. In: Nobre AC, Kastner S, eds. *The Oxford Handbook of Attention*. Oxford, New York: Oxford University Press; 2014:927–971.
21. Gazzaley A, Nobre AC. Top-down modulation: bridging selective attention and working memory. *Trends Cognit Sci*. 2012;16(2):129–135.
22. Gazzaley A. Influence of early attentional modulation on working memory. *Neuropsychologia*. 2011;49(6):1410–1424.
23. Reuter-Lorenz PA, Lustig C. Brain aging: reorganizing discoveries about the aging mind. *Curr Opin Neurobiol*. 2005;15(2):245–251.
24. Reed AE, Carstensen LL. The theory behind the age-related positivity effect. *Front Psychol*. 2012;3:339.
25. Carstensen LL, Isaacowitz DM, Charles ST. Taking time seriously: a theory of socioemotional selectivity. *Am Psychol*. 1999;54(3):165.
26. Cacioppo JT, Berntson GG, Bechara A, Tranel D, Hawkley LC. Could an aging brain contribute to subjective well-being? The value added by a social neuroscience perspective. In: Todorov A, Fiske S, Prentice D, eds. *Social Neuroscience: Toward Understanding the Underpinnings of the Social Mind*. Oxford, New York: Oxford University Press; 2011:249–262.
27. Marsiske M, Lang FB, Baltes PB, Baltes MM. *Selective Optimization with Compensation: Life-Span Perspectives on Successful Human Development*; 1995.
28. Charles ST, Carstensen LL. Social and emotional aging. *Annu Rev Psychol*. 2010;61:383–409.
29. Carstensen LL. The influence of a sense of time on human development. *Science*. 2006;312:1913–1915.
30. Löckenhoff CE, Carstensen LL. Socioemotional selectivity theory, aging, and health: the increasingly delicate balance between regulating emotions and making tough choices. *J Personality*. 2004;72(6):1395–1424.
31. Levine LJ, Edelstein RS. Emotion and memory narrowing: a review and goal-relevance approach. *Cogn Emot*. 2009;23(5):833–875.
32. Carstensen LL, Turan B, Scheibe S, et al. Emotional experience improves with age: evidence based on over 10 years of experience sampling. *Psychol Aging*. 2011;26(1):21–33.
33. Mather M, Canli T, English T, et al. Amygdala responses to emotionally valenced stimuli in older and younger adults. *Psychol Sci*. 2004;15(4):259–263.
34. Berntson GG, Bechara A, Damasio H, Tranel D, Cacioppo JT. Amygdala contribution to selective dimensions of emotion. *Soc Cogn Affect Neurosci*. 2007;2(2):123–129.
35. Carstensen LL, Fung HH, Charles ST. Socioemotional selectivity theory and the regulation of emotion in the second half of the life. *Motiv Emot*. 2003;27:103–123.
36. Treisman AM, Gelade G. A feature-integration theory of attention. *Cogn Psychol*. 1980;12(1):97–136.
37. Wolfe JM. Guided search 2.0 a revised model of visual search. *Psychon Bull Rev*. 1994;1(2):202–238.
38. Desimone R, Duncan J. Neural mechanisms of selective visual attention. *Annu Rev Neurosci*. 1995;18(1):193–222.
39. Vuilleumier P. How brains beware: neural mechanisms of emotional attention. *Trends Cognit Sci*. 2005;9(12):585–594.
40. Pessoa L. On the relationship between emotion and cognition. *Nat Rev Neurosci*. 2008;9(2):148–158.
41. Vuilleumier P, Armony JL, Driver J, Dolan RJ. Effects of attention and emotion on face processing in the human brain: an event-related fMRI study. *Neuron*. 2001;30:829–841.
42. Pessoa L, McKenna M, Gutierrez E, Ungerleider LG. Neural processing of emotional faces requires attention. *Proc Natl Acad Sci USA*. 2002;99(17):11458–11463.
43. Yiend J. The effects of emotion on attention: a review of attentional processing of emotional information. *Cognit Emot*. 2010;24(1):3–47.
44. Mather M, Carstensen LL. Aging and attentional biases for emotional faces. *Psychol Sci*. 2003;14(5):409–415.
45. Isaacowitz DM, Wadlinger HA, Goren D, Wilson HR. Selective preference in visual fixation away from negative images in old age? An eye-tracking study. *Psychol Aging*. 2006;21:40–48.
46. Isaacowitz DM, Allard ES, Murphy NA, Schlangel M. The time course of age-related preferences toward positive and negative stimuli. *J Gerontol B Psychol Sci Soc Sci*. 2009;64B(2):188–192.
47. Sakaki M, Gorlick MA, Mather M. Differential interference effects of negative emotional states on subsequent semantic and perceptual processing. *Emotion*. 2011;11(6):1263–1278.

48. Rosler A, Ulrich C, Billino J, et al. Effects of arousing emotional scenes on the distribution of visuospatial attention: changes with aging and early subcortical vascular dementia. *J Neurol Sci*. 2005;229-230:109–116.
49. Thomas RC, Hasher L. The influence of emotional valence on age differences in early processing and memory. *Psychol Aging*. 2006;21(4):821–825.
50. Knight M, Seymour TL, Gaunt JT, Baker C, Nesmith K, Mather M. Aging and goal-directed emotional attention: distraction reverses emotional biases. *Emotion*. 2007;7:705–714.
51. Allard ES, Isaacowitz DM. Are preferences in emotional processing affected by distraction? Examining the age-related positivity effect in visual fixation within a dual-task paradigm. *Neuropsychol Dev Cogn Section B Aging Neuropsychol Cogn*. 2008;15:725–743.
52. Mather M, Knight M. Goal-directed memory: the role of cognitive control in older adults' emotional memory. *Psychol aging*. 2005;20(4):554–570.
53. Isaacowitz DM, Toner K, Goren D, Wilson HR. Looking while unhappy: mood-congruent gaze in young adults, positive gaze in older adults. *Psychol Sci*. 2008;19(9):848–853.
54. Isaacowitz DM, Toner K, Neupert SD. Use of gaze for real-time mood regulation: effects of age and attentional functioning. *Psychol Aging*. 2009;24:989–994.
55. Larcom MJ, Isaacowitz DM. Rapid emotion regulation after mood induction: age and individual differences. *J Gerontol B Psychol Sci Soc Sci*. 2009;64(6):733–741.
56. Allard ES, Wadlinger HA, Isaacowitz DM. Positive gaze preferences in older adults: assessing the role of cognitive effort with pupil dilation. *Neuropsychol Dev Cogn Section B Aging Neuropsychol Cogn*. 2010;17:296–311.
57. Pessoa L. Emergent processes in cognitive-emotional interactions. *Dialog Clin Neurosci*. 2010;12(4):433–448.
58. Lindquist KA, Wager TD, Kober H, Bliss-Moreau E, Barrett LF. The brain basis of emotion: a meta-analytic review. *Behav Brain Sci*. 2012;35(3):121–143.
59. Wager TD, Kang J, Johnson TD, Nichols TE, Satpute AB, Barrett LF. *A Bayesian Model of Category-Specific Emotional Brain Responses*; 2015.
60. Williams LM, Brown KJ, Palmer D, et al. The mellow years?: neural basis of improving emotional stability over age. *J Neurosci*. 2006;26(24):6422–6430.
61. Mather M. The emotion paradox in the aging brain. *Ann NY Acad Sci*. 2012;1251(1):33–49.
62. Iidaka T, Okada T, Murata T, et al. Age-related differences in the medial temporal lobe responses to emotional faces as revealed by fMRI. *Hippocampus*. 2002;12(3):352–362.
63. St Jacques P, Dolcos F, Cabeza R. Effects of aging on functional connectivity of the amygdala during negative evaluation: a network analysis of fMRI data. *Neurobiol Aging*. 2010;31(2):315–327.
64. Wright CI, Wedig MM, Williams D, Rauch SL, Albert MS. Novel fearful faces activate the amygdala in healthy young and elderly adults. *Neurobiol Aging*. 2006;27(2):361–374.
65. Gunning-Dixon FM, Gur RC, et al. Age-related differences in brain activation during emotional face processing. *Neurobiol Aging*. 2003;24(2):285–295.
66. St Jacques PL, Bessette-Symons B, Cabeza R. Functional neuroimaging studies of aging and emotion: fronto-amygdalar differences during emotional perception and episodic memory. *J Int Neuropsychol Soc JINS*. 2009;15(6):819–825.
67. Leclerc CM, Kensinger EA. Age-related differences in medial prefrontal activation in response to emotional images. *Cogn Affect Behav Neurosci*. 2008;8(2):153–164.
68. Ochsner KN, Gross JJ. The cognitive control of emotion. *Trends Cogn Sci*. 2005;9(5):242–249.
69. Wager TD, Davidson ML, Hughes BL, Lindquist MA, Ochsner KN. Prefrontal-subcortical pathways mediating successful emotion regulation. *Neuron*. 2008;59(6):1037–1050.
70. Fischer H, Sandblom J, Gavazzeni J, Fransson P, Wright CI, Backman L. Age-differential patterns of brain activation during perception of angry faces. *Neurosci Lett*. 2005;386(2):99–104.
71. Gutchess AH, Kensinger EA, Schacter DL. Aging, self-referencing, and medial prefrontal cortex. *Soc Neurosci*. 2007;2(2):117–133.
72. Kensinger EA, Corkin S. Two routes to emotional memory: distinct neural processes for valence and arousal. *Proc Natl Acad Sci USA*. 2004;101(9):3310–3315.
73. Mather M, Sutherland MR. Arousal-biased competition in perception and memory. *Perspect Psychol Sci*. 2011;6(2):114–133.
74. Reed AE, Chan L, Mikels JA. Meta-analysis of the age-related positivity effect: age differences in preferences for positive over negative information. *Psychol Aging*. 2014;29:1–15.
75. Charles ST, Mather M, Carstensen LL. Aging and emotional memory: the forgettable nature of negative images for older adults. *J Exp Psychol Gen*. 2003;132(2):310–324.
76. Kapucu A, Rotello CM, Ready RE, Seidl KN. Response bias in "remembering" emotional stimuli: a new perspective on age differences. *J Exp Psychol Learn Mem Cogn*. 2008;34(3):703–711.
77. Gallo DA, Foster KT, Johnson EL. Elevated false recollection of emotional pictures in young and older adults. *Psychol Aging*. 2009;24:981–988.
78. Kensinger EA, Garoff-Eaton RJ, Schacter DL. Effects of emotion on memory specificity: memory trade-offs elicited by negative visually arousing stimuli. *J Mem Lang*. 2007;56(4):575–591.
79. Emery L, Hess TM. Viewing instructions impact emotional memory differently in older and young adults. *Psychol Aging*. 2008;23(1):2–12.
80. Kensinger EA. Age differences in memory for arousing and nonarousing emotional words. *J Gerontol B Psychol Sci Soc Sci*. 2008;63B:13–18.
81. Kensinger EA, Brierley B, Medford N, Growdon JH, Corkin S. Effects of normal aging and Alzheimer's disease on emotional memory. *Emotion*. 2002;2:118–134.
82. Grühn D, Smith J, Baltes PB. No aging bias favoring memory for positive material: evidence from a heterogeneity-homogeneity list paradigm using emotionally toned words. *Psychol Aging*. 2005;20(4):579–588.
83. Phelps EA. Human emotion and memory: interactions of the amygdala and hippocampal complex. *Curr Opin Neurobiol*. 2004;14(2):198–202.
84. Buchanan TW, Adolphs R. The role of the human amygdala in emotional modulation of long-term declarative memory. In: Moore S, Oaksford M, eds. *Emotional Cognition: From Brain to Behavior*. London, UK: Benjamins; 2002:9–34.
85. LaBar KS, Cabeza R. Cognitive neuroscience of emotional memory. *Nat Rev Neurosci*. 2006;7(1):54–64.
86. Kensinger EA, Schacter DL. Memory and emotion. In: Lewis M, Haviland-Jones JM, Barrett LF, eds. *The Handbook of Emotions*. 3rd ed. New York: Guilford; 2007:601–617.
87. Murty VP, Ritchey M, Adcock RA, LaBar KS. Reprint of: fMRI studies of successful emotional memory encoding: a quantitative meta-analysis. *Neuropsychologia*. 2011;49(4):695–705.
88. Dolcos F, McCarthy G. Brain systems mediating cognitive interference by emotional distraction. *J Neurosci Off J Soc Neurosci*. 2006;26(7):2072–2079.
89. Kensinger EA, Schacter DL. Neural processes supporting young and older adults' emotional memories. *J Cogn Neurosci*. 2008;20(7):1161–1173.
90. Kensinger EA, Garoff-Eaton RJ, Schacter DL. Effects of emotion on memory specificity in young and older adults. *J Gerontol B Psychol Sci Soc Sci*. 2007;62(4):P208–P215.

91. Leclerc CM, Kensinger EA. Neural processing of emotional pictures and words: a comparison of young and older adults. *Dev Neuropsychol*. 2011;36(4):519–538.
92. Addis DR, Leclerc CM, Muscatell KA, Kensinger EA. There are age-related changes in neural connectivity during the encoding of positive, but not negative, information. *Cortex; J Devot Study Nerv Syst Behav*. 2010;46(4):425–433.
93. Murty VP, Sambataro F, Das S, et al. Age-related alterations in simple declarative memory and the effect of negative stimulus valence. *J Cogn Neurosci*. 2009;21(10):1920–1933.
94. Fischer H, Nyberg L, Backman L. Age-related differences in brain regions supporting successful encoding of emotional faces. *Cortex*. 2010;46(4):490–497.
95. Phillips LH, Slessor G, Bailey PE, Henry JD. Older Adults' Perception of Social and Emotional Cues. *Oxf Handb Emot Soc Cogn Probl Solving Adulthood*. 2014;9.
96. Von Hippel W, Henry JD. *Social Cognitive Aging. The Handbook of Social Cognition*. London: Sage; 2012. 390–410.
97. Ruffman T, Henry JD, Livingstone V, Phillips LH. A meta-analytic review of emotion recognition and aging: implications for neuropsychological models of aging. *Neurosci Biobehav Rev*. 2008; 32(4):863–881.
98. Ebner NC, Johnson MK. Age-group differences in interference from young and older emotional faces. *Cogn Emot*. 2010;24(7):1095–1116.
99. Ebner NC, He Y, Johnson MK. Age and emotion affect how we look at a face: visual scan patterns differ for own-age versus other-age emotional faces. *Cogn Emot*. 2011;25(6):983–997.
100. Folster M, Hess U, Werheid K. Facial age affects emotional expression decoding. *Front Psychol*. 2014;5:30.
101. Sze JA, Goodkind MS, Gyurak A, Levenson RW. Aging and emotion recognition: not just a losing matter. *Psychol Aging*. 2012;27(4):940.
102. Krumhuber EG, Kappas A, Manstead ASR. Effects of dynamic aspects of facial expressions: a review. *Emot Rev*. 2013;5(1):41–46.
103. Schultz J, Pilz KS. Natural facial motion enhances cortical responses to faces. *Exp Brain Res*. 2009;194(3):465–475.
104. Krendl AC, Ambady N. Older adults' decoding of emotions: role of dynamic versus static cues and age-related cognitive decline. *Psychol Aging*. 2010;25(4):788–793.
105. Sullivan S, Ruffman T. Social understanding: how does it fare with advancing years? *Brit J Psychol*. 2004;95:1–18.
106. Sullivan S, Ruffman T. Emotion recognition deficits in the elderly. *Int J Neurosci*. 2004;114(3):403–432.
107. Ruffman T, Sullivan S, Dittrich W. Older adults' recognition of bodily and auditory expressions of emotion. *Psychol Aging*. 2009;24(3):614–622.
108. Ruffman T. Ecological validity and age-related change in emotion recognition. *J Nonverbal Behav*. 2011;35(4):297–304.
109. Halberstadt J, Ruffman T, Murray J, Taumoepeau M, Ryan M. Emotion perception explains age-related differences in the perception of social gaffes. *Psychol Aging*. 2011;26(1):133–136.
110. Ruffman T, Murray J, Halberstadt J, Taumoepeau M. Verbosity and emotion recognition in older adults. *Psychol Aging*. 2010;25(2):492–497.
111. Ruffman T, Murray J, Halberstadt J, Vater T. Age-related differences in deception. *Psychol Aging*. 2012;27(3):543.
112. Öhman A, Lundqvist D, Esteves F. The face in the crowd revisited: a threat advantage with schematic stimuli. *J Personality Soc Psychol*. 2001;80(3):381–396.
113. Mather M, Knight MR. Angry faces get noticed quickly: threat detection is not impaired among older adults. *J Gerontol B Psychol Sci Soc Sci*. 2006;61(1):P54–P57.
114. Ruffman T, Ng M, Jenkin T. Older adults respond quickly to angry faces despite labeling difficulty. *J Gerontol B Psychol Sci Soc Sci*. 2009;64(2):171–179.
115. Bailey PE, Henry JD, Nangle MR. Electromyographic evidence for age-related differences in the mimicry of anger. *Psychol Aging*. 2009;24(1):224–229.
116. Isaacowitz DM, Löckenhoff CE, Lane RD, et al. Age differences in recognition of emotion in lexical stimuli and facial expressions. *Psychol Aging*. 2007;22(1):147.
117. Murphy NA, Lehrfeld JM, Isaacowitz DM. Recognition of posed and spontaneous dynamic smiles in young and older adults. *Psychol Aging*. 2010;25(4):811–821.
118. Ekman P, Friesen WV. *The Facial Action Coding System*. Palo Alto, CA: Consulting Psychologists Press; 1978.
119. Ekman P. *Telling Lies: Clues to Deceit in the Marketplace, Politics, and Marriage* [Revised Edition]. NY: WW Norton and Company; 2009.
120. Boraston ZL, Corden B, Miles LK, Skuse DH, Blakemore S-J. Brief report: perception of genuine and posed smiles by individuals with autism. *J Autism Dev Disord*. 2008;38(3):574–580.
121. Slessor G, Miles LK, Bull R, Phillips LH. Age-related changes in detecting happiness: discriminating between enjoyment and non-enjoyment smiles. *Psychol Aging*. 2010;25(1):246–250.
122. Riediger M, Studtmann M, Westphal A, Rauers A, Weber H. No smile like another: adult age differences in identifying emotions that accompany smiles. *Front Psychol*. 2014;5(480).
123. Slessor G, Bailey PE, Rendell PG, Ruffman T, Henry JD, Miles LK. Examining the time course of young and older adults' mimicry of enjoyment and nonenjoyment smiles. *Emotion*. 2014;14(3):532–544.
124. Itier RJ, Batty M. Neural bases of eye and gaze processing: the core of social cognition. *Neurosci Biobehav Rev*. 2009;33(6):843–863.
125. Shepherd SV. Following gaze: gaze-following behavior as a window into social cognition. *Front Integr Neurosci*. 2010;4:5.
126. Adolphs R. THe human amygdala and emotion. *Neuroscientist*. 1999;5:125–137.
127. Ungerleider LG, Mishkin M. Two cortical visual systems. In: Ingle DG, Goodale MA, Mansfield RJQ, eds. *Analysis of Visual Behavior*. Cambridge, MA: MIT Press; 1982.
128. Adolphs R, Tranel D, Damasio AR. Dissociable neural systems for recognizing emotions. *Brain Cogn*. 2003;52(1):61–69.
129. Feinstein JS. Lesion studies of human emotion and feeling. *Curr Opin Neurobiol*. 2013;23(3):304–309.
130. Ebner NC, Johnson MK, Fischer H. Neural mechanisms of reading facial emotions in young and older adults. *Emot Sci*. 2012;3:223.
131. Ebner NC, Johnson MR, Rieckmann A, Durbin KA, Johnson MK, Fischer H. Processing own-age versus other-age faces: neuro-behavioral correlates and effects of emotion. *NeuroImage*. 2013;78:363–371.

SECTION V

BRAIN IMAGING PERSPECTIVES ON THE BASIS OF PROSOCIABILITY

CHAPTER

14

Cultural Neuroscience of Moral Reasoning and Decision-Making

Yi-Yuan Tang[1], *Rongxiang Tang*[2]

[1]Presidential Endowed Chair in Neuroscience, Department of Psychological Sciences, Texas Tech University, Lubbock, TX, USA; [2]Department of Psychology, Washington University, St. Louis, MO, USA

OUTLINE

1. INTRODUCTION

Research in cultural neuroscience examines how cultural and genetic diversity shape the human mind, brain, and behavior using a variety of multimodal methods.[1] Cultural differences in understanding and conceptualizing one's relationships with others may have led to diverse cultural systems for interpreting, thinking, reasoning, and decision-making about the world. Eastern holistic systems of thought rely on connectedness and relations as a primary way of understanding the world, whereas Western analytic systems of thought rely on discreteness as an epistemological way of thinking.[2] Reasoning and decision-making, as important aspects of human high-level cognition, may be influenced by cultural differences. For example, Eastern Asians tend to frame the decision to help as a matter of moral responsibility, whereas Americans are more likely to frame it as one's personal choice.[3]

Previous studies indicate that from attention and cognition to social cognitive processes, neural systems have likewise adapted differently across cultural contexts to facilitate divergent systems of social interactions and relations.[4] In this chapter, we will focus on the cultural differences in high-level cognition (moral reasoning and decision-making), especially between the Chinese and American cultures. We will explore the associated brain networks of cultural variation involved in these processes and will take abstract thinking/reasoning and moral decision-making as examples to demonstrate the neural basis of cultural influences on these processes that involve Eastern and Western cultures. This focus reflects the current growing theoretical and experimental curiosity among behavioral scientists and neuroscientists, as well as the general public.[5] We acknowledge that this chapter provides an incomplete view of the literature on moral reasoning and decision-making.

1.1 Abstract Thinking/Reasoning—Arithmetic Processing

Arithmetic or math processing often involves abstract thinking/reasoning.

One example of a higher cognitive function that may be influenced by holistic vs. analytic thinking styles involves discrete neural patterns of arithmetic processing

Neuroimaging Personality, Social Cognition, and Character
http://dx.doi.org/10.1016/B978-0-12-800935-2.00014-2

in Chinese and Westerners.[5] The universal use of Arabic numbers in mathematics raises the question of whether they are processed the same way in people from different cultures who speak different languages, such as Chinese and English. To address this question, researchers used functional Magnetic Resonance Imaging (fMRI) to scan 12 native Chinese speakers (NCS) and 12 native English speakers (NES) who possessed a college-level education. The participants were instructed to perform four tasks during the scanning (see Figure 1 for examples): (1) the symbol condition: judge the spatial orientation of nonnumerical stimuli in which a triplet of nonsemantic characters or symbols are visually presented either in an upright or in an italic orientation. The task was to decide whether the third symbol had the same orientation as the first two; (2) the number condition: judge the spatial orientation of numerical stimuli, using the same task as in the symbol condition except for using Arabic digits as visual stimuli; (3) the addition condition: determine whether the third digit was equal to the sum of the first two in a triplet of Arabic numbers; and (4) the comparison condition: determine whether the third digit was larger than the larger of the first two in a triplet of Arabic numbers. A baseline condition of matching white and/or gray circular dots was used to control for the motor and nonspecific visual components of the tasks.

Results indicated a different cortical representation of numbers between native Chinese speakers and native English speakers. While the English speakers employed a language process relying on the left perisylvian cortices for mental calculation, such as a simple addition task, the Chinese speakers instead engaged a visuo–premotor association network for the same task (see Figure 1). We further chose two regions of interest, the perisylvian area, including both the Broca (Br) and Wernicke (Wn) areas, and the premotor association area between BA6, BA8, and BA9.[6] Quantitative analyses were conducted by comparing the fMRI signal between the English and Chinese groups. We found the perisylvian activations were significantly larger in the English speakers than in the Chinese speakers (Figure 2(A)). As the arithmetic complexity increased across the four conditions (Symbol < Number < Addition < Comparison), there was a trend toward increased premotor activation in the Chinese speakers, but not in the English speakers (Figure 2(B)). Therefore, there was a double dissociation in brain activation between these two groups during these tasks, supporting clear cultural differences in the processing of numbers. In addition, for both cultural groups, the inferior parietal cortex was activated in the numerical quantity comparison. However, the fMRI connectivity analyses[6] revealed a difference in the brain networks involved in the task for Chinese and English speakers. Two distinct patterns were observed in the functional networks: there was dorsal visuo-pathway dominance

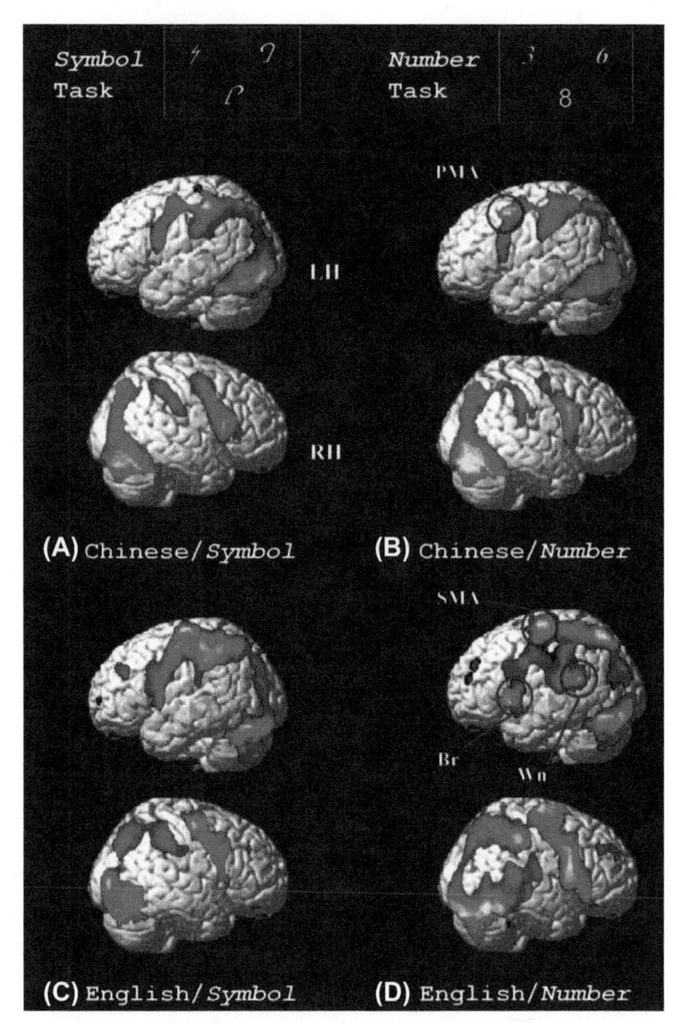

FIGURE 1 **Dissociation in the brain representation of Arabic numbers between native Chinese speakers (NCS) and native English speakers (NES).** (A) During the symbol task in NCS. (B) During the number task in NCS. (C) During the symbol task in NES. (D) During the number task in NES. The task-dependent brain activation was determined by Statistical Parametric Mapping (SPM) by using a liberal threshold ($p < 0.05$) for illustrating a global pattern of the fMRI blood-oxygen-level dependent (BOLD) signal changes. The type-I error of detecting the difference was corrected for the number of resolution elements at each of the activated brain regions defined anatomically by using the SPM add-on toolbox automated anatomical labeling (AAL). The multiple comparison correction is the small volume correction (SVC) procedure implemented in SPM. (A and B) Examples of the visual stimuli used for the symbol task and the number task, respectively, are shown at the top. LH = left hemisphere; RH = right hemisphere; Br = Broca area; Wn = Wernicke area.

(through the parietal–occipital cortex) for the Chinese speakers, but ventral visuo-pathway dominance (through the temporal cortex) for the English speakers.

Our findings have two implications. First, in both Chinese and English speakers, there is cortical dissociation between addition and numerical comparison processing. The addition task seems to be more dependent upon language processing than does the comparison task, which is consistent with the notion that there

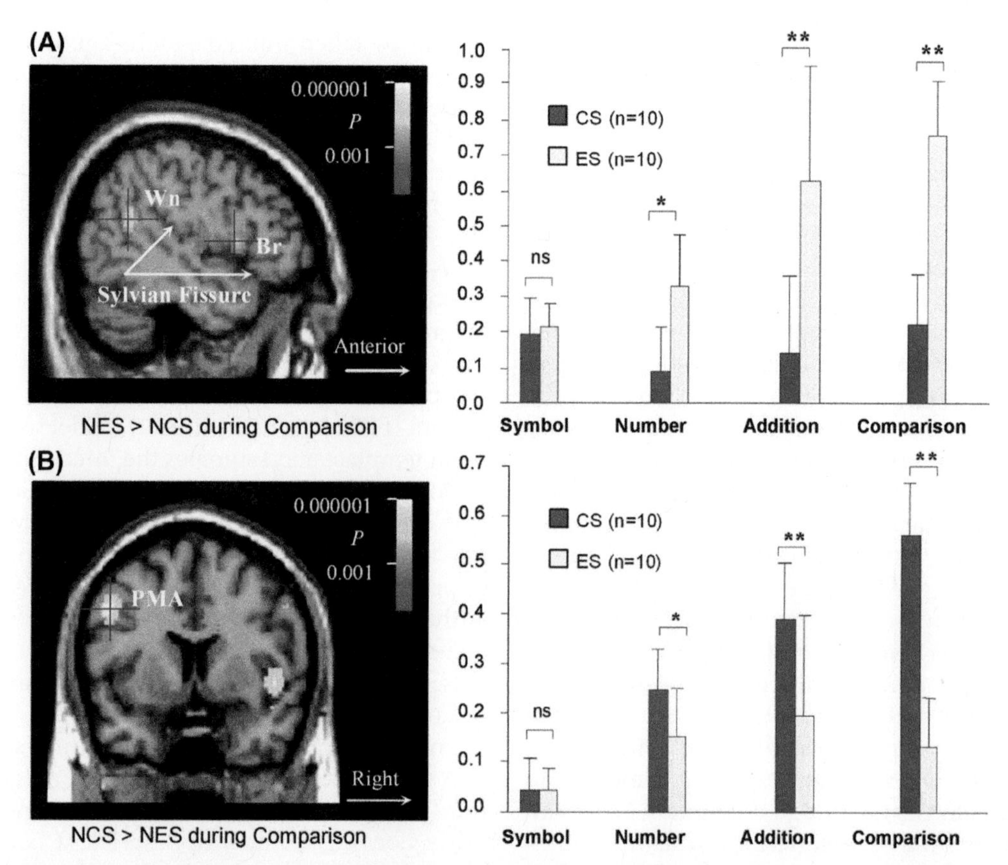

FIGURE 2 **Comparison of the activation intensity between native Chinese speakers (NCS) and native English speakers (NES) in the perisylvian language region (A) and the premotor association area (PMA) (B).** The brain activation maps (left) were determined by contrasting BOLD signals between NCS and NES only during the comparison task, with the NES group showing relative increase of the signal (A, English > Chinese), and the NCS group showing relative increase of the signal (B, Chinese > English). The within-group task-dependent activation was determined by SPM99 by using a threshold ($p < 0.001$, uncorrected) for defining the regions of interest (ROIs) in the perisylvian language region, including both the Broca area (Br) with Talairach coordinates at (−50, 12, 7) and the Wernicke area (Wn) (−57, −59, 16) and in the PMA (−18, 22, 56). For each individual, the fMRI activation index (right) was then determined by integrating the BOLD signal changes in these ROIs for statistical comparisons. Two sample t-tests were used to compare the mean of the activation index for each task. *, $p < 0.05$; **, $p < 0.01$; ns = not significant.

are different neural substrates underlying verbal and numerical processing.[7,8] Second, there are differences in the brain representation of number processing between Chinese and English speakers.

Although differences between the two cultures in terms of language, educational systems, or genetics may contribute to these findings, relatively greater reliance on analytic and formal logical (rather than holistic) styles of cognition among Westerners may have contributed to the greater observed activity in language-related regions among Westerners during the arithmetic task. Future cultural neuroscience research should investigate the consequences of holistic versus analytic forms of thinking on mathematical processing and would help confirm the culturally-variant patterns of neural activity observed during arithmetic tasks. Furthermore, genetics and environmental factors often work together in shaping human cognition and social behavior.[9] Future studies should address how the gene × environment (or experience) interacts to affect high-level cognition in different cultures. For example, if both genes and experience shape human cognitive functions, such as numerical processing in the brain, another challenge will be to understand how different educational systems may influence our core understanding of numbers.[9]

1.2 Moral Reasoning/Decision-Making—Moral Dilemmas

Moral decision-making is a flourishing field of research that has attracted great scholarly attention. A growing literature has tried to uncover the neural mechanisms underlying moral decision-making with a family of moral dilemmas.[10,11] Generally, a moral dilemma is a complex situation that involves a conflict in choosing between two undesirable alternatives, which would evoke the competition between deontological (nonutilitarian) choice and utilitarian response.[3] For example, in a personal moral dilemma (the footbridge dilemma), the only way to save five workers from a runaway trolley is to push a large man off an overpass bridge onto the tracks below. He will die, but his body will stop the trolley from reaching the other five people. This is commonly referred to as the personal moral dilemma, since the participant would be personally required to engage in the harmful act. A corresponding impersonal moral dilemma is the trolley dilemma, in which the only way to save the five workers is to pull a lever redirecting the trolley onto another set of tracks, where it will kill a single worker instead of five workers.[3,10,12] In this scenario, the participant does not directly and physically engage in the harmful act and is actually saving more people. These two different dilemmas usually produce distinct responses from the participants. The deontological response is an aversive emotional response to the harmful act, which would lead to the rejection of a utilitarian response. In contrast, the utilitarian response is to take part in the harmful act, since doing so will maximize good consequences, which would require overcoming the prepotent emotional response.

Although the proposed actions in both personal and impersonal dilemmas would produce similar outcomes, moral judgment in the two dilemma types might be driven by different principles. Previous studies have indicated that most people show agreement with pulling the lever in the trolley dilemma and disagreement with pushing the man in the footbridge dilemma.[11] Neuroimaging results revealed that personal moral dilemmas elicit greater activation in brain regions associated with emotions, whereas impersonal moral dilemmas elicit greater activation in areas associated with problem solving and working memory, suggesting that both cognitive and emotional processes contribute to moral decision-making.[10,11]

For both personal and impersonal ethical dilemmas, each act may lead to certain consequences, and both sides may be right in different senses. If we adopt the utilitarian way of thinking, we would conclude that it is right to kill one instead of five, but it is also right to develop an intuitive rule against the participation in killing others. However, people's moral decisions might be influenced by cultural factors, since people's morals and virtuousness are shaped by culture.[13] Therefore, culture should be taken into consideration when investigating moral decision-making.

Research on the cultural neuroscience of moral decision-making suggests that members of holistic cultures (East Asians) exhibit greater integrative processing of moral dilemmas than do members of analytic cultures (Westerners). In one study (culture × type of dilemma), Wang and colleagues (2014) combined event-related potential (ERP) techniques with standardized low-resolution brain electromagnetic tomography (sLORETA) to study potential cultural variation within the brain.[3] sLORETA is a method that computes images of electric activity from electroencephalogram (EEG) in a realistic head model using the MNI152 brain template and estimates the three-dimensional distribution of current density in voxels.[3] In this study, Chinese and Western college students were recruited to participate in a moral dilemma task that included both personal and impersonal situations. An example of a personal situation is the footbridge dilemma, and an impersonal situation includes the trolley dilemma (see above). Each dilemma was presented in the participants' native language (Chinese or English) and as black text against a gray background on a computer monitor, with a series of three screens. The first two screens described the scenario of a dilemma, and the third posed a question asking whether or not the hypothetical action was morally appropriate. Choosing appropriate options was considered to be utilitarian (taking part in the harmful act since doing so will maximize good consequences—saving five workers in the footbridge dilemma, which requires overcoming the prepotent emotional response), whereas choosing inappropriate options was considered to be nonutilitarian. The dependent variables were the proportion of utilitarian judgments, reaction time, ERP responses, and sLORETA activity.

There was a main effect of the type of dilemma such that participants made a smaller proportion of utilitarian judgments and exhibited longer reaction times in response to personal rather than impersonal dilemmas, with no difference observed between Westerners and Chinese, as shown in Figure 3.

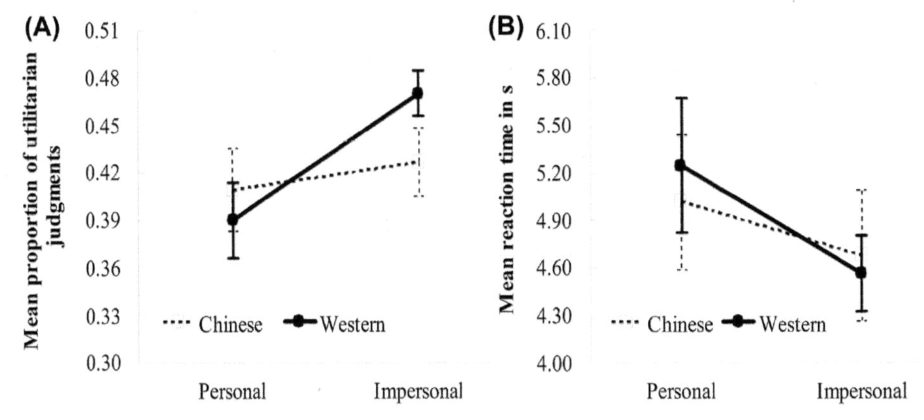

FIGURE 3 (A) Mean proportion of utilitarian judgments by dilemma type for Chinese and Westerners. (B) Mean reaction time by dilemma type for Chinese and Westerners. Error bars indicate standard error of the mean.

The main focus of analysis was the P3 and P260 components (event related potential components elicited in the process of decision-making). Notably, the event-related potential components were significantly different between the two cultural groups. For Westerners, smaller P3 amplitudes were evoked by personal rather than impersonal dilemmas, while for Chinese participants, smaller P260 deflections were elicited by personal rather than impersonal dilemmas. The different ERP components elicited by moral dilemmas may be attributable to cultural differences. It has been widely demonstrated that East Asians and Westerners differ in experience and socialization, which influences cognition and the allocation of attention.[14] Previous research has indicated that P3 is an index of inhibition of task-irrelevant emotional information, with less positive amplitudes for negative stimuli than for neutral stimuli in implicit emotional tasks.[15] The P3 results reported here are consistent with the former findings, suggesting that more negative emotions needed to be inhibited in response to personal (rather than impersonal) dilemmas. Moreover, the P260 component has been suggested to be a combination of a P2 and a P3-like process and has been reported to reflect immediate affective reactions toward options that integrate attention, working memory, and emotional processing.[16] This different cognitive functioning may be related to holistic thinking. Thus, the P260 component may suggest a more integrated process during the solution of moral dilemmas in Chinese compared to Westerners.

The analysis with sLORETA revealed a significantly different pattern of brain activity between Westerners and Chinese. As shown in Figure 4, the main source of both P2 and P3 components for personal dilemmas was the cingulate gyrus, similar to other studies.[17] In contrast,

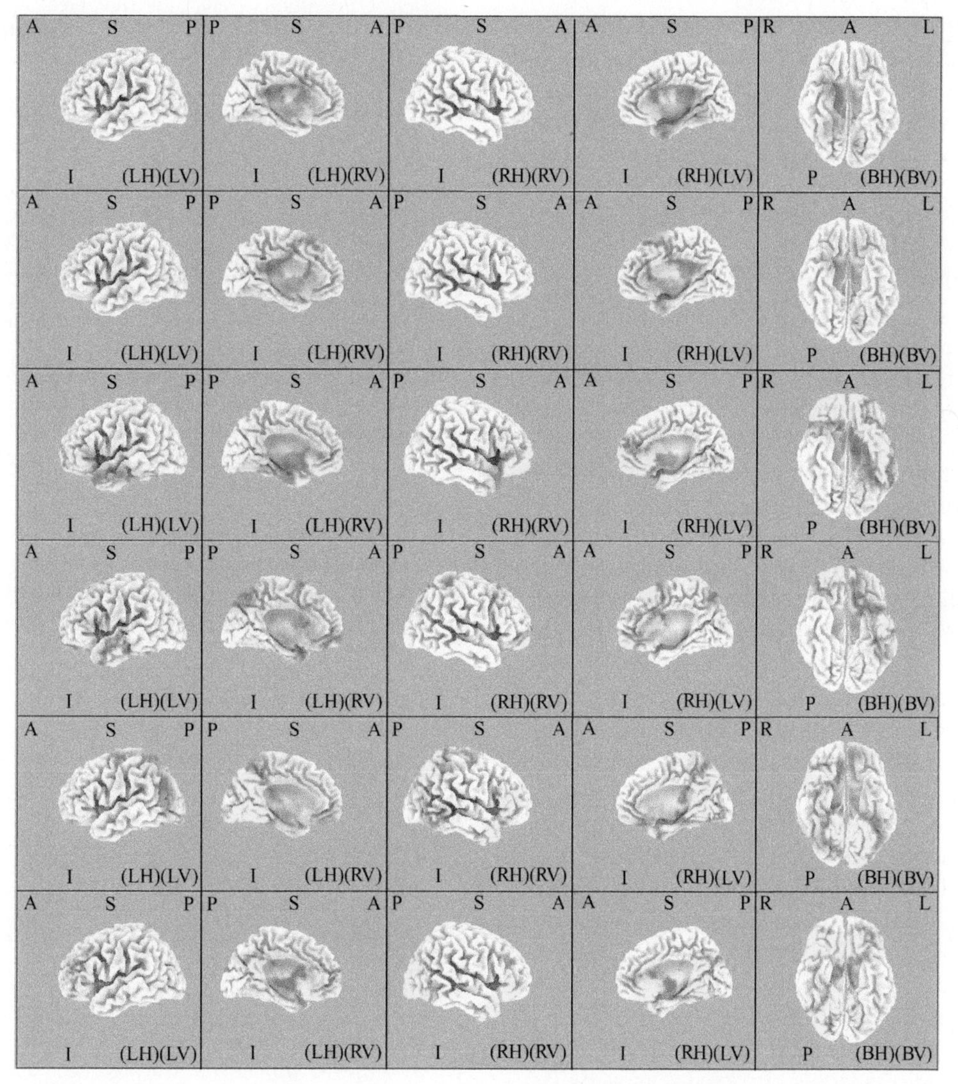

FIGURE 4 **Grand average sLORETA images, derived from voxel-by-voxel *t*-test ($p<0.05$).** First row: grand average P2 for personal dilemmas. Second row: grand average P3 for personal dilemmas. Third row: grand average P2 for impersonal dilemmas. Fourth row: grand average P3 for impersonal dilemmas. Fifth row: grand average P260 for personal dilemmas. Sixth row: grand average P260 for impersonal dilemmas. sLORETA = standardized low-resolution brain electromagnetic tomography. Yellow–red areas indicate the activated areas ($p<0.05$) and maxima are color-coded as yellow. Results are collapsed across cultures.

the main sources of P2 and P3 components for impersonal dilemmas were localized in the medial frontal area and cingulate gyrus (with contributions from several other brain regions, including the temporal and insula areas). But it should be noted that EEG users have reservations concerning the accuracy of localizations using this type of software. These findings are in accordance with brain imaging research, demonstrating that a complex network of brain regions are involved in moral decision-making.[18] Different from the sources of the P2 and P3 components, the P260 component for both dilemma types was mainly activated in areas in the posterior cingulate, parahippocampal gyrus, and cuneus and precuneus cortices, areas considered to be related to emotional processing and evaluation, retrieval of episodic memory representations, attention, detection of salient stimuli, and higher-order cognitive functions.[19–21] It seems that brain areas associated with attention, memory retrieval, and emotional processing were involved in the process of moral decision-making and may involve holistic thinking. Future research examining cultural differences in these processes would benefit from direct manipulations of holistic and analytic thinking.

2. CONCLUDING REMARKS

Although moral decision-making and cultural differences are both major themes in social psychology, the brain mechanisms underlying cross-cultural moral decision-making are still far from being well understood. In this chapter, we exemplify abstract thinking/reasoning and moral decision-making in Eastern (e.g., Chinese) and Western (American) cultures to reveal the underlying brain mechanisms involved, using the event-related potential (ERP) technique and fMRI. These evidences suggest the extensive differences in cognitive processing and moral decision behavior between Eastern and Western cultures. Different brain patterns during the arithmetic processing and the resolution of moral dilemmas were reflected in spatiotemporal cortical activation underlying moral decision-making, and these results, using diverse methods, support a dual-process theory of moral judgment according to which utilitarian moral judgments (favoring the "greater good" over individual rights) are enabled by controlled cognitive processes, while deontological judgments (favoring individual rights) are driven by intuitive emotional responses. Since this chapter focuses on the cross-cultural work on this topic, many studies of neural systems and various facets of moral reasoning have not been cited and discussed here (see examples[22–25]). Future studies will need to address the roles of cognitive control, emotion regulation, and the beliefs in these moral reasoning and decision-making processes under diverse cultural context.

Acknowledgment

This research was supported by the Office of Naval Research.

References

1. Chiao J, Cheon B, Pornpattananangkul N, Mrazek A, Blizinsky K. Cultural neuroscience: progress and promise. *Psychol Inq*. 2013; 24:1–19.
2. Nisbett R, Peng K, Choi I, Norenzayan A. Culture and systems of thought: holistic versus analytic cognition. *Psychol Rev*. 2001;108:291–310.
3. Wang Y, Deng Y, Sui D, Tang YY. Neural correlates of cultural differences in moral decision making: a combined ERP and sLORETA study. *Neuroreport*. 2014;25:110–116.
4. Nisbett RE. *The Geography of Thought: How Asians and Westerners Think Differently and Why*. New York: Free Press. 2003.
5. Tang YY, Zhang W, Chen K, et al. Arithmetic processing in the brain shaped by cultures. *Proc Natl Acad Sci USA*. 2006;103: 10755–10780.
6. Keller S, Crow T, Foundas A, Amunts K, Roberts N. Broca's area: nomenclature, anatomy, typology and asymmetry. *Brain and Lang*. 2009;109:29–48.
7. Dehaene S, Cohen L. Towards an anatomical and functional model of number processing. *Math Cogn*. 1995;1:83–120.
8. Dehaene S, Spelke E, Pinel P, Stanescu R, Tsivkin S. Sources of mathematical thinking: behavioral and brainimaging evidence. *Science*. 1999;284:970–974.
9. Tang YY, Liu Y. Numbers in the cultural brain. *Prog Brain Res*. 2009;178:151–157.
10. Greene JD, Nystrom LE, Engell AD, Darley JM, Cohen JD. The neural bases of cognitive conflict and control in moral judgment. *Neuron*. 2004;44:389–400.
11. Greene JD. The cognitive neuroscience of moral judgment. In: Gazzaniga M, ed. *The Cognitive Neurosciences IV*. Cambridge: MIT Press. 2012.
12. Thomson JJ. *Rights, Restitution, and Risk: Essays, in Moral Theory*. Cambridge: Harvard University Press. 1986.
13. Rozin P. Five potential principles for understanding cultural differences in relation to individual differences. *J Res Pers*. 2003;37: 273–283.
14. Kitayama S, Park J. Cultural neuroscience of the self: understanding the social grounding of the brain. *Soc Cogn Affect Neurosci*. 2010;5:111–129.
15. Yuan J, Zhang Q, Chen A, et al. Are we sensitive to valence differences in emotionally negative stimuli? Electrophysiological evidence from an ERP study. *Neuropsychologia*. 2007;45:2764–2771.
16. Miltner W, Johnson R, Braun C, Larbig W. Somatosensory event-related potentials to painful and non-painful stimuli: effects of attention. *Pain*. 1989;38:303–312.
17. Gajewski PD, Stoerig P, Falkenstein M. ERP–correlates of response selection in a response conflict paradigm. *Brain Res*. 2008;1189: 127–134.
18. Prehn K, Wartenburger I, Mériau K, et al. Individual differences in moral judgment competence influence neural correlates of socio-normative judgments. *Soc Cogn Affect Neurosci*. 2008;3:33–46.
19. Cavanna AE, Trimble MR. The precuneus: a review of its functional anatomy and behavioural correlates. *Brain*. 2006;129:564–583.
20. Schacter D, Addis D. The cognitive neuroscience of constructive memory: remembering the past and imagining the future. *Philos Trans R Soc B: Biol Sci*. 2007;362:773–786.

21. Vogt BA, Vogt L, Laureys S. Cytology and functionally correlated circuits of human posterior cingulate areas. *Neuroimage*. 2006;29:452–466.
22. Yoder KJ, Decety J. The good, the bad, and the just: justice sensitivity predicts neural response during moral evaluation of actions performed by others. *J Neurosci*. 2014;34:4161–4166.
23. Zaki J, López G, Mitchell JP. Activity in ventromedial prefrontal cortex co-varies with revealed social preferences: evidence for person-invariant value. *Soc Cogn Affect Neurosci*. 2014;9:464–469.
24. Young L, Camprodon JA, Hauser M, Pascual-Leone A, Saxe R. Disruption of the right temporoparietal junction with transcranial magnetic stimulation reduces the role of beliefs in moral judgments. *Proc Natl Acad Sci USA*. 2010;107:6753–6758.
25. Koster-Hale J, Saxe R, Dungan J, Young LL. Decoding moral judgments from neural representations of intentions. *Proc Natl Acad Sci USA*. 2013;110:5648–5653.

CHAPTER

15

Empathy

Nicholas M. Thompson, Caroline Di Bernardi Luft, Michael J. Banissy
Department of Psychology, Goldsmiths, University of London, London, UK

OUTLINE

1. INTRODUCTION

The ability to share the feelings of others (empathy) is a vital aspect of human social interaction; it is a core component underpinning our ability to form and maintain interpersonal relationships, to comprehend and predict the behaviors of those around us, and to respond adaptively to the ever-changing demands of complex social situations. Empathizing with somebody allows us to not only understand their perspective, but to appreciate the nature of their experiences on an affective, cognitive, and sensory level. In recent years, this topic has become the focus of a number of studies in social neuroscience. These have included studies related to what neural mechanisms contribute to empathy, how neural correlates of empathy are modulated by individual differences, and how empathy relates to other forms of social evaluation (e.g., sympathy and compassion) within the brain.

In this chapter, we will first discuss what constitutes empathy and paradigms developed to assess empathy. We will then discuss the neural correlates of empathy and factors that mediate empathic brain responses. Throughout, we discuss the role of interindividual differences by focusing on the relationship between variability in empathy traits and the neural correlates of empathy within typical adults, how individual differences in factors related to higher-order control (e.g., expertise) mediate empathic brain responses, and how the empathic brain responds in atypical groups. This chapter concludes with the discussion of a recent model describing the underlying processes through which empathy is accomplished and how it relates to individual differences in empathic outcomes.

2. WHAT IS EMPATHY AND HOW CAN IT BE STUDIED?

The term empathy is common across a variety of disciplines, including psychology, philosophy, and sociology. It is therefore perhaps not surprising that there are a number of definitions for the term. From a neuroscience perspective, empathy is commonly used to refer to circumstances in which individuals experience an affective state that is isomorphic to that of another person. For example, Hoffman[1] defines empathy as "any process where the attended perception of the object's state generates a state in the subject that is more applicable to the object's state or situation than to the subject's own prior state or situation" (also see Refs 2,3 for discussion).

Neuroimaging Personality, Social Cognition, and Character
http://dx.doi.org/10.1016/B978-0-12-800935-2.00015-4

This process can be distinguished from other related sociocognitive processes that involve representing another individual's state. These include mentalizing/theory of mind (ToM), which involves understanding another's intentions and beliefs from cognitively reasoning about their state (for further discussion see Refs 4,5); emotional contagion, which involves experiencing a matching affective state with another, but where that affective state is not explicitly recognized as being experienced by the other (for further discussion see Refs 2,5); imitation/mimicry, which involves representing and sharing action plans (for further discussion see Ref. 5); and sympathy or compassion where, unlike empathy, the affective state elicited does not need to be isomorphic with the state of another person (for further discussion see Ref. 2).

With regards to approaches to studying empathy in the human brain, a useful distinction has recently been made between two lines of research: (1) investigations of what is shared between the object and subject (i.e., the study of empathic outcomes, such as experiencing empathy for pain) and (2) investigations of how empathy is shared (i.e., the study of mechanisms that support empathy).[5] It has been suggested that the majority of studies to date have focused on the study of empathic outcomes, with several studies examining the overlap between brain regions involved in experiencing a particular affective state and observing a similar state in another (see Refs 6,7 for reviews). For example, in the context of empathy for pain, prior brain imaging work suggests that neural regions involved in the self-pain network become active in response to knowledge that a loved one is in pain,[8] and observing pain being inflicted to the body parts of others can also evoke activity in sensorimotor regions involved in experiencing pain to oneself (eg., Refs 9–11).

As noted above, there are a number of cognitive neuroscience studies examining empathic outcomes. These range from studies of sensory aspects of empathy (e.g., Refs 12,13) through to affective aspects (e.g., Refs 8,14). With extant literature providing extensive knowledge regarding the neural mechanisms underpinning nociception, examining neural activation patterns during the perception of pain in others provides a valid and objective means of investigating empathic outcomes.[10] As a result, much of what we have learned about the neural mechanisms involved in what is shared has focused around empathy for pain. Here, two approaches have tended to dominate studies in the literature: (1) pictorial based and (2) cue based investigations.

In pictorial-based investigations, participants are shown visual stimuli (e.g., images and video clips) depicting human body parts (e.g., hands and feet) in various situations suggestive of painful experience (e.g., the insertion of needles), along with neutral stimuli of the same body parts/settings without any accompanying elements indicative of pain.[6] By contrast, in cue-based investigations, participants are not directly exposed to any overt pain stimuli but are instead made aware in real time of the mild painful experience of other individuals who are typically present in the room with them.[6] Evidently there are fundamental differences between pictorial and cue-based approaches; for example, in cue-based investigations, participants have an ongoing interaction with target individuals, and therefore, such studies arguably demonstrate greater ecological validity.[6] Furthermore, in contrast to pictorial-based investigations, during cue-based paradigms, participants do not directly observe any painful somatic consequences, nor are they exposed to any external indicators of the target's painful experience. It is therefore possible that cue-based approaches elicit a greater need for top-down cognitive processing in order for participants to interpret the painful experience of the target individual.[6] We will review findings relating to variability in neural activation patterns across these two approaches in the following section, once we have discussed the proposed mechanisms responsible for mediating empathic outcomes and examined evidence regarding the core brain networks recruited during empathy for pain.

In two seminal cue-based studies,[8,14] target individuals sat alongside participants who were in a functional Magnetic Resonance Imaging (fMRI) scanner. After individualized amplitudes for high (pain) and low (no pain) intensity electrical stimulation were determined, an electrical pain stimulus was delivered to the dorsum of the right hand of either the participant or the target individual across progressive trials. Participants were able to view their own hand and the stimulated hands of targets, and while in the scanner, they were presented with flashes of different colored light (cues), which indicated who would experience the electrical stimulation on each trial, and whether or not this would be painful. Such cue-based procedures can be extremely useful, as they allow researchers to directly compare individual neural activation patterns in response to both first-hand (self) and vicarious (other) pain,[7] thus identifying core regions showing common activation patterns across these two conditions. For example, cue-based studies have provided evidence of activation in anterior insula/fronto-insular cortex (AI/FI) and the anterior cingulate cortex (ACC) in response to first-hand pain and also in response to being made aware of the painful experience of others.[8] Research findings regarding shared activation patterns for self and other pain are discussed in greater detail in the next section of this chapter.

In addition to research investigating empathic outcomes, a number of studies have examined the relationship between interindividual variability in trait empathy and neural correlates of empathic response. Common trait empathy measures include the Interpersonal

Reactivity Index (IRI)[15,16] and Empathy Quotient (EQ).[17] The IRI is a self-report measure of trait empathy, consisting of four subscales: (1) empathic concern, which measures feelings of compassion and sympathy for the observed individual; (2) personal Distress, which measures aversive emotional responses in the observer (e.g., feelings of fear or discomfort at witnessing negative experiences of others); (3) perspective taking, which examines the tendency to think from another perspective; and (4) fantasy, which examines participants' abilities to transpose themselves into fictional situations.

The EQ is another self-report measure of trait empathy, which consists of three subscales: (1) cognitive empathy, which measures intuitive reasoning about others' mental states; (2) emotional reactivity, which measures affective empathic responses that are primarily other-oriented (it correlates with empathic concern on the IRI, but not personal distress); and (3) social skills, which measure how easily individuals engage with others.[17] A variety of studies have associated functional brain activity related to empathic outcomes with these trait measures (e.g., Refs 10,18–23). More recent work has also sought to examine how interindividual differences in brain structure are linked with trait empathy (e.g., Refs 24,25). We discuss these and additional studies investigating the link between trait empathy and functional neural activity, as well as structural brain composition later in this chapter.

3. WHAT IS SHARED AND HOW DOES THIS HAPPEN?

3.1 What Is Shared?

A fundamental feature of the empathic experience is the sharing of representations between self (observer) and other (empathic target).[12] Many models of empathy (see Refs 3,11,26–28 for further detail) assert that the ability to share the internal experience of another individual is mediated (at least in the early stages) by motor-mimicry mechanisms, which automatically trigger a representation of the other's state through vicarious activation of a subset of the brain regions associated with the first-hand experience of the perceived state.[6,8,13,14,29]

Evidence of a direct link between observation and action is well documented,[30–33] and sensory-motor representations (e.g., mirror systems) in the premotor and inferior parietal cortices[34,35] have been proposed as one biological mechanism mediating our ability to simulate the perceived emotional states of others.[8,33,36,37] A number of theories assert that the unconscious and automatic mimicry of others' emotional behaviors (e.g., facial expressions, bodily postures, movements, and vocalizations) triggers activation of the brain regions associated with the observed experience. This vicarious activity then invokes affective and somatic representations of the observed state in the self, thereby inducing an internal experience that is isomorphic to that of the other.[5,28] To ensure clarity, it is important to once again distinguish between emotional contagion and empathy. While both involve the sharing of affective representations between self and other, empathy is defined by an awareness of the extra-personal source of the experienced emotion, which is not present in emotional contagion. This self-other distinction is considered by some models to be an essential criterion for the experience of empathy (e.g., Ref. 5).

Functional neuroimaging investigations provide support for the notion of shared representations during empathy for pain. As noted above, this has been shown in two ways: (1) by demonstrating activation in regions associated with the first-hand experience of pain during the vicarious observation of others in pain[29,38–41] and (2) when making participants aware of the fact that a target individual was experiencing pain, without observing the painful experience (i.e., in cue-based investigations).[8,14] Consistent activation in areas associated with the first-hand experience of pain, most notably the AI/FI and medial regions of the ACC, have been reported in response to the perception of others' pain.[39] Lamm et al.[6] conducted an image and coordinate-based meta-analysis investigation of fMRI studies exploring empathy for pain; they were able to identify a core empathy for pain network that showed reliable activation across different empathy-elicitation procedures (i.e., pictorial and cue based). Consistent with prior research, Lamm et al.[6] concluded that bilateral activation of the AI and the boundary of the anterior/posterior anterior cingulate cortex (aMCC/pACC) represents a core empathy for pain network. These regions showed significant activation during the experience of empathy for pain across all studies examined.

Prior research suggests that the AI and aMCC are part of an interconnected network involved in representing emotions[42–44] and processing the affective and motivational components of pain.[6,45] Reliable activation of these regions in response to the perceived pain of others therefore provides support for the notion that direct affective simulation, made possible by neural systems relating to the individuals' own internal state, plays an important role in empathic outcomes.[2,6,46,47] In addition to empathy for pain, there exists broader evidence of common activation patterns in the AI when experiencing emotions and when observing facial expressions depicting these emotions in others (for example, in relation to disgust).[37]

It is important to note that while there is significant overlap between brain activation patterns during first-hand (self) and vicarious (other) pain, critical differences between these two conditions have been observed. For example, there is a clear posterior-to-anterior gradient within the insula, with more posterior regions (associated with sensory-related components of nociception; e.g., Ref. 11), showing activation in response to first-hand pain but not during empathy for pain.[6] Further differences between self-pain and empathy for pain relate to subtle variations in activity within somatosensory regions implicated in the processing of the sensory dimensions of proprioceptive and nociceptive information.

The term "somatosensory system" typically refers to all brain regions implicated in the processing of somatosensory information, specifically the somatosensory cortices proper along with the insula and ACC.[11] While findings regarding the activation of the AI and ACC during the perception of others' pain provide strong support for the concept of a shared affective experience during empathy for pain,[8,14,48,49] findings regarding whether or not the observer also shares in the more direct sensory discriminative dimensions of the target's painful experience are somewhat less conclusive.

In support of the notion of shared sensorimotor representations during empathy for pain, Osborn and Derbyshire[13] exposed participants to photographs of people suffering injuries (e.g., a broken leg), and found that around one-third of participants reported experiencing pain at the corresponding location on their own body. fMRI results showed significant activation in primary (SI) and secondary (SII) somatosensory regions in individuals who reported these localized pain sensations but not in those who reported experiencing only the affective components. In a broader context, this is consistent with another rare group of individuals whom experience tactile sensations (including pain and touch) when observing tactile events on other's bodies (mirror-touch synesthetes).[50,51] Relative to nonsynesthetes, mirror-touch synesthetes show greater levels of activation in primary (SI) and secondary (SII) somatosensory regions when observing touch to others,[36,52] suggesting that shared sensorimotor representations may contribute to the perception of a range of somatic experiences. Many other studies also support the concept that sharing in the sensory-motor dimensions is important for empathy (e.g., Refs 9,12,53–57; and see Ref. 11 for discussion).

Studies using transcranial magnetic stimulation (TMS) to measure motor evoked potentials (MEPs) from hand muscles that correspond with observed images of pain to another person further support the view that the sensorimotor consequences of observed pain can also be shared (e.g., Ref. 9). Typically in such studies, individuals view stimuli showing needles penetrating body parts of a human model from a first-person perspective, which leads to MEP amplitudes in the participant's corresponding muscle to be suppressed by viewing painful versus neutral control stimuli, implying that the sensorimotor, as well as affective experience of pain (e.g., Refs 8,14) can be empathically shared. It is of note, however, that recent work has questioned the extent to which MEP suppression reflects purely empathic processes versus misidentification of visual information as relating to the observer.[58] Since much of the evidence supporting shared sensorimotor representations during empathy comes from studies examining transcranial magnetic resonance (e.g., Ref. 9) and somatosensory evoked potentials (e.g., Ref. 12), some researchers have questioned whether the inconsistent fMRI findings regarding activation in primary somatosensory areas may be due to a lack of sensitivity with this technique in detecting subtle changes in somatosensory activity.[10]

Cheng et al.[10] used magnetoencephalography (MEG) with a pictorial-based approach to explore activation patterns in the primary somatosensory cortex during the perception of pain in others. Rhythmic oscillatory activity within the alpha (~10 Hz) and beta (~20 Hz) frequency bands (often referred to as the "mu rhythm") recorded over the primary somatosensory cortex is proposed to reflect the functional status of this system, with decreased amplitude indicating activation of sensorimotor systems and increased amplitude indicating an inhibitory idling state.[59,60] Significant suppression of mu rhythm oscillations was observed when participants were attending to pictures depicting both painful and nonpainful situations (relative to baseline). This suppression occurred to a greater degree in response to painful, compared with nonpainful, stimuli. These findings are aligned to prior research demonstrating somatic resonance in response to viewing stimuli depicting the painful and nonpainful touching of others' body parts,[9,12] and further suggests that empathy for pain can indeed influence activity within the primary somatosensory cortex (reflecting sharing of the sensorimotor consequences of the perceived experience).[10]

In addition to the aforementioned variability in the sensitivity of different neuroscience techniques, a further possible explanation for inconsistencies regarding activation within the somatosensory cortices during investigations of empathy for pain relates to the methods and pain stimuli used in such studies. For example, there is evidence to suggest that activation in primary somatosensory regions is more likely when strong visual pain stimuli are used[41,61,62] and when participants are made to focus specifically upon the somatic location to which the pain stimulus is delivered.[63] Furthermore, based on the findings of their meta-analysis, Lamm et al.[6] assert that while the core empathy for pain network (i.e., the AI and ACC) is activated across both pictorial and cue-based investigations, there is evidence of differential recruitment of supplementary brain regions depending on whether a pictorial or cue-based paradigm was adopted. This suggests these approaches might each be underpinned to varying degrees by distinct subcomponents related to empathic outcomes. For example, significant activation in the inferior parietal cortex (supramarginal gyrus and intraparietal sulcus) and ventral premotor areas (inferior frontal gyrus and pars opercularis) was observed in pictorial-based investigations.[6] Given previous findings suggesting a functional role of these regions in computational processes related to action understanding and the prediction of events,[64] the activation recorded in these areas during empathy for pain may reflect predictions regarding the likely consequences of the observed pain stimuli, coupled with the subsequent inferences about the affective state likely to result from such circumstances.[6]

In contrast, cue-based empathy for pain investigations were found to be more strongly associated with brain regions associated with ToM processes (for example, the precuneus, posterior superior temporal cortex, temporoparietal junction (TPJ), the temporal poles, and ventral regions of medial prefrontal cortex (mPFC)).[6,65] These ToM regions are involved not only in the attribution of thoughts to others, but also in self-referential processing.[66] Furthermore, TPJ activity in particular has been shown to be associated with the process of making distinctions between self and other.[67,68] Given likely functional correlates, it has been proposed that a key role of the ToM network during empathy for pain is to enable the utilization of existing knowledge to be able to infer the affective state of the other.[6] In line with this functional interpretation, it is logical that greater activation of brain regions linked to ToM would be observed during cue-based studies, where the individual is heavily reliant on cognitive mentalizing abilities in order to infer the target individual's affective experience. While top-down cognitive processing is likely to be important in cue-based studies, the overt visual pain stimuli presented to participants in pictorial investigations reduces the necessity for such mechanisms and appears to be more likely to engage sensorimotor brain areas.[6]

It is worth noting that Lamm et al.[6] suggest that such activity does not provide definitive evidence that observers share the sensorimotor representations of targets' painful experience, and may simply reflect nonspecific coactivation in response to the observation of others' body parts being touched.

In conclusion, current evidence supports the assertion that empathic outcomes for pain can be mediated by resonance mechanisms, which emulate neural activity in the self-pain matrix.[6] The human mirror system provides a functional bridge between self and other, allowing for direct simulation of the perceived affective and sensorimotor representations in the self (e.g., Refs 5,8,10,14,36). There is also evidence to suggest that the precise nature of what is shared may be moderated to some degree by the experimental paradigm/stimuli.[6,10] Research findings to date indicate that our ability to share in the pain of another is mediated by two primary mechanisms; if we simply know that the individual is experiencing pain (i.e., through the presentation of abstract visual cues), vicarious activation of the AI and ACC induces us to share in the affective components of their experience. If however, we are shown explicit pain stimuli and induced to focus more upon the somatic causes of the other's pain, we may also be able to share more directly in the sensory dimensions of their experience, via recruitment of the primary somatosensory cortices, in addition to the AI and ACC.[6,9,10,12]

3.2 Interindividual Differences in What Is Shared

There is an increasing awareness that stable interindividual differences in levels of empathy exist, and that such differences can have a significant impact on social functioning in multiple domains.[69] Indeed, the notion that some people show more empathic concern than others will no doubt come as little surprise to most of us, but what are the neural substrates of such differences, and how can they be illuminated through neuroscientific investigation? Social neuroscience studies examining this topic have demonstrated that the characteristics of an individual can indeed influence the strength of empathic brain responses (e.g., see Ref. 7 for review); in this section, we review recent research examining how interindividual differences in empathy relate to variations in functional neural activity and structural composition in empathy-related brain areas.

One of the most common means of examining how interindividual differences are associated with different neural activation patterns related to empathic outcomes is by examining the relationship between baseline differences in trait empathy and functional brain activity. Using established measures, such as the IRI,[15,16] EQ,[17] and the Balanced Emotional Empathy Scale,[70] prior research has highlighted positive correlations between various subscales of trait empathy (e.g., personal distress and perspective taking) and activity in core brain regions in the empathy network, such as the insula,[6,8,18,71] ACC,[20,72–74] and sensorimotor cortices.[23]

Higher scores on other trait subscales of empathic disposition, such as empathic concern (IRI),[15,16] have also been shown to predict increased empathy-related activation in a range of regions associated with both affective and cognitive components of empathy, such as the precuneus, inferior parietal cortex, dorsolateral/medial prefrontal cortex,[18] inferior frontal gyrus/premotor cortex,[18–22,75] and superior temporal regions.[18,22] In addition to demonstrating correlations between regional neural activity and trait empathy, there is evidence that individual patterns of activity within the insula during empathic outcomes can reliably predict prosocial behaviors.[76]

Consistent with the findings of fMRI investigations, studies utilizing electrophysiological approaches, such as MEG, have provided further support for a link between trait empathy and neural activation patterns during empathic outcomes. In their aforementioned MEG investigation, Cheng et al.[10] found that mu rhythm suppression (reflecting heightened activation in primary somatosensory regions) in response to pictorial pain stimuli showed a significant correlation with an individual's perspective taking abilities (as measured by the IRI),[15,16] suggesting that subjects with higher levels of trait empathy have a heightened propensity to share the sensorimotor dimensions of another individual's pain.[10]

The anecdotal notion of a heightened empathic response in females relative to males has long existed, and increasing behavioral evidence seems to support this assertion. For example, females often score higher than males on trait empathy measures (e.g., Refs 17,77) and are often superior to males in judging emotions and character traits based upon nonverbal cues, such as facial expressions.[78] Furthermore, disorders characterized by atypical empathic responding, such as psychopathy and conduct disorders, are more common among males.[78]

While findings regarding sex differences in functional neural activation patterns are not entirely conclusive, some fMRI studies have found evidence to suggest that such differences may exist. For example, Schulte-Rüther et al.[79] found that females (relative to matched males) displayed increased neural activation in regions associated with mirror systems during the experience of empathy. Additional support comes from a study by Singer et al.,[14] in which a cue-based empathy for pain paradigm was used to assess whether functional activity in empathy-related brain regions was affected by participants' judgments regarding the "fairness" of the empathic target. While males and females showed similar neural activation in the fronto-insular (FI) and ACC in response to the vicarious pain of "fair" targets, males but not females showed significantly diminished empathy-related brain activity in response to the pain of "unfair" targets. Singer et al.'s[14] finding of a modulation of empathic brain responses in males for "unfair" targets not only provides evidence indicative of sex differences in neural activity related to empathy, but suggests that empathic outcomes may be sensitive to modulation through higher-order cognitive processes (i.e., judgments regarding a target's fairness).

Further support for a moderating role of higher-order processes in empathy comes from studies that have examined the effects of expertise on neural activation patterns of empathic outcomes. In one such study, Cheng et al.[80] used fMRI to compare the neurohemodynamic response to pictorial pain stimuli (images of needles being inserted into different body parts) across two groups of participants: "novices" and "experts." The expert group consisted of physicians with experience administering acupuncture; these individuals were therefore more familiar than the novices with regards to observing the vicarious pain of others caused by needles. For novices, the observation of pain resulted in significant activation in brain regions associated with the processing of the affective and sensory components of pain (including the AI, aMCC, supplementary motor area and somatosensory cortex).[48] The expert group, however, showed no significant signal increase in these regions when observing the painful stimuli. In contrast to the activation in the self-pain matrix observed in novices, experts instead showed increased activation in brain regions associated with ToM and self-other processing (TPJ),[5,67] executive control and emotion regulation (medial and superior prefrontal cortices),[27,48] and memory retrieval (parahippocampal gyrus; e.g., Ref. 81). Furthermore, evidence of significant negative functional connectivity between the mPFC and the insula was observed in experts but not novices. The reported variance in neural activation patterns was also shown to correlate with behavioral measures; self-report ratings of the unpleasantness of the experience were found to be significantly lower in experts relative to novices. Cheng et al.[80] propose that the differences between these two groups is likely a reflection of the differential knowledge and experience of acupuncture across the two groups, which moderated their relative appraisals of the painful situations depicted in the stimuli.

As previously discussed, it is proposed that the activation of the self-pain network, in response to the perception of others' pain, facilitates the process of understanding another individual's experience by representing their perceived experiences in the self. In certain circumstances, however, such automatic mirroring of the affective and sensory experiences of others could be problematic for the observer.[80] For example, surgeons, who must as part of their job regularly be exposed to the pain and suffering of others, would no doubt experience significant anxiety and distress if unable to suppress the self-other overlap of affective and/or sensory representations. Therefore, the ability to regulate empathic outcomes as a result of prior experience could provide a significant adaptive advantage.[80,82] Recent findings also suggest that first-hand familiarity with a particular painful sensation can also affect the extent to which individuals empathically share such painful experiences when observed in others. For example, Derbyshire, Osborn, and Brown[83] found that relative to participants with no such prior experience, participants with past experience of tooth pain induced by eating cold foods showed increased evidence of painful experience in response to viewing images of others experiencing pain when eating a Popsicle.

Additional support for a moderating role of higher-order mechanisms in empathic outcomes comes from studies demonstrating that judgments regarding the empathic target and situational context of the pain can influence empathic brain responses in the AI and aMCC.[14,84] The aforementioned findings suggest that prior knowledge and experience (with either the vicarious observation of such pain in others or direct experience of such pain in the self) is an important factor influencing interindividual variability in empathic outcomes.

In addition to research investigating the relationship between trait empathy and functional activation patterns, a number of studies have examined how interindividual variability in brain structure is associated with differences in levels of trait empathy. The primary structural brain imaging approach, which has been employed by such investigations, is voxel-based morphometry (VBM). VBM is a technique that can accurately detect subtle differences in local cerebral volume based on high resolution structural MRI images.[25,85] Through the use of multiple regression analyses, researchers have used VBM to investigate the relationship between variance in gray matter volume in discrete brain structures and differences in trait empathy.[24,25]

Banissy et al.[24] demonstrated that distinct morphological variations in empathy-related brain regions are predictive of interindividual variability in different subcomponents of trait empathy. Reduced gray matter volume in the left inferior frontal gyrus, left ACC, and left precuneus were shown to be associated with high scores on the empathic concern subscale of the IRI.[15,16] Findings regarding the inferior frontal gyrus (IFG) are supported by prior research demonstrating a relationship between functional neural activity in the IFG and trait measures reflecting affective empathy.[18,20,56,57,71] While the direction of the relationship between brain volume and empathic concern reported by Banissy et al.[24] may seem at odds with the typical "more is better" concept, such negative correlations between gray matter volume and behavioral/trait outcomes are not uncommon.[86] Consistent with prior meta-analyses demonstrating functional neural correlates of cognitive perspective taking in the ACC and dorsolateral prefrontal cortex (DLPFC),[87] Banissy et al.[24] report that the ability to take the perspective of others and place oneself in fictional scenarios was associated with increased local gray matter volume in these areas. Interestingly, higher scores on the empathic concern subscale were associated with decreased gray matter volume in the ACC; this dissociation between gray matter volume and different trait empathic subscales provides strong support for the perspective that empathy is indeed a complex and multidimensional construct.[24]

Additionally, using VBM, Cheng et al.[25] found further confirmation for the notion of sex differences in empathy by finding evidence of structural differences in core areas associated with the human mirror system in males and females. In addition to demonstrating that gray matter volume in the pars opercularis and inferior parietal lobule is positively correlated with trait empathy measures, Cheng et al.[25] found evidence of increased gray matter volume in these areas in females, relative to matched males. The results of structural MRI investigations regarding the core brain networks implicated in empathic outcomes are aligned with those obtained through functional imaging approaches[6] and reveal that morphological differences in core regions in the empathy brain network can, at least in part, account for interindividual variability in trait empathy in normal populations.[24]

Broader interindividual differences in empathy relate to a range of atypical traits associated with empathic outcomes. These include, but are not limited to, autistic spectrum conditions (ASC),[88–90] alexithymia,[5,91,92] and psychopathy,[93,94] each of which has been shown to relate to variations in functional neural activity and/or structural composition in empathic brain regions. We will review evidence regarding the functional and structural abnormalities and possible underlying causes of the empathy deficits observed within these conditions in the final section of this chapter, following the discussion of the mechanistic processes supporting empathy.

In summary, the findings of behavioral, fMRI, and electrophysiological investigations provide evidence to suggest that individual variability in distinct submeasures of trait empathy are predictive of interindividual differences in neural activation patterns during empathic outcomes. Higher scores on trait empathy submeasures have been found to predict stronger activation in empathy-related brain areas, such as precuneus, inferior parietal cortex, dorsolateral/medial prefrontal cortex,[18] inferior frontal gyrus/premotor cortex,[18–22,75] and superior temporal regions.[18,22] Furthermore, MEG investigations provide evidence to suggest that the propensity to empathically share in the sensorimotor components of another's painful experience is predicted by an individual's perspective taking abilities.[10] There is also evidence to suggest that an individual's prior experience of painful situations (with regards to both observing such pain in others and experiencing such pain in oneself)[80,83] and appraisal of the empathic target/context[14,84] can affect the extent to which the individual shares in the other's experience. In addition to demonstrating differences in functional neural activation, there is also evidence that distinct morphological variations in empathy-related brain regions are predictive of interindividual variability in different subcomponents of trait empathy.[24,25] For example, reduced gray matter volume in the left inferior frontal gyrus, left ACC, and left precuneus were found to be associated with higher trait levels of empathic concern.[24]

3.3 How Is It Shared?

Despite the numerous investigations and subsequent progress in explaining what is shared during empathy (i.e., empathic outcomes), relatively fewer studies have explored the mechanisms that support empathic sharing. In this section, we discuss a recent model that proposes a mechanistic framework of the underlying processes necessary for empathy to occur. In the final section of this chapter we then discuss several examples of disorders characterized by atypical empathic outcomes within the context of this empathy model.

The Self-to-Other Model of Empathy (SOME)[5] builds upon previous work[6,95] and proposes a model of the underlying processes involved in representing the affective state of another within the self. Critically, the SOME[5] seeks to address four key questions fundamental to any cognitive model of empathy: (1) how do we know what another is feeling? (2) What is the role of ToM in empathy? (3) How does emotional contagion occur? (4) How do we represent another's emotion once emotional contagion has occurred?

The SOME[5] asserts that empathy relies to varying degrees upon two distinct representational systems: (1) the *ToM system* and (2) the *affective representation system*. The *ToM system* (recruiting the TPJ, mPFC, and precuneus) mediates the process of generating a cognitive representation of the mental state of another individual (e.g., beliefs, knowledge, and goals),[96] whereas the *affective representation system* (recruiting the insular cortex and ACC) is responsible for representing the current affective state of the self through the recruitment of information from autonomic nervous and subcortical emotion systems. It is important to note that while the representation of one's own internal state within the *affective representation system* is typically deemed to be unconscious in nature, it is assumed by the SOME to be accessible by conscious introspection.[5]

For empathy to occur, the individual must first be able to classify the emotional state of the other; Bird and Viding[5] propose that this early stage relies on two distinct input sources, each of which can directly influence the *affective representation system*. These two input systems as proposed by the SOME are the *affective cue classification system* and the *situation understanding system*.

The *affective cue classification system* conducts a low-level categorization of person-related emotion cues (e.g., facial expressions, body language, vocalizations); prior research suggests that these perceptual cues are processed in an implicit manner and automatically feed into the *affective representation system*.[5] The automatic processing of such cues is purported to be accomplished by specialized perceptual systems supported by cortical and subcortical neural networks. For example, activation within the inferior and superior temporal cortices,[97,98] has been linked to the processing of emotional facial expressions but not facial identity.[99]

Bird and Viding[5] propose that the automatic processing of person-related cues by the *affective cue classification system* induces an emotional state in the observer isomorphic to that of the individual (i.e., emotional contagion). This could occur either via a direct link from the *affective cue classification system* to the *affective representation system*, or via an indirect link through mirror systems (located in the premotor and inferior parietal cortices),[34] which trigger automatic mimicry of perceived emotion cues (e.g., facial expressions), thereby influencing the *affective representation system*.

The *situation understanding system* is distinct from the *affective cue classification system* and provides an estimation of an individual's emotional experience based upon deductions regarding the situation in which they are observed.[5] In the absence of any person-related emotional cues, the *situation understanding system* allows the observer to make inferences regarding the emotional state of the other based upon any available contextual cues. For example, in many cultures it is a common convention for males to wear a black tie when attending a funeral; the *situation understanding system* is thought to utilize this learned association to infer that the emotional state of the individual observed in this context is likely to be negative (i.e., black tie = funeral = sad). It is worth noting that the *situation understanding system* is likely to rely upon the observer's stored socioemotional knowledge regarding the contextual stimuli and the likely emotional states associated with such stimuli.[5] Given previous findings demonstrating the involvement of the dorsal and ventral mPFC (see meta-analysis by Ref. 100), and the temporal poles[96] in the storage and application of social script knowledge, the *situation understanding system* is likely supported by these brain areas.[5]

Much like the *affective cue classification system*, the *situation understanding system* can directly influence the observer's *affective representation system*, resulting in emotional contagion. It is important to highlight that low-level processing of emotion cues (person-related and/or situational) within the *affective cue classification system* and the *situation understanding system* can result in the identification of another individual's emotional state (and subsequent emotional contagion) in an implicit manner (i.e., without the need for conscious recognition of these emotions). In other words, while the *ToM system* may be recruited during the processing of situational cues, involvement of the *ToM system* is not a prerequisite for emotional contagion to occur as a result of more low-level processing routes.[5]

In addition to the aforementioned low-level routes to emotion classification, Bird and Viding[5] propose that a more high-level situation understanding route also exists. In contrast to the implicit nature of the low-level processing routes previously discussed, the accomplishment of successful emotion classification via this high-level situation understanding route is dependent upon the involvement of the *ToM system*. Take, for example, a situation in which we are asked how another individual is likely to be feeling; in the absence of any perceptual cues (person-related or situational) from which we could implicitly deduce the likely emotional state of the individual, a high-level situation understanding route would need to be adopted. This would require us to call upon any stored knowledge we might have regarding the current situation of the target individual and the degree to which this is aligned with their goals/intentions, in order to cognitively infer their likely emotional state. While this cognitive process is in itself affectively neutral, the outcome of such an operation is likely to induce a corresponding affective response in the self.[5]

Bird and Viding[5] propose that the *affective cue classification system* and *situation understanding system*, while distinct, can directly influence one another. For example, an individual's appraisal of another's situation may affect how any available person-related emotion cues are interpreted; likewise, any person-related emotion cues may influence the nature of how the other's situation is construed by the observer. Additional involvement of the *ToM system* occurs in instances where there is an overt discrepancy between the available person-related and situational cues. When these two input systems appear to contradict one another, recruitment of the *ToM system* is necessary in order to reconcile the discrepancy by "recalibrating" one of the input systems. The SOME asserts that the *affective cue classification system, situation understanding system*, and *ToM system* can each implicitly influence the *affective representation system*, and the relative influence of each input system is moderated by factors such as attention and motivation.[5]

One key question addressed by the SOME[5] relates to the mechanisms by which each input system can come to influence the affective state of the self. While other accounts have been proposed (e.g., innate emotional facial mimicry),[101] learning-based models arguably provide the most robust and coherent explanation of how each of the input systems proposed by the SOME can influence the *affective representation system* (and thus cause emotional contagion). Such learning-based theories posit a fundamental role of experience in the formation of links between the perceived emotions of others and the associated emotional states within the self.

Based upon the Associative Sequence Learning model of action imitation,[102,103] Heyes and Bird[104] developed a learning-based theory of emotional contagion (see also Ref. 105). This theory asserts that as a child develops, they form bidirectional associative links between perceptual emotion representations (e.g., emotional facial expressions) and the corresponding emotional experience in the self. These associative links can be learned from a variety of sources (see Refs 5,102,103,106 for further detail), and once developed, they can influence the *affective representation system* directly or indirectly via mirror systems, which mediate this perception-emotion link through automatic motor mimicry mechanisms. In much the same way that associations between person-related emotion cues and internal affective states can be learned, associative links between situational emotion cues (e.g., the example of a black tie mentioned previously) and corresponding emotional states in the self could presumably also be formed through experience; the perception of such cues would then be able to automatically trigger the associated affective state within the self.[5]

It is important to highlight that emotional contagion occurring as a result of either the *affective cue classification system* or the *situation understanding system* does not reflect empathy per se; in order for the experience to be defined as empathy, the individual first needs to develop an awareness of the extra-personal source of the resulting affective state. According to the SOME, this critical next step is accomplished via a mechanism referred to as the *self-other switch*. Bird and Viding[5] propose that the state of another individual is represented by the empathizer's perceptual and cognitive systems and not by their affective system. The empathizer's affective system is responsible only for representing the state of the self, and it is through the *self-other switch* mechanism that this internal affective state (resulting from emotional contagion following input from the *affective cue classification system and*/or the *situation understanding system*) is then attributed to the other.

The *self-other switch* biases the input processing systems toward the other, which enables the empathizer to form a cognitive representation of their current emotional state (which was initially represented in the *affective representation system*) and then "tag" this cognitive representation as being relevant to the other and not to the self. This input processing bias focuses the individual's *situation understanding system* upon the situation of the other (rather than the situation of the self), meaning that attentional resources are then targeted toward emotional cues displayed by the other. As a result of this switch, the information feeding into both input systems is now other-focused, and the individual is subsequently able to interpret their current emotional state as resulting from the other; thus empathy is experienced.[5] It has been suggested that the self-focused processing mode is the default setting (at least in Western cultures),[107] and the engagement of the *self-other switch* is most likely the result of a conscious decision to switch to an other-focused mode.[5] While there is no firm consensus regarding the precise neural correlates of this self-other switch mechanism, there is evidence to suggest that the TPJ plays an important role in shifting between self and other processing modes.[68,108]

While the precise mechanisms through which empathy is achieved are still subject to debate, the SOME[5] presents a coherent and compelling account of the processes required for empathy to occur. In addition to providing a detailed overview of the underlying mechanisms involved in representing the affective state of another individual within the self, the SOME provides a framework by which future research can seek to explore the deficits observed in certain disorders characterized by atypical empathic outcomes (e.g., autistic spectrum conditions, alexithymia, and psychopathy) and could have profound implications for the management and treatment of such conditions.

3.4 Empathy in Atypical Individuals

As mentioned at the end of Section 3.2, broader inter-individual differences in empathic outcomes are associated with certain atypical traits. In this final section, we review findings obtained from behavioral and neuroimaging investigations of empathic outcomes in individuals with autistic spectrum conditions (ASC; henceforth autism), alexithymia, and psychopathy. Within the architectural framework proposed by the SOME,[5] we also discuss the possible underlying mechanisms that could mediate the observed empathy impairments typical of each of these conditions.

3.4.1 Autism

Autism is a developmental disorder characterized by impaired social functioning coupled with a restricted repertoire of interests.[109] Atypical empathic responding is commonly observed in autism,[110] and indeed, autistic individuals have in the past been characterized as exhibiting a global empathy deficit.[17] In support of such claims, a wealth of behavioral evidence demonstrates that autistic individuals tend to score lower on trait empathy measures, such as the EQ[18] and IRI,[15,16,111] most notably in relation to subscales measuring the extent to which one is able to take the perspective of others and to share their feelings.[17]

Consistent with such evidence indicative of a ToM deficit in autism, the findings of numerous functional imaging studies suggest that the empathy deficits observed in autistic individuals may be the result of impaired functional activity in ToM brain networks (see Ref. 112 for a review). Research using VBM has also highlighted the presence of structural differences in empathy-related brain regions in autistic individuals relative to matched controls. For example, there is evidence that autism is associated with reduced cortical thickness and gray matter volume in the inferior frontal gyrus[113] and superior temporal sulcus[89]; the latter of which may underpin the cognitive perspective taking impairments commonly observed in autistic individuals.[114]

The SOME[5] asserts that ToM is a prerequisite for empathy, as it enables the formation of a cognitive representation of one's internal affective state, which can then be attributed to the empathic target. Within this framework, a ToM deficit could have a profound effect on empathic outcomes; however, it is likely that the demand placed upon ToM systems in empathy is relatively minor, therefore empathy could be expected to remain relatively intact in autism in all but those cases where such ToM deficits are distinctly severe.[5] With that said, some degree of empathic impairment would be expected in all autistic individuals in instances where inputs from the *ToM system* are required to infer the affective state of another. Furthermore, when there is a discrepancy between the *affective cue classification system* and the *situation understanding system,* the ability to reconcile this incompatibility would likely be impaired in autistic individuals, given the mediating role played by the *ToM system* under such circumstances.[5]

As there is evidence to suggest that autistic individuals show reduced social motivation and attention (e.g., Refs 115,116), it is also possible that such individuals may have less opportunities to form learned associations between perceived emotional cues (situational and person-related) and experienced emotional states in the self. This reduced social motivation (and subsequent delay in developing external-internal emotional associations) could impact the ability of autistic individuals to correctly classify the emotional states of others through the *affective cue classification system* and/or the *situation understanding system;* therefore, there could also be a delayed development of emotional contagion (and therefore empathy) in autism.[5]

While there are some inconsistencies in research findings obtained to date, there is evidence to suggest that mirror systems (e.g., Refs 117,118) and associative learning mechanisms (e.g., Refs 119,120) are intact in autism. Furthermore, there is evidence that the affective representation system also remains intact in autism.[92,121] While deficits in ToM and possible delayed learning of self-other emotion associations in autism may result in impairments in the process of classifying the emotional state of another individual, the systems by which the affective state of the other is represented within the self appear to be intact. Therefore, in situations where the other's state is correctly classified, empathic outcomes would be expected to be relatively unaffected in autistic individuals with small to moderate ToM impairments.[5]

While somewhat more speculative in nature than the aforementioned impairments, there is some support for the notion that autistic individuals may also show atypical function of the *self-other switch* mechanism proposed by the SOME. The SOME[5] asserts that the default processing mode is self-focused; as autistic individuals tend to exhibit an increased self-focused bias (e.g., Ref. 122) and display reduced social motivation and attention (e.g., Refs 115,116), it is possible that their ability to switch to an other-focused processing mode may also be impaired.[5] It has been proposed that the ability to form a cognitive representation of one's internal affective state and then attribute that state to another reduces the degree of distress experienced when perceiving the negative emotions of others.[27] It is therefore likely that an impairment in the *self-other switch* mechanism would result in a heightened level of personal distress (resulting from emotional contagion) in response to the negative emotions displayed by others.[5] Indeed, there is evidence to suggest that autistic individuals do experience a heightened level of distress (relative to typical controls) in response to the perceived distress of others (see Ref. 123 for an overview).

Based upon the SOME framework, the empathic impairments in autism can in fact be relatively mild and are primarily mediated by the degree of ToM impairment exhibited by autistic individuals.[5] This perspective appears to contrast somewhat with the common notion that autism is associated with severe empathy deficits; however, this discrepancy can be accounted for in a number of ways. First, within the SOME,[5] ToM and empathy are operationalized as distinct components (relative to other models where the distinction is often less explicit). Furthermore, a key complication in determining the homogeneity of empathy-related impairments in autism relates to the high comorbidity between autism and alexithymia.[121,124]

3.4.2 *Alexithymia*

Alexithymia is a subclinical cognitive-affective impairment affecting the ability to interpret one's own emotional experiences.[125] Alexithymia is present in approximately 10% of the general population,[126,127] with significantly higher incidence levels within autistic populations (~50%).[121,124] Recent work suggests that comorbidity with alexithymia may be a fundamental factor in some of the social impairments (including empathy and emotion perception) exhibited by autistic individuals.[128] Prior research has demonstrated that, relative to low alexithymia controls, individuals with high alexithymia ratings tend to score lower on trait empathy measures and exhibit reduced activation of the dorsal ACC, DLPFC, and cerebellum during the experience of empathy.[91,92] Silani et al.[92] found evidence of a positive correlation between AI activation during an interoceptive task and the extent to which individuals were able to correctly interpret their own emotional states. Interestingly, this correlation was consistent across autistic individuals and healthy controls, suggesting that the severe empathy deficits associated with autism may be mediated, at least to some extent, by comorbid alexithymia commonly observed with autism. This assertion is supported by another neuroimaging investigation, which found that activity within the self-pain network in response to the perceived pain of others correlated with alexithymia scores but was not associated with the presence of autism per se.[129]

As many of the studies that have identified empathy deficits in alexithymia used cue-based paradigms (which do not require participants to accurately infer the affective states of others based on perceptual cues), the atypical empathic outcomes are presumably not the result of impairments in perceptual input systems.[5] Consistent with this interpretation, a recent pictorial-based investigation found that while alexithymic individuals were able to distinguish between expressions depicting anger and disgust, they were unable to correctly classify these expressions.[128]

Based upon such findings, indicating that individuals with alexithymia are aware of the presence of their own emotions but unable to correctly classify them, Bird and Viding[5] propose that the atypical empathic outcomes observed in alexithymia are likely associated with impairments in the *affective representation system*. Dysfunction of the *affective representation system* would be expected to result in such deficits, as one's ability to form a representation of internal affective states that are accessible to conscious introspection would be impeded. Furthermore, such a deficit in classifying one's own internal affective states would be likely to affect the ability to develop the appropriate learned associations between internal and external emotion cues (which is a critical process in the development of emotional contagion as proposed by learning-based models).[104] Thus, while the *affective cue classification system* may be functioning correctly in alexithymic individuals (allowing for correct classification of perceptual affective cues), dysfunction of the *affective representation system* would result in an inability to form a consciously accessible internal representation of the perceived emotion and to correctly classify external affective cues based upon learned associations with the corresponding internal affective state.[5]

3.4.3 *Psychopathy*

Psychopathy is a disorder characterized by atypical emotional responses and antisocial behaviors.[130] Psychopathy is commonly associated with a reduced empathic response, and this deficit has been highlighted in a number of ways. For example, the willingness of psychopathic individuals to act in an antisocial manner, with little evidence of concern for the impact that such behaviors may have on others,[130] suggests a diminished empathic response. Furthermore, findings indicate that psychopathy is associated with a reduced physiological response in relation to perceptual emotional stimuli[131,132] and when imagining oneself in dangerous/fearful situations.[133,134] Research has also shown that children with psychopathic tendencies score atypically low on measures of emotional contagion in response to stimuli describing the negative emotional experiences of others.[133,135–137]

Neuroimaging research indicates that the atypical empathic outcomes observed in psychopathic individuals may be associated with reduced activity in regions associated with emotional processing, such as the AI and amygdala.[94,138–140] In addition to evidence suggestive of abnormal functional neural activity in psychopathy, studies have also revealed that psychopathy is associated with structural abnormalities in the AI[141] (for example, reduced gray matter volume in this area and also in the amygdala).[93,94] Reduced gray matter volume in the AI and amygdala has similarly been shown to correlate with instances of aggressive behavior and levels of empathy in adolescents with conduct disorder.[93]

As previously discussed, the SOME[5] asserts that emotional contagion is a prerequisite for empathy and develops via learned associations between internal affective states and external cues indicating the corresponding states in others. In addition to the empathy deficits observed in psychopathic individuals, there is evidence that psychopathy is characterized by atypical experiences with regards to one's own emotions.[142,143] Prior research suggests that individuals with psychopathy exhibit abnormal emotional experiences, particularly in relation to distressing emotions such as fear and sadness[144]; within the SOME[5] framework, such atypical functioning could have a significant detrimental effect on the development of emotional contagion (and therefore empathy). If negative emotions and distress are experienced less often (and/or to a lesser extent) by children with psychopathy, it stands to reason that they would have fewer opportunities to learn the appropriate associations between external affective cues and internal affective states, leading to a lack of understanding regarding the distressed states of others—thereby inhibiting emotional contagion and empathy.[5]

While there is robust evidence to support the proposal of a deficit in emotional contagion in individuals with psychopathy, such individuals are not impaired in every aspect of understanding the experiences of others.[5] Psychopathy is often characterized by a willingness and adeptness in manipulating others, typically for personal gain.[145] Such abilities rely heavily on ToM, and indeed, the assertion that ToM abilities remain intact in psychopathic individuals has received support from a number of investigations.[133,137,146,147]

Bird and Viding[5] suggest that while ToM is intact in psychopathy (allowing individuals with psychopathic tendencies to cognitively represent the affective states of others), their reduced opportunities to form associations between internal emotional states and emotional cues exhibited by others means they do not share the affective representations of others' distress, thereby resulting in the empathic deficits characteristic of this condition.

4. SUMMARY

In summary, this chapter has discussed what constitutes empathy and mechanisms that mediate empathic responses. As noted, empathy is a multifaceted construct that relies upon an interaction of a variety of factors that can be importantly influenced through individual differences.

References

1. Hoffman ML. *Empathy and Moral Development: Implications for Caring and Justice*. Cambridge University Press; 2000.
2. deVignemont F, Singer T. The empathic brain: how, when and why? *Trends Cogn Sci*. 2006;10:435–441.
3. Preston SD, de Waal F. Empathy: its ultimate and proximate bases. *Behav Brain Sci*. 2002;25:1–20.
4. Shamay-Tsoory SG. Dynamic functional integration of distinct neural empathy systems. *Soc Cogn Affect Neurosci*. 2014;9:1–2.
5. Bird G, Viding E. The self to other model of empathy: providing a new framework for understanding empathy impairments in psychopathy, autism, and alexithymia. *Neurosci Biobehav Rev*. 2014;47:520–532.
6. Lamm C, Decety J, Singer T. Meta-analytic evidence for common and distinct neural networks associated with directly experienced pain and empathy for pain. *NeuroImage*. 2011;54:2492–2502.
7. Hein G, Singer T. I feel how you feel but not always: the empathic brain and its modulation. *Curr Opin Neurobiol*. 2008;18:153–158.
8. Singer T, Seymour B, O'Doherty JP, Kaube H, Dolan RJ, Frith CD. Empathy for pain involves the affective but not sensory components of pain. *Science*. 2004;303:230–237.
9. Avenanti A, Bueti D, Galati G, Aglioti SM. Transcranial magnetic stimulation highlights the sensorimotor side of empathy for pain. *Nat Neurosci*. 2005;8:955–960.
10. Cheng Y, Yang C-Y, Lin C-P, Lee P-L, Decety J. The perception of pain in others suppresses somatosensory oscillations: a magnetoencephalography study. *NeuroImage*. 2008;40:1833–1840.
11. Keysers C, Kaas JH, Gazzola V. Somatosensation in social perception. *Nat Rev Neurosci*. 2010;11:417–428.
12. Bufalari I, Aprile T, Avenanti A, Di Russo F, Aglioti SM. Empathy for pain and touch in the human somatosensory cortex. *Cereb Cortex*. 2007;17:2553–2561.
13. Osborn J, Derbyshire SW. Pain sensation evoked by observing injury in others. *Pain*. 2010;148:268–274.
14. Singer T, Seymour B, O'Doherty JP, Stephan KE, Dolan RJ, Frith CD. Empathic neural responses are modulated by the perceived fairness of others. *Nature*. 2006;439:466–469.
15. Davis MH. A multidimensional approach to individual differences in empathy. *JSAS Catalog Sel Doc Psychol*. 1980;10:85.
16. Davis MH, Luce C, Kraus SJ. The heritability of characteristics associated with dispositional empathy. *J Pers*. 1994;62:369–391.

17. Baron-Cohen S, Wheelwright S. The Empathy Quotient (EQ). An investigation of adults with Asperger syndrome or high functioning autism, and normal sex differences. *J Autism Dev Disord*. 2004;34:163–175.
18. Chakrabarti B, Bullmore ET, Baron-Cohen S. Empathising with basic emotions: common and discrete neural substrates. *Soc Neurosci*. 2006;1:364–384.
19. Dapretto M, Davies M, Pfeifer J, et al. Understanding emotions in others: mirror neuron dysfunction in children with autism spectrum disorders. *Nat Neurosci*. 2006;9:28–30.
20. Gazzola V, Aziz-Zadeh L, Keysers C. Empathy and the somatotopic auditory mirror system in humans. *Curr Biol*. 2006;16:1824–1829.
21. Nishitani N, Avikainen S, Hari R. Abnormal imitation-related cortical activation sequences in Asperger's syndrome. *Ann Neurol*. 2004;55:558–562.
22. Sculte-Rüther M, Markowitsch HJ, Fink GR, Piefke M. Mirror neuron and theory of mind mechanisms involved in face-to-face interactions: a functional magnetic resonance imaging approach to empathy. *J Cogn Neurosci*. 2007;19:1354–1372.
23. Yang CY, Decety J, Lee S, Chen C, Cheng Y. Gender differences in the mu rhythm during empathy for pain: an electroencephalographic study. *Brain Res*. 2009;1251:176–184.
24. Banissy MJ, Kanai R, Walsh V, Rees G. Inter-individual differences in empathy are reflected in human brain structure. *NeuroImage*. 2012;62:2034–2039.
25. Cheng Y, Chou K-H, Decety J, et al. Sex differences in the neuroanatomy of human mirror-neuron system: a voxel-based morphometric investigation. *Neuroscience*. 2009;158:713–720.
26. Decety J, Jackson PL. The functional architecture of human empathy. *Behav Cogn Neurosci Rev*. 2004;3:71–100.
27. Decety J, Lamm C. Human empathy through the lens of social neuroscience. *Sci World J*. 2006;6:1146–1163.
28. Gallese V, Goldman A. Mirror neurons and the simulation theory of mind-reading. *Trends Cogn Sci*. 1998;2:493–501.
29. Gu X, Han S. Attention and reality constraints on the neural processes of empathy for pain. *NeuroImage*. 2007;36:256–267.
30. Fadiga L, Fogassi L, Pavesi G, Rizzolatti G. Motor facilitation during action observation: a magnetic stimulation study. *J Neurophysiol*. 1995;73:2608–2611.
31. Fadiga L, Craighero L, Destro MF, et al. Language in shadow. *Soc Neurosci*. 2006;1:77–89.
32. Hari R, Forss N, Avikainen S, Kirveskari E, Salenus S, Rizzolatti G. Activation of human primary motor cortex during action observation: a neuromagnetic study. *Proc Natl Acad Sci USA*. 1998;95:15061–15065.
33. Keysers C, Wicker B, Gazzola V, Anton JL, Fogassi L, Gallese V. A touching sight: SII/PV activation during the observation and experience of touch. *Neuron*. 2004;42:335–346.
34. Molenberghs P, Cunnington R, Mattingley JB. Brain regions with mirror properties: a meta-analysis of 125 human fMRI studies. *Neurosci Biobehav Rev*. 2012;36:341–349.
35. Rizzolatti G, Craighero L. The mirror-neuron system. *Annu Rev Neurosci*. 2004;27:169–192.
36. Blakemore SJ, Bristow D, Bird G, Frith C, Ward J. Somatosensory activations during the observation of touch and a case of vision–touch synaesthesia. *Brain*. 2005;128:1571–1583.
37. Wicker B, Keysers C, Plailly J, Royet JP, Gallese V, Rizzolatti G. Both of us disgusted in My insula: the common neural basis of seeing and feeling disgust. *Neuron*. 2003;403:655–664.
38. Botvinick M, Jha AP, Bylsma LM, Fabian SA, Solomon PE, Prkachin KM. Viewing facial expressions of pain engages cortical areas involved in the direct experience of pain. *NeuroImage*. 2005;25:312–319.
39. Jackson PL, Rainville P, Decety J. To what extent do we share the pain of others? Insight from the neural bases of pain empathy. *Pain*. 2006;125:5–9.
40. Lamm C, Nusbaum HC, Meltzoff AN, Decety J. What are you feeling? Using functional magnetic resonance imaging to assess the modulation of sensory and affective responses during empathy for pain. *PLoS One*. 2007;12:e1292.
41. Morrison I, Lloyd D, di Pellegrino G, Roberts N. Vicarious responses to pain in anterior cingulate cortex: is empathy a multisensory issue? *Cogn Affect Behav Neurosci*. 2004;4:270–278.
42. Craig AD. How do you feel? Interoception: the sense of the physiological condition of the body. *Nat Rev Neurosci*. 2002;3:655–666.
43. Craig AD. How do you feel-now? The anterior insula and human awareness. *Nat Rev Neurosci*. 2009;10:59–70.
44. Olsson A, Ochsner KN. The role of social cognition in emotion. *Trends Cogn Sci*. 2008;12:65–71.
45. Price DD. Psychological and neural mechanisms of the affective dimension of pain. *Science*. 2000;288:1769–1772.
46. Decety J, Sommerville JA. Shared representations between self and others: a social cognitive neuroscience view. *Trends Cogn Sci*. 2003;7:527–533.
47. Keysers C, Gazzola V. Towards a unifying neural theory of social cognition. *Prog Brain Res*. 2006;156:379–401.
48. Lamm C, Batson CD, Decety J. The neural substrate of human empathy: effects of perspective taking and cognitive appraisal. *J Cogn Neurosci*. 2007;19:42–58.
49. Jackson P, Meltzoff A, Decety J. How do we perceive the pain of others? A window into the neural processes involved in empathy. *NeuroImage*. 2005;24:771–779.
50. Banissy MJ, Ward J. Mirror-touch synesthesia is linked with empathy. *Nature Neurosci*. 2007;10:815–816.
51. Banissy MJ. Synaesthesia, mirror neurons and mirror-touch. In: Simner J, Hubbard E, eds. *The Oxford Handbook of Synaesthesia*. Oxford: Oxford University Press; 2013.
52. Holle H, Banissy MJ, Ward J. Functional and structural brain differences associated with mirror-touch synaesthesia. *NeuroImage*. 2013;83:1041–1050.
53. Adolphs R. The social brain: neural basis of social knowledge. *Annual Rev Psychol*. 2009;60:693–716.
54. Banissy MJ, Sauter DA, Ward J, Warren E, Walsh V, Scott S. Suppressing sensorimotor activity modulates the discrimination of auditory emotions but not speaker identity. *J Neurosci*. 2010;30:13552–13557.
55. Banissy MJ, Walsh VZ, Muggleton NG. Mirror-touch synaesthesia: a case of faulty self-modelling and insula abnormality. *Cogn Neurosci*. 2011;2:114–115.
56. Hooker CI, Verosky SC, Germine LT, Knight RT, D'Esposito M. Mentalizing about emotion and its relationship to empathy. *Soc Cogn Affect Neurosci*. 2008;3:204–217.
57. Hooker CI, Verosky SC, Germine LT, Knight RT, D'Esposito M. Neural activity during social signal perception correlates with self-reported empathy. *Brain Res*. 2010;1308:110–113.
58. Mahayana IT, Banissy MJ, Chen C, Walsh V, Juan C, Muggleton NG. Motor empathy is a consequence of misattribution of sensory information in observers. *Front Hum Neurosci*. 2014;8:47.
59. Pfurtscheller G, Stancák Jr A, Neuper CH. Event-related synchronization (ERS) in the alpha band - an electrophysiological correlate of cortical idling: a review. *Int J Psychophysiol*. 1996;24:39–46.
60. Pfurtscheller G, Lopes da Silva FH. Event-related EEG/MEG synchronization and desynchronization: basic principles. *Clin Neurophysiol*. 1999;110:1842–1857.
61. Avenanti A, Paluello IM, Bufalari I, Aglioti SM. Stimulus-driven modulation of motor-evoked potentials during observation of others' pain. *NeuroImage*. 2006;32:316–324.
62. Costantini M, Galati G, Romani GL, Aglioti SM. Empathic neural reactivity to noxious stimuli delivered to body parts and non-corporeal objects. *Eur J Neurosci*. 2008;28:1222–1230.
63. Kulkarni B, Bentley DE, Elliott R, et al. Attention to pain localization and unpleasantness discriminates the functions of the medial and lateral pain systems. *Eur J Neurosci*. 2005;21:3133–3142.

64. Schubotz RI. Prediction of external events with our motor system: towards a new framework. *Trends Cogn Sci*. 2007;11:211–218.
65. Van Overwalle F, Baetens K. Understanding others' actions and goals by mirror and mentalizing systems: a meta-analysis. *NeuroImage*. 2009;48:564–584.
66. Schilbach L, Eickhoff SB, Rotarska-Jagiela A, Fink GR, Vogeley K. Minds at rest? Social cognition as the default mode of cognizing and its putative relationship to the "default system" of the brain. *Conscious Cogn*. 2008;17:457–467.
67. Decety J, Lamm C. The role of the right temporoparietal junction in social interaction: how low-level computational processes contribute to meta-cognition. *Neuroscientist*. 2007;13:580–593.
68. Santiesteban I, Banissy MJ, Catmur C, Bird G. Enhancing social ability by stimulating right temporoparietal junction. *Curr Biol*. 2012;22:2274–2277.
69. Zaki J, Ochsner K. The neuroscience of empathy: progress, pitfalls, and promise. *Nat Neurosci*. 2012;15:675–680.
70. Mehrabian A, Epstein N. A measure of emotional empathy. *J Pers*. 1972;40:525–543.
71. Jabbi M, Swart M, Keysers C. Empathy for positive and negative emotions in the gustatory cortex. *NeuroImage*. 2007;34:1744–1753.
72. Montag C, Schuber F, Heinz A, Gallinat J. Prefrontal cortex glutamate correlates with mental perspective-taking. *PLoS One*. 2008;3:e3890.
73. Decety J. The neurodevelopment of empathy in humans. *Dev Neurosci*. 2010;32:257–267.
74. Singer T, Lamm C. The social neuroscience of empathy. *Ann N Y Acad Sci*. 2009;156:81–96.
75. Rankin KP, Gorno-Tempini ML, Allison SC, et al. Structural anatomy of empathy in neurodegenerative disease. *Brain*. 2006;129:2945–2956.
76. Hein G, Silani G, Preuschoff K, Batson CD, Singer T. Neural responses to ingroup and outgroup members' suffering predict individual differences in costly helping. *Neuron*. 2010;68:149–160.
77. Hoffman ML. Sex differences in empathy and related behaviors. *Psychol Bull*. 1977;84:712–722.
78. Chakrabarti B, Baron-Cohen S. Empathizing: neurocognitive developmental mechanisms and individual differences. In: Anders S, Ende G, Junghofer M, Kissler J, Wildgruber D, eds. *Understanding Emotions. Special Issue of Progress in Brain Research*. Vol. 56. Elsevier; 2006:403–417.
79. Schulte-Rüther M, Markowitsch HJ, Shah NJ, Fink FR, Piefke M. Gender differences in brain networks supporting empathy. *NeuroImage*. 2008;42:393–403.
80. Cheng Y, Lin CP, Liu HL, et al. Expertise modulates the perception of pain in others. *Curr Biol*. 2007;17:1708–1713.
81. Takahashi E, Ohki K, Miyashita Y. The role of the parahippocampal gyrus in source memory for external and internal events. *Neuroreport*. 2002;13:1951–1956.
82. Ochsner KN, Gross JJ. The neural architecture of emotion regulation. In: Gross JJ, ed. *Handbook of Emotion Regulation*. New York: The Guilford Press; 2013:87–109.
83. Derbyshire S, Osborn J, Brown S. Feeling the pain of others is associated with self-other confusion and prior pain experience. *Front Human Neurosci*. 2013;7:470.
84. Valentini E. The role of anterior insula and anterior cingulate in empathy for pain. *J Neurophysiol*. 2010;104:584–586.
85. Ashburner J, Friston KJ. Multimodal image coregistration and partitioning: a unified framework. *NeuroImage*. 1997;6:209–217.
86. Kanai R, Dong MY, Bahrami B, Rees G. Distractibility in daily life is reflected in the structure and function of human parietal cortex. *J Neurosci*. 2011;31:6620–6626.
87. Mar RA. The neural basis of social cognition and story comprehension. *Annu Rev Psychol*. 2011;62:103–134.
88. Oberman LM, Hubbard EM, McCleery JP, Altschuler EL, Ramachandran VS, Pineda JA. EEG evidence for mirror neuron dysfunction in autism spectrum disorders. *Cogn Brain Res*. 2005;24: 190–198.
89. Zilbovicius M, Meresse I, Chabane N, Brunelle F, Samson Y, Boddaert N. Autism, the superior temporal sulcus and social perception. *Trends Neurosci*. 2006;29:259–366.
90. Sims TB, Neufeld J, Johnstone T, Chakrabarti B. Autistic traits modulate frontostriatal connectivity during processing of rewarding faces. *Soc Cognitive Affect Neurosci*. 2014;9:2010–2016.
91. Moriguchi Y, Decety J, Ohnishi T, Maeda M, Matsuda H, Komaki G. Empathy and judging other's pain: an fMRI study of alexithymia. *Cereb Cortex*. 2007;17:2223–2234.
92. Silani G, Bird G, Brindley R, Singer T, Frith C, Frith U. Levels of emotional awareness and autism: an fMRI study. *Soc Neurosci*. 2008;3:97–112.
93. Sterzer P, Stadler C, Poustka F, Kleinschmidt A. A structural neural deficit in adolescents with conduct disorder and its association with lack of empathy. *NeuroImage*. 2007;37:335–342.
94. Birbaumer N, Veit R, Lotze M, et al. Deficient fear conditioning in psychopathy: a functional magnetic resonance imaging study. *Arch Gen Psychiatry*. 2005;62:799–805.
95. Blair RJR. Responding to the emotions of others: dissociating forms of empathy through the study of typical and psychiatric populations. *Conscious Cognit*. 2005;14:698–718.
96. Frith U, Frith CD. Development and neurophysiology of mentalizing. *Philos Trans R Soc London Biol Sci*. 2003;358:459–473.
97. Said CP, Moore CD, Engell AD, Todorov A, Haxby JV. Distributed representations of dynamic facial expressions in the superior temporal sulcus. *J Vis*. 2010;10:1–12.
98. Scott SK, Sauter D, McGettigan C. Brain mechanisms for processing perceived emotional vocalizations in humans. *Hand Behav Neurosci*. 2010;19:187–197.
99. Winston JS, Henson RNA, Fine-Goulden MR, Dolan RJ. fMRI-adaptation reveals dissociable neural representations of identity and expression in face perception. *J Neurophysiol*. 2004;92: 1830–1839.
100. Van Overwalle F. Social cognition and the brain: a meta-analysis. *Hum Brain Mapp*. 2009;30:829–858.
101. Meltzoff AN, Gopnik A. The role of imitation in under-standing persons and developing a theory of mind. In: Baron-Cohen S, Tager-Flusberg HE, Cohen DJ, eds. *Understanding Other Minds: Perspectives from Autism*. Oxford University Press; 2013:335–366.
102. Heyes C. Causes and consequences of imitation. *Trends Cogn Sci*. 2001;5:253–261.
103. Heyes C. Where do mirror neurons come from? *Neurosci Biobehav Rev*. 2010;34:575–583.
104. Heyes CM, Bird G. Mirroring, association and the correspondence problem. In: Haggard P, Rossetti Y, Kawato M, eds. *Sensorimotor Foundations of Higher Cognition, Attention and Performance*. Vol. XX. Oxford: Oxford University Press; 2008.
105. Gergely G, Watson JS. The social biofeedback theory of parental affect-mirroring: the development of emotional self-awareness and self-control in infancy. *Int J Psychoanal*. 1996;77:1181–1212.
106. Allen R, Heaton PF. Autism, music, and the therapeutic potential of music in alexithymia. *Music Percept*. 2010;27:251–261.
107. Kitayama S, Park J. Cultural neuroscience of the self: understanding the social grounding of the brain. *Soc Cogn Affect Neurosci*. 2010;5:111–129.
108. Spengler S, von Cramon DY, Brass M. Control of shared representations relies on key processes involved in mental state attribution. *Hum Brain Mapp*. 2009;30:3704–3718.
109. American Psychiatric Association. *Diagnostic and Statistical Manual of Mental Disorders*. 5th ed. Arlington: American Psychiatric Publishing; 2013.
110. Oberman LM, Ramachandran VS. The simulating social mind: the role of the mirror neuron system and simulation in the social and communicative deficits of autism spectrum disorders. *Psychol Bull*. 2007;133:310–327.
111. Lombardo MV, Barnes JL, Wheelwright SJ, Baron-Cohen S. Self-referential cognition and empathy in autism. *PLoS One*. 2007;2:e883.

112. Frith CD, Frith U. The neural basis of mentalizing. *Neuron*. 2006;50:531–534.
113. Hadjikhani N, Joseph RM, Snyder J, Tager-Flusberg H. Anatomical differences in the mirror neuron system and social cognition. *Cereb Cortex*. 2006;16:1276–1282.
114. Frith CD, Frith U. The self and its reputation in autism. *Neuron*. 2008;57:331–332.
115. Dawson G, Webb SJ, McPartland J. Understanding the nature of face processing impairment in autism: insights from behavioral and electrophysiological studies. *Dev Neuropsychol*. 2005;27:403–424.
116. Jones W, Carr K, Klin A. Absence of preferential looking to the eyes of approaching adults predicts level of social disability in 2-year olds with autism spectrum disorder. *Arch Gen Pysch*. 2008;65:946–954.
117. Hamilton AFDC, Brindley RM, Frith U. Imitation and action understanding in autistic spectrum disorders: how valid is the hypothesis of a deficit in the mirror neuron system? *Neuropsychologia*. 2007;45:1859–1868.
118. Cook JL, Bird G. Atypical social modulation of imitation in autism spectrum conditions. *J Autism Dev Disord*. 2012;42:1045–1051.
119. Boucher J, Warrington EK. Memory deficits in early infantile autism: some similarities to the amnesic syndrome. *Br J Psychol*. 1976;67:73–87.
120. Williams DL, Goldstein G, Minshew NJ. The profile of memory function in children with autism. *Neuropsychology*. 2006;20:21.
121. Berthoz S, Hill EL. The validity of using self-reports to assess emotion regulation abilities in adults with autism spectrum disorder. *Eur Psychiatry*. 2005;20:291–298.
122. Kanner L. Autistic disturbances of affective contact. *Nervous Child*. 1943;2:217–250.
123. Smith A. The empathy imbalance hypothesis of autism: a theoretical approach to cognitive and emotional empathy in autistic development. *Psychol Rec*. 2009;59:489–510.
124. Hill E, Berthoz S, Frith U. Brief report: cognitive processing of own emotions in individuals with autistic spectrum disorder and in their relatives. *J Autism Dev Disord*. 2004;34:229–235.
125. Taylor GJ. Alexithymia: concept, measurement, and implications for treatment. *Am J Psychiatry*. 1984;141:725–732.
126. Linden W, Wen F, Paulhus DL. Measuring alexithymia: reliability, validity, and prevalence. In: Butcher J, Spielberger C, eds. *Advances in Personality Assessment*. Hillsdale: Erlbaum; 1995:51–95.
127. Salminen JK, Saarijarvi S, Aarela E, Toikka T, Kauhanen J. Prevalence of alexithymia and its association with sociodemographic variables in the general population of Finland. *J Psychosom Res*. 1999;46:75–82.
128. Cook R, Brewer R, Shah P, Bird G. Alexithymia, not autism, predicts poor recognition of emotional facial expressions. *Psychol Sci*. 2013;24:723–732.
129. Bird G, Silani G, Brindley R, White S, Frith U, Singer T. Empathic brain responses in insula are modulated by levels of alexithymia but not autism. *Brain*. 2010;133:1515–1525.
130. Hare RD, Neumann CS. The PCL-R assessment of psychopathy. In: Patrick CJ, ed. *Handbook of Psychopathy*. Guilford Press; 2007:58–88.
131. Hare RD. Psychopathy and electrodermal responses to nonsignal stimulation. *Biol Psychol*. 1978;6:237–246.
132. Patrick CJ, Bradley MM, Lang PJ. Emotion in the criminal psychopath: startle reflex modulation. *J Abnorm Psychol*. 1993;102:82.
133. Jones AP, Happé FG, Gilbert F, Burnett S, Viding E. Feeling, caring, knowing: different types of empathy deficit in boys with psychopathic tendencies and autism spectrum disorder. *J Child Psychol Psychiatry*. 2010;51:1188–1197.
134. Marsh AA, Finger EC, Schechter JC, Jurkowitz IT, Reid ME, Blair RJR. Adolescents with psychopathic traits report reductions in physiological responses to fear. *J Child Psychol Psychiatry*. 2011;52:834–841.
135. de Wied M, van Boxtel A, Matthys W, Meeus W. Verbal, facial and autonomic responses to empathy-eliciting film clips by disruptive male adolescents with high versus low callous-unemotional traits. *J Abnorm Child Psychol*. 2012;40:211–223.
136. Pardini DA, Lochman JE, Frick PJ. Callous/unemotional traits and socialcognitive processes in adjudicated youths. *J Am Acad Child Adolesc Psychiatry*. 2003;42:364–371.
137. Schwenck C, Mergenthaler J, Keller K, et al. Empathy in children with autism and conduct disorder: group-specific profiles and developmental aspects. *J Child Psychol Psychiatry*. 2012;53:651–659.
138. Singer T. The neuronal basis and ontogeny of empathy and mind reading: review of literature and implications for future research. *Neurosci Biobehav Rev*. 2006;30:855–863.
139. Blair RJ. Fine cuts of empathy and the amygdala: dissociable deficits in psychopathy and autism. *Q J Exp Psychol*. 2008;61:157–170.
140. Sebastian CL, McCrory EJ, Cecil CA, et al. Neural responses to affective and cognitive theory of mind in children with conduct problems and varying levels of callous-unemotional traits. *Arch Gen Psychiatry*. 2012;69:814–822.
141. de Oliveira-Souza R, Hare RD, Bramati IE, et al. Psychopathy as a disorder of the moral brain: fronto-temporo-limbic grey matter reductions demonstrated by voxel-based morphometry. *NeuroImage*. 2008;40:1202–1213.
142. Brook M, Brieman CL, Kosson DS. Emotion processing in Psychopathy Checklist – assessed psychopathy: a review of the literature. *Clin Psychol Rev*. 2013;33:979–995.
143. Steuerwald BL, Kosson DS. Emotional experiences of the psychopath. In: Gacono CB, ed. *The Clinical and Forensic Assessment of the Psychopath: A Practitioner's Guide*. Mahwah: Lawrence Erlbaum Associates Publishers; 2000:111–135.
144. Blair RJR, Budhani S, Colledge E, Scott S. Deafness to fear in boys with psychopathic tendencies. *J Child Psychol Psychiatry*. 2005;46:327–336.
145. Hare RD. *Without Conscience: The Disturbing World of the Psychopaths Among Us*. Guilford Press; 2011.
146. Dolan M, Fullam R. Theory of mind and mentalizing ability in antisocial personality disorders with and without psychopathy. *Psychol Med*. 2004;34:1093–1102.
147. Richell RA, Mitchell DGV, Newman C, Leonard A, Baron-Cohen SE, Blair RJR. Theory of mind and psychopathy: can psychopathic individuals read the "language of the eyes"? *Neuropsychologia*. 2003;41:523–526.

CHAPTER

16

Honesty

Francesca Mameli[1], Giuseppe Sartori[2], Cristina Scarpazza[2], Andrea Zangrossi[3], Pietro Pietrini[3], Manuela Fumagalli[1], Alberto Priori[1,4]

[1]Centro Clinico per la Neurostimolazione, le Neurotecnologie ed i Disordini del Movimento, Fondazione IRCCS Ca' Granda, Ospedale Maggiore Policlinico, Milan, Italy; [2]Università degli Studi di Padova, Padua, Italy; [3]Università degli Studi di Pisa, Pisa, Italy; [4]Università degli Studi di Milano, Milan, Italy

OUTLINE

1. INTRODUCTION

Ample evidence showing that the human brain is innately primed to deceive explains why all societies have shown an interest in deception, right from the earliest written record, and the early appearance of lying in life.[1–3]

According to DePaulo[4] and Vrij,[5] the reasons for lying differ in at least two ways. The first difference depends on whether lying benefits the liar or another person (is self-oriented or other-person oriented), and the second depends on whether lying serves to gain advantages or avoid costs. Self-oriented and other-oriented lies may be told for psychological reasons (to prevent embarrassment or loss of face) or for material reasons (to gain a material advantage or avoid punishment).[4,5] People cheat when a cost-benefit analysis suggests they could gain advantages, as happens in a legal framework.

The cognitive perspective on honesty holds that deception is cognitively more demanding than truth telling.[6] Telling a lie involves one or more of the following mental operations: deciding to lie, withholding the truth, fabricating the lie, monitoring whether the receiver believes the lie, and, if necessary, adjusting the

Neuroimaging Personality, Social Cognition, and Character
http://dx.doi.org/10.1016/B978-0-12-800935-2.00016-6

fabricated story and keeping the lying consistent. These mental operations make lying a cognitively demanding task.[1,3]

This chapter focuses on recent observations regarding the cognitive correlates of deception and proposes a framework for understanding the neural mechanisms that allow people to tell lies. We explore the same topic from a developmental viewpoint, emphasizing the relationship between acquiring the ability to tell a lie and the developing brain. We describe the findings obtained from cognitive and neuroimaging studies on the dishonest brain. Finally, the last sections focus on pathological lying, on the moral brain, and on future research directions in the cognitive neuroscience of deception.

2. WHAT DOES LYING INVOLVE? HOW DOES COGNITIVE NEUROSCIENCE CONTRIBUTE TO LYING?

2.1 Cognitive Studies of Lie Production

The first to underline the role of cognition in deception were Zuckerman, DePaulo, and Rosenthal (1981), who included it among four associated factors: emotion, arousal, control, and cognitive processing.[7] Others later extended the same concept.[8,9] Increasing evidence supports the assumption that lying is more cognitively demanding than truth telling.[7,10]

The role of cognitive load in deception is explained by different theories. For example, *Gombos's theory* suggests that deception comprises two main cognitive processes: control mechanisms for thought (e.g., inhibition of the truth) and active management (analyzing the interlocutor's reactions and adapting their own behavioral features to maintain the lie).[11] The *Activation Decision Construction Action Theory* emphasizes the roles of executive processes, theory of mind, emotions and motivation, thoroughly specifies cognitive processing, and considers the rehearsal of lies.[12–15] The *Working Memory Theoretical Model of Deception* holds that the deception process needs multitasking for liars to plan what to say, to be coherent (without contradicting themselves), to observe listener's reactions, and to modulate their own behavior.[16] Finally, other models, such as the *Sheffield Model*[3] and the *Working Model of Deception*,[17] assume that generating a deceptive response requires inhibition of the prepotent truth response.[18]

Lying therefore requires a greater cognitive load than telling the truth. The conscious decision to lie, inhibiting the truth, and formulating the lie are all considered cognitively demanding tasks,[19] as are continuous monitoring of the information exchange with the interlocutor and feedback assessment.[20]

Evidence showing increased cognitive involvement also comes from several brain imaging studies showing that, in contrast to truth telling, the act of lying involves intense activity in prefrontal brain regions,[21] especially the anterior cingulate cortex (ACC), the dorsolateral prefrontal cortex (DLPFC),[22,23] the ventromedial prefrontal cortex (VMPFC), and superior temporal sulcus,[23] areas that several reviews consider of crucial importance for cognitive control (Figure 1).[1,24]

Further support for the idea that cognition has a major role in deception comes from studies showing a poor ability to deceive when a medical condition or a developmental step undermine prefrontal cortex functionality. Children, for example, whose prefrontal cortex development is not yet complete,[25,26] are less able to deceive than patients with frontal lobe neurodevelopmental and neurodegenerative conditions such as autism, Parkinson's disease,[27] and essential tremor (see Section 5).[28]

2.2 Types of Lies

Some problems related to the study of lying arise because the various kinds of lies differ in cognitive complexity. Investigating cognitive complexity in lying, DePaulo (1996) proposed a taxonomy of lies in everyday life.[4] They distinguished between three main types of lies: outright lies, in which "the information conveyed is completely different from, or contradictory to, the truth";

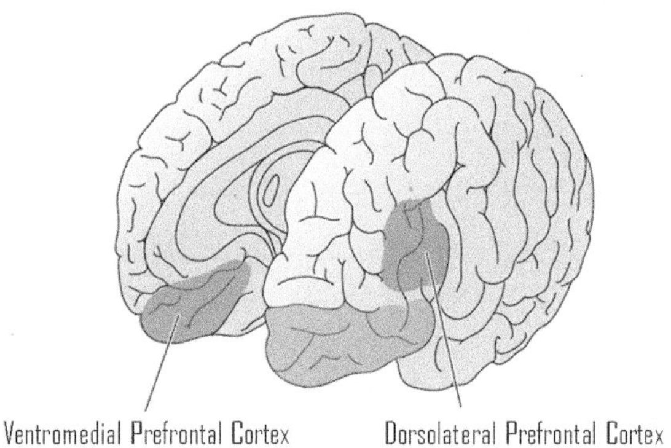

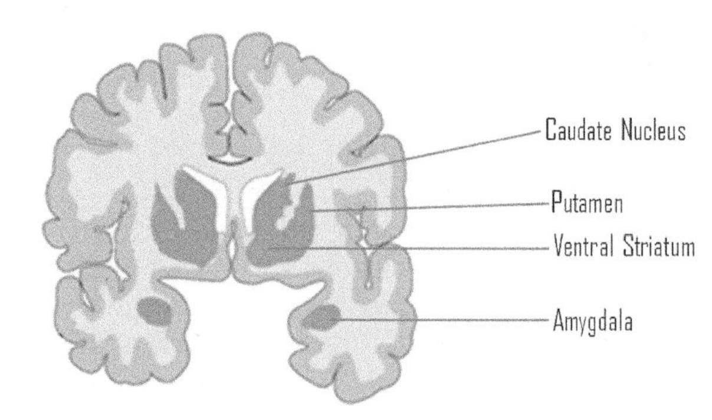

FIGURE 1 **Lateral and medial surface of the cerebral cortex and subcortical brain regions.** All the regions highlighted have been implicated in the existing literature for the neurobiology of deception. GKT, Guilty Knowledge Task; E1, experiment n.1; E2, experiment n.2.

exaggerations, in which "the liars overstate the facts or convey an impression that exceeds the truth"; and subtle lies, in which lying entails "evading or omitting relevant details and telling literal truths that are designed to mislead." This third type also includes behavioral and nonverbal lies.[4] Different types of lies therefore imply a different cognitive load, from the simplest type ("outright lies"), characterized by an information reversal in which the cognitive request focuses on inhibiting the truth and constructing an "alternative truth," to the more complex "subtle lies," such as Ulysses' famous statement "my name is Nobody" in the Odyssey. To build this well-known and complex literary lie, for example, the protagonist had to visualize a future probable scenario in which the lie would have been credible for Polyphemus but detected as false information when conveyed to Ulysses' companions. From the first example to the second, cognitive demand increases (for example, when "theory of mind" requires the liar to have a theory about what the interlocutor is thinking). The most complex lies also require a strength check in several possible alternative scenarios.

3. THE ABILITY TO TELL A LIE: A DEVELOPMENTAL PERSPECTIVE

3.1 How Do We Learn to Lie?

The ability to fabricate lies first appears during a child's early years and is followed by a developmental curve during which most individuals can be considered as "ordinary." Some children, however, especially those with certain neurodevelopmental conditions (i.e., autism spectrum disorder), are less able than others to fabricate lies. Lying is a natural characteristic of normal child development, a universal human behavior or inbred mechanism, observed in children of similar age regardless of culture, upbringing, moral, social, or economic background.

One of the first scientists to take an interest in the dynamics involved in the act of lying was Charles Darwin (1877), who observed and maintained records of his son's development, noting the boy's emerging lie-telling ability since the tender age of two years.[29] What Darwin derived from his studies is that one of the principal driving forces that bring children to lie is to conceal a disobedience or transgression. When children transgress and leave behind evidence, even at a very young age, they show an ability to make untrue statements to cover up their misdeed. Only later in their development do they learn how to use language effectively to strategically support the original lie when further probed to explain evidence, thus managing to make their statements consistent with evidence of their transgression and avoiding implicating themselves.[29]

Darwin's initial interest in the origins of the human's ability for lying was followed by Hartshone and May's (1928) subsequent work with older children.[30] Not until about 50 years later did Lewis, Stanger, and Sullivan (1989) conduct more studies addressing children's deceptive behaviors.[31] Such a longstanding interest in the subject of lying is hardly surprising. Many scholars intuitively realized that to observe and dissect the human impulse to lie can serve as a window into many aspects of children's developing minds, for example, intelligence,[32] theory of mind,[33,34] moral understanding,[35] personality and character formation,[34] and children's competence as witnesses in the courts of law.[36–38]

The method most commonly used in research to assess lie-telling behavior in children is the *Temptation Resistance paradigm*.[39] This procedure consists in placing young children (typically between two to five years of age) in a setting where deception is likely to occur naturally. One such example is being left alone in a room with an exciting toy, which the researcher (before leaving the room) instructs them not to touch or look at. The child is thus subjected to a very tempting situation, very likely to lead them to infringe the experimenter's rule. Upon the researcher's return to the room, the child is asked whether they touched (or looked at) the toy during the researcher's absence. The answer to this question reveals the child's tendency to tell the truth or a lie after committing a transgression.

Studies conducted to date consistently show that children start to display an ability to lie during the preschool years, with about one-third of three-year-olds telling lies. The numbers grow directly proportional with increasing age.[40] For example, the study by Talwar and Lee (2002) showed that most three-year-olds peeked at the toy, and all peekers returned to their original posture either as soon as they finished peeking or when they heard the experimenter opening the door.[37] Others subsequently found that when three- to five-year-olds were left alone in a room and asked not to lift a cup to peek at its contents, five-year-olds tended to peek more than three- and four-year-olds.[41]

These research findings show that children are driven to lie primarily to conceal a transgression, thus confirming Darwin's first conclusion. These children could not be defined as "perfect liars,"[1] however, given that whereas they clearly seem to grasp the concept of "lie," and what "lie" means, they still tend to confuse it with the concept of "moral prohibition."[42] This observation could indicate that even if preschool-aged children can deceive others, they do not necessarily discern deception from the idea of ignoring an interdiction, nor do they fully understand what lying is or how it is done, and therefore cannot make use of effective deception in a social context.[1]

Current knowledge implies that the age at which children lie most is usually between four and six years old, when their ability to support a lie via appropriate body language and acting develops and sharpens.[1] By the time a child reaches school-age, they tend to lie more often, as well as more convincingly than before. As their vocabulary increases, and they begin to understand better how other people think and reason, their fabrications also become more sophisticated.

Most children aged five and younger, and about half of six- to seven-year-olds are incapable of successfully lying strategically when responding to follow-up questions.[26] During these follow-up episodes, they are often caught blurting out the name of the toy, or otherwise betraying the truth, hence conflicting with their initial position that they had not played with or looked at the toy,[26,39,43] and as such, confirming Darwin's second conclusion. The same experiment, conducted with older school-age children provided different results: these children proved able to lie strategically by pretending to ignore the identity of the toy or otherwise offering convincing explanations for their knowledge of the toy.[34] This observation could suggest that alongside their mere ability to tell lies, younger children have an extremely limited aptitude to lie strategically, and they acquire this ability only in the elementary school years. In similar experiments, Talwar and colleagues (2007) found that children's cheating behavior decreases into early adolescence.[34] They used a modified *Temptation Resistance paradigm* to examine 6–11-year-olds' cheating behaviors by asking the children not to peek at the answer to a test question and found that cheating behavior decreased with development. Correspondingly, using a similar *Temptation Resistance paradigm,* Evans and Lee (2011) found that children's cheating behavior decreased between 8 and 16 years of age.[41] Collectively, these studies suggest a developmental trend in cheating behaviors from late childhood into early adolescence, specifically a noticeable decrease.

3.2 Cerebral Structures Mediating Lying in Children

A human being's nervous system develops thanks to the interaction between several concurrent mental processes, some of which are finalized before birth, whereas others continue well into adulthood.[1] The time spans needed for neural and synaptic formation and elimination differ considerably according to the diverse cortical areas involved.[26,44] They initially take place around primary motor and sensory areas and subsequently involve the prefrontal cortex.[26]

Current research findings therefore imply that children (especially preschool-aged children) are capable of lying even though prefrontal cortex development remains incomplete. Some tentatively associate the rudimentary prefrontal control with the fact that children cannot discriminate between skillful or unskillful lying nor can they effectively choose between lying or telling the truth. This speculative, though important, possibility remains to be ascertained in future neurodevelopmental research designed to clarify the role of the prefrontal cortex in deception.

3.3 Cognitive Functions Mediating Lying in Children

Available research on the topic suggests a correlation between a child's ability to tell lies and the said child's Theory of Mind understanding and executive functioning.

3.3.1 *Theory of Mind Understanding*

Studies analyzing the relationship between Theory of Mind (ToM) understanding and lie telling in children led researchers to formulate two hypotheses: the first, ToM1, suggested a relationship between children's lie telling and their first-order belief understanding.[43] Under this hypothesis, successful lying requires a person deliberately to produce or create a false belief in the mind of another. Several studies underline acts of deception, such as lying, as early signs indicating a child's understanding of belief and false belief.[26,33,34] The second hypothesis, ToM2, assumes a relationship between a child's propensity to uphold a lie and their notion of "second-order belief." Research has demonstrated that second-order mental state understanding (e.g., Peter knows that Sally thinks it is raining) rarely becomes apparent until about six years of age and undergoes steady development well into adolescence.[26,34]

Overall, experimental studies have shown that whereas initial false denials sometimes reflect children's first-order belief understanding (ToM1 Hypothesis), the ability to remember and maintain a lie in follow-up questions can relate to a child's faculty to envisage another's beliefs and what the other will infer from any information received by the child (ToM2 Hypothesis).[34,39,43] No studies have yet examined these hypotheses concurrently: one study directly examined the ToM1 Hypothesis with three- and five-year-old children,[43] whereas another study examined the ToM2 Hypothesis in children seven years old and older.[34] Nor have published studies addressed the development of children's lie-telling behavior between three and eight years of age, when ToM understanding undergoes critical changes, or the relationship between children's first-order and second-order belief understanding and their actual lie-telling behavior.

3.3.2 *Executive Functioning*

Lying involves various frontal skills, including set shifting, working memory, inhibition process (or the act of shutting out the truthful response) and conflict monitoring ("conflict" here meaning the conflict between the automatic truthful response and the lie response).[45,46] These cognitive functions entail "high-functioning frontal skills" and take place mainly in the ACC, a frontal brain area involved in action-monitoring and resolving problems with conflicting responses.[22] In practice, to maintain their lies, children must inhibit those thoughts and statements that are contrary to their lie and would otherwise betray their infraction, while keeping in memory the contents of their lie.

Prompted by early observations by Carlson and colleagues (1998), others noted that children of preschool age who encountered difficulty with executive functioning tasks (and especially tasks requiring a high level of inhibitory control) also appeared to face difficulties with the act of physical deception (i.e., pointing).[47] Although Carlson and colleagues (1998) did not specifically study lie-telling behavior, their results seem to imply that children may also have difficulties with lying when they lack advanced executive functioning skills, particularly inhibitory control and working memory. No study has been designed to test this hypothesis directly.

Collectively, these findings clearly indicate that the ability to lie develops early in children, and with increasing age, cognitive skill, and prefrontal cortex development, they later become skillful liars.[26] Whereas much of the foregoing research centers on cognitive ability and suggestibility in children,[48] less interest focuses on children's lying and truth-telling behavior. Research to date has yielded inconsistent findings. The discrepancy predominantly stems from the complex testing materials used that confused the children's linguistic and mnemonic skills with their lying and truth-telling competence. These methodological drawbacks led researchers to underestimate children's lying and truth-telling behaviors. Further research is needed to study, in a straightforward and ecologic manner, the relationships between other cognitive abilities and the development of children's lie-telling abilities.

4. NEUROIMAGING STUDIES OF DECEPTION

The development of neuroimaging techniques has enabled researchers to measure directly brain activity associated with various cognitive functions, the neural correlates that subtend cognition, emotion, and behavior.[49–51] To do so, they compared brain activity during an experimental task and a control task designed to test only the cognitive feature under study.

The initial neuroimaging studies hypothesized that, compared with tasks generating truthful responses, tasks generating lying responses would be associated with greater fMRI BOLD activity in the DLPFC.[21] Moreover, the concomitant inhibition of relatively preponderant truthful responses would be associated with greater activation in the ventrolateral prefrontal cortex (VLPFC), brain regions known to be involved in response inhibition.[52]

Neuroimaging studies have provided previously unobtainable insights into how cognitive load differs during the various types of lying, specifying minimal load when subjects simply deny a fact that actually happened but a high load when they fabricate complex lies.[53]

The first set of neuroimaging studies used a modified version of the Guilty Knowledge Task, a questioning technique that has been extensively used in the forensic field. In the studies conducted by Langleben and colleagues,[22,54] the first researcher to use this paradigm in an fMRI experiment, participants were given a playing card (ace of hearts) and were instructed to deny their possession of the card when later queried. Once in the scanner, participants were shown a series of playing cards and were asked whether they had each card. They responded truthfully to all the cards except the ace of hearts. The fMRI results showed greater BOLD activation in the ACC, superior frontal gyrus, and anterior parietal cortex during deceptive, rather than during truthful, responses (Table 1). This experimental procedure provided only minimal insight into deception as a psychological process because the deceptive response was confounded by recognition memory effects for the relevant playing card.

Subsequently, several neuroimaging studies have used different deception tasks, asking participants about autobiographical information or specific personal experiences or both, and prompting them to respond incorrectly to a subset of the questions.[53,55] For example, in the study conducted by Spence and colleagues (2008), participants had to recount past personal events that they would typically intend to conceal because the experience was embarrassing.[56] In other studies, the authors focused on deception in relation to recently completed action events,[21,57] or to complex and structured situations, such as in the mock crime tasks, in which the subjects were required to really steal an object (a watch or a ring) and subsequently lie about it.[58] Finally, subjects could be required to lie about a piece of news recently acquired through passive experience. For instance, they had to lie about an object's location[59] or to lie about the positive or negative valence of the emotion expressed by observed faces.[60]

Despite the different experimental protocols used across previous studies, the area invariably associated with deception is the frontal executive system.[1,6] Frequently activated areas include the DLPFC, the VMPFC, the anterior prefrontal cortex (aPFC), and the ACC (Figure 1). In all these paradigms, researchers compared brain activation during a truth condition (participants

TABLE 1 Neuroimaging Studies on Deception. Studies Are Clustered According to the Paradigm Used and Were Selected If They Reported a Whole Brain Analysis

First Author	Year	Task
MODIFIED GKT PARADIGM		
Langleben et al.[54]	2002	Modified GKT paradigm
Langleben et al.[22]	2005	Modified GKT paradigm
Phan et al.[105]	2005	Modified GKT paradigm
Monteleone et al.[106]	2008	Modified GKT paradigm
Ganis et al.[107]	2011	GKT paradigm
AUTOBIOGRAPHIC INFORMATION		
Lee et al.[36]	2002 (E2)	Autobiographic memory task
Ganis et al.[53]	2003	Questions about work experiences
Spence et al.[3]	2004	Vocal response to auditory stimuli
Nunez et al.[108]	2005 (E2)	Answer to autobiographic questions
Abe et al.[55]	2007	Answer to autobiographic questions
Spence et al.[56]	2008	Answer to autobiographic questions
Ganis et al.[2]	2009 (E1)	Respond to self-related statement
Ding et al.[109]	2012	Lie about own identity
Markewka et al.[110]	2012 (E2)	Answer to autobiographic questions
RECENT ACTION OR EVENT		
Spence et al.[21]	2001	Answer to questions
Kozel et al.[58]	2005	Mock crime
Abe et al.[111]	2006	Look at video depicting events
Kroliczak et al.[112]	2007	Grasping task
Fullam et al.[57]	2009	Judging 36 acts
Baumgartner et al.[102]	2009	Trust game
Kozel et al.[113]	2009	Mock sabotage crime
Kozel et al.[114]	2009	Mock sabotage crime
Mohamed et al.	2011	Mock shooting situation
RECENT KNOWLEDGE		
Lee et al.[36]	2002 (E1)	Digit memory task
Kozel et al.[59]	2004	Lie about object's location
Kozel et al.[115]	2004	Lie about object's location
Nunez et al.[108]	2005 (E1)	Answer to general knowledge questions
Abe et al.[116]	2008	Word recognition
Browndyke et al.[117]	2008	Purposefully malingered and normal recognition memory performances
Bhatt et al.[118]	2009	Unfamiliar face recognition
		Familiar face recognition
Ganis et al.[2]	2009	Respond to other related statement
Lee et al.[119]	2009	Word recognition

TABLE 1 Neuroimaging Studies on Deception. Studies Are Clustered According to the Paradigm Used and Were Selected If They Reported a Whole Brain Analysis—cont'd

First Author	Year	Task
RECENT KNOWLEDGE		
Lee et al.[60]	2010	Valence evaluation of IAPS stimuli
Sip et al.[120]	2010	Interactive game
Ito et al.[62]	2011	New/old judgment
McPherson et al.[121]	2011	Forced two-choice task involving tones and words
Ito et al.[122]	2012	Living/nonliving evaluation
Kireev et al.[123]	2012	Judging arrows directions
Markewka et al.[110]	2012 (E1)	General knowledge questions

GKT, Guilty Knowledge Task; E1, Experiment 1 described in the original paper; E2, Experiment 2 described in the original paper.

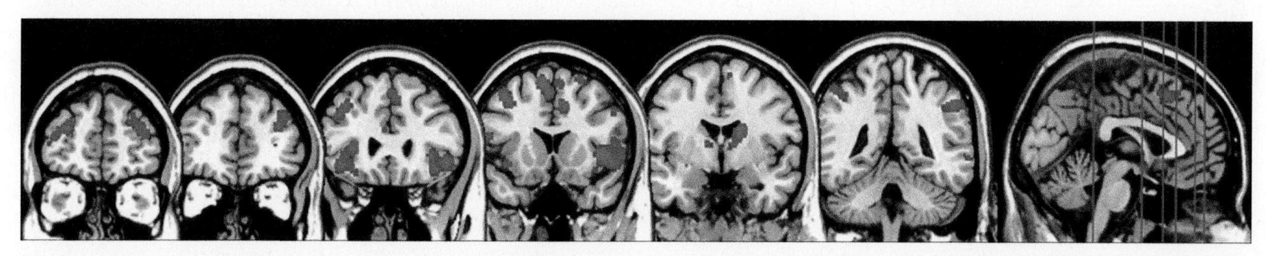

FIGURE 2 **Brain regions that were more active during the deception task than during the control task.** Results are shown in coronal sections. Sections 1 and 2 show the activation of the bilateral DLPFC. Section 3 shows the activation of the anterior insula, in the DLPFC and in the medial PFC. Section 4 shows the activation in the insular cortex (more pronounced in the right hemisphere) in the DLPFC and in the ACC. Section 5 shows the significant activation in the right caudate. Section 6 shows the bilateral activation in the inferior parietal lobule.

responded correctly to the presented stimuli) with brain activation during a deception condition (participants intentionally responded incorrectly to stimuli). Despite these similarities, the specific methodology used to generate a deceptive response varied across studies and involved differences in context, motivation, spontaneity of behavior, and response modalities. Given this variability and the difficulty in replicating results, the findings varied widely across studies.

To assess the results more objectively for interstudy consistency, we included neuroimaging studies of deception published to date in a coordinate-based Activation Likelihood Estimation (ALE) meta-analysis.[61] This approach aims to identify areas in which reported coordinates converge across experiments at a higher rate than expected from a random spatial association. We included only studies that used fMRI, analyzed whole-brain fMRI data, analyzed group data for healthy participants, used a statistical analysis that clearly distinguished locations showing greater activation for deceptive than for a control task, and reported results on a standardized coordinate space. These criteria identified 35 studies as eligible for inclusion in the meta-analysis (Table 1). Together, these studies reported 471 activation foci obtained from 48 individual experiments (with a "study" referring to a paper, and "experiment" referring to an individual contrast reported in this paper), representing regions showing significantly greater activation in the deception task than in the control task. The ALE meta-analysis identified several brain regions that showed greater activation during deception than during a control task (Figure 2). These areas included the inferior and middle frontal gyrus, together referred to as the DLPFC, the anterior insula, the right nucleus caudatus, the ACC, and the VMPFC.

Confirming the DLPFC as a key brain region involved in deception, in 35 of the 48 experiments included in our meta-analysis, ALE identified DLPFC activation during the deception task. Previous investigators proposed that the DLPFC might be responsible for the executive aspects of deception, regardless of the task used.[62] For instance, in an experiment investigating the neural correlate of deception in a memory task for emotional and neutral events, fMRI showed DLPFC activation during the deceptive response in both the emotional and neutral scenarios, i.e., regardless of the emotional valence of the memory content.[62] The various brain areas activated during deceptive tasks seem to have specific roles in deception. Whereas the ACC is thought to be involved in suppressing the default truth response, the DLPFC is thought to be involved in producing a credible substitution (the lie). The anterior insula intervenes in controlling visceral responses that usually accompany deceptive behavior. The VMPFC probably intervenes in processing emotional stimuli, a brain process that is active during deceptive behavior as it is in other real-life situations. The functional role of the nucleus caudatus in deceptive behavior is still unclear, and calls for further research. However, it has been suggested that the higher activation of the caudate might reflect the performance monitoring.[36] Indeed, a previous research[64] found that the caudate activation correlates with different measures of task inhibition. Supporting this hypothesis, Amos and coworkers (2000)[64] reported increased errors in performance when striatal activations were reduced, as observed, for instance, in diseases of the striatal system (i.e., Parkinson's and Huntington disease). These findings suggest an important role for the caudate in the inhibition of the usual response when a person is trying to deceive.

4.1 Is Deception an Isolated Function?

Evidence from neuroimaging studies implicating the prefrontal cortex and nearby regions in deception supports the hypothesis that lying involves the executive control system. Executive control refers to a set of higher order cognitive processes that allow for flexible changes in thought and behavior in response to changing cognitive and environmental contexts.[65]

In their study on complex executive functions, Miyake (2000) suggested conceptualizing executive control as comprising at least three different processes: working memory, task switching, and inhibitory control.[66] Others suggest that all three processes contribute to deception to the extent that deceit requires: keeping the truth in mind while formulating a truthful response (working memory), suppressing a truthful response (inhibitory control), and switching between truthful and deceptive responses (task switching).[3]

Executive control during deception has been experimentally explored in a meta-analysis conducted by Christ and colleagues (2009).[24] The authors identified the brain regions involved in deception and evaluated whether different regions implicated in deception were engaged in different aspects of executive control. To do so, they contrasted the deception meta-analysis with additional meta-analyses generated on studies focusing on working memory, inhibitory control, and task switching. The results showed a significant overlap among all the three executive control maps in portions of the bilateral VLPFC, the left DLPFC, the left ACC, and the left posterior parietal cortex (Figure 3).[24]

A new finding that provided a greater insight into the brain areas modulating deception was that the executive control meta-analysis and the deception meta-analysis overlapped in the bilateral anterior insula, the left inferior frontal gyrus, the left middle frontal gyrus, the right ACC, and the right intraparietal sulcus.[24] The overlap between regions involved in deception and those underlying executive control suggests that the aforementioned cognitive processes may indeed contribute to the psychological phenomenon of deception. The results showed that frontal regions mediate task set maintenance and implement moment-by-moment cognitive control.[67] The insula specializes in control of visceral functions and representation of the body's interoceptive state.[68] Early observations already showed that deceptive responses often lead to visceral responses (i.e., changes in blood pressure, heart rate, and breathing rate).[69] Finally, in none of the three executive control meta-analyses did deception-related activation in parietal lobules overlap, suggesting that their involvement in deception may be related to neurocognitive processes other than working memory, inhibitory control, or task switching. Parietal lobule activation in deception might play a role in maintaining attention to the environmental context so as to detect and respond appropriately when instances requiring deception arise.[24]

A literature overview on neuroimaging studies of deception underlines considerable agreement on which brain areas are more active during lying than during truth telling. One of the main obstacles to applying the foregoing reported findings in the real world is that lies examined in laboratory settings differ distinctly from those we would customarily try to detect in ecological settings. Although researchers have been concerned with real-world effectiveness and have presented ecologically valid tasks, such as the "mock crime" task, experimental paradigms still differ in many important ways from the situations in which lie detection would be used in the real world. In laboratory studies, subjects lie because they are instructed to lie, and they lie about matters with little or no personal relevance in highly constrained and contrived situations. In addition, the familiarity of the information being concealed and the level of emotion associated with it are typically much lower in laboratory studies than in real life. Real-world deception is likely to be highly emotional and personally relevant. In fMRI studies, emotion has been shown to alter the neural circuitry of memory,[70] inhibition and cognitive control,[71] and working memory interference,[72] all of which are processes that underlie brain activation during deceptive responses. Hence, the currently available neuroimaging studies on deception are only a first step toward studying such a complex cognitive process.

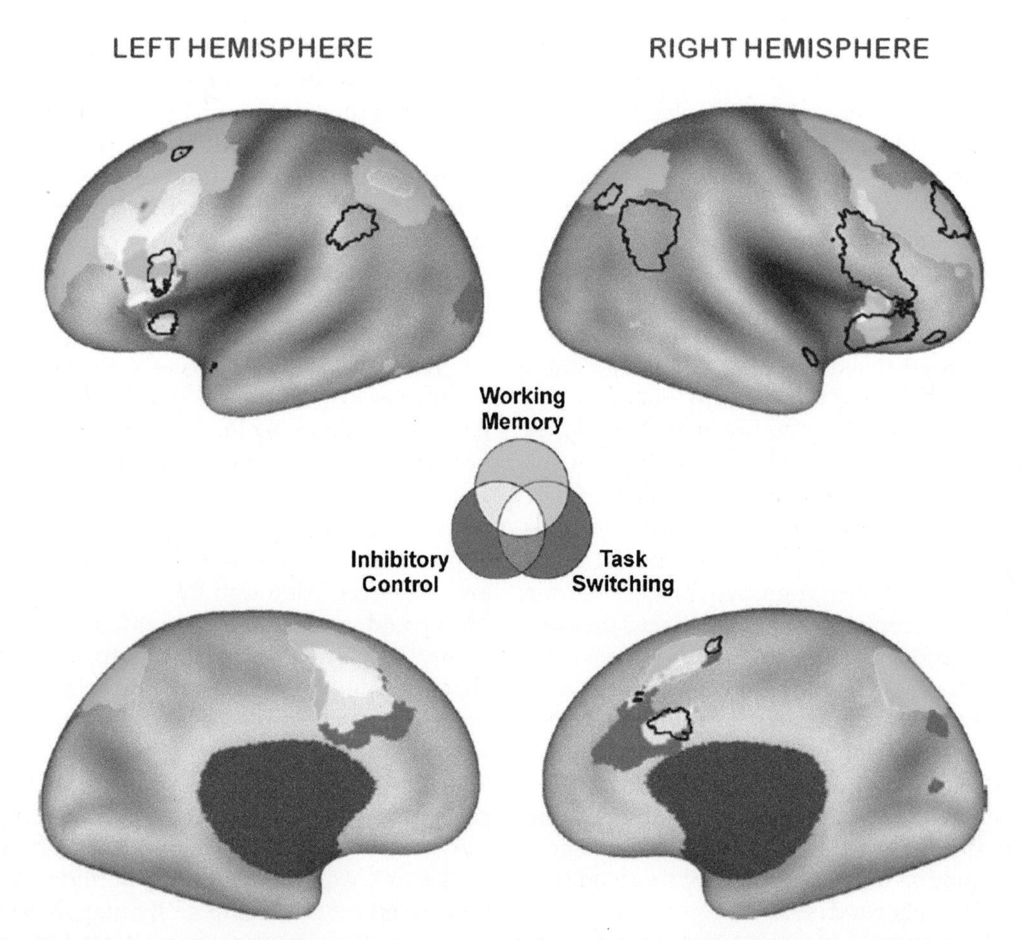

FIGURE 3 **Results of the executive functions components meta-analyses and deception meta-analyses (black borders).** Results of the working memory (green), inhibitory control (red), task switching (blue), and deception (black borders). The intersection between executive functions is located in the right ACC (by Christ et al., 2009).

5. BRAIN STIMULATION STUDIES OF DECEPTION

In recent years, transcranial direct current stimulation (tDCS) and transcranial magnetic stimulation (TMS) have been used to examine causal effects in the brain areas mediating deceit or truth telling.[73] Both procedures have been used to assess the validity of brain imaging findings and to attempt to establish a direct correlation between activity in a cortical region and deceptive behavior by transiently inhibiting cortical excitability.[73]

5.1 Studies with Transcranial Direct Current Stimulation

Transcranial direct current stimulation (tDCS) is a noninvasive technique able to manipulate brain neuroplasticity and modulate cortical function by delivering weak direct currents.[74] It induces long-term potentiation (LTP)-like synaptic changes that facilitate cortical excitability, neuroplasticity, and learning.[75,76] Anodal tDCS increases, whereas cathodal tDCS decreases, excitability in the stimulated areas.[74]

As earlier research predicted, tDCS can effectively balance cognitive functions, such as memory, attention, learning, and language, with motor functions within healthy and in clinical populations.[77] The first systematic study describing tDCS-modulated lying[46] showed that tDCS (1.5 mA for 10 min) applied on the DLPFC (F3/F4 according to the 10–20 Electroencephalogram System) interferes with deceptive responses. Truthful and deceptive abilities were tested with a Guilty Knowledge Task using a computer-controlled exercise. All participants were asked to perform tasks before and immediately after anodal and cathodal tDCS and sham sessions (*offline procedure*). The results showed that focal changes in DLPFC excitability after anodal tDCS can increase deceptive responses in speed and efficiency. This finding explains how anodal tDCS leads to an "anodal block" and therefore impairs inhibitory function in the DLPFC.

To interfere with deceptive responses, Karim and colleagues (2010) modulated excitability in the aPFC.[78]

In a double-blind, repeated-measures design, subjects received cathodal, anodal, or sham tDCS (1mA for 13min) during a role play interrogation (*online procedure*) using a modified version of the Guilty Knowledge Task. The results showed that focal changes in aPFC excitability induced by cathodal tDCS failed to impair deceptive behavior but rather facilitated it, as measured by faster reaction times, a decreased sympathetic skin conductance response, decreased feelings of guilt, and an increased behavioral pattern during skillful lying. The investigators proposed that cathodal tDCS, a technique known to suppress cortical excitability, relieved the moral conflict brought about by the act of deception—as represented in the aPFC—thus facilitating deception.

In a double-blind, randomized, sham-controlled study conducted in our laboratory, Mameli and colleagues (2010) investigated whether bilateral DLPFC–tDCS specifically influences cognitive processing for general knowledge deception or also influences the construct of personal information deception.[45] To do so, we administered a modified Guilty Knowledge Task during a tDCS session (*online procedure*). We found that after anodal tDCS (2mA for 15min) applied on the DLPFC, reaction times decreased significantly but did so only for lies involving general knowledge. These findings show that tDCS specifically modulates deceptive responses for general information deception, leaving those on personal information unchanged.

Finally, in a pseudo-randomized study and sham-controlled study, Fecteau et al. (2013) applied tDCS over the DLPFC to interfere with untruthful answers in a variety of contexts (including deceiving on guilt-free questions about daily activities, generating previously memorized lies about past experiences, and producing spontaneous lies about past experiences) and response modalities (verbal and motor responses).[79] The experimental design included three electrode arrangements: right anodal/left cathodal, right cathodal/left anodal, and sham stimulation (2mA for 20min). Participants completed tasks before and immediately after the tDCS session (*offline procedure*). The results showed that real, but not sham, tDCS over the DLPFC reduces response latency for untruthful over truthful answers across various contexts and response modalities. Also, hemispheric laterality differs according to the deception context.

5.2 Studies with Transcranial Magnetic Stimulation

TMS with targeted magnetic fields that temporarily disrupt neural processing in the focal area can alter brain activity in a specific cortical region.[73] By measuring small but functionally important changes in behavior, this technique therefore allows researchers to study functioning in a certain brain area in relation to an existing behavior. Repetitive TMS (rTMS) alters cortical excitability either increasing or decreasing it, depending on the stimulation variables. Stimulation at about 1Hz typically induces inhibition and stimulation and higher intensities elicits excitation.[80] For example, in their first study Lo and coworkers (2003) delivered TMS to the left and right motor cortices (at 110% of resting motor threshold) and recorded motor-evoked potentials (MEPs) before the question and immediately after questions that elicited a false response.[81] The task required subjects to give "yes" or "no" answers to questions involving simple lies or truths ("Are you a man?" or "Is the moon round?") or complex lies or truths ("How old are you?" or "How many legs does a dog have?"). The results suggested that generating a false response was associated with increased excitability in the corticospinal tract, perhaps reflecting increased motor readiness or a general arousal effect. Although this finding demonstrated that TMS could be used to detect differences in cortical excitability between deceptive and truthful responses, the study did not examine the neural substrate for deceptive processing by testing whether TMS could alter deceptive responding.

To test whether asymmetric functions in the prefrontal cortex influence spontaneous propensity to lying, Karton and colleagues conducted three experiments.[80,82,83] In the first experiment,[80] they applied rTMS to the left and right DLPFC during a "spontaneous lie task." After applying stimulation, they conducted behavioral testing in which the participating subjects could freely choose whether to lie or tell the truth. They applied rTMS to the left and right DLPFC and left and right parietal cortex (PC) as a control area. Subjects underwent two experimental sessions. In one session, before the lie task, they received a 1Hz rTMS train over the left or right DLPFC and in the other session over the same hemisphere of the PC. The lie task consisted in naming the color of the red and blue discs appearing in quasi-random order on a computer monitor. Participants were instructed to name the color correctly or just lie about it, naming the other color that was not presented in this trial, while being free to choose whether to lie. When subjects could name the color of the presented objects correctly or incorrectly at their free will, stimulation succeeded in manipulating the tendency to stick to truthful answers. Right-hemisphere DLPFC disruption decreased lying compared with the control condition, and left-hemisphere DLPFC disruption increased the propensity to lie.

In the second experiment, Karton and colleagues[83] tested whether by applying an opposite rTMS effect—facilitating the DLPFC—they could reverse the propensity to be more truthful to the propensity to produce relatively more untruthful responses during right-DLPFC stimulation and vice versa. Comparatively, they also tested this hypothesis for the left DLPFC and

developing children to cover up their initial lie; that is, children with ASD had difficulty exercising semantic leakage control (the ability to maintain consistency between their initial lie and subsequent statements).[97]

Other neurological conditions in which patients have difficulty in formulating lies include the movement disorders Parkinson's disease (PD) and essential tremor (ET).

Seeking information on the ability to tell a lie in patients with neurological conditions who have cognitive impairments in executive dysfunction, Abe and colleagues (2009) examined deception in patients with PD.[27] Others have described patients with PD as "honest" in that they tend not to deceive others.[98] Questioning these earlier conclusions, Abe (2009) assumed that, rather than choosing not to tell lies, patients with PD may actually have difficulty lying owing to cognitive deficits resulting from pathological changes in certain brain regions.[27] Their results showed that patients with PD had greater difficulty than healthy controls in making deceptive responses in an experimental novel lie task (a recognition memory test during which the participants were asked to tell the truth or a lie on material previously displayed).

Critically, resting-state 18F-fluorodeoxyglucose positron-emission tomography (PET) showed that this difficulty correlated significantly with decreased metabolic rates in the left dorsolateral and right anterior prefrontal cortices. These findings demonstrate that the supposed honesty found in patients with PD has a neurobiological basis, and they provide neuropsychological evidence of the brain mechanisms crucial for human deceptive behavior.[27]

Finally, in a study conducted in our laboratory, we investigated whether deceptive responses are also impaired in patients with ET, compared with PD patients and with a healthy control group.[28] Truthful and deceptive responses were evaluated using a computer-controlled procedure, a simplified version of the Guilty Knowledge Task.[46] Our results showed that patients with ET are less accurate than healthy subjects in producing deception (Figure 4). As expected in accordance with a previous report, the Guilty Knowledge Task disclosed a similar difficulty in producing deceptive responses in patients with PD.[27] These findings suggest that difficulty in lying is an aspecific cognitive feature in movement disorders characterized by fronto-subcortical circuit dysfunction, such as PD and ET. Collectively, findings in patients with PD and ET[27,28] confirm that prefrontal brain regions contribute critically to deception.

7. NEUROSCIENCE OF HONESTY AND MORAL JUDGMENT

Current research on deception and its neural correlates mainly involves subjects who were instructed to lie, so their behavior was not genuinely dishonest. What

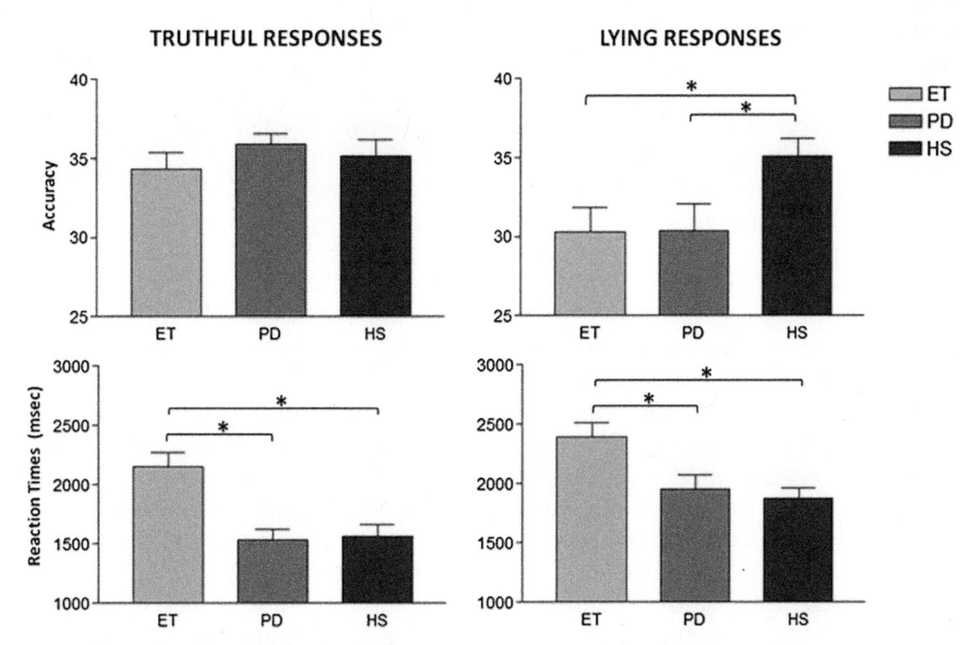

FIGURE 4 **Guilty Knowledge Task results in patients with essential tremor and Parkinson disease.** Accuracy (upper panel) and reaction times (RTs) (lower panel) for true (left) and lying (right) responses in the deception task (Guilty Knowledge Task). The histograms show mean task accuracy and RTs. Error bars are the standard error of the mean. *Significant difference (analysis of variance, $p<0.05$). ET, essential tremor; HS, healthy subjects; PD, Parkinson's disease (by Mameli et al., 2013).

does neuroscience know about the intention to lie? What happens in our brain when we judge dishonest behavior? In this section, we report studies that investigated the neural processes underlying deception from two different perspectives: the intention to lie and how to judge a dishonest behavior.

When the intention to lie emerges spontaneously, the cognitive, motivational, and emotional patterns involved differ from those underlying the lie itself. In an fMRI study, Greene and Paxton[99] (2009) recruited 35 healthy subjects undergoing a random computerized coin-flip task: in the "opportunity" condition, subjects made their prediction (head or tail) privately and received a reward based on their self-reported accuracy, affording them the opportunity to cheat; in the "no opportunity" condition, subjects recorded their predictions in advance, thus denying them the opportunity to cheat by lying about their accuracy. The investigators presented the experiments to subjects as a study on paranormal abilities to predict the future. They tested two competing hypotheses concerning the cognitive features of honesty, the Will hypothesis and the Grace hypothesis. The Will hypothesis postulates that honesty results from active resistance to temptation, a response controlled by cognitive processes that enable subjects to delay gratification. Conversely, the Grace hypothesis suggests that honesty results from an absence of temptation, behavior being determined by the presence or absence of automatic processes. In their fMRI study, Greene and Paxton found that behavioral performances differed for honest and dishonest subjects: reaction times in the opportunity loss condition were significantly slower in dishonest than in honest participants. fMRI detected increased activity in the control network (ACC/SMA, DLPFC, VLPFC, dorsomedial prefrontal cortex [DMPFC], right parietal lobe) in the dishonest group during the opportunity loss trials, whereas control network activity in the honest group remained unchanged. In honest subjects, lying activated the bilateral VLPFC. Behavioral and neuroimaging data support the Grace hypothesis, suggesting that honest decisions reflect the absence of temptation: honest subjects did not engage additional controlled cognitive processes when they chose to behave honestly and nor did their reaction times imply that they took longer than dishonest subjects to forgo opportunities for dishonest gain. Conversely, individuals who behaved dishonestly exhibited increased activity in control-related regions of the prefrontal cortex, both when choosing to behave dishonestly and on occasions when they refrained from dishonesty.

The results obtained in the study by Greene and Paxton[101] (2009) prompted another fMRI study[100] that used the same coin-flip task and investigated whether the Grace hypothesis is also associated with weak responses to anticipated rewards. In this study, 28 subjects performed the coin-flip task and a Monetary Incentive Delay (MID) task, during which they experienced brief delays before claiming monetary rewards of variable value. Because the nucleus accumbens specifically responds to anticipated reward, the investigators hypothesized that this brain structure responds also to dishonest behavior. Behavioral performances replicated the results obtained in the study by Greene and Paxton (2009). The fMRI BOLD signal during the tasks showed significantly higher nucleus accumbens activity during high-reward trials than during low-reward trials in the MID task. Nucleus accumbens responses also correlated positively with the frequency of dishonest behavior in the coin-flip task. Subjects with large nucleus accumbens responses to anticipated reward in the MID task required additional cognitive control (specifically in the bilateral DLPFC) to forgo available rewards in the coin-flip task. The investigators suggested that neural responses to reward are important cognitive and neurobiological determinants of dishonesty.

In the study by Abe and colleagues (2014),[101] during fMRI scanning, subjects read scenarios concerning events that could happen in real life and were asked to decide whether to tell a lie in relation to these events. Three types of scenarios were presented: harmful stories, in which the subject's decision to tell a lie served to harm the target; helpful stories, in which the subject's decision to tell a lie served to help the target; and control stories, in which the subject's decision was neither honest nor dishonest. Results showed that the subjects made more dishonest decisions in helpful stories than in harmful stories. fMRI data showed that deciding to tell the truth or a lie in the harmful and helpful stories, in contrast to control stories, elicited activity in several brain areas, including the left DLPFC. Dishonest decisions in harmful stories, in contrast to honest decisions in harmful stories, were associated with activity in the right temporo-parietal junction and the right medial frontal cortex. No regional activity discriminated between the honest and dishonest decisions made in helpful stories. These findings suggest that the neural basis for dishonest decisions changes according to whether lying serves to harm or to help the target.

In an fMRI study aiming to investigate brain correlates of the decision to break or to keep a promise, Baumgartner[102] (2009) used a deception paradigm that resembles an economic exchange situation: when the task began, subjects had to make a promise decision indicating whether they always, mostly, sometimes, or never planned to be trustworthy. After the promise, the exchange took place and subjects could decide whether they wanted to keep or break the promise. Three different processes were studied: promising, anticipating the effect of promise on the exchange partner's decision to trust, and decision-making, during which subjects had to decide whether to keep or break the promise. According to task performance, and specifically according to whether subjects kept or did not keep their promise, they were subdivided into two groups: honest and dishonest. Brain imaging data showed highly specific brain activation patterns that distinguished subjects who broke a promise and those who kept a promise. During a promise decision, in dishonest subjects (unlike honest ones), fMRI showed increased activation in the ACC and the fronto-insular cortex, demonstrating that subjects who behave dishonestly already form their intent to break the promise during the promise stage. During anticipation, when no decision related to the dishonest or honest act had to be made, fMRI in dishonest subjects showed increased brain activation in the right anterior insula and right inferior frontal gyrus. Because these brain areas are associated with uncertainty and anticipation of a stressful and emotional situation, the investigators suggested that increased activation in these areas anticipates a dishonest act. Finally, in honest subjects, the decision stage is related to ventral striatum BOLD activation, which reflects inhibition of the impulse to answer honestly.

In another study on the propensity for deceptive behavior with resting EEG,[103] 50 healthy subjects performed the same task described in Baumgartner et al.[102] (2009), and a 64-electrode EEG was recorded at rest. Baseline EEG activation showed that the higher the baseline activation level in the anterior insula, the lower was the individuals' propensity for deceptive behavior. Baumgartner et al. hypothesized that a high baseline activation level in the anterior insula might predispose individuals to be honest, given that a hyperactive emotional system makes a deceptive act too stressful and bothersome.

Only one study investigated the neural bases underlying choices to behave honestly or dishonestly.[104] This study aimed to investigate the neural correlates of moral judgment in anti- and prosocial lying during fMRI scanning. The subjects read scenarios similar to those used by Abe (2014)[101] concerning events that could happen in real life and that end with dishonest or honest protagonist acts. Two types of scenario were presented: harmful stories, in which the protagonist's dishonesty would harm the listener and the protagonist would gain some type of benefit, and helpful stories, in which the protagonist's dishonesty would help the listener and the listener would gain some type of benefit. Each story could have an honest or a dishonest end. Subjects had to judge whether the protagonist's act was morally appropriate. Neuroimaging results showed that antisocial lying was judged to be morally inappropriate and led to increased BOLD activity in the right VMPFC, the right middle frontal gyrus, the right precuneus/posterior cingulate gyrus, the left posterior cingulate gyrus, and the bilateral temporo-parietal junction. Prosocial lying was judged to be morally appropriate and led to increased BOLD activity in the right middle temporal gyrus, the right supramarginal gyrus, and the left middle cingulate gyrus. These findings suggest that the cognitive and neural processes modulating the moral judgment of lying differ according to whether the lie serves to harm or benefit listeners.

In conclusion, these neuroimaging studies disclose several features that help characterize the intention to lie. Dishonesty arises through a complex neural circuit that includes brain control network areas (ACC/SMA, DLPFC, VLPFC, DMPFC, and the right parietal lobe),[99] reward-related brain structures (nucleus accumbens),[100] areas related to inhibition of the impulse to answer honestly (ventral striatum),[102] and areas associated with uncertainty and anticipation of a stressful and emotional situation (right anterior insula and right inferior gyrus).[102,103] The neural structures that modulate dishonest decisions depends on whether lying serves to harm or benefit: the bilateral temporo-parietal junction, the right medial frontal cortex, the right middle frontal gyrus, the right precuneus/posterior cingulate gyrus, and the left posterior cingulate gyrus are specifically involved in dishonest decisions in harmful stories.[101,104]

8. CONCLUSION

The past decade has witnessed increasing research outlining a cognitive neuroscience of deception.[1] Findings from neuroimaging, neurostimulation, and clinical studies converge on the conclusion that the prefrontal cortex plays a pivotal role in human honest behavior.[6,28,45,46,73] From a biological perspective, our findings and those of others suggest that deception engages the higher centers in the human brain and places certain demands upon the cognitive abilities of the individual who is lying.

From a psychiatric perspective, several researchers have attempted to clarify the neural correlates of pathological lying, and gradually accumulating evidence implicates specific neural correlates, such as changes in prefrontal volume.

Finally, studies on neural basis of honest behavior show that the neural circuits in the limbic and paralimbic systems include several brain control network areas (the ACC/SMA, DLPFC, VLPFC, DMPFC, and right parietal lobe).

To allow the cognitive neuroscience of deception to make further progress, future research needs to develop new experimental approaches based upon multiple techniques, including neuroimaging and neurostimulation.

8.1 Limitations in this Field of Research

Studies in this field need to be expanded and intensified to deepen knowledge of the neural correlates of deception. Information is lacking to show how multiple brain areas, such as the frontal cortex and subcortical areas, interact with each other in deceptive behavior. In some contexts, cognitive and neural processes may integrate in a decision to tell a lie, whereas in more demanding, difficult situations, the same processes might conflict.[6,28,45,46,73]

Equally important, no research has yet clarified the conditions that facilitate or inhibit people's tendencies to be honest or dishonest in a given social context. For example, deception can be facilitated by the amount of possible gain and inhibited by the risk of being revealed. Clarifying the facilitating and inhibitory factors for deception and their neural correlates would help to understand how the brain processes decisions about whether to tell a lie in a complex social interaction.

References

1. Abe N. How the brain shapes deception: an integrated review of the literature. *Neuroscientist*. 2011;17(5):560–574.
2. Ganis G, Morris RR, Kosslyn SM. Neural processes underlying self- and other-related lies: an individual difference approach using fMRI. *Soc Neurosci*. 2009;4(6):539–553.
3. Spence SA, Hunter MD, Farrow TF, et al. A cognitive neurobiological account of deception: evidence from functional neuroimaging. *Philos Trans R Soc Lond B Biol Sci*. 2004;359(1451):1755–1762.
4. DePaulo BM, Kashy DA, Kirkendol SE, Wyer MM, Epstein JA. Lying in everyday life. *J Pers Soc Psychol*. 1996;70(5):979–995.
5. Vrij A. *Detecting Lies and Deceit: Pitfalls and Opportunities*. Chichester: John Wiley and Sons Ltd; 2007.
6. Abe N. The neurobiology of deception: evidence from neuroimaging and loss-of-function studies. *Curr Opin Neurol*. 2009a;22(6):594–600.
7. Zuckerman M, DePaulo BM, Rosenthal R. Verbal and nonverbal communication of deception. In: Berkowitz L, ed. *Advances in Experimental Social Psychology*. New York: Academic Press; 1981;14:1–57.
8. Buller DB, Burgoon JK. Interpersonal deception theory. *Commun Theory*. 1996;6:203–242.
9. Ekman P. *Telling Lies: Clues to Deceit in the Market-Place, Politics and Marriage*. New York: Norton; 1985.
10. Vrij A, Fisher R, Mann S, Leal S. Detecting deception by manipulating cognitive load. *Trends Cogn Sci*. 2006;10(4):141–142.
11. Gombos VA. The cognition of deception: the role of executive processes in producing lies. *Genet Soc Gen Psychol Monogr*. 2006;132(3):197–214.
12. Walczyk JJ, Harris LL, Duck TK, Mulay D. A social-cognitive framework for understanding serious lies: activation—decision—construction—action theory. *New Ideas Psychol*. 2014;34:22–36.
13. Walczyk JJ, Mahoney KT, Doverspike D, Griffith-Ross DA. Cognitive lie detection: response time and consistency of answers as cues to deception. *J Bus Psychol*. 2009;24:33–49.
14. Walczyk JJ, Roper KS, Seemann E, Humphrey AM. Cognitive mechanisms underlying lying to questions: response time as a cue to deception. *Appl Cogn Psychol*. 2003;17:755–774.
15. Walczyk JJ, Schwartz JP, Clifton R, Adams B, Wei M, Zha P. Lying person-to-person about life events: a cognitive framework for lie detection. *Pers Psychol*. 2005;58:141–170.
16. Sporer SL, Schwandt B. Paraverbal indicators of deception: a meta-analytic synthesis. *Appl Cogn Psychol*. 2006;20:421–446.
17. Vendemia JM, Buzan RF, Simon-Dack SL. Reaction time of motor responses in two-stimulus paradigms involving deception and congruity with varying levels of difficulty. *Behav Neurol*. 2005;16(1):25–36.
18. Debey E, De Houwer J, Verschuere B. Lying relies on the truth. *Cognition*. 2014;132(3):324–334.
19. Vrij A. Interviewing to detect deception. *Eur Psychol*. 2014;19(3):184–194.
20. Sip KE, Roepstorff A, McGregor W, Frith CD. Detecting deception: the scope and limits. *Trends Cogn Sci*. 2007;12(2):48–53.
21. Spence SA, Farrow TF, Herford AE, Wilkinson ID, Zheng Y, Woodruff PW. Behavioural and functional anatomical correlates of deception in humans. *Neuroreport*. 2001;12(13):2849–2853.
22. Langleben DD, Loughead JW, Bilker WB, et al. Telling truth from lie in individual subjects with fast event-related fMRI. *Hum Brain Mapp*. 2005;26(4):262–272.
23. Lieberman MD. Social cognitive neuroscience: a review of core processes. *Annu Rev Psychol*. 2007;58:259–289.
24. Christ SE, Van Essen DC, Watson JM, Brubaker LE, McDermott KB. The contributions of prefrontal cortex and executive control to deception: evidence from activation likelihood estimate meta-analyses. *Cereb Cortex*. 2009;19(7):1557–1566.
25. Hala S, Russell J. Executive control within strategic deception: a window on early cognitive development? *J Exp Child Psychol*. 2001;80(2):112–141.
26. Talwar V, Lee K. Social and cognitive correlates of children's lying behavior. *Child Dev*. 2008;79(4):866–881.
27. Abe N, Fujii T, Hirayama K, et al. Do parkinsonian patients have trouble telling lies? The neurobiological basis of deceptive behaviour. *Brain*. 2009b;132(Pt 5):1386–1395.
28. Mameli F, Tomasini E, Scelzo E, et al. Lies tell the truth about cognitive dysfunction in essential tremor: an experimental deception study with the guilty knowledge task. *J Neurol Neurosurg Psychiatry*. 2013;84(9):1008–1013.
29. Darwin C. A biographical sketch of an infant. *Mind*. 1877;2:285–294.
30. Hartshone H, May MS. Studies in the Nature of Character. *Studies in Deceit*. New York: Macmillian; 1928;1.
31. Lewis M, Sullivan MW, Stanger C. Deception in 3-year-olds. *Dev Psychol*. 1989;25:439–443.
32. Binet A. Psychology of prestigitation. In: *Annual Report of the Board of Regents of Smithsonian Institution 1896*; 1894:555–571.
33. Sodian B. The development of deception in young children. *Br J Dev Psychol*. 1991;9:173–188.
34. Talwar V, Gordon HM, Lee K. Lying in the elementary school years: verbal deception and its relation to second-order belief understanding. *Dev Psychol*. 2007;43(3):804–810.
35. Piaget J. *The Moral Judgment of the Child*. New York: The Free Press; 1932.
36. Lee TM, Liu HL, Tan LH, et al. Lie detection by functional magnetic resonance imaging. *Hum Brain Mapp*. 2002;15(3):157–164.
37. Talwar V, Lee K, Bala N, Lindsay RC. Children's conceptual knowledge of lying and its relation to their actual behaviors: implications for court competence examinations. *Law Hum Behav*. 2002;26(4):395–415.
38. Talwar V, Lee K, Bala N, Lindsay RC. Children's lie-telling to conceal a parent's transgression: legal implications. *Law Hum Behav*. 2004;28(4):411–435.
39. Talwar V, Lee K. The development of lying of conceal a trasgression: children's control of expressive behavior during verbal deception. *Int J Behav Dev*. 2002a;26:436–444.
40. Xu F, Bao X, Fu G, Talwar V, Lee K. Lying and truth-telling in children: from concept to action. *Child Dev*. 2010;81(2):581–596.
41. Evans AD, Xu F, Lee K. When all signs point to you: lies told in the face of evidence. *Dev Psychol*. 2011;47(1):39–49.
42. Peterson CA, Peterson JL, Seeto D. Developmental changes in ideas about lying. *Child Dev*. 1983;54:1529–1535.
43. Polak A, Harris PL. Deception by young children following noncompliance. *Dev Psychol*. 1999;35(2):561–568.
44. Huttenlocher PR, Dabholkar AS. Regional differences in synaptogenesis in human cerebral cortex. *J Comp Neurol*. 1997;387(2):167–178.

45. Mameli F, Mrakic-Sposta S, Vergari M, et al. Dorsolateral prefrontal cortex specifically processes general - but not personal—knowledge deception: multiple brain networks for lying. *Behav Brain Res*. 2010;211(2):164–168.
46. Priori A, Mameli F, Cogiamanian F, et al. Lie-specific involvement of dorsolateral prefrontal cortex in deception. *Cereb Cortex*. 2008;18(2):451–455.
47. Carlson SM, Moses LJ, Hix HR. The role of inhibitory processes in young children's difficulties with deception and false belief. *Child Dev*. 1998;69(3):672–691.
48. Bussey K, Grinbeek EJ. Children's conceptions of lying and truth-telling: implication for child witnesses. *Leg Criminol Psychol*. 2000;5:187–199.
49. Pietrini P. Toward a biochemistry of mind? *Am J Psychiatry*. 2003;160(11):1907–1908.
50. Pietrini P, Alexander GE, Furey ML, Hampel H, Guazzelli M. The neurometabolic landscape of cognitive decline: in vivo studies with positron emission tomography in Alzheimer's disease. *Int J Psychophysiol*. 2000;37(1):87–98.
51. Pietrini P, Furey ML, Guazzelli M. In vivo biochemistry of the brain in understanding human cognition and emotions: towards a molecular psychology. *Brain Res Bull*. 1999;50(5–6):417–418.
52. Starkstein SE, Robinson RG. Mechanism of disinhibition after brain lesions. *J Nerv Ment Dis*. 1997;185(2):108–114.
53. Ganis G, Kosslyn SM, Stose S, Thompson WL, Yurgelun-Todd DA. Neural correlates of different types of deception: an fMRI investigation. *Cereb Cortex*. 2003;13(8):830–836.
54. Langleben DD, Schroeder L, Maldjian JA, et al. Brain activity during simulated deception: an event-related functional magnetic resonance study. *Neuroimage*. 2002;15(3):727–732.
55. Abe N, Suzuki M, Mori E, Itoh M, Fujii T. Deceiving others: distinct neural responses of the prefrontal cortex and amygdala in simple fabrication and deception with social interactions. *J Cogn Neurosci*. 2007;19(2):287–295.
56. Spence SA, Kaylor-Hughes C, Farrow TF, Wilkinson ID. Speaking of secrets and lies: the contribution of ventrolateral prefrontal cortex to vocal deception. *Neuroimage*. 2008;40(3):1411–1418.
57. Fullam RS, McKie S, Dolan MC. Psychopathic traits and deception: functional magnetic resonance imaging study. *Br J Psychiatry*. 2009;194(3):229–235.
58. Kozel FA, Johnson KA, Mu Q, Grenesko EL, Laken SJ, George MS. Detecting deception using functional magnetic resonance imaging. *Biol Psychiatry*. 2005;58(8):605–613.
59. Kozel FA, Padgett TM, George MS. A replication study of the neural correlates of deception. *Behav Neurosci*. 2004;118(4):852–856.
60. Lee TM, Lee TM, Raine A, Chan CC. Lying about the valence of affective pictures: an fMRI study. *PLoS One*. 2010;5(8):e12291.
61. Turkeltaub PE, Eickhoff SB, Laird AR, Fox M, Wiener M, Fox P. Minimizing within-experiment and within-group effects in activation likelihood estimation meta-analyses. *Hum Brain Mapp*. 2012;33(1):1–13.
62. Ito A, Abe N, Fujii T, et al. The role of the dorsolateral prefrontal cortex in deception when remembering neutral and emotional events. *Neurosci Res*. 2011;69(2):121–128.
63. Semrud-Clikeman M, Steingard RJ, Filipek P, Biederman J, Bekken K, Renshaw PF. Using MRI to examine brain-behavior relationships in males with attention deficit disorder with hyperactivity. *J Am Acad Child Adolesc Psychiatry*. 2000;39(4):477–484.
64. Amos A. A computational model of information processing in the frontal cortex and basal ganglia. *J Cogn Neurosci*. 2000;12(3):505–519.
65. Stuss DT. Biological and psychological development of executive functions. *Brain Cogn*. 1992;20(1):8–23.
66. Miyake A, Friedman NP, Emerson MJ, Witzki AH, Howerter A, Wager TD. The unity and diversity of executive functions and their contributions to complex "Frontal Lobe" tasks: a latent variable analysis. *Cogn Psychol*. 2000;41(1):49–100.
67. Dosenbach NU, Visscher KM, Palmer ED, et al. A core system for the implementation of task sets. *Neuron*. 2006;50(5):799–812.
68. Craig AD. Interoception: the sense of the physiological condition of the body. *Curr Opin Neurobiol*. 2003;13(4):500–505.
69. Cutrow RJ, Parks A, Lucas N, Thomas K. The objective use of multiple physiological indices in the detection of deception. *Psychophysiology*. 1972;9(6):578–588.
70. Phelps EA. Emotion and cognition: insights from studies of the human amygdala. *Annu Rev Psychol*. 2006;57:27–53.
71. Bush G, Luu P, Posner MI. Cognitive and emotional influences in anterior cingulate cortex. *Trends Cogn Sci*. 2000;4(6):215–222.
72. Levens SM, Phelps EA. Insula and orbital frontal cortex activity underlying emotion interference resolution in working memory. *J Cogn Neurosci*. 2010;22(12):2790–2803.
73. Luber B, Fisher C, Appelbaum PS, Ploesser M, Lisanby SH. Non-invasive brain stimulation in the detection of deception: scientific challenges and ethical consequences. *Behav Sci Law*. 2009;27(2):191–208.
74. Priori A. Brain polarization in humans: a reappraisal of an old tool for prolonged non-invasive modulation of brain excitability. *Clin Neurophysiol*. 2003;114(4):589–595.
75. Nitsche MA, Cohen LG, Wassermann EM, et al. Transcranial direct current stimulation: state of the art 2008. *Brain Stimul*. 2008;1(3):206–223.
76. Nitsche MA, Paulus W. Excitability changes induced in the human motor cortex by weak transcranial direct current stimulation. *J Physiol*. 2000;527(Pt 3):633–639.
77. Leal S, Vrij A. The occurence of eye blinks during a guilty knowledge test. *Psychol Crime Law*. 2010;16:349–357.
78. Karim AA, Schneider M, Lotze M, et al. The truth about lying: inhibition of the anterior prefrontal cortex improves deceptive behavior. *Cereb Cortex*. 2010;20(1):205–213.
79. Fecteau S, Boggio P, Fregni F, Pascual-Leone A. Modulation of untruthful responses with non-invasive brain stimulation. *Front Psychiatry*. 2013;3:97.
80. Karton I, Bachmann T. Effect of prefrontal transcranial magnetic stimulation on spontaneous truth-telling. *Behav Brain Res*. 2011;225(1):209–214.
81. Lo YL, Fook-Chong S, Tan EK. Increased cortical excitability in human deception. *Neuroreport*. 2003;14(7):1021–1024.
82. Karton I, Palu A, Joks K, Bachmann T. Deception rate in a "lying game": different effects of excitatory repetitive transcranial magnetic stimulation of right and left dorsolateral prefrontal cortex not found with inhibitory stimulation. *Neurosci Lett*. 2014b;583:21–25.
83. Karton I, Rinne JM, Bachmann T. Facilitating the right but not left DLPFC by TMS decreases truthfulness of object-naming responses. *Behav Brain Res*. 2014a;271:89–93.
84. Verschuere B, Schuhmann T, Sack AT. Does the inferior frontal sulcus play a functional role in deception? A neuronavigated theta-burst transcranial magnetic stimulation study. *Front Hum Neurosci*. 2012;6:284.
85. Dike CC, Baranoski M, Griffith EE. Pathological lying revisited. *J Am Acad Psychiatry Law*. 2005;33(3):342–349.
86. Healy W, Healy MT. *Pathological Lying, Accusation, and Swindling: A Study in Forensin Psychology*. Boston: Little, Brown, and Co.; 1915.
87. King BH, Ford CV. Pseudologia fantastica. *Acta Psychiatr Scand*. 1988;77(1):1–6.
88. Deutsch H. On the pathological lie (pseudologia phantastica). *J Am Acad Psychoanal*. 1982;10(3):369–386.
89. Yang Y, Raine A, Lencz T, Bihrle S, Lacasse L, Colletti P. Prefrontal white matter in pathological liars. *Br J Psychiatry*. 2005;187:320–325.
90. Hare RD, Hart SD, Harpur TJ. Psychopathy and the DSM-IV criteria for antisocial personality disorder. *J Abnorm Psychol*. 1991;100(3):391–398.

91. American Psychiatric Association. *Diagnostic and Statistical Manual of Mental Disorders* (DSM IV). 4th ed. Washington: APA; 1994.
92. Baron-Cohen S, Tager-Flusberg H, Cohen DJ. *Understanding Other Minds: Perspectives from Autism*. New York: Oxford University Press; 1994.
93. Leekam SR, Prior M. Can autistic children distinguish lies from jokes? A second look at second-order belief attribution. *J Child Psychol Psychiatry*. 1994;35(5):901–915.
94. Baron-Cohen S. Out of sight or out of mind? Another look at deception in autism. *J Child Psychol Psychiatry*. 1992;33(7):1141–1155.
95. Sodian B, Frith U. Deception and sabotage in autistic, retarded and normal children. *J Child Psychol Psychiatry*. 1992;33(3):591–605.
96. Yirmiya N, Solomonica-Levi D, Shulman C, Pilowsky T. Theory of mind abilities in individuals with autism, Down syndrome, and mental retardation of unknown etiology: the role of age and intelligence. *J Child Psychol Psychiatry*. 1996;37(8):1003–1014.
97. Li AS, Kelley EA, Evans AD, Lee K. Exploring the ability to deceive in children with autism spectrum disorders. *J Autism Dev Disord*. 2011;41(2):185–195.
98. Menza M. The personality associated with Parkinson's disease. *Curr Psychiatry Rep*. 2000;2(5):421–426.
99. Greene JD, Paxton JM. Patterns of neural activity associated with honest and dishonest moral decisions. *Proc Natl Acad Sci USA*. 2009;106(30):12506–12511.
100. Abe N, Greene JD. Response to anticipated reward in the nucleus accumbens predicts behavior in an independent test of honesty. *J Neurosci*. 2014;34(32):10564–10572.
101. Abe N, Fujii T, Ito A, et al. The neural basis of dishonest decisions that serve to harm or help the target. *Brain Cogn*. 2014;90:41–49.
102. Baumgartner T, Fischbacher U, Feierabend A, Lutz K, Fehr E. The neural circuitry of a broken promise. *Neuron*. 2009;64(5):756–770.
103. Baumgartner T, Gianotti LR, Knoch D. Who is honest and why: baseline activation in anterior insula predicts inter-individual differences in deceptive behavior. *Biol Psychol*. 2013;94(1):192–197.
104. Hayashi A, Abe N, Ueno A, et al. Neural correlates of forgiveness for moral transgressions involving deception. *Brain Res*. 2010;1332:90–99.
105. Phan KL, Magalhaes A, Ziemlewicz TJ, Fitzgerald DA, Green C, Smith W. Neural correlates of telling lies: a functional magnetic resonance imaging study at 4 Tesla. *Acad Radiol*. 2005;12(2):164–172.
106. Monteleone GT, Phan KL, Nusbaum HC, et al. Detection of deception using fMRI: better than chance, but well below perfection. *Soc Neurosci*. 2009;4(6):528–538.
107. Ganis G, Rosenfeld JP, Meixner J, Kievit RA, Schendan HE. Lying in the scanner: covert countermeasures disrupt deception detection by functional magnetic resonance imaging. *Neuroimage*. 2011;55(1):312–319.
108. Nunez JM, Casey BJ, Egner T, Hare T, Hirsch J. Intentional false responding shares neural substrates with response conflict and cognitive control. *Neuroimage*. 2005;25(1):267–277.
109. Ding XP, Du X, Lei D, Hu CS, Fu G, Chen G. The neural correlates of identity faking and concealment: an FMRI study. *PLoS One*. 2012;7(11):e48639.
110. Marchewka A, Jednorog K, Falkiewicz M, Szeszkowski W, Grabowska A, Szatkowska I. Sex, lies and fMRI—gender differences in neural basis of deception. *PLoS One*. 2012;7(8):e43076.
111. Abe N, Suzuki M, Tsukiura T, et al. Dissociable roles of prefrontal and anterior cingulate cortices in deception. *Cereb Cortex*. 2006;16(2):192–199.
112. Kroliczak G, Cavina-Pratesi C, Goodman DA, Culham JC. What does the brain do when you fake it? An FMRI study of pantomimed and real grasping. *J Neurophysiol*. 2007;97(3):2410–2422.
113. Kozel FA, Johnson KA, Grenesko EL, et al. Functional MRI detection of deception after committing a mock sabotage crime. *J Forensic Sci*. 2009;54(1):220–231.
114. Kozel FA, Laken SJ, Johnson KA, et al. Replication of Functional MRI Detection of Deception. *Open Forensic Sci J*. 2009;2:6–11.
115. Kozel FA, Revell LJ, Lorberbaum JP, et al. A pilot study of functional magnetic resonance imaging brain correlates of deception in healthy young men. *J Neuropsychiatry Clin Neurosci*. 2004;16(3):295–305.
116. Abe N, Okuda J, Suzuki M, et al. Neural correlates of true memory, false memory, and deception. *Cereb Cortex*. 2008;18(12):2811–2819.
117. Browndyke JN, Paskavitz J, Sweet LH, et al. Neuroanatomical correlates of malingered memory impairment: event-related fMRI of deception on a recognition memory task. *Brain Inj*. 2008;22(6):481–489.
118. Bhatt S, Mbwana J, Adeyemo A, Sawyer A, Hailu A, Vanmeter J. Lying about facial recognition: an fMRI study. *Brain Cogn*. 2009;69(2):382–390.
119. Lee TM, Au RK, Liu HL, Ting KH, Huang CM, Chan CC. Are errors differentiable from deceptive responses when feigning memory impairment? An fMRI study. *Brain Cogn*. 2009;69(2):406–412.
120. Sip KE, Lynge M, Wallentin M, McGregor WB, Frith CD, Roepstorff A. The production and detection of deception in an interactive game. *Neuropsychologia*. 2010;48(12):3619–3626.
121. McPherson B, McMahon K, Wilson W, Copland D. "I know you can hear me": neural correlates of feigned hearing loss. *Hum Brain Mapp*. 2011;33(8):1964–1972.
122. Ito A, Abe N, Fujii T, et al. The contribution of the dorsolateral prefrontal cortex to the preparation for deception and truth-telling. *Brain Res*. 2012;1464:43–52.
123. Kireev MV, Korotov AD, Medvedev CV. fMRI study of deliberate deception. *Fiziol Cheloveka*. 2012;38(1):41–50.

SECTION VI

BRAIN IMAGING AND SOCIETY

CHAPTER

17

The Henchman's Brain: Neuropsychological Implications of Authoritarianism and Prejudice

Kelsey Warner[1,3], *Daniel Tranel*[1,2], *Erik Asp*[3,4]

[1]Department of Neurology, University of Iowa, Iowa City, IA, USA; [2]Department of Psychology, University of Iowa, Iowa City, IA, USA; [3]Department of Psychiatry, University of Iowa, Iowa City, IA, USA; [4]Department of Psychology, Hamline University, St. Paul, MN, USA

OUTLINE

1. INTRODUCTION

From my childhood, obedience was something I could not get out of my system. When I entered the armed services at the age of 27, I found being obedient not a bit more difficult than it had been during my life to that point. It was unthinkable that I would not follow orders…A life predicated on being obedient and taking orders is a very comfortable life indeed. Living in such a way reduces to a minimum one's own need to think. *Otto Adolf Eichmann (1962)*[1]

Otto Adolf Eichmann was the ideal subordinate for Adolf Hitler in Nazi Germany. The greatest virtue to Eichmann was to follow orders, regardless of circumstance or personal aversion. This ruthless devotion to authority enabled him to orchestrate the mass deportation of Jews to ghettos and extermination camps during World War II. Nazi Heinrich Müller suggested, "If we had fifty Eichmanns, we would have won the war."[2] Upon reflection of the atrocities of the Holocaust, many have questioned what kind of person could do such inhuman and cruel acts. What factors enable an individual to obey authorities and harm others? Arguably some of the most important, infamous, and controversial studies in the history of psychology attempted to understand how people could commit horrific acts of aggression against others, such as the Holocaust. This chapter will review the classic and contemporary work on authoritarianism and prejudice toward outgroups. We will also bring recent neuropsychological and neuroimaging evidence to bear on the issue, arguing that certain neural regions are critical for resistance to authoritarian persuasion. Our review aims to understand the mind and brain of henchmen like Eichmann, those individuals who easily follow an authority's orders and are intent on harming others.

Eichmann was hanged for his crimes against humanity five days after Stanley Milgram[3] completed his first obedience study.[4] In a seminal and controversial research design, Milgram had an experimenter act as an authority figure to a naïve participant and a confederate. The participant was given the role of a "teacher" and was required to administer ostensibly real electric shocks to

Neuroimaging Personality, Social Cognition, and Character
http://dx.doi.org/10.1016/B978-0-12-800935-2.00017-8

the "learner" (the confederate). Each time the learner produced an incorrect response, the teacher needed to administer greater levels of shocks to the learner. After several trials, it became apparent to the participant that the learner was in distress. Appeals to the experimenter were met with a series of prods used to continue the experiment. Remarkably, most of the participants succumbed to the pressure of the experimenter: 62.5% of the subjects delivered the maximum amount of 450 volts (labeled XXX) and 80% gave shocks after the learner screamed, "Let me out of here! My heart's bothering me. Let me out of here!...Get me out of here! I've had enough. I won't be in the experiment anymore."[3] Milgram emphasized the critical role of situational factors in the likelihood to obey directions from an experimenter that ostensibly produced harm toward another human being.[4] Proximity of the victim or the experimenter, institutional context, and the validity of the authority all influenced obedience. For instance, in the "touch-proximity" experiment, the learner was seated directly next to the teacher. At the 150-volt level, the learner demanded to be left free and refused to place his hand on the shock plate. The experimenter then ordered the teacher to force the learner's hand down onto the plate. Milgram's rather banal conclusion was that this condition reduced obedience relative to the original experiment, as only 30% of the subjects delivered the maximum shock amount. However, the most disquieting and unsettling fact from this setup is the simple percentage of subjects that forcibly produced ostensible pain and even death from the authority of the experimenter. Even under no explicit threat from the experimenter, almost one in every three people would continue the experiment when the learner was slumped over and unresponsive. Who are these highly obedient individuals? What are the specific traits or characteristics that overlap between these individuals and real henchmen like Otto Eichmann? Are these individuals, to some degree, modern day henchmen?

2. THE AUTHORITARIAN PROFILE

Although Milgram highlighted some case examples of the individuals in his study, he did not examine potential common traits in those that obey and those that do not. During the same period, several researchers at the University of California–Berkeley published the highly-influential book *The Authoritarian Personality*[5] that attempted to delineate the traits and dispositions of individuals that would submit to and be aggressive for a perceived authority figure. Early work was marred by poor psychometrics (e.g., the F scale)[6] and political controversies (e.g., right-wing authoritarianism versus left-wing authoritarianism); however, research on authoritarianism has matured with highly reliable and valid scales, as well as convincing studies demonstrating that authoritarianism in the general population is relatively nonexistent on the left side of the political spectrum.[7]

More specifically, a sample of over 2500 individuals was surveyed using a reliable left-wing authoritarianism scale that was structurally similar to the right-wing authoritarianism scale. Not a single left-wing authoritarian could be identified in the general population sample.[7] More recent research has suggested that left-wing authoritarianism may exist, but only in a small and specific minority of political activists in extremist parties, such as anarchists.[8] These authors emphasized that they did not expect to (nor did they) find left-wing authoritarians among the population of "ordinary citizens" nor among individuals from mainstream political parties—as is commonly found in right-wing authoritarianism. To be sure, political beliefs that are currently ascribed as left-wing or right-wing have fluctuated under either rubric across time. For example, right-wingers tended to have more egalitarian beliefs in the US in the 1860s, whereas left-wingers tend to have more today. Thus, it is likely that the set of political beliefs currently deemed "right-wing" are not inherently associated with high authoritarianism. As the political beliefs recognized as "right-wing" shift, authoritarians may not shift with them. It is also important to emphasize that while the majority of authoritarians hold conservative political views, there are many individuals who hold right-wing political views that are not authoritarians.[7] That is, while authoritarianism strongly predicts one's political views, conservativism and authoritarianism are independent constructs.

This discussion focuses on the characteristics of authoritarianism that exist across a large spectrum of the population; thus, here we will focus on what has been traditionally termed right-wing authoritarianism. To reiterate, in this chapter, the general term "authoritarianism" will refer to the more generalizable traits present in the broad population of so-called "right-wing authoritarians."

Robert Altemeyer[7] has defined authoritarianism as a covariation of three attitudinal clusters: (1) authoritarian submission, a high degree of submission to perceived authorities in society; (2) authoritarian aggression, a general aggressiveness toward others that is perceived to be sanctioned by authorities; and (3) conventionalism, a high degree of adherence to social conventions that are perceived to be endorsed by authorities. Authoritarianism is an individual difference variable developed on the idea that some people need little situational pressure to submit to authority and attack others, while others need significantly more. Indeed, in a design similar to Milgram's obedience studies, authoritarianism measures strongly predicted which "teachers" would give the highest levels of shocks to "learners" ($r = 0.43$).[6]

Authoritarians also tend to show indifference to government injustices directed against unconventional groups, little interest in protecting human rights, a general punitiveness against persons convicted of crimes, increased sexual aggressiveness, increased acceptance of immoral actions committed by authorities, and increased support for military attack.[7,9,10] Thus, Altemeyer's construct of authoritarianism has high external validity.

Another useful trait of a henchman is negative attitudes toward individuals outside their in-group. Considering the case example of Eichmann and his social attitudes, it may be unsurprising that measures of authoritarianism are highly correlated with prejudice toward outgroups.[7,11–13] Obedience to authority and prejudice against Jews were cornerstones of Nazi indoctrination tactics.[14] However, even outside of these extreme "brainwashing" programs, authoritarianism predicts negative attitudes toward almost every "outgroup": ethnic minorities, homosexual individuals, women, the homeless, criminals, drug dealers, prostitutes, and atheists.[7,15–18] Stemming from Gordon Allport's[19] seminal book, *The Nature of Prejudice*, contemporary theories of social attitudes see prejudice as a tension between an automatic, unintentional stereotyping and a secondary, controlled compensation based on egalitarian beliefs.[20] In this model, if egalitarian beliefs are absent or cognitive control is disrupted, individuals will show increased prejudicial attitudes and behaviors toward outgroups. More education and increased endorsement of egalitarian values is associated with decreased prejudice.[21,22] In addition, poor executive control/inhibitory ability has been associated with increased prejudice and stereotyped judgments.[20,23,24]

Some studies have found specific cognitive functioning differences between authoritarians and nonauthoritarians. However, the correlation between authoritarianism and general intelligence is relatively weak to nonexistent[7,25] (but see Altemeyer's book for a discussion), suggesting that one can be relatively high in IQ and can also be high in authoritarianism. Rather authoritarians display a unique profile of beliefs and cognitive abnormalities that obstructs independent, skeptical thought. First, authoritarianism strongly correlates with religious fundamentalism and general dogmatism.[13,15] They tend to have high religious beliefs that are held with an immutable, unjustified certainty. Of course, this does not mean that authoritarians hold *all* dogmatic beliefs they are exposed to, nor does it imply that nonauthoritarians do not endorse some dogmatisms. Rather, on average, authoritarians generally tend to hold more dogmatic attitudes. This strong correlation suggests that authoritarians rely on authorities to provide their beliefs for them and, importantly, tend to be less likely to counter these beliefs with independent thought.[7] As Eichmann mentioned (introductory quote), he did not need to think for himself, only to believe the statements and follow the orders he was given. Of course, this may not be a cognitive problem, per se, but a motivational one. However, authoritarians also tend to show highly compartmentalized beliefs to antithetical statements in situations without a motivational component.[7] Authoritarians are more likely to agree with both the statements, "If human beings were really honest with each other, there would be a lot more anger and hostility in the world" and "If human beings were really honest with each other, there would be more sympathy and friendship in the world" than nonauthoritarians."[a] Thus, even when contradictory ideas are presented within minutes of each other, authoritarians fail to notice the discrepancy and do not change their beliefs to be consonant with one another. They tend to think with a "forked mind" and are particularly swayed by slogans and propaganda.[7] Finally, authoritarians are particularly poor at recognizing decidedly false inferences.[7,26] The evidence suggests authoritarians have an increased bias toward believing information: when a statement is true, they will tend to think it's true; however, when a statement is false, they will also tend to think it's true.[b]

Not only do authoritarians tend to believe contradictory ideas, but they also endorse contradictory principles. Their judgment justifications tend to ignore alternative viewpoints as they often employ double standards.[7] Authoritarianism negatively correlates with empathy and perspective taking.[27] As a result, authoritarians tend to be egocentric and relatively blind to the concerns and welfare of others. Moreover, individuals high in authoritarianism also tend to carry around less guilt than nonauthoritarians.[7] Altemeyer suggests that their low guilt might be attributable to an increased ease in expunging moral transgressions using religious prayer and confession. However, it is also possible that they simply experience reduced social emotions, such as empathy and guilt, and this leads to their egocentric behavior and callous attitude toward others. Future research should address this important distinction.

Thus far, we have highlighted many attributes of authoritarians: their behavior, social attitudes, cognitive functioning, and even their affect. This particular constellation of psychological tendencies is common among authoritarians from the general population; it is not confined to individuals who display extremist behaviors in response to "brainwashing" efforts (as with

[a] In this study, the antithetical statements were presented on two separate pages in the same research session.

[b] We pick up a discussion of authoritarianism implications for the mechanisms of belief and disbelief below.

Otto Eichmann). Indeed, perhaps the combination of an authoritarian profile (commonly seen in the general population) and focused, consistent persuasion techniques against outgroups result in the atrocious behavior of Nazis like Eichmann. The general population authoritarian profile is most likely derived from several nonpersuasion-based factors: genetics, parental rearing style, experiences with authoritarian punishment, and experiences with outgroups all influence the probability to which one will be high in authoritarianism.[7,28] However, it is also possible neural functioning may be related to authoritarian attitudes and behaviors. Given the unique psychological profile of quotidian authoritarians, the next section investigates whether a particular neural dysfunction could account for many of these psychological tendencies.

3. AUTHORITARIANISM TRAITS IN PATIENTS WITH VENTROMEDIAL PREFRONTAL CORTEX DAMAGE

The prefrontal cortex is often considered the brain region responsible for what makes us "who we are." Since the seminal observations of human lesion patient Phineas Gage, it has been known that damage to the prefrontal cortex can profoundly alter personality.[29] More recent research with patients who have damage to the ventromedial prefrontal cortex (vmPFC) has revealed an interesting profile of cognitive, affective, and behavioral tendencies that are strongly reminiscent of individuals high in authoritarianism. First, it is important to point out what is preserved following damage to the vmPFC. Most patients have normal language abilities, visuospatial function, and reading performance. Performances on general intelligence measures (such as the Wechsler Adult Intelligence Scale) are often in the normal to superior range.[30] However, the patients may have deficits in the so-called executive functions: planning, decision-making, inhibition, and attention.[31–33] Thus, while these patients can have executive function deficits, it is important to note that basic intellectual and cognitive capacities in these patients remain intact.[34] In addition, patients with vmPFC damage have problems properly regulating their emotions. In a particularly compelling bilateral vmPFC case, patient "EVR" revealed a profound inability to express emotion about complex personal and social situations, often leading to disadvantageous real-world social behavior.[30,35] EVR has diminished emotional responsivity, blunted affect, and has particular problems evoking social emotions, such as empathy and guilt.[30,36] He displays restricted emotions (i.e., low emotional expressivity) accompanied by sporadic inappropriate emotional outbursts.[37] Although intelligent and easy to talk to, he cannot hold down a job and has difficulties maintaining relationships. Indeed, EVR and other prefrontal patients often show increased aggression toward others.[38,39]

Patients with vmPFC damage also evince deficits in decision-making and moral reasoning. The emotional deficits present in these patients have an effect on their ability to make normal decisions on the Iowa Gambling Task, a computerized card game that simulates real-life decision-making.[40] This has been associated with a failure to activate somatic signals as indexed by skin conductance response.[36,41] Relative to comparison groups, these patients also show impaired moral reasoning. For example, patients with vmPFC damage judge attempted harm (e.g., attempting, but failing to poison someone) as more morally permissible than *accidentally* harming someone (e.g., accidentally poisoning someone, leading to their death).[42] These results run counter to the findings in healthy age-matched adults and individuals with damage outside the region of the vmPFC. These comparison groups judge attempted harm as *less* morally permissible than accidental harm. In another study, patients with vmPFC damage were more likely than comparison groups to endorse high-conflict personal moral dilemmas, for example, smothering your own baby to save the lives of others. Patients with vmPFC damage showed normal judgments for impersonal moral scenarios, such as putting false information on a resume to look more impressive, as well as for nonmoral scenarios, such as deciding whether to take a bus or train to get to a meeting on time. The authors refer to the distinct endorsement of high-conflict personal scenarios as more utilitarian in that these patients elect to maximize aggregate welfare. Thus, generally patients with damage to the vmPFC region will endorse actions that many consider to be moral violations. Of final note, it has been shown that these patients also tend to be more punitive toward others in an economic game in which they are slighted.[43]

Interestingly, the characteristics of patients with vmPFC damage and healthy individuals high in authoritarianism show considerable overlap. Both have reduced empathy and guilt, increased punitive judgments, increased endorsement of harmful actions, and increased egocentric behavior. Moreover, virtual simulations of Milgram's[3] obedience paradigm have shown vmPFC activations in a functional Magnetic Resonance Imaging (fMRI) study and increased autonomic responses when healthy participants see the virtual "learner" in pain.[44,45] Of course, the characteristics in patients with vmPFC damage are acquired from brain damage, whereas the healthy authoritarians profile is likely derived from genetic factors and early environmental conditions. Could damage to the vmPFC actually produce the authoritarianism personality? Is it possible that a lesion to the vmPFC might create an individual that is geared to submit to authorities and attack others? To answer these questions, a deeper investigation of the cognitive tendencies in these patients was necessary, along with a neuropsychological model that may account for their pattern of beliefs and behavior.

As mentioned above, healthy individuals high in authoritarianism tend to show an increased belief contradiction. They are less likely to notice and correct two mutually exclusive ideas. Thus, they are less likely to have cognitive dissonance,[46] and they tend to compartmentalize their beliefs. Patients with vmPFC damage also show difficulties integrating mutually exclusive beliefs. They are often prone to pathological confabulation, wherein they truly believe their (sometimes florid) assertions, even though contradictory evidence to these assertions is salient and obvious.[47] Clinical observations have also associated a general credulity with patients with vmPFC damage, which could be due to a deficit in the ability to compare and correct discrepant beliefs.[48] Using this clinical data and the hypothesis of an *acquired* authoritarian personality in these patients, the False Tagging Theory (FTT)—a neuropsychological model of belief and doubt[31]—was developed. This model attempts to unify prefrontal cortex functioning and may offer some interesting insights into healthy individuals high in authoritarianism and prejudicial beliefs.

4. MECHANISMS OF BELIEF AND DOUBT: THE FALSE TAGGING THEORY

The central tenets of the FTT include the following: (1) belief occurs in two stages: (a) mental representation (i.e., the existence of meaningful information in a mental system) and (b) mental assessment (i.e., the acceptance or rejection of such information); (2) all ideas that are represented are believed during the initial representation stage, but a secondary psychological process can produce doubt after assessment; (3) the initially believed representation of the idea must be "tagged" to indicate falsehood, thereby generating doubt; (4) the prefrontal cortex is vital for the "false tag" in the assessment component of belief; and (5) the "false tags" are affective in nature.[30,31,49–51] The FTT's core tenets rest on basic belief principles outlined by Baruch Spinoza. In Spinoza's view, disbelief is merely a deliberate revision of belief; thus, comprehension and initial acceptance are the same process. This can be contrasted with René Descartes' (i.e., the Cartesian) model of belief, which suggests that the comprehension of meaningful information precedes the act of both acceptance and rejection.[49] The FTT employs a Spinozan framework and suggests that mental representations are initially believed, and a secondary, psychological analysis produces disbelief. The FTT argues that the prefrontal cortex is a critical hub in a network of brain regions that mediates this secondary disbelief (or doubt).

Intuitively, the Cartesian model of belief seems to be the more likely process by which we believe information. Introspective experience suggests that we logically weigh positive and negative evidence to believe or disbelieve an idea.[51] However, several convincing psychological experiments have shown support for the Spinozan model. For example, in the Phony Man Experiment, participants were shown smiling faces and were informed either before or after each presentation that the face was expressing either true or false happiness.[52] On some trials, participants' processing of the face was interrupted by having them quickly perform an unrelated tone-discrimination task. Participants were once again presented the original faces and asked to determine whether each was expressing true or false happiness. In regards to the Spinozan and FTT belief and doubt models, interruption should cause participants to mistake false ideas for true ones, but not vice versa. Results indicate that interruption had no effect on the correct identification of true faces, but significantly reduced correct identification of false faces. Thus, participants seem to have initially represented each face as expressing true happiness, and then attempted to alter that representation when the face expressed false happiness. Therefore, this experiment demonstrated a dissociation between belief and disbelief, but no dissociation between comprehension and belief.

More evidence on propositional knowledge stems from another experiment conducted by Gilbert and colleagues.[50] Participants read criminal vignettes and determined appropriate prison terms for each perpetrator. The crime stories contained explicitly labeled true information and false information. True or false statements were denoted by color: white statements were true while red statements were false. Individuals who underwent resource depletion during the reading of the crime vignettes (i.e., by pressing a button in response to a noise while the information was presented) were more likely to accept the false information as true, but were not more likely to accept true information as false. These increased false-as-true errors correlated with their criminal sentencing judgments. When the ostensibly false statements exacerbated the crime in the stories, resource depletion increased the criminal sentences. When the false statements mitigated the severity of the crime, resource depletion decreased the criminal sentences. Thus, resource depletion acts to increase credulity to the explicitly labeled false information. This lends additional support toward the Spinozan model of doubt, as resource depletion prevented the disbelief of information that was simultaneously comprehended and believed.

The FTT posits that the prefrontal cortex mediates "false tagging" or falsification to postrolandic association cortices. In this model, prefrontal cortex damage from strokes or tumor resections should result in a "doubt deficit" whereby an individual has increased credulity, or tendency toward belief.[31] The idea that the prefrontal cortex is critical for "false tagging" novel information that is compulsorily, initially believed is based on several lines of evidence. First, these patients display dispositional or personality patterns that are consistent with a "doubt deficit": overconfidence, boastfulness, grandiosity, obstinacy, and egocentricity.[31,33] These personality patterns combined with clinical observations suggesting increased credulity in patients with vmPFC damage led to the design of an empirical study examining belief and doubt within this patient population. Specifically, patients with vmPFC damage, brain damaged comparison patients (i.e., patients with damage outside the vmPFC region), and healthy age-matched adults were provided with a series of advertisements that had been deemed deceptive by the Federal Trade Commission.[53] Consistent with the prediction of a doubt deficit in patients with vmPFC damage, results showed that this group was more credulous to the misleading ads than the comparison groups. Patients with vmPFC damage also presented with increased intention to purchase the products showcased in the ads. Increased credulity in these patients was found even when the deceptive ads contained a disclaimer rebutting the misleading claim, suggesting that skepticism is generally lower in these individuals. Indeed, these findings were not due to differences in general cognitive functioning, such as intelligence, memory, or reading ability. The site of the lesion was the only consistent factor related to credulity. In addition, the authors were interested in whether patients with vmPFC damage would have "forked minds," or an increased compartmentalization in their beliefs. The FTT predicts that these patients should believe many propositions and perceptions that are inconsistent with their extant knowledge, but fail to compare and falsify discordant ideas with one another. Using the same stimuli in Altemeyer's[7] authoritarian self-contradiction study,[c] it was found that patients with vmPFC damage had increased compartmentalization to their beliefs than brain damaged comparison patients and healthy adults.[54]

Again, the similarity between the psychological profile of patients with vmPFC damage and healthy authoritarians should be noted. Authoritarians tend to believe superficially appealing slogans, and patients with vmPFC damage are credulous to ostentatiously misleading ads. These patients, along with healthy authoritarians, both have an increased bias to believe information that is labeled as false. Moreover, they both have high belief self-contradiction, or an increased probability of believing conflicting ideas. These results, combined with the affective and behavioral evidence, persuasively argue that damage to the vmPFC may indeed create a profile of increased authoritarianism. However, to confirm these suspicions, a more direct assessment of authoritarianism and related attitudes in patients with vmPFC damage was essential.

5. AUTHORITARIANISM ATTITUDES IN PATIENTS WITH vmPFC DAMAGE

> Dogmatism [is] a dead give-away that the person doesn't know why he believes what he believes. ***Robert Altemeyer***

To examine authoritarianism and related attitudes, patients with vmPFC damage (see Figure 1), brain damaged comparison patients, medical comparison patients (individuals who had undergone a life-threatening but nonneurological medical event), and healthy adults from the general population were provided with scales measuring authoritarianism, religious fundamentalism, religious behaviors, specific religious beliefs, and prejudicial attitudes.[55] It was theorized that patients with vmPFC damage would show high levels of authoritarianism, religious fundamentalism, and prejudice toward outgroups. In line with these predictions, patients with damage to the vmPFC had the highest scores on scales of authoritarianism (e.g., greater endorsements of statements like, "Our country will be great if we honor the ways of our forefathers, do what the authorities tell us to do, and get rid of the 'rotten apples' that are ruining everything") and religious fundamentalism (e.g., "God has given humanity a complete, unfailing guide to happiness and salvation, which must be totally followed") relative to the comparison groups.[55] It could be argued that the increase in authoritarianism is simply a product of high religiosity in these patients (i.e., an individual difference variable which is completely unrelated to their brain damage). However, when items on the authoritarianism scale that explicitly mentioned topics of religion were removed, the authoritarianism differences between the groups survived.[55] Moreover, the results could not be accounted for by differences in general cognitive functioning, demographic variables, religious affiliation, religious upbringing, or religious service attendance. Neither an aversive medical event, per se, nor brain damage, per se, led to the high levels of religious beliefs in patients with vmPFC damage.

This research suggests that if individuals with vmPFC lesions have a deficit in the ability to "tag" incoming information as false, they may rely on authorities to provide information for them, leading them to hold beliefs more consistent with authority figures. Thus, the authoritarian characteristics we see in patients with vmPFC

[c] See above for a discussion.

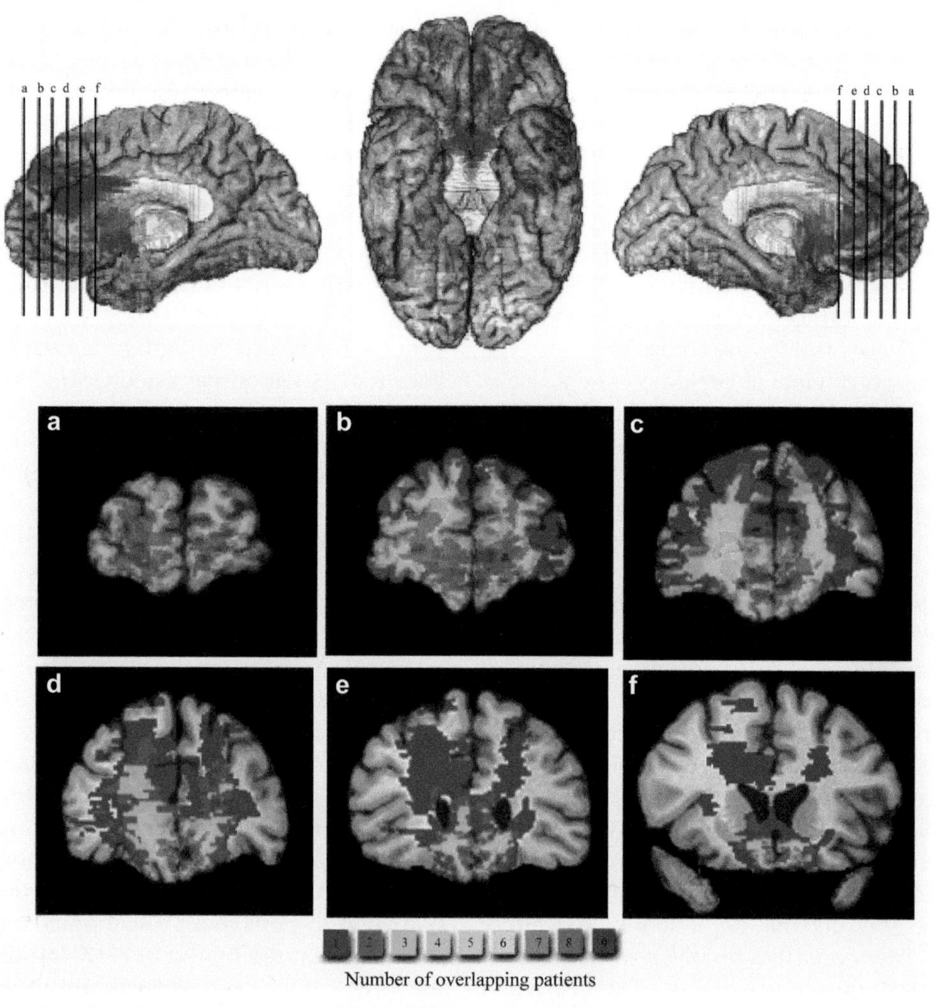

FIGURE 1 Lesion overlap of vmPFC patients. Lesions of the vmPFC patients displayed in mesial and coronal slices. The color bar indicates the number of overlapping lesions at each voxel.

damage may stem from an underlying "false tagging" dysfunction. Damage to this region is associated with the personality profile of authoritarianism across several domains (affective, cognitive, behavioral, and attitudinal). It is hypothesized that these characteristics may be the product of a decreased ability to doubt or falsify information. Rather than having a reduced motivational desire to think independently (see introductory quote by Eichmann), we argue that patients with vmPFC damage have a reduced ability to reject authoritarian direction. This general deficit in the ability to reject propositions coupled with a decreased emotional aversion toward harming others[56] suggests damage to the vmPFC may create the ideal henchman: an individual with high submissiveness to authorities and high aggressiveness toward others. Indeed, Milgram's[3] case description of Mr Bruno Batta, who displayed extreme submissiveness as he unemotionally forced the learner's hand on the shock plate in the touch-proximity experiment, contains striking parallels to patients with vmPFC damage characterized by blunted emotions[33] and stagnant autonomic responses to provocative social stimuli.[36] Behavioral paradigms in these patients measuring tractability toward authorities and aggression against others would help solidify these findings. Future research will need to address behavioral outcomes in authoritarian situations.

6. NEURAL CORRELATES OF PREJUDICE

A prejudice, unlike a simple misconception, is actively resistant to all evidence that would unseat it. *Gordon Allport*

As mentioned above, authoritarianism is strongly associated with explicit prejudice toward outgroups.[13] In the examination of authoritarian attitudes in lesion patients, it was found that patients with vmPFC damage also displayed high explicit prejudice toward ethnic minorities and homosexual individuals.[54] Thus, when patients with vmPFC damage, brain damaged comparison patients, and healthy age-matched adult participants are given prejudicial statements, such as "Many minorities are spoiled; if they really wanted to improve their lives, they would get jobs and get off welfare" and "Homosexuals should be forced to take whatever treatments science can come up with to make them normal," only the patients with vmPFC damage tend to show increased prejudicial endorsement. This is additional evidence that patients with damage to this region are less likely to doubt authoritarian ideals and the general social milieu. These findings are compatible with the Allportian contemporary view of prejudice: prejudice is a failure of a cognitive control process to compensate for automatic, unintentional stereotyping. Research suggests that executive processes are recruited in order to suppress the prejudice behavior and stereotypes that come to mind unintentionally and automatically.[20,23,57] The FTT argues that "false tagging" can account many executive processes, including general inhibition, cognitive switching, planning, decision-making, attentional focusing, and working-memory maintenance, that are dependent on the prefrontal cortex.[58] Thus, damage to the prefrontal cortex should lead to decreased inhibition of automatic stereotypes, which result in higher prejudice toward outgroups.

Several neuroimaging studies in healthy individuals support this finding. In an fMRI study, researchers investigated whether differences in racial bias among white participants predict the depletion of executive resources during later contact with black individuals.[24] In this experiment, white participants were provided with sets of unfamiliar black faces, and brain activity was assessed. Racial bias predicted activity in the prefrontal cortex in response to the stimuli shown during the task. Individuals then had an interracial interaction and afterward were given the Stroop interference task (a neuropsychological test requiring executive control). Results showed that activity in the prefrontal cortex during the fMRI task predicted Stroop interference and mediated the relationship between racial bias and Stroop interference. This research supports the idea that executive function resource depletion can occur via interracial contact. The prefrontal cortex critically mediates both cognitive control during interracial contact and the Stroop task. It also supports the idea that prejudice is increased when cognitive control and the prefrontal cortex are compromised.

In addition, individuals who show the ability to take the perspective of and have empathy toward others reveal reduced prejudice compared to those who do not.[28,59,60] As described above, patients with prefrontal damage have known deficits in empathy and perspective taking.[33] Thus, both from a strictly cognitive perspective and a social-affective angle, the prefrontal cortex is considered a critical mediator for lower prejudicial beliefs.

Other studies examined the roles of the prefrontal cortex (as a mediator of cognitive control) and the amygdala (as a mediator for racially-induced fear). In one fMRI study, white egalitarian-motivated participants were shown black and white faces at fast or slow speeds in the scanner.[61] To create more of a racially negative stereotypic environment, participants listened to violent rap music in the background. In other conditions, participants either listened to no music or death metal. Results showed that only the violent rap music condition showed amygdala activation for black faces, and this activation persisted during slow exposure. The amygdala response positively covaried with activation in a region of the prefrontal cortex often associated with cognitive control. The authors concluded that while white individuals are successful at controlling an initial arousal reaction (amygdala response) to a black target in a neural context, this arousal response is not downregulated in the presence of negative stereotypical cues.

One of the most heavily used tasks that putatively measures only the implicit or automatic prejudicial component is the implicit association test (IAT).[62] The IAT purports to measure the strength of association between concepts, such as white and black, and attributes, such as good and bad. For instance, in white individuals that show no race preference on explicit measures, there is a strong preference for positive stereotypes of white faces rather than black faces.[63] Authoritarianism strongly correlates with both racial and homosexual implicit prejudice as measured by the IAT.[64,65] Implicit measures, such as the IAT, have been consistently associated with amygdala activation.[66,67] While one might predict prefrontal cortex structural integrity to have no effect on implicit measures of prejudice, several studies have shown that damage to the prefrontal cortex affects implicit stereotyping.[68,69] Indeed, patients with lesions to the vmPFC have shown increased stereotypical attitudes on the IAT.[69] This result and other studies have suggested that the IAT is a rather poor measure of implicit attitudes in isolation.[70,71] The IAT indices likely reflect both automatic and controlled components (the latter involving some prefrontal cortex mediation).[71] Nevertheless, the extant evidence implicates a critical role for the prefrontal cortex in the mitigation of involuntary, believed prejudicial attitudes and stereotypes.

7. CONCLUSION

> Punishment may make us obey the orders we are given, but at best it will only teach an obedience to authority, not a self-control which enhances our self-respect. ***Bruno Bettelheim***

This chapter has reviewed psychological and neuropsychological evidence on authoritarianism and prejudice, attributes that are commonly associated with obedience to authority. It is clear that tendencies toward authoritarian attitudes and prejudicial beliefs are the culmination of environmental and genetic factors; however, we provide research suggesting that circumscribed damage to the ventromedial prefrontal cortex may act to *create* authoritarian individuals. On a battery of cognitive and psychometric tests, patients with lesions to this region show a profile consistent with submissiveness to authoritarian commands intent on harming others and aggressiveness in the name of authority, mirroring the profile of healthy authoritarians.[7] Furthermore, patients with vmPFC damage also present marked prejudicial beliefs toward ethnic and homosexual minorities. Neuroimaging studies complement these neuropsychological findings and provide evidence that the vmPFC and amygdala are critical structures involved in inhibiting and facilitating attitudes toward outgroups. These results beg more questions: what do these findings mean on a broader societal level? How can this information be interpreted within the general population of authoritarian individuals?

Certainly these findings do *not* mean that individuals who are high on scales of authoritarianism or religious fundamentalism have damage to the ventromedial prefrontal cortex.[55] We consider authoritarianism and religiosity to be multidetermined, with several factors beyond brain integrity leading to one's particular ensemble of authoritarian and religious beliefs. Indeed, it is improbable that even extreme cases of obedience to authority, such as Otto Eichmann, can be explained by neurological injury. As much as it may seem fitting, we cannot retrospectively assign brain damage to Eichmann or any other individual who holds authoritarian beliefs. The more restricted claim to be made is that damage to the vmPFC may act to *increase the likelihood* that an individual holds authoritarian and religious beliefs. To reiterate, it appears probable that damage to the vmPFC promotes the authoritarian psychological profile. That being said, we would be remiss not to mention the case of Dr Robert Ley, another Nazi henchman who authorized, directed, and participated in crimes against humanity. Ley committed suicide on October 25, 1945, and his autopsy showed a "long-standing degenerative process of the frontal lobes."[72] This is likely a curious coincidence, in that brain damage of Nazi officers probably played an insignificant role during the Nazi scourge. However, one could speculate that in some individuals, vmPFC dysfunction could enable an authoritarian mind-set that may be selected for by military hierarchy.

Further studies should try to address whether patients with vmPFC damage are more obedient to authorities when instructed to hurt others. However, given the obvious ethical implications of such a design (especially when considering this vulnerable and valuable subject population), any potential study will have considerable methodological constraints in direct examination of the question.

Despite this, a significant developmental implication garnered from this research is that children may be especially susceptible to belief and subsequently vulnerable to prejudicial and authoritarian attitudes.[73] Prejudicial attitudes are often implicitly learned in childhood, and exposure to authoritarian-style parenting methods may act to increase exposure to authoritarian principles.[7,74] This environment, coupled with the notion that the prefrontal cortex is still developing in childhood,[75] may place these children at a higher probability of becoming authoritarians in adulthood. On the other end of the developmental spectrum, older adults may be at a higher risk of authoritarian and prejudicial beliefs as their prefrontal cortex integrity declines with age.[76] Further research is warranted in both of these developmental populations to help illuminate potential mechanisms behind the development of the authoritarian profile.

While authoritarianism has often been viewed as a negative factor in society (largely from its correlation to prejudice of outgroup members), it also has been shown to correlate with high levels of in-group cooperation.[77] Increased adherence of commands and instruction can also be good for a society and individuals (assuming that the commands are adaptive for the group). To be sure, adherence to authoritarian instruction often circumvents disadvantageous, painful, or even deadly trial-and-error learning (e.g., *"Don't stick a knife into an electric outlet"*). Moreover, authoritarianism may lead to reduced mental distress,[78] and thus enable increased positive affect in one's life. Authoritarianism is certainly not a uniformly negative individual difference variable. Rather, the context (and more specifically, the quality of the instruction to the individual and society) determines the beneficence or malfeasance of authoritarianism.

Stanley Milgram certainly did not anticipate the degree to which individuals would be willing to shock innocent participants at extreme levels in his experimental task.[3] Indeed, it is likely that Milgram's findings were at least partly attributed to the generalized cultural acceptance of obedience to authority in the era of his study. However, a recent (partial) replication of Milgram's study demonstrated no significant differences between the percentages of participants who continued to administer electric shocks to the "learner" than in Milgram's study.[79] The author of this study concluded that although societal attitudes on obedience may have changed, this has not had an effect on obedience to authority over the past 45 years. This suggests that authoritarianism may be less dependent on environmental and situational factors and more driven by biological mechanisms. These neural and biological underpinnings may play a critical role in identifying henchmen-like individuals that are prone to committing aggressive acts in the name of authority.

References

1. Eichmann OA. *Adolf Eichmann: Memoirs*. Jewish Virtual Library; 1999.
2. Robinson J. *And the Crooked Shall Be Made Straight: The Eichmann Trial, the Jewish Catastrophe, and Hannah Arendt's Narrative*. Philadelphia: Jewish Publication Society; 1965.
3. Milgram S. *Obedience to Authority*. New York: Harper and Row Publishers; 1974.
4. Benjamin LT, Simpson JA. The power of the situation: the impact of Milgram's obedience studies on personality and social psychology. *Am Psychol*. 2009;64:12–19.
5. Adorno TW, Frenkel-Brunswik E, Levinson DJ, Sanford RN. *The Authoritarian Personality*. New York: Harper and Row Publishers; 1950.
6. Altemeyer B. *Right-wing Authoritarianism*. Manitoba: University of Manitoba Press; 1981.
7. Altemeyer B. *The Authoritarian Specter*. Cambridge: Harvard University Press; 1996.
8. Van Hiel A, Duriez B, Kossowska M. The presence of left-wing authoritarianism in western Europe and its relationship with conservative ideology. *Political Psychol*. 2006;27:769–793.
9. McFarland SG. On the eve of war: authoritarianism, social dominance and American students' attitudes toward attacking Iraq. *Pers Soc Psychol Bull*. 2005;31:360–367.
10. Walker WD, Rowe RC, Quinsey VL. Authoritarianism and sexual aggression. *J Pers Soc Psychol*. 1993;65:1036–1045.
11. Sibley CG, Duckitt J. Personality and prejudice: a meta-analysis and theoretical review. *Pers Soc Psychol Rev*. 2008;12:248–279.
12. Johnson MK, LaBouff JP, Rowatt WC, Patock-Reckham JA, Carlisle RD. Facets of right-wing authoritarianism mediate the relationship between religious fundamentalism and attitudes toward Arabs and African-Americans. *J Sci Study Relig*. 2012;51:128–142.
13. Rowatt WC, Shen MJ, LaBouff JP, Gonzalez A. Religious fundamentalism, right-wing authoritarianism, and prejudice. In: Paloutzian RF, Park CL, eds. *Handbook of the Psychology of Religion and Spirituality*. 2nd ed. New York: Guilford Press; 2013:457–475.
14. Voigtlander N, Voth H. Nazi indoctrination and anti-Semitic beliefs in Germany. *Proc Natl Acad Sci*. 2015.
15. Altemeyer B, Hunsberger B. Authoritarianism, religious fundamentalism, quest, and prejudice. *Int J Psychol Relig*. 1992;2:113–133.
16. Christopher AN, Mull MS. Conservative ideology and ambivalent sexism. *Psychol Women Q*. 2006;30:223–230.
17. Peterson BE, Doty RM, Winter DG. Authoritarianism and attitudes toward contemporary social issues. *Pers Soc Psychol Bull*. 1993;19:174–184.
18. Asbrock F, Sibley CG, Duckitt J. Right-wing authoritarianism and social dominance orientation and the dimensions of generalized prejudice: A longitudinal test. *Eur J Pers*. 2010;24:324–340.
19. Allport GW. *The Nature of Prejudice*. Cambridge: Addison-Wesley; 1954.
20. Payne BK. Conceptualizing control in social cognition: how executive functioning modulates the expression of automatic stereotyping. *J Pers Soc Psychol*. 2005;89:488–503.
21. Biernat M, Vescio TK, Theno SA, Crandall CS. Values and prejudice: toward understanding the impact of American values on outgroup attitudes. In: Seligman C, Olson JM, Zanna MP, eds. *The Psychology of Values: The Ontario Symposium*. Hillsdale: Erlbaum; 1996.
22. Wagner U, Zick A. The relation of formal education to ethnic prejudice: its reliability, validity, and explanation. *Eur J Soc Psychol*. 1995;25:41–56.
23. von Hippel W, Silver LA, Lynch ME. Stereotyping against your will: the role of inhibitory ability in stereotyping and prejudice among the elderly. *Pers Soc Psychol Bull*. 2000;26:523–532.
24. Richeson JA, Shelton JN. When prejudice does not pay: effects of interracial contact on executive function. *Psychol Sci*. 2003;14:287–290.
25. Heaven P, Ciarrochi J, Leeson P. Cognitive ability, right-wing authoritarianism, and social dominance orientation: a five-year longitudinal study amongst adolescents. *Intelligence*. 2011;39:15–21.
26. Wegmann MF. *Information Processing Deficits of the Authoritarian Mind*. Santa Barbara: Psychology, The Fielding Institute; 1992.
27. Backstrom M, Bjorklund F. Structural modeling of generalized prejudice. *J Individ Differ*. 2007;8:10–17.
28. McFarland SG. Authoritarianism, social dominance, and other roots of generalized prejudice. *Political Psychol*. 2010;31:453–477.
29. Harlow HF. Recovery from the passage of an iron bar through the head. *Hist Psychiatry*. 1993;4:274–281.
30. Damasio A. *Decartes' Error: Emotion, Reason and the Human Brain*. New York: Grosset/Putnam; 1994.
31. Asp EW, Tranel D. False tagging theory: toward a unitary account of prefrontal cortex function. In: Stuss DT, Knight RT, eds. *Principles of Frontal Lobe Function*. 2nd ed. New York: Oxford University Press; 2013:383–416.
32. Stuss DT, Benson DF. Neuropsychological studies of the frontal lobes. *Psychol Bull*. 1984;95(1):3–28.
33. Damasio AR, Anderson SW, Tranel D. The frontal lobes. In: Heilman KM, Valenstein E, eds. *Clinical Neuropsychology*. 5th ed. New York: Oxford University Press; 2012:417–465.
34. Tranel D, Anderson SW, Benton A. Development of the concept of "executive function" and its relationship to the frontal lobes. In: Boller F, Grafman J, eds. *Handbook of Neuropsychology*. Vol. 9. Amsterdam: Elsevier; 1994:125–148.
35. Tranel D. Emotion, decision making, and the ventromedial prefrontal cortex. In: Stuss DT, Knight RT, eds. *Principles of Frontal Lobe Function*. New York: Oxford University Press; 2002:338–353.
36. Damasio AR, Tranel D, Damasio H. Individuals with sociopathic behavior caused by frontal damage fail to respond autonomically to social stimuli. *Behav Brain Res*. 1990;41(2):81–94.
37. Barrash J, Asp EW, Markon K, Manzel K, Anderson SW, Tranel D. Dimensions of personality distrubance after focal brain damage: investigation with the Iowa Scales of Personality Change. *J Clin Exp Neuropsychol*. 2011;33:833–852.
38. Giancola PR. Evidence for dorsolateral and orbital prefrontal cortical involvement in the expression of aggressive behavior. *Aggressive Behavior*. 1995;21(6):431–450.
39. Grafman J, Schwab K, Warden D, Pridgen A, Brown HR, Salazar AM. Frontal lobe injuries, violence, and aggression. *Neurology*. May 1, 1996;46(5):1231.
40. Bechara A, Tranel D, Damasio H. Characterization of the decision-making deficit of patients with ventromedial prefrontal cortex lesions. *Brain*. 2000;123:2189–2202.
41. Bechara A, Tranel D, Damasio H, Damasio AR. Failure to respond autonomically to anticipated future outcomes following damage to prefrontal cortex. *Cereb Cortex*. 1996;6(2):215–225.
42. Young L, Bechara A, Tranel D, Damasio H, Hauser M, Damasio A. Damage to ventromedial prefrontal cortex impairs judgment of harmful intent. *Neuron*. 2010;65:845–851.
43. Koenigs M, Tranel D. Irrational economic decision-making after ventromedial prefrontal damage: evidence from the ultimatum game. *J Neurosci*. 2007;27:951–956.
44. Cheetham M, Pedroni AF, Antley A, Slater M, Jäncke L. Virtual milgram: empathic concern or personal distress? Evidence from functional MRI and dispositional measures. *Front Hum Neurosci*. 2009;3:1–13.
45. Slater M, Antley A, Davison A, et al. A virtual reprise of the Stanley Milgram obedience experiments. *PLoS One*. 2006;1(1):e39.
46. Festinger L. *A Theory of Cognitive Dissonance*. Stanford: Stanford University Press; 1957.
47. Gilboa A, Moscovitch M. The cognitive neuroscience of confabulation: A review and a model. In: Baddeley AD, Kopelman MD, Wilson BA, eds. *Handbook of Memory Disorders*. 2nd ed. Chichester: John Wiley; 2002:315–342.

48. Berlyne N. Confabulation. *Br J Psychiatry*. 1972;120:31–39.
49. Gilbert DT. How mental systems believe. *Am Psychol*. 1991;46: 107–119.
50. Gilbert DT, Tafarodi RW, Malone PS. You can't not believe everything you read. *J Pers Soc Psychol*. 1993;65:221–233.
51. Mandelbaum E. Thinking is believing. *Inquiry*. 2014;57:55–96.
52. Gilbert DT, Krull DS, Malone PS. Unbelieving the unbelievable: some problems in the rejection of false information. *J Pers Soc Psychol*. 1990;59:601–613.
53. Asp EW, Manzel K, Koestner B, Cole CA, Denburg NL, Tranel D. A neuropsychological test of belief and doubt: damage to ventromedial prefrontal cortex increases credulity for misleading advertising. *Front Neurosci*. 2012;6:100.
54. Asp EW. *A Neuroanatomical Investigation of Belief and Doubt*. Iowa City: Neurology, University of Iowa; 2012.
55. Asp EW, Ramchandran K, Tranel D. Authoritarianism, religious fundamentalism, and the human prefrontal cortex. *Neuropsychology*. 2012;26:414–421.
56. Koenigs M, Young L, Adolphs R, et al. Damage to the prefrontal cortex increases utilitarian moral judgements. *Nature*. 2007;446:908–911.
57. Devine PG. Stereotypes and prejudice: their automatic and controlled components. *J Pers Soc Psychol*. 1989;56:5–18.
58. Asp EW, Manzel K, Koestner B, Denburg NL, Tranel D. Benefit of the doubt: a new view of the role of the prefrontal cortex in executive functioning and decision making. *Front Neurosci*. 2013;7(86).
59. Hodson G, Busseri MA. Bright minds and dark attitudes: lower cognitive ability predicts greater prejudice through right-wing ideology and low intergroup contact. *Psychol Sci*. 2012;23:187–195.
60. Hodson G, Hogg SM, MacInnis CC. The role of "dark personalities" (narcissism, machiavellianism, psychopathy), big five personality factors, and ideology in explaining prejudice. *J Res Pers*. 2009;43:686–690.
61. Forbes CE, Cox CL, Schmader T, Ryan L. Negative stereotype activation alters interaction between neural correlates of arousal, inhibition and cognitive control. *SCAN*. 2012;7:771–781.
62. Greenwald AG, McGhee DE, Schwartz JLK. Measuring individual differences in implicit cognition: the implicit association test. *J Pers Soc Psychol*. 1998;74:1464–1480.
63. Ames DL, Jenkins AC, Banaji MR, Mitchell JP. Harvesting implicit group attitudes and beliefs from a demonstration web site. *Group Dyn*. 2002;6:101–115.
64. Jonathan E. The influence of religious fundamentalism, right-wing authoritarianism, and Christian orthodoxy on explicit and implicit measures of attitudes toward homosexuals. *Int J Psychol Relig*. 2008;18:316–329.
65. Rowatt WC, Franklin LM. Christian orthodoxy, religious fundamentalism, and right-wing authoritarianism as predictors of implicit racial prejudice. *Int J Psychol Relig*. 2004;14:125–138.
66. Phelps EA, O'Connor KJ, Cunningham WA, et al. Performance on indirect measures of race evaluation predicts amygdala activation. *J Cogn Neurosci*. 2000;12:729–738.
67. Kubota JT, Banaji MR, Phelps EA. The neuroscience of race. *Nat Neurosci*. 2012;15:940–948.
68. Milne E, Grafman J. Ventromedial prefrontal cortex lesions in humans eliminate implicit gender stereotyping. *J Neurosci*. 2001;21: 1–6.
69. Gozzi M, Raymont V, Solomon J, Koenigs M, Grafman J. Dissociable effects of prefrontal and anterior temporal cortical lesions on stereotypical gender attitudes. *Neuropsychologia*. 2009;47: 2125–2132.
70. Mandelbaum E. *Attitude, Inference, Association: On the Propositional Structure of Implicit Bias*. Philpapers: Philosophical Research Online; 2014:1–36.
71. Bartholow BD. Event-related brain potentials and the role of cognitive control in implicit race bias. In: Derks E, Scheepers D, Ellemers I, eds. *The Neuroscience of Prejudice*. London: Taylor and Francis; 2013:190–208.
72. Zillmer E, Harrower M, Ritzler BA, Archer RP, eds. *The Quest for the Nazi Personality: A Psychological Investigation of Nazi War Criminals*. Hillsdale: Lawrence Erlbaum Associates; 1995. Weiner IB, ed. The LEA Series in Personality and Clinical Psychology.
73. Doyle AB, Aboud FE. A longitudinal study of white children's racial prejudice as a social-cognitive development. *Merrill-Palmer Q: J Dev Psychol*. 1995;41:209–228.
74. Duckitt J. A dual process cognitive-motivational theory of ideology and prejudice. *Adv Exp Soc Psychol*. 2001;33:41–113.
75. Diamond A. Normal development of prefrontal cortex from birth to young adulthood: cognitive functions, anatomy, and biochemistry. In: Stuss DT, Knight RT, eds. *Principles of Frontal Lobe Function*. New York: Oxford University Press; 2002:466–503.
76. West RL. An application of prefrontal cortex function theory to cognitive aging. *Psychol Bull*. 1996;120:272–292.
77. Kessler T, Cohrs JC. The evolution of authoritarian processes: fostering cooperation in large-scale groups. *Group Dyn: Theory Res Pract*. 2008;12:73–84.
78. Van Hiel A, De Clercq B. Authoritarianism is good for you: right-wing authoritarianism as a buffering factor for mental distress. *Eur J Pers*. 2009;23:33–50.
79. Burger JM. Replicating Milgram: would people still obey today? *Am Psychol*. 2009;64:1–11.

C H A P T E R

18

The Neural Mechanisms of Prejudice Intervention

Keith B. Senholzi[1], *Jennifer T. Kubota*[2,3]

[1]Harvard Medical School, Boston, MA, USA; [2]Department of Psychology, University of Chicago, Chicago, IL, USA; [3]Center for the Study of Race, Politics, and Culture, University of Chicago, Chicago, IL, USA

O U T L I N E

We view the world through a social lens that colors our environment with categorical labels, providing information about, among many things, people's age, gender, and race. This lens ultimately lays the foundation for how we perceive the world and its organization, including where we live, how we make social connections, the education we receive, our healthcare, the jobs we take on, and how we go about managing our finances. Perhaps most prominently, this lens affects how we see and are seen by others. Although processing social category information may serve as an important and positive function by providing an efficient means to think about those around us, it can also have deleterious effects. Social categorization can result in the application of inaccurate stereotypes and the perpetuation of intergroup conflict.[1]

The purpose of this chapter is to integrate across the behavioral science and neuroimaging literature on prejudice in an effort to elucidate the mechanisms of prejudice intervention from which scientists can derive innovative theoretical insights for future research. We will focus our overview and analysis primarily on racial prejudice directed toward Blacks in the United States, not because other types of prejudice do not exist, but primarily due to the unfortunate lack of available data involving other types of prejudice and groups (see the Discussion section for suggestions regarding potentially fruitful avenues for future research relating to this concern).

We will highlight the effectiveness of functional magnetic resonance imaging (fMRI) in illuminating the underlying neural substrates of prejudice. The work we review implicates a network of brain regions related to prejudice, namely those involved in person perception and emotion processing—the amygdala, fusiform face area (FFA), and medial prefrontal cortex (mPFC)—and regulation—the dorsolateral prefrontal cortex (dlPFC), anterior cingulate cortex (ACC), and orbitofrontal cortex (OFC). In addition, we present emergent evidence

Neuroimaging Personality, Social Cognition, and Character
http://dx.doi.org/10.1016/B978-0-12-800935-2.00018-X

for successful behavioral interventions that influence race processing across this network, with the majority of interventions explored to date altering self-regulatory processing.

Our chapter will feature answers to three questions and the theoretical implications of each for understanding the mechanisms of prejudice. We ask (and attempt to answer) the following: (1) How is prejudice defined and measured?; (2) Does a network of brain areas exist that is reliably associated with prejudice?; and (3) Are neural responses during race perception and prejudice expression malleable?[a]

1. HOW IS PREJUDICE DEFINED AND MEASURED?

Prejudice has taken on many definitions, but for the purposes of this chapter, we regard it as any attitude or emotion toward a member or members of a group that directly or indirectly indicates some negativity or antipathy toward that group.[2] This characterization highlights the way in which one's social group can ultimately dictate how they are perceived and responded to. As such, one of the fundamental questions in stereotyping and prejudice research is how people extract and use information about the social groups to which targets of our perception belong. This emphasis is reflected in the field as a whole, in which a remarkable amount of research is devoted to understanding processes involved in social categorization and group identification. Of the vast array of categorical person dimensions, scholars have primarily focused on three: race, gender, and age.[3–5] This is likely because each of these distinctions is readily observed from our visual appearance and each is incredibly relevant for many social judgments. Because we typically have highly accessible beliefs associated with groups that fall within these three dimensions, attending to them is also socially expedient: doing so allows perceivers to quickly and efficiently draw inferences about an individual in exchange for only minimal cognitive effort.[4]

Although group-based associations may not be veridical or even necessarily relevant to a perceiver's situation, they nonetheless readily come to mind.[6–8] Unfortunately, these types of associations oftentimes manifest themselves as prejudices. Once social category labels are applied to someone, they can guide how a perceiver gets to know that person.[9] For example, after a perceiver ascribes a social category label to an individual target of perception, information that is inconsistent with stereotypes about that target's social group tends to be forgotten or explained away as unusual in nature when perceivers do not have the time, mental resources, or motivation to attend to or encode the counterstereotypic information.[10,11,171,172] Pernicious associations can affect a wide range of behaviors, including subtle aspects of interactions such as nonverbal behavior (e.g., smiling or eye contact, interaction proximity), and also outright discrimination.[12]

When most people think of prejudice, notions of harsh discriminatory acts come to mind. As such, prejudice has traditionally been characterized as *explicit* in nature. Explicit prejudice refers to negative attitudes based on group membership that are consciously endorsed and subject to deliberate control in their expression.[13] This characterization connotes a high degree of intentionality in the expression of prejudice. Because of this, explicitly asking people how they feel about certain groups or members of groups, and thus acquiring their introspective reports, has been the most prevalent way to measure prejudice. Implicit prejudice, by contrast, corresponds to prejudice that (typically) lacks self-reflective access and is unintentionally triggered[14,15] (c.f. Ref. 16). For these reasons, implicit prejudice is measured by performance on cognitive tasks that do not require introspection. It is thought that implicit associations are derived from affective and cognitive knowledge stored in memory acquired from years of exposure to cultural associations regarding members of social groups. These associations slowly emerge over time and unintentionally affect how we perceive and behave toward others. As a result of these well-learned associations, even individuals who are explicitly egalitarian may at times unintentionally act in prejudicial ways.[102]

A clear distinction between trends in explicit versus implicit prejudice is possible when considering research on racial bias in the United States. On the one hand, explicit prejudice against Blacks in the US has become increasingly attenuated and is at an all-time low.[17] Conversely, current implicit anti-Black prejudice in the US is ubiquitous.[18] These findings highlight the fact that implicit versus explicit racial prejudice may not necessarily cohere in a way that one might expect; in fact, they may show very low to nonexistent relations.[16,18–23]

[a] Race perception (how we come to identify the race of another person), race attitudes (the evaluations we have about a racial group), and prejudice expression (or a prejudicial behavior) are separable processes. For example, an individual can visually perceive race, but the act of race identification may not necessarily result in prejudice expression. These processes can feed off one another such that race perception can activate a host of evaluative associations based on group membership, and these associations can influence behavior. In this chapter, we will often discuss these processes together, but readers should keep in mind that the interventions discussed may not apply equally well to all of these processes. For example, interventions that affect prejudice expression may not affect race processing.

Despite a probable lack of coherence between implicit and explicit prejudice, implicit prejudice is a rather robust phenomenon that holds implications for real-world behavior. A multitude of studies show the existence of implicit prejudice across multiple domains (e.g., gender, age, sports teams, etc.), and individual differences in implicit prejudice are predictive of discriminatory behavior. For example, several studies find that implicit prejudice predicts less friendly nonverbal behavior in intergroup interactions,[13,24,25] as well as biased judgments in social impression formation[26] and mock hiring decisions.[27] In a meta-analysis of studies employing the implicit association test (IAT), a task devised to measure implicit associations between various stimuli and evaluative associations,[28] implicit prejudice was *more* predictive of behaviors and judgments than explicit prejudice.[29] The predictive utility of implicit prejudice as it relates to real-world behavior is apparent even at the societal level: Payne et al.[30] showed that in the 2008 presidential election, voters who were higher in implicit prejudice associating Blacks with unpleasantness were either less likely to vote for Barrack Obama or rather more likely to abstain from voting altogether.[31] The careful reader may wonder whether anti-Black or pro-Black implicit biases had a stronger impact on decisions to vote for Obama. The researchers tested this question by treating each as a separate predictor in the model. Across three studies, increasingly anti-Black attitudes predicted a lower likelihood of voting for Obama, whereas increasingly pro-Black attitudes predicted a greater likelihood of voting for him. Although these findings partially highlight the important implications pro- and anti-Black implicit biases may hold, it is also important to note that because respondents in these studies (much like many other studies on implicit race bias) exhibited an anti-Black bias *on average*, the net effect was a disadvantage for Barack Obama.

The findings from Payne et al.[30] showing the relationship between implicit prejudice and presidential voting behavior serves as a strong representation of the critical relationship between implicit prejudice and population-level outcomes. Payne and colleagues' results indicate that implicit attitudes, despite their seemingly unconscious and unintentional nature, represent a genuine and powerful roadblock to prejudice-reduction efforts at both the individual and the population levels. However, this is just one (albeit important) depiction of the relationship between prejudice and a behavior. It is important to consider a wider range of outcomes because the systematic valuation and beliefs about some groups can add to systemic oppression while the privileged position of other groups reinforces their dominance.[32] Therefore, these and similar effects may be far-reaching and not isolated to one particular situation.

Importantly, group-based biases are not exclusive to readily apparent, deeply familiar social categories such as race, gender, and age; biased person perception is also influenced by social factors that include ostensibly incidental group membership.[33–37] As an illustration of the almost inconceivable way in which this effect can unfold, the arbitrary assignment of a person to a distinct and objectively meaningless novel group is sufficient to create intergroup biases in which members of the perceiver's own group are preferentially favored.[38] These minimal group effects emerge implicitly,[39,40] and even modulate neural responses to faces within 200ms,[41] implying that they occur with some degree of automaticity.

The above studies provide robust evidence that group membership is an important factor in our daily lives and affects how others within our environment are perceived and responded to. Members of one's own social group, or their in-group, are afforded preferential attention and treatment, whereas members of the out-group are processed, on average, in a more superficial manner and treated more poorly than one's in-group members. The magnitude of the consequences of these perceptions and behaviors can be large and can oftentimes result in prejudiced behavior, and such biases even extend to arbitrary groups with which a perceiver has little or no experience.

Many social neuroscientists, much like behavioral social psychologists, are interested in studying intergroup relations. However, they do so with a different and complementary methodological toolbox that allows for the integration of convergent evidence from an exploration of the neural systems to help understand the mechanisms that underlie these phenomena. Neuroscientists correlate neural activation within these systems with measured behaviors—such as stereotyping, prejudice, and discriminatory behaviors—to gain a better understanding of the function of these brain regions in intergroup relations. To elucidate the biological processes of intergroup dynamics, social neuroscientists have increasingly turned to functional magnetic resonance imaging (fMRI).[42–44] fMRI measures brain activity by assessing the local oxygenation of neural tissues (i.e., blood-oxygenation-level dependent, or BOLD signal, which is used as a proxy measure of neuronal activity with the assumption that activated neurons increase consumption of oxygenated blood). This type of measurement is well-suited for quantifying the mechanisms of prejudice, providing insight into the psychological variables that give rise to prejudice, as well as allowing for an implicit measure of mechanism (i.e., individuals do not self-report their brain activity) that can be used to predict attitudes and behavior.

We review the relevant fMRI research on race below, first briefly considering more general functions of each brain region, and provide an overview of how each area contributes to prejudice. We then propose what we believe to be a useful framework (a network of brain areas) within which neuroscientists interested in studying intergroup biases may test relevant theory and make novel predictions to inform our understanding, not only of the mechanisms underlying prejudice, but also the possibilities for effective intervention efforts.

2. DOES A NETWORK OF BRAIN AREAS EXIST THAT IS RELIABLY ASSOCIATED WITH PREJUDICE EXHIBITION?

Converging fMRI evidence suggests that race perception and prejudice evoke a network of activity in regions involved in person perception and emotion processing, including the amygdala, fusiform face area (FFA),[b] and medial prefrontal cortex (mPFC), as well as regions involved in regulation, including dorsolateral prefrontal cortex (dlPFC), anterior cingulate cortex (ACC), and orbitofrontal cortex (OFC).[42–45] We review and discuss recent research using fMRI to investigate race perception and prejudice below.

3. RACE PERCEPTION AND PREJUDICE

Recent reviews have extensively covered the amygdala's role in race perception and prejudice.[42,44,45] The amygdala encompasses a group of nuclei in the anterior temporal lobe that has vast subcortical and cortical connections.[46,47] As such, it is involved in a host of psychological processes that vary from emotional experience, to attention, to memory. Despite these varied functions, amygdala nuclei are most typically associated with the automatic processing of emotional stimuli, particularly with respect to salient emotional stimuli and fear conditioning.[48–50,168] For example, the amygdala is critical for the acquisition, storage, and expression of associative threat and fear.[46,49] It is also involved in the processing of stimuli that have an acquired emotional significance due to previous experience, and it plays an active role in sensitivity to salient environmental cues.[47,51–54]

Taken together, these findings suggest that the amygdala responds broadly to emotional salience[47,54] of *both* negatively and positively valenced stimuli.[55,56] Given these functions, it is perhaps not surprising that the amygdala is critical for the acquisition of affective associations learned within one's social environment, such as those involved in the learning of social category associations.[57] A large body of fMRI research on prejudice implies that the expression of social group biases may share the neural circuits important for fear learning,[58] signifying the amygdala's potentially critical role in this domain.

A large proportion of empirical work on the amygdala's role in social perception and evaluation demonstrates greater amygdala activity when viewing social out-group, rather than in-group, faces.[59–67] For example, White perceivers show increased amygdala activity to Black faces, even in the absence of conscious awareness.[60] The original interpretation from race perception studies was that out-group members evoke threat and consequently increase amygdala reactivity.[62] However, this interpretation has recently been questioned[42,44] as a result of inconsistencies in data, with some studies failing to report greater mean-level amygdala activity when viewing Black versus White faces for White Americans,[63,68–71] and others finding that Black participants show either greater amygdala activity when viewing racial in-group faces[72] or out-group faces.[62]

Intergroup amygdala research suggests that race perception and prejudice may not show *invariant* effects on amygdala activation but must instead be understood within the context of the environment, perceiver, and task. Amygdala findings also support the idea that group-based amygdala differences are, in part, a function of underlying cultural associations and may be less sensitive to in-group/out-group distinctions. For example, a Black perceiver may still hold Black-danger implicit cultural associations, and so amygdala activity in this case may be greater to images of the perceiver's *in-group*.[72] Therefore, it is not necessarily the case that in the presence of an out-group target, amygdala activity will increase.

Although the interpretation that the amygdala is sensitive to cultural associations of racial groups coheres with much of the existing psychological and neuroscience literature on race perception and prejudice, the amygdala is involved in the processing of many types of salient stimuli, not just those that are negatively valenced. This points toward a need to more stringently define the parameters that affect amygdala activity in prejudice.[42] For example, the amygdala responds to novel or ambiguous stimuli,[73] as well as to extreme negatively and positively valenced images.[55,56,74,75] Prior research also shows amygdala activity to cohere with activity in brain regions involved in motivational salience.[47,76] The amygdala may therefore function in part to inform a perceiver about what is important in the environment and then facilitate modulation of appropriate perceptual and attentional processes to respond to the salient stimulus.[47,76] This implies that the interpretation of the

[b] The FFA has largely been explored in the context of face perception. Although this region may be important for decoding and encoding person identity, there is little research that implicates the FFA in prejudice. Future research should clarify the role of the FFA in both prejudice and discrimination based on social group membership.

amygdala's role in prejudice is likely more complex than originally thought and highlights the need for a model of the computational components of amygdala reactivity and a more nuanced determination of the predictive power of these components for discriminatory behavior.

Researchers are actively moving beyond basic race perception studies to explore the relationship between amygdala activity and evaluations. Earlier, we highlighted that negative implicit associations about a social group are predictive of discriminatory behavior.[29] Researchers examining race perception have been interested in elucidating the relationship between neural responses to racial out-group members and implicit attitudes. In an initial demonstration of this relationship, Phelps et al.[68] had participants view pictures of Black and White faces while measuring fMRI and correlated its activity with implicit race bias. Findings showed that the greater the amygdala activity difference to Black compared with White faces, the greater an individual's implicit (anti-Black, pro-White) race bias. However, when the amygdala is damaged, patients still display IAT (pro-White) race bias, implying that implicit associations are not strictly amygdala dependent.[77] In addition, there is no correlation between amygdala activation and explicit race attitudes.[58,60,63,68] Findings such as these support a more complex neural model of prejudice whereby these attitudes are not *singularly* determined, but instead involve a network of brain regions and a larger set of psychological processes.

4. RACE PERCEPTION AND EVALUATION BEYOND THE AMYGDALA

Clues about social group membership are often readily apparent when viewing only a face. As such, face perception is an important aspect of understanding how social identity is processed and recognized. A bevy of research from social and cognitive psychology has provided strong evidence for what is known as the "cross-race effect": individuals are faster and more accurate at remembering and recognizing faces of racial in-group, rather than out-group, members.[78–81] One potential reason for this is because out-group faces are thought to be processed primarily at the category level (e.g., racial group) at the expense of encoding individuating information, and it is perhaps evolutionarily advantageous to more deeply encode in-group members.[82,83] A prime neural candidate for differentiating in- versus out-group faces is the fusiform face area (FFA) in the ventral occipito-temporal cortex, as it is consistently implicated in the recognition of faces and face identity.[84–86]

The cross-race effect and findings showing face sensitivity in the FFA led Golby et al.[87] to assess how this brain region relates to the in-group recognition advantage, the assumption being that the processing of in-group faces may be more nuanced and fine-grained than the processing of out-group members. This differential activity could theoretically lead to better recognition of in-group faces, mirroring the behavioral cross-race effect. In the study, White participants were asked to remember pictures of unfamiliar Black and White faces and non-face objects (i.e., antique radios). Behaviorally, the cross-race effect was replicated—participants showed superior memory for in-group White faces. Moreover, FFA activity was heightened when participants viewed same-race faces compared with other-race faces, and the in-group/out-group activation difference in the left hemisphere was correlated with the in-group memory advantage.[65] Golby et al.[87] reasoned that out-group members were not encoded at the individual level to an equivalent extent as in-group members, as reflected by the lesser FFA activity. This more superficial encoding may relate to poorer memory for out-group members, and future research should attempt to clarify whether the FFA is necessary to produce the in-group memory advantage.

A recent investigation using multivoxel pattern analysis (MVPA) to determine if fMRI activation patterns can predict race from face stimuli showed a much more nuanced relationship between neural activity generated in the FFA and race perception.[88] The researchers successfully predicted the race of faces using FFA activity, but only for those who were higher in implicit pro-White bias.[89,90] This finding has multiple implications. Firstly, greater bias decreases the similarity of FFA representations of race, implying that stronger race bias may be associated with larger differences in the perceptual experience of Black and White faces. That is, those who are higher in implicit race bias may show more preferential or, in this case, individuated processing, of racial in-group faces, whereas those who are lower in implicit bias may experience the perception of in- and out-group faces in a more similar manner. This finding also supports a model whereby cultural associations may drive differences in FFA activation, given that cultural associations are thought to in part drive implicit attitudes. This implies that one's culture can shape the way even seemingly basic perceptual processing of social group membership is carried out.

Critical ways in which in-group members are distinct from out-group members are that individual perceivers typically have more experience and contact with in-group members and often assume that in-group members are more similar to them than out-group members.[91–95] Research examining neural correlates of self-processing suggest that thinking about one's own personality traits or the traits of a familiar but unrelated person (e.g., a famous actor) is linked to activity in the middle mPFC,[96] as compared to thinking about the personality of a dissimilar person.[97–99] By contrast, thinking about a dissimilar other results in heightened dorsal (d) mPFC activity. Recent research has applied these findings to further delineate how the mPFC distinguishes between in-group and out-group biases outside of the domain of racial prejudice. One study had participants think about the opinions and preferences of a person who had a similar or dissimilar political affiliation to their own.[98] The prediction was that more politically identified participants would process the similar person as an in-group member and thus show heightened activity in brain areas, such as the mPFC, that have been linked to self-referential processing. This prediction was made for members of each separate political party, with the expectation that mPFC activity would increase in response to the perceiver's respective political group members as compared to political out-group members. Findings showed that considering the mental state of a member of one's own political party led to activity in the ventral (v) mPFC, whereas considering the mental state of a member of the other political party lead to heightened activity in the dmPFC. Interestingly, individuals who more strongly identified with their respective political group on an implicit measure showed heightened vmPFC activity to politically similar others and less dmPFC activity to politically dissimilar others. The interpretation of these findings is that similar others, relative to dissimilar others, are processed in a way that is closer to how the self is processed. Consistent with this hypothesis, Harris and Fiske[101] showed that when participants viewed members of social out-groups that typically arouse feelings of contempt, such as drug users, less vmPFC activation occurred. If people are more able, motivated, or willing to think deeply about the thoughts and feelings of people with whom they share strong group membership, out-group members may not (on average) receive this preferential processing. However, it is also important to keep in mind that like other brain areas, the mPFC underlies a varied set of psychological processes, so one cannot always assume that self-referential processing has taken place simply due to its activation.

These findings suggest a possible differentiation in the neural correlates of in-group versus out-group perception that are relevant for the exhibition of prejudice and implicate specific subregions of the mPFC in social evaluations that distinguish between those that are categorized as similar to the self versus those that are unfamiliar or dissimilar to the self. The cited research is outside of the domain of race, and as such, the implications of this research for understanding prejudice are speculative, but some have implied that this type of in-group versus out-group processing differentiation may even form the basis of prejudice.[100] Future research should seek to clarify the role of mPFC in racial prejudice.

The reviewed research thus far suggests that the amygdala, FFA (with the caveat that this may be specific to race perception), and the mPFC constitute a network that supports the representation of social group membership and evaluation. However, our perception of social category membership and our underlying evaluative associations are just one piece of the puzzle that aids in our understanding of the expression of prejudice. Although individuals readily notice social category information and are typically deeply familiar with the cultural prejudices and stereotypes that are associated with certain groups, these facts do not necessarily result in prejudicial behavior. In fact, individuals may have strong egalitarian motives that drive their behaviors in intergroup contexts. We next turn to the role of self-regulation in prejudice.

5. SELF-REGULATION AND PREJUDICE

Research on the neural regions involved in prejudice has focused on areas associated with response conflict detection and performance monitoring. This is because responding in a prejudiced manner sometimes results in a conflict between implicit associations and explicit egalitarian goals. As such, many individuals who possess chronic egalitarian goals spontaneously bring online neural mechanisms to diminish implicit race bias.[102,103] This results in a conflict between biased associations and intentional response goals, resulting in activation of the ACC and dlPFC. Both of these regions contribute to executive function and self-regulation, with the dlPFC involved in top-down goal maintenance and emotion regulation,[104,105] and regions of ACC associated with response-related selection,[106] conflict detection,[107,108] and inhibition of prepotent responses.[109,110] This leads to the prediction that the dlPFC and ACC are involved in overcoming expressions of prejudice.[60,61,64] These two regions may work in concert, with the ACC detecting response conflict and the dlPFC engaging regulatory mechanisms to resolve the conflict.[109,111] Additionally, some perceivers may be more likely to recruit regulatory resources. For example, perceivers who hold more egalitarian beliefs may either recruit more self-regulatory resources to successfully decrease their chances of expressing prejudice or these perceivers may require less self-regulatory resources because they may have become more efficient at recruiting executive functions and/or their underlying evaluative associations may have changed. In this way, motivations to control racial prejudice may be chronically activated. A variety of studies have shown both ACC and dlPFC activation in response to simply viewing out-group versus in-group faces.[63,68,69,112,113] Engagement of these regions when passively viewing out-group faces may serve a preemptive function by recruiting the regulatory resources necessary for overcoming a prejudicial response.

Although dlPFC and ACC activation is a typical response for perceivers viewing out-group versus in-group faces, it appears that activation in these regions does not necessarily occur automatically. It is more likely that individuals must first detect the potential for racial bias in order to bring these resources online. For example, Cunningham et al.[60] found attenuation of amygdala responses to Black faces presented when participants were aware they were viewing them (supraliminally), compared with Black faces presented when participants were unaware of their presentation (subliminally). Attenuation of amygdala activation correlated with activation of both the dlPFC and ACC.[63,69,114,115] Therefore, neural mechanisms may be in place that serve to regulate racially biased responding, but only if a perceiver is aware of the potential for responding in a prejudiced manner. Studies of this nature implicate a model of prejudice regulation such that the ACC monitors for conflict between explicit intentions and implicit attitudes, and the dlPFC is brought online to control unwanted, implicit expression of racial biases.[66]

Of course, regulatory resources such as these are not always available and are sensitive to contextual changes. For example, stress and fatigue can diminish executive functions[116,117] and resource depletion exacerbates race bias, perhaps influencing the effectiveness of interventions.[118,119] Moreover, situational factors can shape the egalitarian goals of perceivers and can modulate ACC and dlPFC race-based activity. For example, Krill and Platek[120] found that being excluded by in-group partners enhanced ACC activation, as compared with social exclusion by out-group partners, implying greater conflict induced by own-race exclusion. Additionally, Forbes et al.[114] found that listening to stereotypical music (rap compared with heavy metal) increased amygdala and dlPFC activation to Black versus White faces. These findings in concert point to the fragility of and variance in self-regulation in intergroup contexts.

It is important to note that the studies reviewed above typically involved relatively simple perceptual judgments in which race was not directly task-relevant (e.g., whether the facial stimulus was presented to the right or left of fixation). Mere exposure to a stigmatized racial out-group may activate some degree of behavioral regulation,[121] but these studies do not provide compelling evidence regarding precisely how areas involved in behavior regulation engage in response to racial cues during more complex tasks that themselves present a regulatory challenge. Moreover, these studies do not provide direct evidence that refraining from a prejudicial response is the result of dlPFC and ACC functions. Research aimed at providing more insight into the role of prejudice regulation in more complex, real-world scenarios will prove fruitful in elucidating the mechanisms underlying its control.

The dlPFC and ACC are most likely not alone in regulating intergroup responding. The OFC inhabits the ventral (bottom) surface of the frontal part of the brain and is implicated in a variety of processes, but with respect to situations involving intergroup prejudice, it is theorized to be involved in the evaluation of the relative appropriateness of one's responses, activating both to receiving rewards and avoiding punishments.[121,122] Thus, the OFC appears to be more generally involved in current subjective evaluation. Given this association, the OFC may be a prime candidate for facilitating regulation of perceivers' evaluations of targets when there is a potential to respond with prejudice.[123] For example, recent research shows that OFC activity is associated with perceivers' judgments about the potential to become friends with out-group Black individuals.[124] Furthermore, given the OFC's reciprocal connections with the amygdala, it may play a critical role in modulating amygdala activity elicited by exposure to racial out-group faces if a perceiver's initial prejudiced response conflicts with explicit, overarching egalitarian motives.[61,125–128] In this context, the OFC may function much like the dlPFC and ACC reviewed above. It will be fruitful to clarify the independent and interactive roles of the OFC, dlPFC, and ACC in integrating motives with behavior in an intergroup context. Moreover, OFC activity is associated with perceivers' preferences for members of their own experimentally manipulated minimal groups, independent of target race.[34] These findings suggest that the OFC may play a broader role in social evaluation, one that extends beyond the realm of racial prejudice.

To the extent that the amygdala relays information regarding expected outcomes following the perception of a target that elicits a prejudiced response, and the OFC integrates social motives with behaviors to represent the current state of the perceiver, the dense reciprocal connections between amygdala and OFC allow for a comparison of expected rewards and punishments (e.g., social shaming) with current experience (i.e., the feeling of prejudice). Support for this idea comes from research demonstrating OFC activations following value-based expectancy violations[129] and the inability of patients with OFC damage to update representations when predictions and outcomes are incongruent.[130,131] Thus, whereas subcortical systems, such as the amygdala, provide a low-resolution estimate of likely outcomes, regions of the OFC may be involved in integrating such output with current experience, allowing the current context to dictate how social evaluation is shaped.[125,126,132,133,134] This function of the OFC is important to consider in the domain of prejudice. It appears that the OFC functions in such situations to modulate prejudiced responses to a target so that the evaluative response coheres with the current context. Theoretically, this could result in entirely different patterns of results, depending upon whether the experimental context is one in which social expectancies are salient versus when social expectancies are minimal, or if the environment is discouraging versus encouraging of biased responses (e.g., private versus public response conditions[135]). Future research should vary the experimental context within which intergroup processing is measured to provide a more nuanced gauge of OFC response variability in social perception and evaluation.

The regions reviewed above are interesting to consider in the domain of prejudice, given their functional roles in emotional responding and learning, motivational salience, cognitive control, and the experience and expectation of reward and evaluation (Figure 1).

Their anatomical connections provide more reason to consider each of their roles in tandem, moving from a modulatory to a network exploration of function. Research on the neural mechanisms underlying intergroup processing provides insight into prejudice and discrimination, highlighting the basic psychological processes involved. By integrating cognitive and affective science and psychology with our social psychological knowledge of intergroup processing, we may be more likely to identify points of intervention. Translation of this knowledge into prejudice interventions requires a better understanding of not only the acquisition, storage, and expression of prejudice, but also the mechanisms that diminish prejudice.

6. ARE NEURAL RESPONSES DURING PREJUDICE EXPRESSION MALLEABLE?

Given that prejudice appears to be a rather robust and easily induced phenomenon, the fact that it can be implicit in nature, and also that a network of neural regions is reliably activated during intergroup perception and prejudice expression, should we resign ourselves to the belief that prejudice is necessary or inevitable? As behavioral and brain sciences have progressed, a more flexible view of social categorization has emerged, with dominant theory suggesting that person perception is a

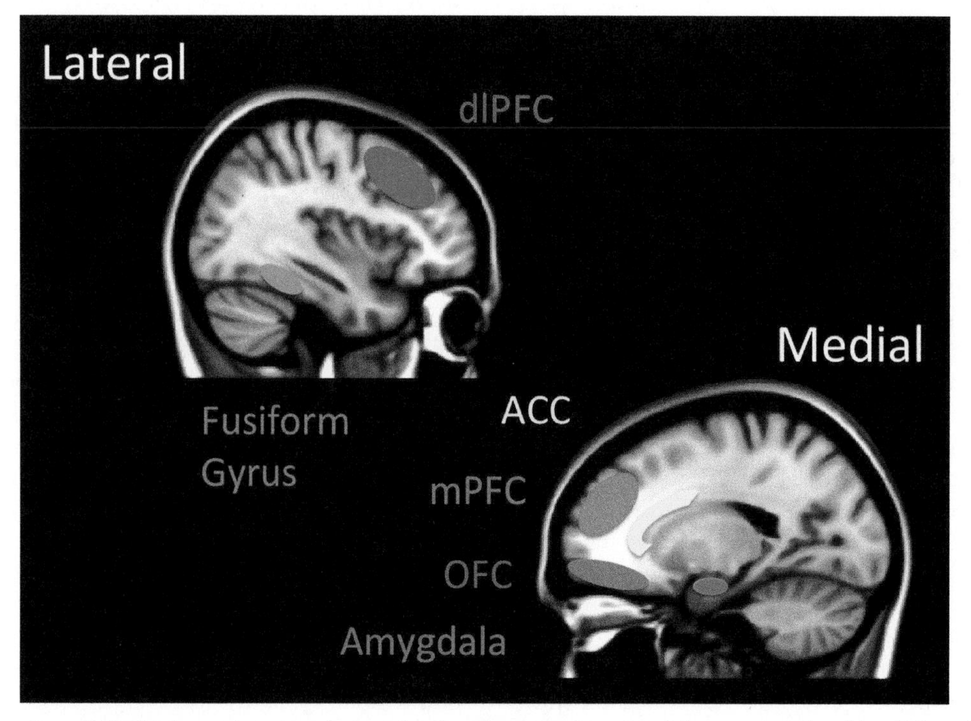

FIGURE 1 The regions of the brain most commonly associated with processing of social group membership. Although these regions are involved in a number of processes, in this chapter, we highlight their theoretical contribution to racial prejudice. We propose a network of regions implicated in race processing and prejudice that includes areas important for race perception and evaluation (amygdala, FFA, and mPFC) and areas important for self-regulation (dlPFC, ACC, and OFC). The amygdala (medial view) is implicated in learning about and detecting salient things in our environments and plays a role in fear learning and expression. The fusiform face area (FFA lateral view), located in the fusiform gyrus, is thought to extract physical information from faces to distinguish among individuals. The medial prefrontal cortex (mPFC; medial view) is commonly activated when thinking about one's self and similar others compared with dissimilar others. Together, the amygdala, FFA, and mPFC seem to support the perception and evaluation of racial out-group members. The dorsolateral prefrontal cortex (dlPFC) and anterior cingulate cortex (ACC) are implicated in self-regulation and are important for the top-down goal maintenance and emotion regulation. The orbitofrontal cortex (OFC) is involved in decision making, emotion regulation, and reward expectation and may be important for integrating societal group norms and internalized motives. The common activation of this network of regions when people think about the feelings, thoughts, and intentions of individuals from different social groups indicates that intergroup processing involves a variety of complex psychological processes.

dynamic process[135] and that racial prejudice is not necessarily innate or inevitable.[136,137] In fact, an effective organizational framework for prejudice intervention has been proposed. Racial prejudice is thought to contain an associative component (Automatic Prejudiced Association: Social Group X=Bad), as well as a control component (Stable Egalitarian Goal: Social Group X≠Bad).[138] With this framework in mind, the most effective and enduring prejudice reduction techniques likely focus on altering both the automatic associative component (Social Group X=Good) as well as the initiated self-regulatory component (i.e., reinforce egalitarian goals). Social neuroscientists have recently begun to explore the neural mechanisms underlying intervention and how they relate to current cognitive and affective neuroscience models of self- and emotion regulation.[44] Despite these efforts, many unanswered questions remain. We briefly review the little that is known about how the malleability of intergroup perception and prejudice is reflected in the brain. Due to the prevalence of implicit negative racial associations as opposed to explicit racial prejudice in the US, we focus our review on implicit prejudice interventions.

6.1 Counterstereotypic Imagining

Counter stereotypic imagining is a strategy that provides perceivers with concrete examples of individuals who do not conform to common stereotypes[139] or prejudices.[140,141] These examples can range from those who are famous or familiar (e.g., Barack Obama) or unknown and unfamiliar (e.g., a Black professor). This technique gives perceivers a counterexample that they may not otherwise encounter in their daily lives and aims to reinforce the recognition that applying overgeneralized evaluations to every individual in a social group is a flawed method. In an fMRI study that sought to understand the neural mechanisms associated with this strategy, perceivers were presented with familiar positive Black and White Americans.[68] Unlike conditions in which unfamiliar individuals were presented, individual differences in amygdala activity between Black and White familiar/positive exemplars was not predictive of implicit pro-White/anti-Black bias. Recent work further delineates the amygdala's role in counterstereotypic imagining, showing that when depictions of White and Black individuals are shown to violate stereotypic norms (e.g., a White individual in a negative role and a Black individual in a positive role) amygdala activity is heightened relative to norm-consistent behaviors.[71] This finding provides additional support for the prediction that counterstereotypic imagining alters group-based amygdala processing and implies that amygdala activation may not be race- or group-specific, but instead is likely sensitive to novel and/or salient stimuli. Moreover, this work highlights the potential importance of forming counterattitudinal associations as a means of dynamically shaping intergroup processing.

6.2 Perspective Taking

Perspective taking is a prejudice-reduction technique whereby perceivers are encouraged to think about the world from the vantage point of out-group members.[141] This strategy affords individuals the opportunity to understand how similar they are to out-group members and reinforces that it is important to think about the intentions and situation of out-group members, rather than relying on group-based assumptions.[143] Critically, perspective taking decreases stereotyping and increases empathy.[142] One possible way to attenuate differential patterns of activity observed in the neuroimaging race literature on in-group/out-group perception is to increase out-group empathy via perspective taking.[101,145–150] As highlighted above, thinking of others' internal mental states activates the mPFC,[142,143] and considering the personality traits of familiar others, as compared with dissimilar others, also increases mPFC activity.[97–99] The mechanism posited to account for these effects is that accessibility of the self-concept and self-other overlap in mental representations increases during perspective taking, resulting in diminished differences in mPFC activity between out-group and in-group members, and also prompting in-group-relevant processing and less stereotyping.[98,141,145] These findings are relevant when considering a recent finding that shows that the neural representation of race is impacted by racial identification, or the extent to which an individual identifies with their own racial group.[113] It is perhaps the case that individuals who identify less with their racial in-group view out-group members as more similar to themselves. To date, it is unclear how perspective taking shapes associative intergroup learning and self-regulatory intergroup processing. However, it is possible that having a goal to take the perspective of another person may engage greater attention to the target and self-referential processes that may relate to mPFC activity.

6.3 Individuation

Individuation is a strategy that requires a perceiver to put forth time and cognitive effort to process others.[3,5] This strategy encourages perceivers to learn and consider personal information about out-group members or expectancy-violating information.[171] As such, this strategy gives perceivers the ability to associate out-group members with personal information, rather than relying on group-based associations.[152] Those who are similar to a perceiver, such as in-group members, are oftentimes spontaneously individuated.[5] Moreover, similar to perspective taking, researchers have observed heightened mPFC activity when a perceiver's goal is to form an individuated impression.[144] Much research from social psychology and social cognition supports the notion that an individuation goal during an encounter can influence intergroup processing.[151] In line with this, social neuroscience work has shown that when participants' goals are to think about Black and White individuals' preferences while viewing pictures of their faces, no race-based amygdala differences emerge.[67,72] Therefore, a goal to individuate out-group targets may moderate the neural systems underlying group-based processing and perhaps increase the likelihood of more nuanced representations of out-group members.[152] Future research should clarify the neural mechanism(s) underlying individuation and also consider the extent to which individuation training employing a single target generalizes to individuated processing of other out-group members.

6.4 Contact

Positive interactions with out-group members are afforded via increased intergroup contact.[92,153] These types of interactions can serve to fend off negative group stereotypes and reduce any uncertainties that can arise from novel intergroup interactions. A key question in race processing is whether negative group-based associations are innate or if they rather develop due to exposure to cultural associations.[154] Recent fMRI work seeking to answer this question hypothesized that amygdala differences due to race are culturally acquired and likely emerge over time.[155] Results showed that race-based amygdala differences materialize during adolescence (around the age of 16) but are nonexistent in early childhood (around age 4). Moreover, greater interracial contact during adolescence attenuates amygdala responses to Black versus White familiar faces (Figure 2).[59] However, it is important to keep in mind that simple exposure to out-group members may not be sufficient to reduce biases in amygdala activity due to race, whereas *quality* of contact may prove to be a more successful determinant. As an illustration of this, the number of romantic out-group partners one has is negatively associated with fear extinction learning for out-group members.[154] In other words, whereas out-group members may be associated with negativity at the mean level, increased close contact attenuates this response.

6.5 Prejudice Replacement

Prejudice replacement refers to the strategy of learning to identify and replace prejudicial responses with nonprejudicial responses. For example, this could take the form of replacing the "Asians are cold" stereotype with "Asians are warm."[156] Although often described as an explicit strategy, prejudice replacement is akin in some respects to fear extinction learning. Fear extinction to in-group members is much easier than that for out-group members, suggesting that it may be difficult to restructure overly learned associations underlying out-group members, particularly in the long term.[154] Recent research has extended prejudice replacement research by capitalizing on the known relationship between mechanisms involved in fear learning and expression and the learning and expression of racial attitudes, predicting that it may be possible to abolish implicit bias expression by way of pharmacological interventions that target emotional

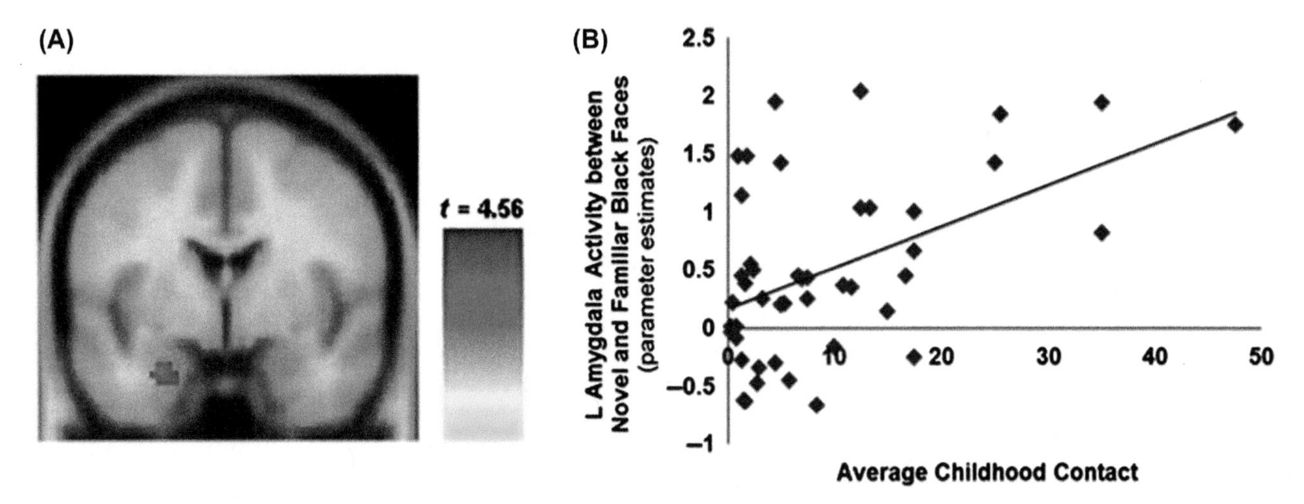

FIGURE 2 Panel A represents a whole-brain regression analysis exploring how childhood intergroup contact relates to amygdala activity to Black novel faces versus Black familiar faces. Cloutier, Li, and Correll[59] observed greater left amygdala activity to novel Black faces compared with familiar Black faces for individuals with greater intergroup childhood contact. Panel B is a scatterplot of the relationship within the left amygdala. This research represents a remarkable advancement in our understanding of how early childhood intergroup contact can shape neural responses to race even years later. Future research should also consider the interplay between quantity and quality of intergroup contact in race processing. *Panels A and B modified, with permission, from Ref. 59 © (2014) Massachusetts Institute of Technology Press.*

memory and perception.[44] An especially promising pharmacological intervention is the use of propranolol, a β adrenergic receptor antagonist that impairs memory consolidation and reconsolidation in humans.[157–160] A single dose of propranolol, relative to a placebo control, effectively diminishes implicit race bias.[161] Reconsolidation research in humans has been explored in the context of newly acquired fear associations; therefore, the mechanism by which propranolol diminishes overly learned, implicit group-based attitudes that have a complex associative structure is unknown. Although pharmacological intervention is perhaps an extreme intervention prospect for prejudice reduction, research of this type may shed further light on the psychological mechanisms underlying prejudice intervention.

The reviewed research on prejudice intervention begins to outline potential mechanisms of prejudice reduction and raises the critical question of whether there may be a common set of psychological factors that underscore these interventions that can be identified through fMRI. What mechanisms result in the most robust and reliable changes in neural activity and discriminatory behavior? Do changes within these regions predict decreases in discrimination? fMRI research in this domain has begun to address these questions. The reviewed research proposes that the most successful and lasting prejudice reduction techniques target the associative components, bolster the activated self-regulatory component, and bring processing inline with similar others (Table 1) as we observe changes both in amygdala and prefrontal activity. To date, we have not observed a change in FFA activation when manipulating these intervention techniques. However, that does not rule out a role for the FFA or other regions in implicit prejudice reduction. For example, when individuals participate in a minimal group task with mixed race participants, FFA activity is similar for in-group members of various races.[34]

From the reviewed research, we can infer that many of these interventions influence the regions involved in forming and expressing our group-based associations, but also seem to influence areas that both regulate and integrate that information into decisions. We can also infer that this combination results in the successful attenuation of implicit biased responding, but support for this assumption is rather limited. Additionally, it remains unclear the relative importance of each of these neural regions in reducing implicit prejudice. Future work should consider under what circumstances these interventions are successful, what exact psychological mechanism(s) are altering implicit prejudice, whether these changes are lasting, and how these interventions affect real-world discriminatory behaviors. It is also unclear to what degree simply shaping race perception versus changing underlying group-based associations and/or activating motivations is impactful for reducing implicit prejudice. What is clear is that we have just begun to delve into prejudice intervention neuroscience research and future work should strive to manipulate the factors in addition to measuring individual differences that relate to these interventions. The reviewed strategies are also strikingly similar to techniques used in emotion regulation and fear attenuation and imply that researchers may benefit from borrowing theory and methods from these literature in an effort to discover more useful information about how to intervene in implicit prejudice. In doing so, however, it will be important for researchers to fully consider feasibility and external validity of these interventions as they apply to public policy recommendations. Additionally, despite the great potential for these types of programs to succeed in combating prejudice, it will also be important to fully consider the potential risks inherent in adopting them.

7. CONCLUSIONS AND CONSIDERATIONS

The understanding of social group processing and evaluation is invaluable, as it gives scholars an insight into the mechanisms involved in reducing discrimination. Social neuroscientists since the 1990s have deepened our understanding of prejudice intervention, but ultimately researchers are seeking to push this frontier further by bridging the gap between laboratory brain science and real-world behavior during interactions and judgments that hold more realistic social consequences. In other words, increased attention has justifiably been directed to the neural and psychological correlates of real-world social decision making.[170] We do not operate in a vacuum in our everyday lives, nor should the social processes that we study in the lab. We have only begun to scratch the surface in this domain, but there is much to be optimistic about.

One promising approach to modeling how group membership influences social decision making uses neuroeconomics as a theoretical and methodological framework to bridge brain and behavioral science. Neuroeconomics is highly interdisciplinary in nature, as it combines economic paradigms, computational modeling, and neuroscience.[162] A study by Stanley et al.[166] (2011) was one of the first investigations to adapt a neuroeconomic approach to intergroup decision making, exploring the relationship between implicit bias and economic decisions to trust a partner.[167] In this study, IAT was correlated with decisions to trust. Individuals with pro-White bias invested more money with White than Black partners. In a follow-up examination, greater investment in White compared with Black partners correlated with activity in the striatum, a brain region implicated in valuation.[58] This supports

TABLE 1 Neuroimaging Studies Exploring Implicit Prejudice Intervention

Intervention	Strategy description	Study	Findings	MNI (X, Y, Z)	Region
1. Prejudice replacement	Identifying prejudices, labeling them, and replacing them with nonprejudicial responses	None	None	None	None
2. Counterstereotypic imaging	Imagining examples of out-group members who counter held stereotypes	Schreiber and Iacoboni[71]	Greater activity to norm violating Black and White targets.	Anatomically defined for right and left: Harvard–Oxford subcortical structural atlas	Amygdala
				−2, 48, −7	mPFC
				16, −74, −10 (R) −26, −58, −14 (L)	Fusiform
				36, −80, 12 (R) −30, −100, 6 (L)	Middle occipital
				18, −56, 14 (R)	Posterior cingulate
		Phelps et al.[68]	No longer a relationship between IAT performance and race differences in amygdala activity when viewing positive famous Black and White faces	31.7, −5, 12.2 (R) −17.6, −5, −10.8 (L) (Talairach)	Amygdala
3. Individuation	Viewing others according to their personal, rather than stereotypic, characteristics	Wheeler et al.[67]	When making an individuated preference judgment no longer observe race differences in amygdala activity	−20, −10, −14 (L) (Talairach)	Amygdala
4. Perspective taking	Adopting the perspective of an out-group member	None	None	None	None
5. Contact	Increasing exposure to out-group members	Cloutier, Li, and Correll[59]	Greater activity to familiar Black than familiar White faces	−21, −102, −9 (left)	Inferior occipital cortex
			Greater activity to familiar Black than unfamiliar Black faces	39, −63, 39 (R) −33, −66, 42 (L)	Inferior parietal lobe
				−3, −30, 30	Posterior cingulate gyrus
				−9, −69, 30	Precuneus
				−45, 18, 36 (L)	Middle frontal gyrus
				−39, 45, 6 (L)	Inferior frontal gyrus

Studies included in this table represent only research where these intervention techniques were manipulated. This table highlights the gaps in this research and emphasizes the need for more fMRI implicit prejudice intervention research. From this small body of research, it appears that these implicit prejudice interventions shape neural activity across a wide range of both subcortical and cortical regions.

a model whereby action values are integrated with evaluative associations in the amygdala via the striatum.[164]

Stanley et al.[58] study highlight's another promising avenue for future social neuroscience research: a broader consideration of the under-examined brain structures that may be involved in social category processes. Research relevant for the striatum's role in prejudice is scant, but its role in intergroup biases has been shown in a scenario in which perceivers have no experience with the out-group (i.e., it is experimentally manipulated).[34] In this study, participants were randomly assigned to a novel mixed-race team without a history of contact or conflict with an out-group team. Subsequently, participants memorized the team membership of various faces, and these faces were then presented during fMRI scanning. Heightened ventral striatum activation to in-group, rather than out-group, faces occurred, and this activation correlated with self-reported preferences for in-group (versus out-group) members. The results from this study support social psychological posits that without a history of prejudiced responses toward the out-group or preexisting stereotypical associations, one's in-group may be motivationally primary. These findings expand upon the neural model of prejudice and indicate that intergroup action values may also modulate intergroup discrimination. Findings such as these that involve a wider network of brain structures will likely emerge more in the future, as the scope of fMRI and intergroup neuroscience research grows. It will be increasingly important to consider the activation of the entire network and relationships among regions to gain a richer understanding of intergroup relations.

Financial decisions represent only a subset of real-world decisions that are impacted by social category information. For example, race can also influence legal decision making,[165] and social neuroscientists have begun to explore how neural processing of race influences judicial behaviors. Specifically, race differences in BOLD responses are shown to correlate with discrimination damage awards for Black victims,[115] with dlPFC and parietal cortex increases relating to damages awarded. These studies provide a model for future intergroup neuroscience research to continue to bridge our laboratory research with socially consequential discriminatory behavior. Research that investigates the real-world domains where implicit prejudice is more or less likely to occur will broaden our understanding of intergroup behavior. Moreover, implicit prejudice intervention research should continue to expand into other domains in which discrimination can occur, such as education, employment, and health care.

There is a clear need for more work outlining the psychological and neural factors of implicit prejudice intervention. Initial studies have focused on basic-level phenomena or behavioral processes that exacerbate implicit prejudice. More recent work is extending beyond exploring only the factors that produce and exacerbate racial bias to understand the mechanisms of implicit prejudice mitigation. With these efforts, a more detailed picture of the underlying psychological and neural mechanisms of prejudice intervention will emerge. Additionally, it is not enough to demonstrate diminished prejudice in the lab, and research should aim to extend our understanding of how the neural and behavioral correlates of prejudice intervention predict real-world decreases in discrimination. A possible next step in this area is to compare the impact of intervention strategies and explore how effective these strategies are across contexts to provide a better understanding of which strategy is most effective, and in what types of situations.

To date, the majority of the prejudice literature examines responses to Black and White race categories in US participants, and as a result, this chapter primarily concentrated on this literature. It is important that future neuroscience work on race include a variety of racial groups across cultures to facilitate a more complete understanding of stereotyping and prejudice and the steps that can be taken to diminish discrimination. Moreover, researchers interested in social group membership should broadly sample both in terms of stimuli and participants when exploring discrimination interventions to make more informed policy recommendations. By combining affective, social, cognitive, and economic neuroscience approaches and insights with decision tasks reflecting socially relevant consequences, we will obtain a better understanding of how our implicit biases may, or may not, impact the choices we make[169]. Additionally, we highlighted some studies that employed the use of novel groups (e.g., Ref. 34). Studies such as these allow for the examination of biases in social processes that may generalize across social groups, those that may not be confounded with preexisting attitudes and cultural associations, and provide insight into the basic set of psychological and neural processes that underlie discrimination. We view these types of studies as critical to understanding the perception and processing of social categories more generally, and as such, an important foundation from which more nuanced examinations may be conducted.

References

1. Brewer MB, Feinstein ASH. Dual processes in the cognitive representation of persons and social categories. In: Chaiken S, Trope Y, eds. *Dual Process Theories in Social Psychology*. New York: The Guilford Press; 1999:255–270.
2. Brown R. *Prejudice: Its Social Psychology*. John Wiley & Sons; 2011.
3. Brewer MB. A dual process model of impression formation. In: 1st ed. Srull T, Wyer R, eds. *Advances in Social Cognition*. vol. 1. Hillsdale, NJ: Lawrence Erlbaum Associates; 1988:1–36.
4. Bruner JS. On perceptual readiness. *Psychol Rev*. 1957;64(2): 123–152. http://dx.doi.org/10.1037/h0043805.
5. Fiske S, Neuberg SL. A continuum of impression formation, from category-based to individuating processes: Influences of information and motivation on attention and interpretation. In: 23rd ed. Zanna MP, ed. *Advances in Experimental Social Psychology*. vol. 23. New York: Academic Press; 1990:1–74.
6. Tajfel H. Cognitive aspects of prejudice. *J Biosoc Sci*. 1969;1(S1): 173–191. http://dx.doi.org/10.1017/S0021932000023336.
7. Bodenhausen GV, Macrae CN, Garst J. Stereotypes in thought and deed: social-cognitive origins of intergroup discrimination. In: Sedikides C, Schopler J, Insko CA, eds. *Intergroup Cognition and Intergroup Behavior*. vol. xiv. Mahwah, NJ: Lawrence Erlbaum Associates Publishers; 1998:311–335.

8. Bodenhausen GV, Macrae CN, Sherman JS. On the dialectics of discrimination: dual processes in social stereotyping. In: Chaiken S, Trope Y, eds. *Dual-Process Theories in Social Psychology*. vol. xiii. New York: Guilford Press; 1999:271–290.
9. Macrae CN, Bodenhausen GV, Milne AB, Jetten J. Out of mind but back in sight: stereotypes on the rebound. *J Pers Soc Psychol.* 1994;67(5):808–817. http://dx.doi.org/10.1037/0022-3514.67.5.808.
10. Cohen CE. Person categories and social perception: testing some boundaries of the processing effect of prior knowledge. *J Pers Soc Psychol.* 1981;40(3):441–452. http://dx.doi.org/10.1037/0022-3514.40.3.441.
11. Stangor C. Stereotype accessibility and information processing. *Personal Soc Psychol Bull.* 1988;14(4):694–708. http://dx.doi.org/10.1177/0146167288144005.
12. Dovidio JF, Kawakami K, Johnson C, Johnson B, Howard A. On the nature of prejudice: automatic and controlled processes. *J Exp Soc Psychol.* 1997;33:510–540. http://dx.doi.org/10.1006/jesp.1997.1331.
13. Fazio RH, Jackson JR, Dunton BC, Williams CJ. Variability in automatic activation as an unobtrusive measure of racial attitudes: a bona fide pipeline? *J Pers Soc Psychol.* 1995;69(6):1013–1027. http://dx.doi.org/10.1037/0022-3514.69.6.1013.
14. Fazio RH, Olson MA. Implicit measures in social cognition research: their meaning and uses. *Annu Rev Psychol.* 2003;54:297–327. http://dx.doi.org/10.1146/annurev.psych.54.101601.145225.
15. Gawronski B, Bodenhausen GV. Associative and propositional processes in evaluation: an integrative review of implicit and explicit attitude change. *Psychol Bull.* 2006;132(5):692–731. http://dx.doi.org/10.1037/0033-2909.132.5.692.
16. Hahn A, Judd CM, Hirsh HK, Blair IV. Awareness of implicit attitudes. *J Exp Soc Psychol.* 2014;143(3):1369–1392. http://dx.doi.org/10.1037/a0035028.
17. Schuman H, Steeh C, Bobo L, Krysan M. *Racial Attitudes in America: Trends and Interpretations.* Cambridge, MA: Harvard University Press; 1997.
18. Nosek BA, Smyth FL, Hansen JJ, et al. Pervasiveness and correlates of implicit attitudes and stereotypes. *Eur Rev Soc Psychol.* 2007;18(1):36–88. http://dx.doi.org/10.1080/10463280701489053.
19. Nosek BA. Moderators of the relationship between implicit and explicit evaluation. 2005;134(4):565–584. http://dx.doi.org/10.1037/0096-3445.134.4.565.
20. Nosek BA, Hansen JJ. Personalizing the Implicit Association Test increases explicit evaluation of target concepts. *Eur J Psychol Assess.* 2008;24(4):226–236. http://dx.doi.org/10.1027/1015-5759.24.4.226.
21. Blair IV. Implicit stereotypes and prejudice. In: Moskowitz GB, ed. *Cognitive Social Psychology: The Princeton Symposium on the Legacy and Future of Social Cognition.* Mahwah, NJ: Lawrence Erlbaum Associates, Inc; 2001:359–374.
22. Hofmann W, Gschwendner T, Schmitt M. On implicit-explicit consistency: the moderating role of individual differences in awareness and adjustment. *Eur J Pers.* 2005;19(1):25–49. http://dx.doi.org/10.1002/per.537.
23. Hofmann W, Gawronski B, Gschwendner T, Le H, Schmitt M. A meta-analysis on the correlation between the implicit association test and explicit self-report measures. *Personal Soc Psychol Bull.* 2005;31(10):1369–1385. http://dx.doi.org/10.1177/0146167205275613.
24. Dovidio JF, Kawakami K, Gaertner SL. Implicit and explicit prejudice and interracial interaction. *J Pers Soc Psychol.* 2002;82(1):62–68. http://dx.doi.org/10.1037//0022-3514.82.1.62.
25. McConnell AR, Leibold JM. Relations among the implicit association test, discriminatory behavior, and explicit measures of racial attitudes. *J Exp Soc Psychol.* 2001;37(5):435–442. http://dx.doi.org/10.1006/jesp.2000.1470.
26. Lambert AJ, Payne BK, Ramsey S, Shaffer LM. On the predictive validity of implicit attitude measures: the moderating effect of perceived group variability. *J Exp Soc Psychol.* 2005;41(2):114–128. http://dx.doi.org/10.1016/j.jesp.2004.06.006.
27. Ziegert JC, Hanges PJ. Employment discrimination: the role of implicit attitudes, motivation, and a climate for racial bias. *J Appl Psychol.* 2005;90(3):553–562. http://dx.doi.org/10.1037/0021-9010.90.3.553.
28. Greenwald AG, McGhee DE, Schwartz JLK. Measuring individual differences in implicit cognition: the implicit association test. *J Pers Soc Psychol.* 1998;74(6):1464–1480. http://dx.doi.org/10.1037/0022-3514.74.6.1464.
29. Greenwald AG, Poehlman TA, Uhlmann E, Banaji MR. Understanding and using the Implicit Association Test: III. Meta-analysis of predictive validity. *J Pers Soc Psychol.* 2009;97(1):17–41. http://dx.doi.org/10.1037/a0015575.
30. Payne BK, Krosnick JA, Pasek J, Lelkes Y, Akhtar O, Tompson T. Implicit and explicit prejudice in the 2008 American presidential election. *J Exp Soc Psychol.* 2010;46(2):367–374. http://dx.doi.org/10.1016/j.jesp.2009.11.001.
31. Greenwald AG, Smith CT, Sriram N, Bar-Anan Y, Nosek BA. Implicit race attitudes predicted vote in the 2008 U.S. presidential election. *Anal Soc Issues Public Policy.* 2009;9(1):241–253. http://dx.doi.org/10.1111/j.1530-2415.2009.01195.x.
32. Hardin CD, Banaji MR. The nature of implicit prejudice: implications for personal and public policy. In: Shafir E, ed. *Policy Implications of Behavioral Research.* 2013:13–30.
33. Hugenberg K, Corneille O. Holistic processing is tuned for in-group faces. *Cogn Sci.* 2009;33(6):1173–1181. http://dx.doi.org/10.1111/j.1551-6709.2009.01048.x.
34. Van Bavel JJ, Packer DJ, Cunningham WA. The neural substrates of in-group bias: a functional magnetic resonance imaging investigation. *Psychol Sci.* 2008;19(11):1131–1139. http://dx.doi.org/10.1111/j.1467-9280.2008.02214.x.
35. Van Bavel JJ, Packer DJ, Cunningham WA. Modulation of the fusiform face area following minimal exposure to motivationally relevant faces: evidence of in-group enhancement (not out-group disregard). *J Cogn Neurosci.* 2011;23(11):3343–3354. http://dx.doi.org/10.1162/jocn_a_00016.
36. Young S, Bernstein MJ, Hugenberg K. When do own-group biases in face recognition occur? Encoding versus post-encoding. *Soc Cogn.* 2010;28(2):240–250. http://dx.doi.org/10.1521/soco.2010.28.2.240.
37. Young SG, Hugenberg K. Mere social categorization modulates identification of facial expressions of emotion. *J Pers Soc Psychol.* 2010;99(6):964–977. http://dx.doi.org/10.1037/a0020400.
38. Tajfel H, Billig M, Bundy R, Flament C. Social categorization and intergroup behaviour. *Eur J Soc Psychol.* 1971;1:149–178.
39. Ashburn-Nardo L, Voils CI, Monteith MJ. Implicit associations as the seeds of intergroup bias: how easily do they take root? *J Pers Soc Psychol.* 2001;81(5):789–799. http://dx.doi.org/10.1037/0022-3514.81.5.789.
40. Otten S, Wentura D. About the impact of automaticity in the minimal group paradigm: evidence from affective priming tasks. *Eur J Soc Psychol.* 1999;29(8):1049–1071. http://dx.doi.org/10.1002/(SICI)1099-0992(199912)29:8<1049::AID-EJSP985>3.0.CO;2-Q.
41. Ratner KG, Amodio DM. Seeing "us vs. them": minimal group effects on the neural encoding of faces. *J Exp Soc Psychol.* 2013;49(2):298–301. http://dx.doi.org/10.1016/j.jesp.2012.10.017.
42. Chekroud AM, Everett JAC, Bridge H, Hewstone M. A review of neuroimaging studies of race-related prejudice: does amygdala response reflect threat? *Front Hum Neurosci.* 2014;8:179. http://dx.doi.org/10.3389/fnhum.2014.00179.
43. Cikara M, Van Bavel JJ. The neuroscience of intergroup relations an integrative review. *Perspect Psychol Sci.* 2014;9(3):245–274. http://dx.doi.org/10.1177/1745691614527464.

44. Kubota JT, Banaji MR, Phelps EA. The neuroscience of race. *Nat Neurosci*. 2012;15:940–948.
45. Amodio DM. The neuroscience of prejudice and stereotyping. *Nat Rev Neurosci*. 2014. http://dx.doi.org/10.1038/nrn3800.
46. LeDoux J. The amygdala. *Curr Biol*. 2007;17(20):R868–R874. http://dx.doi.org/10.1016/j.cub.2007.08.005.
47. Cunningham WA, Brosch T. Motivational salience amygdala tuning from traits, needs, values, and goals. *Curr Dir Psychol Sci*. 2012;21(1):54–59. http://dx.doi.org/10.1177/0963721411430832.
48. Davis M, Whalen PJ. The amygdala: vigilance and emotion. *Mol Psychiatry*. 2001;6(1):13–34. http://dx.doi.org/10.1038/sj.mp.4000812.
49. Phelps EA, LeDoux JE. Contributions of the amygdala to emotion processing: from animal models to human behavior. *Neuron*. 2005;48(2):175–187. http://dx.doi.org/10.1016/j.neuron.2005.09.025.
50. Whalen PJ. Fear, vigilance, and ambiguity: initial neuroimaging studies of the human amygdala. *Society*. 1998;7(1948):177–188. http://dx.doi.org/10.1111/1467-8721.ep10836912.
51. Fitzgerald DA, Angstadt M, Jelsone LM, Nathan PJ, Phan KL. Beyond threat: amygdala reactivity across multiple expressions of facial affect. *Neuroimage*. 2006;30(4):1441–1448. http://dx.doi.org/10.1016/j.neuroimage.2005.11.003.
52. Fudge JL, Emiliano AB. The extended amygdala and the dopamine system: another piece of the dopamine puzzle. *J Neuropsychiatry Clin Neurosci*. 2003;15(3):306–316. http://dx.doi.org/10.1176/appi.neuropsych.15.3.306.
53. Santos A, Mier D, Kirsch P, Meyer-Lindenberg A. Evidence for a general face salience signal in human amygdala. *Neuroimage*. 2011;54(4):3111–3116. http://dx.doi.org/10.1016/j.neuroimage.2010.11.024.
54. Whalen PJ, Shin LM, McInerney SC, Håkan F, Wright CI, Rauch SL. A functional MRI study of human amygdala responses to facial expressions of fear versus anger. *Emotion*. 2001;1(1):70–83. http://dx.doi.org/10.1037/1528-3542.1.1.70.
55. Breiter HC, Etcoff NL, Whalen PJ, et al. Response and habituation of the human amygdala during visual processing of facial expression. *Neuron*. 1996;17(5):875–887. http://dx.doi.org/10.1016/S0896-6273(00)80219-6.
56. Hennenlotter A, Schroeder U, Erhard P, et al. A common neural basis for receptive and expressive communication of pleasant facial affect. *Neuroimage*. 2005;26(2):581–591. http://dx.doi.org/10.1016/j.neuroimage.2005.01.057.
57. Olsson A, Nearing KI, Phelps EA. Learning fears by observing others: the neural systems of social fear transmission. *Soc Cogn Affect Neurosci*. 2007;2(1):3–11. http://dx.doi.org/10.1093/scan/nsm005.
58. Stanley DA, Sokol-Hessner P, Fareri DS, et al. Race and reputation: perceived racial group trustworthiness influences the neural correlates of trust decisions. *Philos Trans R Soc London Ser B Biol Sci*. 2012;367(1589):744–753. http://dx.doi.org/10.1098/rstb.2011.0300.
59. Cloutier J, Li T, Correll J. The impact of childhood experience on amygdala response to perceptually familiar Black and White faces. *J Cogn Neurosci*. 2014;26(9):1992–2004. http://dx.doi.org/10.1162/jocn_a_00605.
60. Cunningham WA, Johnson MK, Raye CL, Gatenby JC, Gore JC, Banaji MR. Separable neural components in the processing of black and white faces. *Psychol Sci*. 2004;15(12):806–813. http://dx.doi.org/10.1111/j.0956-7976.2004.00760.x.
61. Forbes CE, Cox CL, Schmader T, Ryan L. Negative stereotype activation alters interaction between neural correlates of arousal, inhibition and cognitive control. *Soc Cogn Affect Neurosci*. 2011. http://dx.doi.org/10.1093/scan/nsr052.
62. Hart AJ, Whalen PJ, Shin LM, McInerney SC, Fischer H, Rauch SL. Differential response in the human amygdala to racial outgroup versus ingroup face stimuli. *Neuroreport*. 2000;11(11):2351–2355. http://dx.doi.org/10.1097/00001756-200008030-00004.
63. Krill AL, Platek SM. In-group and out-group membership mediates anterior cingulate activation to social exclusion. *Front Evol Neurosci*. 2009;1:1. http://dx.doi.org/10.1016/j.brainres.2009.05.076.
64. Richeson JA, Baird AA, Gordon HL, et al. An fMRI investigation of the impact of interracial contact on executive function. *Nat Neurosci*. 2003;6(12):1323–1328. http://dx.doi.org/10.1038/nn1156.
65. Ronquillo J, Denson TF, Lickel B, Lu Z-L, Nandy A, Maddox KB. The effects of skin tone on race-related amygdala activity: an fMRI investigation. *Soc Cogn Affect Neurosci*. 2007;2(1):39–44. http://dx.doi.org/10.1093/scan/nsl043.
66. Stanley D, Phelps EA, Banaji MR. The neural basis of implicit attitudes. *Curr Dir Psychol Sci*. 2008;17(2):164–170. http://dx.doi.org/10.1111/j.1467-8721.2008.00568.x.
67. Wheeler ME, Fiske ST. Controlling racial prejudice: social-cognitive goals affect amygdala and stereotype activation. *Psychol Sci*. 2005;16(1):56–63. http://dx.doi.org/10.1111/j.0956-7976.2005.00780.x.
68. Phelps EA, O'Connor KJ, Cunningham WA, et al. Performance on indirect measures of race evaluation predicts amygdala activation. *J Cogn Neurosci*. 2000;12(5):729–738. http://dx.doi.org/10.1162/089892900562552.
69. Richeson JA, Baird AA, Gordon HL, et al. An fMRI examination of the impact of interracial contact on executive function. *Nat Neurosci*. 2003;6:1323–1328. http://dx.doi.org/10.1038/nn1156.
70. Richeson JA, Todd AR, Trawalter S, Baird AA. Eye-gaze direction modulates race-related amygdala activity. *Gr Process Intergr Relations*. 2008;11(2):233–246. http://dx.doi.org/10.1177/1368430207088040.
71. Schreiber D, Iacoboni M. Huxtables on the brain: an fMRI study of race and norm violation. *Polit Psychol*. 2012;33(3):313–330. http://dx.doi.org/10.1111/j.1467-9221.2012.00879.x.
72. Lieberman MD, Hariri A, Jarcho JM, Eisenberger NI, Bookheimer SY. An fMRI investigation of race-related amygdala activity in African-American and Caucasian-American individuals. *Nat Neurosci*. 2005;8(6):720–722. http://dx.doi.org/10.1038/nn1465.
73. Blackford JU, Buckholtz JW, Avery SN, Zald DH. A unique role for the human amygdala in novelty detection. *Neuroimage*. 2010;50(3):1188–1193. http://dx.doi.org/10.1016/j.neuroimage.2009.12.083.
74. Cunningham WA, Van Bavel JJ, Johnsen IR. Affective flexibility: evaluative processing goals shape amygdala activity. *Psychol Sci*. 2008;19(2):152–160. http://dx.doi.org/10.1111/j.1467-9280.2008.02061.x.
75. Todorov A, Engell AD. The role of the amygdala in implicit evaluation of emotionally neutral faces. *Soc Cogn Affect Neurosci*. 2008;3(4):303–312. http://dx.doi.org/10.1093/scan/nsn033.
76. Vuilleumier P, Brosch T. Interactions of emotion and attention in perception. In: Gazzaniga MS, ed. *The Cognitive Neurosciences*. vol. IV. Cambridge, MA: MIT Press; 2009:925–934.
77. Phelps EA, Cannistraci CJ, Cunningham WA. Intact performance on an indirect measure of race bias following amygdala damage. *Neuropsychologia*. 2003;41(2):203–208. http://dx.doi.org/10.1016/S0028-3932(02)00150-1.
78. Malpass RS, Kravitz J. Recognition for faces of own and other race. *J Pers Soc Psychol*. 1969;13(4):330–334. http://dx.doi.org/10.1037/h0028434.
79. Brigham JC, Barkowitz P. Do "They all look alike?" The effect of race, sex, experience, and attitudes on the ability to recognize faces. *J Appl Soc Psychol*. 1978;8(4):306–318. http://dx.doi.org/10.1111/j.1559-1816.1978.tb00786.x.
80. Brigham JC, Malpass RS. The role of experience and contact in the recognition of faces of own- and other-race persons. *J Soc Issues*. 1985;41(3):139–155. http://dx.doi.org/10.1111/j.1540-4560.1985.tb01133.x.

81. Chance JE, Goldstein AG. The other-race effect and eyewitness identification. In: Sporer SL, Malpass RS, Koehnken G, eds. *Psychological Issues in Eyewitness Identification*. Hillsdale, NJ: Lawrence Erlbaum Associates, Inc; 1996:153–176.
82. Ostrom TM, Carpenter SL, Sedikides C, Li F. Differential processing of in-group and out-group information. *J Pers Soc Psychol*. 1993;64(1):21–34. http://dx.doi.org/10.1037/0022-3514.64.1.21.
83. Senholzi KB, Ito TA. Structural face encoding: How task affects the N170's sensitivity to race. *Soc Cogn Affect Neurosci*. 2012;8(8): 937–942. http://dx.doi.org/10.1093/scan/nss091.
84. Haxby JV, Horwitz B, Ungerleider LG, Maisog JM, Pietrini P, Grady CL. The functional organization of human extrastriate cortex: a PET-rCBF study of selective attention to faces and locations. *J Neurosci*. 1994;14:6336–6353. Available at: http://psycnet.apa.org/index.cfm?fa=search.displayRecord&uid=1995-20408-001.
85. Kanwisher N, McDermott J, Chun MM. The fusiform face area: a module in human extrastriate cortex specialized for the perception of faces. *J Neurosci*. 1997;17(11):4302–4311. Available at: http://www.ncbi.nlm.nih.gov/pubmed/9151747.
86. Rossion B, Schiltz C, Crommelinck M. The functionally defined right occipital and fusiform face areas discriminate novel from visually familiar faces. *Neuroimage*. 2003;19(3):877–883. http://dx.doi.org/10.1016/S1053-8119(03)00105-8.
87. Golby AJ, Gabrieli JDE, Chiao JY, Eberhardt JL. Differential responses in the fusiform region to same-race and other-race faces. *Nat Neurosci*. 2001;4(8):845–850. Available at: http://dx.doi.org/10.1038/90565.
88. Brosch T, Bar-David E, Phelps EA. Implicit race bias decreases the similarity of neural representations of Black and White faces. *Psychol Sci*. 2013;24(2):160–166. http://dx.doi.org/10.1177/0956797612451465.
89. Ratner KG, Kaul C, Van Bavel JJ. Is race erased? Decoding race from patterns of neural activity when skin color is not diagnostic of group boundaries. *Soc Cogn Affect Neurosci*. 2012:1–6. http://dx.doi.org/10.1093/scan/nss063.
90. Contreras JM, Schirmer J, Banaji MR, Mitchell JP. Common brain regions with distinct patterns of neural responses during mentalizing about groups and individuals. *J Cogn Neurosci*. 2013;25(9):1406–1417. http://dx.doi.org/10.1162/jocn_a_00403.
91. Allport G. *The Nature of Prejudice*. Reading, MA: Addison-Wesley; 1954.
92. Pettigrew TF. Intergroup contact theory. *Annu Rev Psychol*. 1998; 49(1):65–85. http://dx.doi.org/10.1146/annurev.psych.49.1.65.
93. Allen VL, Wilder DA. Categorization, belief similarity, and intergroup discrimination. *J Personal Soc Psychol*. 1975;32(6):971–977. http://dx.doi.org/10.1037/0022-3514.32.6.971.
94. Brown R, Abrams D. The effects of intergroup similarity and goal interdependence on intergroup attitudes and task performance. *J Exp Soc Psychol*. 1986;22(1):78–92. http://dx.doi.org/10.1016/0022-1031(86)90041-7.
95. Turner JC, Hogg MA, Oakes PJ, Reicher SD, Wetherell MS. *Rediscovering the Social Group: A Self-Categorization Theory*. Oxford: Basil Blackwell; 1987.
96. Kelley WM, Macrae CN, Wyland CL, Caglar S, Inati S, Heatherton TF. Finding the self? An event-related fMRI study. *J Cogn Neurosci*. 2002;14:785–794. http://dx.doi.org/10.1162/08989290260138672.
97. Gobbini MI, Leibenluft E, Santiago N, Haxby JV. Social and emotional attachment in the neural representation of faces. *Neuroimage*. 2004;22(4):1628–1635. http://dx.doi.org/10.1016/j.neuroimage.2004.03.049.
98. Mitchell JP, Macrae CN, Banaji MR. Dissociable medial prefrontal contributions to judgments of similar and dissimilar others. *Neuron*. 2006;50(4):655–663. http://dx.doi.org/10.1016/j.neuron.2006.03.040.
99. Heatherton TF, Wyland CL, Macrae CN, Demos KE, Denny BT, Kelley WM. Medial prefrontal activity differentiates self from close others. *Soc Cogn Affect Neurosci*. 2006;1(1):18–25. http://dx.doi.org/10.1093/scan/nsl001.
100. Qiu J. Peering into the root of prejudice. *Nat Rev Neurosci*. 2006;7:508–509. http://dx.doi.org/10.1038/nrn1959.
101. Harris LT, Fiske ST. Dehumanizing the lowest of the low: neuroimaging responses to extreme out-groups. *Psychol Sci*. 2006;17(10): 847–853. http://dx.doi.org/10.1111/j.1467-9280.2006.01793.x.
102. Devine PG. Stereotypes and prejudice: their automatic and controlled components. *J Personal Soc Psychol*. 1989;56(1):5–18. http://dx.doi.org/10.1037/0022-3514.56.1.5.
103. Norton MI, Mason MF, Vandello JA, Biga A, Dyer R. An fMRI investigation of racial paralysis. *Soc Cogn Affect Neurosci*. 2013; 8(4):387–393. http://dx.doi.org/10.1093/scan/nss010.
104. Delgado MR, Nearing KI, LeDoux JE, Phelps EA. Neural circuitry underlying the regulation of conditioned fear and its relation to extinction. *Neuron*. 2008;59(5):829–838. http://dx.doi.org/10.1016/j.neuron.2008.06.029.
105. Ochsner KN, Bunge SA, Gross JJ, Gabrieli JDE. Rethinking feelings: an FMRI study of the cognitive regulation of emotion. *J Cogn Neurosci*. 2002;14(8):1215–1229. http://dx.doi.org/10.1162/089892902760807212.
106. Milham MP, Banich MT, Webb A, et al. The relative involvement of anterior cingulate and prefrontal cortex in attentional control depends on nature of conflict. *Cogn Brain Res*. 2001;12(3):467–473. http://dx.doi.org/10.1016/S0926-6410(01)00076-3.
107. Nieuwenhuis S, Yeung N, Van Den Wildenberg W, Ridderinkhof KR. Electrophysiological correlates of anterior cingulate function in a go/no-go task: effects of response conflict and trial type frequency. *Cogn Affect Behav Neurosci*. 2003;3(1):17–26. http://dx.doi.org/10.3758/CABN.3.1.17.
108. Van Veen V, Carter CS. The timing of action-monitoring processes in the anterior cingulate cortex. *J Cogn Neurosci*. 2002;14(4): 593–602. http://dx.doi.org/10.1162/08989290260045837.
109. Botvinick MM, Braver TS, Barch DM, Carter CS, Cohen JD. Conflict monitoring and cognitive control. *Psychol Rev*. 2001;108(3): 624–652. http://dx.doi.org/10.1037/0033-295x.108.3.624.
110. Swainson R, Cunnington R, Jackson GM, et al. Cognitive control mechanisms revealed by ERP and fMRI: evidence from repeated task-switching. *J Cogn Neurosci*. 2003;15(6):785–799. http://dx.doi.org/10.1162/089892903322370717.
111. Carter CS, van Veen V. Anterior cingulate cortex and conflict detection: an update of theory and data. *Cogn Affect Behav Neurosci*. 2007;7(4):367–379. http://dx.doi.org/10.3758/CABN.7.4.367.
112. Beer JS, Stallen M, Lombardo MV, Gonsalkorale K, Cunningham WA, Sherman JW. The Quadruple Process model approach to examining the neural underpinnings of prejudice. *Neuroimage*. 2008;43(4):775–783. http://dx.doi.org/10.1016/j.neuroimage.2008.08.033.
113. Mathur VA, Harada T, Chiao JY. Racial identification modulates default network activity for same and other races. *Hum Brain Mapp*. 2012;33(8):1883–1893. http://dx.doi.org/10.1002/hbm.21330.
114. Forbes CE, Cox CL, Schmader T, Ryan L. Negative stereotype activation alters interaction between neural correlates of arousal, inhibition and cognitive control. *Soc Cogn Affect Neurosci*. 2011. http://dx.doi.org/10.1093/scan/nsr052.
115. Korn HA, Johnson MA, Chun MM. Neurolaw: differential brain activity for Black and White faces predicts damage awards in hypothetical employment discrimination cases. *Soc Neurosci*. 2011:1–12. http://dx.doi.org/10.1080/17470919.2011.631739.
116. Arnsten AFT. Stress signalling pathways that impair prefrontal cortex structure and function. *Nat Rev Neurosci*. 2009;10(6): 410–422. http://dx.doi.org/10.1038/nrn2648.

117. Kubota JT, Mojdehbakhsh R, Raio CM, Brosch T, Uleman JS, Phelps EA. Stressing the person: legal and everyday person attributions under stress. *Biol Psychol*. 2014;103:117–124. http://dx.doi.org/10.1016/j.biopsycho.2014.07.020.
118. Payne BK. Conceptualizing control in social cognition: how executive functioning modulates the expression of automatic stereotyping. *J Pers Soc Psychol*. 2005;89(4):488–503. http://dx.doi.org/10.1037/0022-3514.89.4.488.
119. Richeson JA, Shelton JN. When prejudice does not pay: effects of interracial contact on executive function. *Psychol Sci*. 2003;14(3):287–290. http://dx.doi.org/10.1111/1467-9280.03437.
120. Krill A, Platek SM. In-group and out-group membership mediates anterior cingulate activation to social exclusion. *Front Evol Neurosci*. April 2009;1:1. http://dx.doi.org/10.3389/neuro.18.001.2009.
121. Amodio DM, Harmon-Jones E, Devine PG. Individual differences in the activation and control of affective race bias as assessed by startle eyeblink response and self-report. *J Pers Soc Psychol*. 2003;84(4):738–753. http://dx.doi.org/10.1037/0022-3514.84.4.738.
122. Cunningham WA, Kesek A, Mowrer SM. Distinct orbitofrontal regions encode stimulus and choice valuation. *J Cogn Neurosci*. 2009;21(10):1956–1966. http://dx.doi.org/10.1162/jocn.2008.21148.
123. Kim H, Shimojo S, O'Doherty JP. Is avoiding an aversive outcome rewarding? Neural substrates of avoidance learning in the human brain. *PLoS Biol*. 2006;4(8):e233. http://dx.doi.org/10.1371/journal.pbio.0040233.
124. Forbes CE, Grafman J. The role of the human prefrontal cortex in social cognition and moral judgment. *Annu Rev Neurosci*. 2010;33:299–324. http://dx.doi.org/10.1146/annurev-neuro-060909-153230.
125. Gilbert SJ, Swencionis JK, Amodio DM. Evaluative vs. trait representation in intergroup social judgments: distinct roles of anterior temporal lobe and prefrontal cortex. *Neuropsychologia*. 2012;50(14):3600–3611.http://dx.doi.org/10.1016/j.neuropsychologia.2012.09.002.
126. Beer JS, Heerey EA, Keltner D, Scabini D, Knight RT. The regulatory function of self-conscious emotion: insights from patients with orbitofrontal damage. *J Pers Soc Psychol*. 2003;85(4):594–604. http://dx.doi.org/10.1037/0022-3514.85.4.594.
127. Blair RJR. The roles of orbital frontal cortex in the modulation of antisocial behavior. *Brain Cogn*. 2004;55(1):198–208. http://dx.doi.org/10.1016/S0278-2626(03)00276-8.
128. Elliott R, Deakin B. Role of the orbitofrontal cortex in reinforcement processing and inhibitory control: evidence from functional magnetic resonance imaging studies in healthy human subjects. In: 65th ed. Bradley RJ, Harris RA, Jenner P, eds. *International Review of Neurobiology*. vol. 65. San Diego, CA: Gulf Professional Publishing; 2005:89–116.
129. Cunningham WA, Zelazo PD. Attitudes and evaluations: a social cognitive neuroscience perspective. *Trends Cogn Sci*. 2007;11(3):97–104. http://dx.doi.org/10.1016/j.tics.2006.12.005.
130. Nobre AC, Coull JT, Frith CD, Mesulam MM. Orbitofrontal cortex is activated during breaches of expectation in tasks of visual attention. *Nat Neurosci*. 1999;2(1):11–12. http://dx.doi.org/10.1038/4513.
131. Rolls ET. The functions of the orbitofrontal cortex. *Brain Cogn*. 2004;55(1):11–29.http://dx.doi.org/10.1016/S0278-2626(03)00277-X.
132. Fellows LK, Farah MJ. Ventromedial frontal cortex mediates affective shifting in humans: evidence from a reversal learning paradigm. *Brain*. 2003;126(8):1830–1837. http://dx.doi.org/10.1093/brain/awg180.
133. Rolls ET. The orbitofrontal cortex and reward. *Cereb Cortex*. 2000;10(3):284–294. http://dx.doi.org/10.1093/cercor/10.3.284.
134. Rolls ET, Hornak J, Wade D, McGrath J. Emotion-related learning in patients with social and emotional changes associated with frontal lobe damage. *J Neurol Neurosurg Psychiatry*. 1994;57(12):1518–1524. http://dx.doi.org/10.1136/jnnp.57.12.1518.
135. Amodio DM, Kubota JT, Harmon-Jones E, Devine PG. Alternative mechanisms for regulating racial responses according to internal vs. external cues. *Soc Cogn Affect Neurosci*. 2006;1(1):26–36. http://dx.doi.org/10.1093/scan/nsl002.
136. Freeman JB, Ambady N. A dynamic interactive theory of person construal. *Psychol Rev*. 2011;118(2):247–279. http://dx.doi.org/10.1037/a0022327.
137. Kubota JT, Banaji MR, Phelps EA. The neuroscience of race. *Nat Neurosci*. 2012;15:940–948. http://dx.doi.org/10.1038/nn.3136.
138. Dasgupta N. Mechanisms underlying the malleability of implicit prejudice and stereotypes: the role of automaticity and cognitive control. In: Nelson TD, ed. *Handbook of Stereotyping, Prejudice, and Discrimination*. New York: Psychology Press; 2009:267–284.
139. Payne BK. Prejudice and perception: the role of automatic and controlled processes in misperceiving a weapon. *J Personal Soc Psychol*. 2001;81(2):181–192. http://dx.doi.org/10.1037/0022-3514.81.2.181.
140. Blair IV, Ma JE, Lenton AP. Imagining stereotypes away: the moderation of implicit stereotypes through mental imagery. *J Pers Soc Psychol*. 2001;81(5):828–841. http://dx.doi.org/10.1037/0022-3514.81.5.828.
141. Kubota JT, Ito TA. The role of expression and race in weapons identification. *Emotion*. 2014;14(6):1115–1124. http://dx.doi.org/10.1037/a0038214.
142. Galinsky AD, Moskowitz GB. Perspective-taking: decreasing stereotype expression, stereotype accessibility, and in-group favoritism. *J Pers Soc Psychol*. 2000;78(4):708–724. http://dx.doi.org/10.1037/0022-3514.78.4.708.
143. Kubota JT. *The neural correlates of categorical and individuated impressions*; 2010. 3433360. Available at: http://gradworks.umi.com/34/33/3433360.html.
144. Ames DL, Jenkins AC, Banaji MR, Mitchell JP. Taking another person's perspective increases self-referential neural processing. *Psychol Sci*. 2008;19(7):642–644. http://dx.doi.org/10.1111/j.1467-9280.2008.02135.x.
145. Rilling JK, Dagenais JE, Goldsmith DR, Glenn AL, Pagnoni G. Social cognitive neural networks during in-group and out-group interactions. *Neuroimage*. 2008;41(4):1447–1461. http://dx.doi.org/10.1016/j.neuroimage.2008.03.044.
146. Mathur VA, Harada T, Lipke T, Chiao JY. Neural basis of extraordinary empathy and altruistic motivation. *Neuroimage*. 2010;51(4):1468–1475. http://dx.doi.org/10.1016/j.neuroimage.2010.03.025.
147. Xu X, Zuo X, Wang X, Han S. Do you feel my pain? Racial group membership modulates empathic neural responses. *J Neurosci*. 2009;29(26):8525–8529. http://dx.doi.org/10.1523/JNEUROSCI.2418-09.2009.
148. Azevedo RT, Macaluso E, Avenanti A, Santangelo V, Cazzato V, Aglioti SM. Their pain is not our pain: brain and autonomic correlates of empathic resonance with the pain of same and different race individuals. *Hum Brain Mapp*. 2013;34(12):1–14. http://dx.doi.org/10.1002/hbm.22133.
149. Losin EAR, Cross KA, Iacoboni M, Dapretto M. Neural processing of race during imitation: self-similarity versus social status. *Hum Brain Mapp*. April 2013;35(4). http://dx.doi.org/10.1002/hbm.22287.
150. Harris LT, Todorov A, Fiske ST. Attributions on the brain: neuro-imaging dispositional inferences, beyond theory of mind. *Neuroimage*. 2005;28:763–769. http://dx.doi.org/10.1016/j.neuroimage.2005.05.021.
151. Frith U, Frith CD. Development and neurophysiology of mentalizing. *Phil Trans R Soc Lond B*. 2003;358:459–473. http://dx.doi.org/10.1098/rstb.2002.1218.
152. Kubota JT, Senholzi KB. Knowing you beyond race: the importance of individual feature encoding in the other-race effect. *Front Hum Neurosci*. 2011;5:1–2. http://dx.doi.org/10.3389/fnhum.2011.00033.

153. Mitchell JP, Macrae CN, Banaji MR. Encoding-specific effects of social cognition on the neural correlates of subsequent memory. *J Neurosci*. 2004;26:4912–4917. http://dx.doi.org/10.1523/JNEUROSCI.0481-04.2004.
154. Blair IV, Judd CM, Sadler MS, Jenkins C. The role of Afrocentric features in person perception: judging by features and categories. *J Pers Soc Psychol*. 2002;83(1):5–25. http://dx.doi.org/10.1037/0022-3514.83.1.5.
155. Greer TM, Vendemia JMC, Stancil M. Neural correlates of race-related social evaluations for African Americans and white Americans. *Neuropsychology*. 2012;26(6):704–712. http://dx.doi.org/10.1037/a0030035.
156. Pettigrew TF, Tropp LR. A meta-analytic test of intergroup contact theory. *J Pers Soc Psychol*. 2006;90(5):751–783. http://dx.doi.org/10.1037/0022-3514.90.5.751.
157. Olsson A, Ebert JP, Banaji MR, Phelps EA. The role of social groups in the persistence of learned fear. *Science*. 2005;309(5735):785–787. http://dx.doi.org/10.1126/science.1113551.
158. Telzer EH, Humphreys KL, Shapiro M, Tottenham N. Amygdala sensitivity to race is not present in childhood but emerges over adolescence. *J Cogn Neurosci*. 2013;25(2):234–244. http://dx.doi.org/10.1162/jocn_a_00311.
159. Monteith MJ. Self-regulation of prejudiced responses: implications for progress in prejudice-reduction efforts. *J Personal Soc Psychol*. 1993;65(3):469–485. http://dx.doi.org/10.1037/0022-3514.65.3.469.
160. Schwabe L, Nader K, Wolf OT, Beaudry T, Pruessner JC. Neural signature of reconsolidation impairments by propranolol in humans. *Biol Psychiatry*. 2012;71(4):380–386. http://dx.doi.org/10.1016/j.biopsych.2011.10.028.
161. McGaugh JL. Memory–a century of consolidation. *Science*. 2000;287(5451):248–251. http://dx.doi.org/10.1126/science.287.5451.248.s.
162. Nader K, Schafe GE, Le Doux JE. Fear memories require protein synthesis in the amygdala for reconsolidation after retrieval. *Nature*. 2000;406(6797):722–726. http://dx.doi.org/10.1038/35021052.
163. Cahill L, Prins B, Weber M, McGaugh JL. β Adrenergic activation and memory for emotional events. *Nature*. 1994;371(6499):702–704. http://dx.doi.org/10.1038/371702a0.
164. Terbeck S, Kahane G, McTavish S, Savulescu J, Cowen P, Hewstone M. Propranolol reduces implicit negative racial bias. *Psychopharmacology (Berl)*. 2012;222(3):419–424. http://dx.doi.org/10.1007/s00213-012-2657-5.
165. Glimcher PW, Fehr E. *Neuroeconomics: Decision Making and the Brain*. 2nd ed. San Diego, CA: Elsevier; 2013.
166. Stanley DA, Sokol-Hessner P, Banaji MR, Phelps EA. Implicit race attitudes predict trustworthiness judgments and economic trust decisions. *Proc Natl Acad Sci*. 2011;108(19):7710–7715. http://dx.doi.org/10.1073/pnas.1014345108.
167. Kubota JT, Li J, Bar-David E, Banaji MR, Phelps EA. The price of racial bias: intergroup negotiations in the ultimatum game. *Psychol Sci*. 2013;24(12):2498–2504. http://dx.doi.org/10.1177/0956797613496435.
168. Delgado MR, Rita LJ, LeDoux JE, Phelps EA. Avoiding negative outcomes: tracking the mechanisms of avoidance learning in humans during fear conditioning. *Front Behav Neurosci*. 2009;3:33. http://dx.doi.org/10.3389/neuro.08.033.2009.
169. Lane KA, Kang J, Banaji MR. Implicit social cognition and law. *Annu Rev Law Soc Sci*. 2007;3(1):427–451. http://dx.doi.org/10.1146/annurev.lawsocsci.3.081806.112748.
170. Ito TA, Senholzi KB. Us versus them: understanding the process of race perception with event-related brain potentials. *Vis Cogn*. 2013;21(9–10):1096–1120. http://dx.doi.org/10.1080/13506285.2013.821430.
171. Cloutier J, Gabrieli JD, O'Young D, Ambady N. An fMRI study of violations of social expectations: when people are not who we expect them to be. *NeuroImage*. 2011;57(2):583–588. http://dx.doi.org/10.1016/j.neuroimage.2011.04.051.
172. Macrae CN, Bodenhausen GV, Schloerscheidt AM, Milne AB. Tales of the unexpected: executive function and person perception. *J Pers Soc Psychol*. 1999;76(2):200–213. http://dx.doi.org/10.1037/0022-3514.76.2.200.

CHAPTER

19

Political Neuroscience

Ingrid J. Haas

Department of Political Science and Center for Brain, Biology, and Behavior, University of Nebraska-Lincoln, Lincoln, NE, USA

OUTLINE

1. POLITICAL NEUROSCIENCE

The field of political science has traditionally had close ties to disciplines like economics, history, and sociology. While political science has always been somewhat interdisciplinary in nature, in recent years this interdisciplinary approach has expanded to include biology, psychology, and neuroscience. This interest in the human sciences has led to the development of new subfields within political science, including biopolitics, political psychology, and political neuroscience (also called neuropolitics). What these new subfields have in common is an interest in individual human behavior and decision-making as an approach to understanding political behavior. While political science has traditionally focused on understanding politics in the aggregate, new methods and techniques are improving our ability to understand political behavior at the individual level and consider how individual differences in information processing may give rise to political behavior that is observed at the mass level.

While political science, psychology, and neuroscience have fairly distinct intellectual histories, it makes sense to combine them. While some political scientists think about politics as a special type of human behavior, and some psychologists dismiss the study of politics as too applied, a case can be made for the idea that politics and psychology share significant overlap. From the perspective of human evolution and the development of social behavior, it seems clear that social and political behavior have been historically intertwined.[1,2] Just as the brain evolved to deal with larger and larger social groups, it became necessary to consider how those groups should be governed. From this perspective, it seems obvious to suggest that political behavior can be understood through the lens of human psychology, biology, and neuroscience.

Neuroimaging Personality, Social Cognition, and Character
http://dx.doi.org/10.1016/B978-0-12-800935-2.00019-1

As with any interdisciplinary or multidisciplinary approach, there are a number of challenges for researchers working in this area. While it has become increasingly clear that recent advances in social and cognitive neuroscience will help improve our understanding of political behavior, there are significant challenges that arise when trying to engage in this type of multilevel analysis—it is not always easy to translate what happens at the neural level into much more abstract notions about how society functions.

In this chapter, I will outline the contributions that political neuroscience has made thus far and discuss areas where political neuroscience may have the most to contribute moving forward. The chapter will focus on four important questions within political psychology and discuss the role for neuroimaging work within these areas: (1) political attitudes and evaluation, (2) social cognition and politics, (3) emotion and politics, and (4) individual differences in political behavior. Given that political neuroscience is in its infancy, the discussion of work in this area will be supplemented with relevant work from social and cognitive neuroscience, as well as social and political psychology more broadly. I think political scientists and neuroscientists can benefit from firmly grounding their ideas in social psychological theory, and social psychologists can benefit from an increasing understanding of brain function, as well as increased consideration of the role of context. After reviewing the current state of the literature in political neuroscience, I will offer some suggestions for future work.

2. A BRIEF OVERVIEW OF NEUROIMAGING METHODS

Many of the chapters in this volume will no doubt discuss neuroimaging methodology in extensive detail, but it is worth providing a brief overview here of neuroimaging methods used by political neuroscientists, especially for those readers who may be new to the field. The growth of methods like fMRI has been exponential since 1990.[3] The use of these methods to study questions in political neuroscience has followed suit, although I think it is fair to say that political neuroscience is not quite as well established at this point. It is still possible to create a (relatively short) list of all the political neuroscience studies that have been conducted with structural Magnetic Resonance Imaging (MRI) or functional neuroimaging (fMRI) or electroencephalography (EEG).

There have been multiple attempts to describe how political science and cognitive neuroscience might be able to learn from one another, perhaps dating back to a special issue of the journal *Political Psychology* published in 2003.[4–6] Although acceptance of this idea has grown, some political scientists remain skeptical of the idea that adopting neuroimaging methods will strengthen the field. More recently, people have called for balance on this issue—it is important to not overstate the claims we can make based on methods like functional Magnetic Resonance Imaging (fMRI) but also important not to be overly dismissive and to be aware that this field is still in its infancy.[7] These methods are unlikely to replace traditional methods in political science research, but they may be used by some political scientists (though additional training or through collaboration) to help supplement our understanding of important questions in the field.

Given that this volume is focused on the role of neuroimaging, my goal here is to discuss neuroimaging research as it relates to political science. With that in mind, I will focus primarily on research using structural and functional imaging. It is important to note that a lot of research in the domain of biopolitics has also begun to investigate these issues using methods from psychophysiology, but that is beyond the scope of this chapter. For readers who are less familiar with these methods, I will offer a brief overview below. It is important to be aware that different methods have different strengths and weaknesses, and may be amenable to different types of research questions.

2.1 Structural Magnetic Resonance Imaging

Magnetic Resonance Imaging (MRI) is a technology that allows researchers to take three-dimensional images of the brain. This technology allows for much more detailed resolution than older imaging technologies like X-ray, CT scan, or positron emission technology (PET) scan. For example, a 3 Tesla MRI can produce a three-dimensional image of the brain at a resolution of 1×1×1 mm. Structural MRI can be used to examine and compare the size and composition of different subregions within the brain across individuals. In relation to politics, this technique is most useful for individual difference analyses where different brain regions or structures are compared across individuals as a function of some personality trait or characteristic. There are additional methods available for structural imaging, such as diffusion tensor imaging (DTI), which allows for measurement of anatomical connections between brain regions. To my knowledge, no studies have used DTI yet to study political differences so I will not discuss that here.

2.2 Functional Magnetic Resonance Imaging

Magnetic resonance imaging can also be used for research on brain function. This type of research uses a different strategy for image collection. While structural imaging produces high-resolution images of the brain, functional imaging typically produces lower resolution images focused not on structure, but on changes in brain function as a product of some experimental task. It is important to realize that fMRI relies on *indirect* measurement of brain activation.[8,9] This measurement of brain function is dependent on what is known as the blood-oxygen-level dependent (BOLD) signal. The basic assumption of fMRI is that measurement of increased cerebral blood flow using the BOLD signal can serve as a proxy for increased neural activity. While this idea is relatively well-established at this point, questions remain about exactly what the BOLD response represents. Neuroscientists have argued that the BOLD response is more closely tied to input and processing within a brain region and less related to output or "spiking" from that region.[8,9]

There are some important caveats to be aware of related to the interpretation of fMRI data. First, any signal produced is a relative (not absolute) measure of activation. As you will see in the description of political neuroscience work in the sections that follow, typically any results from an fMRI study are discussed in terms of contrast effects (i.e., observing greater activation for one condition compared to a baseline, control, or second experimental condition). Second, analysis of fMRI data is highly dependent on assumptions about the BOLD signal—data is typically analyzed using the hemodynamic response function (HRF), which assumes that the BOLD signal peaks about four to six seconds after stimulus presentation and then returns to baseline.[9] Finally, fMRI data is correlational—you can show support for a relationship between BOLD activation and stimuli or task demands, but it is difficult to establish causal relationships. Simply showing that brain activation is related to some task does not establish that region is necessary for the task, and researchers who study social or political behavior need to be aware that some of the brain activation they observe may be due to basic cognitive processes (e.g., viewing a stimulus, task switching demands). Careful experimental design can help ameliorate some of these concerns related to interpretation, but it is worth being cautious when interpreting fMRI data.

2.3 Electroencephalography

There is an often discussed trade-off when discussing fMRI versus electroencephalography (EEG) and other neuroimaging methods. While fMRI provides superior spatial resolution, EEG provides superior temporal resolution. This is sometimes mistaken to mean that fMRI provides no information about timecourse, which is not exactly true. Data from fMRI allow for examination of the timecourse on a slower scale—in terms of seconds as opposed to milliseconds with EEG. It is also important to realize that EEG and fMRI are not measuring the same thing—EEG measures electrical activity, whereas fMRI is dependent on the BOLD signal, as discussed above.

The primary strength of EEG is its ability to measure electrophysiological responses in a matter of milliseconds. EEG uses a net of electrodes that is placed on the scalp and measure surface-level electrical activity. The primary difficulty with EEG is determining where those electrical signals are coming from. Although methods for source localization have become significantly more refined over the years, the spatial resolution from EEG cannot compete with spatial resolution obtained through fMRI and it can be difficult to obtain a signal from subcortical regions (e.g., amygdala) using EEG. Most of the work in political neuroscience that I will discuss in this chapter has used structural or functional MRI, but there have also been some studies in this area using EEG, which I will describe below.

3. IMPORTANT QUESTIONS IN POLITICAL NEUROSCIENCE

In the sections that follow, I will provide an overview of political neuroscience research related to four important questions within political psychology: (1) how do people evaluate political information?, (2) how do people think about and process politically similar and dissimilar others?, (3) what is the impact of emotion on politics?, and (4) how do individual differences influence political thought and behavior? For each section, I will provide an overview of the relevant research in political neuroscience, placing this work in the context of social and political psychology, as well as social and cognitive neuroscience. The existing work in political neuroscience becomes easier to interpret if it is placed in context, and a broader understanding of research in these related fields may help to illustrate gaps in the existing work and help generate ideas for future research.

4. POLITICAL ATTITUDES AND EVALUATION

Social psychologists have been interested in the properties of attitude structure, function, and change for quite some time. Traditional work in this area has relied most heavily on self-reported attitudes, using measurement tools such as Likert scales or semantic differentials, and these were among the first measurement tools to be adopted by political scientists interested in studying attitudes. More recently, there has been a burgeoning interest in implicit or indirect attitude measurement, typically relying on cognitive response latency measures such as the Implicit Association Test (IAT)[10] or the Affective Misattribution Procedure (AMP).[11,12] Along with these shifts in methodology have been related shifts in attitude theory. Given that we now have two different categories of measurement techniques, researchers have speculated about what these two types of measures may represent.

In recent years, there has been a lot of debate within social psychology regarding what are known as dual processing and dual systems models of attitudes. Without getting too carried away with the details here, researchers have argued back and forth about whether we should consider implicit and explicit attitudes to be two different types of attitudes (presumably existing as two different types of attitudes stored in the brain or related to different types of memory or processing)[13–15] versus a focus on implicit and explicit *measures* tapping into the same attitudinal information (but sometimes producing different "attitudes" at least in terms of measurement outcome).[16,17] This debate over *dual system* versus *dual process* models of attitudes can potentially be informed by greater understanding of the neural underpinnings of evaluation.

In recent years, social cognitive neuroscience has weighed in on this question by suggesting that we should focus not on different types of attitudes per se, but on the process of evaluation. In a recent model of attitudes and evaluation informed by neurobiology, the Iterative Reprocessing (IR) Model, Cunningham and Zelazo proposed that the process of evaluation should be understood as unfolding over time.[16,18] Attitudinal information is stored in memory and accessed as needed to provide evaluations relevant to the current situation or context. This process is iterative, meaning that attitudes can be updated as new information is received externally or additional information is accessed internally. So, different "attitudes" may result from implicit versus explicit measures capturing the current evaluation at different time points or after varying degrees of information processing, updating, and reorganization.

The IR Model proposes a network of brain regions that are likely implicated in the process of evaluation, including subcortical regions like the amygdala, insula, and hypothalamus, regions that allow for additional processing, like the anterior cingulate and orbitofrontal cortex, as well as regions likely involved in higher-order processing, such as areas of prefrontal cortex (PFC—dorsolateral PFC, ventrolateral PFC, and rostrolateral PFC).[16,18] Many of these regions have also been implicated in studies of political neuroscience (see Figure 1 for a visual representation).[19] These regions allow for the integration of sensory information with affective knowledge and are thought to combine to produce an evaluation in any given situation.[16,18] Importantly, the IR Model can explain potential differences in implicit versus explicit measurement of attitudes by inserting the concept of time. Initial responses are likely to be somewhat rapid and automatic (amygdala, insula), while later responses allow time for regions involved in more

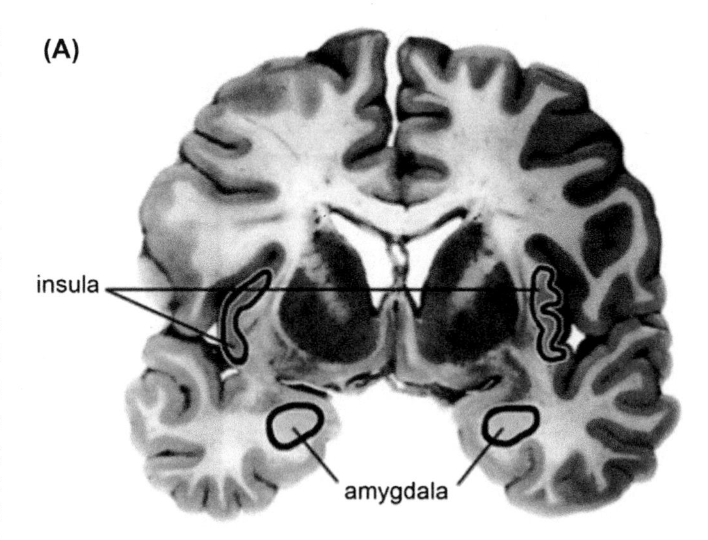

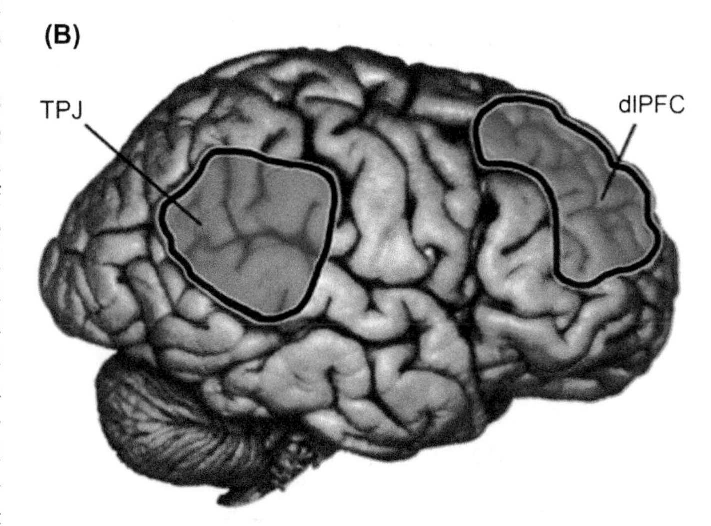

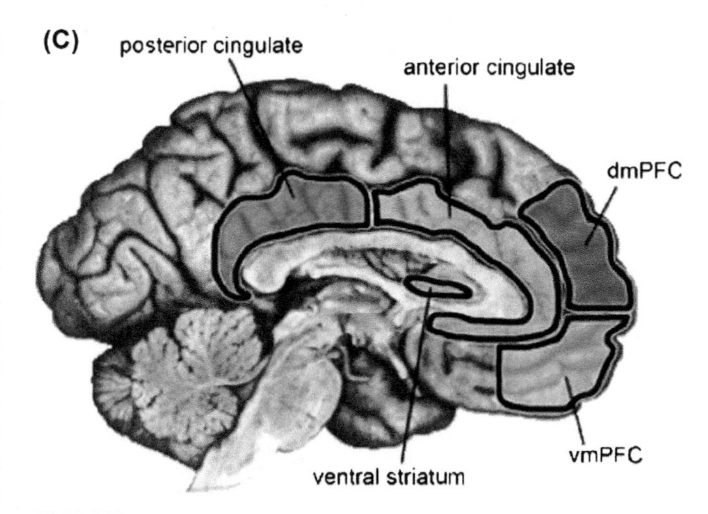

FIGURE 1 Brain regions and structures that are most commonly observed in studies of political neuroscience: views of (A) coronal, (B) sagittal, and (C) midsagittal planes. *Reprinted with permission from Ref. 19.*

reflective processing to become involved (anterior cingulate cortex, dorsolateral PFC, rostrolateral PFC, and ventrolateral PFC).

Importantly, recent work in social neuroscience has suggested that we may need to further reevaluate how we think about attitudes. While the distinction between automatic versus controlled processes has been generative for research, it may not be reflective of how the brain actually works. While some processes do happen more rapidly than others, even rapid evaluative processes (e.g., amygdala) are subject to guidance and input from what we would typically think about as more "controlled" systems. Consistent with this view, we have shown that amygdala activation is sensitive to evaluative goals.[20] Given that people are constantly monitoring the environment for stimuli and events that are goal congruent or incongruent, it makes sense that even early signals in the brain are sensitive to this information as well.

4.1 The Psychology of Political Attitudes

Political psychology research on political attitudes and evaluation has focused on similar issues in the context of politics. Many of these debates in political psychology have mirrored debates in social psychology—examining the role of online versus memory-based evaluation, automatic versus controlled processing, and the automaticity of affect. A number of papers, for example, have shown support for the "Hot Cognition Hypothesis," or the idea that political attitudes are affectively charged and that this affective response is automatically activated upon encountering a stimulus.[21–23] This theory is somewhat reminiscent of the now infamous debate between Zajonc and Lazarus about the primacy of affect,[24,25] with Lodge and colleagues coming down on the side of Zajonc—arguing for the primacy of affect. Most of the neuroimaging work in political neuroscience related to attitudes and evaluation has focused on the evaluation of political policies or candidates, both in terms of rapid, automatic processes and slower, more controlled processes.

4.2 Political Evaluation

The amygdala has been implicated in multiple aspects of evaluative processing, including attitudes toward political candidates. One early study in political neuroscience scanned participants while they completed an Implicit Association Test (IAT) using names and faces of well-known Democratic (e.g., Hillary Clinton, Al Gore) and Republican politicians (e.g., Condoleezza Rice, Ronald Reagan).[26] Results showed that the amygdala responded to familiar faces, and the strength of the amygdala response was related to the strength of the affective response (positive or negative) toward both parties and candidates. This study also found fusiform gyrus activation for familiar faces and prefrontal cortex activation for various aspects of the task. While the authors conclude that these results are consistent with a dual processing view of attitudes, I think that is somewhat difficult to discern based on the results of this one study. The prefrontal cortex activation, for example, could have been related to activation of stereotypic knowledge, as they suggest, or more generally related to the cognitive demands of completing a task like the IAT.

More recently, researchers have examined the amygdala response to political candidates across cultures, showing that the amygdala responded more strongly to faces of candidates that participants chose to vote for in samples from both the United States and Japan.[27] However, the results from this study also showed that overall, the amygdala responded more strongly to cultural outgroup members than ingroup members. As with the previous study, the interpretation of these results is somewhat dependent on current theories of the amygdala's role in evaluative processing.

Historically, the amygdala was thought to respond primarily to negative or fear-inducing stimuli,[28–30] but more recent work has shown that the amygdala also responds to positive,[31–33] arousing,[34] uncertain,[35] or motivationally relevant[20,36,37] stimuli. Given this ambiguity in terms of amygdala function, the only thing we can really conclude here is that participants probably viewed these candidates as relevant in some way and, given that participants chose to vote for them, we might assume that the amygdala response is related to positive evaluations. However, future work might be able to examine situations where this does or does not hold. From a political science perspective, we know that people sometimes vote because they are excited about their preferred candidate. However, it is also possible that people vote because they are opposed to the candidate from the other party. Given that the amygdala responds to emotional intensity or relevance, it could be the case that this pattern of responding would differ depending on some of these contextual or individual difference variables, but that is speculative at this point.

Other researchers have focused on examining what happens when people view faces of candidates they dislike. For example, Kaplan and colleagues showed participants pictures of the 2004 Presidential candidates' faces (i.e., George W. Bush, John Kerry, Ralph Nader) during fMRI, demonstrating that when people were viewing opposition candidates, there was greater activation in dorsolateral PFC and anterior cingulate.[38] The authors suggest that this activation may be consistent with the idea that participants were engaging cognitive control processes to regulate the negative responses, but because this was a passive viewing task, there is no direct evidence for this hypothesis. Future work could try to test this more directly.

More recent work has shown that this activation in insula and anterior cingulate to disliked candidates is also predictive of election outcomes.[39] Real-world candidates (selected from recent Congressional and Governor races) for whom participants showed greater activation in these areas were more likely to lose elections. Importantly, these researchers also showed that negative trait perceptions (i.e., threat) related to this activation were more predictive of election outcomes than positive trait perceptions, suggesting that there may be an overall negativity bias in candidate evaluation. While this is consistent with a variety of work showing evidence for a general negativity bias in evaluation,[40,41] it is worth noting that this study relied on rapid trait evaluations of candidates. Given more time, it is possible that positive information may have played a greater role in evaluation. In other words, it is possible that part of the reason for the negativity bias is that negative information has some degree of primacy in terms of temporal processing. Indeed, there do seem to be differences in terms of negative versus positive perceptions of ambiguous stimuli leading to activation in more ventral versus dorsal regions within the amygdala.[42] So, while it is possible that initial, negative perceptions may help drive perceptions of political candidates, it is worth considering the role of both ambivalence and time allowing for correction or modification of these early evaluative responses in future work.

One remaining question here is whether there are differences between active versus passive viewing of political candidates, perhaps being representative of the distinction between automatic versus controlled processing or implicit versus explicit attitudes. Although, it is important to remember that this distinction is likely an oversimplification of how the brain processes information.[43] To examine the role of attention, participants in a recent study viewed pictures of German politicians during fMRI and either attended to the pictures or viewed the pictures while completing a demanding visual fixation task.[44] Politicians represented either the Christian Democratic Union Party or the Social Democratic Party. The researchers could then examine whether brain function differed during active versus passive viewing, and as a function of candidates' party affiliation. When politicians were unattended, they found that regardless of party affiliation, preference for candidates was related to activation in ventral striatum, a region often implicated in reward processing. Preferences for candidates of one's preferred party was related to processing in additional regions, including insula and cingulate cortex (both anterior and posterior regions). This work is consistent with theories in political psychology such as the "Hot Cognition Hypothesis," suggesting that people have a tendency to automatically process political information (i.e., faces of political candidates).

4.3 Motivated Reasoning

Another important question in political psychology has been to examine the role of motivated reasoning. Classic research on attitudes in social psychology, for example, showed that people were able to distort incoming information to be consistent with their prior beliefs, whether they were in favor of or opposed to the death penalty.[45] Kunda offered a theoretical overview of motivated reasoning, suggesting that people experience conflict between the motivation for accuracy versus the motivation to reach desired conclusions.[46] More recently, researchers have begun to examine the neural underpinnings of these effects.

In another early study in political neuroscience, Westen and colleagues examined motivated reasoning by asking participants to make judgments during fMRI about information that was threatening to either their own candidate, an opposing candidate, or a neutral control target.[47] Data was collected from "committed partisans" during the 2004 U.S. Presidential election. The study found behavioral evidence for motivated reasoning, showing that Democrats were more likely to perceive contradictions from George W. Bush and vice versa, Republicans were more likely to perceive contradictions from John Kerry. When participants were engaged in motivated reasoning, a network of regions was involved including ventromedial PFC, anterior and posterior cingulate cortex, insula, and lateral orbitofrontal cortex. Given that many of these regions (e.g., insula, ACC, lateral OFC) are involved in processing negative information, detecting conflict, and integrating cognitive and affective responses,[48–51] this overall pattern of results seems consistent with what we would expect when people are engaged in motivated reasoning. Interestingly, the study also found that when participants were given the opportunity to resolve this inconsistency, activation in ventral striatum was observed. Given that the ventral striatum has often been implicated in reward processing,[52,53] this is consistent with the view that resolving inconsistency may be experienced as rewarding.

4.4 Attitude Change

While research on motivated reasoning has shown that people often maintain preexisting attitudes through biased reasoning processes, it is important to note that attitudes can and do change in response to new information. Indeed, this was an important part of Kunda's theory—people are also motivated by a desire for accuracy and may abandon desired conclusions if the counter-attitudinal evidence is overwhelming.[46] Attitude change is more likely to occur when people do not hold strong preferences to begin with, or when attitudes are weaker or less certain.[54] Attitudes can become stronger—a positive attitude might become even more positive, or they can change valence—a positive attitude might become a negative attitude, for example. Importantly, given our understanding of attitudes, these positive and negative processes appear to be independent of one another.[55] So, thinking of an attitude purely as positive versus negative is likely an oversimplification. Attitudes can contain some degree of positive information, negative information, both (ambivalence), or neither (apathy).

Researchers have begun to examine what leads to both positive and negative political attitude change using fMRI. It appears that different regions of the prefrontal cortex may be implicated in positive versus negative change. One study showed that the dorsolateral prefrontal cortex response to negative campaign videos predicted attitude change in a negative direction—lower ratings of a political candidate.[56] It has been suggested that dorsolateral prefrontal cortex is involved in cognitive control and reprocessing of evaluative information,[16,18,57] consistent with the idea that it may be involved in attitude change, especially for more abstracted information. The Kato study also showed a relationship between medial prefrontal cortex and increased ratings of candidates.[56] This may be consistent with a larger body of work, suggesting that medial prefrontal cortex is implicated in Theory of Mind[58]—imagining the mental states of others—and is more likely to occur when thinking about people we like (as opposed to those we dislike).[59] That work will be described in more detail below.

5. SOCIAL COGNITION AND POLITICS

The term social cognition means different things to different people, but for the purposes of this chapter I will focus primarily on discussing issues related to Theory of Mind or mentalizing. The overarching question guiding research in this area has been to determine how people think about and make decisions about the behavior of other people.

Research on Theory of Mind has focused on understanding how people think about or "mentalize" about the minds of others. Engaging in this process allows individuals to think about the causes of others' behavior and anticipate how others will respond to them. At this point, the neural underpinnings of Theory of Mind are fairly well established. Neuroimaging work has primarily focused on the role of the medial prefrontal cortex and temporoparietal junction.[58,60] Interestingly, recent work has shown that people may deploy this resource somewhat strategically. Kozak and colleagues showed, for example, that people were more likely to engage in Theory of Mind processing for liked versus disliked others.[59]

In the political context, social cognition research has been applied primarily to understanding how people think about political candidates or politicians, although there are also examples of work examining how people think about politically similar or dissimilar others. This work has primarily focused on the role of the medial prefrontal cortex. Mitchell and colleagues hypothesized that people would be more likely to mentalize about politically similar versus dissimilar others.[61] During fMRI, participants were presented with other people who either held similar or dissimilar political views. They found that ventromedial prefrontal cortex activation was greater for similar others, whereas dorsal medial prefrontal cortex activation was greater for dissimilar others. This is consistent with the behavioral work showing that people engage in mentalizing somewhat selectively.[59]

Additional brain regions have been implicated in perspective taking, aside from the medial prefrontal cortex. In an fMRI study conducted immediately prior to the 2008 election, Falk and colleagues showed that taking the perspective of a same party candidate resulted in greater poster cingulate cortex activation, whereas taking the perspective of an opposing candidate led to activation in bilateral temporoparietal junction and insula.[62] While many of these regions have been implicated in social cognition processes, it is still somewhat unclear why different regions within this network would respond differentially to ingroup versus outgroup targets. Given that there have only been two studies (to my knowledge) on the political neuroscience of perspective taking and Theory of Mind, this area is ripe for future investigation.

6. EMOTION AND POLITICS

Political science has had a long relationship with fields like economics that focus more on rational choice models of decision-making. But with increasing interest in psychology and neuroscience, some political scientists have expressed a growing interest in understanding both biases in human information processing and the role of emotion. It is worth noting that it is not necessarily the case that emotion leads to biased reasoning. It certainly can, but it is important to realize that emotion probably exists because it was adaptive. The amygdala's role in fear detection, for example, likely evolved because it was adaptive for both human and nonhuman animals to prioritize their response to threatening stimuli.[37] Disgust is another emotion that is often described in terms of its evolutionary origins.[63] Most of the recent work on emotion in social psychology and cognitive neuroscience has focused on trying to understand the structure and function of emotion. Political scientists are interested in emotion insofar as it helps to explain political behavior.

So, the focus for political scientists is more on how emotion influences political evaluation and decision making.

6.1 Social Psychological Models of Emotion

Social psychologists have offered many different theories of emotion over the years, with one potential starting point being the basic emotion models. These models attempted to specify a set of emotions that were both basic and universal—the list typically included the basic six: happiness, sadness, fear, anger, disgust, and surprise.[64,65] While there is some evidence that the basic emotions map on to facial expressions, the evidence for unique neural substrates for each of these emotions is rather limited, perhaps even nonexistent.[66,67] A more recent approach to emotion informed by work in affective neuroscience suggests that psychological constructivism is likely the more appropriate approach.[66] From this perspective, there is a network of brain regions involved in social and affective processes across emotion categories. Given that the basic emotion models may have limited explanatory power, social psychologists have proposed other models of emotion over the years—most can be considered either dimensional or appraisal models of emotion.

Both dimensional and appraisal models have evolved and have become more complicated over time, especially given recent work in affective neuroscience, but I will summarize the basic arguments here. The dimensional models of emotion typically attempt to simplify emotions into two or more dimensions. The most popular version being Russell's Circumplex Model, which includes both valence (ranging from positive to negative) and arousal (ranging from low to high).[68] These dimensions are orthogonal, so emotions can be positive or negative and high in arousal (e.g., fear, excitement) or low in arousal (e.g., contentment, sadness). From this perspective, it is important to consider both valence and arousal when examining the role of emotions—to see if the impact of a given emotion is due to its valence, arousal, or some interaction of the two.

Appraisal models of emotion have focused more directly on the cognitive processes that give rise to the experience of emotions—suggesting that emotions are constructed to help people deal with specific types of situations and respond appropriately. Appraisal models have taken different forms over the years, but typically suggest that people have some sort of physiological or affective response, which then gets interpreted in light of the current situation or context and then gets labeled as a specific emotion.[25] This view is probably most consistent with the psychological constructivism approach advanced by scholars like Lisa Feldman Barrett.[66,67] Indeed, some of the newer appraisal models are directly informed by research on affective and computational neuroscience.[69]

Importantly, the appraisal models have helped to elucidate the idea that emotions, which may look similar in dimensional models, may actually have distinct behavioral outcomes. For example, fear and anger are both high-arousal negative emotions but have been shown to lead to different outcomes in relation to risky behavior.[70] In general, fear is more likely to lead to avoidance behavior, whereas anger is more likely to lead to approach behavior.[71] Research on emotion and politics has only just begun to incorporate and adapt these models of emotion to increase our understanding of political behavior.

6.2 Models of Emotion and Politics

Political scientists have begun to incorporate models of emotion into their research, but I think it is fair to say that there is still much work to be done in this area. Early work in the area of emotion and politics focused on positive versus negative emotions, largely assuming that all negative emotions should lead to similar outcomes. However, consistent with dimensional or appraisal models of emotion discussed above, the work in political psychology has increasingly shown that emotions cannot necessarily be collapsed into a simple valence dimension, and even emotions that look similar in terms of a dimensional analysis—fear and anger, for example—lead to different outcomes in political behavior.[72,73] There have been some attempts to synthesize the current state of research on emotion and politics, and interested readers may want to review edited volumes such as *The Affect Effect* for more background.[74] I will give a brief overview of some of the recent work in political science below.

Research on emotion and politics has shown that positive and negative emotions have differential impact on politics, but also that different negative emotions may have different outcomes. For example, building on appraisal models of emotion, Huddy and colleagues have attempted to distinguish among anxiety, anger, and threat, showing that each of these emotions has distinct connections to foreign policy preferences.[75,76] Other work has examined how these emotions are connected to candidate appraisals and voting behavior, showing that emotions like hope and anger may be most likely to lead to voter mobilization.[77] In contrast, emotions like anxiety may be more likely to lead people to abstain from voting or reconsider their options and switch sides.[78] At this point, there is only one major theory of emotion and politics that has been put forth, and I will outline that theory below.

One of the primary theories in political science of how emotion influences politics is the Affective Intelligence Theory (AIT).[78,79] Marcus and colleagues have argued for this theory on the basis of neurobiological models of affect, suggesting that there is an important distinction to be made between behavior in familiar versus unfamiliar contexts. When objects or situations are familiar, people can rely on preexisting attitudes and beliefs to guide behavior either toward (for appetitive stimuli) or away (for aversive stimuli) from that stimulus. When situations are unfamiliar, AIT argues that people will experience greater anxiety and engage in additional information search in order to gain the information necessary to direct behavior in that situation. From this perspective, political scientists have shown support for the idea that anxiety leads to increased attention and political learning,[80] consideration of opposing viewpoints, and willingness to compromise.[81] While this may be consistent with some of the research on uncertainty in social psychology and cognitive neuroscience, it is largely inconsistent with the large body of work linking uncertainty (and anxiety) to threat, suggesting the interplay of these affective states may be more complicated than initially assumed.

From a motivational perspective, uncertainty can serve as a signal that we are lacking enough information to deal with situational challenges appropriately.[82] While informational uncertainty may sometimes lead to an increase in epistemic motives, other types of uncertainty (e.g., personal uncertainty) may be inherently more threatening and lead to the opposite response—increased closed-mindedness.[83] One way to examine this question is to treat uncertainty and threat as distinct conceptual variables (see Figure 2). Threat can be understood as the potential for harm, whereas uncertainty is often more ambiguous and context-dependent, signaling any lack of information or clarity on some issue.[84,85] From this perspective, uncertainty and threat interact to produce distinct affective states that may then lead to distinct behavioral outcomes.

In some of our work, we hypothesized that the effects of uncertainty on political tolerance would differ as a function of context—namely, whether the uncertainty was associated with threat or not (see Figure 3).[84] In an experimental context, we found that threat moderated the impact of uncertainty on political tolerance. Uncertainty increased tolerance in a neutral or positive context, but decreased tolerance in a threatening context. In a more recent study, I found support for a similar pattern of results with respect to support for compromise—uncertainty increased support for compromise in a relatively neutral or positive context, but had no effect on support for compromise in a more threatening context.[86] We are currently investigating the neural underpinnings of this interaction, both in terms of basic affective processes and political information processing. Preliminary analyses have shown that the amygdala, for example, responds differentially to uncertainty associated with negative versus positive affective information.[85]

I have included a section about emotion and politics here not because there is a dearth of neuroimaging

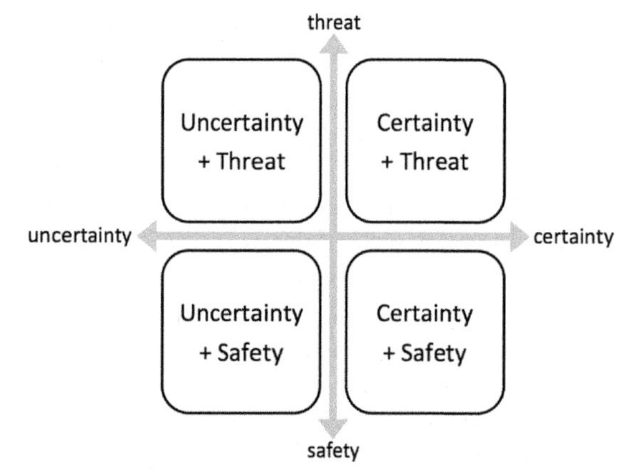

FIGURE 2 Conceptual relationship between uncertainty and threat. *Reprinted with permission from Ref. 84.*

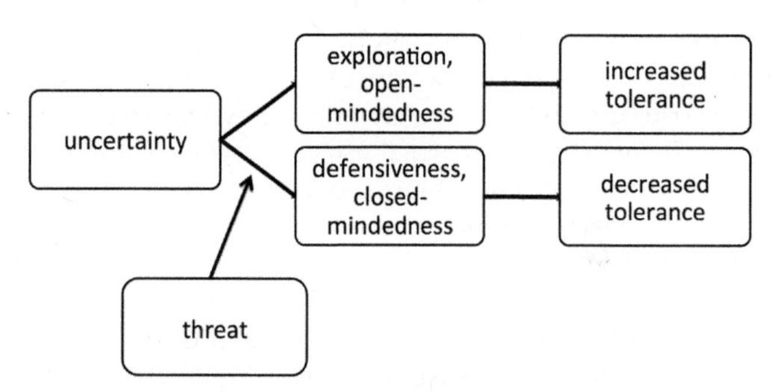

FIGURE 3 Hypothesized relationship between uncertainty and political tolerance as a function of threat. *Reprinted with permission from Ref. 84.*

research on the topic, but because I think there should be. This is an area of research where neuroimaging methods may be useful, and there is plenty of opportunity in terms of available research questions. It is important to realize that the future goals of this research are unlikely to be quite so simple as mapping the neural pathways that differentiate basic emotional responses. I think a more useful approach, at least in terms of the implications for politics, will be to focus on examining how different emotional experiences change political information processing. While existing research in this area is limited, there has been a fair amount of work examining individual differences in response to affective stimuli, primarily between political liberals versus conservatives. In the following section, I will provide an overview of political neuroscience research on individual differences.

7. INDIVIDUAL DIFFERENCES IN POLITICAL BEHAVIOR

One of the questions that fMRI is well-suited for is asking whether brain function differs across different types of people. Work in political neuroscience has taken advantage of this to examine differences in social and affective processing as a function of individual differences in political ideology, values, and political interest and expertise.

7.1 Ideological Differences

Ideological differences has been one of the primary topics guiding neuroimaging work on politics. The growth of political neuroscience studies on this topic is likely related to increased interest in both psychological and biological differences between liberals and conservatives. In 2003, Jost and colleagues published a meta-analysis arguing that conservatism has motivational underpinnings, primarily related to differential response to negative or threatening information.[87] Recent work in the area of biopolitics has largely corroborated this idea, showing that conservatives and liberals differ in their physiological response to negatively valenced emotional stimuli.[88–92] The relevant neuroimaging work has used both structural and functional neuroimaging to examine the neural underpinnings of these psychological and biological differences between liberals and conservatives.

7.1.1 Brain Structure

Given the growing evidence of a link between biology, genetics, and ideological differences, one possibility is that liberalism versus conservatism is related to differences in brain structure. Consistent with this idea, one study has shown structural brain differences between liberals and conservatives. Kanai and colleagues used structural MRI to examine gray matter volume, finding that liberals had increased volume in the anterior cingulate cortex, whereas conservatives had increased volume in the right amygdala, left insula, and right entorhinal cortex (see Figure 4).[93] While this work supports the hypothesis that liberals and conservatives have different brain structure, there are multiple ways to interpret this data and I think some caution is warranted, given that this is just one study.

First, it is probably necessary to consider whether or not these structural differences are indicative of functional brain differences. It is not necessarily the case that having more gray matter in a region means you will find significant differences in function, but it does mean that these brain regions are probably worth examining in subsequent studies of brain function. While this is purely speculative at this point, the amygdala finding from the Kanai study could be consistent with behavioral work suggesting that conservatives are more sensitive to threat or negativity than liberals.[87,89,94] However, it is important to note that the relationship between amygdala activation and fear or negativity is not a one-to-one mapping. While the amygdala has been implicated in fear, it has also been implicated in responses to positive or arousing information, as discussed earlier in this chapter.[31–33] More recent theorizing about amygdala function has suggested that the amygdala response is not specific to positive or negative information per se, but that the amygdala may respond to any information that is motivationally relevant for the individual.[95]

Second, structural differences are also difficult to interpret because there is sometimes an assumption that they are an indicator of biological or genetic differences in brain growth or development (suggesting the influence of nature over nuture). An alternative, or perhaps complementary, viewpoint is that brain structure is influenced not just by biology and genetics, but also by psychology and behavior. For example, some recent longitudinal work has shown that a stress-reduction intervention actually reduced both reported stress level and gray matter volume in the right basolateral amygdala.[96] So, having a larger amygdala might increase sensitivity to stress, but a decrease in stress may also decrease the size of the amygdala. Jost and colleagues have labeled this the "chicken-and-egg" problem in political neuroscience.[19,89] It could be the case that genetic differences shape brain structure in a way that gives rise to ideological differences, or it could be the case that people adopt certain patterns of behavior and that behavior then shapes brain structure. In reality, brain structure is probably determined by some combination of these two perspectives. The take-home point is that we need to be careful about assuming causal relationships from studies of brain structure. Regardless of the causal pathways, the Kanai study[93] suggested to researchers that these regions (e.g., the amygdala and anterior cingulate) are

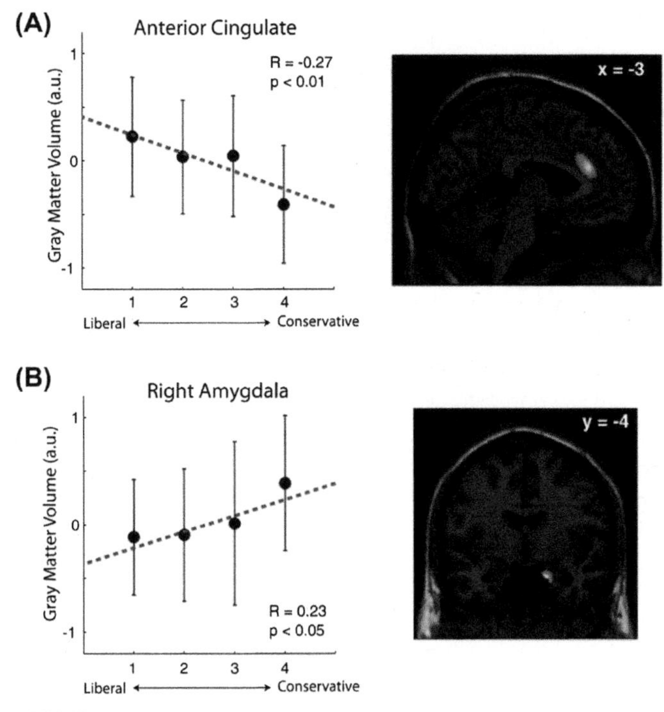

FIGURE 4 Individual differences in political attitudes and brain structure. *Reprinted with permission from Ref. 93.*

worth examining in future studies of ideology, brain structure, and brain function.

7.1.2 Brain Function

Research on whether or not brain function differs between liberals and conservatives is still relatively limited, but the existing research is consistent with the idea that there may be processing differences in regions implicated in social and affective processing—such as the amygdala, insula, and anterior cingulate. One recent study showed that during a decision-making task, Democrats showed greater activation in the left insula and Republicans showed greater activation in the right amygdala.[97] Interestingly, while brain function appeared to differ, the decisions participants made in this study did not. In other words, it could be the case that these differences are due, at least in part, to differences in the decision-making *process* as opposed to the outcomes. This conclusion should be treated as tentative at this point, given that this is just one study and other research has shown that there may be differences in both process and outcomes. Using EEG, Amodio and colleagues found that liberals demonstrated greater conflict-related activity and were more accurate on a go/no-go task.[98] The conflict-related activity on no-go trials was localized to the dorsal anterior cingulate cortex, which is one of the regions where structural differences between liberals and conservatives have been found.[93] Taken together, these studies suggest that there may be functional differences between liberals and conservatives in some of the regions where structural differences were previously found—the amygdala, insula, and anterior cingulate.

A lot of the recent behavioral work or work in biopolitics has focused on ideological differences in emotional responding, suggesting that liberals and conservatives may differ in their response to negatively-valenced emotional stimuli.[88,89] Only one study (that I am aware of) has examined this question using neuroimaging. Ahn and colleagues had participants engage in a passive viewing task during fMRI using stimuli (emotional images) from the International Affective Picture System (IAPS).[99] Consistent with behavioral work showing that conservatives are more responsive to disgust,[91,100] conservatives in this study appeared to be more sensitive to disgusting images (especially those images related to animal mutilation). Liberals showed greater activation than conservatives to disgusting stimuli mainly in the insula, whereas for conservatives, there was greater activation for this contrast (disgust>neutral) in a whole network of regions, including the amygdala, hippocampus, basal ganglia, thalamus, anterior cingulate, dorsolateral prefrontal cortex, and ventromedial prefrontal cortex. While this is consistent with the view that conservatives may be more sensitive to disgusting stimuli than liberals, it is difficult to conclude what processes each of these regions may have been engaged in, given that this was a passive viewing task. Future work may want to examine differences between liberals and conservatives using different tasks that require additional processing demands (e.g., emotion regulation). It could be the case that liberals and conservatives respond differently to disgusting images, or they might differentially engage emotion regulation strategies to cope with the disgust response.

While a growing body of work has suggested that conservatives may be more sensitive to negative information than liberals,[87,99] one alternative perspective is that conservatives are more sensitive to arousal (and not negativity, per se).[89,101] From the perspective of dimensional models of emotion (see earlier section on Emotion and Politics), it is important to consider both the valence of emotions and their level of arousal. If researchers are comparing high-arousal negative emotions to low-arousal positive emotions, any observed differences could be due either to valence (negative-positive) or arousal (low-high). It is important to note that the Ahn study did not examine differences between negative versus positive emotions directly, and the positive images used in the study appear to be lower in arousal (e.g., animals, babies) than the negative images used (e.g., snakes, violence, mutilation).[99] So, the question about whether conservatives are more responsive to negative or highly arousing stimuli, even when positively valenced, remains.[101] To examine this question directly, researchers will need to compare emotional responses to positive and negative stimuli while controlling for, equating, or manipulating arousal.

7.2 Values

An alternative approach to understanding ideological differences in brain function has been to examine differences in neural processing related to specific values that might underlie ideological differences. This research builds on a long tradition in psychology and political science of studying values as a way to understand political attitudes.[102] Political ideology has been most closely tied to values related to preference for tradition versus change and views about equality.[103] Neuroimaging work has attempted to examine the brain activity that might underlie political decision-making related to these values, although this work has been largely exploratory.

Zamboni and colleagues had participants evaluate political beliefs during fMRI and examined the relationship between evaluative processing and values—individualism, conservatism, and radicalism.[104] They found differentiation on the basis of these values: individualism was related to activity in the ventromedial and dorsomedial prefrontal cortex and temporoparietal junction, conservatism was related to the dorsolateral prefrontal cortex, and radicalism was related to the ventral striatum and posterior cingulate. While this research does suggest that values may be related to differential processing of political statements, it is still unclear why these differences are occurring or how they map on to differences in the process of decision-making. The authors do speculate about what this activation might mean—perhaps individualism-related prefrontal cortex activation signals self-referential processing, dorsolateral prefrontal cortex activation might be related to additional evaluative processing for liberals versus conservatives, and radicalism might be related to greater emotional responses to these statements.[104] It is important to note, however, that these assumptions are based on reverse inference and may or may not be corroborated by future work. Additional work will be needed to test these assumptions directly.

A second study on this topic focused on the role of egalitarianism, showing that egalitarian preferences during a decision-making task were related to activation in the insula and ventromedial PFC.[105] The insula has been implicated in a number of studies related to emotion and empathy, or just more general integration of bodily states, so this connection between insula activity and egalitarianism seems plausible. But again, more work will be needed to clearly examine the underlying mechanisms here.

7.3 Political Interest and Expertise

Another important individual difference that has often been discussed by political scientists is the distinction between political experts versus novices. The classic research in political science often discussed differences between the mass public versus political elites, often arguing that political elites were really the only individuals with something that resembled a political ideology.[106] This question has also been examined through the use of neuroimaging methods, looking primarily at differences between political experts versus novices or people who are highly interested in politics versus uninterested.

Interestingly, while the impact of expertise on brain function has often been discussed in a cold, cognitive sense, purely in relation to knowledge, it may be the case that there is a motivational component as well. In other words, some people may actually be motivated to learn or read about politics and may experience that activity as rewarding. Consistent with this view, Gozzi and colleagues showed that individuals with a strong interest in politics experience greater activation in the amygdala and ventral striatum (putamen) when reading statements they agreed with.[107] While we need to be careful about making reverse inferences here, this is consistent with the idea that viewing these positive statements may be rewarding for individuals interested in politics. Interestingly, this may help explain why some political junkies literally cannot stop reading about or watching the news.

8. DIRECTIONS FOR FUTURE RESEARCH

The field of political neuroscience is relatively young, and research in this area is fairly limited at this point. While this makes it difficult to generalize and draw broader conclusions based on the work, I think we do have enough at this point to say that neuroimaging has the potential to inform research in political psychology and biopolitics. The challenge for researchers then becomes figuring out how we should move forward in this interdisciplinary area. Below, I offer some suggestions for researchers currently working in the field of political neuroscience, or those who are interested in getting involved in this work.

Most, if not all, of the political neuroscience studies described here have been exploratory in nature. There is nothing inherently wrong with this, given that exploratory research is often the first stage of a larger research program—gathering initial data on important questions can help clarify theory and generate hypotheses for subsequent research. However, I do think we are getting closer to the point where we can move out of the initial exploratory phase and into the hypothesis-testing phase.

The main problem with exploratory fMRI studies is that they are difficult to interpret, for a number of reasons. As discussed at the outset of this chapter, fMRI data is correlational. It is difficult (if not impossible) to demonstrate causal relationships using fMRI data. The bigger issue with these exploratory studies is that interpretation has largely relied on reverse inference and assumptions about the processes that participants were engaged in during the task. Now that some of this initial work has been conducted, we can start to develop more nuanced hypotheses about how people are processing political information and why we are observing brain activation in specific regions or networks. Given that we have specific hypotheses about brain mechanisms, we can carefully design experimental tasks to test these mechanisms by directly manipulating or measuring variables of interest. Only through careful experimentation can we determine, for example, why the amygdala is responding to political candidates, how the amygdala is involved in political evaluation, how this differs across individuals, and what the boundary conditions are that limit or constrain these effects.

It would be impossible to overemphasize the importance of careful research design in the fMRI environment. In an ideal world, any political neuroscience study using fMRI would include the following: theoretical background, specific hypotheses about brain function, behavioral pilot data, careful experimental design with multiple conditions, task data during scanning (as well as relevant postscan ratings), measurement of relevant individual differences, and great care taken when drawing conclusions not to overgeneralize or rely on unwarranted reverse inferences about brain function. Other scholars have offered similar suggestions for those working in the field of political neuroscience,[7] as well as those working in social and affective neuroscience more generally.[108]

It is worth noting here that there is a relatively steep learning curve when it comes to learning neuroimaging methods such as fMRI, so collaboration among interdisciplinary teams should be encouraged. Political scientists interested in working in the area of political neuroscience will likely benefit from working with social and cognitive psychologists, who have expertise in experimental design, as well as social and cognitive neuroscientists, who have expertise in conducting research within the MRI context and dealing with unique considerations for both experimental design and data analysis. Psychologists and neuroscientists interested in politics can benefit from working with political scientists, who have expertise in the role of context and are more well-versed in issues related to external validity. Many fMRI studies, especially early studies using this methodology, have relied on very small sample sizes, typically with convenience samples. I think future studies in political neuroscience will probably want to move toward collecting data from larger, more diverse samples and attempt to use that data to predict real-world outcomes, in addition to performance on experimental tasks. These are ambitious goals and might be difficult for any lone scientist to master but will likely be easier to tackle in larger, interdisciplinary teams.

There is one very good reason why fMRI studies have typically relied on small sample sizes—cost. Given the expense attached to doing research using fMRI (as of this writing, an hour of scan time typically costs around $500 or $600 at universities in the US), the challenge for researchers becomes trying to figure out if and when fMRI will be a valuable method to add to their toolbox. Many questions in political science may not require the use of methods like fMRI, and researchers should not feel obligated to use these methods if they are not directly relevant to the questions they are interested in. fMRI is most likely to be a useful tool when researchers have ideas about processes or mechanisms that give rise to observable political behavior and are interested in testing whether or not those mechanisms are tenable given neurobiological structure and function.

9. CONCLUSION

Neuroimaging cannot replace traditional methods and measurement techniques in political science, but it can serve as a useful technique for examining whether or not theories about political behavior are biologically plausible. Social psychology has demonstrated over the years that people have many assumptions about human behavior that are not always supported at the behavioral level, and the same is likely true for what neuroimaging can show us at the neural level. For example, neuroscience work has already led many scientists to question some of the traditional assumptions that have guided research on social behavior in recent decades—that emotion and cognition are distinct processes, or that implicit versus explicit attitudes are categorically distinct and stored separately in memory—and it will likely lead us to question many other assumptions about human behavior. We have really only just begun trying to use neuroimaging methods to understand how people make social and political decisions at the neural level, and it will likely be the case that, in the years to come, we will continue to learn that our current theories of human social and political behavior are incomplete.

In sum, while research in political neuroscience has increased rapidly in recent years, we still have a long way to go before we have a clear picture of the neural mechanisms that underlie political evaluation, cognition, and decision-making. While this means there are still a number of challenges for researchers working in this area in terms of methods, design, and interpretation, it also means it is an exciting time for anyone interested in the subfield of political neuroscience, either as a participant or an observer. It is an area of research where there is still much left to explore.

References

1. McDermott R. The case for increasing dialogue between political science and neuroscience. *Political Res Q*. 2009;62(3):571–583.
2. Lopez AC, McDermott R. Adaptation, heritability, and the emergence of evolutionary political science. *Political Psychol*. 2012;33(3):343–362.
3. Friston KJ. Modalities, modes, and models in functional neuroimaging. *Science*. October 16, 2009;326(5951):399–403.
4. Cacioppo JT, Visser PS. Political psychology and social neuroscience: strange bedfellows or comrades in arms? *Political Psychol*. 2003;24(4):647–656.
5. Lieberman MD, Schreiber D, Ochsner KN. Is political cognition like riding a bicycle? How cognitive neuroscience can inform research on political thinking. *Political Psychol*. 2003;24:681–704.
6. Raichle ME. Social neuroscience: a role for brain imaging. *Political Psychol*. 2003;24:759–764.
7. Theodoridis AG, Nelson AJ. Of BOLD claims and excessive fears: a call for caution and patience regarding political neuroscience. *Political Psychol*. 2012;33(1):27–43.
8. Logothetis N, Pauls J, Augath M, Trinath T, Oeltermann A. Neurophysiological investigation of the basis of the fMRI signal. *Nature*. 2001;412(6843):150–157.
9. Logothetis NK, Wandell BA. Interpreting the BOLD signal. *Annu Rev Physiol*. 2004;66:735–769.
10. Greenwald AG, McGhee DE, Schwartz JLK. Measuring individual differences in implicit cognition: the implicit association test. *J Pers Soc Psychol*. 1998;74:1464–1480.
11. Payne BK, Cheng CM, Govorun O, Stewart BD. An inkblot for attitudes: affect misattribution as implicit measurement. *J Pers Soc Psychol*. September 2005;89(3):277–293.
12. Murphy ST, Zajonc RB. Affect, cognition, and awareness: affective priming with optimal and suboptimal stimulus exposures. *J Pers Soc Psychol*. 1993;64(5):723–739.

13. Gawronski B, Bodenhausen GV. Unraveling the processes underlying evaluation: attitudes from the perspective of the APE model. *Soc Cogn*. 2007;25:687–717.
14. DeCoster J, Banner MJ, Smith ER, Semin GR. On the inexplicability of the implicit: differences in the information provided by implicit and explicit tests. *Soc Cogn*. 2006;24:5–21.
15. Deutsch R, Strack F. Duality models in social psychology: from dual processes to interacting systems. *Psychol Inquiry*. 2006;17:166–172.
16. Cunningham WA, Zelazo PD, Packer DJ, Van Bavel JJ. The iterative reprocessing model: a multilevel framework for attitudes and evaluation. *Soc Cogn*. 2007;25:736–760.
17. Fazio RH. Attitudes as object-evaluation associations of varying strength. *Soc Cogn*. 2007;25:603–637.
18. Cunningham WA, Zelazo PD. Attitudes and evaluations: a social cognitive neuroscience perspective. *Trends Cogn Sci*. 2007; 11:97–104.
19. Jost JT, Nam HH, Amodio DM, Van Bavel JJ. Political neuroscience: the beginning of a beautiful friendship. *Adv Political Psychol*. 2014;35:3–42.
20. Cunningham WA, Van Bavel JJ, Johnsen IR. Affective flexibility: evaluative processing goals shape amygdala activity. *Psychol Sci*. 2008;19:152–160.
21. Burdein I, Lodge M, Taber C. Experiments on the automaticity of political beliefs and attitudes. *Political Psychol*. 2006;27(3): 359–371.
22. Lodge M, Taber CS. The automaticity of affect for political leaders, groups, and issues: an experimental test of the hot cognition hypothesis. *Political Psychol*. 2005;26:455–482.
23. Morris JP, Squires NK, Taber CS, Lodge M. Activation of political attitudes: a psychophysiological examination of the hot cognition hypothesis. *Political Psychol*. 2003;24(4):727–745.
24. Zajonc RB. On the primacy of affect. *Am Psychol*. 1984;39:117–123.
25. Lazarus RS. Cognition and motivation in emotion. *Am Psychol*. 1991;46:352–367.
26. Knutson KM, Wood JN, Spampinato MV, Grafman J. Politics on the brain: an fMRI investigation. *Soc Neurosci*. 2006;1:25–40.
27. Rule NO, Freeman JB, Moran JM, Gabrieli JD, Adams Jr RB, Ambady N. Voting behavior is reflected in amygdala response across cultures. *Soc Cogn Affect Neurosci*. June 2010;5(2–3): 349–355.
28. LeDoux JE, Cicchetti P, Xagoraris A, Romanski LM. The lateral amygdaloid nucleus: sensory interface of the amygdala in fear conditioning. *J Neurosci*. 1990;10:1062–1069.
29. Davis M. The role of the amygdala in fear and anxiety. *Annu Rev Neurosci*. 1992;15:353–375.
30. Adolphs R, Tranel D, Damasio H, Damasio AR. Fear and the human amygdala. *J Neurosci*. 1995;15:5879–5891.
31. Garavan H, Pendergrass JC, Ross TJ, Stein EA, Risinger RC. Amygdala response to both positive and negatively valenced stimuli. *Neuroreport*. 2001;12:2779–2783.
32. Hamann SB, Ely TD, Hoffman JM, Kilts CD. Ecstasy and agony: activation of the human amygdala in positive and negative emotion. *Psychol Sci*. 2002;13:135–141.
33. Hamann SB, Mao H. Positive and negative emotional verbal stimuli elicit activity in the left amygdala. *Neuroreport*. 2002;13:15–19.
34. Adolphs R, Russell JA, Tranel D. A role for the human amygdala in recognizing emotional arousal from unpleasant stimuli. *Psychol Sci*. 1999;10:167–171.
35. Whalen PJ. The uncertainty of it all. *Trends Cogn Sci*. December 2007;11:499–500.
36. Ousdal OT, Jensen J, Server A, Hariri AR, Nakstad PH, Andreassen OA. The human amygdala is involved in general behavioral relevance detection: evidence from an event-related functional magnetic resonance imaging Go-NoGo task. *Neuroscience*. October 15, 2008;156(3):450–455.
37. Sander D, Grafman J, Zalla T. The human amygdala: an evolved system for relevance detection. *Rev Neurosci*. 2003;14:303–316.
38. Kaplan JT, Freedman J, Iacoboni M. Us versus them: political attitudes and party affiliation influence neural responses to faces of presidential candidates. *Neuropsychologia*. 2007;45:55–64.
39. Spezio ML, Rangel A, Alvarez RM, et al. A neural basis for the effect of candidate appearance on election outcomes. *Soc Cogn Affect Neurosci*. December 2008;3(4):344–352.
40. Ito TA, Larsen JT, Smith NK, Cacioppo JT. Negative information weighs more heavily on the brain: the negativity bias in evaluative categorizations. *J Pers Soc Psychol*. 1998;75:887–900.
41. Fazio RH, Eiser JR, Shook NJ. Attitude formation through exploration: valence asymmetries. *J Pers Soc Psychol*. 2004;87:293–311.
42. Neta M, Whalen PJ. The primacy of negative interpretations when resolving the valence of ambiguous facial expressions. *Psychol Sci*. July 2010;21(7):901–907.
43. Van Bavel JJ, Xiao YJ, Cunningham WA. Evaluation is a dynamic process: moving beyond dual system models. *Soc Pers Psychol Compass*. 2012;6:438–454.
44. Tusche A, Kahnt T, Wisniewski D, Haynes JD. Automatic processing of political preferences in the human brain. *Neuroimage*. May 15, 2013;72:174–182.
45. Lord CG, Ross L, Lepper MR. Biased assimilation and attitude polarization: the effects of prior theories on subsequently considered evidence. *J Pers Soc Psychol*. 1979;37:2098–2109.
46. Kunda Z. The case for motivated reasoning. *Psychol Bull*. 1990;108:480–498.
47. Westen D, Blagov PS, Harenski K, Kilts C, Hamann S. Neural bases of motivated reasoning: an fMRI study of emotional constraints on partisan political judgment in the 2004 U.S. presidential election. *J Cogn Neurosci*. 2006;18:1947–1958.
48. Botvinick MM, Braver TS, Barch DM, Carter CS, Cohen JD. Conflict monitoring and cognitive control. *Psychol Rev*. 2001;108(3):624–652.
49. Gu X, Liu X, Van Dam NT, Hof PR, Fan J. Cognition-emotion integration in the anterior insular cortex. *Cereb Cortex*. January 2013;23(1):20–27.
50. Cunningham WA, Johnsen IR, Waggoner AS. Orbitofrontal cortex provides cross-modal valuation of self-generated stimuli. *Soc Cogn Affect Neurosci*. 2011;6:286–293.
51. Kringelbach ML, Rolls ET. The functional neuroanatomy of the human orbitofrontal cortex: evidence from neuroimaging and neuropsychology. *Prog Neurobiol*. 2004;72:341–372.
52. Grahn JA, Parkinson JA, Owen AM. The cognitive functions of the caudate nucleus. *Prog Neurobiol*. November 2008;86(3):141–155.
53. Kuhn S, Gallinat J. The neural correlates of subjective pleasantness. *Neuroimage*. May 15, 2012;61(1):289–294.
54. Petty RE, Krosnick JA, eds. *Attitude Strength: Antecedents and Consequences*. Mahwah: Erlbaum Associates; 1995.
55. Cacioppo JT, Berntson GG. Relationship between attitudes and evaluative space: a critical review, with emphasis on the separability of positive and negative substrates. *Psychol Bull*. 1994;115:401–423.
56. Kato J, Ide H, Kabashima I, Kadota H, Takano K, Kansaku K. Neural correlates of attitude change following positive and negative advertisements. *Front Behav Neurosci*. 2009;3:1–13.
57. O'Reilly RC, Noelle DC, Braver TS, Cohen JD. Prefrontal cortex and dynamic categorization tasks: representational organization and neuromodulatory control. *Cereb Cortex*. 2002;12:246–257.
58. Frith U, Frith CD. Development and neurophysiology of mentalizing. *Philos Trans R Soc Lond B Biol Sci*. March 29, 2003;358(1431): 459–473.
59. Kozak MN, Marsh AA, Wegner DM. What do I think you're doing? Action identification and mind attribution. *J Pers Soc Psychol*. 2006;90(4):543–555.
60. Saxe R, Kanwisher N. People thinking about thinking people: the role of the temporo-parietal junction in "theory of mind". *Neuroimage*. 2003;19:1835–1842.

61. Mitchell JP, Macrae CN, Banaji MR. Dissociable medial prefrontal contributions to judgments of similar and dissimilar others. *Neuron*. 2006;50:655–663.
62. Falk EB, Spunt RP, Lieberman MD. Ascribing beliefs to ingroup and outgroup political candidates: neural correlates of perspective-taking, issue importance and days until the election. *Philos Trans R Soc Lond B Biol Sci*. 2012;367(1589):731–743.
63. Rozin P, Haidt J. The domains of disgust and their origins: contrasting biological and cultural evolutionary accounts. *Trends Cogn Sci*. 2013;17(8):367–368.
64. Ekman P. Are there basic emotions? *Psychol Rev*. 1992;99:550–553.
65. Ekman P. An argument for basic emotions. *Cogn Emotion*. 1992; 6(3/4):169–200.
66. Lindquist KA, Wager TD, Kober H, Bliss-Moreau E, Barrett LF. The brain basis of emotion: a meta-analytic review. *Behav Brain Sci*. June 2012;35(3):121–143.
67. Barrett LF. Are emotions natural kinds? *Perspect Psychol Sci*. 2006;1:28–58.
68. Russell J. A circumplex model of affect. *J Pers Soc Psychol*. 1980; 39:1161–1178.
69. Scherer KR. Emotions are emergent processes: they require a dynamic computational architecture. *Philos Trans R Soc Lond B Biol Sci*. December 12, 2009;364(1535):3459–3474.
70. Lerner JS, Keltner D. Fear, anger, and risk. *J Pers Soc Psychol*. 2001;81(1):146–159.
71. Harmon-Jones E. Clarifying the emotive functions of asymmetrical frontal cortical activity. *Psychophysiology*. 2003;40:838–848.
72. Lerner JS, Gonzalez RM, Small DA, Fischhoff B. Effects of fear and anger on perceived risks of terrorism: a national field experiment. *Psychol Sci*. 2003;14(2):144–150.
73. Skitka LJ, Bauman CW, Aramovich NP, Morgan GS. Confrontational and preventative policy responses to terrorism: anger wants a fight and fear wants "them" to go away. *Basic Appl Soc Psychol*. 2006;28:375–384.
74. Neuman WR, Marcus GE, Crigler A, MacKuen M, eds. *The Affect Effect: Dynamics of Emotion in Political Thinking and Behavior*. Chicago: University of Chicago Press; 2007.
75. Huddy L, Feldman S, Cassese E. On the distinct political effects of anxiety and anger. In: Neuman WR, Marcus GE, Crigler AN, MacKuen M, eds. *The Affect Effect: Dynamics of Emotion in Political Thinking and Behavior*. Chicago: University of Chicago Press; 2007:202–230.
76. Huddy L, Feldman S, Taber C. Threat, anxiety, and support of antiterrorism policies. *Am J Political Sci*. 2005;49:593–608.
77. Just MR, Crigler AN, Belt TL. Don't give up hope: emotions, candidate appraisals, and votes. In: Neuman WR, Marcus GE, Crigler AN, MacKuen M, eds. *The Affect Effect: Dynamics of Emotion in Political Thinking and Behavior*. Chicago: University of Chicago Press; 2007.
78. MacKuen M, Marcus GE, Neuman WR, Keele L. The third way: the theory of affective intelligence and American democracy. In: Neuman WR, Marcus GE, Crigler AN, MacKuen M, eds. *The Affect Effect: Dynamics of Emotion in Political Thinking and Behavior*. Chicago: University of Chicago Press; 2007:124–151.
79. Marcus GE, Neuman WR, MacKuen M. *Affective Intelligence and Political Judgment*. Chicago: University of Chicago Press; 2000.
80. Marcus GE, MacKuen MB. Anxiety, enthusiasm, and the vote: the emotional underpinnings of learning and involvement during presidential campaigns. *Am Political Sci Rev*. 1993;87: 672–685.
81. MacKuen M, Wolak J, Keele L, Marcus GE. Civic engagements: resolute partisanship or reflective deliberation. *Am J Political Sci*. 2010;54:440–458.
82. Hirsh JB, Mar RA, Peterson JB. Psychological entropy: a framework for understanding uncertainty-related anxiety. *Psychol Rev*. January 16, 2012;119(2):304.
83. McGregor I, Nash K, Mann N, Phills CE. Anxious uncertainty and reactive approach motivation (RAM). *J Pers Soc Psychol*. 2010;99:133–147.
84. Haas IJ, Cunningham WA. The uncertainty paradox: perceived threat moderates the effect of uncertainty on political tolerance. *Political Psychol*. 2014;35(2):291–302.
85. Haas IJ. *The Context-Dependent Nature of Uncertainty: Responses to Uncertainty are Moderated by the Presence or Absence of Threat* [Dissertation]. Columbus: Department of Psychology, The Ohio State University; 2012.
86. Haas IJ. *Uncertainty, Threat, and Support for Political Compromise during the 2013 U.S. Government Shutdown*; Manuscript submitted for publication. 2015.
87. Jost JT, Glaser J, Kruglanski AW, Sulloway FJ. Political conservatism as motivated social cognition. *Psychol Bull*. 2003;129:339–375.
88. Hibbing JR, Smith KB, Peterson JC, Feher B. The deeper sources of political conflict: evidence from the psychological, cognitive, and neuro-sciences. *Trends Cogn Sci*. March 2014;18(3):111–113.
89. Hibbing JR, Smith KB, Alford JR. Differences in negativity bias underlie variations in political ideology. *Behav Brain Sci*. June 2014;37(3):297–307.
90. Oxley DR, Smith KB, Alford JR, et al. Political attitudes vary with physiological traits. *Science*. 2008;321:1667–1670.
91. Smith KB, Oxley D, Hibbing MV, Alford JR, Hibbing JR. Disgust sensitivity and the neurophysiology of left-right political orientations. *PLoS One*. 2011;6(10):e25552.
92. Dodd MD, Balzer A, Jacobs CM, Gruszczynski MW, Smith KB, Hibbing JR. The political left rolls with the good and the political right confronts the bad: connecting physiology and cognition to preferences. *Philos Trans R Soc Lond B Biol Sci*. March 5, 2012;367(1589):640–649.
93. Kanai R, Feilden T, Firth C, Rees G. Political orientations are correlated with brain structure in young adults. *Curr Biol*. 2011;21(8):677–680.
94. Shook NJ, Fazio RH. Political ideology, exploration of novel stimuli, and attitude formation. *J Exp Soc Psychol*. 2009;45(4): 995–998.
95. Cunningham WA, Brosch T. Motivational salience: amygdala tuning from traits, needs, values, and goals. *Curr Dir Psychol Sci*. 2012;21:54–59.
96. Holzel BK, Carmody J, Evans KC, et al. Stress reduction correlates with structural changes in the amygdala. *Soc Cogn Affect Neurosci*. March 2010;5(1):11–17.
97. Schreiber D, Fonzo G, Simmons AN, et al. Red brain, blue brain: evaluative processes differ in Democrats and Republicans. *PLoS One*. 2013;8(2):e52970.
98. Amodio DM, Jost JT, Master SL, Yee CM. Neurocognitive correlates of liberalism and conservatism. *Nat Neurosci*. 2007;10:1246–1247.
99. Ahn W-Y, Kishida Kenneth T, Gu X, et al. Nonpolitical images evoke neural predictors of political ideology. *Curr Biol*. 2014; 24(22):2693–2699.
100. Inbar Y, Pizarro DA, Bloom P. Conservatives are more easily disgusted than liberals. *Cogn Emotion*. 2009;23(4):714–725.
101. Tritt SM, Inzlicht M, Peterson JB. Preliminary support for a generalized arousal model of political conservatism. *PLoS One*. 2013;8(12):e83333.
102. Schwartz SH, Bilsky W. Toward a theory of the universal content and structure of values: extensions and cross-cultural replications. *J Pers Soc Psychol*. 1990;24:1–91.
103. Jost JT. The end of the end of ideology. *Am Psychol*. 2006;61:651–670.
104. Zamboni G, Gozzi M, Krueger F, Duhamel J-R, Sirigu A, Grafman J. Individualism, conservatism, and radicalism as criteria for processing political beliefs: a parametric fMRI study. *Soc Neurosci*. October 1, 2009;4(5):367–383.
105. Dawes CT, Loewen PJ, Schreiber D, et al. Neural basis of egalitarian behavior. *Proc Natl Acad Sci USA*. April 24, 2012;109(17): 6479–6483.

106. Converse PE. The nature of belief systems in mass publics. In: Apter DE, ed. *Ideology and Discontent*. New York: Free Press; 1964:206–261.
107. Gozzi M, Zamboni G, Krueger F, Grafman J. Interest in politics modulates neural activity in the amygdala and ventral striatum. *Hum Brain Mapp*. November 2010;31(11):1763–1771.
108. Berkman ET, Cunningham WA, Lieberman MD. Research methods in social and affective neuroscience. In: Reis HT, Judd CM, eds. *Handbook of Research Methods in Personality and Social Psychology*. 2nd ed. New York: Cammbridge University Press; 2014:123–158.

CHAPTER

20

Science in Society: Neuroscience and Lay Understandings of Self and Identity

Cliodhna O'Connor

Department of Psychology, Maynooth University, Maynooth Co Kildare, Ireland

OUTLINE

1. INTRODUCTION

The dying days of the twentieth century were witness to an invigoration of the brain sciences, as governments worldwide declared that the 1990s would go down in official history as the "decade of the brain." The symbolic and material investments that accompanied such pronouncements propelled a dramatic expansion of both the volume of neuroscientific research produced and of the range of phenomena to which neuroscientific explanations were applied. In this period of disciplinary growth, the subfield of social neuroscience, which strives to elucidate the biological underpinnings of social cognition and behavior, gained particular momentum.[1,2] This reflects the growing ambition of neuroscientific research agendas, whose gaze increasingly extends beyond basic neurocognitive processes to the complex dynamics of personal, social, and cultural worlds. Topics previously the preserve of the humanities and social sciences, such as religion, love, art, crime, and politics, are now standard targets of neuroscientific investigation.[3]

Due to the increasingly direct social relevance of neuroscientific data, much rhetoric about the rise of neuroscience has framed it not merely as a matter of *scientific* progress, but as a development with profound cultural significance. In particular, many have conjectured that day-to-day representations of the human self, spirit, or soul are being progressively reconfigured in neurochemical terms. Such claims frequently emanate from within the neuroscientific community itself; for instance, when the European Union's (EU) Human Brain Project was launched, its promotional materials asserted that its scientific endeavors would "undoubtedly…have a deep impact on our deepest felt convictions—in particular our concepts of personhood, free will and personal responsibility, the way we see ourselves as persons, personally responsible for our actions."[4] Moreover, many sociologists, ethicists, and philosophers have trained their lens on neuroscience's position in contemporary society, to conclude that subjective experience is increasingly filtered through such constructs as "neurochemical selves,"[5] "cerebral subjects,"[6] and "brainhood."[7] Until

Neuroimaging Personality, Social Cognition, and Character
http://dx.doi.org/10.1016/B978-0-12-800935-2.00020-8

recently, however, such debates were limited to largely speculative analyses, due to a paucity of empirical research documenting how neuroscientific concepts are resonating with the lay public.[8]

Fortunately, a growing body of social scientific research has taken up the challenge of tracking what "happens" to neuroscience knowledge as it travels through the public sphere and what marks it leaves on the individuals and communities it meets along the way. This chapter reviews the existing literature on how neuroscience has influenced everyday understandings of self and identity. It also presents original data from an interview study exploring how members of the public engage with contemporary brain research. Such research affords valuable insight into how neuroscience assimilates into subjective experience, how it intermingles with existing values and beliefs, and the shifts in self-conceptions and social relations that may ensue as a result.

2. THE NEUROLOGIZATION OF FOLK PSYCHOLOGY: WHY DOES IT MATTER?

If neuroscience is affecting common-sense understandings of human activity and identity, why might this matter? First, people's "folk psychology," or implicit understanding of how minds work, guides how they interpret and respond to behavior. As such, if neuroscience changes folk psychologies, it might prompt a corresponding shift in social interactions. For example, some have speculated that the increasing prominence afforded to neuroscientific explanations of behavior will gradually erode people's belief in free will.[9–11] If this prospect materializes, it could have profound social implications. Correlational and experimental research has linked disbelief in free will to an increased proclivity for aggression,[12] a reduced inclination toward helpfulness,[12] and dishonest behavior.[13] A heightened belief in free will, on the other hand, supports empathy toward the disadvantaged people and increased commitment to equality and social mobility.[14] Another example of potential neuroscientific influence on folk psychology relates to attributions of responsibility: conceivably, materialistic accounts of behavior could make individuals seem less personally culpable for their harmful actions and less commendable for their acts of virtue. Thus, via their influence on folk psychology, neuroscientific concepts may have tangible effects on the common-sense understandings on which social institutions and interactions are based.

Second, neuroscience's resonance with folk concepts of personhood matters because it fuels the appropriation of neuroscience to serve a diverse range of professional, pragmatic, and ideological interests. Large portions of the political classes have recently developed a conviction in the relevance of neuroscientific findings for public policy, in areas as diverse as education, pornography, national security, unemployment, and financial behavior.[8,15–17] The enthusiasm with which political actors have embraced neuroscientific justifications has led many to question whether neuroscience's newfound political utility owes more to its rhetorical power than its evidential value.[18,19] This worry is also elicited by uses of neuroscience in the commercial sphere, where a belief that neuroscience reveals something important about people's "true" selves has ignited an explosion of interest in applications of neuroscientific technology to such services as "neuromarketing," "brain-training," and "lie detection."[20,21]

A further field in which neuroscientific concepts are applied to practical purposes is the law, whose practitioners have devoted considerable resources to examining the implications of neuroscientific findings for legal principles of responsibility and intentionality.[22–24] Speculation on a criminal biological "type" is not new, extending back to Cesare Lombroso's nineteenth century search for the anatomical characteristics of the "born criminal."[5] Such thinking subsided in the later decades of the twentieth century due to unease with the eugenics ideas it had subsumed. With the rise of the neurosciences, however, speculation about the biological basis of antisocial behavior has enjoyed a resurgence. Neuroimaging technologies are increasingly finding their way into courtrooms to support arguments that accused criminals could not control their antisocial impulses.[25–27] Neuroimaging technologies have also been utilized in personal injury cases to "prove" the presence of debilitating pain[28] and, despite scientific misgivings, have been held up as a means of detecting deception. For example, in 2008, an Indian court found a woman guilty of murder largely on the basis of an electroencephalogram (EEG) "lie detector" test that most experts would deem invalid.[29]

Thus, the ways neuroscience resonates with common-sense understandings of personhood can spark both micro and macro social effects, modulating day-to-day thought and behavior, as well as large-scale legal, political, and economic dynamics. This means that neuroscience research is not merely *describing* processes of emotion or social cognition; in its circulation through the public sphere, it may be actively *shaping* the very processes it seeks to explain. This process is captured by the philosopher Ian Hacking's concept of a "looping effect," which posits a constant circle of mutual influence between the understandings of science, individuals, and society.[30] The social or pragmatic concerns of a particular society establish certain questions as interesting or important (e.g., why are certain children disruptive in classrooms?): scientific research pursues these questions and offers some conclusion (e.g., they possess a particular constellation of neuropsychological characteristics, captured under the diagnostic category of attention-deficit/hyperactivity disorder (ADHD)); individual behavior, emotion, and identity shifts in line with this new knowledge (e.g., as a child is ascribed a diagnostic label and associated medical, psychotherapeutic, or educational intervention); and the collective understandings of society evolve and posit new questions for science to address (e.g., how can children with ADHD be best supported in achieving effective learning?).

A socially responsible science should be cognizant of these feedback loops and engage in critical scrutiny of the ways in which it influences and is influenced by the cultural, political, and ideological environments of which it is part. As such, it is apposite for neuroscientists to reflect on the wider cultural processes that their research outputs may be setting in motion. Social scientific research is an invaluable resource in this enterprise, helping to ensure that debates about neuroscience's sociocultural implications are realistic and evidence-based rather than anecdotal. What follows surveys the social scientific research that has tracked whether and how neuroscientific concepts are propelling shifts in common-sense understandings of self and others.

3. NEUROSCIENCE IN THE MEDIA

The popular media are the vehicle by which neuroscientific findings reach the general public, and as such, it is useful to look at how press coverage construes neuroscience's implications for understandings of human personhood. A study of the British press conducted by O'Connor, Rees and Joffe found that neuroscience coverage increased dramatically between 2000 and 2010.[31] This large dataset of media articles revealed numerous ways in which neuroscience was drawn into articulations of self and identity. Most notably, the brain was universally lauded as the wellspring of human potential and cast as a valuable resource that required careful nurturing from its owner. This laid the foundations for a constant appeal for "brain optimization," whereby the media transfigured neuroscientific findings into a regime of lifestyle choices that individuals should undertake in order to enhance their neurocognitive capacity. According to this media theme, a responsible citizen should carefully modulate their (or their child's) nutritional, cognitive, and behavioral activity in accordance with the advice of neuro-experts, in order to maximize their neurocognitive productivity. This trend closely resonates with deeply entrenched ideologies in Western culture, which valorize individual self-reliance, self-improvement, and self-discipline.[32–34] The displacement of these values onto the site of the brain suggests that neuroscientific concepts are interpreted through and harnessed to reinforce prevailing cultural ideals of individualism.

The media's absorption of neuroscientific concepts into abiding cultures of identity was further evident in another key trend detected by this analysis: the use of neuroscience to underline differences between categories of people. The media were exceedingly enthusiastic about the prospect that the many intergroup divisions that exist in our societies—relating to such variables as gender, sexuality, morality, psychopathology, personality, and political attitudes—can ultimately be traced to different social categories' possession of distinct "types" of brains. Much of this material revolved centrally around explaining patterns of behavior deemed pathological or abnormal; attributing phenomena such as criminality, homosexuality, or addiction to disordered brains served to symbolically distance these populations from the "normal" majority. This media content was often deterministic in tone, reproducing long-established stereotypes, reinforcing sharp divisions between social categories, and constituting social groups as wholly internally homogeneous. For example, the media often deploy neuroscientific research on sex differences to legitimize traditional gender stereotypes and role divisions.[35]

The results of the O'Connor et al. analysis chime with another media analysis by Racine, Waldman, Rosenberg, and Illes.[36] This analysis found that a key facet of popular coverage of functional Magnetic Resonance Imaging (fMRI) research was a trend the researchers termed *neuroessentialism*, in which the brain is construed as synonymous with the more global concepts of person, self, or soul. In their data, the brain often stood as the grammatical subject of a sentence, with the causal power of self-propelling neurochemical processes effacing conventional notions of individual agency. These reductionist media tendencies were rarely tempered by critical reflection on the conceptual or methodological robustness of the research in question. The researchers concluded that in the popular press, neuroscience is afforded a level of authority in determining "who we are" that is not merited by the current sophistication of neuroscientific theories and techniques.

However, an article by Whiteley proposes that Racine et al.'s study was insufficiently attuned to the rhetorical contexts of media articles.[37] Whiteley suggests that the identified instances of neuroessentialism do not necessarily indicate a serious neuroscientific colonization of ordinary concepts of personhood but rather may reflect instances of irony, humor, or metaphor. She also questions the proposition that critique of neuroscience is rare in popular contexts, noting that critique can be expressed through many discursive forms beyond explicit, reasoned argument. In an analysis aimed at documenting the nature of critical engagements with neuroimaging, Whiteley applied principles of discourse analysis to 249 newspaper, magazine, and online articles that suggested that neuroimaging research could enlighten human nature. This analysis revealed ample occasions where neuroimaging evidence was questioned or rejected, particularly when the research topic was one on which the writer claimed personal familiarity (e.g., gender relations or adolescence). This resistance was selective, however: when the writer agreed with the purported implications of neuroimaging research, its authority tended to be endorsed.

Whiteley's analysis highlights the limitations of evaluating neuroscience's social implications solely in terms of its literal depiction in the media. Textual information can convey rhetorical meaning that extends beyond its literal content, and audiences' readings of that information can be diverse and unpredictable. With this in mind, it is important to look beyond media analyses to also examine how neuroscientific ideas resonate in everyday social and personal realities.

4. NEUROSCIENCE IN PERSONAL LIVES

Most research exploring how neuroscience assimilates into personal and social identities has focused on populations affected by a psychological or neurological disorder, for whom neuroscientific information is assumed to be particularly directly relevant. This research has detected numerous examples of communities for whom neuroscientific ideas are profoundly important for self-conception and social presentation. One such case is the burgeoning "neurodiversity movement," which deliberately harnesses neuroscientific frameworks to cultivate positive social identities. This campaign, spearheaded by the autism community, propagates an interpretation of developmental disorders (e.g., autism spectrum disorders) as simply alternative ways of being that are equally legitimate as "neurotypicality."[7,38] Similar logic has been detected in the self-concepts of individuals with developmental disorders, who can adopt neuroscientific language to represent themselves as subject to unique, "hard-wired" challenges and abilities.[38–41] Singh observes that children with ADHD conceptualize the self-brain relationship as mutable and context-dependent, with the brain most causally implicated in the context of misbehavior.[42] This indicates that while neurobiology does not form an immutable, hegemonic framework of self-understanding, brain attributions can be deployed instrumentally within specific psychosocial contexts. Thus, for groups diagnosed with particular psychiatric conditions, neurobiological explanations of their thoughts and feelings are sometimes psychologically and socially functional, with their endorsement serving identity-protective ends.

Research has also explored the reception of neuroscientific information by the *families* of those diagnosed with psychiatric or psychological disorders. Feinstein suggests that a child's diagnosis with autism stimulates a progressive, dynamic engagement with neuroscience in which scientific knowledge is mingled with ordinary, everyday meanings.[43] Much of the discourse celebrating the prospect of neurogenetic explanations of disorders, such as autism and ADHD, has focused on their potential to obviate the parental blame that these conditions have traditionally invited, exemplified in the mid-twentieth century "refrigerator mother" theory of autism and schizophrenia. Singh's interviews with mothers of boys with ADHD found them to endorse the notion that biological explanations refuted parental culpability: in the mothers' narratives, the time of diagnosis marked the point at which they were absolved of blame for their child's disruptive behavior.[44] However, Singh's analysis ultimately concludes that despite mothers' explicit renunciation of culpability, clinical diagnosis had reconstituted rather than expunged mother-blame. For example, mothers' knowledge that their son's bad behavior was biologically caused provoked shame when they felt anger or frustration toward him. Similar findings are reported in an analysis of interviews with relatives of individuals with schizophrenia.[45] Relatives repeatedly invoked biogenetic causation to repulse blame that might otherwise be directed toward them or other family members, with siblings particularly motivated to protect their mothers from blame. However, they continued to search for things that family members could have done that "triggered" the emergence of the disorder.

Thus, research indicates that for many people affected by psychological disorders, neuroscience forms an emotionally meaningful and pragmatically useful resource for identity construction. However, much less is known about how neuroscience has resonated in the self-concepts of those who have not been alerted to neuroscientific information via clinical diagnosis. Pickersgill, Cunningham–Burley, and Martin incorporated both clinical and nonclinical populations (e.g., teachers, clerics, and neuroscientists themselves) into a focus group study examining how neuroscience relates to ordinary subjective experience.[46] Their analysis suggested that for these participants, the brain is characterized by "mundane significance": people appreciated its importance when speaking in the abstract, but brain information rarely struck them as relevant to their day-to-day lives. Similarly, Choudhury, McKinney, and Merten's investigation of how adolescents engage with concepts of the "teenage brain" found that while teenagers stated that this field of science was objectively important, they also saw it as boring and personally irrelevant.[47] In neither of these studies was biology foremost in people's understandings of self or others: behavior was primarily seen as resulting from interactions between material circumstances, previous experiences, and social relationships.

Although the Pickersgill et al. and Choudhury et al. studies provide valuable insight into the uneven take-up of neuroscience beyond clinical populations, their designs have a number of restrictions. In particular, neither provides a clear test of whether neuroscience has registered with members of the lay public who do not have some preexisting investment in neuroscience. Pickersgill et al.'s sample was composed of neuroscientists, patients, or members of professions that the researchers saw as relevant to brain research, while Choudhury et al. concentrated exclusively on adolescents' responses to the idea of a "teen brain." There remains a marked absence of research on how members of the public at large, rather than people for whom neuroscience has been designated specifically relevant, engage with ideas about the brain.

5. A NEW SOURCE OF EVIDENCE: INTERVIEWS WITH THE BRITISH PUBLIC

To address this empirical void, researchers in University College London recently initiated a study to explore how lay publics, who maintain no specific investment in the neuroscience field, engage with neuroscientific information within their day-to-day lives. Interviews were conducted with 48 London residents, who were purposively selected to ensure an equal distribution of age, gender, and tabloid/broadsheet newspaper readership (which in the British context operates as a proxy for socioeconomic status, since broadsheets are typically associated with a middle-class and tabloids with a working-class readership). Ethical approval was obtained from University College London and interviews took place between May and October 2012. Participants were not told the topic of the research in advance of the interview.

In accordance with the Grid Elaboration Method,[48] an interview technique for eliciting people's naturalistic, subjectively valid patterns of understanding, respondents were first presented with a grid of four empty boxes and asked to write or draw the first four ideas that came to mind on hearing the phrase "brain research." Figure 1 displays an example of a completed grid. The resulting set of free associations functioned as a stimulus for the subsequent verbal interview, which encouraged people to expand on their subjective responses to the topic of brain research. This nondirective approach, which involved minimal intercession by the interviewer, facilitated a naturalistic glimpse into the chains of association people spontaneously mobilized in conceptualizing the topic.

All interviews (average duration 34 min) were recorded, transcribed, and analyzed using the ATLAS.ti qualitative analysis software. A procedure of thematic analysis was deployed, which involved systematically coding all text using a coding frame that was validated through interrater reliability testing. ATLAS.ti's functions for detecting co-occurrence and sequencing patterns were then utilized to explore the relationships between codes and identify the key themes that traversed the data.

A detailed account of the full results of this analysis has been presented elsewhere.[49,50] Briefly, the analysis found that this lay sample was far less aware of contemporary neuroscience than much social commentary on neuroscience has assumed. Most individuals who were interviewed struggled to recall any prior encounter with neuroscientific information, and those associations that were summoned related primarily to the medical domain. Despite the documented preponderance of neuroscientific concepts and imagery in the popular press, for these interviewees the scalpel of the brain surgeon remained a more salient symbol than the vividly-colored fMRI scan. However, the analysis did reveal a number of specific contexts where neuroscientific concepts had more deeply penetrated common-sense understanding, most notably in the very consensual representation of mental illness as a brain disease and in the notion that brain function can be improved by adapting one's lifestyle choices. This suggests that neuroscience can indeed influence lay people's everyday understanding, but in selective, partial, and contingent ways.

With this in mind, the remainder of this paper considers the light this interview data throws on the debate about whether and how neuroscience is influencing "folk psychology." Do people spontaneously use neuroscientific concepts to understand themselves, and do they use them to understand others? Do people ascribe causal significance to the brain in explaining individual characteristics, abilities, and life outcomes? What follows mobilizes the British interview data to shed light on these questions.

5.1 Was the Brain Seen as Determining Behavior?

In the interviews, neural processes were frequently ascribed a causal role in conceptualizing human behavior. Observed differences in surface traits, such as personality and intelligence, were explicitly attributed to differences in people's brains.

> You know, people say you're brainy because people are more intelligent than others and some people are just naturally intelligent. So obviously their brain must work in a different way.
> ***Female, tabloid-reader, aged 38–57***

> I'm sure somebody who has a, let's say an overly happy excitable person, their brain may look very similar to a depressed person's brain in terms of the structure, but the way it's, the way people are using the structure I guess could be different.
> ***Male, broadsheet-reader, aged 18–37***

What are the first four ideas that come to mind when you hear the term 'brain research'?

Please express what comes to mind using words and/or drawings, putting each idea in a single box. There are no correct or incorrect answers, I am simply interested in what you associate with brain research.

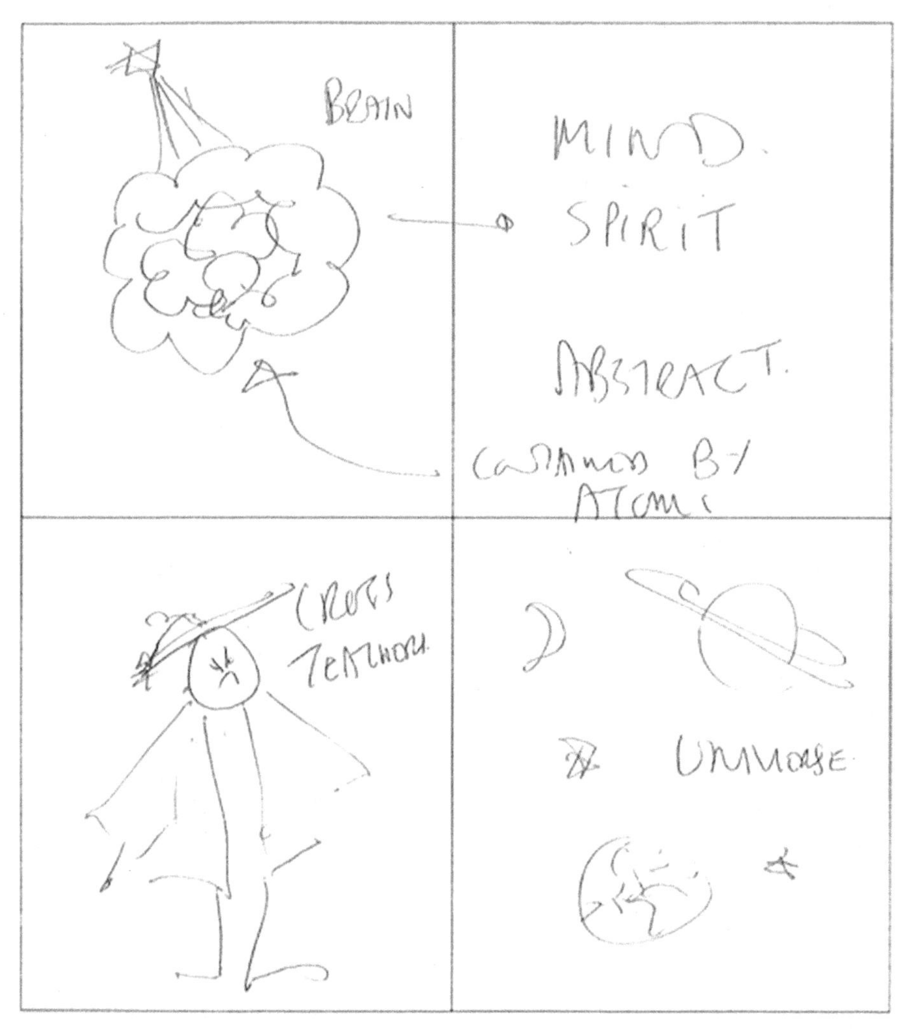

FIGURE 1 **Example of a completed free association grid, produced by a 58-year-old female broadsheet reader.** This participant first drew an image of a brain, which she linked with arrows to the second box to convey the idea that in the brain, abstract phenomena such as "mind" and "spirit" are "constrained by atoms." She went on to draw a picture of a "cross teacher," which captured her sense that scientific knowledge is controlled and disseminated in authoritarian ways. In her final box, she turned to thinking about the importance of the brain in human life, suggesting that its implications are as vast as the "universe."

> With brains no two people are the same. And so therefore it is the brain that creates who you are and makes you different and makes you respond in a different way and react in a different way. *Female, broadsheet-reader, aged 38–57*

However, these neurobiological attributions did not preclude the acknowledgment of additional causal forces. Reference to environmental factors in individual development also occurred frequently in the data, with the family constituting a particularly salient locus of environmental influence. Most people did not see neurobiological and environmental causality as contradictory, instead endorsing a biology–environment interaction. Although very few made specific reference to the scientific concept of neuroplasticity, many intuitively grasped the principle that experience modulates brain structure and function.

> So obviously we're predisposed to, you know, emotions, the way we think, the way we feel. There must be a certain pattern that's sort of imprinted in there to start off with and the way you learn and the way you take stuff in as you grow. You grow one way, you grow another way, it must, it must all be like that. There must be a starting point of like being hard-wired in the brain. But then as you learn, whether you're learning at school, whether you're learning through life. It must take you in different directions. *Male, broadsheet-reader, aged 38–57*

Additionally, assertions of neurological determination of behavior were often intermingled with endorsements of individual agency. For example, one-quarter of interviewees introduced a metaphor that compared the brain to aspects of electrical or mechanical systems. The brain was variously described as a "hub," "control room," "engine room," "battery," "IT center," "motor," "mighty powerhouse," "centrifugal force," "starter motor," "great electrical center," and "central processor." On the surface, these metaphors, which condensed the source of human vitality into the single site of the brain, could be interpreted as reflecting a deterministic conception of the brain as an all-controlling power. However, many of the mechanisms to which the brain was compared, such as computers or batteries, were not framed as self-sufficient automatons but rather objects of instrumental value, which are used by individuals to achieve intended outcomes.

> And it is up to you but you have got to, you have got to tell the brain and you've got to find the brain, the part of the brain that's going to react. That's how I see it. It's all a bit like a computer. I see it like a computer, that you're the one that's operating it so if you make a mistake, it's not the computer's fault, it's you. ***Female, broadsheet-reader, aged 58–77***

Such metaphors therefore constituted the brain as a tool that individuals could willfully exploit to achieve desired ends. The brain coordinated human activity, but the biochemical directions that it issued were subject to intentional control. Thus, literal descriptions of the brain as all-commanding did not necessarily bypass notions of conscious control or individual autonomy.

> Brain is not really in control of it. We ask him to control. It's resting there. He works hard. And your eyes or your hands or whatever, you know, they send signal to the brain. But at the moment brain is not doing anything, brain automatically don't do it, you've got to think with your eyes and go to brain, then it reacts. Brain is not reacting on its own. Although it's sitting there, but just like electricity, there is electricity there, if you need it you just plug it and then it comes, things start working. ***Male, tabloid-reader, aged 58–77***

> Well it's there for us, isn't it, to be, to be used. Our brain is everything about us. We need our brain. If we haven't got a brain then we can't do anything. Our brain tells us what to do. ***Female, broadsheet-reader, aged 18–37***

In sum, drafting the brain into conceptualizations of human behavior did not impose complete materialism or determinism. Rather, lay thinking was characterized by holistic, multidimensional explanations of behavior, in which attributions to the brain directly intermeshed with endorsements of environmental influence and individual agency.

5.2 Did People Use Neuroscientific Ideas to Understand Themselves?

Just over one-third of participants spontaneously related the brain to their own traits or characteristic thought patterns. In imprinting their individuality on their brain, these people revealed a sense of ownership or identification with *"the way my brain works."* Here, the phrase *"my brain"* operated as shorthand for the cognitive attributes that the participant saw as uniquely self-characteristic. This trend represented one of the rare points in the data at which participants directly incorporated the physical brain into self-conception.

> The way my brain works, literally my train of thought is always speeding forwards. Sometimes I've got to try and slow myself down or write things down. I'll think of an idea and all of a sudden, thump, I've worked it through twenty stages in a few seconds! ***Male, broadsheet-reader, aged 38–57***

> I think there's different types of intelligence and I think that's okay. Like I'm not really an academic person and I don't think my brain works like that and I don't think it will ever work like that. ***Male, broadsheet-reader, aged 18–37***

Fifteen interviewees volunteered statements resembling "neuroessentialism," which condensed the entirety of personhood into the brain. These overtly philosophical musings directly equated concepts like "spirit," "soul," and "essence" with the material brain.

> I think the brain defines who you are. So that any research or any meddling or… is really unwrapping and unfolding and revealing something about the personality and the person and the character of that person and the very nature of that person and the very, the very essence of that person. […] Well it's, it's you. It's not your body, it's you, it's your personality, it's who you are, your spirit, your character. ***Female, broadsheet-reader, aged 38–57***

> Yeah, well the brain is what makes a person, gives them their essence I suppose. ***Female, tabloid-reader, aged 38–57***

However, commitment to such sentiments often faltered under further reflection. Several participants expressed discomfort with the idea of an entirely material self, and in contemplating, it became mired in a form of existential anxiety. Some disclosed that they purposely avoided thinking about the topic for this reason.

> No, 'cause then you've got the thing of is the brain the soul, do you believe in the soul, is the soul winging away as the brain… That's a difficult one. I'm not too sure about that kind of thing at the minute. Really not too sure. That's something that I think we all choose not to think about too much as well. ***Female, broadsheet-reader, aged 58–77***

> You can, it's very reductive, isn't it. So it's reducing yourself to just a series of impulses and electrical, you know electrical impulses and you're one big, you know, biological circuit board.

> Or the brain is connected to sort of muscles which are just again sort of series of, you know, contracting fibres and… So that's all quite, so I suppose it's sort of where does it end, you know. 'Cause we like to think of ourselves as being quite important and special. ***Male, broadsheet-reader, aged 18–37***

One participant, who worked in environmental science, offered a particularly lucid articulation of the inconsistency between abstract belief in material personhood and immediate self-understanding. This individual identified as scientifically-minded and on a conscious level fully endorsed a materialistic view of the mind. However, he made an explicit separation between his *"theoretical"* beliefs and his day-to-day thinking, asserting that it is existentially impossible to maintain a purely materialistic view in ordinary life. This conviction was premised on his positioning of materialism and personal autonomy as mutually exclusive principles. He rejected materialistic thinking in his day-to-day life because he believed that to accept it would necessitate sacrificing his sense of personal control and attendant feelings of achievement, which he imagined would be *"doing yourself a disservice."* He framed this in explicitly emotional terms, characterizing materialistic views of oneself as *"sad," "nihilistic," "isolating,"* and *"cold."* This participant painted the retention of what he ultimately saw as the fiction of his free will as an emotional imperative, necessary to sustain his ability to function normally in society. This example illustrates how people's willingness to endorse materialism on an abstract level teetered when it breached their concrete, immediate self-understanding.

> You can think about it like that, you know, when I'm speaking about it consciously, but in your day-to-day making decisions, that kind of thing, you have to forget about that, otherwise it would be a bit nihilistic and sad. […] at the end, it was always going to happen through this weird cascade of chemical activity – I don't like that very much. I don't know, I kind of do like it but I don't like it, if that makes sense. I like it theoretically but, you know, when you're in that moment looking at the things you've achieved I think it's hard to separate the two. ***Male, tabloid-reader, aged 18–37***

Thus, although numerous participants drew a connection between the material brain and more ephemeral ideas of selfhood when speaking in the abstract, there was little indication that this understanding pervaded their day-to-day experience. Very few recounted specific examples of previously encountering neuroscientific information that had affected their self-conception. Further, it is worth noting that while some participants nominally linked individuality to the organ of the brain, they did not allude to particular neurological processes, structures, or chemicals. The role of the "neuro" in self-understanding was confined to a basic understanding that cognition "happens" in the brain, rather than any specific knowledge of contemporary neuroscientific concepts.

5.3 Did People Use Neuroscientific Ideas to Understand Others?

Though participants intermittently invoked the notion of brain difference to explain interindividual variation in personality or intellectual ability, the neurobiological domain was not a key reference point for understanding one's immediate social circle or "people like me." However, the data revealed that a particular point at which respondents turned to the brain for explanation was when confronted with individuals who seemed unusual or strange. Unusual behavior was experienced as intuitively incomprehensible, and the mystification this produced was resolved by enlisting a brain explanation. For example, one woman expressed bewilderment at an acquaintance's perpetually benevolent disposition, which she saw as so extraordinary that the only possible explanation was an atypical brain. Another person described encountering a man acting bizarrely on the street and drawing the conclusion that his brain must function irregularly.

> Like it was very strange. Like just shouting at people and to himself and talking to himself non-stop. It was just, it was very, it was very very strange the way he behaved and you wouldn't do that unless, I'm sure there was something wrong with his brain. I'm definitely sure. Because you wouldn't speak like that. ***Female, broadsheet-reader, aged 18–37***

The invocation of the brain to understand abnormal "others" was exemplified in discussion of violent criminality, which occurred in one-third of the interviews. This usually centered upon the extreme offences of mass murder, terrorism, or pedophilia and was often personified by individuals notorious for their evil acts, such as Adolf Hitler or Anders Behring Breivik (whose trial for the murder of 77 people in Norway was ongoing at the time of interviews). In contemplating instances of criminal atrocity, participants produced a stream of "why" questions, conveying a sense of complete bewilderment.

> I mean, you know, look at people like Adolf Hitler. Why did he think the way he did? Why did he do what he did? You know. So I'm fascinated by that. You know, these people are, created so many – they were powerful but they were very cruel and evil. Why is one person more evil than the next? You know, why do some people commit murder and others that are just normal? […] I'm just trying to think as an intelligent person, you know, 'cause I'm, I'm baffled by it all. You know, sometimes I think, why do they do that? You know, why did they, why create that? Why did they, what are they up to? You know, why do they do these things? ***Male, broadsheet-reader, aged 38–57***

This gulf in understanding was strongly emotionally tinged. The confusion provoked by confrontation with alien mentalities was evidently experienced by some as distressing.

> So you know, just the thought of entertaining ideas about, reading up about killing somebody, for me is just terrifying. You know what I mean, like. I'd be like, oh my God. But people must, I mean, I don't know, they must do that, right, they must be like – I just don't know how their brain would work, you know? *Female, broadsheet-reader, aged 18–37*

To abate this discomfort, participants attempted to articulate some explanation of why people commit such acts. Scanning their conceptual registers for a viable cause of such unfathomable attributes, participants ultimately alighted on the concept of "dysfunctional brain." Respondents did not articulate any explicit, logical rationale for this explanation; rather, the attribution seemed to flow from an intuitive sense that this deviance *must* be reflected biologically.

> You know, people who do like terrible things. You must think, well there must be something in, there must be something to do with their brain that's made them do that because a normal person wouldn't be able to do, you know, really kind of horrible things. So it must be to do with something, something to do with the brain that makes them like that. *Female, broadsheet-reader, aged 18–37*

> Well I'd say that, you know like you've had these terrorists and all that. You know, some of them believe that if they go onto a bus and kill themself and a thousand, or how many hundred people with them, that they're going to go to some lovely place somewhere. Now to me nobody with a normal brain would speak like that or would think like that. *Female, tabloid-reader, aged 58–77*

Attribution to the brain seemed to satisfy respondents' need to explain such behavior. They did not feel obliged to probe deeper into precisely how neurobiological processes could produce such behavior; the concept of "different brain" was sufficient to resolve the psychological tension elicited by encountering incomprehensible behavior.

> Say you had a mad axeman, right. Here's the normal brain. Here's the mad axeman's brain. Now see this bit, it is more, more active. And that is the reason, they're saying that this is the reason why he is like he is. *Male, tabloid-reader, aged 58–77*

As well as abating psychic discomfort, attribution of antisocial behavior to the brain had the additional consequence of reinforcing intergroup divides. It often involved a level of essentialism, with those who committed such acts constituted as intrinsically and irrevocably evil. This instituted firm boundaries between social categories: people were either normal or wholly evil, with no acknowledgment of potential areas of ambiguity between these poles. Certain people were simply born to be "bad."

> I think there's got to be something in you to do that. An evilness or sadness or something. I believe that that person is born with that bad seed. I genuinely believe that. *Male, broadsheet-reader, aged 38–57*

> Like people who go around killing people. That's right to them, they think that's fine. So there's something in the brain that's clicked and gone this is, this is okay to be like this. I think it's, you can't change. It just runs. You can't sort of go, 'I don't want to think like that anymore.' *Male, tabloid-reader, aged 18–37*

Such quotes articulate an understanding that biogenetic fate impels antisocial behavior. Interestingly, however, only one participant implied that this deterministic biological causality would diminish legal or moral responsibility for destructive behavior. All other respondents who touched on the issue held fast to the principle of personal responsibility, which for them remained commensurate with biological causation.

> But seeing a human being as a, as a body with a brain, you can't say that, it's like nature versus nurture and why is somebody a criminal, you can't take somebody's fault away because they've killed someone 'cause the brain told you to. 'Cause I think that's stupid. I think that's when it starts crossing the line of, oh it's not my fault, it's my brain's fault. So [laughs] yes, that could cross the line of what we call insanity but I personally think that you are in control of your, your actions. *Female, broadsheet-reader, aged 58–77*

In summary, the data suggested that invoking the notion of "different brain" functioned to resolve the discomfort elicited by encountering radically abnormal behavior. In the process, it naturalized the social and symbolic divides separating "bad people" from the normal majority. This did not, however, compel society to relinquish the prerogative to hold depraved individuals morally accountable for their wrongdoing.

5.4 Summary of Interview Findings

This interview data confirms that when pressed to speculate about the brain, people spontaneously relate it to concepts of personhood, self, and identity. Respondents directly attributed their own and others' behavior to biology. Discussion often evolved to incorporate wider, explicitly philosophical concerns: participants instinctively felt that brain research would have relevance for notions of self, spirit, and soul. However, when the surrounding context of such statements was scrutinized, it was clear that they did not reflect a comprehensive materialization of ordinary understandings of personhood. Some participants actively resisted neuroessentialist ideas, unnerved or unconvinced by these scientific conceptualizations. Further, even those who were comfortable accepting biological determination of personal traits refrained from positioning the brain as paramount. When given space to elaborate on their understandings, respondents revealed complex explanatory networks in which neurobiological causation, environmental influence, and individual intentionality occupied equally valid, interlocking positions. For example, participants would attribute an individual's level of intelligence directly to their brain characteristics, but on reflecting further would attribute these neural resources to the personal effort they expended in education, which was in turn attributed to the person's upbringing and cultural values and expectations. Thus, while the brain was positioned as the proximal source of intelligent cognition, it was ultimately a medium for the more fundamental causes of culture and individual will.

These dynamics shifted somewhat as conversation moved beyond the parameters of "normal" interindividual variation to mentalities imbued with a sense of abnormality and "otherness." Here, attention to environmental or other nonbiological causality dramatically subsided: participants were strongly invested in attributing deviance to an essential biological aberrancy. Many commentators on contemporary neuroscience have speculated that knowledge about the biological roots of antisocial behavior will have profound implications for the criminal justice system, undermining the principle of legal responsibility. In this data, however, respondents rejected outright the suggestion that biological causality of criminal behavior was incompatible with the ascription of moral responsibility. Indeed, it is possible that attributing criminality to brain difference may foster *more* punitive attitudes, because the construction of criminals as an irredeemably bad "other species" may attenuate empathy or identification with criminal populations.

This interview data bolsters the emerging empirical consensus that when neuroscientific concepts breach registers of common sense, they do not drive out prevailing modes of understanding personhood.[16,42,51,52] Overtly contradictory ideas can coexist independently, preferentially evoked in different discursive contexts, or can indeed directly interact to form complex, multifaceted explanatory networks. An important contribution of this research is therefore to highlight that, due to the multivalent nature of common-sense knowledge, neuroscience does not assimilate into society in linear, predictable ways. The extensive public attention to neuroscience means that neuroscientists may indeed be contributing toward cultural shifts in how people conceive of themselves and others. However, the direction of these shifts can only be discerned through empirically-informed analysis that remains fully alive to the complexities of both common-sense cognition and real-world social environments.

6. HOW SHOULD NEUROSCIENTISTS RESPOND?

It is clear that in contemporary society, neuroscience is interpreted and applied in many diverse ways. These appropriations of neuroscientific knowledge are sometimes driven by neuroscientists themselves, who are working within an academic environment that increasingly prizes research impact. The desire to catch media attention and emphasize the real-world relevance of their research may encourage scientists to make unwarranted leaps between data and interpretation. For instance, an analysis of media coverage of a high-profile study of sex differences found that some of the most egregious features of media coverage, which presented the research as a validation of traditional gender stereotypes, were supported by quotes that the researchers themselves provided in their informal communications with journalists.[35] This highlights the importance of remaining alert to the cultural biases and ideologies with which a particular piece of research might resonate. If there is a possibility that scientific findings could be illegitimately recruited to serve a particular sociopolitical agenda, scientists must take care to avoid fueling this and could indeed take preemptive steps to disclaim certain interpretations of the research. In devising their research programs, scientists should devote careful consideration to the sociocultural processes that their research might set in motion and develop a dissemination plan that is sensitive to those.

It would be wise not to leave these measures solely to the responsibility of individual researchers, whose ability to recognize and address problems may be limited by resource constraints and their closeness to their own research. If the neuroscience community is genuinely committed to scrutinizing the social repercussions of its research, structures that facilitate this must be established at the institutional level. Ethics committees and institutional review boards should broaden their purview beyond a prospective study's immediate influence on its participants, to the more profound and long-term ripple effects that may occur as science interacts with sociocultural contexts. The social scientific literature on neuroscience in society offers a rich and evidence-based source of insight into the potential promises and perils of a particular piece of research. While these need not necessarily be the ultimate arbiter of whether ethical approval is granted or refused, mandating attention to these issues would cultivate a more holistic view among scientists and ensure they are alerted to controversies that might arise as their research program develops. Establishing systems that make active consideration of a study's social implications a routine part of project planning would therefore benefit both science and society.

It is important to note, however, that the social evolution of scientific ideas is not a process that can or should be controlled solely by scientists themselves, no matter how considered their public engagement strategy. Often these developments are not instigated by scientific experts, but reflect more bottom-up transformations, whereby social actors read into neuroscientific concepts a certain significance for their own personal, social, or professional interests. Upon leaving the laboratory, neuroscientific concepts can take on a "life of their own" and develop implications that were not foreseen by their creators.

This sparks unease among many in the scientific community, who have taken to newspapers and blogs to protest at what they see as a booming industry in "neuro-nonsense," "neuro-bollocks," or "neuro-trash." However, the susceptibility of neuroscience research to reinterpretation in wider society is not necessarily a wholly negative attribute. Rather, it is a testament to the societal pertinence of neuroscientific research programs, which are clearly tapping culturally important questions. This represents an opportunity rather than a threat for neuroscientists, whose invitation into these dynamic social debates promises valuable returns in intellectual energy and public support. Undoubtedly, some appropriations of neuroscientific information might occur that aggrieve the scientific community; and if scientists deem a certain interpretation wrong or damaging, they should not shy away from attempts to convince the wider public of its relative demerits. However, it should be remembered that while scientific experts often have privileged insight into the factual *correctness* of scientific claims, they have no exclusive authority to arbitrate their cultural *meaning*. It is inevitable that in a pluralistic society, the same information will connote different things to different people. The scientific community should be prepared to engage with these differences of perspective fully and respectfully, appreciating that lay populations can have legitimate and valuable reservoirs of local expertise they can bring to bear on scientific debates.[53,54]

7. CONCLUSION

A socially responsible science requires sensitivity to the social contexts in which it is mobilized and the social effects it incites therein. Increasingly, neuroscience's influences on wider society are mediated by how neuroscientific concepts interact with cultural understandings of personhood. To ensure its research is conducted and communicated in a socially responsible manner, the neuroscience community must cultivate awareness of these processes. Neuroscientists should be prepared to engage with the perspectives of lay communities, both through direct public outreach activities and through acquaintance with the social science literature that tracks neuroscience's route through particular social contexts. Doing so will ensure that the continued development of neuroscience's scientific and public profile will be informed by a sound understanding of the role it occupies in modern society.

References

1. Adolphs R. Cognitive neuroscience of human social behaviour. *Nat Rev Neurosci*. 2003;4:65–178.
2. Illes J, Kirschen MP, Gabrieli JDE. From neuroimaging to neuroethics. *Nat Neurosci*. 2003;6:205.
3. Frazzetto G, Anker S. Neuroculture. *Nat Rev Neurosci*. 2009;10: 815–821.
4. Human Brain Project. *Ethics*. Retrieved 17 May, 2013, from http://www.humanbrainproject.eu/ethics.html.
5. Rose N. *The Politics of Life Itself: Biomedicine, Power, and Subjectivity in the Twenty-First Century*. Princeton: Princeton University Press; 2007.
6. Ortega F. The cerebral subject and the challenge of neurodiversity. *Biosocieties*. 2009;4:425–445.
7. Vidal F. Brainhood, anthropological figure of modernity. *Hist Hum Sci*. 2009;22:5–36.
8. O'Connor C, Joffe H. How has neuroscience affected lay understandings of personhood? A review of the evidence. *Public Underst Sci*. 2013;22:254–268.
9. Churchland PM. *The Engine of Reason, the Seat of the Soul: A Philosophical Journey into the Brain*. Cambridge: MIT Press; 1995.
10. Crick F. *The Astonishing Hypothesis: The Scientific Search for the Soul*. New York: Touchstone; 1995.
11. Green J, Cohen J. For the law, neuroscience changes nothing and everything. *Philos Trans R Soc Lond B Biol Sci*. 2004;359:1775–1785.
12. Baumeister RF, Masicampo EJ, DeWall CN. Prosocial benefits of feeling free: disbelief in free will increases aggression and reduces helpfulness. *Pers Soc Psychol Bull*. 2009;35:260–268.
13. Vohs KD, Schooler JW. The value of believing in free will. *Psychol Sci*. 2008;19:49–54.
14. Vonasch AJ, Baumeister RF. Implications of free will beliefs for basic theory and societal benefit: critique and implications for social psychology. *Br J Soc Psychol*. 2012;52:219–227.
15. O'Connor C, Joffe H. Media representations of early human development: protecting, feeding and loving the developing brain. *Soc Sci Med*. 2013;97:297–306.
16. Pickersgill M. The social life of the brain: neuroscience in society. *Curr Sociol*. 2013;61:322–340.
17. Choudhury S, Gold I, Kirmayer LJ. From brain image to the Bush doctrine: critical neuroscience and the political uses of neurotechnology. *AJOB Neurosci*. 2010;1:17–19.
18. Macvarish J, Lee E, Lowe P. The "first three years" movement and the infant brain: a review of critiques. *Sociol Compass*. 2014;8:792–804.
19. Seymour B, Vlaev I. Can, and should, behavioural neuroscience influence public policy? *Trends Cogn Sci*. 2012;16:449–451.
20. Abi-Rached JM. The implications of the new brain sciences. *EMBO Rep*. 2008;9:1158–1162.
21. Chancellor B, Chatterjee A. Brain branding: when neuroscience and commerce collide. *AJOB Neurosci*. 2011;2:18–27.
22. Kulynych J. Psychiatric neuroimaging evidence: a high-tech crystal ball? *Stanford Law Rev*. 1997;49:1249–1270.
23. Schweitzer NJ, Saks MJ. Neuroimage evidence and the insanity defense. *Behav Sci Law*. 2011;29:592–607.
24. Walsh C. Youth justice and neuroscience: a dual-use dilemma. *Br J Criminol*. 2011;51:21–39.
25. Farisco M, Petrini C. The impact of neuroscience and genetics on the law: a recent Italian case. *Neuroethics*. 2012;5:317–319.
26. Hughes V. Science in court: head case. *Nature*. 2010;464:340–342.
27. Mobbs D, Lau HC, Jones OD, Frith CD. Law, responsibility, and the brain. *PLoS Biol*. 2007;5:e103.
28. Reardon A. Neuroscience in court: the painful truth. *Nature*. 2015;518:474–476.
29. Saini A. The brain police: judging murder with an MRI. *Wired*. 27 May, 2009. Retrieved 21 August 2015, from http://www.wired.co.uk/magazine/archive/2009/06/features/guilty.

30. Hacking I. The looping effects of human kinds. In: Sperber D, Premack D, Premack AJ, eds. *Causal Cognition: A Multidisciplinary Debate*. Oxford: Oxford University Press; 1995:351–383.
31. O'Connor C, Rees G, Joffe H. Neuroscience in the public sphere. *Neuron*. 2012;74:220–226.
32. Becker D, Marecek J. Dreaming the American dream: individualism and positive psychology. *Soc Pers Psychol Compass*. 2008;2:1767–1780.
33. Joffe H, Staerklé C. The centrality of the self-control ethos in Western aspersions regarding outgroups: a social representational approach to stereotype content. *Cult Psychol*. 2007;13:395–418.
34. Sampson E. The debate on individualism: indigenous psychologies of the individual and their role in personal and societal functioning. *Am Psychol*. 1988;43:15–22.
35. O'Connor C, Joffe H. Gender on the brain: a case study of science communication in the new media context. *PLoS One*. 2014;9:e110830.
36. Racine E, Waldman S, Rosenberg J, Illes J. Contemporary neuroscience in the media. *Soc Sci Med*. 2010;71:725–733.
37. Whiteley L. Resisting the revelatory scanner? Critical engagements with fMRI in popular media. *Biosocieties*. 2012;7:245–272.
38. Fein E. Innocent machines: Asperger's syndrome and the neurostructural self. In: Pickersgill M, van Keulen I, eds. *Sociological Reflections on the Neurosciences*. Bingley: Emerald; 2011:27–49.
39. Ortega F, Choudhury S. 'Wired up differently': autism, adolescence and the politics of neurological identities. *Subjectivity*. 2011;4:323–345.
40. Rapp R. A child surrounds this brain: the future of neurological difference according to scientists, parents and diagnosed young adults. In: Pickersgill M, van Keulen I, eds. *Sociological Reflections on the Neurosciences*. Bingley: Emerald; 2011:3–26.
41. Singh I. A disorder of anger and aggression: children's perspectives on attention deficit/hyperactivity disorder in the UK. *Soc Sci Med*. 2011;73:889–896.
42. Singh I. Brain talk: power and negotiation in children's discourse about self, brain and behaviour. *Sociol Health Illn*. 2013;35:813–827.
43. Feinstein NW. Making sense of autism: progressive engagement with science among parents of young, recently diagnosed autistic children. *Public Underst Sci*. 2014;23:592–609.
44. Singh I. Doing their jobs: mothering with Ritalin in a culture of mother-blame. *Soc Sci Med*. 2004;59:1193–1205.
45. Callard F, Rose D, Hanif EL, et al. Holding blame at bay? "Gene talk" in family members' accounts of schizophrenia aetiology. *Biosocieties*. 2012;7:273–293.
46. Pickersgill M, Cunningham-Burley S, Martin P. Constituting neurologic subjects: neuroscience, subjectivity and the mundane significance of the brain. *Subjectivity*. 2011;4:346–365.
47. Choudhury S, McKinney KA, Merten M. Rebelling against the brain: public engagement with the "neurological adolescent". *Soc Sci Med*. 2012;74:565–573.
48. Joffe H, Elsey J. Free association in psychology and the grid elaboration method. *Rev Gen Psychol*. 2014;18:173–185.
49. O'Connor C, Joffe H. Social representations of brain research: exploring public (dis)engagement with contemporary neuroscience. *Sci Commun*. 2014;36:617–645.
50. O'Connor C, Joffe H. How the public engages with brain optimization: the media-mind relationship. *Sci Technol Hum Values*. 2015;40:712–743.
51. Bröer C, Heerings M. Neurobiology in public and private discourse: the case of adults with ADHD. *Sociol Health Illn*. 2013;35:49–65.
52. Meurk C, Carter A, Hall W, Lucke J. Public understandings of addiction: where do neurobiological explanations fit? *Neuroethics*. 2014;7:51–62.
53. Wynne B. Misunderstood misunderstanding: social identities and public uptake of science. *Public Underst Sci*. 1992;1:281–304.
54. Wynne B. Public uptake of science: a case for institutional reflexivity. *Public Underst Sci*. 1993;2:321–337.

CHAPTER

21

Toward a Neuroscience of Wisdom

Patrick B. Williams, Howard C. Nusbaum

The University of Chicago, Chicago, IL, USA

OUTLINE

1. TOWARD A NEUROSCIENCE OF WISDOM

Wisdom is a quality of human nature that has been discussed extensively throughout history, perhaps most notably by Aristotle. In modern times, however, despite being considered a pinnacle of human cognition, there has been little public discourse about wisdom or its importance in human enterprise and even less scientific study of wisdom, although in recent years this has been increasing. Furthermore, much of the scientific study of wisdom has focused on describing the components of wisdom and its association (or lack thereof) with age and not on wisdom as a unified construct or how wisdom may be cultivated in life, although this too has been changing in recent research. In general, wise decisions and actions go beyond being smart, clever, or knowledgeable—being wise requires the quality of *prudent* judgment based on reflection of the reasons and values underlying one's own and others' thoughts, motivations, and behaviors. Aristotle defined one kind of wisdom as involving practical decisions that lead to human flourishing (phronêsis) or well-being, grounding wisdom in a more prosocial notion of human well-being in terms of seeking the highest human good. On this view, which is in line with modern philosophical and psychological descriptions, wisdom integrates a balance of cognitive and social expertise and knowledge.

In contrast to descriptions arising from ancient western philosophy, early eastern descriptions of wisdom, originating primarily from India and China, emphasized emotional balance. While there have been distinct differences in how wisdom was characterized between eastern and western cultures, there is also significant overlap in descriptions of wisdom, including aspects of prosocial consideration, gaining an understanding of oneself and others through careful reflection, value relativism, and tolerance.

The empirical study of wisdom began in earnest in the 1970s when psychologists began to make inquiries into the skills and dispositions that contribute to successful aging. Accordingly, the initial research on the association of age and wisdom has been driven by folk psychological intuitions that wisdom comes with age.[1] This makes intuitive sense, as each day that a person lives provides opportunities to gain and learn from experience, and these experiences may facilitate wisdom.[2] It appears then that age may be necessary but certainly not sufficient for wisdom. Emotional regulation and reappraisal are two characteristics that tend to improve with age and may account for

Neuroimaging Personality, Social Cognition, and Character
http://dx.doi.org/10.1016/B978-0-12-800935-2.00021-X

age-related increases in wisdom.[3] Furthermore, research indicates that everyday practical wisdom increases with experience,[4] which suggests that perhaps instead of age being necessary for wisdom, age may serve as a proxy for experience. We need to better understand the kinds of experiences that lead to wisdom. As wisdom research has developed beyond the scope of associations between wise reasoning and aging, investigation has shifted focus from the nature of context-general wisdom[5] to focus on the practical applications of wisdom and wise reasoning to complex everyday situations.

Prior to the relatively recent emergence of multiple psychological theories, definitions, and descriptions of wisdom, the study of wisdom was largely the province of philosophy and religion. Today, wisdom research is increasing as researchers across a range of disciplines seek to understand positive human characteristics related to well-being and how such characteristics may be cultivated. Wisdom is inherently difficult to define, as reflected in the numerous psychological definitions that currently exist in wisdom literature, though this hardly makes wisdom not worth studying and reflects similar difficulties that existed in early intelligence research.[6] While varied across research labs and modes of measurement, there are several commonalities among definitions, such as the need for a large base of pragmatic knowledge gained from life experience, self and other reflectiveness, and prosocial attitudes and behaviors. At the end of a research project supported by the John Templeton Foundation, referred to as the Defining Wisdom Project,[7] a group of scholars and scientists proposed the following as a definition:

> We distinguish wisdom from intelligence, cleverness, knowledge, and expertise. Wisdom requires moral grounding, but is not identical to it (i.e., wisdom must be moral but morality need not be wise). Wisdom can be observed in individual or collective wise action or counsel. Action or counsel is perceived as wise when a successful outcome is obtained in situations involving risk, uncertainty, and the welfare of the group. (We recognize that understanding the definition of a successful outcome is a substantial problem on its own.) Wisdom flexibly integrates cognitive, affective, and social considerations, but can be studied profitably by understanding its constituent elements. Because of the fundamentally multifaceted nature of wisdom, interdisciplinary discourse is extremely useful in advancing the research.

1.1 Roots of Wisdom Research

By Socrates' account, wisdom was an awareness of and humility toward one's own knowledge and its limitations. Aristotle further defined wisdom as an intellectual virtue of harmony between plan and action, without regret. He made a distinction between general wisdom, as it pertains to the knowledge of a god-like entity, and practical wisdom, which is gained through everyday experience and insights taken from one's own life. In the Aristotelian view, wisdom is treated as an integrated trait reserved for persons who follow a virtuous development into *male* adulthood, though contemporary views tend to represent wisdom as a multicomponent characteristic distributed across the general population and not restricted to one or another gender.

Concepts of wisdom go far back into Eastern traditions as well, as ancient practices like Buddhism view it as one part of a series of attributes, trainable by contemplation, toward a path of enlightenment. Eastern philosophy tends to emphasize wisdom as having a strong component of emotional stability, compared to the strong emphasis on knowledge and cognition in Western philosophy. However, significant overlap between Eastern and Western philosophy exists in defining wisdom as including components of compassion, altruism, and insight. Present-day psychological models of wisdom are varied but have in common a core of attributes influenced by these ancient roots, including an extensive knowledge of the world and how it works, social expertise based in empathy, compassion, prosocial behavior toward others, and decision-making based on insights gained from reflection on oneself and others.[8] By viewing wisdom as a synthesis of existing personal characteristics situated in specific but varying contexts and situations, early neuroscientific research into these characteristics, contexts, situations, and their interactions can help move forward an understanding of the neural bases for wisdom and wise decision-making.

1.2 Modern Wisdom Research

Ancient philosophers described the nature and function of wisdom, and in recent years, philosophers such as Tiberius[9] have carried on this tradition, while contributing further to describe the process by which wisdom may be practiced and developed. Tiberius describes wisdom as knowledge and reasoning based on practical reflection on the reasons behind decision-making in situations that involve multiple conflicting values, in which wise reasoning should lead to the best possible outcome for the largest number of people. This description of wisdom is a process model that allows for the development of wisdom through life experience. Wise reflection takes into account one's own values and perspectives, as well as the values and perspectives of others affected in a particular situation or context. Determining the most appropriate or wise action relies on knowing what matters in a particular situation, presumably based on knowledge gained from life experience, awareness of the limitations of that knowledge, and based on sensitivity to one's own and others' emotions. In this way, wisdom may increase with experience, though the extent to which this relationship exists depends on the types of experiences that occur, the perspective taken toward knowledge gained, and the ability to reflect on and tolerate multiple conflicting points of view. Wisdom as practical reflection on the values and reasoning behind beliefs fits well with psychological models of wisdom that have developed over the latter part of the twentieth and the beginning of the twenty-first century.

Following initial work by Clayton and colleagues to understand wisdom as it may relate to aging, the effort to measure and describe wisdom systematically was initially led by the Berlin wisdom group, which conceptualized wisdom as a sort of expert pragmatic knowledge system based largely in cognitive processes that develop with age. The Berlin group describes a multicomponent model of wisdom using data taken from responses to vignettes describing possible real-life social scenarios involving other people. This model includes five interacting parts including the following: (a) a rich and practical factual knowledge of the world and its complexities; (b) a rich procedural knowledge of strategies to solve problems related to life; (c) life span contextualization, or the ability to understand the varying contexts and temporal relationships of life; (d) a relativistic point of view, in which one has an understanding of individual differences in goals and values; and (e) a comfort with uncertainty and the ability to manage it. Researchers in this group have described wisdom largely by collecting from typical people and from individuals deemed to be wise based on responses to a "life-review problem." Success in resolving such problems is determined by trained researchers who independently judge responses based on the Berlin model of wisdom.

Wisdom is not pragmatic knowledge alone, but also relies on balancing interpersonal, intrapersonal, and extrapersonal interests in order to come to a solution that strikes a balance between adapting to the existing environment, changing this environment, and selecting a new environment.[10] In the context of group social affiliation and leadership, wisdom exists as a synthesis of intelligence—the ability to successfully adapt to the environment—and creativity—the ability to produce high quality, novel, and appropriate solutions for the task at hand.[11] Intelligence in this context refers to pragmatic intelligence, influenced by personal and social experiences, as opposed to abstract reasoning abilities, such as crystallized intelligence and working memory.[4] Wisdom incorporates intelligence and creativity while maintaining its role as a unique attribute, as a wise person must make and carry out decisions on a balance between a need for change, requiring creativity, and a need to maintain the stability of existing environmental and social structures, requiring intelligence.[12] By this model, a wise person would necessarily be both intelligent and creative, but an intelligent or creative person would not necessarily be wise. Successful leaders, for example, require wisdom and the disposition to carry out wise decisions, in addition to intelligence and creativity, to be successful in the long term. Consider that a leader who is intelligent in a relevant domain, and creative in their approach to problem solving, would require pragmatism and interpersonal sensitivity to ensure that wise decisions are carried out and are not only self-serving, but also take into account the greatest common good. These prominent models of wisdom suggest that the neuroscientific study of inter and intrapersonal intelligence, mental flexibility in the form of creativity, value relativism, and ambiguity tolerance, are good candidates for investigation in the study of the neurobiological bases of wisdom.

Personal wisdom varies largely across individuals and grows through life experience but can be described as the use of certain types of pragmatic reasoning skills that are prosocial and help to navigate and resolve important life challenges. Like the concept of multiple intelligences,[13] wisdom may be conceptualized as a single construct with separable but overlapping dimensions.[8] The cognitive dimension of wisdom is similar in part to Baltes' definition of wisdom as a deep pragmatic knowledge of life,[14] or to Socrates' concept of epistemic humility—to be aware of and to acknowledge the limits of what one knows, an acknowledgment of the positive and negative aspects of human nature, and a willingness to work within the context of inherently limited and often ambiguous knowledge. Diminished emotional self-centeredness and a deeper understanding of others' affairs characterize the affective dimension of wisdom, which is marked by compassion and overlaps with Aristotle's concept of wisdom as, by definition, being tied with a virtuous disposition. The reflective dimension of wisdom is characterized by an increased accuracy in perceiving reality brought about by looking at it from many different points of view. Because it is associated with gaining a deep understanding of life and because genuine feelings of sympathy and compassion require perspective taking, reflective wisdom is said to be the most crucial dimension of wisdom, facilitating the cultivation of cognitive and affective wisdom.

1.3 Neurobiology of Wisdom

As wisdom is manifested in human thought and behavior within specific contexts and environments, it will be illuminating to understand the neurobiological basis of wisdom if it exists as a part of human psychology. One seminal overview[15] of wisdom research outlines a broad set of brain regions associated in the processing of information related to characteristics aligned with the components of wisdom described above. Meeks and Jeste point out in their analysis that wisdom is a unique psychological characteristic and not merely a convenient label for a collection of desirable traits. In this synthesis of the various extant models and conceptualizations of wisdom, the authors classified wisdom across six categories and mapped each domain to neurobiological substrates in order to facilitate future research into the neurobiological bases of wisdom.

Wisdom by this model is a stable but malleable attribute broken down to include: (1) prosocial attitudes and behavior, (2) social decision-making (i.e., pragmatic knowledge of life), (3) emotional homeostasis, (4) reflection and self-understanding, (5) value relativism, and (6) acknowledgment of and dealing effectively with uncertainty and ambiguity. The brain regions associated with these categories include frontal and parietal regions related to intelligence and reasoning,[16] as well as cingulate and subcortical regions associated with affect and reward. The inclusion of cingulate and subcortical regions[17,18] points to the importance of emotional self-regulatory strategies[19,20] that bring about the emotional homeostasis needed for wise decision-making. More specifically, frontal and prefrontal regions interact with cingulate and subcortical emotional regions to downregulate emotion in contexts that require reason and prudent use of intelligence. While neuroscientific research into the components of wisdom—such as self and other reflection, moral reasoning, prosocial attitudes and behavior, and emotional homeostasis—indicate a large variety of unique and networked regions of activation,[21] some (such as the medial prefrontal and cingulate cortex) appear particularly important for bringing together cognitive strategies and emotional regulation for the goal of wise reasoning.

If wisdom can be decomposed into constituent cognitive and affective components, future investigations of how these components develop and change in the specific context of wisdom and wise decision-making can provide insight into mechanisms by which wisdom is cultivated and how it affects reasoning and decision-making. Though the work to date has looked at how the different components have been studied largely outside of this context, the overall pattern is that wisdom arises as higher-order cognitive regions, such as the prefrontal cortex, work to regulate immediate reward and emotional processing in striatal and cingulate cortex structures. *Prosocial attitudes and behaviors*, for example, are associated with the putative mirror neuron system—frontal and prefrontal brain regions that show the same pattern of activation during both motor performance and observation—and cortical regions that show activation in response to simulating the mental states of others, a process related to Theory of Mind.[22] It is important to note in describing associations between attitudes, behaviors, and associations of brain regions in functional Magnetic Resonance Imaging (fMRI)—given the nature of such research—findings presented here are merely potential avenues to explore the neural underpinnings of wisdom components and may not represent direct mechanisms for the representation of wisdom in the brain.

Associations between *pragmatic decision-making* and neural activity in the medial prefrontal cortex overlap with that of prosocial attitudes and behaviors. In a study among business students who viewed moral and nonmoral narratives,[31] viewing moral narratives evoked greater activation in medial prefrontal cortex, as well as in the posterior central sulcus and superior temporal sulcus, than did nonmoral narratives. Wise reasoning often requires self-regulation of instinctual impulses in the pursuit of a greater good. As such, *emotional homeostasis* is brought about by a coordination of functional activity between prefrontal control and emotional regions in the cingulate cortex. Specifically, the dorsal anterior cingulate cortex (dACC)—believed to detect conflict between automatic emotional responses and more socially acceptable responses—coordinates with the lateral PFC—believed to coordinate responses in working memory that are perceived to be socially advantageous.[23] *Self and other reflectiveness* also relies on activation of lateral PFC regions, as studies of "Theory of Mind" suggest that this region allows one to inhibit one's own point of view, in order to take on the perspective of others.[24] Furthermore, patients with lesions to this region are highly self-focused and exhibit difficulty in interpreting the social cues of others.[25,26] The neural bases of *value relativism* are somewhat similar to those described for other wisdom-related characteristics. Automatic amygdala activation in response to the depiction of other races and ethnicities—depictions in which effort to overcome prejudice may contradict automatic emotional responses—is mediated by dACC regions that detect such conflict and lateral PFC regions that regulate reaction.[27,28] A thorough understanding of these networks of regions, their individual characteristics, and cross-regional interaction is important for the development of a neurobiologically informed model of wisdom and wise decision-making.

While outlining how different regions of the brain are associated with components of wisdom does a great deal to help in understanding how it is represented in the brain, there is currently a lack of understanding of wisdom and its neurobiological foundation as a unified construct. If wisdom represents the regular joint action of automatic processing in regions like the amygdala, regions that are controlled upstream by higher-level cortical prefrontal and cingulate regions, and that wisdom develops as a practice of practical reflection contexts and situations of ambiguous information and interests, we would expect that there is an underlying network structure, albeit dynamically modified, that would relate to wisdom. The development of such a view requires emotional regulation abilities to manage negative emotions and stress during reasoning and decision-making that may arise as one takes in personal and vicarious values in coming to an appropriate decision. As such, wisdom may be reflected neurobiologically by an increased functional and structural connectivity of the brain regions described above, implicated in social pragmatic decision-making, value relativism, emotional homeostasis, prosocial attitudes and behaviors, and the skill and disposition to reflect on one's own and others' beliefs.

1.4 Practical Wisdom

As described, it is commonly understood that wisdom encompasses thoughtful or pragmatic decision-making, compassion, prosocial goals, moral judgment, and insight into personal and interpersonal problem solving via practical reflection. As described by Tiberius,[29] wisdom depends on a process in which possible decisions or choices are evaluated in the context of specific value commitments, when those value commitments are grounded in the virtues. Wise decisions are made by being able to flexibly shift perspectives (as in taking another person's perspective or the perspective of another culture) and comparing the value commitments for other perspectives for making decisions. As such, wise decisions depend on epistemic humility (recognizing the importance of other value commitments, knowledge, and perspectives than one's own), on reflection (being able to think analytically about value commitments and perspectives, as well as engage others in discourse about these), perseverance and the willingness to engage in intellectual struggle to deal with difficult problems or choices, and cognitive creativity to seek solutions that may not be apparent. These are all very high-level psychological constructs that are rooted in the flexible use of attention, working memory, long-term memory, reasoning and problem solving, sophistication of language use and knowledge, social interaction and understanding, and emotional reasoning. Given that there is not a single language region, memory region, or knowledge region in the brain, understanding the neurobiological bases of wisdom will depend on the interaction of complex neural networks.

Wisdom can therefore be characterized as a complex psychological process that is related to interactions of higher-order processing in the cortex, emotional activity in cingulate cortices, and reward processing in striatal regions of the brain, as well as insula processing related to homeostatic regulation and, along with other regions such as the amygdala, sensitivity to risks and negative outcomes. To the degree that wisdom is considered the successful integration of thought and affect during decision-making, the discussion of potential neuroscientific models of wisdom is related to the way individuals make moral decisions based on a pragmatic knowledge of life and how such decisions are rooted in empathy, compassion, and altruism. Given the need in wise reasoning to take on multiple conflicting points of view, emotional regulation is a critical component of wisdom, and such regulation is again rooted in the interaction of frontal, cingulate, and reward processing regions of the brain. Of particular interest is how individuals use self-reflection and other reflection and social interaction to overcome individual emotional reactions to stress and anxiety—owing to the uncertainty and ambiguity that exists in complex real-life problem solving scenarios, where personal values and those of others may come into conflict. Following a review of moral sensitivity and decision-making, prosocial attitudes and behavior, self-regulation and emotional homeostasis, self/other reflectiveness, and value relativism, this chapter will put forward ideas on how wisdom may be developed across the lifespan and how the components of wisdom may be synthesized into a neurobiologically unified construct.

1.5 Tying Practical Wisdom Philosophy to Psychobiology

After the dust settles following a particularly challenging situation, decisions are often judged as wise when they are shown to lead to the largest benefit for the greatest number of people over the long-term. Wise reasoning then depends on thought processes that take into account multiple points-of-view and an understanding of the larger third-person perspective. Reducing uncertainty and confusion in conflict resolution may require what philosopher Valery Tiberius[29] refers to as practical reflection. Such reflection must take into account not only the facts as they pertain to the context and situation, but also one's personal values and the possibly conflicting values of others. Taking into account the values of others is not just a matter of knowing what is important to other people—it also depends on feeling the impact of those value commitments. In order to realize that impact, it is important to be able to adopt someone else's perspective. As a result, wise reflection depends on flexibility in shifting perspective. In this way, reflection and perspective taking are central to wise reasoning, and both rely largely on the cognitive awareness of facts and contingencies, as well as on the subjective sense of possible affective outcomes of decision-making. This description incorporates psychological constructs of wisdom by requiring a strong pragmatic knowledge base, the ability to reflect on one's own values and the values of others, and the disposition to carry out reasoned decision-making with regard for social goods. Further, wisdom and wise reflection suggests a process by which wisdom could be practiced and cultivated.

In situations that require wisdom, decisions are based on value judgments. Values differ between and within groups of people. Therefore, people have some level of consideration that their values are justified. However, as Tiberius points out,[29] sometimes this is just a sense that values are justifiable, which means that people sometimes accept what others around them do (perhaps through culture) and that there exists some justification that could be recovered in some fashion. Sometimes, justification simply occurs at a gut level through some intuitive sense from cultural exposure, but value justification can also be based on reflection from internal considerations, as well as from discussions with others. In general, this view suggests that we generally take a perspective in which one set of value commitments holds and for a reflective person, these value commitments frame the decision process. A wise person can go beyond this process, flexibly shifting perspectives to adopt or consider other value commitments than their own, and the value commitments of the wise person in some perspectives are grounded in virtues such as generosity or kindness. This does not necessarily mean that wise reasoning through practical reflection leads to objectively correct conclusions, but such reflection leads to decisions that are based on what is known and on a mental simulation of how outcomes of wise reasoning will play into one's own interests, as well as the interests of others.

Because the wise person takes into consideration the outcomes and prospects of his or herself, as well as those of others, wise decisions require moral sensitivity to understand the relativity of values across individuals. Such an understanding can facilitate outcomes of wise reasoning that lead to the greater good of all those involved in a particular situation or context. Moral sensitivity has been studied extensively using psychological and neuroscientific methods, the results of which characterize it as being paramount for decision-making that leads to the greatest benefit for self, others, and society as a whole.

1.6 Moral Reasoning

Wise reasoning relies on the ability to frame reasoning in the context of moral values not only from one's own point of view, but also from the perspective of other people and groups. The ability to use moral value commitments relies both on moral sensitivity and the disposition to carry out moral decisions. To balance decision-making in such a way as to account for personal goals and goals for a larger group or society as a whole, moral sensitivity and decision-making relies on autobiographical memories of past decision-making during moral conflicts, as well as on perspective taking to gain insight into possible outcomes of decision-making for others. Flexibly using one's own and others' perspectives and goals relies on both cognitive and affective systems in cortical and subcortical regions of the brain related to affect, intention, memory, perspective taking, and decision-making.

Moral sensitivity refers to the awareness of how different individuals or groups of individuals may be affected by the outcome of a decision based on a particular issue.[30] Given the overlap of personal experience and taking into account the perspective of others in making wise and moral decisions, it follows that neural regions underpinning moral sensitivity are those associated with autobiographical memory retrieval and social perspective taking processes.[31] Moral sensitivity is itself a prerequisite for ethical decision-making, which is central to wise reasoning about human social behavior.

Functional magnetic resonance imaging, in which the metabolic activity of neurons and glial cells is used as a proxy for activation of large regions of neural tissue, points to three main brain regions as important for moral sensitivity.[31] These regions include the medial prefrontal cortex (MPFC), the posterior cingulate cortex (PCC), and the posterior superior temporal sulcus (pSTS). The MPFC has been implicated in both implicit and explicit moral decision-making[32–34] and is tied to self-monitoring behavior and self-referential processing. In contrast to self-referential processing, the MPFC with the temporoparietal junction (TPJ) is also associated with theorizing about others' states of mind, also known as mentalizing or Theory of Mind (ToM).[35] This implies that moral sensitivity may be a perspective taking process that involves both self-knowledge, as well as other knowledge and reflections.

The PCC has been linked to emotional evaluations of the appropriateness of responses to personal moral dilemmas[32] and may serve as an interface between emotion and cognition.[36] It is possible that the PCC facilitates moral sensitivity by integrating emotional, cognitive, and affective memories of past moral conflicts. The pSTS and adjacent TPJ contribute to moral sensitivity by facilitating the integration of one's personal point of view with the point of view of others, which facilitates empathic emotions such as guilt and compassion.[37]

Because it facilitates moral and ethical decision-making, moral sensitivity is an antecedent of wisdom, and its understanding may contribute to a greater discourse about wisdom in national and international enterprise. Its training may facilitate greater wisdom, particularly given that postconventional reasoning—principles of moral reasoning in which social good is placed above personal or selfish motives—has been shown to plateau during professional development unless ethics interventions are present. Professional enterprise may be considered a largely moral enterprise, and decisions taken in business that affect many partners and associates across a range of situations and contexts require wise reasoning, making it a critical environment for the practice of wise reasoning. Professional educational research suggests that moral reasoning may be trained through the implementation of ethics interventions in professional education settings.[38] It is possible that the use of such interventions could generally impact wisdom, and future interdisciplinary research is needed to investigate this possibility.

1.7 Prosocial Attitudes and Behavior

As has been described, wisdom requires the processing of personal values and perspectives in coordination with the points of views and possibly conflicting values and perspectives of others. It follows that wise reasoning requires empathy and compassion in coordinating personal thoughts and behaviors with those of other people within a particular context or situation. Empathy is a feeling or sense of sameness between one's emotions and those experienced by others. Empathy may lead to prosocial behavior, but it could also cause the kind of distress that leads to disengagement from the empathy-inducing circumstance. Wise reasoning must strike a balance between dampening empathic distress while increasing empathic concern, so that feelings of vicarious distress do not interfere with social reasoning processes that may facilitate prosocial behavior.

Empathy can be broken down into cognitive and affective components, reflecting differences in the mental understanding and affective mirroring of others' thoughts, feelings, and intentions. The first step of cognitive empathy is in distinguishing one's own feelings, thoughts, and intentions from those of others. Following this cognitive distinction, one may imagine how another person feels or believes in a manner that will not overly tax one's own emotional state. With this imagined model of others' states of mind, it is then possible to extrapolate future thoughts, feelings, intentions, and behaviors. Though, to the degree that this mental model is built on one's existing understanding of the world and its workings, predictions will have varying degrees of accuracy. As an interaction of lower-level emotional processes in the automatic mirroring of others' emotions, and higher-level cognitive control of these reactions, empathy recruits affective regions in subcortical, as well as more highly evolved medial prefrontal regions that are implicated in the coordination of emotion and behavior.

Social situations involving multiple persons or groups—within uncertain and ambiguous contexts and in which the thoughts, feelings, and intentions of others in respect to the self may be very different—can give rise to negative emotional responses. For wise reasoning to take place, negative feelings in situations of ambiguity must be regulated and reflected upon, so that the most appropriate actions can be executed. Given the previous description of moral sensitivity as taking into account personal and vicarious points of view, the cognitive and emotional processes underlying empathy share neurophysiological bases with processes of moral reasoning.[39] In a meta-analysis of almost 80 studies, Seitz and colleagues[40] found that the MPFC, for example, plays a prominent role in multiple separable components of empathy. These components relate to cognitions, emotions, and intentions to act triggered by internal and external states of introspection and mentalizing. Furthermore, these processes are coordinated via functional connectivity between the MPFC and the anterior cingulate cortex (ACC), as internal thoughts and emotions and external circumstances may come into conflict during social interaction.

Mentalizing refers to the ability to identify and comprehend the beliefs, intentions, traits, and emotions of oneself and others,[41] and core regions subserving these functions include the MPFC and the TPJ. Mentalizing is more than the mental simulation of others' thoughts and feelings and requires some conceptual knowledge of how the mind works.[35] Recent research has shown that mentalizing skills are stronger among individuals from interdependent cultures than among people from individualistic cultures, and these differences are likely due to a decreased self-centeredness and increased focus on the importance of others' thoughts, feelings, and motivations.[42] While this does not speak to wisdom directly, these findings imply that wisdom may be manifested in different ways across different cultures.

Empathy may be central for the self-control of behavior, as it is thought to allow the subjective experience of others' mental and affective states,[43] which in turn informs present and future social behavior. Evidence suggests that the accomplishment of cognitive functions related to empathy is brought about by large-scale cortical networks through nodes of convergence that link info from different and varying sources. These nodes include the temporal pole, which gives access to knowledge of past experience; the superior temporal sulcus, which provides information about observed behavior; and MPFC, which serves to link cognitive information to basic emotion[44] and is recruited preferentially by mentalizing processes, but only very rarely for nonsocial executive functioning such as working memory and attention.[45] The social nature of wise reasoning and the differential use of the MPFC between social and nonsocial processing reflects the difference between wisdom and more commonly studied processes, such as intelligence and attention.

Humans are social animals, and virtually all actions, from external behaviors to internal thoughts and desires, occur in response to others.[46] As such, wise reasoning in social settings relies on empathy (which one may use to understand the feelings of others), compassion, or a feeling that motivates helping behavior. Vital to the development and sense of empathy is an awareness of others' emotions, as well as emotional regulation to maintain a self-other affective distinction.

As such, empathy is a complex form of psychological inference in which memory, knowledge, and reasoning are combined to yield insights into the thoughts and feelings of others.[47] Put another way, empathy incorporates a cognitive component of knowing what another person is feeling, an affective component of feeling what another person is feeling, and a behavioral component of intending to respond compassionately to the distress of others. However, not all empathic responses are cognitive, affective, and intentional, as sociopaths and patients with lesions to the MPFC may be capable of cognitive, but not affective, empathy, and even in the case that one feels and knows another's emotional state, this does not necessitate action to alleviate such vicarious distress. Furthermore, one may be moved to compassionate responding without affectively resonating with their pain.

To reason wisely, it is not enough to simply reflect the emotions of others and predict future outcomes based on these predictions. It is also important to reflect on one's own model of how the world works and how others work within it. This reflects the idea that an understanding of how the world and people within it work is never complete or accurate, and that only by careful reflection on one's own and others' internal states and external circumstances can one come to more accurately predict and prepare for the future. Such reflection on one's own internal state in comparison to others—in either case, these emotions may be dissonant—takes emotional self-regulation.

1.8 Emotional Homeostasis and Impulse Control

Emotional homeostasis refers to emotional stability in the face of uncertainty,[48] which is considered in the Ardelt Three-Dimensional Wisdom Scale to be a part of the affective dimension of wisdom.[49] As it takes place by an interaction of cognitive and affective processes, emotional homeostasis is associated with functional changes in lateral and medial prefrontal cortices, as well as in orbital frontal cortices and the amygdala.[15] Emotional self-regulation is associated with old age, such that positive affect increases across the lifespan.[50] Neuroscientific findings suggest that emotional regulation strategies change over the lifespan, as individuals in later life seek emotional homeostasis through cognitive regulation, regulate negative affect, and feel good in the present moment.[51]

It can be argued that those who become wiser in old age are those who can successfully regulate emotional responding in complex personal and social situations. Like other components of wisdom, such as moral reasoning, emotional homeostasis requires an interaction of both cognitive and emotional abilities, which is reflected in the interplay of corresponding neural substrates. Specifically, emotional regulation is largely understood as a cognitive regulation of emotional appraisal and response through a dampening of emotional and impulsive responses in the ventral striatum by the prefrontal cortex.[52–54] Evidence suggests that emotional regulation strategies are used without deficit across the lifespan, though the mechanisms by which these regulation strategies are carried out changes with old age.

One way in which negative feelings may be regulated in social conflicts, like those associated with wise reasoning, is through a strategy of cognitive reappraisal. Cognitive reappraisal refers to a flexible regulatory strategy that draws on cognitive control and executive functioning to reframe stimuli or situations within the environment to change their meaning and emotional valence.[55] These functions and processes are associated with activation within the PFC and the posterior parietal cortex.[32] Recent research suggests that both young and older adults use cognitive reappraisal strategies, despite evidence of age-related declines in PFC volume.[56] However, reappraisal in old age appears to rely even more on PFC activation than in youth, particularly at points of peak emotional experience. Additionally, integration of PFC with other cognitive regulation regions, such as the ACC, is as robust in old age as in young adulthood.[57] This suggests that how people maintain emotional homeostasis changes over the lifespan, without degrading as with other cognitive functions.

Older and young adults similarly recruit the ACC when using cognitive reappraisal strategies, but the network of functional connectivity between this region and others is what appears to set older adults apart from youth.[57] The ACC plays an important role in emotional homeostasis, as its activation is associated with conflict detection. Functional connectivity in this circumstance refers to the network of regions that are similarly active with the ACC during moments of cognitive reappraisal. While both youth and older adults show ACC and PFC activation during cognitive reappraisal, unlike young adults, elders recruit regions associated with the inhibition of negative emotional reactivity in the lateral portion of the orbital frontal cortex. Younger adults recruit dorsolateral and dorsomedial cortices, which are associated with the manipulation of reappraisal in working memory and the monitoring of the success of cognitive reappraisal. As the affective dimension of wisdom is described, not only as an increase in positive emotions in the face of uncertainty, but also as a decrease in negative emotions,[8] that older adults differentially recruit regions associated with the inhibition of negative affect compared to youths is in line with this view of wisdom. Experience with life problems and adversity across the lifespan may also lead to changes in the way that problems are represented in memory, such that older adults' working memory is not as taxed in situations requiring wise reasoning.

1.9 Practice and Wisdom

As described in this chapter, wisdom develops with life experience, though the type of experience and reactivity of the individual in context is critical to the cultivation of wisdom. Wise reasoning seems to be invoked by the ability to reflect on one's own and other's thoughts, beliefs, and actions, by empathy, compassion, and prosocial behavior, and by epistemic humility—the Socratic notion *I know one thing: that I know nothing*. If wisdom may be increased by some intervention, this may occur by targeting specific components of wisdom that are thought malleable, such as perspective taking, compassion, and prosocial behavior. There are gender and cultural differences in perspective taking, which may inform ways in which perspective taking may be targeted for improvement, and recent research indicates that certain forms of contemplative practice may train key components of wisdom, such as compassion. Complementary to this research, our lab has shown that certain structured practices, such as meditation and even ballet, are associated with increases in wisdom. While research into the cultivation of wisdom by practice is only beginning, it holds great promise for the development of curriculum for the intentional teaching of wisdom.

The ability to take others' perspectives is vital to reflecting on others' points of view, which contributes to wise reasoning. Perspective taking is often linked to ToM, or the ability to take on the mental states of others to understand their emotions, motivations, and frames of mind, and how these differ from our own.[58] An extensive literature shows that these abilities develop during the preschool ages of three to five years old,[59,60] and ToM and related skills continue to develop throughout life. While ToM skills develop similarly in all typically developing children, interdependent East Asian cultures tend to have increased abilities in ToM compared to individualistic cultures, such as those in Western countries. Further, gender differences in ToM that are evidenced in Western cultures, whereby females score higher on tests of mentalizing compared to males, are not as strong among East Asian cultures.[61] Understanding the mechanisms by which these differences occur can help to inform future interventions aimed at increasing perspective taking and facilitating the reflective nature of wisdom.

Evidence that an interdependent point of view may lead to better perspective taking comes from research in which either Chinese or American participants were asked to follow instructions on how to move objects in a grid, following a director's instructions.[42] Chinese participants in this study made fewer critical mistakes—mistakes that required perspective taking to avoid. By careful analysis of decisions and reaction times, the authors were able to determine that increased perspective taking abilities stemmed from a focusing of attention toward others and away from the self. Given the Chinese culture of interdependence, increased perspective taking comes about because the self is defined by its relationship with others, increasing the importance of and focus on the role that others play, as well as of their actions, knowledge, and needs. Kessler and colleagues[61] have additionally found that Westerners are slower than East Asians at tasks involving embodied perspective taking and show stronger gender differences in speed and depth of mental perspective taking. Taken together, these results suggest that individuals with a stronger ability to take on the thoughts, emotions, and motivations of others, rather than simply imagining their visual point of view are better perspective-takers and have increased associated social skills. It is possible that such increases are also associated with increased wise decision-making, though continued research must be conducted into whether this is the case.

One method by which perspective taking may be enhanced is by structured meditation practice. Participants trained in a secularized compassion meditation program (cognitive-based compassion training; CBCT) show increased empathic accuracy in the Reading the Mind in the Eyes Task (RMET), compared to an active control group.[62] The RMET involves identifying the affect expressed in a series of black-and white-pictures of eyes, requiring the cognitive simulation of others' feelings from minimal visual cues, and performance has been shown to improve with the experimental administration of oxytocin,[63] a hormone associated with social bonding. Mascaro and colleagues claim that meditation could act as a behavioral intervention to enhance empathy by defending against deleterious nervous system responses during stressful or adversarial situations. If the meditation experience leads to increased understanding of others' emotions, which in turn increases interpersonal and prosocial interactions, this could indicate that it increases wisdom over a period of sustained practice.

Interventions based in the practice of loving-kindness meditation have also been used in recent years to facilitate increased compassionate feelings and responding. Recent interest in the experimental manipulation of compassion for the easement of suffering and promotion of prosocial behavior have deep roots in ancient contemplative practices linked to the cultivation of wisdom and cessation of suffering.[64] A steadily growing body of literature is building the case that brief compassion training may increase positive interpersonal skills and behavior, skills that are important for wisdom and wise reasoning. Further, these changes in interpersonal skills have been tied to changes in different networks of activation in the brain related to affiliation and perception of pain.[65]

Some of the earliest studies of compassion provide evidence of increased wisdom-related skills and dispositions, such as increased support and implicit positive evaluation of others following compassion-state induction brought about by a brief loving-kindness meditation.[60,66] In a study of loving-kindness meditation, participants who practiced only a brief meditation showed increased connectivity to and positive regard for others, compared to controls.[67] In a more recent study, in which participants played a social Simon task, which measures the integration of their own and others' perspectives in decision-making, practicing Buddhists showed greater self-other integration than nonreligious controls.[67] Moreover, in a study that used magnetic resonance imaging (MRI) to measure cortical gray matter volume in expert meditators compared to novices, experts had greater gray matter volume in regions of the brain associated with affective regulation.[68] This suggests that increased social-connectedness, and subsequently increased compassion and empathy, are related to long-lasting effects of meditation practice over time.

More recently, compassion training has been associated with decreased negative affect in response to videos depicting others in distress—videos that previously elicited increased negative affect following a similar program to train empathy in the same participants.[65] The authors of this study used behavioral and neurobiological measures to suggest that compassion increases the ability to cope with distress not by suppression of negative emotions in response to suffering, but by the generation and strengthening of positive affect. This is in line with the concept of the wise individual as one who does not push away negative emotions, but instead sees things from a larger perspective in order to cope with a situation appropriately. Compassion training has further been associated with increased prosocial behavior in both laboratory-based and in real-world settings.[68,69] This recent body of research holds promise for the development of programs to cultivate wisdom through compassion, possibly through long-term changes in regions of the brain associated with empathy, feelings of affiliation with others, and prosocial behavior.

Research from our own lab (in review) indicates that meditation and certain structured practices, such as ballet, are associated with increased cognitive, reflective, and compassionate dimensions of wisdom, as measured by Ardelt's Three-Dimensional Wisdom Scale (3DWS[8]). The association between wisdom and meditation is not surprising, given the above-described findings and historical associations between meditation and the development of wisdom in Buddhist and Taoist traditions.[70] The association between wisdom and ballet experience is more surprising, for while ballet requires great physical and mental self-regulation to excel to high levels of expertise, as a practice, it is not typically associated with wisdom or wise reasoning.

Meditation is generally associated with characteristics that have also been identified as essential components of wisdom, such as regulation of attention, self-control, and interpersonal understanding.[15,62,71] Wisdom incorporates several interrelated characteristics that seem linked to meditation. For example, greater wisdom is associated with increased prosocial behavior, based around both self-reflection and compassion.[8,72] Furthermore, the openness, curiosity, and acceptance of experience found in meditation is similar to several components of wisdom, including tolerance and value relativism, the development of pragmatic knowledge of life, and the ability to effectively deal with uncertainty and ambiguity.[72,73]

Though the practice comes in many forms, meditation often involves the cultivation of characteristics belonging to a state of mindful awareness. These characteristics break down into self-regulation of attention and an investigative awareness characterized by openness, curiosity, and acceptance of experience.[74] This state of open awareness is thought to lead eventually to the acceptance of experiences that may otherwise cause stress, thereby allowing meditation practitioners to navigate life with lower anxiety and increased cognitive capacity in the present moment. In a study of focused-breathing meditation, for example, trained participants viewed both neutral and negative pictures significantly less negatively and were more willing to view optional negative pictures than were control participants,[75] suggesting lowered stress responses to negative stimuli. Research into mental training associated with meditation has been linked to lowered anxiety,[76] and lowered anxiety can in turn free up valuable cognitive resources like working memory (a short-term memory system involved in cognitive control). When working memory resources are disrupted by anxiety, performance can suffer,[77] but interventions that reduce worrying and decrease anxiety have been shown to boost cognitive performance.[78] Meditation practice may play a similar intervening role, in that lowering anxiety frees up mental resources, creating a reflective mental space that promotes wise decision-making.

The ability to deal successfully with hardship correlates with an increase in psychological health for elders identified as wise and may be a prerequisite for the development of wisdom.[79] By improving psychological health, it is possible that practicing meditation helps people deal with hardship in a more successful and wise manner. Meditation practice in general, and mindfulness practice in particular, is associated with improvements in psychological well-being, as evidenced by improvements across a variety of psychiatric disorders, such as depression, anxiety, and addiction.[80] Regular meditation practice then may provide psychologically healthy individuals with resources to handle a challenge rather than viewing it as a threat. It is also true that many people are attracted to begin meditation as a means to deal with hardship, which suggests the possibility of a self-selection bias toward meditation favoring those who develop wisdom. Future research utilizing interventions among meditation for naïve individuals is needed to further understand whether meditation experience leads directly to the development of wisdom and over what time frame.

1.10 Conclusion

While wisdom has been a topic of discussion and debate among philosophers for thousands of years, wisdom research did not begin in earnest until the second half of the twentieth century, led by researchers interested in its association with successful aging.[1,14,81] Such research gained momentum near the beginning of the twenty-first century, as a consensus grew toward a unified definition of wisdom, or at least wisdom as it exists within individuals. Of particular interest in the context of this chapter is the development of research into the neurobiological underpinnings of personal wisdom.

Based on a study of wisdom experts,[82] there is now some agreement over what constitutes the characteristics of wisdom. In a development that promotes the unification of wisdom as a construct in the psychological literature, these characteristics largely fall into three categories that align with the cognitive, reflective, and affective dimensions of wisdom developed by Monica Ardelt[8] for the Three-Dimensional Wisdom Model. A rich knowledge of life, skills in social cognition, tolerance of ambiguity, and acceptance of uncertainty, for example, map onto a cognitive dimension; a sense of justice and fairness, self-insight, and tolerance of differences among others belong to a reflective dimension; and characteristics that compose the affective dimension include empathy and social cooperation. In terms of relating wisdom to neurobiology, five of the characteristics described by wisdom experts also appear in a review of the neurobiological underpinnings of wisdom.[15]

Wisdom characteristics, notably prosocial attitudes and behaviors, social decision-making and pragmatic knowledge of life, emotional homeostasis, and self and other reflection, are facilitated by the coordinated effort of higher-order processing systems in the prefrontal and temporal cortices, the anterior cingulate cortex, and the ability of these regions to regulate otherwise automatic processing in deeper brain regions, such as the amygdala and ventral striatum, that are associated with fear, reward, and punishment. As noted by Tiberius,[29] it is possible that one may cultivate wisdom by practicing self and other reflection, but it is unclear at this time, at least from the perspective of psychological science, whether such practices lead to changes in wisdom, and complimentarily, long-term changes in the neural substrates of wisdom characteristics: more empirical research is needed. As an advance in wisdom research along these lines, we have shown that one form of structured practice, meditation, which tends to focus on self and other reflection, is associated with increased wisdom. However, future research is needed to better understand this relationship, and through what mechanism—such as by increased practice reflecting on oneself or others—meditation may affect wisdom.

Prior to the wisdom research of the latter part of the twentieth and the beginning of the twenty-first century, wisdom was largely ignored by science and was little spoken of outside of these research paradigms, perhaps because of the somewhat mysterious and fuzzy nature of wisdom as a topic. By understanding the underpinnings of wisdom, both psychologically as well as neurobiologically, it may be possible to develop interventions or classroom curricula that cultivate wise reasoning. As a nation that seeks to cultivate entrepreneurship and innovation, intelligence and creativity are prized characteristics. However, as has been described in this chapter, successful aging takes something beyond intelligence and creativity, specifically the wisdom to know when to apply intelligence and when to apply creativity in a prudent and prosocial manner.

Acknowledgment

Preparation of this chapter was supported by the John Templeton Foundation.

References

1. Clayton VP, Birren JE. The development of wisdom across the life span: a reexamination of an ancient topic. In: Baltes PB, Brim Jr OG, eds. *Life-Span Development and Behavior*. New York: Academic Press; 1980:103–135.
2. Baltes PB, Staudinger UM. Wisdom: a metaheuristic to orchestrate mind and virtue towards excellence. *Am Psychol*. 2000;55:122–136.
3. Sullivan S, Mikels JA, Carstensen LL. You never lose the ages you've been: affective perspective taking in older adults. *Psychol Aging*. 2010;25:229–234.
4. Grossmann I, Na J, Varnum MEW, Kitayama S, Nisbett RE. A route to well-being: intelligence versus wise reasoning. *J Exp Psychol Gen*. 2012;142:944–953.
5. Staudinger UM, Smith J, Baltes PB. Wisdom-related knowledge in a life review task: age differences and the role of professional specialization. *Psychol Aging*. 1992;7:271–281.
6. Sternberg RJ. Personal wisdom in the balance. In: Ferrari M, Weststrate NM, eds. *The Scientific Study of Personal Wisdom*. New York: Springer; 2013:53–74.
7. Wisdom Research. The University of Chicago. http://wisdomresearch.org/; Updated February 10, 2015. Accessed 20.02.14.
8. Ardelt M. Empirical assessment of a three-dimensional wisdom scale. *Res Aging*. 2003;25:275–324.
9. Tiberius V. In defense of reflection. *Philos Issues*. 2013;23:223–243.
10. Sternberg RJ. A balance theory of wisdom. *Rev Gen Psychol*. 1998;2:347–365.
11. Sternberg RJ. A systems model of leadership: WICS. *Am Psychol*. 2007;62:34–42.

12. Sternberg RJ. Why schools should teach for wisdom: the balance theory of wisdom in educational settings. *Educ Psychol*. 2001; 36:227–245.
13. Gardner H. *Frames of Mind: The Theory of Multiple Intelligences*. New York: Basic Books; 1983.
14. Baltes PB, Smith J. Toward a psychology of wisdom and its ontogenesis. In: Sternberg RJ, ed. *Wisdom: Its Nature, Origins, and Development*. New York: Cambridge University Press; 1990:87–120.
15. Meeks TW, Jeste DW. Neurobiology of Wisdom: a literature overview. *Arch Gen Psychiatry*. 2009;66:355–365.
16. Jung RE, Haier RJ. The parieto-frontal integration theory (P-FIT) of intelligence: converging neuroimaging evidence. *Behav Brain Sci*. 2007;30:135–154.
17. Kim SH, Hamann S. Neural correlates of positive and negative emotion regulation. *J Cogn Neurosci*. 2007;19:776–798.
18. McClure SM, Laibson DI, Loewenstein G, Cohen JD. Separate neural systems value immediate and delayed monetary rewards. *Science*. 2004;15:503–507.
19. Leiberman MD, Eisenberger NI, Crockett MJ, Tom SM, Pfeifer JH, Way BM. Putting feelings into words: affect labeling disrupts amygdala activity in response to affective stimuli. *Psychol Sci*. 2007;18:421–428.
20. Hariri AR, Bookheimer SY, Mazziotta JC. Modulating emotional responses: effects of neocortical network on the limbic system. *Neuroreport*. 2000;11:43–48.
21. Sanders JD, Jeste DV. Neurobiological basis of personal wisdom. In: Ferrari M, Weststrate NM, eds. *The Scientific Study of Personal Wisdom*. New York: Springer; 2013:99–112.
22. Perner J, Lang B. Development of theory of mind and executive control. *Trends Cogn Sci*. 1999;3:337–344.
23. Congdon E, Canli T. The endophenotype of impulsivity: reaching consilience through behavioral, genetic, and neuroimaging approaches. *Behav Cogn Neurosci Rev*. 2005;4:262–281.
24. Satpute AB, Lieberman MD. Integrating automatic and controlled processes into neurocognitive models of social cognition. *Brain Res*. 2006;1079:86–97.
25. Samson D, Apperly IA, Humphreys GW. Error analyses reveal contrasting deficits in "theory of mind": neuropsychological evidence from a 3-option false belief task. *Neuropsychologia*. 2007;45:2561–2569.
26. Samson D, Apperly IA, Kathirgamanathan U, Humphreys GW. Seeing it my way: a case of a selective deficit in inhibiting self-perspective. *Brain*. 2005;128:1102–1111.
27. Amodio DM, Harmon-Jones E, Devine PG, Curtin JJ, Hartley SL, Covert AE. Neural signals for the detection of unintentional race bias. *Psychol Sci*. 2004;15:88–93.
28. Cunningham WA, Johnson MK, Raye CL, Chris GJ, Gore JC, Banaji MR. Separable neural components in the processing of black and white faces. *Psychol Sci*. 2004;15:806–813.
29. Tiberius V. *The Reflective Life: Living Wisely with our Limits*. New York: Oxford University Press; 2008.
30. Rest JR. Background: theory and research. In: Rest JR, Narvaez D, eds. *Moral Development in the Professions: Psychology and Applied Ethics*. Vol. 1. Hillsdale: Erlbaum; 1994:26.
31. Robertson D, Snarey J, Ousley O, et al. The neural processing of moral sensitivity to issues of justice and care. *Neuropsychologia*. 2007;45:755–766.
32. Greene JD, Sommerville RB, Nystrom LE, Darley JM, Cohen JD. An fMRI investigation of emotional engagement in moral judgment. *Science*. 2001;293:2105–2108.
33. Heekeren HR, Wartenberger I, Schmidt H, Schwintowski HP, Villringer A. An fMRI study of simple ethical decision-making. *Neuroreport*. 2003;14:1215–1219.
34. Moll J, Oliveira-Souza RD, Eslinger PJ, et al. The neural correlates of moral sensitivity: a functional magnetic resonance imaging investigation of basic and moral emotions. *J Neurosci*. 2002;22:2730–2736.
35. Saxe R, Wexler A. Making sense of another mind: the role of the right temporo-parietal junction. *Neuropsychologia*. 2005;43:1391–1399.
36. Bush G, Luu P, Posner MI. Cognitive and emotional influences in anterior cingulate cortex. *Trends Cogn Sci*. 2000;4:212–215.
37. Moll J, Oliveira-Souza R, Garrido GJ, et al. The self as a moral agent: linking the neural bases of social agency and moral sensitivity. *Soc Neurosci*. 2007;2:336–352.
38. Bebeau MJ. The defining issues test and the four component model: contributions to professional education. *J Moral Educ*. 2002;31:271–295.
39. Seitz RJ, Schäfer R, Scherfeld D, et al. Valuating other people's emotional face expression: a combined functional magnetic resonance imaging and electroencephalography study. *Neuroscience*. 2008;152:713–722.
40. Seitz RJ, Nickel J, Azari NP. Functional modularity of the medial prefrontal cortex: involvement in human empathy. *Neuropsychology*. 2006;20:743–751.
41. Amodio DM, Frith CD. Meeting of minds: the medial frontal cortex and social cognition. *Nat Rev Neurosci*. 2006;7:268–277.
42. Wu S, Keysar B. The effect of culture on perspective taking. *Psychol Sci*. 2014;18:600–606.
43. Seitz RJ, Nickel J, Azari NP. Reflection on oneself and others: evidence for functional modularity of the medial frontal cortex. *NeuroImage*. 2005;26:276.
44 Frith U, Frith CD. Development and neurophysiology of mentalizing. *Philos Trans R Soc B Biol Sci*. 2003;358:459–473.
45. Van Overwalle F. Social cognition and the brain: a meta-analysis. *Hum Brain Mapp*. 2009;30:829–858.
46. Batson CD. Self-report ratings of empathic emotion. In: Eisenberg N, Strayer J, eds. *Empathy and its Development*. New York: Cambridge University Press; 1990:356–360.
47. Ickes W. *Empathic Accuracy*. New York: The Guilford Press; 1997.
48. Brugman GM. Wisdom and aging. In: Birren JE, Schaie KW, eds. *Handbook of the Psychology of Aging*. 6th ed. Burlington: Elsevier Academic Press; 2005:445–469.
49. Ardelt M, Achenbaum WA, Oh H. The paradoxical nature of personal wisdom and its relation to human development in the reflective, cognitive, and affective domains. In: Ferrari M, Weststrate NM, eds. *The Scientific Study of Personal Wisdom*. New York: Springer; 2013:265–295.
50. Carstenson LL, Mikels JA. At the intersection of emotion and cognition. *Curr Dir Psychol Sci*. 2005;14:117–121.
51. Carstensen LL, Isaacowitz DM, Charles ST. Taking time seriously: a theory of socioemotional selectivity. *Am Psychol*. 1999;54:165–181.
52. Mischel W, Underwood B. Instrumental ideation in delay of gratification. *Child Dev*. 1974;45:1083–1088.
53. Mischel W, Ayduk O, Berman MG, et al. "Willpower" over the life span: decomposing self-regulation. *Soc Cogn Affect Neurosci*. 2011;6:252–256.
54. Casey BJ, Somerville LH, Gotlib IH, et al. Behavioral and neural correlates of delay of gratification 40 years later. *Proc Natl Acad Sci*. 2011;108:14998–15003.
55. Ochsner KN, Gross JJ. The cognitive control of emotion. *Trends Cogn Sci*. 2005;9:242–249.
56. Allard ES, Kensinger EA. Age-related differences in neural recruitment during the use of cognitive reappraisal and selective attention as emotion regulation strategies. *Front Psychol*. 2014;5:296.
57. Allard ES, Kensinger EA. Age-related differences in functional connectivity during cognitive emotion regulation. *J Gerontol B Psychol Sci Soc Sci*. 2014;69:852–860.
58. Premack D, Woodruff G. Does the chimpanzee have a theory of mind? *Behav Brain Sci*. 1978;1:515–526.
59. Bartsch K, Wellman HM. *Children Talk about the Mind*. New York: Oxford University Press; 1995.
60. Flavell JH. Cognitive development: children's knowledge about the mind. *Annu Rev Psychol*. 1999;50:21–45.

61. Kessler K, Cao L, O'Shea J, Wang H. A cross-culture, cross-gender comparison of perspective taking mechanisms. *Proc R Soc B Biol Sci*. 2014;10:1–9.
62. Mascaro JS, Rilling JK, Negi L, Raison CL. Compassion meditation enhances empathic accuracy and related neural activity. *Soc Cogn Affect Neurosci*. 2013;8:48–55.
63. Domes G, Heinrichs M, Michel A, Berger C, Herpertz SC. Oxytocin improves "mind-reading" in humans. *Biol Psychiatry*. 2007;61:731–733.
64. Bajracharya R, Bajracharya B. *Buddhist Wisdom and Compassion*. Naxal, Kathmandu, Nepal: New Nepal Press; 2009.
65. Klimecki OM, Leiberg S, Ricard M, Singer T. Differential pattern of functional brain plasticity after compassion and empathy training. *Soc Cogn Affect Neurosci*. 2013;9:873–879.
66. Hutcherson CA, Seppala EM, Gross JJ. Loving-kindness meditation increases social connectedness. *Emotion*. 2008;8:720–724.
67. Colzato LS, Zech H, Hommel B, Verdonschot R, van den Wildenberg WPM, Hsieh S. Loving-kindness brings loving-kindness: the impact of Buddhism on cognitive self-other integration. *Psychon Bull Rev*. 2012;19:541–545.
68. Leiberg S, Klimecki O, Singer T. Short-term compassion training increases prosocial behavior in a newly developed prosocial game. *PLoS One*. 2011;6:e17798.
69. Condon P, Desbordes G, Miller WB, DeSteno D. Meditation increases compassionate responses to suffering. *Psychol Sci*. 2013; 24:2125–2127.
70. Bodhidharma. *The Zen Teaching of Bodhidharma* [Red Pine, Trans.]. New York: North Point Press; 1987.
71. Tang YY, Ma Y, Wang J, et al. Short-term meditation training improves attention and self-regulation. *Proc Natl Acad Sci USA*. 2007;104:17152–17156.
72. Brugman GM. *Wisdom and Aging*. Amsterdam: Elsevier; 2006.
73. Baltes PB, Smith J, Staudinger UM. Wisdom and successful aging. *Nebr Symp Motiv*. 1991;39:123–167.
74. Bishop SR, Lau M, Shapiro S, et al. Mindfulness: a proposed operational definition. *Clin Psychol Sci Pract*. 2004;11:230–241.
75. Arch JJ, Craske MG. Mechanisms of mindfulness: emotion regulation following a focused breathing induction. *Behav Res Ther*. 2006;44:1849–1858.
76. Chen KW, Berger CC, Manheimer E, et al. Meditative therapies for reducing anxiety: a systematic review and meta-analysis of randomized controlled trials. *Depress Anxiety*. 2012;29:545–562.
77. Beilock SL, Carr TH. When high-powered people fail: working memory and "choking under pressure" in math. *Psychol Sci*. 2005;16:101–105.
78. Ramirez G, Beilock SL. Writing about testing worries boosts exam performance in the classroom. *Science*. 2011;331:211–213.
79. Ardelt M. Social crisis and individual growth: the long-term effects of the great depression. *J Aging Stud*. 1998;12:291–314.
80. Hofmann SG, Sawyer AT, Witt AA, Oh D. The effect of mindfulness-based therapy on anxiety and depression: a meta-analytic review. *J Consult Clin Psychol*. 2010;78:169–183.
81. Clayton VP. Erikson's theory of human development as it applies to the aged: wisdom as contradictory cognition. *Hum Dev*. 1975;18:119–128.
82. Jeste DV, Ardelt M, Blazer D, Kraemer HC, Vaillant G, Meeks TW. Expert consensus on characteristics of wisdom: a Delphi method study. *Gerontologist*. 2010;50:668–680.

Index

Note: Page numbers followed by "f" indicate figures and "t" indicate tables.

C

D

G

H

I

O

P

R

S

T